Gustav Mann (1836-1916). A Hanoverian, he was first a Kew gardener and then a plant collector (the 'Kew Collector' of Sir Richard Burton). Mann was the first collector on Mt Cameroon. He visited three times during 1861 and 1862, staying several months in total and collecting many hundreds of specimens, though concentrating on the upland flora at the instruction of Kew's director, Sir William Hooker.

The Plants of Mt Cameroon

– A Conservation Checklist –

Compiled and edited by

Stuart Cable and Martin Cheek

Collaborating Institutions:

Royal Botanic Gardens, Kew
Mount Cameroon Project, Limbe

Published by the Royal Botanic Gardens, Kew

First Published 1998

Printed by Whitstable Litho Printers Ltd., Whitstable, Kent

Typeset by Maureen Bradford

Cover design by Jeff Sayer, Media Resources, R.B.G., Kew

ISBN 1 900347 57 1

Front cover: the characteristic bole of *Desbordesia glaucescens* (Engl.) Tiegh. in the Bonadikombe-Bimbia (Mabeta-Moliwe) forest in the eastern foothills of Mt Cameroon, October 1995. Photo: Andrew McRobb © Mount Cameroon Project and R.B.G., Kew.

Rear cover: Landsat™ image of Mt Cameroon © Mt Cameroon Project - GTZ. Showing plant specimen collecting densities at minute intervals. Produced by Justin Moat R.B.G., Kew.

CONTENTS

PREFACE

FOREST CONSERVATION AND USE ON MOUNT CAMEROON

Joseph Besong

Director, Mount Cameroon Project

The more it is proclaimed how rich in plant diversity Mount Cameroon is, the more the questions ring: What does it benefit the local people? Are they involved, is it in their interests - all this work going on? The answers are clearly 'YES'. Now there is proof for all these declarations. The checklist of plants on Mount Cameroon is an important step in the bid to know, use and conserve the riches of the Mountain. MCP's mission of "maintaining the richness and diversity of the plants and animals in the Mount Cameroon region by helping people and institutions to manage the forests wisely so as to obtain their benefits for ever" is certainly attainable, thanks to the existence and use of this checklist.

Scientists will use it, the local people also. Administrators will be interested by it, schoolchildren, women herbalists. As many get to know that this list exists, they will begin to appreciate not just the immensity of the task of using the forest wisely but also the risk of abusing it. It can be lost - and species which appear on the list may cease to exist, so that the list becomes an annal of history. Herein lies our challenge today - all of us who read the checklist. As for MCP and all with a heart, not only for the forest, but also for the people:- the work ahead is only greater - to add new discoveries to the list, to refine the details for example, concerning rarity, endemicity, threat of extinction, change of name or status etc. This work shall surely always go on, thanks to the herbaria in Limbe, Yaounde, Kew etc. and the staff all conscientious and applied to keep abreast with our very own rate of effecting change as we impact on the forest ecosystems.

But this work and especially the result of this work, the knowledge gathered - this checklist and subsequent revisions of it - must be put to good use. We must act wisely and use the forests as people who know their importance and the danger of living one day without them. If this is never allowed to happen, then the checklist will have served its purpose.

ACKNOWLEDGEMENTS

First and foremost, we wish to thank the West and North Africa Department of the British Government's Department of International Development (DFID, formerly ODA) for sponsoring, in connection with the Mount Cameroon Project (DKP) this book and the work which led to it. John Gibb, head of the department, and his predecessor John Gilbert, their staff, particularly John Howarth, Peter Feinson and Jalal Farooqui, have all provided support over the years. Ian Napier, former Forest Adviser to the Mount Cameroon Project for the West and North Africa Department gave support in funding Karen Sidwell to set up at R.B.G., Kew the initial specimen database which eventually led to this book. Peter Wood, Ian Napier's successor, approved the botanical inventories on Mt Cameroon which led to the discovery and rediscovery of so many of the species listed here. His successor, John Hudson, recommended funding for SC to work at R.B.G., Kew over several years on the naming and databasing of these. Graham Chaplin supported the extension of this work and its publication on recommendation from Glyn Davies, former DFID Team Leader at the Mount Cameroon Project.

Since 1993, Earthwatch Europe with the support of DG VIII of the European Commission have sponsored us to extend fieldwork on the mountain in collaboration with the Mount Cameroon Project. We are grateful to all the numerous Earthwatch fellows and volunteers that gave their time free to help us in the forests of Cameroon particularly Leona Cobham, Polly Riemann, Anne Dillen, Eric Aguilar, Eleanor Jones, Rebecca Pearson-Gee, Bob Schilling, Victor Martin, William Green, Sue Williams, Morgan Seiffert, Jane Solomons, Sizwe Cawe, Tim Holden, Jane Hill, Rachel Scott, Beatrice Khayota, Kim Wood, Arlene Golembiewski, Menassie Gashaw, Maanda Ligavha, Penelope Furneaux, Margaret and Marc Faucher, David Laidley, John Lopes, Kenneth Norris, Russel Dekalb, Leonard Sullivan, Gail McAleese, Kevin FitzGerald, Lorissa Boyd, Judith Randall, Lori Furda, and also to the people of the villages that have hosted us: Dikulu, Upper Boando, Likombe, Bakingili, Njonji and Enyenge. From head office, in Oxford, Sally Moyes was hugely helpful in providing advice on how to run our first expeditions, and joined us on our very first term. Gill Barker, Robert Barrington and Julian Laird have also been of immense assistance. We would also like to thank Simon Martyn, Sean Doolan, Andrew Mitchell and Pamela Mackney.

In Cameroon we are extremely grateful to Joseph Besong, Director of the Mount Cameroon Project, and to all his staff, past and present, who have done so much over the years to organize and execute specimen collection, databasing, identification and despatch. Peguy Tchoutou, head of the Limbe herbarium, has been particularly helpful. We also thank his team, especially Juliana Nning and Maurice Betafo. Together with Emma Crowe they have shouldered the unglamorous and often repetitive work of specimen drying and sorting, as well as managing the entering of specimen collection data. Ekema Ndumbe, Nouhou Ndam, Brendan Jaff, Charles Tekwe, Guillaume Akogo, Jean-Marie Mbani and François Nguembock are all past or present Mount Cameroon Project officers from the Department of Forestry who have been prominent in leading plant collecting teams in the forests of Mt Cameroon. James Acworth, British Government Technical Co-operation Officer, and his predecessors Joe Watts and Jos Wheatley have also shared this task. Often they have camped out for weeks on end in inclement conditions in order to further our knowledge of the wild plants of the mountain.

Just as numerous people have assisted with plant collection in the forests of Mt Cameroon in the last few years, so have many persons from around the world been involved in the identification of the resultant specimens. The following list of persons is generated from the BRAHMS database. It enumerates all who have been credited with naming the specimens cited in this checklist, whether a single specimen, or a thousand. The ninety-seven people listed here are from 12 countries. They include numerous World specialists as well as many students and technicians. We sincerely thank them all, and their institutions, for working, often without remuneration, towards producing the names which are used in this book.

Wheatley J., Ndam N., Friis I., Sonke B., Lane P., Ekema N., Elad M., Upson T., Bygrave P., Harris D., Mbani J., Mezili P., Groves M., Koufani A., Nkongmeneck B., de Wilde J., Sosef M., Tekwe C., Breteler F., Lemmens R., Jongkind C., Bos J., Burtt B., Vollesen K., Watts J., Cope T., Sidwell K., Jaff B., Polhill R., Faruk A., Onana J., Leeuwenberg A., Goyder D., Bidgood S., Paton A., Radcliffe-Smith A., Prance G., Jeffrey C., Verdcourt B., Huynh K., Simpson D., Asonganyi J., Boyce P., Dransfield J., Lock M., Mbugua P., Wilkin P., De Block P., Bridson D., Townsend C., Ambrose B., Cribb P., Zapfack L., Akogo M., Baker W., Gosline G., von Rege I., Hunt L., Pennington T., Faden R., Besina M., Mackinder B., Lewis G., Achoundong G., Goldblatt P., Sunderland T., Verboom A., Acworth J., Tchouto P., Thomas D., Edwards P., Rico L., Schrire B., Nning J., Hepper N., Frodin D., Dekker A., Beentje H., Meve U., Benl G., Dawson S., Hind N., Marsden J., Brodie S., Sands M., Thomas S., Banks H., Atkins S., Biggs N., Omino E., Wieringa J., Sebsebe D., Poulsen A., Johns R., Lye K.

We thank Denis Filer of the Oxford Forestry Institute, inventor and developer of BRAHMS, the database from which this conservation checklist has been produced. Without his constant and unstinting help over the years, this work might still be waiting to see the light of day.

William Hawthorne provided the ecological guild ratings and, with Peguy Tchouto, the star ratings that are incorporated, with modification, in this book.

George Gosline wrote the macro for formatting the checklist from the BRAHMS database and has provided database support and good advice since 1995.

Many of the specimens cited here were named, on a preliminary basis, at the National Herbarium of Cameroon (YA) by Mt Cameroon Project staff. We thank Dr Satabie, recently retired Director of the National Herbarium, and his staff, for their assistance in this process.

Charlotte Lusty and Harriet Gillett of WCMC Cambridge provided listings of Red Data taxa for Cameroon from their tree database, and also from their all species database. They also provided useful discussion on IUCN Red Data ratings. Wendy Strahm, Alan Hamilton, Henk Beentje and Steve Davies provided information on Red Data and checklist matters in Africa.

For commenting on drafts and help in book production, we are indebted to our colleagues at Kew, Eimear Nic Lughadha, Diane Bridson and Susie Dickerson.

Finally we thank Nigel Hepper, who played such an important part in 1986 in starting what we now know as the Mount Cameroon Project. Since its inception in 1988 until his retirement, he has championed the cause of Limbe. The genesis of this book was his suggestion to Ian Napier (ODA Forestry Adviser) at the Atlantik Beach Hotel in Limbe, that the records of plant species known from the mountain be collated, perhaps as a computer database.

NEW NAMES

The following are published in this volume for the first time:

OCHNACEAE: ***Campylospermum monticola*** (Gilg) Cheek   p 92

RUBIACEAE: ***Cuviera wernhamii*** Cheek   p 106

Geology, Geomorphology, Soils and Climate.

These factors play a large part in deciding the vegetation. The following short notes are taken largely from Hawkins and Brunt (1965) and Courade (1974), but include some personal observations.

Geology & Geomorphology (Map 1).

A fault, running nearly NE-SW between two major African plates, is the origin of Mt Cameroon and the mountainous areas nearby that lie in a band 50-100 km wide along that fault. 40 Km to the SW of Mt Cameroon is the mountain-island of Bioko (formerly Macias Nguema or Fernando Po), geologically and botanically nearly a twin of Mt Cameroon. Moving NW from Mt Cameroon, 100 km along the fault, is Mt Kupe, followed by the Manengouba massif, the Bamboutos Mts, the Bamenda Highlands and then, further inland, yet more mountain ranges. These mountains are formed largely of igneous material, both volcanic and plutonic. Three main periods of volcanic activity and one of plutonic uplift have been reviewed by Courade (1974).

The age of the oldest volcanic period has been suggested as the Cretaceous, but this is controversial, and the Tertiary has been preferred by Courade (1974). The "lower black series" of volcanic activity gave rise to the basalts, sometimes porphyritic, which formed the heart of Mt Cameroon, Mt Etinde and presumably the larger part of the western, eastern and northern foothills. Hexagonal columns of basalt believed to date from this period are exposed on the coast between Mabeta and Dikulu. The second period of volcanic activity "the middle white series" gave rise to much of the Bamenda Highlands, but appears to have contributed little to the Mt Cameroon area. This was followed by a plutonic period in which, for example, the main bulk of Mt Kupe was formed by the uplift of granites and syenites. Mt Etinde is indicated as being of the same material as Mt Kupe on the geological map accompanying Courade (1974), but this seems to be an error, in conflict with the accompanying memoir.

The third period of volcanic activity "the upper black series", took place in the Quaternary. Basalts forming the perimeter of Mt Cameroon are exposed e.g. in the seasonal river bed above Njonji at about 300 m alt. Crater lakes arising from volcanic explosions are present on the southern edge of the massif at Debundscha and Njonji and also on the northern edge, namely Lakes Mbwamdong and Barombi Koto. They are also believed to date from the third period. More lava flows have followed in the Quaternary, building the bulk of the main massif which appears from the surface to be formed largely of recent volcanic scoria and cinders, increased steadily by additions from eruptions of lava which still occur at roughly 20 year intervals. In this century, lava flows have arisen from different vents on several aspects of the mountain, between the 3,000 and 4,000 m contours. Flows tend to be no more than one or two kilometres across and do not always reach the lowest slopes. There have been five major eruptions so far this century. In 1909 an eruption produced a lava flow that descended towards the town of Muyuka in the northeast; in 1922 a lava flow descended from a crater at about 3000 m alt. towards the southwest, entering the sea at Bibundi, destroying the village; no lava was produced by the eruption in 1954; the 1959 eruption produced lava that flowed east, near Ekona descending to about the 800 m contour but stopping before the village was reached; in 1982 a lava flow descended southwards, west of Mann's Spring. Above the treeline the mountain slopes seem to be made of a succession of lava flows of different ages and differing stages of recolonization, interspersed with cinder cones and craters. Numerous features around the mountain derive from these recent episodes of volcanic activities. The absence of settlement over most of the main massif is due to the free draining volcanic lava that drains rapidly and eliminates permanent surface water supplies but instead feeds the numerous springs that well-up from the foot of the mountain. Some lava tubes are hollow and on collapsing, have left radial gullies, sometimes destroying roads that are built over them. Volcanic dust, turning to mud with rainfall, has produced unstable slopes and land slides on the western slopes of the massif. The islands in Ambas Bay, e.g. the `Pirate Islands', are remnants of sea-eroded lava flows and eroded fragments form the black beaches around the coast near Limbe.

At 4095 m high, Mt Cameroon, locally known as Fako, is by far the tallest mountain, and the only active volcano, in West-Central Africa. One has to travel about 1800 km eastwards to the Virunga Mts of eastern Zaire before encountering mountains of comparable height, or active volcanoes. However, there is an active volcano nearer to Mt Cameroon in the Cape Verde Islands.

Mt Cameroon is classified as an 'Hawai'ian Type' volcano. Its massif sprawls over an enormous area, about 45 km along its longest axis, and 28 km across its narrowest. The gentle slopes of its rounded convexity are characteristic of this volcano type. By contrast, the geologically older Mt Etinde or 'Little Mt Cameroon', the 1700 m high volcanically extinct subpeak that arises from the southern slopes of the main massif, has an angular, sharp peak formed from three or four steep, blade-like ridges arising from a narrow base.

Sedimentary rocks are scarce around Mt Cameroon. Tertiary sedimentary deposits are known in parts of the eastern foothills and there are small patches of recent alluvial deposits at Idenau and in Mabeta-Moliwe.

Soils (Map 2).

Hawkins and Brunt (1965) classify the soils of Mt Cameroon using fertility classes one to four, as follows:
first class or excellent; second class or good; third class or average and fourth class or mediocre.

The soils of the main massif of Mt Cameroon, thanks to their recent volcanic origin, are first class for agriculture, excepting Etinde, the peak, and recent lava flows which at most have poorly evolved mineral soils. The western and northern foothills are described as having good, second class soils composed of a mix of young and old volcanic soils. Parts of the eastern foothill areas as at Onge and Mokoko, have soils overlain by white sand. Mabeta-Moliwe, and other parts of the eastern foothills are classified as third class, or average, being composed of old lateritic soils. Boulders, up to a metre diameter occur in the eastern foothills, as at Mabeta-Moliwe.

Climate (Map 3).

Whilst there is some rain every month on Mt Cameroon, most falls in the 8-9 month wet season between April and November. There are no months of the year with less than 50 mm precipitation at Debundscha, two months at Tiko and three months at the Bai estate. There are dramatic variations around the mountain in the annual rainfall figures. Most of the mountain receives between 3 m and 10 m of rain per annum. The wettest place is the SW, where Cape Debundscha receives 10-15 m per annum. Even in the drier season, the road always seems to be wet when passing through Debundscha. Together with an area in Hawai'i and another in Assam, NE India, Cape Debundscha is believed to be one of the three wettest places in the world. The driest area is the rainshadowed NE, corresponding with part of the S. Bakundu reserve, where only 2 m per annum is recorded. Rainfall also tends to decrease with altitude on the mountain, the upper part of the main massif only receiving 3 m per annum, while the topmost part, around the peak, receives only 2 m of rain per annum.

Snow is rare on the summit (4095 m alt.) of Mt Cameroon, though frost is not uncommon. The mean annual temperature is 4 °C, rising at Buea (915 m alt.) to 20 °C and at Tiko (near sea-level) to about 28 °C. Relative humidity varies at Tiko from 65 degrees in the driest month, January, to 86 degrees in August, the wettest month.

At sea-level, there is often a gentle breeze at the coast, but generally air movements are sluggish. Hurricanes are unknown. At the beginning of the wet season however, the opening frontal storms can be violent and this is the main season for natural tree fall.

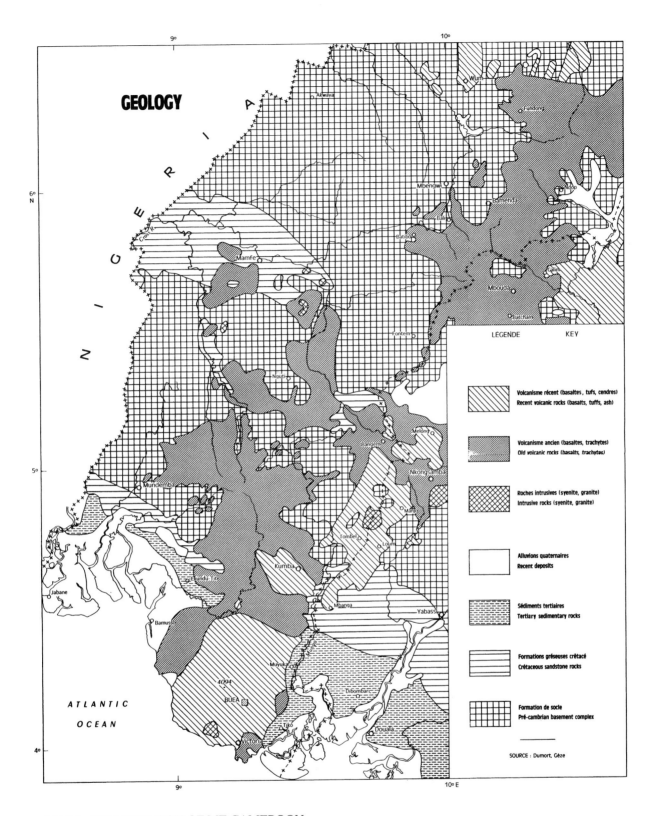

GEOLOGY

LEGENDE   KEY

Volcanisme récent (basaltes, tufs, cendres)
Recent volcanic rocks (basalts, tuffs, ash)

Volcanisme ancien (basaltes, trachytes)
Old volcanic rocks (basalts, trachytes)

Roches intrusives (syenite, granite)
Intrusive rocks (syenite, granite)

Alluvions quaternaires
Recent deposits

Sédiments tertiaires
Tertiary sedimentary rocks

Formations gréseuses crétacé
Crétaceous sandstone rocks

Formation de socle
Pré-cambrian basement complex

SOURCE : Dumort, Gèze

MAP 1. THE GEOLOGY OF MT CAMEROON
Reproduced with the permission of ORSTOM from Courade (1974).

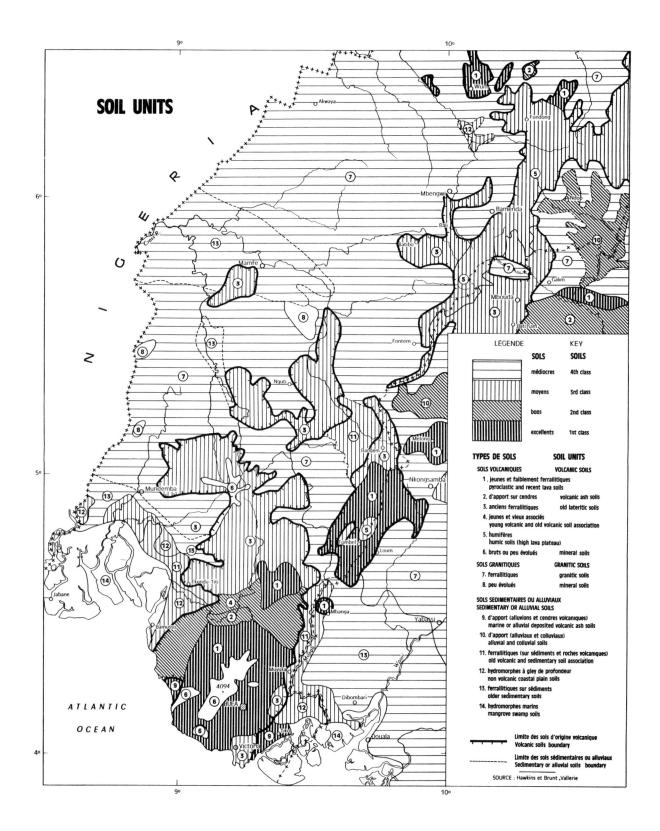

SOIL UNITS

LÉGENDE   KEY

SOLS	SOILS
médiocres	4th class
moyens	3rd class
bons	2nd class
excellents	1st class

TYPES DE SOLS   **SOIL UNITS**

SOLS VOLCANIQUES   VOLCANIC SOILS

1. jeunes et faiblement ferrallitiques
   pyroclastic and recent lava soils

2. d'apport sur cendres   volcanic ash soils

3. anciens ferrallitiques   old lateritic soils

4. jeunes et vieux associés
   young volcanic and old volcanic soil association

5. humifères
   humic soils (high lava plateau)

6. bruts ou peu évolués   mineral soils

SOLS GRANITIQUES   GRANITIC SOILS

7. ferrallitiques   granitic soils

8. peu évolués   mineral soils

SOLS SEDIMENTAIRES OU ALLUVIAUX
SEDIMENTARY OR ALLUVIAL SOILS

9. d'apport (alluvions et cendres volcaniques)
   marine or alluvial deposited volcanic ash soils

10. d'apport (alluviaux et colluviaux)
    alluvial and colluvial soils

11. ferrallitiques (sur sédiments et roches volcaniques)
    old volcanic and sedimentary soil association

12. hydromorphes à gley de profondeur
    non volcanic coastal plain soils

13. ferrallitiques sur sédiments
    older sedimentary soils

14. hydromorphes marins
    mangrove swamp soils

Limite des sols d'origine volcanique
Volcanic soils boundary

Limite des sols sédimentaires ou alluviaux
Sedimentary or alluvial soils boundary

SOURCE : Hawkins et Brunt ,Vallerie

MAP 2. THE SOILS OF MT CAMEROON
Reproduced with the permission of ORSTOM from Courade (1974).

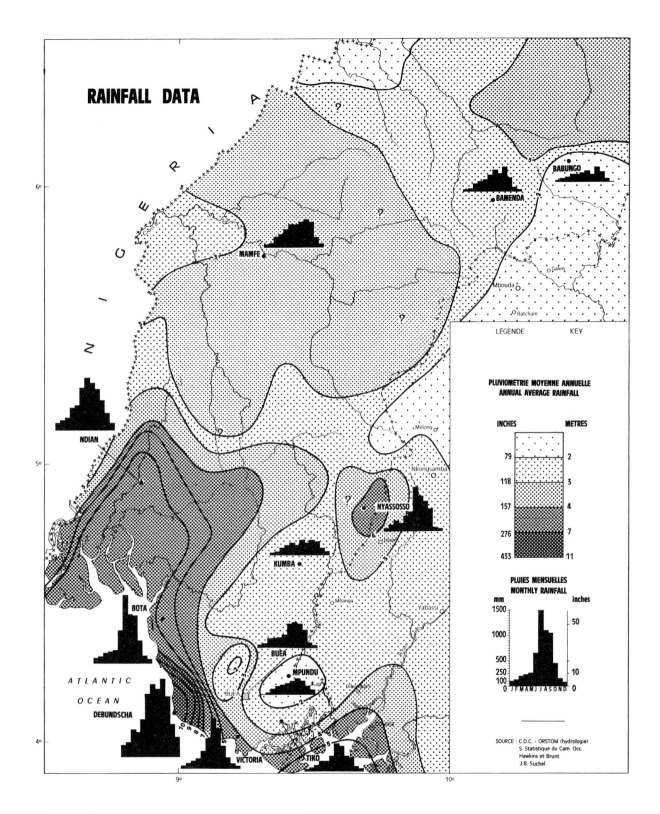

MAP 3. THE CLIMATE OF MT CAMEROON
Reproduced with the permission of ORSTOM from Courade (1974).

The Checklist Area (Map 4.)

The Mount Cameroon area, for the purpose of this checklist, is taken as roughly ovate in outline, the point orientated to the northeast. It encompasses the main massif, about half the total area, and the foothills to the northwest (Onge and Mokoko Forests), those to the east (including Mabeta-Moliwe, also known as Bimbia-Bonadikombo) and the more extensive and less steep foothill area to the north, the Southern Bakundu Forest Reserve. Moving clockwise, we take the southern boundary as the coast, beginning in the mangroves at Tiko in the southeast corner of the checklist area, and proceeding westwards for about 90 km, past the eastern foothills, Limbe, and onwards to Bota, Debundscha and Idenau to Njangassu below Thump Mount. Moving inland, northeastwards from Njangassu we take the road that passes, through Boa, Bonjare, Dikome and Illoani to Mbonge as the boundary since this road divides the western foothills from the Boa Plain. From Mbonge we take as the boundary the road to Kumba, that runs slightly north of east. Kumba is the most northerly point in the checklist area. From it our boundary heads south, along the road (parallel to the Mungo River nearby to the east) to Muyuka on the main massif. From Muyuka we rejoin Tiko by the Mpundu road so as to include the northern part of the eastern foothills.

This boundary encompasses an area of approximately 2700 km^2, but the species listed in this checklist almost all come from the protected areas (some only proposed) that together comprise only about a third of this area, i.e. about 1000 km^2, as follows: the Bambuko Forest Reserve (266.8 km^2), Southern Bakundu Forest Reserve (194.3 km^2), Mabeta-Moliwe or Bimbia-Bonadikombo (c. 36 km^2), the Etinde Reserve (c. 360 km^2), Onge (c. 100 km^2) and the Mokoko River Forest Reserve (90.7 km^2). Two further protected areas occur just outside the checklist area: the Meme River Forest Reserve (51.8 km^2) and the Mouyouka-Kompina Forest Reserve (50 km^2, but only 20 km^2 still intact). Since both are botanically unknown and geographically peripheral, we exclude them here. The areas given are taken from Cheek and Thomas (1994).

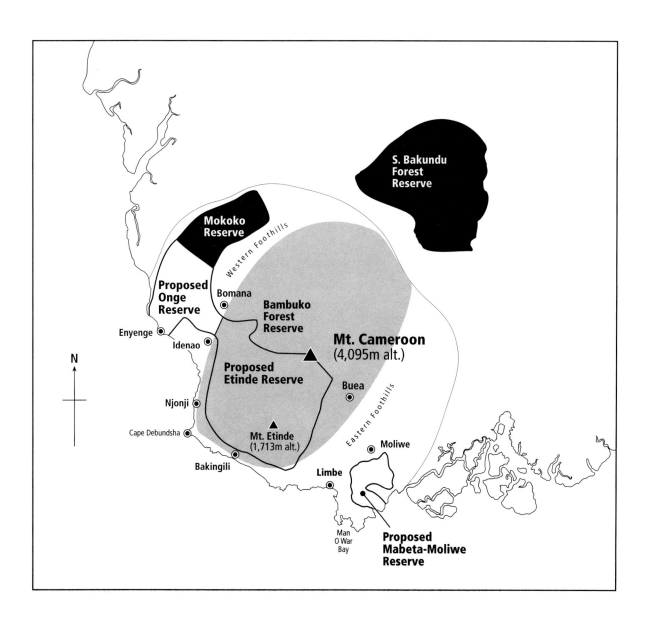

MAP 4. THE CHECKLIST AREA
Reproduced from Cheek et al. (1996).

Vegetation (Map 5).

It is beyond the scope of this work to provide a detailed analysis of the highly varied vegetation of Mt Cameroon. We aim here only to give a brief overview as a background to the checklist. Numerous papers and reports have been published on different facets of the vegetation of the mountain. There are 17 pages on the vegetation of the proposed Etinde reserve alone in Thomas & Cheek (1992).

As reference for our notes, we take the map published by Maley & Brenac (1998), which has the virtue of showing the vegetation of the whole checklist area clearly and concisely on one page. We are most grateful to them, and to ORSTOM, for permitting this map to be reproduced here. Their mapping units are modified from those of Letouzey (1985). It is to be hoped in a future edition that a map based on satellite image analysis will be available.

Mapping unit 1. "Evergreen Biafrean Forest with numerous Caesalpinaceae". This unit accounts for the single most important vegetation type, in terms of species diversity and area, in the checklist area. There are three disjunct areas, each with their own characteristics and each with unique species. Firstly, that of the S. Bakundu F.R. to the north, secondly the western foothills forests of Onge and Mokoko (the least disturbed and also the largest contiguous area, reviewed by Thomas (1993a and 1994)), thirdly the eastern foothills, of which Mabeta-Moliwe (Bimbia Bonadikombo) is the largest of the remnants that remain, reviewed by Cheek (1992).

Mapping units 2 & 3. "Evergreen transitional to littoral Atlantic Forest poor in Caesalpinaceae". This is also lowland rainforest, and largely occupies the coastal strip between Limbe and Idenau. Much has been cleared for Oil Palm plantation and the remainder is under threat. *Trichoscypha* sp. nov., for example is unique to this forest. It has been reviewed in Thomas & Cheek (1992).

Mapping unit 4. "Evergreen Littoral Atlantic Forest with rare Caesalpinaceae". This vegetation type occurs on the Boa Plain, west of the western foothills, and so does not occur in the checklist area.

Mapping unit 5. "Semi-Deciduous Forest". Characterised by deciduous species such as *Triplochiton scleroxylon*, this forest is found in the drier rain-shadowed area immediately to the north of the main massif.

Mapping unit 6. "Grass Savanna with a few trees". Four small patches of this are shown to the north of the main massif, in the area where the Bambuko, Mokoko and S. Bakundu forests converge. *Borassus aethiopum*, a palm, characterizes these savannas.

Mapping unit 7. "Marantaceae open canopy forest (altitude ca. 500-2000 m)". This occurs as a peripheral belt from the west side of Mt Etinde around to the northern part of the main massif, increasing in altitudinal amplitude as one moves west. We refer to this as *Hypselodelphys* thicket (e.g. Thomas & Cheek 1992). It appears as a dense thicket of Maranataceae and herbs 1.5-3 m tall, much favoured by those elephants that survive on the mountain, though relatively poor in plant species. Much, if not most of the Bambuko forest is believed to be composed of this vegetation.

Mapping units 8 & 9. "Montane Rain Forests". This unit girdles the main massif from an altitude of about 800 m to 1900-2500 m. We recognize (Thomas & Cheek 1992) three types of forest within this unit:

1. Montane Forest (1800-2500 m alt.) characterized by *Schefflera abyssinica, Schefflera mannii, Prunus africana, Xymalos monospora, Hypericum revolutum, Clausena anisata* and *Nuxia congesta.*

2. Submontane Forest, Closed Canopy. Localized to Mt. Etinde and occurring between 800-1700 m alt., this vegetation is characterized by the presence of species not seen in the other submontane vegetation type, e.g. Sapotaceae, *Garcinia smeathmannii* and *Pseudagrostistachys africana.* The strongest affinities are thought to be with submontane forest of the Rumpi Hills and the Bakossi Mts (outside our checklist area).

3. Submontane forest, Discontinuous Canopy. This occupies the same altitudinal range as the closed canopy forest, but is more extensive, almost girdling the mountain. It is characterized by numerous gaps and contains large swards of monocarpic Acanthaceae including *Mimulopsis solmsii, Acanthopale decempedale* and *Oreacanthus mannii.* Trees include *Polyscias fulva, Alangium chinense, Psydrax dunlapii, Cyathea manniana, Oncoba lophocarpa, Oncoba ovalis* and *Xylopia africana.* Large stranglers are common, particularly at higher altitudes.

Mapping unit 10. "Montane grassland". This vegetation type occurs from 1900-2500 m up to the summit at 4095 m. We have recognized three types of vegetation within it (Thomas & Cheek 1992).

1. Montane scrub. This occurs at the boundary of forest and grassland, but is poorly developed, possibly due to fire pressure. Equated by some authors with the Ericaceae belt of East African mountains, but poor in that family in both species diversity and in abundance, with only *Erica mannii, Erica tenuicaulis* and *Agauria salicifolia. Hypericum roeperianum, Gnidia glauca, Maesa lanceolata* and *Myrica arborea* are also noticeable.

2. Montane grassland or prairie. Occurring up to about 2800 m, it is richest in species at lower altitudes, particularly at the transition with forest, or if present, scrub.

3. Subalpine grassland or prairie. Much poorer in species diversity and with plants not forming a complete sward, this type occurs above c. 2800 m and has been characterized by the absence of *Loudeta simplex* (Thomas & Cheek 1992). Otherwise it intergrades with montane grassland. *Veronica mannii* and *Pentaschistis mannii* are examples of species restricted to this vegetation type.

The absence of *Arundinaria alpina, Podocarpus latifolius, Ternstroemia polystachya* and *Protea elliotii* are remarkable since they occur on other mountains in Cameroon of lower altitude.

Mapping unit 11. "Mangrove". This vegetation covers extensive areas to the west and east of the mountain, but is poorly represented on the exposed coast in the checklist area. However, small portions can be found in the south eastern corner and there is one small, largely pristine patch in the Mabeta-Moliwe area, near Dikulu.
Details of this vegetation can be found in Cheek (1992).

Mapping unit 12. "Recent lava flow with pioneer forest". This is placed on the 1922 lava flow which reaches the sea at Bibundi. A review of the regeneration on this lava flow, based on the work of the "Njonji 1995" expedition from R.B.G., Kew, is in press. Currently, *Lannea welwitschii*, *Hymenodictyon biafranum* and *Syzygium guineense* dominate this lava flow at sea-level. Other lava flows have not been studied.

Mapping unit 13. "Plantations, villages and various anthropic vegetation". Most plantations are lowland apart from Tole Tea at Saxenhof. Extensive plantations of Rubber, Oil Palm and Bananas occur, particularly on the eastern and southern slopes. Small scale agriculture, including farm fallow is most prevalent on the eastern side. Threats to natural vegetation are discussed in the Red Data chapter.

Mapping reference L1, L2 and L3. These refer to the crater lakes, of which there are five in the checklist area, all botanically unknown.

Littoral vegetation is referred to in Cheek (1992). *Sesuvium portulacastrum, Machaerium lunatum, Phoenix reclinata, Ipomoea pes-caprae, Calophyllum inophyllum* and *Lonchocarpus sericeus* are species common in this type. No Zosteraceae are known.

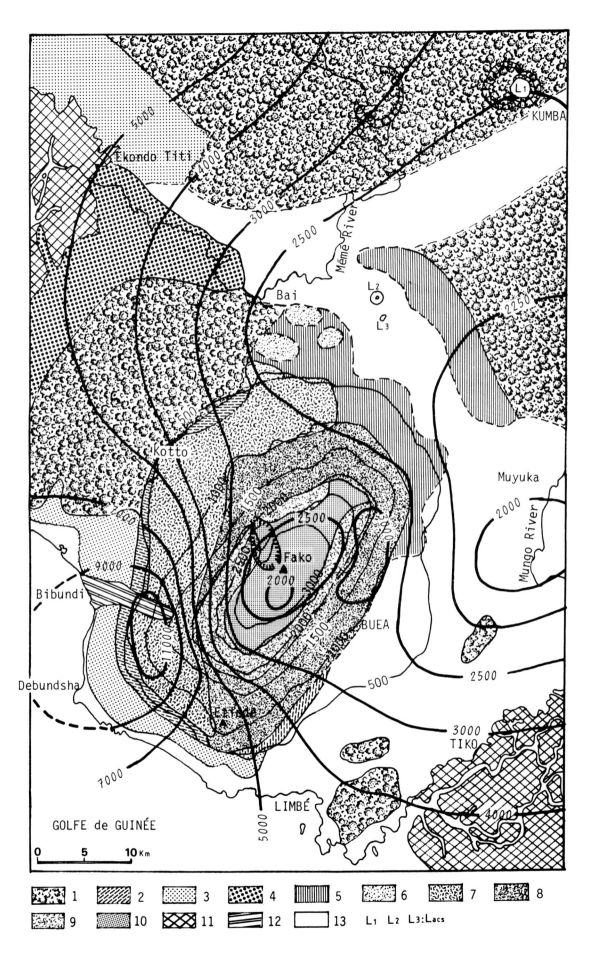

MAP 5. THE VEGETATION OF MT CAMEROON
Reproduced from Maley & Brenac (1998) with permission. The mapping units are referred to in the text.

Phytogeography

A detailed analysis of the phytogeography of Mt Cameroon is beyond the scope of this work. However, analysis of the 1860 taxa which have been scored in the database for phytogeography, using the classification described in 'Read this First!', gives the following breakdown:

Phytochorion	% Species
Guineo-Congolian	43%
Lower Guinea	16%
'Western Cameroon Uplands'	5%
Afromontane	10%
Mt Cameroon	2.6%
Other (e.g. Tropical Africa or Cosmopolitan)	23.4%

The term 'Western Cameroon Uplands' we use for the largely upland area that extends from Bioko in the south to the Bamenda Highlands of Western Cameroon and includes the Oban Hills and Obudu Plateau areas of S.E. Nigeria. The Guineo-Congolian area contributes the largest single slice of the species of Mt Cameroon, but the contributions from the Afromontane and Lower Guinea phytochoria are also significant, though lower than might be expected. Clearly these figures may change if all the species on the mountain are scored.

A surprising number of the higher altitude species from Mt Cameroon are common to temperate areas of the northern hemisphere. Examples include *Callitriche stagnalis*, *Radiola linoides*, *Juncus capitatus*, *Sibthorpia europaea*, *Anagallis minima*, *Stellaria media*, *Poa annua*, *Aira caryophyllea*, *Vulpia bromoides* and *Cardamine hirsuta*. Many of these may have arrived via Ethiopia and the East African mountains, where some are known to occur today. However, it is likely that some of these high altitude species, or their ancestors, migrated to Mt Cameroon directly from Europe. A likely example is *Succisa trichotocephala*. The genus *Succisa* is Eurasian and is not found elsewhere in Africa. *Anagallis minima* is represented by numerous specimens throughout Europe and North America, but is known in Africa only from one specimen in Ethiopia and another from Mt Cameroon.

STATISTICAL SUMMARY

The total number of indigenous and naturalized species listed here for Mount Cameroon is 2435. This figure is conservative since it excludes specimens cited as e.g. *'Garcinia cf. smeathmannii'* if this name follows specimens cited with certainty as *'Garcinia smeathmannii'*. It also excludes e.g. *'Garcinia* sp.' Further collections and research on those specimens cited as *'Garcinia cf. smeathmannii'* may either show that they belong to a species not already cited in the checklist, or else may unite them with *'Garcinia smeathmannii'*. All of the species listed here are based on cited specimens, with the exception of a few extremely well-known and conspicuous species that appear to have remained somehow unvouchered (e.g. *Tridax procumbens, Triplochiton scleroxylon* and *Borassus aethiopum*). As time goes on, the total number of species for the checklist area is likely to rise as 'cf.' species are resolved, and as specimens representing species new to Mt Cameroon are gathered from undercollected parts of the checklist area. Several large parts of the Mt Cameroon area, particularly the northern part of the massif, are still almost unknown botanically (see collecting density map on rear cover) or very poorly covered. It is possible to speculate that as many as two hundred species could yet be added to the total listed here. Some habitats such as the *Borassus* savanna and the crater lakes, appear not to have been botanically investigated yet. This reflects the concentration of effort over the last six years in investigating the most species-rich, most threatened, and until now the most poorly known of habitats, those of the evergreen forest on the southern half of the mountain and its foothills.

Very few checklists are available for areas of similar size in the Gulf of Guinea area, making comparison of total species numbers difficult. We were not able to trace checklists for Gabon or Nigeria. However, figures are available for Bioko (Exell 1944) and for the Korup Project area (Thomas 1993b):

Checklist site	Area	No. Species
Bioko	2018 km2	842
Korup Project Area	2510 km2	1693
Mt Cameroon Area	2700 km2	2435

The lower figure for Bioko, the mountain-island only 32 km from Mt Cameroon, reflects the clearance of the lower altitude forest for agriculture before the indigenous species could be fully enumerated. Fernandes Casas (*pers. comm.*) has since discovered many new additions to the flora of Bioko, but even so, Bioko remains a poor relation to Mt Cameroon in terms of species diversity. The figure for the Korup Project area more nearly approaches that of Mt Cameroon. However, although Korup contains large areas of highly species-diverse forest it appears not to have the habitat diversity or altitudinal range seen within the Mt Cameroon area, and this perhaps explains the lower figure for species diversity. It is likely that the Korup figure will increase as more botanical inventory work is done in the poorly known northern and eastern parts (Cheek & Cable 1997).

This checklist is largely produced from a BRAHMS database containing 9,600 records of determined specimens from Mt Cameroon. Of these several hundred derive from incomplete records extracted from FWTA which were based on collections made between 1861 and the 1960s; up to 500 derive from the 1993 Oxford Univ. Expedition (Baker 1995), a similar number date from the R.B.G. Kew 'Njonji' expedition of 1995 and over a thousand derive from Earthwatch Europe-funded R.B.G., Kew botanical inventories conducted between 1993-

1995 (Cheek & Cable 1997), while the remainder derive from DFID-funded inventory work conducted between 1991-1995. Numerous additional records remain to be added. For example, we list here only two specimens (*Mann* 1312 and *Banks* 68) of the near endemic *Veronica mannii* for Mt Cameroon, whereas there are 25 specimens in the Kew Herbarium. We have estimated that there may be as many as 10,000 historical specimens from western Cameroon in the Kew Herbarium, of which very few are on our database. We hope that for the second edition of this checklist we will be able to incorporate data captured from these historical specimens.

Final confirmations and determinations of the specimens in this checklist were made for 57 of the families, about half the total, by almost as many specialists (some researchers specialize in more than one family). These family specialists are based at 13 institutions, principally R.B.G., Kew (31), Wageningen (9) and then the following: E(3), O (3), ETH (2), YA (2), and then the following, one each: US, G, NLH, BG, BM, NY and MSTR. Specialists are credited in the checklist at the beginning of their family treatment. The remaining families were principally identified by SC beginning in 1994, with determinations in 1992 by MC and Karen Sidwell.

The breakdown of species by major group is as follows:

Dicotyledons	1757
Monocotyledons	479
Gymnosperms	2
Ferns and Fern Allies	197

The species on the mountain represent 117 families of flowering plants (dicotyledons and monocotyledons). The most important plant families in terms of numbers of species are as follows:

1	Rubiaceae	261
2	Orchidaceae	147
3	Leguminosae	136
4	Gramineae	110
5	Euphorbiaceae	94
6	Acanthaceae	72
7	Compositae	67
8	Apocynaceae	55
9	Annonaceae	45
10	Moraceae	43

The predominance of the Rubiaceae, as shown by this list, is even more pronounced in the field. In the southern forests, the proportion of individual plants in a given area from this family is much higher than their proportion of the total species on the mountain (c. 11%) would lead one to expect.

Several wetland families, such as Eriocaulaceae, Xyridaceae and Droseraceae (all of which occur, e.g. on Mt Oku) are notable for their absence from Mt. Cameroon as are certain aquatic families e.g. Nymphaeaceae, Ceratophyllaceae and Alismataceae. Owing to the free-draining volcanic substrate, wetland habitats are scarce, but careful searching on the eastern slopes might yet locate boggy flushes. It is quite possible that the crater lakes might yield several aquatic families not yet recorded for the mountain.

The most important genera in terms of numbers of species are as follows:

	Genus and Family	No. Species	Predominant Lifeform
1	Bulbophyllum (Orchidaceae)	41	Epiphytic herb
2	Asplenium (Aspleniaceae)	38	Epiphytic herb
3	Psychotria (Rubiaceae)	37	Shrub
4	Cola (Sterculiaceae)	24	Tree
5	Ficus (Moraceae)	23	Tree
6	Cyperus (Cyperaceae)	23	Terrestrial herb
7	Salacia (Celastraceae)	22	Liane
8	Diospyros (Ebenaceae)	22	Tree
9	Polystachya (Orchidaceae)	22	Epiphytic herb
10	Begonia (Begoniaceae)	22	Epiphytic herb
11	Strychnos (Loganiaceae)	20	Liane
12	Pavetta (Rubiaceae)	19	Shrub
13	Aframomum (Zingiberaceae)	17	Terrestrial herb
14	Culcasia (Araceae)	17	Liane
15	Impatiens (Balsaminaceae)	15	Terrestrial herb
16	Dichapetalum (Dichapetalaceae)	15	Liane
17	Dracaena (Dracaenaceae)	13	Shrub
18	Pteris (Pteridaceae)	13	Terrestrial herb
19	Trichoscypha (Anacardiaceae)	12	Tree
20	Campylospermum (Ochnaceae)	12	Shrub

It is clear that the most diverse genera are usually understorey trees and shrubs or epiphytic herbs. In contrast, analysis of the lifeform data in the database as a whole shows that terrestrial herbs account for almost a third of the total diversity at species level. Together, understorey trees and shrubs account for 28% of the total diversity at species level while the epiphytic herbs, so prominent among the major genera, actually account for only 10% of the total species diversity. Note that shrubs are defined as woody plants generally less than 3 m high at maturity, or multi-stemmed if taller than 3 m. Note also that not all records on the database include lifeform data and that a single species may be scored as exhibiting more than one lifeform. Thus the total of the %s slightly exceeds 100.

Large trees (>15 m tall)	11%
Understorey trees	18%
Shrubs (multistemmed or <3 m tall)	10%
Stranglers	c. 0.5%
Woody Climbers	13%
Herbaceous climbers	5%
Terrestrial herbs	32%
Epiphytic herbs	10%
Hemi-epiphytes	<0.5%
Lithophytes	<0.5%
Rheophytes	<0.5%

| Saprophytes | c. 0.5% |
| Parasites | c. 0.5% |

For statistics on endemic and conservation priority species, e.g. IUCN and "star" ratings, see the Red Data chapter. For those on Phytogeography, see there.

BIBLIOGRAPHY

Adams, C.D. (1957). Observations on the fern flora of Fernando Po. J. Ecol. 45: 479-494.

Alston, A.H.G. (1956). New Africa Ferns. Bot. Soc. Brot. II. 30: 16.

Baker, W. et al. (1995). Mabeta-Moliwe 1993. An Oxford University Expedition to Cameroon. Cyclostyled report. 24 pp.

Barthlott, W., Lauer, W. & Placke, A. (1996). Global distribution of species diversity in vascular plants: towards a world map of phytodiversity. Erkunde Band 50: 317-328 (with supplement and figure).

Brummitt, R.K. (1992). Vascular Plant Families & Genera. (1992). Royal Botanic Gardens, Kew.

Brummitt, R.K., & Powell, C.E. (1992). Authors of plant names. Royal Botanic Gardens, Kew.

Cable, S. & Cheek, M. (1996). Provisional plant species checklist for Etinde. R.B.G. Kew. Report to ODA/MCP. Cyclostyled. 99 pp.

Cheek, M. (1992). A Botanical Inventory of the Mabeta-Moliwe Forest. Report to ODA/MCP. R.B.G., Kew. Cyclostyled. pp. 22, 112.

Cheek, M. (1994). Rx from the rain forest. Earthwatch Magazine. March–April 1994: 10-11.

Cheek, M. (1994). The Mount Cameroon Project. Kew Scientist, Issue 6: 5.

Cheek, M. (1996). The Mount Cameroon Protected area. The World of Plants, 131: 353. Asahi Shimbun, Japan.

Cheek, M. & Cable, S. (1996). Mount Cameroon: a biodiversity conservation joint project between the ODA and the Cameroon Forestry Department. In Jermy, Long, Sands, Stork, Winser (Eds.). Biodiversity Assessment: a guide to good practice. Vol. I. p. 136.

Cheek, M. & Cable, S. (1996). Progress in the compilation of plant species checklist for Mount Kupe, Cameroon. R.B.G. Kew. Cyclostyled. 5 pp.

Cheek, M. & Cable, S. (1997). Plant inventory for conservation management: the Kew-Earthwatch programme in Western Cameroon, 1993-96. pp. 29-38 in Doolan, S. (ed.) African Rainforest and the Conservation of Biodiversity. Proceedings of the Limbe Conference. Earthwatch Europe, Oxford. pp 170.

Cheek, M., Cable, S., Hepper, F.N., Ndam, N. & Watts, J. (1996). Mapping Plant Diversity on Mount Cameroon. pp. 110-120 in L.J.G. van der Maesen et al. (eds). The Biodiversity of African Plants. Kluwer Academic. Dordrecht.

Cheek, M. & Hepper, F.N. (1994). Progress on the Mt. Cameroon rainforest genetic resources project. Wildlife Conservation in West Africa. Symposium Proceedings, Nigerian Field Society, pp 15-19.

Cheek, M. & Ndam, N. (1996). Saprophytic plants of Mount Cameroon. pp. 612-617 in L.J.G. van der Maesen et al. (eds). The Biodiversity of African Plants. Kluwer Academic. Dordrecht.

Cheek, M. & Thomas, D. W. (1994). Mount Cameroon, pp. 163-166 in Davis, S.D., Heywood, V.H. & Hamilton, A.C. (eds), Centres of Plant Diversity, I. WWF & IUCN.

Courade, G. (1974). Commentaire des cartes. Atlas Regional. Ouest I. ORSTOM, Yaounde.

Davis, S.D. et al. (1986). Plants in danger: What do we know? IUCN, pp 55-57.

Dundas, J. (1947). Victoria Botanic Gardens. Farm and Forest. Jul.-Dec. pp 86-87.

Fairchild, D. (1928). Two expeditions after living plants. The Scientific Monthly. pp 97-127.

Fitton, J.G., Kilburn, C.R.J., Thirlwall, M.F. & Hughes, D.J. (1983). 1982 eruption of Mount Cameroon, West Africa. Nature 306. pp 327-328.

Freeth, S. (1992). The deadly cloud hanging over Cameroon. New Scientist, pp 23-27. August.

Giresse, P., Maley, J. & Brenac, P. (1994). Late quaternary palaeoenvironments in the Lake Barombi Mbo (West Cameroon) deduced from pollen and carbon isotopes of organic matter. Palaeogeography, Palaeoclimatology & Palaeoecology. Elsevier Science B.V. 107: 65-78.

Haig, E.F.G. (1937). The Cameroon Mountain, a general conspectus. The Nigerian Field. 6 (3): 118-128. & (4): 172-182.

Hall, J.B. (1973). Vegetational zones on the Southern slopes of Mount Cameroon. Vegetation. 27 (1-3): 49-69.

Hall, J.B. (1995). Botanical research needs and options relating to the Mount Cameroon Project. Cyclostyled, 80 pp.

Hawkins, P. & Brunt, M. (1965). The Soils and Ecology of West Cameroon. 2 vols. FAO, Rome.

Hawthorne, W.D. (1998). A Database (TREMA) of Mount Cameroon Project's botanical samples, with brief comments on the data. Report to DFID/MCP. Cyclostyled 59 pp & Appendices.

Hepper, F.N. (1987). Limbe (Victoria) Botanical Gardens, SW. Cameroon. Proceedings on the International Association of Botanical Gardens, 10th General Meeting, Frankfurt. pp 77-85.

Hepper, F.N. (1989). Report on the Limbe Botanic Garden Herbarium. Report to ODA/MCP. Cyclostyled, 4 pp.

Hepper, F.N. (1994). The centenary of Limbe Botanic Garden, SW. Cameroon. 1892-1992. The Nigerian Field. pp 8.

Holmgren, P.K., Holmgren, N.H. & Barnett, L.C. (1990). Index Herbariorum. Eighth Ed. New York Botanical Garden. 693 pp.

Hooker, J.D. (1864). On the plants of the temperate regions of Cameroons Mountains and Islands in the Bight of Benin collected by Gustav Mann. J. Linn. Soc. (Bot.) 7: 171-240.

Keay, R.W.J. (1955). Montane Vegetation and Flora in the British Cameroons. Proc. Linn. Soc. London. 165: 140-143.

Keay, R.W.J. (1959). Lowland vegetation on the 1922 lava flow, Cameroons Mountain. J. Ecology 47: 25-29.

Keay, R. W. J. & Hepper, F. N., eds. (1954-1972). Flora of West Tropical Africa, 2nd ed., 3 vols. Crown Agents, London.

Lack, H.W. (1997). Damals in Afrika. J. Museum 11(2): 80-81.

Lebrun, J-P. & Stork, A.L. (1991-1997). Enumeration des Plantes a Fleurs d'Afrique Tropicale. 4 vols. Ville de Geneve, Geneva.

Letouzey, R. (1985). Notice de la carte phytogéographique du Cameroun. Encyl. Biol. 49; 508.

Mace, G.M. & Stuart, S.N. (1994). IUCN Red List Categories. IUCN, Gland.

Maley, J. (1997). Middle to late Holocene changes in Tropical Africa and other continents: Palaeomonsoon and sea surface temperature variations. Palaeoenvironnments & Palynologie (CNRS/ISEM & ORSTOM). NATO ASI series, Vol. 149: 611-640

Maley, J. & Brenac, P. (1998). Vegetation dynamics, palaeoenvironments and climatic changes in the forests of western Cameroon during the last 28,000 years B.P. Review of Paleobotany and Palynology. 99: 157-187.

Meve, U. (in press). *Tylophora anomala* (Asclepiadacaeae)- a cytologically anomalous species. In: Smets, E., Ronse Decraene, L. P. & Robbrecht, E. (eds), Proceedings, 13th Symp. Morph. Anat. & Systematics, Leiden. Opera Bot. Belg.

Morton, J.K. (1953). Notes on Cameroons Commelinas. J. Linn. Soc. Bot. 55: 318-319

Morton, J.K. (1961). The upland flora of West Africa-their composition, distribution and significance in relation to climate changes. Extrait des comptes rendus de la IVe réunion plénière de l' AETFAT (Lisbonne et Coïmbre, 1-23 Sept. 1960).

Morton, J.K. (1972). Phytogeography of the West Africa Mountains. in Valentine, D.H. (ed.). Taxonomy, Phytogeography and Evolution. Academic Press, London. pp 221-236.

Morton, J.K. (1993). Chromosome numbers and polyploidy in the flora of Cameroons Mountain. Opera Bot. 121: 159-172.

Ndam, N. (1991?). Ecosystem of Africa's Highlands: Case Study of Central and West Africa Montane Forests. Unpublished Ms. 16 pp.

Ndam, N., Acworth, J., Tchouto, P., Healey, J. & Hall, J. (1997). Mt Cameroon Case Study: review of past and present methods used for biodiversity assessment and forest management. Case study presented at the DFID Forest Research Programme's Rapid Biodiversity Assessment workshop, Heriot-Watt University, Edinburgh 23-26th September. Cyclostyled.

Reynaud, I. & Maley, J. (1994). Histoire récente d'ine formation forestière de Sud-Ouest Cameroon à partir de l'analyse pollinique. C. R. Acad. Sci. Paris, Sciences de la vie/Life sciences, 317: 575-580.

Richards, P.W. (1963). Ecological notes on West African vegetation. J. Ecology 51 (2):123-149.

Richards, P.W. (1963). The upland forests of Cameroons Mountain. J. Ecology 51 (3): 529-554.

Ruxton. F.H. (1922). Volcanic eruptions on the Cameroons Mountain. London Topographical Society, pp 135-140.

Sosef, M.S.M. (1994). Refuge Begonias. PhD. thesis. Wageningen. 306 pp

Stoffelen, P., Cheek, M., Bridson, D. & Robbrecht, E. (1997). A New Species of *Coffea (Rubiaceae)* and notes on Mount Kupe (Cameroon). Kew Bull. 52: 989-994.

Thomas, D.W. (1984). Botanical exploration of the evergreen forests of SW Cameroon. Cyclostyled. 21 pp.

Thomas, D.W. (1986). Vegetation in the Montane Forests of Cameroon. pp. 20-27 in Stuart, S.N. (Ed.) Conservation of Cameroon Montane Forests. International Council of Bird Preservation. Cambridge, 263 pp.

Thomas, D.W. (1993a). Suitability of the Wonge river-Mokoko river area for inclusion in the GEF-Cameroon Program. Cyclostyled.

Thomas, D.W. (1993b). Korup Project Plant List (all species). Revised July 1993. Cyclostyled. 55 pp.

Thomas, D.W. (1994). Vegetation and conservation of the Mokoko River Forest Reserve, Cameroon. Report to ODA. Cyclostyled, 70 pp.

Thomas, D.W. & Cheek, M. (1992). Vegetation and plant species on the south side of Mount Cameroon in the proposed Etinde reserve, pp. 8-37 (with checklist appendix pp.7) in report on Limbe Gardens Conservation Project. R.B.G. Kew/Govt. Cameroon/ODA. Cyclostyled.

Watts, J. & Akogo, G.M. (1994). Biodiversity assessment and developments towards participatory forest management of Mount Cameroon. Commonwealth Forestry Review. 73 (4): 221-230.

White, F. (1983). Long-Distance Dispersal and the Origins of the Afromontane Flora. Sonderbd. Naturwiss. Ver. Hamburg. 7: 87-116.

Wood, N.P. (1980). An inventory of the Forest Resources of the Boa Plains. Report to the Cameroon Development Corporation. Microfiche Document. 47 pp.

A RED DATA LIST FOR MOUNT CAMEROON

This Red Data list is the first, we believe, for an African mountain, and the first, so far as we know to accompany a botanical checklist. It started from a candidate list produced from previous lists of endemic species and rarities for the area (Cheek 1992, Thomas & Cheek 1992, Cable & Cheek 1996) and from a subset of the W. Cameroon and Cross-River State BRAHMS database selected by Stuart Cable on the criterion of taxa of Mt Cameroon with highly restricted distributions. In addition, Orchids were assessed for inclusion from Sanford's unpublished manuscript for the Flore du Cameroun and Legumes from a list of Cameroonian endemic taxa extracted from the ILDIS database by Barbara Mackinder. All these taxa were then investigated more intensively by herbarium and library research before being included in this chapter or rejected.

CRITERIA FOR INCLUSION OF TAXA

Taxa (species, subspecies and varieties) given first priority for inclusion here are those that are strictly endemic to Mt Cameroon (including the foothill areas) or those that are nearly endemic, i.e. occurring on Mt Cameroon and one or two other areas, usually Bioko or the Bamenda Highlands, especially if the taxa are generally rare or if at the localities concerned their habitats are restricted or threatened.

We have excluded, for example, those species widespread and relatively common in Ethiopian and East African mountains but in West Africa known only from Mt Cameroon, e.g. *Brownleea parviflora* or *Aristea ecklonii,* since this would greatly increase the size of the list. Such taxa might be included in a future edition.

We have also excluded many relatively widespread species that occur on Mt Cameroon even though they are listed in the two recent global WCMC species treatments. Space and time were not sufficient to include these species, but we hope to refer to them in more detail in a future edition. There is no doubt that the species included here are of higher conservation priority.

The 1997 IUCN red list of threatened plants includes 87 taxa which occur in Cameroon (Lusty *pers. comm.*). Of these 87, 42 are recorded from Mt Cameroon in our checklist, of which 28 are included in our red data list for the mountain (out of a total of 116 taxa). The 14 species that we do not include, together with their IUCN ratings (pre Mace & Stuart 1994) are as follows: *Amanoa strobilacea* (I), *Anthospermum cameroonense* (V), *Bulbophyllum teretifolium* (V), *Bulbophyllum lupulinum* (V), *Craibia atlantica* (V), *Culcasia panduriformis* (V), *Dorstenia poinsettifolia* var. *angularis* (V), *Hesperanthera petitana* (I), *Justicia preussii* (I), *Medusandra richardsiana* (V), *Oricia trifoliolata* (I), *Peucedanum camerunensis* (I), *Pseudagrostistachys africana* (V), and *Warneckea memecyloides* (I).

The Cameroon subset from the World List of Threatened Trees (Lusty *pers. comm.*) lists 56 species, of which 21 are found in our checklist. 16 of these 21 do not occur in the IUCN red list. Only one occurs on our red data list. The 16 are as follows: *Autranella congolensis* (CR A1cd), *Baphia leptostemma* var. *gracilipes* (VU A1c, B1+2c), *Cola semecarpophylla* (VU A1c, B1+2c), *Cordia platythyrsa* (VU A1d), *Daniellia oblonga* (NE), *Diospyros crassiflora* (EN A1d), *Sterculia oblonga* (VU A1c), *Garcinia staudtii* (VU A1c, B1+2c), *Hemandradenia mannii* (LR nt), *Loesnera talbotii* (VU A1c, B1+2c), *Millettia macrophylla* (VU A1c, B1+2c),

Neoboutonia mannii (VU A1c, B1+2c), *Sorindeia mildbraedii* (NE), *Trichoscypha manniii* (VU A1c, B1+2c), *Trichoscypha preussii* (NE), *Uvariastrum zenkeri* (VU A1c, B1+2c).

The total of Red Data list species, from our list (116), together with the supplement from the 1997 IUCN list (14) and the World Tree list (16), thus reaches 146 taxa for Mount Cameroon.

IUCN RED DATA CATEGORIES: A GLOBAL APPROACH

IUCN assignments of the level of threat to taxa (Mace & Stuart 1994) must be geared to a particular geographical level. In this Red Data treatment, we have assigned IUCN categories on the basis of the global distributions of the taxa concerned, not just from a Cameroonian, nor indeed, a Mt Cameroonian perspective. As an example, we can again take *Brownleea parviflora* or *Aristea ecklonii*. Even if these were on the brink of extinction on Mt Cameroon, we would not include them in this list, though if this list were written only from the perspective of Mt Cameroon or West-Central Africa, this species would then receive the highest possible level of threat. However, the extinction of such a species from Mt Cameroon, although a loss for Mt Cameroon, Cameroon and West-Central Africa, would hardly affect the viability of the main population of the species in eastern Africa. In the end, this chapter is written from a global, rather than a Cameroonian perspective of conservation importance, since we feel that the plants of Mt Cameroon are of global importance.

IUCN CATEGORIES: APPARENT ANOMALIES

It may surprise some that certain lowland species endemic to the foothills of Mt Cameroon and found there in several lowland forests, e.g. *Psychotria sp. nov. 1* are given a higher IUCN rating here than more narrowly endemic species confined to much smaller areas higher on the mountain, e.g. *Isoglossa nervosa* restricted to a single site at Mann's Spring. This reflects the fact that the lowland forest habitat around Mt Cameroon is much more highly threatened than the montane habitat.

HERBARIUM SPECIMENS AS A PRIMARY SOURCE

The primary source for this Red Data list is plant specimens, principally those at the Kew herbarium, supplemented by personal observations of the taxa concerned (where available) on the mountain. All candidate species were assessed against specimens and the plant specialist for the relevant group was consulted where possible. For each taxon a range is given, in most cases including the number of historic collections (as an index of rarity) for each area in which the species occurs. Historic collections are defined here as those collected between 1860 and 1988, the last date being the beginning of the DFID-funded conservation project on Mt Cameroon. Royal Botanic Gardens, Kew is fortunate in having the largest and most comprehensive collection of specimens in the world for Mt Cameroon and possibly for Tropical Africa as a whole. All the most important of the historic collectors on the mountain (such as Mann, Johnston, Preuss, Schlechter, Maitland, Morton, Letouzey & Thomas) have left a set of duplicates at Kew. In most cases this is the top set, or the only surviving set. Of course there are gaps in the representation at Kew, and to fill these, the literature has been checked as far as possible, and some specimens have been borrowed from Paris and Berlin. Non-botanists may be interested to note that botanical literature is only a secondary source, albeit useful. Botanical literature is always founded on specimens, if it is to be of any use. Botanical literature always lags behind specimens. As soon as a checklist, Flora or monograph is published, it is the case that new collections are made which provide evidence to extend the range or reduce the rarity of the taxa included. Much data is only available from specimens and not from literature. For example, *Cyphostemma mannii* is recorded in both FWTA and Flore du Cameroun, as restricted to Mt Cameroon and Bioko. However, scrutiny of the herbarium shows that specimens of this species are also known from Kivu, in montane eastern Zaire, a fact published nowhere, since this genus has not yet been treated in the Flora of Central Africa. This extension of range for this species alters greatly its conservation priority (from a global perspective) on Mt Cameroon.

THE SPECIES TREATMENTS

Taxa are presented in alphabetical order, as in the checklist: species within genera, genera within species and families within the larger groupings of, respectively, Dicotyledons, Monocotyledons and ferns and Fern Allies.

For each taxon included, we have used the correct name (in the sense of the International Code of Botanical Nomenclature), following the most up-to-date and authoritative treatment or specialist determination (cited in full in the main checklist). The correct name is followed by an IUCN rating which indicates the category of threat and the justification, using IUCN criteria, for giving that rating, following Mace & Stuart (1994). In brief, the categories assigned here to the species included in this red data list are: "extinct", "threatened" (with extinction) or of "lower risk" (of extinction). Most of the species here fall under one of the three threatened categories: "critically endangered" (CR), "endangered" (EN) or "vulnerable" (VU). All of the "lower risk" taxa included here fall under the "near threatened" subdivision, so are listed as LR nt. A note on the range of the taxon follows the IUCN category. The main body of the taxon treatments takes the form of notes given on the detailed distribution, discovery and rarity of the taxon. It is here that notes are given on whether the taxon was rediscovered, or as was often the case, not rediscovered, in the detailed field surveys of 1991-94. Finally, there are sections on the habitat, threats, and management suggestions.

Many of the taxa included are new species, discovered in the course of the recent inventory work and as yet undescribed. These are only given full treatment if work on the description and formal publication is advanced. Some other probable new species which may, after further verification, also be formally published, are referred to only briefly in order to flag their existence. They may appear in more detail in subsequent editions of this Red Data list if further work shows that they warrant this.

'DIFFICULT' GENERA

Several genera have their major centres of diversity in the Mt Cameroon region. Some, such as *Cola, Begonia* and *Drypetes* are well worked out, with a firm foundation on which to work thanks to decades of first-class work by their respective specialists. In these cases, it is easy to elucidate which taxa, if any, are new, narrowly endemic or rare, i.e. of importance for inclusion in a Red Data list such as this. In other genera, such as *Trichoscypha, Leptonychia, Campylospermum* and *Psychotria*, the foundations are less solid, older, or lacking entirely. In these cases we have either elected to include only particularly distinct new taxa (*Trichoscypha* and *Psychotria*), or we have entirely omitted taxa from this Red List (*Leptonychia* and *Campylospermum*) since the basis for assessment seems so incomplete. There is no immediate prospect for correcting this difficulty with the last named genera. However for others in this category, new work will soon be at hand that will allow their resolution for taxonomic, and hence for conservation-assessment purposes. Work on the speciose tree and shrub genus *Rinorea* is almost completed by Dr Gaston Achoundong at Yaounde. Dr David Harris at R.B.G., Edinburgh has begun a two-year programme working out the species of *Aframomum*. When this work is available, it is most likely that taxa from these genera will appear in future Red Data works for the mountain.

ENDEMIC PLANTS

The 49 strictly endemic taxa (species, subspecies and varieties) of Mt Cameroon, i.e., those that occur nowhere else but this mountain, are listed below, grouped by habitat. There are no strictly endemic plant genera on Mt Cameroon, but Mt Cameroon is the only locality in Africa for two genera (*Oxygyne* and *Gastrodia*) otherwise confined to Tropical Asia/Australasia. Palaeoendemics (relictual taxa such as the species of *Stelechantha* and *Synsepalum*) are generally confined to lower or mid-altitudes. Neoendemics (recently evolved taxa of rapidly evolving, often speciose genera) appear more numerous than palaeoendemics.

Montane grassland endemics
These four taxa, for the most part, do not occur near the forest edge (for which see the grassland/forest endemics listed below), but occur in the grassland proper, mostly between 2100-3500 m alt. It is notable that no strictly endemic taxa occur at the summit (4095 m).

Silene biafrae (Caryophyllaceae)
Bulbostylis densa var. cameroonensis (Cyperaceae)
Hypseochloa cameroonensis (Gramineae)
Habenaria obovata (Orchidaceae)

Montane grassland/forest ecotone
The boundary between montane grassland and forest contains many species not found elsewhere on the mountain, apart from in gaps in the upper montane forest. Numerous temperate herb genera are more or less restricted to this belt, and not found in the forest shade, nor in the open grassland e.g. *Rumex, Geranium, Viola* and *Sibthorpia*. The following five species all seem to occupy this niche. The last two are epiphytic on montane trees at the forest edge but may yet be found deeper in the forest.

Isoglossa nervosa (Acanthaceae)
Myosotis sp. nr. vestergrenii (Boraginaceae)
Helichyrum biafranum (Compositae)
Angraecopsis cryptantha (Orchidaceae)
Genyorchis macrantha (Orchidaceae)

Submontane and montane forest
These 11 species do not occur in the species-poor upper montane forest which extends up to 2500 m alt., but in the 800-1800 m alt. belt, roughly corresponding with 'Cloud Forest'. Six species, a little more than half the total, are epiphytic.

Impatiens grandisepala (Balsaminaceae)
Impatiens sp. nov. 1 (Balsaminaceae)
Impatiens sp. nov. 2 (Balsaminaceae)
Oxygyne sp. nov. (Burmanniaceae)

Plectranthus dissitiflorus (Labiatae)
Afrardisia oligantha (Myrsinaceae)
Bulbophyllum modicum (Orchidaceae)
Disperis kamerunensis (Orchidaceae)
Polystachya albescens subsp. angustifolia (Orchidaceae)
Xiphopteris villosissima var. laticellulata (Fern)
Pteris ekemae (Fern)

Lowland forest
It is the lowland forest, (below c. 800 m alt.) of the foothill areas such as Bambuko, S. Bakundu, Mabeta-Moliwe (Bimbia-Bonadikombo), Onge and Mokoko, together with the forest of the coastal strip from Limbe-Idenau, that holds by far the largest part of Mt Cameroon's strictly endemic species, i.e. the highest priorities for conservation. Most of these species are found between sea-level and about 400 m alt. Of the 29 species listed below, 17 are newly discovered as a result of recent botanical inventory work and await publication as new species. The total number of lowland forest endemic species is likely to increase in the next edition, since it is likely that more undescribed endemic species will be identified and published from the specimens known from the mountain: not all "species novae" of the checklist are included here.

Trichoscypha bijuga (Anacardiaceae)
Trichoscypha camerunensis (Anacardiaceae)
Trichoscypha sp. nov. (Anacardiaceae)
Ancistrocladus sp. nov. (Ancistrocladaceae)
Isolona sp. nov. (Annonaceae)
Piptostigma sp. nov. (Annonaceae)
Polyceratocarpus sp. nov. (Annonaceae)
Begonia quadrialata subsp. dusenii (Begoniaceae)
Oxygyne triandra (Burmanniaceae)
Salacia sp. nov. 1 (Celastraceae)
Salacia sp. nov. 2 (Celastraceae)
Sibangea sp. nov. (Euphorbiaceae)
Beilschmiedia preussii (Lauraceae)
Crudia bibundina (Leguminosae-Caesalpinoideae)
Ardisia schlechteri (Myrsinaceae)
Eugenia kameruniana (Myrtaceae)
Gastrodia africana (Orchidaceae)
Genyorchis platybulbon (Orchidaceae)
Psychotria sp. nov. 1 aff. dorotheae (Rubiaceae)
Psychotria sp. nov. 2 aff. bidentata (Rubiaceae)
Rutidea sp. nov. (Rubiaceae)
Sacosperma sp. nov. (Rubiaceae)
Stelechantha sp. nov. (Rubiaceae)
Synsepalum brenanii (Sapotaceae)
Cola praeacuta (Sterculiaceae)
Cola sp. nov. aff. flavo-velutina (Sterculiaceae)
Cola sp. nov.1 aff. philipi-jonesii (Sterculiaceae)
Asplenium sp. 8 (Fern)
Asplenium sp. 9 (Fern)

Of the 49 strictly endemic taxa listed above for Mt Cameroon, nine have not been seen for over 60 years. This despite botanical investigation on the mountain in each intervening decade and intensive inventories over many parts of the mountain in the last ten years. Although it cannot be ruled out that some, perhaps all of these nine survive in some forgotten corners of the mountain, they may well be extinct. We list them here together with the dates on which they were last collected. Repeated efforts since 1991 to rediscover *Oxygyne triandra*, type of the genus, have failed. These efforts include those of a dedicated Oxford Undergraduate expedition in 1993. The other eight species have only been identified as of importance recently, and as yet, no special effort is thought to have been made to refind them. We recommend that this be rectified. The following nine species are the rarest of the rare and, should they ever be refound, would require protection.

Oxygyne triandra (Burmanniaceae) 1905
Plectranthus dissitiflorus (Labiatae) 1891
Beilschmiedia preussii (Lauraceae) c. 1891
Crudia bibundina (Leguminosae-Caesalpinoideae) 1928
Ardisia oligantha (Myrsinaceae) 1906
Ardisia schlechteri (Myrsinaceae) c. 1905
Eugenia kameruniana (Myrtaceae) c. 1905
Disperis kamerunensis (Orchidaceae) 1930
Gastrodia africana (Orchidaceae) c. 1890

Eight of the above nine species occurred, and may still occur, in lowland forest around the foot of the mountain. Most specimens of the nine species are a legacy from the days when German botanists were active on the mountain at the turn of the century. All of the strictly endemic species known from the collections of the earliest collector, Gustav Mann, in 1861 and 1862 have been rediscovered. Endemic taxa collected by Mann are not listed amongst those that are possibly extinct because he collected mostly in the upper altitudes and the main threat is in the forest at lower altitudes.

Sixty years is an arbitrary figure, although the one generally used for making assumptions about whether a species is extinct or not. However, several other strictly endemic species have also not been found, despite botanical surveys in the last ten years. These are also a cause for concern:

Impatiens grandisepala (Balsaminaceae) 1979
Myosotis sp. nr. vestergrenii (Boraginaceae) 1952
Helichrysum biafranum (Compositae) 1952
Hypseochloa cameroonensis (Gramineae)
Bulbophyllum modicum (Orchidaceae) 1965
Genyorchis platybulbon (Orchidaceae) c. 1950
Synsepalum brenanii (Sapotaceae) 1948
Xiphopteris villosissima var. laticellulata (Fern) 1975
Pteris preussii (Fern) 1975

The threats to individual species are generally those to the plant communities or habitats in which the species grow, and the threats vary widely on the mountain. We have chosen, for convenience, to discuss threats on an altitudinal basis, in three tranches: lowland habitats, i.e. sea-level to c. 800 m alt.; mid-elevation habitats, i.e. c. 800–2200 m alt. and upland habitats, i.e. c. 2200–4100 m alt.

Lowland habitat around Mount Cameroon, particularly the evergreen forest in the foothill areas of Bambuko (only partly evergreen forest), Mokoko, Onge, the coastal strip between Idenau and Limbe, S. Bakundu and Mabeta-Moliwe (also known as Bimbia-Bonadikombo), is undoubtedly richest in rare endemic species. This habitat is also the most highly threatened as well as being the habitat that has suffered most from destruction in the past, e.g. most destructively, conversion to plantations. Most of the endemic species of Mt Cameroon that are rated by us as possibly extinct (e.g. *Oxygyne triandra, Ardisia schlechteri, A. oligantha, Crudia bipindina* and *Gastrodia africana*) are from this habitat. If these species are indeed extinct, their extinction can be attributed to habitat destruction. Fortunately, some other species thought to be endemic to Mt Cameroon (e.g. Thomas & Cheek, 1992), such as *Afrothismia pachyantha* and *Amorphophallus preussii*, have since been discovered at Mt Kupe, so their possible extinction on Mt Cameroon has not resulted in their global extinction, as is feared to be the case in the previous examples.

Although the lowland area around Mt Cameroon is mostly evergreen forest on old volcanic rocks, the species composition varies widely from place to place, depending presumably on variations in soil type, rainfall and phytogeographic boundaries (the flora on the west side of the mountain differs significantly from that on the east). On the west side soils are overlain by sand, particularly in the Mokoko area. Small sedimentary pockets occur in both east and west. The coastal strip is based on recent volcanic rock. A littoral community (with *Ipomoea pes-caprae, Calophyllum inophyllum, Phoenix reclinata* etc.) is found along the seaward edge. No sea grasses have been found, perhaps due to the turbidity of the water or to the severity of wave action. Rainfall varies from 10-15 m at Debundscha, and 3-4 m on the seaward side generally, to 2 m in the rainshadow to the northeast, where semi-deciduous forest is present in the S. Bakundu Forest. Species diversity in the lowland area is enhanced by patches of mangrove (Mabeta-Moliwe), freshwater swamp forest (Mabeta-Moliwe and Onge), recent lava-flow (e.g. coastal strip at Bibundi and at Ekona north of Buea), *Hypselodelphys* thicket at upper altitudes of the lowland zone above Njonji and extending westwards, and *Borassus* palm savanna in the Mokoko area. Whilst none of these habitats is particularly speciose, they do contain many taxa not found in lowland evergreen forest and the ecotone at the boundaries between habitats is believed to be important for the survival of some scarce species, e.g. *Oxystigma mannii* at the boundary of mangrove and lowland forest in Mabeta-Moliwe.

We now review threats to locations of conservation interest around the mountain, beginning in the north, and moving anticlockwise.

The Bambuko Forest Reserve, on the north and northwestern flank of the main massif, is mostly *Hypselodelphys* thicket with a large percentage of *Pennisetum purpureum* (known locally as 'Elephant bush', with patches of forest and savanna, including some *Borassus* savannah (Tchouto *pers. comm.*). No report has been found of any survey conducted there apart from the unpublished photographs and notes of Keay, and although it has been visited by Keay and Thomas, it remains without a botanical inventory. However, Tchouto of the Mt

Cameroon Project is currently conducting a land use and NTFP survey in the area (July 1998). *Uvariodendrum fuscum*, a near endemic, was reported as "very common in this area" by Keay in 1958 (Keay in FHI 37485). The lower slopes are now extensively cultivated for market gardening by businesses from Douala and the produce is sent to Douala and exported to Gabon. Further clearance of natural vegetation is a considerable threat. Farming is the main activity in the area and with the significant recent rises in prices for cocoa, small holder plantations are expanding. These plantations bring together large numbers of people as plantation workers who also exert secondary pressure on natural resources. Encroachment by commercial and subsistence farmers has led to the conversion of the north-eastern part of the reserve to farmlands. Furthermore, the construction of an all-season good road (from Idenau to Muyuka) has permitted the easy transportation of foods, cash crops, timber and other forest products to markets. Consequently, farm expansion is likely to continue and additional pressure will be put on the forest (Tchouto *pers. comm.*).

The Mokoko and Onge "western forests" areas are the main areas of forest in the western foothills. The areas of less hilly terrain are being threatened by encroachment from subsistence and small scale commercial agriculturalists. These will soon be displaced into the western forests as a "knock on" effect of imminent plantation expansion in the Boa Plains area (Tchouto *pers. comm.*). Mokoko is gazetted as a forest reserve. Environmental Impact Assessments were executed in these areas in 1997. Together these forests today constitute the most intact and extensive foothill forest of Mt Cameroon and contain several species, e.g. of *Cola*, not found elsewhere in the world. These forests show clear phytogeographic affinities with the southern Korup (e.g. presence of *Oubanguia alata*, *Cola semecarpophylla*, *Belonophora sp. nov.* and high diversity in Caesalpinoideae and Chrysobalanaceae). They also have species in common, though far fewer, with forests immediately to the South of the Sanaga. These particular species do not occur in the Mt Cameroon coastal strip or in the eastern and northern foothills, e.g. *Cola letouzeyi* and *Brazzeia soyauxii*, for both of which Onge-Mokoko is the western limit. These forests are still relatively poorly known botanically and further inventory is likely to yield a larger number of species than is known at present.

The coastal strip between Idenau and Limbe has mostly been denuded of its natural vegetation below the 200 m contour. It has been replaced by *Elaeis guineensis* (oil palm) plantation feeding the mill at Idenau. A strip of natural vegetation still reaches the coast at the 1922 lava flow at Bibundi which is still too inhospitable for oil palms. Surviving forest in the coastal strip above the 200 m contour is the main site for e.g. *Trichoscypha sp. nov.* and *Solenostemon sp. nov.?* and is the last and best hope for refinding the long-lost species *Crudia bibundina, Ardisia schlechteri* and *Gastrodia africana*. Land currently leased to the largest state-owned plantation company (CDC: the Cameroon Development Corporation) covers an area up to 800 m altitude on the West Coast. Unless this lease is reviewed, it is possible that in future, development of new crops, or change in the global commodities market will lead to the expansion of plantation agriculture well above the 200-300 m maximum that has so far been achieved with oil palm. Such expansion would remove all the West Coast lowland forest, which appears to be unique both in the Mt Cameroon region and globally.

A few scraps of natural vegetation, a littoral community, occur along and above the beach, but this has been and will continue to be steadily destroyed by building and touristic developments along the tarmacadamed road from Limbe to Idenau. An oil refinery occupies part of this area.

The whole of the eastern flank of the main massif, up to about 800 m altitude, is now bereft of natural vegetation, it is generally believed. However, a survey to locate any persisting fragments of natural vegetation might well prove worthwhile. Perhaps 80-90% of the eastern foothills are converted to plantations and other agricultural operations, including farm fallow.

The Mabeta-Moliwe (Bimbia-Bonadikombo) forest is the most extensive and least disturbed of the surviving eastern foothill fragments, and probably the constitutes the last refuge for the species that once occurred to the east of the mountain. The forest along the one hour drive to Douala in the east from Limbe is now entirely replaced by *Musa, Hevea* and *Elaeis* plantation. Mabeta-Moliwe is immediately threatened to the west by extensive, sometimes illegal, small-farmer plots run by people in Limbe, immediately adjacent. More serious is the fact that almost the entire forest stands on land reserved for plantation expansion. Several species are restricted to this forest. Mabeta-Moliwe is believed to be the only forest on the mountain with direct sea-frontage (lowland rainforest overhanging the Atlantic) and Mangrove. Prospects for its future survival, however, seem bleak.

The Southern Bakundu Forest is gazetted as a production forest reserve, although it was partly inventoried by the Cambridge Univ. expedition of 1948, and further investigated botanically in the 1950s after a reconnaissance by Keay in 1951. Official Forestry plant collectors (e.g. most famously Onochie, Daramola, Latilo) were sent by Keay (Keay, unpublished notes) from the Forest Herbarium Ibabdan. It has not been inventoried in the life of the current DFID project, since it has never been included in its remit. Over the last decade illegal logging, and in the south, Nigerian yam farmers, are believed to have taken their toll. In the last few years, an ITTO experimental silvicultural project has been executed. This has been aimed at suppression of non-timber species to accelerate growth of timber species. Fortunately for the forest's rare plants, this programme appears to have been implemented only on a pilot basis. Some replanting (species unknown) and research on NTFP s is taking place. More importantly this project is trying to resecure the boundaries (now marked with *Tectona grandis*, Teak) and remove encroaching farmers (Tchouto *pers. comm.*). These activities can only help conserve this species-rich and historically important forest and must be welcomed. This forest is the type locality of *Medusandra richardsiana* (type species of the *Medusandraceae*, discovered by Brenan with the Cambridge expedition in 1948), and the only known site of species such as *Synsepalum brenanii*.

Mid elevation habitat (c. 800–2200 m alt.) is confined to the main massif (including Mt Etinde), of the mountain. It consists mostly of submontane and montane forest in the southwestern, eastern and southern sides of the mountain, and *Hypselodelphys thicket* with some forest and *Borassus* savanna in the northwestern and northern sides, i.e. principally in the Bambuko area. A substantial area of this habitat was covered by the Sept-Oct 1992 Etinde botanical inventory which also extended into the coastal zone below and the montane grassland zone above. The area of the mid elevation habitat covered by this survey was the Mt Etinde area, the southwestern flank of the main massif from Etinde to Idenau, and the southeastern flank from Mt Etinde to Buea.

It is not clear whether the market gardens in Bambuko extend up to the 800 m contour. Generally, the northwestern and northern flanks are unpopulated by villages, apart from some scattered settlements along the perimeter road at the foot of the main massif, and there seems to be little threat to natural habitat. These extensive areas, having fallen outside the remit of the Mount Cameroon Project, remain very poorly known botanically. Extensive areas have not yet been visited by botanical collectors.

On the eastern side of the mountain, the villages of the Bakweri extend up into the submontane area in a line at about the 800 m contour, culminating in the former Capital of Cameroon, Buea. Around and below these villages, there is little or no undisturbed natural vegetation, but only farmbush. In the Tole and Saxenhof areas, extensive tea plantations produce most of the Cameroon crop of this product. Above the 800 m contour, the submontane forest is largely converted to cocoyam (the local staple) farms up to about 1000 m altitude. Above that, farms, felling and heavy disturbance occur in patches as high as 1800 m altitude. At the southern end of the main massif is the subpeak of Etinde, or "Small Mount Cameroon" (c. 1700 m) which geologically, has little connection with the main mountain, being much older, non-volcanic, granitic, with steep, angular ridges and submontane/montane forest of different composition to that on the main massif, for example, with conspicuous Sapotaceae (see Thomas and Cheek 1992). Villages here extend up to about 500 m alt., e.g. Upper Boando and Ekonjo, both with very small permanent populations and with relatively little impact on the adjoining forest compared with the Mapanja to Ekona Lelu line of villages on the eastern side. The southwestern slopes of the main massif, from Etinde to Idenau, are populated only just above sea-level and farming does not reach the 800 m. However, although it is feared that plantations, far more destructive of natural vegetation, are being considered for extension to the 1000 m contour. Such an extension would endanger the best preserved area of submontane and montane forest on the mountain.

The upper elevation habitats (c. 2200–4100 m) are montane and subalpine grassland. This area is remarkably free of human disturbance, the principal activity of man in these habitats being to hunt game. Cultivation and grazing by cattle, sheep and goats are absent. There has been some concern that annual fires set by hunters to clear paths of overgrown grass and possibly to drive game, and to facilitate honey collection might threaten the survival of some plant species, and reduce the extent of the forest at higher altitudes. The tree-line varies greatly in altitude. Generally, in the southern corner it lies about the 2200 m contour, though tongues of forest extend up the mountain to 2400 m alt. and possibly to 2500 m alt. in the area around the hunting settlement of Wikile, west of Mann's Spring. Above Bokwango, east of Etinde, the grassland dips to only 1900 m alt. Above Buea it also drops to a lower altitude, c. 1400 m alt. Photographs taken by Keay in conjunction with his notes, show that in the Bambuko area, tongues of forest extend only up to c. 1650 m alt. On African mountains at this latitude, forest can be expected to reach up to about 3000 m, as on Mt Oku, c. 250 km to the northeast,. There is no doubt that the treeline is depressed on Mt Cameroon, and that this is caused by fire. However, fire is a natural phenomenon, brought about by lightning strikes in the dry season and, at roughly 20 year intervals, by volcanic eruptions on this mountain. Between December and April the grassland vegetation is usually tinder-dry and highly flammable. This is, no doubt, accentuated by the exceedingly free draining volcanic soils which allow no persistent streams on the mountain above about eight hundred metres elevation. This has contributed to the depression of the tree line. As suggested by their persistent underground rootstocks many, if not most of the grassland herbs appear to be fire-adapted (e.g. *Succisa trichotocephala, Cyanotis barbata, Cynoglossum amplifolium* and *Crassula vaginata)*. They may well require fire for their survival, to remove the thick dense growth 45-60 cm thick of *Gramineae* culms that blanket the lower reaches of the montane grassland by the end of the wet season. Morton (*pers. comm.*) reports that he has observed in March and April, at the end of the dry season, a flush of flowering after fires in what he considers to be a pyrophytic community. *Hypoxis camerooniana,* another grassland species, is undoubtedly a pyrophytic species according to the *Hypoxis* specialist Nordal (*pers. comm.*).

DICOTYLEDONS

ACANTHACEAE

Afrofittonia silvestris Lindau
VU A1c+2c
Range: Cross River State Nigeria (five collections), Bioko and S.W. Cameroon (c. three pre-1988 collections).
In Cameroon known only in Korup (where locally abundant) apart from Mt Cameroon.
The inventories of the 1990s yielded a further nine specimens of this species from Mt Cameroon, from all lowland forest areas, except from Mokoko. It seems relatively abundant in Onge.
A monotypic genus, its closest relative *Fittonia albivenis* (Lindl. ex Veitch) Brummitt is native to South America (Columbia, Peru & Ecuador) and is valued for amenity horticulture and used for treating migraine.
Habitat: lowland evergreen forest, often in damp, shady spots
Threats: forest clearance
Management suggestions: none

Isoglossa nervosa C.B.Clarke
VU D2
Range: Mt Cameroon (one pre-1988 collection).
First collected in Dec. 1862 at 7,000 ft. by Mann (type collection). Not recollected until 130 years later (*Thomas* 9333). Last collection known from W. of Mann Spring in 1993 (*Cheek* 5355). Only known from these three collections.
Habitat: montane forest 2000-2500 m.
Threats: unknown
Management suggestions: investigate range. Is this species confined to Mann Spring or is it widespread in the montane forest? What is the population size? Is regeneration occurring? Could this be a fourth mass-flowering, long-cycle Acanthaceae on the mountain (this would help explain the long interval before recollecting). Potential threats to this species need be evaluated. Confusion with *Isoglossa glandulifera* (lower altitudes, widespread) should be avoided.

Sclerochiton preussii (Lindau) C.B.Clarke
EN B1+2e
Range: Obudu Plateau (one collection), Mt Cameroon (one pre-1988 collection), Mt Kupe.
First collected "near Buea" in Nov. 1891. Rediscovered on Mt Cameroon 101 years later in the saddle of Mt Etinde-Mt Cameroon (*Wheatley* 596). Recently found in the 1960s at Obudu (1960s) and Mt Kupe (1992, 1995). Only six collections known, two of which on Mt Cameroon.
Habitat: submontane/montane forest c. 1100-1300 m.
Threats: unknown on Mt Cameroon, but stems cut for trap "springing sticks" on Mt Kupe (*pers. obs.*).

Management suggestions: data needed on range and population size of this species. Evaluation of threat posed by exploitation for trapping advisable.

AMARANTHACEAE

Achyranthes talbotii Hutch. & Dalziel
EN A2c, B1+2c
Range: Cross River State Nigeria (one pre-1988 collection) and Mt Cameroon (three pre-1988 collections).
Apart from the type collection in Nigeria, all other collections of this species are from rivers in the north (S. Bakundu) and west (Onge) of Mt Cameroon. About seven new collections are known. The proposed conversion of the western forests to plantation, is likely to threaten the plant communities of the rivers that drain the area, such as the Onge (one of the main habitats for this species) bringing more silt and fertilizer into the water, for example.
Habitat: fast-flowing rivers, growing half submerged, rooting among boulders (*pers. obs.*)
Threats: logging
Management suggestions: none.

ANACARDIACEAE

Trichoscypha bijuga Engl.
CR A1c, B 1+2abc
Range: Bioko (extinct?) and Mt Cameroon (probably now strictly endemic, one pre-1988 collection).
Although discovered on Bioko, not recollected there this century and possibly extinct there due to clearance of almost all forest below 1000 m alt. On the mainland known from Likomba (where now probably extinct) and Mabeta-Moliwe where numerous specimens (c. 25) were collected in 1992.
Habitat: lowland evergreen forest
Threats: forest clearance for agriculture, particularly plantations
Management suggestions: protection measures are needed.

Trichoscypha camerunensis Engl.
CR A2ac, B1+2abc
Range: Mt Cameroon (one pre-1988 collection).
Known only from three collections: S. Bakundu area (*Preuss* 99, type) and the Etinde coastal strip (*Kwangue* 102 and *Tchouto* 134).
Habitat: lowland evergreen (or semi-deciduous ?) forest.
Threats: forest clearance for agriculture, particularly plantations. Forestry management practices in removal of "weed" species.
Management suggestions: plants need to be pin-pointed and the population assessed. Propagation for reintroduction may be advisable.

Trichoscypha sp. nov.
CR A1c+2ac,
Range: strictly endemic to Mt Cameroon
Known only from six collections mostly from the coastal strip between Idenau and Limbe, (Etinde) but also from Onge, mostly made in 1992 and 1993. Relatively common inside its range. Not known from Mabeta-Moliwe, nor Mokoko.

Habitat: lowland evergreen forest.

Threats: forest clearance for agriculture, particularly plantations.

Management suggestions: habitat protection.

ANCISTROCLADACEAE

Ancistrocladus letestui Pellegr.

VU B1+2a

Range: S. Cameroon & Gabon, outlier on Mt Cameroon.

Habitat: lowland evergreen forest.

Threats: forest clearance for timber and agriculture.

Management suggestions: known only from a single sterile specimen from Mokoko collected in 1994. Recollection to confirm identity needed.

Ancistrocladus sp. nov.

CR A1c+2c, C1+2a

Range: Mt Cameroon.

Discovered in 1992. Known from only c. six plants in Mabeta-Moliwe (*Cheek et al.* 3436, 3915 and *Baker* 325). Accepted as new species by Gereau (MO, monographing genus). Sterile specimens from Rumpi Hills and Douala-Edea may prove conspecific if flowering material can be obtained.

Habitat: lowland evergreen forest.

Threats: clearance for agriculture.

Management suggestions: protection of known plants, propagation for reintroduction.

ANNONACEAE

Artabotrys sp. aff. rhopalocarpus Le Thomas (*Cheek, Elad & Ndumbe* 3470, Mabeta Peninsula, July 1992; *Wheatley* 644 from the coastal strip may also belong here.).

This taxon may merit inclusion in the next edition of this Red Data list. Ideally, more collections are needed before it is evaluated further.

Boutiquea platypetala (Engler & Diels) Le Thomas

EN A1c+2c

Range: endemic to coastal Cameroon and Bioko.

This monotypic genus is close to, and may yet be united with *Neostenanthera* Exell. It was erected by Le Thomas (Adansonia Ser. 2, 5: 531 (1966)) and is known from 8 specimens, but only five localities. The type is from Bipindi and it is also known from Kribi and Korup, besides Bioko. On Mt Cameroon, it is known only from Nkeng 69 (June 1992, Idenau). A shrub or tree 3-5(-10) m tall, this species has conspicuous flowers and fruits, and is not easily overlooked when fertile. Consequently, the sparse and scattered collections probably indicate rarity, rather than undercollection.

Habitat in lowland evergreen forest.

Threats: clearance of forest for timber extraction, followed by agriculture and particularly plantations. Plantations in the Idenau area are believed to be scheduled for expansion soon. Lowland forest is probably the most heavily threatened vegetation type on Mt Cameroon.

Management suggestions: an attempt to be made to rediscover this tree. Is it a singleton or are there scattered individuals in this area? Incorporation of some individuals in a heavily protected area might be investigated.

Isolona sp. nov.
CR A1c+2c
Range: Mt Cameroon
This small to medium sized tree is known from two collections, *Watts* 687 (Bambuko F.R.) and *Wheatley* 501 (Etinde). More collections needed in order to describe. Identified as new and to be described by Stuart Cable.
Habitat: lowland evergreen forest
Threats: forest clearance for timber and agriculture.
Management suggestions: more information on the range, number of individuals, and regeneration of the population of this species is required before suggestions on management can be made.

Piptostigma sp. nov. aff. glabrescens Oliv.
CR A1c+2c
Range: Mt Cameroon
Identified as new, to be described by Stuart Cable, this cauliflorous, medium-sized tree was discovered during the inventories of the western forests in 1993-94. It is known from three collections in Onge and one in Mokoko (see checklist).
Habitat: lowland evergreen forest
Threats: forest clearance for timber and agriculture.
Management suggestions: more information on the range, number of individuals, and regeneration of the population of this species is required before suggestions on management can be made.

Polyceratocarpus sp. nov.
CR A1c+2c
Range: Mt Cameroon
Identified as new and to be described by Stuart Cable when more material is available, this species is a slender tree to 12 m tall, known only from the Mokoko/Bambuko boundary (*Watts* 627, March 1993, Mundongo).
Habitat: lowland evergreen forest; 320 m alt.
Threats: forest clearance for timber and agriculture.
Management suggestions: more information on the range, number of individuals, and regeneration of the population of this species is required before suggestions on management can be made.

Uvariodendron fuscum (Benth.) R.E.Fries
LR nt
AGIN-JA (Asaka Lang.) young leaves used for soup (*Cheek* 5145).
Range: Bioko (two collections) and Mt Cameroon (four pre-1988 collections)
This lowland to submontane tree appears widespread around the mountain. Keay (in FHI 37485) reports it in Feb. 1958 at the eastern boundary of the Bambuko F.R. as "very common in this area and extending from 3-5,000' alt.". It is also known from Buea "above upper farms, fairly frequent, 4,000' " (*Maitland* 453 in March 1929) and from the 1992 inventory at Mabeta-Moliwe, e.g. plots TD-1 (W693) and TD 12 (W 343). In the coastal strip, it has been collected above Small Koto (*Thomas* 4469, March 1985), and at Onge descends to near sea-level near Enyenge (*Cheek* 5145, Oct. 1993).
The collections on Bioko are both historic (*Mann* 308, and *Mildbraed* 6428) and there is no evidence of a recent sighting.

Care should be taken not to confuse this species with its congeners, e.g. *U. giganteum, U. mirabile* or *U. connivens.*

Habitat: submontane, less usually lowland evergreen forest; sea-level to 1500 m alt.

Threats: there is no evidence that traditional use is a threat to this species, but this should be investigated. Although *U. fuscum* is restricted in its range to Mt Cameroon and Bioko, within its range, certainly on Mt Cameroon, it seems widely spread and is common at some locations. Compared with many other species with such a range, it seems in no immediate danger of extinction.

Management suggestions: in view of the lack of immediate threat to this species, protection measures seem unnecessary at the moment. Monitoring of populations is advised.

ARISTOLOCHIACEAE

Pararistolochia preussii (Engl.) Hutch. & Dalziel
CR A1c+2c C2a
Range: Mt Cameroon (two pre-1988 collection) and S. Province, Cameroon (2 collections).
Historic material of this genus at Kew is on loan to Gonzalez in New York, so we must rely largely on his comments and on Keay's notes in FWTA. This species was only known from a leafless specimen, *Preuss* 108 (type, assumed destroyed at Berlin), until a collection attributed to this species by Keay was found in the S. Bakundu area (*Brenan* 9484, collected in March 1948). Recently this species was recollected in Mabeta-Moliwe, on the southern base line, at 400 m from the origin (*Sunderland* 1197, fl. 6 April 1992 and Wheatley 137, fl. 9 April 1992). According to Gonzalez (*pers. comm.*) a further specimen of this species is known from the Kribi-Campo area (*de Wilde* 8011, WAG) and Poncy in Adansonia 17: 484, 1978 lists in addition a collection from Ebolowa: *Letouzey* 10053 (P). This species is undoubtedly very rare. More work is needed on delimitation between this species and *P. ceropegioides*, also on Mt Cameroon and also rare.
Habitat: disturbed lowland forest.
Threats: forest clearance. All known sites of this species are under pressure from timber extraction or forest clearance.
Management suggestions: this ornamental species is a prime candidate for cultivation and multiplication at Limbe Botanic Garden, with a view to reintroduction to the wild at secure sites.

Several collections from submontane forest of what appears to be a new orange-flowered species of *Pararistolochia* have been made on Mt Cameroon and Mt Kupe (*P. sp. aff. ceropegioides* (S.Moore) Hutch. & Dalziel of the checklist), but more work is needed to confirm this.

ASCLEPIADACEAE

Neoschumannia kamerunensis Schltr.
CR A1c+2c, B1+2c
Range: Mt Cameroon (one pre-1988 collection), Mt Kupe, Ivory Coast (one collection) and C.A.R. (one collection).
Discovered at Man O War bay near Mabeta-Moliwe in 1899 by Schlechter, it has not yet been rediscovered there. One specimen was located in the 1960 s in Ivory Coast, but it was not until March 1995 that an Earthwatch sponsored expedition rediscovered this extremely rare vine above Likombe, in farmbush. Subsequently, one of the original discoverers, Etuge (the other

being Meve, an Asclepiad specialist) found a population of seven plants at Mt Kupe in forest at the edge of a farm. Since Oct. 1995 the area at Mt Kupe has been rented from the local farmer in order to protect the plant. Finally, David Harris discovered this species in the Zangha-Zangha reserve in southern C.A.R. in 1996 (references in checklist).

Habitat: lowland to submontane forest, withstanding and perhaps benefiting some disturbance; sea-level to 1000 m alt.

Threats: although known to withstand some disturbance, presumably being able to regenerate after cutting, there is no doubt that intensive agriculture and tree clearance would destroy this rare species.

Management suggestions: the now well-known plant above Likombe should be protected by agreement with the local landowner. A search for more individuals in the same area was carried out without success in 1995. This species may be a candidate for propagation at Limbe Botanic Garden for reintroduction to a protected, managed area or areas.

Rhynchostigma racemosum Benth.

VU A1c

Range: widespread, but rare Bioko (4 collections), Mt Cameroon (one pre-1988 collection, possibly now extinct), Bakossi Mts (1), Zaire (1), Rwanda (3), Burundi (4) and Uganda (1).

A monotypic genus, related to *Secamone*, not rediscovered on Mt Cameroon since the type collection (*Mann* 1273, 4,500' in Feb.). The lowest altitudinal record is from the edge of Lake Edip (Bakossi Mts) where a single plant was seen twining on a shrub 2 m tall, and flowering abundantly in Feb. 1998 (*pers obs*).

Habitat: montane forest 1200-2400 m alt.

Threats: forest clearance for Agriculture.

Management suggestions: possibly extinct on Mt Cameroon, but further searching may yet result in its rediscovery.

Tylophora anomala N.E.Br.

LR nt

Range: East and South-East Africa, Mt Cameroon

Discovered on Mt Cameroon in March 1995 by Meve on an "Earthwatch" expedition at four sites above Likombe, the Mt Cameroon population is genetically distinct (decatetraploid cytotype, 2n+ 132-154) from the numerous E. African populations (hexaploid cytotype, 2n= 66). There are slight floral differences also, but not enough for taxonomic recognition (Meve, in press).

Habitat in montane forest; 1300m, 1450 m, 1600m and 1750 m altitudes

Threats: unknown.

Management suggestions: this species is not such a high priority for action as are the threatened strictly endemic species, for example. However, it would be useful to obtain data on the range of this species on the mountain, and investigate whether it faces any threat.

Tylophora cameroonica N.E.Br.

LR nt

Range: S. W. Cameroon (two pre-1988 collections), Kivu (1 collection) and Uganda (2 collections).

First collected in "Rio del Rey, Camaroons" by Johnson before June 1887 (but the notes are poor and the habitat and altitude of Rio Del Rey are unlikely. Johnson also collected on Mt Cameroon and this may be the real origin of his specimen)), this species was described by Brown in 1895. It was first reliably found on Mt Cameroon on 14 Feb. 1927, near Buea

(*Dalziel* 8236) and rediscovered many decades later (*Tchouto* 60). Curiously, the first collection of *T. cameroonica* from outside of Cameroon was made by Maitland (*Maitland* 1073, May 1928, Makerere, Kampala) a year before he became superintendent at the Victoria Botanic garden at Mt Cameroon. It is also known from Uganda by another collection (*Chandler* 2789, April 1939) and by one from Kivu (*Troupin* 3719, June 1957).

Habitat: submontane forest 850-1200 m alt.

Threats: forest clearance

Management suggestions: an attempt should be made to refind and safeguard the site of *Tchouto* 60. Propagation should be considered. The rarity of this plant on the mountain is suggested by the fact that Meve, a noted Asclepiad specialist, apparently did not find this plant during his time in the area.

Tylophora urceolata Meve

VU B1+2c

Range: Mt Cameroon, Bioko and the Uluguru Mts of Tanzania

This species was discovered on Mt Cameroon on 23rd Feb. 1995 by Meve, NW of and above Likombe at 1050 m alt. (*Meve* 901) and described as new by him in 1996. Known from only four collections, two in Tanzania and one each in Bioko and Mt Cameroon.

Habitat in montane forest, 1000-1600 m alt.

Threats: creation of new cocoyam farms in the submontane belt might threaten this species on Mt Cameroon.

Management suggestions: this species is not such a high priority for action as are the threatened strictly endemic species. However, it would be useful to refind the plant from which *Meve* 910 was collected and attempt to propagate it with a view to reintroduction. Data on the range of this species on the mountain would be useful for conservation management.

BALSAMINACEAE

Impatiens grandisepala Grey-Wilson

DD

Range: strictly endemic to Mt Cameroon (one pre-1988 collection).

Known only from a single collection (*Satabié* 303, YA, P) made in June 1979 from Efolofo, 30 km NW of Muyuka on the N. side of the mountain. "*Impatiens grandisepala* is an extraordinary species known only from a single gathering in the Cameroun. The most obvious features of this plant are the very large and conspicuous lateral sepals which droop downwards, partially or wholly concealing the lower sepal and spur, a characteristic quite unlike any other African species of *Impatiens*." (Grey-Wilson in Impatiens of Africa 1980: 216)

Habitat: epiphytic in deeply shaded montane forest; 1150 m alt.

Threats: forest clearance.

Management suggestions: this species should be refound in the wet season with the help of Dr Satabié (recently retired head of the National Herbarium) and the state and extent of the population assessed.

Impatiens sp. nov. 1

VU D2

Range: strictly endemic to Mt Cameroon.

First collected by Upson from Mt Etinde in 1990, recollected there, and in the saddle with Fako, several times in 1992. This species is related to *Impatiens hians* var. *bipindensis* but has smaller, flame-red flowers, with distinctive spurs. The stems of this species are triangular and grow along the top surfaces of horizontal, moss-covered branches. Like the next species, it seems to flower only in the wet season and so for that reason has probably been overlooked until recently. In 1998, similar plants were seen in the Bakossi Mts, but these may constitute a third new species of *Impatiens*.

Habitat: epiphytic in montane forest 1200-1700 m

Threats: unknown.

Management suggestions: monitoring of population.

Impatiens sp. nov. 2.

VU D2

Range: strictly endemic to Mt Cameroon.

First collected near summit of Mt Etinde by Tim Upson (*Upson* 40) on 2nd Oct. 1990, then on 10th July 1992 (*Cheek & Tchouto* 3600). Three other collections have been made since (e.g. *Thomas, Koufani & Tchouto* 9542, 24 October 1992) and probably also *Cheek* 5539 above Njonji. This species, related to *Impatiens grandisepala*, is common as an epiphyte on the peak and ridges of Etinde, but not known elsewhere apart from the Njoni collection. Colour photographs of it have been published (*Cheek* 1994, 1996). It is unique amongst African epiphytic species of this genus in having tubers (spherical or ellipsoid). Several erect branches 30-60 cm tall grow from the moss-embedded main tuber, and bear highly attractive orange-red and yellow flowers. This plant grows in the forks of stunted trees only 3 m from the ground near the peak of Etinde, but the Njonji collection was taken from the upper part of a tall fallen tree.

Habitat: (submontane-)montane forest, (700-) 1500-1700 m alt.

Threats: none known

Management suggestions: monitoring of population.

BEGONIACEAE

Begonia oxyanthera Warb. (including *B. jussiaecarpa* Warb. of FWTA)

LR nt

Range: S.W. Cameroon (including Mt Cameroon- two pre-1988 collections) and Bioko, possibly S.E. Nigeria.

This species, of. sect. *Tetraphila* A.DC. is closely related to *B. preussii* and can be regarded as an altitudinal vicariant (information on these species from Marc Sosef and Hans De Wilde, Wageningen). The earliest collection known, but not the type, of the accepted name, is *Dusen* 427 "Kamerun Gebirge 1500 m". Two recent collections of this species are known from the mountain, one from the Idenau-Limbe area, the other from between Mt Etinde and Buea. (*Thomas* 9324 and 9478).

Habitat: submontane and montane forest 1200-1800 m alt.

Threats: forest clearance for agriculture.

Management suggestions: none.

Begonia preussii Warb. (including *Begonia sessilanthera* Warb.).

EN A1c+2c.

Range: Bioko, S.W. Cameroon (including Mt Cameroon- one pre-1988 collection) and S.E. Nigeria.

Closely related to *B. oxyanthera* (q.v.). this species is known on Mt Cameroon from *Preuss* 1261, collected between Victoria and Bimbia. No other collection of this species appears to have been made this century of this species from the mountain and it may be locally extinct. It is to be hoped that it persists at other sites in its range.

Habitat: lowland evergreen forest.

Threats: clearance for agriculture, especially plantations.

Management suggestions: none.

Begonia quadrialata subsp. *dusenii* (Warb.) Sosef

CR A1c+2c

Range: strictly endemic to Mt Cameroon (four pre-1988 collections), apparently between Bakingili and Njonji.

Sosef (1994: 194), in his monumental revision of the 73 yellow-flowered African "refuge begonias", lists four collections of this taxon, including the type (*Dusen* 90, probably collected in 1891/2). *Watts* 501 is the only collection of this species made in the 1990's inventories. He recollected it in Oct. 1992, at Njonji Lake, alt. 500 m, on a small tree, at about 2 m from the ground. Only two of the earlier collections have detailed locality data, and both are from Bakingili, so it appears that species may be restricted to a belt between 500-800 m alt. between Bakingili and Njonji.

This taxon is one of seven yellow-flowered Begonias on the mountain, one of largest conglomerations in Africa and perhaps indicative (see Sosef 1994) of its importance as a refuge area.

Habitat: on tree trunks or wet rocks in deeply shaded undergrowth of forest; 500- 800 m alt.

Threats: forest disturbance for timber extraction and clearance for agriculture, particularly plantations.

Management suggestions: this subspecies should be refound, and its range and population re-assessed. Care should be taken not to confuse it with the other six yellow-flowered *Begonia* taxa on the mountain!

BORAGINACEAE

Myosotis sp. nov. (*M. sp. nr vestergrenii* Stroh of FWTA)

VU D2

Range: strictly endemic to Mt Cameroon

Myosotis vestergrenii is a species of montane Africa. It is known from Ethiopia, Uganda, Kenya and Tanzania, also from S. Africa. In the account for Boraginaceae in FTEA (from which the foregoing range was taken), Verdcourt does not include Cameroon in the distribution for this species, but says "A closely allied species occurs on Mount Cameroon: Milne Redhead (adnot.) is convinced it is distinct". The Mount Cameroon plant seems larger, lanker, and more branched than the eastern plants, but microscopic study is required if this plant is to be confirmed as a new species and then published as such.

Mann appears to have overlooked this species. It was first collected (*Johnston* 72) in Dec. 1886 "forest, 8,000'", then "Ukele Camp, 8,000'" in Feb. 1932 (*Maitland* 1342). *Brenan* 9520 (March 1948, nr Mann's Spring) bears the most detailed habitat data: "very local, on margin of forest with *Geranium simense*. Herb with trailing, ascending stems to 1' ". The last collection known seems to be that made in Dec. 1958 (*Morton* K875) "several miles W of Mann Spring in woodland, c. 9,000' ". It is remarkable that this species was not collected in the 1992-94 inventories, but it may have been passed over by collectors confusing it with the superficially similar and abundant *Cynoglossum lanceolatum*.

Habitat at boundary of montane grassland with forest; alt. 2400-2700 m

Threats: unknown
Management suggestions: the range and frequency of this species needs to be determined.

CAMPANULACEAE

Lobelia columnaris Hook. f.
LR nt
Range: Bioko, Mt Cameroon, Mt Oku and the Bamenda Highlands, Chappal Wadi, Mambila Plateau.
This Giant Lobelia, with 2 m spikes of blue flowers, is the only representative in West Africa of a group for which East African mountains are famous. It thrives at the montane forest: grassland ecotone. Although there are numerous collections from Mt. Cameroon, it proved extremely rare during the 1992 survey of Etinde, when only one or two specimens were seen. In contrast, it is abundant in many localities at Mt Oku (*pers. obs.* Nov. 1996), where it is gregarious, 10-50 plants being found at a site.
Habitat: boundary between montane forest & grassland; c. 2200 m alt.
Threats: unknown.
Management suggestions: none.

Wahlenbergia ramosissima (Hemsley) Thulin subsp. *ramosissima*
VU B1+2bc
Range: Mt Cameroon (three collections), Bamenda Highlands, Chappal Wadi, Mambila Plateau (one collection each).
A delicate, blue-flowered annual herb, only 8-10 cm tall, this species was first collected on the mountain in Dec. 1861 (*Mann* 1333, "7,000' ", lectotype). The third historic collection was made in Dec. 1930 (*Maitland* 1185, "Nyanga Camp, 5,000' "). In 1992 it was twice recollected between Mann's Spring and Bokwangwo Hut 3 (*Tchouto* 269, *Cheek* 3678).
Habitat: montane grassland: forest ecotone, between grass tussocks with *Radiola (pers. obs.)*; 1500-2600 m alt.
Threats: unknown.
Management suggestions: none.

CARYOPHYLLACEAE

Silene biafrae Hook.f.
VU D2
Range: Mt Cameroon.
This erect, probably annual herb, 20-30(-50) cm high, with 1-7 brownish pink flowers, is common at about 3,000 m alt., scattered thinly, between Mann's Spring and Bokwango Hut 3 (*pers. obs.*, *Cheek* 3661 and *Thomas* 9326). It favours sparse grassy areas, often on cinder slopes. Its closest relative may be *S. burchellii* Otth., a species of montane eastern Africa including E. Zaire, Rwanda and Burundi. However, *S. lynesii* Norman is less geographically distant, occurring in the Tibesti as well as in S. Sudan and is also noted to occur on volcanic soils. 10 historic specimens are known, the earliest being the type, *Mann* 2034, collected at Mann's Spring, 8-10,000 ft in November 1862. It has also been collected above Buea at Huts 2 and 3, and so appears to be distributed around a good swathe of the mountain, perhaps around its whole circumference.
Habitat: lava and cinder slopes, grassland; 2700-3600 m alt.

Threats: none known.

Management suggestions: none. This species appears in no immediate danger of extinction.

CELASTRACEAE

Salacia fimbrisepala Loes.
CR A1c+2c
Range: Mt Cameroon (one pre-1988 collection) and Kwahu, Ghana.
Apparently known from only two collections, one in each part of its range (FWTA 1: 632). The Mt Cameroon specimen is cited as being *Lehmbach* 228 from Buea, probably collected before 1910. No modern specimen is known from Cameroon. This species is a shrub or small tree, flowering in April, with waxy, yellow flowers. The lacerated sepal margin of this species is unusual in the genus and the source of the specific epithet. Kew material of this species is on loan to WAG so these notes are based solely on FWTA.
Habitat: "flowing stream" (presumably in submontane forest); 1000 m alt.
Threats: forest clearance.
Management suggestions: since this species has not been recollected recently there is cause for concern regarding its survival, especially since Buea, which is given as the locality, has lost much of its natural vegetation.

Salacia sp. nov. 1 (*sp. A* of FWTA 1: 632).
CR A1c+2c
Range: Mt Cameroon (one pre-1988 collection).
A cauliflorous treelet 2-5 m tall with purple flowers, fruits and leaves 30 cm long, unusual in this genus of lianas. Related to a Gabonese taxon attributed to a variety of *Salacia loloensis*. First collected in 1948 (*Brenan* 9590, Victoria, close to coast). 15 collections were made in Mabeta-Moliwe in 1992-1993 and others made since from Mokoko.
Habitat: lowland evergreen forest, sometimes disturbed; 0-100 m alt.
Threats: clearance of forest for plantation agriculture.
Management suggestions: this species appears relatively common in the coastal sector of Mabeta-Moliwe and also in Mokoko but is likely to be exterminated if extensive clear-felling and conversion for plantations takes place.

Salacia sp. nov. 2
CR A1c+2c
Range: Mt Cameroon.
A liana unusual in its large flowers with fimbriate petals, known from only two collections in Mabeta-Moliwe, the first made at the end of May 1992 in the Mabeta peninsula by Wheatley (*Wheatley* 260). In the absence of flowers, this species is easily confused with the numerous other lianescent species of this genus in the lowland forests of the Mt Cameroon foothills. Lianas are generally poorly sampled in botanical inventories, especially canopy-flowerers such as this seems to be. The low number of collections of this taxon may be attributed to this fact. A concerted search at the correct season may lead to the discovery of many more individuals and extend the known range of this species.
Habitat: lowland evergreen forest.
Threats: forest clearance.
Management suggestions: skilled tree climbers will be helpful in refinding this species.

CHRYSOBALANACEAE

Dactyladenia mannii (Oliv.) Prance & F.White
CR A1c+2c
Range: Bioko (possibly extinct) and Mt Cameroon (three pre-1988 collections).
The type collection, and the only record from Bioko (*Mann* 1427) refers to a "climbing shrub 15-20". However, Maitland records it (*Maitland* 467, March 1929, nr Buea 2800') as "a small tree of 30 ft on edge of forest"). Two other historic collections are also from above Buea (*Maitland* s.n.) and from Kumba (*Staudt* 513), but the only modern collection (*Wheatley* 790) is from the Onge forest. Known from a total of only 5 collections.
Habitat: lowland and submontane forest, up to 800 m alt.
Threats: forest clearance for housing, timber and agriculture.
Management suggestions: since all of the sites from which this species is known are under heavy threat, it is advisable to assess the surviving population of this species and develop a means of protecting this species.

Magnistipula cuneatifolia Haum.
CR A1c+2c
Range: Mt Cameroon, Abang Mbang, Gabon.
This tree (to 80 cm diam.) is known from only three sites, N. Gabon (*Le Testu* 9376, type), Abong Mbang (*Letouzey* 3961) and a plot voucher specimen from Mabeta-Moliwe (W 831, transect B, plot -1, June 1992). This tree seems extremely rare though fairly widespread.
Habitat: lowland evergreen forest.
Threats: forest clearance for timber extraction and agriculture.
Management suggestions: it should be feasible to retrace this plant at the plot in which it was found in Mabeta-Moliwe (TB-1), and thence to acquire the information necessary to protect this species at that site.

COMPOSITAE

Crassocephalum bougheyanum C.D.Adams
LR nt
Range: Bioko (three collections), Mt Cameroon (14 pre-1988 collections) and the Bamboutos Mts (one collection).
Segregated from *C. vitellinum* (Benth.) S. Moore in 1957 by Adams, *C. bougheyanum* is a larger plant, 90-180 cm tall, with a more restricted distribution. *C. montuosum* is another closely related species and care should be taken to avoid confusion in identification. On Mt Cameroon, *C. bougheyanum* has been collected most frequently from above Buea (at Hut 1) and from near Mann's Spring. It is curious that no recent collections have been made. Perhaps this reflects the "weedy" look of this species!
Habitat: clearings and at the edge of montane forest; 1500-2500 m alt.
Threats: forest clearance.
Management suggestions: none.

Helichrysum biafranum Hook.f.
VU D2
Range: Mt Cameroon.
Known from only two collections. The type, *Mann* 1934 *pro parte* (a mixed collection) is recorded as "10' tall, 7,000', Dec. 1862" and is probably somewhere in the vicinity of Mann's

Spring, since his letters to Hooker in the Kew archives show that he was camped there at that time. The second collection, *Boughey* in GC 12656), was made on 11 Dec. 1952: "Woody herb in second tongue of forest from Mann's Spring". Recent inventories have not found this plant, but since they have not been conducted in this habitat in December (when *H. biafranum* seems to flower) insofar as is known, the species may have been overlooked.

Habitat: edge of montane forest; 2100-2200 m alt.

Threats: unknown, but a lava flow over Mann's Spring could destroy this species, if it is as localized as it seems.

Management suggestions: high priority should be given to finding this species, preferably searching in December, in the forest: grassland ecotone. An attempt to investigate the range and numbers of individuals of this species (is it really restricted to only Mann's Spring?) is important for conservation planning. If the species does occur at this ecotone, it is to be hoped that its range extends to cover a larger part of the grassland-forest perimeter than just that near Mann's Spring.

Helichrysum cameroonense Hutch. & Dalziel

LR nt

Range: Mt Cameroon (10 pre-1988 collections), Mt Oku (1 collection), Bafut-Ngemba (1 collection) and Chappal Waddi (3 collections). Apparently absent from Bioko.

This species can be distinguished from the similar *H. mannii* by its greater height (to 1.5 m tall, not 50 cm), leaves aromatic and viscid-hairy (rather than subscabrid above), involucral bracts bright golden yellow (not silvery white). *H. cameroonense* appears to be a biennial. In the first year it produces a rosette resembling a small *Dendrosenecio.* In the second year, after vertical growth, the stem produces in a dense mass about 50 capitula, each about 2.5 cm diam. These large plants are highly conspicuous and in many areas relatively common scattered in thin grassland at about 2500 m alt., e.g. above Mann's Spring. Several collections were made in 1992 by Thomas and in 1993 by Cheek. This was one of the first large angiosperm species to recolonize the 1982 lava-flow at 2200 m alt (*pers. obs.* 1993).

Habitat: montane grassland, particularly on old lava flows; 1900-3000 m alt.

Threats: none known.

Management suggestions: none.

Helichrysum mannii Hook.f.

LR nt

Range: Bioko (two collections) and Mt Cameroon (11 pre-1988 collections).

The characters distinguishing this species from the similar *H. cameroonense* are listed there. This pair may be altitudinal vicariants. Certainly *H. mannii* occurs at much higher altitudes than *H. cameroonense*, e.g. *Dundas* in FHI 20362: "within 50 feet of the mountain top". The lower altitudinal limit of *H. mannii* is not clear from the data available. Mann certainly confused the two species. Vegetatively they may be almost indistinguishable although the differences in indumentum should help discriminate between them. The absence of specimens of *H. mannii* in the 1992 inventory probably indicates the lack of coverage of altitudes above 3000 m. Hannah Banks (Lava-flow expedition 1995) collected this species above Buea "three-quarters of the way to Hut 3, in rocky places".

Habitat: montane grassland, particularly on cinders at high altitude; to 4000 m alt.

Threats: none known.

Management suggestions: none.

Mikaniopsis maitlandii C.D.Adams

LR nt

Range: Bioko (one collection), Mt Cameroon (five pre-1988 collections) and Chappal Waddi (two collections).

Closely related to *M. tedliei* (Oliv. & Hiern) C.D.Adams (also of Mt Cameroon, but extending to Upper Guinea, Kivu and Rwanda) and also closely related to *M. paniculata* Milne-Redh. of Ogoja and São Tome, this taxon was only segregated by Adams in 1957. A rambling climber to 20 feet tall, its flowers are dull yellow and inconspicuous. It appears not to have been collected during the last decade.

Habitat: montane forest gaps, sometimes with *Mimulopsis solmsii*; 1000-2200 m alt.

Threats: none known.

Management suggestions: an attempt should be made to refind this species and evaluate the population size and range.

DICHAPETALACEAE

Dichapetalum sp. nov.? (*von Rege* 62a, Transect B, Mabeta-Moliwe, August 1993)

Good flowering material is required of this plant, described as a tree 1 m tall, and having pilose stems (both unusual in this genus) before this can be confirmed as a new species by Dr Breteler. If this is done, it may rate inclusion in a future addition of this Red Data list.

DIPSACACEAE

Succisa trichotocephala Baksay

LR nt

Range: Mt Cameroon (10 pre-1988 collections), Bamboutos Mts (two collections) and Mt Oku (one collection).

This montane scabious is the only species of *Succisa* in Tropical Africa, its congeners being Eurasian. In the montane grassland between Mapanja and Mann's Spring, it can be found thinly scattered in the grassland within a few hundred metres from the forest (*Sidwell* 57, *Thomas* 9325, *pers. obs.* October 1992 and October 1993). This species appears to be a perennial, with a subterranean perennating rootstock that withstands fire. It produces biennial rosettes that flower in their second year, though more observations are needed to confirm this. This is a conspicuous species in flower, plants bearing several laxly erect inflorescences 60-90 cm tall, each carrying 3-8 white capitula 3-4 cm diam. Observations on Mt Oku, above Oku-Elak in October 1996, showed that this species survived, though not abundantly, and was flowering in fired grassland that was being cattle grazed. Cattle seem to avoid eating *S. trichotocephala*. On Kilum, dense populations were seen in *Gnidia glauca* woodland and in a *Pteridium*-grassland island inside montane forest. The latter site yielded an estimated 400-500 plants in about a hectare.

24 collections of this species are held at Kew, of which 11 are not available for study, being loaned to WU in 1992. Hooker treated *Mann* 1309 (Jan. 1862, 10,500 ft.) the first specimen known of this species, as identical to the European 'Devil's Bit Scabious', *Succisa pratensis*. *Succisa trichotocephala* was only described as a separate species in 1952, before which the African plants were only varietally distinguished (in the 1930s) as *Succisa pratensis* var. *kamerunensis* B.L. Burtt.

Habitat: montane grassland 2000-2800 m alt.

Threats: absence of intermittent fire may threaten the existence of this species.

Management suggestions: this species appears in no immediate danger of extinction, being relatively common within its range. Nonetheless, if surveys at the montane forest: grassland ecotone are being conducted, the opportunity should be taken to collect data on the population size and range of this species.

EUPHORBIACEAE

Drypetes tessmanniana (Pax) Pax & K.Hoffm.
CR A1c+2c
Range: Rio Muni and Mt Cameroon (one pre-1988 collection; possibly extinct).
The type specimen of this species, *Tessmann* 996, was collected in Rio Muni in 1908/09 and had not been recollected until a specimen believed to be conspecific (*Mildbraed* 10600) was collected at Likomba in Nov. 1928. This species appears to be a shrub, 2 m tall, so should not be easily overlooked in flower or fruit and the fact that there no further collections have been made (e.g. in the Mabeta-Moliwe inventory of 1992) suggests that *D. tessmanniana* is genuinely rare. It may be that the species is extinct in Cameroon, in view of forest destruction in the Likomba area over the last 60 years. Heavy logging of lowland forest in Rio Muni is reported at the moment.
Habitat: lowland evergreen rainforest.
Threats: forest clearance
Management suggestions: if further inventory work is conducted in the eastern foothills, this species should be looked out for.

Sibangea sp. nov. aff. similis (Hutch.) Radcl.-Sm.
CR A1c+2c
Range: Mt Cameroon.
Known from only three collections, all gathered in April 1992 from Mabeta-Moliwe (Transect C, 6256 m, *Wheatley* 218; Transect B, 6890 m, *Watts* 171 and Transect C, 5580 m, *Wheatley* 208). Work on description of this species is complete, and no further collections are needed, though a fruiting collection would help complete knowledge of this species. This is a tree of 4-8 m tall. *Thomas 6823* (YA) from Korup bears superficial similarities and may prove conspecific when opportunity allows a detailed comparison.
Habitat in disturbed lowland evergreen forest; 80-100 m alt.
Threats: clearance for plantation agriculture.
Management suggestions: a survey of the population and range of this species would help development of a management plan for its survival.

Drypetes sp. nov. aff. pellegrini Leandri (*Cheek, Elad & Ndumbe* 3571, Mabeta Peninsula).
Further collections, especially flowering collections, are required of this species if it is to be fully worked out. This is a tree c. 20 m tall, known only in fruit (July 1992). This taxon may warrant detailed treatment in a future edition of this Red Data List.

FLACOURTIACEAE

Oncoba lophocarpa Oliv.
LR nt
Range: Mt Cameroon (five pre-1988 collections) and Mt Kupe (1 collection) and the Bakossi Mts.

A tree 9-15 m tall with a tendency to flagelliflory, the type collection is from Mt Cameroon (*Mann* 2162). Two recent (post-1988) collections are known. This is an infrequent, if conspicuous member of the montane forest of Mt Kupe and the Bakossi Mts (*pers. obs.*).
Habitat: submontane and montane forest (600-) 1200-1900 m alt.
Threats: forest clearance for agriculture (Mt Cameroon).
Management suggestions: none.
Oncoba ovalis Oliv.
LR nt
Range: Mt Cameroon (two pre-1988 collections), Mt Kupe (one collection) and the Bakossi Mts.
This tree, which can flower when as small as 2 m tall, yet reach 8-10 m tall, may also occur in Nigeria (Ikom. *Keay* in FHI 2848 "*Camptostylis* sp. nov. aff. *ovalis*"). It can be locally dominant, as in parts of the saddle between Fako and Mt Etinde, or in montane forest near Kodmin (Bakossi Mts.). Two collections were made on Mt Cameroon during the inventories of the 1990 s.
Habitat: submontane/montane forest; 1000-1400 m alt.
Threats: clearance of forest for agriculture.
Management suggestions: none.

HOPLESTIGMATACEAE

Hoplestigma pierreanum Gilg
CR A1c+2c
Range: coastal Cameroon evergreen forest, including Mt Cameroon (one pre-1988 collection). This tree, one of only two species in this endemic Tropical African family, is endemic to Cameroon and only known from the syntype collections of *Zenker* (2575, 3650, 2632 & 3383) at Bipindi in 1903-08 (no recent collections from this area have been seen) and from the foothills of Mt Cameroon. The first collection from the mountain was that from Likomba (*Mildbraed* 10774, Dec. 1928; no natural forest now survives there, it is believed), just east of Mabeta-Moliwe, where it should be looked for. The second collection (rather poor, only a few fallen leaves) is from the western foothills, in the Mokoko forest (*Thomas* 10022). The second species of the family is more widespread (Gabon to Cote D'Ivoire) but is still extremely rare. It is distinguished by its smaller, subglabrous and more papery leaves from *H. pierranum*. Several trees of *H. klaineanum* Pierre are scattered in the Korup and share the distinctive *Cordia*-like slash and characteristic bony endocarps of the sister species.
Habitat: lowland evergreen coastal forest at or near sea-level.
Threats: forest clearance for timber and agriculture, especially plantations.
Management suggestions: in view of the extreme rarity of this species and the evolutionary isolation of the genus, it is worth considerable effort to refind it. The best starting point would be Mokoko, where the most recent collection has been made. It is recommended that the site where the original collection was made be pinpointed.

LABIATAE

Plectranthus dissitiflorus (Guerke) J.K.Morton
CR A1c
Range: Mt Cameroon (one pre-1988 collection; possibly extinct).
Known from a single specimen, the type (*Preuss* 1055), FWTA 2: 460 lists the collecting locality as "Buea to Bimbia", i.e. anywhere between 800 m alt. on the east flank of the main

massif (Buea) and sea-level (Bimbia). However, this is contradicted by the protologue which reads "in Buschwald zwischen Buea und Mimbia, 940 m. 9 Oct. 1891". This narrows down the search area considerably.

The specimens at Kew are duplicates of the holotype (presumably destroyed at Berlin).

This species is a herb one metre tall, the leaves are membranous and unusually in the genus, trullate and deeply crenate-sinuate. The flowers are large (2 cm long) and blue. Related to the "Coleus" of commerce, and possibly with horticultural potential.

Habitat: submontane forest; 940 m alt.

Threats: forest clearance for agriculture and habitation.

Management suggestions: the vicinity of Buea should be searched for this species.

Solenostemon sp. nov.?

Several specimens (*Cable* 216, 227, *Cheek* 5563 and *Williams* 52) made on the Earthwatch funded inventory above Njonji in 1993 proved to be a probable new species closely related, and perhaps conspecific, with a taxon in Sierra Leone. More work is needed to resolve this question. It is possible that the Njonji taxon may appear in a future edition of this red list.

LAURACEAE

Beilschmiedia preussii Engler
CR A1c+2c
Range: Mt Cameroon (one pre-1988 collection; possibly extinct).

Only known from the type collection made a century ago (*Preuss* 1272, forest between Victoria and Bimbia), this species has been accepted by all subsequent workers on the genus in Africa. The classical locality for this species equates closely with the Mabeta-Moliwe area from which 10 unidentified specimens of this genus were obtained in 1992/93. Eight of these are sterile plot vouchers from the 1992 inventory and so are unlikely to be identified further, but two, collected in 1993 on the Oxford University expedition, (*Baker* 226 and *Hunt* 14) are believed to be fertile and should be re-evaluated to see if either match the characteristics of *B. preussii*. There is no duplicate of the type at Kew, but a plate is available in FDC. The genus *Beilschmiedia* lacks an active specialist for the African species, and with its numerous species (55 in Tropical Africa in the last revision in 1948), and its literature in German and French, together with the diagnostic characters being held in the minute flowers, it is a daunting genus for the generalist to essay.

Habitat: lowland evergreen forest.

Threats: forest clearance for small-holder agriculture, housing, plantations of *Hevea* and *Elaeis*.

Management suggestions: re-evaluation of the Oxford University specimens against the FDC account is a priority. Searching for fertile *Beilschmeidia* in the Mabeta-Moliwe area in March should follow. Given that no other species is recorded from that area, but that at least 10 specimens of the genus are known from Mabeta-Moliwe, there must be a fair chance that this species persists. If this is confirmed, population data should be gathered to enable formulation of a management plan for this species.

LEGUMINOSAE

Anthonotha leptorrhachis (Harms) J.Léonard
CR A1c+2c
Range: Cameroon coastal forest, including Mt Cameroon (two pre-1988 collections).

This distinctive tree is known from two collections in Bipindi (including the type, *Zenker* 2445, made in 1902). It is also known from Eseka, so appears widely spread in the evergreen coastal forest of Cameroon and may yet be found in adjoining Equatorial Guinea or Nigeria. Our earliest record from Mt Cameroon is in August 1951 from S. Bakundu: *Olorunfemi* in FHI 30737 (tree 50', 3' girth). *Mambo & Thomas* 82 is from Barombi Koto (June 1986) and it is also known from Onge (e.g. *Thomas* 9779). Several collections were made in Mabeta-Moliwe (e.g. plot voucher W1161) in 1992, usually near rivers.

Habitat: lowland coastal evergreen forest, particularly near rivers.

Threats: forest clearance for agriculture.

Management suggestions: none.

Crudia bibundina Harms (*Crudia sp. A.* of FWTA (1: 467) 1958 pro parte).

CR A1c+2c

Range: Mt Cameroon (two pre-1988 collections).

Named after Bibundi, the locality for the type collection (*Mildbraed* 10645, Nov. 1928), this species as treated in Lock (Legumes of Africa, 1989) is only known from one other collection, made on 14 Feb. 1927, *Dalziel* 8246, collected "between Victoria and Buea". It appears not to have been recollected by the 1992-94 inventories and may now be extinct. However, as far as we are aware, no effort has yet been made to look for this tree. For want of tree climbers, tree species have been undercollected in recent surveys, so this species may simply have been overlooked.

Habitat: lowland (but possibly submontane), evergreen forest (available data imprecise).

Threats: forest clearance for agriculture, particularly plantations.

Management suggestions: rediscovery of this possibly extinct species is necessary if its population biology and habitat requirements are to be discovered.

Microberlinia bisulcata A.Chev.

CR A1c+2c.

Range: S.W. Cameroon, including Mt Cameroon (two pre-1988 collections).

This large (to 40 m tall) tree is most famous as one of three species forming groves in the southern Korup N.P. However, it also occurs around Mt Cameroon, in the western (Mokoko: *Ndam* 1175 and *Mbani* 334, both c. 1994) and northern (S. Bakundu, two collections) foothills. It is not generally realised that this species is restricted to SW Cameroon and that records of it occurring elsewhere are spurious.

Habitat: lowland rainforest, usually on sandy soils in flat areas.

Threats: forest clearance.

Management suggestions: habitat protection.

Ormocarpum klainei Tisser. (*Ormocarpum sp. A* of FWTA)

Range: Mt Cameroon (one pre-1988 collection, possibly extinct) and Gabon (two collections).

CR A1c

This shrub was collected on Mt Cameroon by Maitland below (c. 450 m alt.) Buea in January 1931. It has not been collected since. The only other collections of this species are from Gabon (*Klaine* 1829 and *Soyaux* 321), both dating from the last century. Forest on the eastern slopes of Mt Cameroon has virtually ceased to exist, apart from at Mabeta-Moliwe, and this species may be extinct in Cameroon. The best chance of this species surviving is in Gabon where habitat destruction is believed to have been less extensive than in its Cameroonian locality. However, there have been no recent collections of *O. klainei* from that country.

The Maitland specimen appears to have first been linked with the Gabonese material by Jan Gillett.

Habitat: lowland forest; c. 450 m alt.

Threats: clearance for agriculture and housing.

Management suggestions: it is worth conducting a survey for any surviving patches of natural vegetation in the lowland belt of the eastern slopes. If this fails to yield more plants of this species, Mabeta-Moliwe may be the best hope for refinding this species on Mt Cameroon.

Plagiosiphon longitubus (Harms) J.Léonard
CR A1c+2c

Range: Cameroon coastal forest, including Mt Cameroon (one pre-1988 collection).

Discovered in Bipinde in 1911 (*Zenker* 4063, type), this species is also known from Sangmelima and Campo, and appears widespread, though rare, in Cameroon coastal forest. It may yet be found in Equatorial Guinea, and even Nigeria. The first record from Mt Cameroon is *Keay* s.n., from Banga, S. Bakundu (1951). Recent records are from Onge (*Ndam* 749) and Mokoko (*Ndam* 1249, *Thomas* 10048).

Habitat in lowland evergreen forest.

Threats: forest clearance for agriculture.

Management suggestions: none.

MYRSINACEAE

Ardisia etindensis Taton
CR A1c+2c

Range: Mt Cameroon (one pre-1988 collection) & Eseka, Cameroon (one collection).

Known only from two collections, more may yet be discovered under the name *A. cf. staudtii* of this checklist. Taton's work has not yet been incorporated at Kew and so unfortunately was not used in the naming of the material of this genus. The only specimen thus far known on Mt Cameroon is the type, *Letouzey* 14962, from Etinde, SW slopes, near Boando, a specimen c. 50 cm tall.

Habitat: lowland/submontane forest; c. 600 m alt.

Threats: forest clearance for agriculture.

Management suggestions: the Mt Cameroon specimens of this genus need to be evaluated against Taton's revision to establish if any more collections of this taxon can be found. An attempt should then be made to evaluate the population of this species in the field.

Ardisia oligantha (Gilg & Schellenb.) Taton
CR A1c+2c

Range: Mt Cameroon (one pre-1988 collection; possibly extinct).

Whilst several names in *Ardisia* of Gilg and Schellenberg have been reduced to synonymy, *A. oligantha* and *A. schlechteri* have not. They are accepted by both the latest reviewers of African *Ardisia*, i.e. de Wit in 1957, and more recently by Taton, and so must be accepted as good species, despite the fact that they are only known from the types, now presumed destroyed. The only specimen of this species known was *Weberbauer* 48. Detailed locality data is unavailable to us; in this respect, the protologue states only "am Kamerunberg (*Weberbauer* n. 48...Bluh 1906)".

Habitat: unknown, probably montane or submontane forest.

Threats: unknown, in absence of locality data, but probably forest clearance.

Management suggestions: Weberbauer's itinerary should be researched. Existing *Ardisia* specimens from Mt Cameroon should be reassessed against the criteria for this species listed by Taton, e.g. presence of mucronate sepals and petals.

Ardisia schlechteri Gilg
Range: Mt Cameroon (one pre-1988 collection; possibly extinct).
CR A1c+2c
Taton rejects Hepper's conjecture (FWTA 2: 31) that *Olorunfemi* in FHI 30692 might be considered of this species. This leaves *Ardisia schlechteri* in the same position as *A. oligantha*, i.e. known only from the destroyed type, albeit with more locality data. *Schlechter* 12417 was collected at Bibundi.
Habitat: presumably lowland evergreen forest, near sea-level.
Threats: lava-flows, clearance for agriculture and settlement.
Management suggestions: as for *Ardisia oligantha*.

Embelia mildbraedii Gilg & Schellenb. (*E. sp. aff. welwitschii* (Hiern) K. Schum. of FWTA 2: 32).
LR nt
Range: Mt Cameroon (2 pre-1988 collections) and Bamenda Highlands (1 collection).
The type collection was made at Mann's Spring (*Mildbraed* 3452) presumably about 1928. *Maitland* 692 (neotype) was recollected between Buea and Masaka camp. Four recent collections (*Tchouto* 377, *Tekwe* 189, 320, and *Thomas* 9458) are probably this species, so it appears to be relatively common. This is a genus which awaits comprehensive revision.
Habitat: montane forest 1800-2300 m alt.
Threats: none known.
Management suggestions: this species, though rare, appears widespread in the montane forest of Mt Cameroon and in no immediate danger of extinction.

MYRTACEAE

Eugenia kameruniana Engl.
CR A1c
Range: Mt Cameroon (one pre-1988 collection; possibly extinct).
The taxonomic status of this species is uncertain. Keay in FWTA 1: 238 lists this as 'an imperfectly known species'. In Engler's protologue, published in 1899, he gives only "*Dusen* 9" by way of specimen data and it is not certain that this derives from Mt Cameroon: research on Dusen's itinerary is needed. Keay gives credibility to this species by uniting with it *E. hankeana* Winkler as a later synonym, so adding a second specimen, *Winkler* 1110, collected between Debundscha and Bibundi. Both specimens are assumed destroyed at Berlin, and there are no duplicates at Kew. In the 1992-94 inventories, no specimens were attributed to this species, and it may be extinct.

Another rare species on Mt Cameroon, *Eugenia kalbreyeri* Engl. & Brehmer has also not been rediscovered in Cameroon, the only collection being the type (*Kalbreyer* 157 "Shrub 10 ft. Fl. white, axillary. Open spaces, elevation 2000'. Mapanja 3/77"). However, it is known from several countries in Upper Guinea so we do not include it on the present Red Data list, though it is evidently very rare on Mt Cameroon, and also possibly extinct in Cameroon.

Three *Eugenia* taxa discovered in Mabeta-Moliwe in 1992 (see checklist) may prove to be new to science. However, it cannot yet be ruled out that one of these is conspecific with the lost *E. kameruniana*.

Habitat: probably in lowland evergreen rainforest; c. 200 m alt.

Threats: probably forest clearance for agriculture.

Management suggestions: some attempt should be made to rediscover this species on Mt Cameroon. A search for Myrtaceae between Bibundi and Debundscha would be worthwhile, as would research on Dusen's itinerary to identify the locality of the type specimen.

PIPERACEAE

Peperomia kamerunana C.DC.
LR nt
Range: Bioko (two collections) and Mt Cameroon (five pre-1988 collections).

Brenan 9371 (Mann's Spring, 2245 m alt., 25 March 1949): "locally plentiful epiphyte in moss layer on tree trunks in mist forest. Herb with rather fleshy creeping, pubescent stems. Leaves rather fleshy, midgreen, not glossy above, mottled purplish green beneath. Spikes green."

It is curious that this epiphytic species has not been recollected in recent inventories, and also that Mann did not collect it in 1861-62. *Migeod* 84 is from above Buea, so the species seems to be widespread on the mountain. It may simply have been overlooked in view of its inconspicuous green flowers. *Maitland* s.n. and 1174 are from "Tonga" on Mt Cameroon (6700 ft and 7600 ft). We have not traced this locality on the mountain.

Habitat: upper montane forest, epiphytic; 1800-2300 m alt.

Threats: none known.

Management suggestions: if further surveys are conducted in montane forest on the mountain, this species should be sought. Care should be taken to discriminate it from the seven other species of the genus on the mountain, e.g. the larger and more common *P. vulcanica*. *Peperomia kamerunana* can be recognized by its broadly obovate leaves only 1-1.5 cm long.

Peperomia thomeana C.DC. (including *P. vaccinifolia* C.DC.).
LR nt
Range: São Tome (three collections), Bioko (one collection), Mt Cameroon (five pre-1988 collections) and the Bamenda Highlands (three collections).

This species appears not to have been recollected in recent surveys on Mt Cameroon. The same comments apply to this species as to *P. kamerunana* above. Both species seem to fill similar niches but with slightly different altitudinal ranges.

Habitat: epiphytic in montane forest 1460-2100 m alt.

Threats: none known.

Management suggestions: the same comments made above for *P. kamerunana* apply here. *P. thomeana* can be distinguished from other species on the mountain by its rhombic to narrowly elliptic leaves with a distinctly notched apex.

POLYGALACEAE

Polygala tenuicaulis Hook. f. subsp. *tenuicaulis*
VU D2
Range: Mt Cameroon (seven pre-1988 collections) and Bamboutos/Bamenda Highlands (three collections)

This subspecies overlaps in the Bamboutos and Bamenda Highlands with *P. tenuicaulis* subsp. *tayloriana* Paiva which extends to Mambila Plateau, and Chappal Waddi. The latter generally has larger flowers, c. 8 mm long compared with those of subsp. *tenuicaulis* on Mt Cameroon (5 or 6 mm long). The first collections are those of Mann on Mt Cameroon near Mann's Spring (*Mann* 1282, Dec. 1861 and 1982, Nov. 1862) and subsequent collections on the mountain have been made in the same area, and not from above Buea, so its range may be limited. The collections of the typical subspecies from Bamenda mostly have the larger flower size of the northern subspecies and Paiva (monographer of the genus) has determined them with a question mark.

On Mt Cameroon this plant can be locally common and gregarious, forming little patches of about 30 cm X 30 cm. The plants have two or three slender, unbranched stems c. 30 cm tall, of a purplish green hue, and purple flowers. Anthocyanin-free mutants are seen intermixed on occasion (*pers. obs.* 1992, 93). These have yellow flowers and yellowish green stems. It is not clear whether this species is an annual, or perennates from an underground rootstock.

Habitat: montane grassland, within a few hundred metres of the forest boundary, but not at the forest edge; 2100-2400 m alt.

Threats: this subspecies may require fire for regeneration.

Management suggestions: a survey of the range of this subspecies on the mountain is advisable. Rootstock presence should be investigated since this has implications for the fire ecology of the taxon.

ROSACEAE

Prunus africana (Hook.f.) Kalkman
Range: montane Africa (to East and South Africa) and Madagascar.

This species is one of about ten Pan-African montane tree species (including e.g. *Agauria salicifolia, Ilex mitis* and *Myrica arborea)* and is not remotely in danger of extinction, so long as some montane forest survives somewhere within its enormous range. Locally it can be very common. However, on Mt Cameroon as with some other areas within the range of this species, many trees have died as a result of girdling caused by bark removal. The bark from the trees on Mt Cameroon is transported to the Plantecam factory at Mutengene where it is extracted to produce a powder for export to a company in France. A great deal of attention, and funding has been paid by International Conservation organizations to investigate and address this harvest and perhaps for this reason the species has received a high conservation rating. In the opinion of this author, it merits a "Lower Risk, near threatened" at best.

Habitat: montane forest, usually at about 1800-2200 m alt.

Threats: harvesting of bark for the European medicinal market.

Management suggestions: the Mt Cameroon Project has already conducted a detailed survey of this species on the mountain and has been energetic in developing methods for propagating and replanting it.

RUBIACEAE

Belonophora sp. nov.
CR A1c+ 2c
Range: Mt Cameroon and Korup N.P.
This species only came to light during the identifications made after the Onge inventory of October 1993. Five specimens are known from Onge (see checklist) and initially the species was thought to be endemic to that forest. However, a further specimen, from the northern

Korup came to light at Kew whilst the species was being described. SW Cameroon is the centre of diversity for this small genus of shrubs and trees, for which a revision is being prepared at Kew, including a description of this new species.
Habitat: lowland rainforest.

Threats: clearance for plantation agriculture.
Management suggestions: habitat protection.

Psychotria sp. nov. 1 aff. dorotheae Wernham
CR A1c+ 2c
Range: Mt Cameroon
This species first came to light in 1992. A conspicuous treelet 2-6 m tall, with leaves c. 40 X 20 cm, abundant collections were made in Mabeta-Moliwe, particularly at the southern end of transect *B. Jaff* 149, *Watts* 127, *Ndam* 748, *von Rege* 68, *Cheek* 5709, 5761, *Cable* 593, *Tchouto* 945 and *Baker* 272 all represent this taxon. This species also occurs in the Onge Forest (*Akogo* 94), but strangely appears much rarer there.
Habitat: lowland evergreen forest, sometimes disturbed; 0-100 m alt.
Threats: clearance of forest for plantation agriculture.
Management suggestions: this species appears relatively common in the coastal sector of Mabeta-Moliwe but is likely to be exterminated if clear-felling and conversion for plantations takes place.

Psychotria sp. nov. 2 aff. bidentata (Thunb. ex Roem. & K.Schult.) Hiern
CR A1c+ 2c
Range: Mt Cameroon (one pre-1988 collection).
A shrub to 2 m tall, this species was first collected at Victoria in April 1899 (*Schlechter* 12374). More recently it is known from the southern baseline and NE corner of Mabeta-Moliwe (*Watts* 110, plot voucher W932).
Habitat: lowland evergreen forest.
Threats: clearance for plantation agriculture.
Management suggestions: this species is less commonly collected so appears much scarcer than the foregoing. If refound it is a candidate for cultivation and propagation at the Limbe botanic garden.

Rutidea sp. nov.aff. hispida Hiern
CR A1c+ 2c
Range: Mt Cameroon.
This vine is known from a single collection (*Sunderland* 1012) from Mabeta-Moliwe
Habitat: lowland evergreen forest.
Threats: forest clearance for plantation agriculture.
Management suggestions: a concerted effort should be made to refind and evaluate the population size of this species.

Sacosperma sp. nov.
CR A1c+ 2c
Range: Mt Cameroon
Initially known from a single specimen (*Cheek* 3577) collected along the strand near Mabeta in July 1992, several collections (1993 and 1995) have since been made in Mabeta-Moliwe and at least four plants of this climber are now known over a range of about 5 km (*pers. obs.*).

Habitat: lowland evergreen forest.

Threats: clearance for plantation agriculture.

Management suggestions: as for *Rutidea* sp. nov. (above).

Stelechantha sp. nov

CR A1c+ 2c.

Range: Mt Cameroon

This blue-flowered, cauliflorous treelet first came to light in the inventory of Mabeta-Moliwe in June 1992 (*Wheatley* 298). A further collection (*Tchouto* 1246) was found at Barombi, Mokoko F.R. in June 1994. A third collection from just outside our area (*Nemba et al.* 613, mile 16 Kumba-Mamfe Rd. collected in July 1987) was discovered in the Kew Herbarium by Sally Dawson, who is describing the species.

This species, even within its range, seems very rare given the few collections known despite the intensive collecting over recent years in the forests where it occurs. As a treelet 1-2 m tall, with spectacular flowers, it is not likely to be overlooked if present in an area.

There are four species of *Stelechantha* known: the present, one in Angola, another in S. Cameroon, and a third in Upper Guinea (Guineé Bissau to Guineé-Conakry). All are very rare species of evergreen forest.

Habitat: lowland evergreen forest; c. 200 m alt.

Threats: forest clearance for agriculture, especially plantations.

Management suggestions: rediscovery and protection of this species is advised. A survey of the population would be useful.

SAPOTACEAE

Synsepalum brenanii (Heine) T.D.Penn.

CR A1c+ 2c

Range: strictly endemic to Mt Cameroon (two pre-1988 collections; possibly extinct).

Known only from two collections near Banga in the S. Bakundu F.R. Brenan, for whom the species is named, gives the following notes which may help to refind it, if it survives: "Frequent in undergrowth of high forest. Small tree to about 15' high. Trunk to c. 3" diam., cauliflorous and flowers on twigs up to those of second year. Young shoots brown-lanate, older purplish brown. Leaves subcoriaceous, deep green and glossy above, lateral nerve slightly impressed, pale green beneath. Petals and calyx brownish. Flowers dirty greenish-cream" (*Brenan* 9273, March 1948). It was recollected in the same area in March 1956 (*Binuyo & Daramola* in FHI 35589) where it was described as a "shrub to 6 ft high".

There is considerable interest in the allied species *S. dulcificium* 'the miraculous berry' which has the property to depress appetite and make sour and salty subjects taste sweet, due to the glycoprotein miraculin. It is not known whether the fruits of *S. brenanii* have the same property.

Habitat: lowland forest

Threats: the S. Bakundu F.R. has suffered from illegal timber extraction in recent decades. Nigerian yam farmers are reported to be cultivating the southern part. See notes in introduction on threats in S. Bakundu F.R.

Management suggestions: if this species can be refound, the forestry department authorities at Kumba, who manage this forest, should be notified and involved in any protection measures taken.

SCROPHULARIACEAE

Veronica mannii Hook. f.

Bioko (four collections), Mt Cameroon (24 pre-1988 collections), Mt Oku (two recent collections).

First collected in December 1860: "on the very top of Clarence Peak, Fernando Po" (*Mann* 604), this species is an erect herb 10-45 cm tall arising from a weak, woody underground rootstock. Most of the collections have been made above Buea, between Hut 2 and Hut 3, but Morton records it from a location 3 miles from Mann's Spring so it is not completely limited to the area above Buea. Descriptions of flower colour vary from "vivid royal blue" to "violet" and it has generally been collected in flower from December to April. No recent specimens are known, apart from *Banks* 68 (Nov. 1995), but the 1992 Etinde inventory did not extend sufficiently high in altitude to reach the range of this species, nor was it conducted during its flowering season. Recently (Nov. 1996, *pers. obs.*) this species has been discovered on Mt Oku.

Habitat: montane grassland; (2700-)3000-4000 m alt.

Threats: none known.

Management suggestions: none.

SCYTOPETALACEAE

Rhaptopetalum sp. nov.

CR A1c+ 2c

Range: Mt Cameroon (two pre-1988 collections) and Rio Del Rey (one collection).

Although Letouzey in Flore du Cameroun 20: 177 (1978) treated *R. pachyphyllum* as a disjunct species occurring in Rio Muni-Gabon and SW Cameroon, he listed sufficient differences between the plants of these two disjunct areas to recognize two species. The SW Cameroon taxon was first collected in the S.Bakundu reserve (*Binuyo & Daramola* in FHI 35094 (fr. Jan.), 35577 (fl. Feb.)), then 20 km S. of Mundemba (*Letouzey* 15158). During the survey of the Boa Plains during 1997, about 15 plot vouchers were made of what are probably this species.

Habitat: lowland evergreen forest.

Threats: conversion of forest to agricultural land,

Management suggestions: the Boa Plain area should be revisited to obtain fertile material to confirm the tentative identification of the plot vouchers. If confirmed, the Boa Plains area, on the evidence of the number of vouchers, seems densely populated with this species and might be considered as a focus for its conservation management. If the identification is not confirmed, S. Bakundu should be searched in the Banga and Bopo-Pete areas (these localities are given on the specimens cited) to refind this species.

STERCULIACEAE

Cola nigerica Brenan & Keay

CR A1c+ 2c

Range: Mt Cameroon (1 pre-1988 collection) & Nigeria (c. 5 collections)

This 8 m tall tree is one of the simple-leaved, heterophyllous, cauliflorous, species of the genus. The stipules are c. 0.5 cm broad and caducous; the flowers are white or cream and broadly campanulate, succeeded by fruits with 2-3 subglobular, red, drying matt, fruitlets. *Cola nigerica* is extremely rare in each of the five disjunct areas in its range. For example in the 1992 survey of Mabeta-Moliwe, while 22 specimens of *Cola flavo-velutina* were recorded, only a single specimen of *Cola nigerica* was found (*Watts* 244 at TD 6250 m, 24th April 1992). The only other specimens known on the mountain are *Keay* in FHI 37412 (Bambuko F.R.), and *Tchouto* 694 (Onge).

Habitat: lowland evergreen forest.

Threats: All the areas in which *Cola nigerica* occurs(ed) have been cleared or are under threat of forest clearance and cultivation.

Management suggestions: marking and protecting individual trees might be a useful stopgap before making detailed population surveys.

Cola praeacuta Brenan & Keay

CR A1c + 2c

Range: Mt Cameroon (one pre-1988 collection).

A tree or shrub 3-7 m tall, *Cola praeacuta* is only known from about 10 collections in the northern and western foothills, and in the coastal strip. It is unknown from Mabeta-Moliwe. It seems relatively common in the western foothills and this area seems most suited for developing a plan for the conservation of this species.

Habitat: lowland evergreen forest.

Threats: forest clearance for agriculture.

Management suggestions: see following species.

Cola sp. nov. aff. flavo-velutina

CR A1c + 2c

Range: Mt Cameroon.

First collected in the western foothills in March 1993 (*Watts* 670 from Mundongo "Near MC 03"), the only other collection is that from near Bonjari village, made in May 1994 (*Ndam* 1160 "Plot M 11-2"). This shrub, about 2 m tall, is known only from fruiting collections.

Habitat: lowland evergreen forest.

Threats: clearance for agriculture, particularly plantations.

Management suggestions: see following species.

Cola sp. nov.1 aff. philipii-jonesii

CR A1c + 2c

Range: Mount Cameroon.

Known only from the Etinde coastal strip and the western foothills (Mokoko forest). This shrub is distinct in its shortly petioled, homophyllous leaves which dry a distinctive metallic grey. *Kwangue* 97 (June 1992, Batoke, Mile 9, c. 200 m alt.) is the first record of this species, followed by *Kwangue* 132 (Nov. 1992, between Upper Boando and Etome, c. 700 m alt.). The third and final record, *Watts* 1169 is from Ekombe-Mofako (May 1994).

Habitat: lowland evergreen forest.

Threats: clearance for agriculture, particularly plantations.

Management suggestions: a survey of the population of this species is advisable, as are measures to protect what remains from further destruction.

Cola sp. nov. 2 aff. philippi-jonesii

CR A1c + 2c

Range: Mount Cameroon and the southern sector of the Korup Forest (3 recent collections).

On Mt Cameroon known only from the western foothills (4 recent collections).

This is a shrub or small tree 2-8 m tall, unusual in the genus in its subopposite, homphyllous leaves.

Wheatley 792 (March 1993) is the earliest record of this species ("Onge, Likengi, 3.5 hr S on Wonge R."). Later that year it was collected nearby (*Tchouto* 658, Sept. 1993) and then near Mount Thump (*Thomas* 9869) and at the Bomana-Koto junction (*Ndam* 797). The only record from the Mokoko forest is adjacent to the Onge area: *Ndam* 1240 (May 1994, "Boa, transect to plot 14").

Habitat: lowland evergreen forest.

Threats: clearance for agriculture, particularly plantations.

Management suggestions: a survey of the population of this species is advisable, as are measures to protect what remains from further destruction.

VERBENACEAE

Vitex sp. nov. aff. grandifolia (*Cheek, Elad, P. Ndumbe* 3586, coast of the peninsula between Dikulu and Mabeta, near Elephant River, July 1992). Flowering material of this species is required before it can be described as new by Dr Verdcourt and considered for inclusion in a future edition of this list. The original tree (20-30 m tall) was revisited in December 1993 (*Cheek & P. Ndumbe* 3960), but was then sterile and in poor condition.

MONOCOTYLEDONS

ANTHERICACEAE

Chlorophytum petrophilum K. Krause
CR A1c+2c
Range: SW and S Cameroon and Bipindi
Known only from four collections *fide* Nordal & Poulsen *pers. comm.* The type collection is from Bipindi, but the other collections are from SW Province. Discovered on Mt Cameroon in Nov. 1993 (*Cheek* 5451)

Habitat: lowland evergreen forest.

Threats: conversion of forest to oil-palm plantation.

Management suggestions: site protection.

ARACEAE

Amorphophallus preussii (Engl.) N.E.Br.
EN A1c+2c
Range: Mt Cameroon (one pre-1988 collection), Mt Kupe and the Bakossi Mts.

This species has not been recollected on Mt Cameroon this century and may be extinct there. The original description, published in 1892, appears to be based on two specimens from Buea at 3000 ft alt., *Preuss* 588 and *Lehmbach* 127. Flowers are recorded in December. Lately this species has been rediscovered at Mt Kupe where it is locally abundant (*Cable* 697, 867 and 1200 all in January and February 1995, in flower) and it has also been seen in flower in the Bakossi Mts (Cheek, *pers. obs.* 1998). At one site at Mt Kupe, Cable records this species as being gregarious, with numerous individuals scattered over a rock face.

Habitat: rocky areas in montane forest 850-1600 m alt.

Threats: none are known.

Management suggestions: rocky areas and cliff faces around Buea should be searched during the flowering time (December-February) in order to refind this species on Mount Cameroon.

BURMANNIACEAE

Afrothismia pachyantha Schltr.
CR A1c+2c, B1+2abcde, C2b, D.
Range: Mt Cameroon (possibly extinct) and Mt Kupe
First collected near Moliwe in 1905 (*Schlechter* 15789) and not rediscovered since on Mt Cameroon, despite searches. Rediscovered on Mt Kupe in Oct. 1995 where population protected and regularly monitored.
Habitat: lowland evergreen forest
Threats: forest clearance for agriculture and plantations.
Management suggestions: continued searches for this species on Mt Cameroon are suggested.

Afrothismia winkleri (Engl.) Schltr.
CR A1c+2c.
Range: Gulf of Guinea
First collected near Moliwe in 1905 (*Schlechter* 15788) and despite searches, not rediscovered since on Mt Cameroon, (apart from a collection by Winkler at about the same time). Discovered on Mt Kupe in May 1996 where population protected and regularly monitored.
Habitat: lowland evergreen forest
Threats: forest clearance for agriculture and plantations.
Management suggestions: continued searches for this species on Mt Cameroon are advised.

Oxygyne triandra Schltr.
CR A1c+2c
Range: Mt Cameroon.
First collected near Moliwe in 1905 (*Schlechter* 15790) and not rediscovered since on Mt Cameroon, despite searches. The Oxford University expedition to Mt Cameroon (July-Sept. 1993) had the rediscovery of this species, together with *Afrothismia* as a key objective.
Habitat: lowland evergreen forest
Threats: forest clearance for agriculture and plantations.
Management suggestions: continued searches for this species on Mt Cameroon.

Oxygyne sp. nov.
CR D
Range: Mt Cameroon
Discovered in Oct. 1992 (*Cheek et al.* 3816) in the saddle of Etinde and not seen since. Known only from a single patch of two square metres where four plants were found
Habitat: montane forest with *Oncoba ovalis*, c. 1300 m alt.
Threats: highly vulnerable to small scale disturbance in view of the diminutive single population. A camp-fire or clearing created at its site could push the species to extinction.
Management suggestions: revisit original site, demarcate and designate as protected after establishing ownership rights with neighbouring villagers.

COMMELINACEAE

Palisota preussiana K. Schum. ex C.B.Cl.
VU D2
Range: Bioko (one collection) and Mt Cameroon (two pre-1988 collections).
An erect herb with an aerial stem c. 90 cm tall, narrowly elliptic leaves 25-30 x 7 cm long, and a diminutive spike c. 8-10 x 2-3 cm bearing mauve flowers in October-December. Five new collections, mostly from the eastern slopes, were made of this species in the 1990s, so it appears in no danger of immediate extinction.
Habitat: submontane/montane forest; c. 1200 m alt.
Threats: clearance of forest for cocoyam farms.
Management suggestions: this species should be refound and its range and population density assessed.

CYPERACEAE

Bulbostylis densa (Wall.) Hand.-Mazz. var. *cameroonensis* (C.B.Cl.) Hooper
VU D2
Range: Mt Cameroon (two pre-1988 collections).
This diminutive sedge, probably an annual, was only known from the types (*Mann* 1360b, 2093b) until it was recollected in the 1992 inventory (*Thomas* 9407), 130 years late, in the same general area of the mountain, near Mann's Spring. If it is confined to a small area, it may be vulnerable to lava flows. However, it is an inconspicuous plant, so may also have been overlooked.
Care must be taken not to confuse this plant with the sympatric typical variety. Hooper in FWTA notes " Differing from all the other species in Africa in having spikelets on short pedicels producing a compact inflorescence and glumes with a distinctly excurrent nerve".
Habitat: montane grassland 1800-3000 m alt.
Threats: none known
Management suggestions: it is suspected that this species may be an annual, growing between tussocks, but this needs confirmation. More data on the range and density of individuals of this variety on the mountain are needed.

Bulbostylis schoenoides (Kunth) C.B.Cl. subsp. *erratica* (Hook.f.) Lye
LR nt
Range: Mt Cameroon (c. nine pre-1988 collections), Bamenda Highlands, Chappal Waddi and a specimen from the Loma Mts (Sierra Leone).
Most of the 16 specimens studied of this species are from Mt Cameroon. This distinctive sedge is known both from above Buea and from Mann's Spring. The fertile specimens studied usually showed carbonization of the older, outer culms, indicating that these plants had withstood grassland fires. Fire may be desirable for this species in order to reduce competition from other species, or to trigger flowering. The first specimen collected was made by Mann (*Mann* 1344, January 1862) above the spring named for him. The most recent specimens date from the 1992 'Etinde' inventory (*Thomas* 9276, 9397). The outlier in Sierra Leone is *Morton* in SL 3563.
Habitat: montane grassland; 2400-3300 m alt.
Threats: annual burning may favour this species.
Management suggestions: not the highest priority for conservation, since it appears widespread and relatively common on the mountain and extends to the Bamenda Highlands. If

surveys of rare plants in the montane grassland are being conducted however, this species should be included. More data on the density of individuals in the population, and the range of the species on the mountain would be useful.

Carex preussii K.Schum.

LR nt

Range: Mt Cameroon (two pre-1988 collections), Bamenda Highlands, including Mt Oku (7 collections) & Mambila Plateau (1 collection).

This species was erroneously identified as *C. simensis* A. Rich. by Clarke, and so overlooked by him as a new species, although Mann had collected it as early as Dec. 1862 (*Mann* 2099). Schumann described *C. preussii* 35 years later. Mann gives the altitude for this collection as 9-10,000 ft, presumably an error, since this would take it above its habitat in montane forest gaps and edges. Most collections where altitudinal data is given indicate a concentration in the area of 2100 m alt. Unusually, many more specimens of this species are available from the Bamenda Highlands than from Mt Cameroon, the reverse of the usual situation.

The species was not recollected on the mountain until March 1948 (*Brenan & Onochie* 9538), at 2440 m alt, near Mann's Spring. The third collection, from the same area, is *Thomas* 9372.

According to the data available *Carex preussii* is limited to the Mann's Spring area on the mountain, and is not recorded from above Buea. Although restricted in distribution on Mt Cameroon, it appears widespread, perhaps common, in its habitat in the Bamenda Highlands.

Habitat: in gaps in montane forest; 2100-2400 m alt.

Threats: none known on Mt Cameroon, but clearance of montane forest in the Bamenda Highlands is a major threat.

Management suggestions: none.

DRACAENACEAE

Dracaena bueana Engl.

DD

Range: Mt Cameroon (one pre-1988 collection, possibly extinct) and Ghana (one collection).

This species commemorates the town of Buea, capital of the Bakweri, and was described from the conservatory in Berlin, based on a flowering specimen (*Deistel* 461) collected at Buea in Feb. 1900, reportedly growing in a hedge at c. 1000 m alt. *Dracaena bueana* has not been identified from the mountain since and its taxonomic status is not yet resolved. Bos, the authority on the genus, had no knowledge of this plant until 1984, when Leeuwenberg brought a specimen from Ghana, which at that time was judged to be either *D. fragrans* or *D. arborea*: they cannot be distinguished when sterile. However, as the plant developed, the leaves became twice as long as those of *D. arborea*, so it was suspected of being a new species. Eventually it flowered and Bos identified it as Engler's *D. bueana*. Its reproductive characters are intermediate between the two tree species mentioned above, and it may be of hybrid origin, however, the plant is fertile, and produces fruits. Living seedlings were distributed by Bos to a number of Botanic Gardens, including Limbe and Ghana (Bos *pers. comm.*). Collections of fertile tree *Dracaena* are needed if we are to determine whether *Dracaena bueana* persists on Mt Cameroon, and whether also *D. arborea* does. *D. fragrans* (more often shrub-like than tree-like) is already known from the mountain, as is *D. mannii*, another arborescent species, though from only a single specimen (*Thomas* 4507). There is no doubt that the difficulty in making good herbarium specimens of tree *Dracaena* has deterred most plant collectors from collecting what is all too easily assumed to be "*Dracaena arborea*". Only the gathering of further fertile specimens will enable us to sort out the tree *Dracaena* on the mountain and establish which are rare and in need of protection. In the Bakossi mountains, tree *Dracaena* are propagated and are planted in lines as barriers to livestock, thus indirectly conserving the species concerned (*pers. obs.*), a case of conservation through cultivation!

Habitat: "Buea, im hohen lichten Wald, im Buschwald und an freien Plätzen, ungefähr um 1000 m, wo sie oft vereinzelt emporragt" (*Deistel* 461, quoted by Engler in the species protologue in Engl. Jahrb. lix. Beibl. 131, 20 (1924)).

Threats: unknown.

Management suggestions: further research into the survival of this species is needed as outlined above.

Several species of *Dracaena* on Mt Cameroon are extremely rare and may be included in future editions of this Red Data list, e.g. *D. bicolor* Hook. f. The species of *Dracaena* on the mountain are numerous, but poorly worked out and more collections, particularly fertile ones, are needed to secure certain identifications in this genus of great potential as ecological indicators (the forest understorey species) and of considerable horticultural utility.

GRAMINEAE

Agrostis mannii (Hook.f.) Stapf
LR nt
Range: Bioko (six collections), Mt Cameroon (15 pre-1988 collections), Cameroon Highlands (one collection each from Bamboutos & Oku))
A common, tufted, silvery grass 1-2' tall. The number of collections suggests that this species is relatively abundant in its habitat on Mt Cameroon. Two further collections were made on Mt Cameroon in 1992 (*Thomas* 9322, 9387).
Habitat: montane grassland, sometimes at forest edge; 1850-3800 m alt.
Threats: unknown.
Management suggestions: none.

Helicotrichion mannii (Pilger) C.E.Hubb.
LR nt
Range: Bioko (four collections) and Mt Cameroon (10 pre-1988 collections).
The most recent collections on the mountain are, in 1992, *Tchouto* 382 and, in 1995 (Kew ASO expedition), *Marsden* 36.
Habitat: clearings in forest and deep soils in montane grassland; 1700-3000 m alt.
Threats: none known.
Management suggestions: none

Hypseochloa cameroonensis C.E.Hubb.
VU D2
Range: Mt Cameroon (six pre-1988 collections).
First collected (*Mildbraed* 10881) in 1928 (overlooked by Mann, surprisingly), and known from a total of only six collections, this is the rarest and the only incontestably strictly endemic grass on the mountain. It is a diminutive annual 15-18 cm tall. The other species of *Hypseochloa*, *H. matengoensis* C.E. Hubb., is known from a single collection in grassland at Songea, S. Tanzania. It is curious that *Hypseochloa cameroonensis* was not found in the inventories of the 1990s.

Habitat: montane grassland; (1000-)2100- 2800 m. Frequent between tufts of high grass, 2800 m (*Mildbraed*), frequent in grassland strewn with boulders. Mostly collected between Hut 2 and Hut 3 above Buea.

Threats: unknown

Management suggestions: more ecological, range and populational data are needed on this scarce endemic, but it is possible that, as an annual of the lusher lower altitude grassland it needs open areas for establishment, and regular firing of grassland may be found to favour the survival of this species.

Pentaschistis mannii C.E. Hubb.

LR nt

Range: Mt Cameroon (20 pre-1988 collections).

This species is now generally considered not be distinct from *P. pictigluma* (Steud.) Pilger, a species of montane Rwanda, Sudan (Imatongs) and Ethiopia (Linder *pers. comm.*). Indeed at Kew, *P. mannii* is not now recognized! It is notable that this plant is not found on neighbouring mountains, such as Bioko or Mount Oku, nor is any other *Pentaschistis* although those mountains have sufficient altitude to sustain this species. First collected on Mt Cameroon by Mann in Jan 1862 (type), 20 collections are now known from Mt Cameroon and most botanists ascending to the peak (where it is one of only six flowering plant species) collect it. The absence of recent collections owes to the fact that altitudes above 3000 m. were not covered in the 1992 inventory.

Habitat: upper montane grassland; 2700-4000 m alt. "summit with mosses in tussocks on bare volcanic ash" Keay.

Threats: none known.

Management suggestions: none known, perhaps unnecessary at this juncture.

ORCHIDACEAE

This account is based on screening a copy at Kew of Sanford's unpublished account of Orchids for Flore du Cameroun, due to be published in Yaounde in 1998/9 in three parts after scientific modification in Paris by Szlachetko and Olszewski. Sanford does not accept the work by Vermeulen on *Bulbophyllum*, in which many species accepted in FWTA are reduced to varieties. Since Sanford is not alone in this respect, we have followed his nomenclature here.

Angraecopsis cryptantha P. J. Cribb

VU D2

Range: Mt Cameroon

Described as a new species only recently (Cribb, 1996), this species is known only from the type collection made during an ODA funded botanical inventory of Etinde: *Thomas* 9443, collected on 11th Oct. 1992 "Likombe to Mann's Spring". It is remarkable that Mann apparently missed this species despite being in this location at the same season for a lengthy period.

Habitat: montane mist forest dominated by *Agauria salicifolia* and *Schefflera* on old lava flow, adjacent to grassland; 1900-2000 m alt.

Threats: none known.

Management suggestions: the site should be revisited to see if this species can be rediscovered. Numbers of individuals present should be recorded, and possible threats, if any, evaluated.

Angraecopsis tridens (Lindl.) Schltr.

EN A1c+2c

Range: Mt Cameroon (three pre-1988 collections), Bioko (four collections), Manengouba and Bamboutos Mts (one collection each).

Originally discovered on Bioko (*Mann* 646, type), this species was first collected on Mt Cameroon at Buea, 1250 m (*Preuss* 965), later by Gregory and then in 1965, by Sanford. The only collection of this diminutive epiphyte during the inventories of the 1990s was *Wheatley* 632 in October 1992 from 1400 m on the N face of the main peak of Etinde.

Habitat: epiphyte of montane forest 1250-1500 m alt.

Threats: none known.

Management suggestions: an attempt to rediscover this species should be followed by an evaluation of its population in terms of size, regeneration levels, and threats.

Barombia gracillima (Kraenzl.) Schltr.

LR nt

Range: Gabon (two collections), Mt Cameroon, Mbalomayo and the Bakossi Mts (one collection each).

The combination in *Aerangis*, necessary since *Barombia* is no longer maintained, has not yet been published as far as we know. Discovered at Barombi Mbo (*Preuss* 459) to the extreme north of our area, it has not yet been located in the Mount Cameroon area (hence it is not included in the checklist), but is included here since it is on the edge. It has also been found in the nearby Bakossi's at Supé (*Letouzey* 14361). This species is spectacular in bloom, with 4-5 white flowers each 7-8 cm across on each spike. Since it is conspicuous in flower, the sparsity of collections suggests that it has not been overlooked by collectors, but is actually highly rare.

Habitat: epiphytic in lowland forest.

Management suggestions: an attempt to rediscover this species should be followed by an evaluation of its population in terms of size, regeneration levels, and threats. Much of the forest around this crater lake survives due to the steepness of the walls being inconvenient *inter alia* for timber extraction and agriculture.

Bulbophyllum filiforme Kraenzl.

CR A1c+2c

Range: Mt Cameroon (two pre-1988 collections), Korup and Niger Delta (two collections).

Described from a collection made between Bimbia and Limbe (*Preuss* 1242, 20 April 1984), it was recollected later at Man O War Bay (*Schlechter* 15759). In March 1995 it was rediscovered (three collections) at Mabeta-Moliwe, i.e. in the same general area as the original collections, on an Earthwatch funded expedition led by Stuart Cable. Treated as a variety of *B. resupinatum* Ridl. by Vermeulen

Habitat: lowland evergreen forest, as an epiphyte.

Threats: forest clearance for agriculture, particularly plantations.

Management suggestions: effort should be made to rediscover this species and safeguard it.

Bulbophyllum gravidum Lindl.

VU D2

Range: endemic to Bioko and Mt Cameroon (one pre-1988 collection).

First discovered by Mann at Bioko, later (*Mann* 2126) at Mt Cameroon. Treated as a variety of *B. cochleatum* Lindl. by Vermeulen.

Habitat: epiphytic in montane forest; 1500 m alt.

Threats: on Mt Cameroon, past and future forest clearing for cocoyam farms pose a moderate threat.

Management suggestions: an attempt to rediscover this species should be followed by an evaluation of its population in terms of size, regeneration levels, and threats.

Bulbophyllum modicum Summerh.

EN A1c+2c

Range: Mt Cameroon (three pre-1988 collections).

Published in 1957, this species is known only from three collections, all from Buea (*Gregory* 193, *Keay* FHI 25439 and *Sanford* 630/65). A fourth possible collection from Bambui (Bamenda) is dubious since sterile. Treated in synonymy of the widespread *B. josephii* by Vermeulen.

Habitat: epiphytic in montane forest; 900-1200 m.

Threats: possibly forest clearance for agriculture, firewood and housing, all believed to be at their most intense in this vegetation type in the Buea area.

Management: this species merits effort to rediscover it and to assess population size and threats. It is possible, from the collection data available, that it may be restricted to the edge of the large town of Buea, and so threatened by urban expansion and proximity. Searches for the plants and promotion of the rarity of this species to those unwittingly threatening it should be considered.

Bulbophyllum porphyrostachys Summerh.

LR nt

Range: S. Nigeria (Okumu, Sapoba and Usonigbe F.R.s, and Calabar: at least five collections), Mt Cameroon (two pre-1988 collections) and Congo-Brazzaville (one collection).

Known on Mt Cameroon from only two collections (*Rosevear* 60/37 and *Keay* in FHI 28133), it is debatable whether this species qualifies as sufficiently rare for this list in view of its widespread distribution. Accepted as a species by Vermeulen.

Habitat as an epiphyte (but also on the 1922 lava flow as a terrestrial).

Threats: unknown

Management suggestions: an attempt to rediscover this species should be followed by an evaluation of its population in terms of size, regeneration levels, and threats. An exhaustive survey of the 1922 lava flow in 1995 did not turn up this species.

Diaphananthe bueae (Schltr.) Schltr.

EN A1c+2c

Range: endemic to Mt Cameroon (two pre-1988 collections) and the Cameroon Highlands (two collections).

As the specific epithet suggests, this species was first collected at Buea (*Deistel* s.n., type, collected 24 July 1905, "auf der altern Rinde hoher Baumen in gesellscaft anderer Orchideen. Walden in d. Ungebung Buea") and rediscovered there over 40 years later (*Gregory* 153). More recently it has been found near Mt Oku (*Letouzey* 8889 and *Mbenkum* 354). A possible record of this species from Ivory Coast (*Perez-Vera* 725, in fruit Nov. 1974) needs support from flowering material before it is confirmed.

Habitat: submontane forest as an epiphyte; 1000 m alt.

Threats: all localities known are believed to be under pressure for forest clearance for agricultural, firewood collection and (Buea) urban expansion.

Management suggestions: see under *Bulbophyllum modicum* which may be sympatric.

Disperis kamerunensis Schltr.

EN B1+2a

Range: Mt Cameroon (two pre-1988 collections; possibly extinct).

Known from two collections only. Not seen since 1930. First collected near Buea (*Preuss* 609, 28th September 1891, 1100-1200 m), it was later recollected at Musake Camp (locality unknown) about 1930 (*Maitland* 106 "in the forest." 1800 m alt.). It is curious that this species was not recollected in 1992 during the "Etinde" inventory, since this was conducted in the same habitat as that in which the type collection was made, at the same time of year. It may be that the species is localized to the Buea area, which was not covered so thoroughly as other areas in 1992.

Habitat in montane and submontane forest, terrestrial; 1100-1800 m alt.

Threats: clearance of forest for agriculture.

Management suggestions: an attempt to rediscover this species should be followed by an evaluation of its population in terms of size, regeneration levels, and threats. The identification of the whereabouts of "Musake Camp" would allow a potential second locality for the species to be searched and a better idea of the natural range (possibly former) of this species.

Gastrodia africana Kraenzl.

CR A1c+2c

Range: Mt Cameroon (one pre-1988 collection; now extinct?) and ?? Mt Kala.

Apparently overlooked by Summerhayes in FWTA, this appears to be the westernmost and only native African species of a genus of about 35 species of mycotrophic, sometimes edible and medicinal orchids, mostly found in S.E. Asia and tropical Australia. Collected by Dusen at Mt Cameroon between "Love and Ndiva" in the month of April (*Dusen* 397, type), it is unclear where these localities are or were. They are not referred to in Letouzey's gazetteer of 1968. However, if the letters v in these words are converted to b (they can be interchanged in German) then the localities are rendered Lobe and Ndibe, recognizable as a major plantation area immediately to the north of Mt Cameroon. The type is believed destroyed in Berlin. Sanford refers to this species a collection from Mt Kala (*Letouzey* 9509), but with a great deal of doubt and discussion.

Habitat: unknown, but presumably lowland forest.

Threats: unknown, but if lowland forest, then forest clearance at low altitude poses a major threat to this species, if it is not already extinct.

Management suggestions: research is needed on old place names of Mt Cameroon and on the itinerary in Cameroon of Dusen, so that the localities he mentions for this species can be traced. Only then can an attempt be made to rediscover *G. africana*.

Genyorchis macrantha Summerh.

VU D2

Range: Mt Cameroon (one pre-1988 collection).

Known from a single collection made in April 1948 at Mann's Spring (*Brenan & Richards* 9570), this species was not rediscovered until March 1988 (*Nemba et al.* 938, forest edge, northeast slope, radio station road. "the largest flowers so far recorded in the genus" (Summerhayes in Kew Bull. 11: 124 (1957)).

Habitat: epiphytic on tree trunk at margin of mist forest; 2260 m alt.

Threats: unknown.

Management suggestions: see under *Angraeopsis cryptantha* with which it is probably sympatric

Genyorchis platybulbon Schltr.

CR A1c+2c

Range: Mt Cameroon (two pre-1988 collections).

Sanford's MS lacks the page referring to this species, so we include it on the basis of its listing in FWTA (1968: 242) as being known from only two collections, the first from Moliwe (*Stammler* s.n., type), the second from Victoria (unknown collector in FHI 11155). A third collection from Douala-Edea (*Thomas* 311, February 1978) awaits confirmation, but may well extend the range of this species.

Habitat: lowland evergreen forest, epiphytic.

Threats: forest clearance for agriculture, particularly plantations, and urban spread.

Management suggestions: effort should be made to rediscover this species and safeguard it.

Habenaria microceras Hook. f.

LR nt

Range: Bioko (one collection), Mt Cameroon (three pre-1988 collections) and the Bamenda Highlands (two collections).

The three collections on Mt Cameroon are *Mann* 2116 (type), *Johnston* 31, 32 ("8,000 ft. 12/86"), and *Preuss* 967, ("Buea 2500 m, 24/9/91").The two collections in the Bamenda Highlands are *Savoury* UCI 462, *Bambuluwe & Letouzey* 14271, Mt Nseshele. Rediscovered in October 1992 on Mt Cameroon, when two collections were made near Bokwangwo Hut 3: *Thomas* 9381 at 2400 m alt. and *Cheek & Tchouto* 3641, the latter at a study of the forest: grassland ecotone at about the same altitude.

Habitat: montane grassland, sometimes at the ecotone with forest (rarely epiphytic); 2000-3050 m alt.

Threats: unknown

Management suggestions: a survey to estimate the range, density and threats (if any) to the population on Mt Cameroon seems worthwhile.

Habenaria obovata Summerh.

VU D2

Range: Mt Cameroon (five pre-1988 collections).

The five historical collections known are *Johnston* 29 (type, "comm. 12/86"), *Mary Kingsley* s.n. (c. 1890 "comm. iii. 96"), *Preuss* 980 (24 November 1891), *Maitland* 804 ("very frequent" December 1929) and *Boughey* s.n. (September 1954). "A remarkable species without any near relative known to me." (Summerhayes, in the protologue of this species). Recently rediscovered (*Sidwell* 39) in grassland at 2820 m alt. in early October 1992 above Mann's Spring "reasonably common in grassland, seen down to around 2400 m alt." The most recent collection (*Banks* 104) was collected in early October 1995 between Huts 2 & 3 above Buea.

Habitat: terrestrial in montane grassland; 2150-3050 m alt.

Threats: unknown.

Management suggestions: a survey to estimate the range, density and threats (if any) to the population on Mt Cameroon seems worthwhile.

Liparis goodyeroides Schltr.

CR A1c+2c

Range: Nigeria (one collection) and W. Cameroon (three collections, one of which on Mt Cameroon).

The type collection of this species is from Moliwe (*Stammler* s.n., 1900) and the only other collections known are south of Ngu at the Plain of Mbaw (*Brunt* 479), Takamanda (*Thomas et al.* 30 April 1987) and, in Nigeria, the Niger Estuary (?) (*Barter* 2029).

Habitat: terrestrial in evergreen forest usually of foothills.

Threats: forest clearance for agriculture (particularly plantations) and firewood.

Management suggestions: although this species was not rediscovered in the botanical inventory of Mabeta-Moliwe in 1992, nor in 1993, it may not yet be extinct on Mt Cameroon, and efforts should be made to rediscover it in Mabeta-Moliwe.

Polystachya albescens Ridl. subsp. *angustifolia* (Summerh.) Summerh.

EN A1c+2c

Range: Mt Cameroon (four pre-1988 collections).

Discovered by Miss Gregory in 1947, the four collections from which epiphyte is known, are all from near Buea: (*Gregory* 165, 284; *Ejiofor* in FHI 25340 and *Keay* in FHI 22411). The last two specimens were grown from material obtained from Miss Gregory. Her notes (*Gregory* 165, 3rd July, are as follows: "Buea, upper part of golf course, tree. 20 feet from ground, very dense growth reminiscent of mistletoe. Some pendulous stems three feet long." The only recent collection is *Tekwe* 107, who may be only the second person to have collected this subspecies in the wild.

Habitat: submontane forest, epiphytic; c. 900 m alt.

Threats: possibly forest clearance for agriculture, firewood and housing.

Management: this species merits effort to rediscover it and to assess population size and threats. Charles Tekwe's assistance will be required. It is possible, from the collection data available, that it may be restricted to the edge of the large town of Buea, and so threatened by urban expansion and proximity. Searches for the plants and the promotion of the rarity of this species to those unwittingly threatening it should be considered.

Polystachya bicalcarata Kraenzl.

EN A1c+2c

Range: Bioko (one collection), Mt Cameroon (five pre-1988 collections) and the Bamboutos Mts (one collection).

Of the six Cameroonian collections, only one (*Jacques-Félix* 5439, Bamboutos Mts), is not from Mt Cameroon. The five collections known on the mountain are: *Deistel* 62 C (type), 79; *Maitland* 730, *Dundas* in FHI 15303 and *Sanford* 488/65, all with Buea given as locality, where known. *Thomas* 9178 (1992) is the only recent collection of the species from the mountain.

Habitat: submontane and montane forest, epiphytic; 950-1500 m alt.

Threats: see *Liparis goodyeroides*

Management suggestions: see *Liparis goodyeroides*

Polystachya superposita Reichb. f.

EN A1c+2c

Range: Bioko (one collection) and Mt Cameroon (two pre-1988 collections).

Discovered on Mt Cameroon (*Mann* 2125, type, "Nov. 1862, 5,000 ft.") this species is otherwise only known on the African mainland from a single collection (*Deistel* s.n.) near Buea.

Habitat: submontane and montane forest, epiphytic; c. 900-1500 m. alt.

Threats: see *Liparis goodyeroides*

Management suggestions: see *Liparis goodyeroides*

Polystachya victoriae Kraenzl.
CR A1c+2c
Range: coastal Gabon (one collection) and Cameroon (six collections, two of which from Mt Cameroon).
First described (*Diestel* 190, type, February 1899) from Victoria, it was recollected there by Simon at the Botanic Garden (*Simon* 14, 22 February 1909 "Victoria in der Nahe des Bot. Gartens, auf hohem Baum"). Otherwise it is also known from Eseka, Edea, Kribi, Ebolowa and Bipindi
Habitat: lowland evergreen forest; epiphytic
Threat: forest clearance for agriculture, particularly plantations.
Management suggestions: forest protection.

ZINGIBERACEAE

This family, particularly *Aframomum*, the most species-rich genus, is being actively revised in Africa by several workers. Dr Harris (Edinburgh) is dealing with Cameroon species and has advised us that while it is likely that several species on Mt Cameroon may prove to be rare, *Aframomum* sp. A and *Renealmia albo-rosea* of FWTA, should not be included here (both of which appeared on our candidate red-data list). It is notable that of the 18 species of *Aframomum* referred to in the checklist, only six can be given scientific names at present.

FERNS

Asplenium adamsii Alston
LR nt
Range: Mt Cameroon (three pre-1988 collections), East Africa (four collections).
Published in 1956, this species is known from only three collections, *Adams* 1271, 1278 and 1284 on Mt Cameroon. It is notable that this species has not been rediscovered since it was described, even by the 1992-4 surveys. This may indicate that populations are small, the plants inconspicuous and that they are highly specialized in their habitat requirement. No specific locality data is available, but from the habitat data it is likely that the specimens were gathered either between hut 1 and 2 or near Mann's Spring.
The East African collections date from the 1960s: *Faden* 69/155 and *Polhill* 12026 from the Aberdare Mts of Kenya, and *Gilbert* 3467 and *Vesey-Fitzgerald* 5632 from Mt Meru, Tanzania.
Habitat: rocks at lower limits of montane grassland, c. 2600-2700 m alt.
Threats: unknown.
Management suggestions: effort should be made to rediscover this apparently very rare species and elucidate basic data on range, population size, ecological requirements and threats, if any.

Asplenium sp. 8. (of R. Johns).
CR A1c+2c
Range: Mt Cameroon.
One of 14 new African species of simple-fronded 'birds nest' *Asplenium* identified in recent years by Prof. Johns. Collected in March 1995 (*M. Groves* 230) along the old road to Bimbia north of the village of Dikulu and so far known only from this specimen.
Habitat: epiphyte on *Elaeis* in secondary forest along the shore.

Threats: forest clearance for agriculture and timber extraction.

Management suggestions: sampling of simple-fronded *Asplenium* in the Mount Cameroon area should be intensified so that the ranges of the six species (two apparently endemic to Mt Cameroon) of this group on the mountain are better known. These species may prove amenable to cultivation at the Limbe botanic garden.

Asplenium sp. 9 (of R. Johns)

CR A1c + 2c

Range: Mt Cameroon

Known only from a single collection (*Cheek* 5576, Nov. 1993) made above Njonji on the path to Lake Njonji.

Habitat: epiphytic on trunk 1.5 m above the ground in lowland evergreen forest; 300 m alt.

Threats: forest clearance for plantations.

Management suggestions: see notes for *Asplenium sp.* 8 (above).

Pteris ekemae Benl

CR A1c + 2c

Range: Mt Cameroon (two pre-1988 collections).

Discovered with the assistance of the 'old man' of the Mount Cameroon Project, Ekema Ndumbe for whom it was named by Benl. It is known from only two collections: *Benl* 75/49 and *Benl* 75/69, both made above and to the NW of Mapanja. Care should be taken not to confuse this species with *P. manniana* Mett. ex Kuhn (syn. *P. camerooniana* Kuhn) and *P. quadriaurita* Retz. subsp. *togoensis* (Hieron.) Schelpe (syn. *P. togoensis* Hieron.), both of which are reported by Benl as growing abundantly in the neighbourhood of *Pteris ekemae*. Diagnostic characters of the latter are the rows of 2-3 areoles formed by vein anastomoses, the slightly winged rhachis and the tufted fronds.

Habitat: "moist and shady habitats in the evergreen rain forest on the southern slope of Cameroon Mt (Benl 1976: 150); 1100-1250 m alt."

Threats: unknown, but clearance for new coco-yam farms may pose a threat.

Management suggestions: the assistance of Ekema Ndumbe should be sought in refinding this species. Is this species highly localized or does its range extend around the whole or part of the periphery of the mountain? This is one question that should be addressed before attempts are made to manage this species.

Xiphopteris villosissima (Hook.f) Alston var. *laticellulata* Benl

EN A1c + 2c

Range: Mt Cameroon (two pre-1988 collections).

Published in 1976 (see *Pteris ekemae*), this variety was collected above Buea at 1200 m alt. in 1970/71 (*Benl* 70-71/27). An epiphyte, it was not recollected in the 1992 inventory, but Buea was not the focus of activity at that time.

Habitat: epiphyte in montane forest; 1200 m alt.

Threats: unknown, but firewood collection for Buea may prove a threat.

Management suggestions: those working above Buea should take the opportunity to rediscover this variety.

READ THIS FIRST!

Before using this checklist, the following explanatory notes to the conventions and format used should be read.

There is no index. Names of taxa are organized alphabetically: species within genus, genus within family, and family within the groups Dicotyledons, Monocotyledons, Gymnosperms, Fern allies and Ferns, respectively. The families and genera accepted here follow Vascular Plant Families and Genera (Brummitt 1992), excepting in the *Ochnaceae*, where *Campylospermum* Tiegh., *Rhabdophyllum* Farron and *Idertia* Tiegh. are recognized in place of *Gomphia* Schreb., and in *Cyperaceae*, where we follow Lye in sinking *Pycreus* Pal. and *Kyllinga* Rottb. into *Cyperus* L.

The species names adopted (in bold type) follow the most recently published research that is available and acceptable. In many cases the specimens cited, after preliminary naming in Cameroon (see acknowledgements) had their names confirmed or determined by a family specialist, who is named together with the standard abbreviation for their herbarium (following Index Herbariorum, Holmgren et al. 1990) at the head of the account, below the family name. The same specialist or specialists have also checked the species data (see below) under each entry, compiled by SC and edited by MC. Specimens in the remaining families were mostly checked or named at RBG, Kew by SC, beginning in 1994, apart from the c. 2300 specimens from the 1992 Mabeta-Moliwe inventory named by Karen Sidwell and MC, together with Terry Sunderland, Adam Faruk and numerous specialists (Cheek, 1992). Abbreviations of the authors of species names follow Authors of Plant Names (Brummitt & Powell 1992).

A reference, or references, are given under each species name to a place where that name has been used recently and authoritatively, and where a description, diagnostic keys and more information than can be provided here is available. In most cases this reference is to F.W.T.A. (The Flora of West Tropical Africa, Keay & Hepper 1954-1972), but in other cases we refer to Flore du Cameroun or Flore du Gabon, or to a research paper, or rarely to the place where the name was validated. Names added during editing generally lack references. It is important to note that the reference given is not usually the protologue as is common practice in many botanical works. Each reference has been seen by SC.

Synonyms, in italics, with references, are listed below the correct species names. About 25 per cent of the species names accepted here do not appear in FWTA although Mt Cameroon does fall (just) within the FWTA area. This is because these species are new additions to the FWTA area, being newly discovered at Mt Cameroon, or because the names used in FWTA for the species concerned have been superseded by research over recent decades. New names were provided by specialists and were also discovered while checking names of specimens at K. The Enumeration des Plantes a Fleurs D'Afrique Tropicale (Lebrun & Stork 1991-1997) was also of great assistance in checking for the latest nomenclature. Several well known montane species in particular, as a result of continental-wide revisions of the genera in which they fall, have suffered name changes. Amongst the heathers, *Philippia mannii* is now *Erica mannii* and *Blaeria mannii* is *Erica tenuicaulis*, as result of research in South Africa that shows that neither *Philippia* nor *Blaeria* can be maintained as distinct from *Erica*. What was referred to as *Hypericum lanceolatum* on Mt Cameroon in all previous works should now be called *Hypericum revolutum*, and similarly *Hypericum riparium* is a synonym of *Hypericum roeperianum. Geranium ocellatum* is now *Geranium mascatense* and *Geranium simense* is now *Geranium arabicum*.

Not all names listed here are straightforward binomials with authorities. A generic name followed by "sp." generally indicates that the specimens cited are inadequate to name to species. These specimens are often sterile plot voucher specimens. While sterile specimens with adequate collection data can usually be named to species, though often with considerable difficulty in a poorly known flora, for some genera, such as *Tabernaemontana*, this is impossible. A generic name followed by "cf." e.g. *Garcinia cf. smeathmannii*, indicates that the specimens cited below the name should be compared with, in this case *Garcinia smeathmannii*. This is an indication of doubt, suggesting that the material in question is closely related to the species named, but may be something different. It is often the case that more complete material is needed to resolve the question. The use of "sp. nov." or "sp. nov. aff." as in *Cola sp. nov. aff. philipi-jonesii* is a firm statement that the specimens concerned are considered to be new species to science, in this case, a new species related to *Cola philipi-jonesii*.

The species data given has been compiled by SC. It includes a brief habit description, e.g. "herb" or "straggling shrub to 2 m." which has been taken from the specimens cited. This is followed for many species by a note on the inferred ecological guild, e.g. "pioneer", "light demander" (non pioneers) or "shade-bearer" which has been provided by William Hawthorne largely on this basis of his work in Ghana (Hawthorne 1995). Habitat information is taken from the specimens cited and is simplified to e.g. "forest", "grassland" or "forest-grassland transition". It is important to read this information in correlation with the altitudinal records for the species in order to determine, if the habitat is, for example forest, whether the species is predominantly of lowland, submontane or montane forest.

The distribution of species by country, is largely taken from FWTA and so is not necessarily definitive. Countries are cited in FWTA fashion, from West to East, then North to South. Note that "Guinea" refers to Guinea-Conakry and that "Congo" is Congo-Brazzaville, as opposed to Zaire (or Congo-Kinshasha). Equatorial Guinea is for our purposes referred to either as "Rio Muni" (the mainland part) or "Bioko" (the island).

The chorological term most commonly used here, "Guineo-Congolian" refers to species occurring in both the Guinean forest area (approximately Senegal to Gabon) and the Congolian area (principally Zaire and bordering rainforest areas of C.A.R. and Uganda). "Guinea" as a chorological unit is broken down further to "Upper Guinea" (Senegal to Ghana) and "Lower Guinea" (Nigeria to Gabon). More widely spread species are generally cited as "Tropical Africa", i.e. more widespread than Guineo-Congolian but being confined to the area North of the Tropic of Capricorn and south of the Sahara. "Afromontane" refers to species restricted to those parts of Africa predominantly above about 1500 m alt. "Western Cameroon Uplands (montane)" is a subdivision of this unit used by us for species restricted to montane Bioko, the western Cameroon mountains and adjoining montane parts of Nigeria. Other terms used, e.g. "Tropical and southern Africa"; "Palaeotropical" and "Cosmopolitan" are self-explanatory.

"Star" refers to the star-rating system developed by William Hawthorne during rapid botanical inventory surveys in Ghana (Hawthorne 1995). His Genetic Heat Index is a mechanism for assessing the conservation value of an area, on the basis of the score accumulated from the points assigned to species in that area. Points are assigned to species in roughly inverse proportion to their geographical range on a degree square basis. Thus "black star" species are considered to be those needing urgent conservation attention, occurring in only one or two degree squares, and score 27 points in the Hawthorne system. Gold star species are fairly rare locally or internationally, occur in 4-12 degree squares, and score 9 points. Blue star species are those widespread in the country, but rare locally, or vice-versa, occur in 12 -37 degree squares and score three points. Scarlet, Red and Pink star species are different categories of common, but exploited species, occurring in 24-56 degree squares, and score two points. Green star species are common, of no particular concern for conservation, occur in up to 130 degree squares and score zero points. The star ratings used in this checklist are modified and supplemented by SC from those made in 1997 by Peguy Tchouto and William Hawthorne for Mt Cameroon species.

The altitudinal range given for each species is generated from the BRAHMS Mt Cameroon specimen data base, and shows the number of specimens recorded from each 200 m interval; likewise, flowering and fruiting periods have been generated in the same way and show the number of specimens in flower or fruit in the months indicated. Specimens captured from FWTA lack this data.

Specimens are cited for each of the protected areas around or on the mountain. Protected areas are cited in alphabetical, not geographical order. The areas cited are illustrated in Map 4, with the exception that specimens from the former Etinde area have been subdivided to follow recent changes at the Mount Cameroon Project. "Eastern slopes" is the area from Mt. Etinde in the south, extending northeastward to Buea, whilst "Etinde" is now confined to the coastal area from Mt Etinde northwestwards to Idenau. Botanical inventories made in the last ten years have been made with reference to these areas and it has been relatively easy to place specimens in an area. Many older specimens lack precise location data e.g. "Cameroons Mt" and so are given as "unlocated". Older specimens that do have more precise location data were often only captured onto the database from the abbreviated FWTA citation which excludes this data. Hence these specimens are also cited as "unlocated". Some older specimens have been "shoehorned" into some of the protected areas. For example, the Mildbraed specimens collected in 1928 from Likomba a few kilometres east of Mabeta Moliwe have been cited as from that area. Likewise, some specimens from Kumba collected by Duncan Thomas in the 1980 s have been shoehorned, so to speak, into the S. Bakundu area. Due to space constraints, it was not practicable to included more detailed specimen data than has been presented here. This can be found on the specimens themselves and on the specimen database.

Within each protected area specimens are cited in alphabetical order of the collectors name. The specimens cited are located at the Herbarium in the Limbe Botanic Garden (SCA), with duplicates at the National Herbarium of Cameroon at Yaounde (YA) and at the Herbarium of the Royal Botanic Gardens, Kew (K). Where more than three duplicates were collected, the fourth collection can be found at the herbarium of the specialist concerned in naming the family concerned, and thereafter specimens are placed at Wageningen (WAG), Missouri (MO), Brussels (BR) and Paris (P). Pre-1990 specimens cited are generally at the Royal Botanic Gardens, Kew.

DICOTYLEDONS

ACANTHACEAE

K. Vollesen (K)

Acanthopale decempedalis C.B.Clarke
F.W.T.A. 2: 398 (1963).

Habit: herb. Habitat: forest-grassland transition. Distribution: SE Nigeria, Bioko and W. Cameroon. Chorology: Western Cameroon Uplands (montane). Star: gold. Alt. range: 600-800m (1), 1601-1800m (1), 2001-2200m (1), 2201-2400m (1). Flowering: Sep (1), Oct (1), Nov (2). Specimens: - *Eastern slopes:* Cheek M. 5345; Preuss P. R. 947; Tchouto P. 324. *Etinde:* Cheek M. 5594; Thomas D. W. 9184

Acanthus montanus (Nees) T.Anderson
F.W.T.A. 2: 410 (1963); Fl. Gabon 13: 112 (1966).

Habit: shrub to 2 m. Habitat: forest. Distribution: Benin to Bioko, Cameroon, Gabon, Congo (Kinshasa), Angola and CAR. Chorology: Guineo-Congolian. Star: green. Alt. range: 1-200m (1), 201-400m (1), 601-800m (1). Flowering: Nov (1), Dec (1). Specimens: - *Etinde:* Lighava M. 17; Maitland T. D. FHI 7749. *Mabeta-Moliwe:* Plot Voucher W 1206. *Mokoko River F.R.:* Fraser P. 464. *Onge:* Harris D. J. 3793

Adhatoda buchholzii (Lindau) S.Moore
F.W.T.A. 2: 422 (1963).

Habit: woody climber. Habitat: forest. Distribution: SE Nigeria, Cameroon and Gabon. Chorology: Lower Guinea. Star: blue. Alt. range: 200-400m (1), 401-600m (1). Flowering: Nov (1), Dec (1). Specimens: - *Etinde:* Cable S. 295; Cheek M. 5474

Adhatoda camerunensis Heine
F.W.T.A. 2: 422 (1963).

Habit: shrub to 2 m. Habitat: forest. Distribution: Cameroon. Chorology: Lower Guinea. Star: gold. Flowering: Jan (1). Fruiting: Jan (1). Specimens: - *Bambuko F.R.:* Keay R. W. J. FHI 37391

Afrofittonia silvestris Lindau
F.W.T.A. 2: 417 (1963).

Habit: trailing herb. Habitat: forest, sometimes on rocks. Distribution: SE Nigeria, Bioko and SW Cameroon. Chorology: Western Cameroon. Star: black. Alt. range: 1-200m (3), 201-400m (2), 401-600m (1). Flowering: Sep (1), Oct (3), Nov (2). Fruiting: Oct (1). Specimens: - *Etinde:* Tchouto P. 471; Wheatley J. I. 537. *Mabeta-Moliwe:* Plot Voucher W 732, W 1152. *Onge:* Cheek M. 5146; Tchouto P. 866; Thomas D. W. 9846; Watts J. 919

Ascotheca paucinervia (T.Anderson ex C.B.Clarke) Heine
Fl. Gabon 13: 210 (1966).
Rungia paucinervia (T.Anderson ex C.B.Clarke) Heine, F.W.T.A. 2: 429 (1963).

Habit: herb. Habitat: forest. Distribution: SE Nigeria, Cameroon and Gabon. Chorology: Lower Guinea. Star: blue. Alt. range: 200-400m (1). Flowering: Mar (1). Specimens: - *Mokoko River F.R.:* Tchouto P. 523

Asystasia decipiens Heine
F.W.T.A. 2: 413 (1963); Fl. Gabon 13: 136 (1966).

Habit: herb. Guild: pioneer. Habitat: forest. Distribution: Sierra Leone to Cameroon, Gabon and CAR. Chorology: Guineo-Congolian. Star: green. Alt. range: 200-400m (3), 401-600m (1), 601-800m (1), 801-1000m (1). Flowering: Oct (3), Nov (1), Dec (2). Specimens: - *Etinde:* Cable S. 382; Kwangue A. 129; Lighava M. 15. *Onge:* Ndam N. 641; Tchouto P. 894; Watts J. 822

Asystasia gangetica Lindau
F.W.T.A. 2: 413 (1963); Fl. Gabon 13: 134 (1966).
Asystasia calycina Benth., F.W.T.A. 2: 413 (1963).

Habit: herb. Guild: pioneer. Habitat: forest and farmbush. Distribution: palaeotropical. Star: excluded. Alt. range: 1-200m (5). Flowering: Mar (1), May (2), Aug (1), Dec (1). Specimens: - *Mabeta-Moliwe:* Baker W. J. 345; Cable S. 562, 1450; Ndam N. 532; Sunderland T. C. H. 1356

Asystasia lindaviana Hutch. & Dalziel
Fl. Gabon 13: 138 (1966).
Filetia africana Lindau, F.W.T.A. 2: 417 (1963).

Habit: shrub to 1 m. Habitat: forest. Distribution: Cameroon and Gabon. Chorology: Lower Guinea. Star: blue. Alt. range: 1-200m (1). Flowering: Mar (2). Specimens: - *Etinde:* Nkeng P. 45. *Southern Bakundu F.R.:* Brenan J. P. M. 9412

Asystasia macrophylla (T.Anderson) Lindau
F.W.T.A. 2: 412 (1963); Fl. Gabon 13: 127 (1966).

Habit: shrub to 3 m. Habitat: forest. Distribution: SE Nigeria, Bioko, Cameroon and Gabon. Chorology: Lower Guinea. Star: blue. Alt. range: 1-200m (4), 201-400m (1), 401-600m (1). Flowering: Feb (1), Mar (5), May (1). Specimens: - *Bambuko F.R.:* Watts J. 610. *Etinde:* Khayota B. 550. *Mabeta-Moliwe:* Cable S. 1452, 1475; Ndam N. 529; Wheatley J. I. 37. *No locality:* Nkeng P. 11. *Southern Bakundu F.R.:* Olorunfemi J. FHI 30545

Asystasia vogeliana Benth.
F.W.T.A. 2: 412 (1963); Fl. Gabon 13: 130 (1966).

Habit: shrub to 1.5 m. Guild: pioneer. Habitat: forest and stream banks. Distribution: Sierra Leone to Bioko, Cameroon, Gabon, Congo (Kinshasa) and CAR. Chorology: Guineo-Congolian. Star: green. Alt. range: 1-200m (3), 201-400m (3), 401-600m (2), 601-800m (1), 801-1000m (1), 1001-1200m (1). Flowering: Jan (2), Feb (2), May (1), Nov (3), Dec (3). Specimens: - *Eastern slopes:* Cable S. 1325; Tekwe C. 9. *Etinde:* Cable S. 279; Cheek M. 5590, 5841, 5862; Mbani J. M. 9; Sunderland T. C. H. 1076. *Mabeta-Moliwe:* Cable S. 521; Nguembock F. 37; Sunderland T. C. H. 1302. *No locality:* Mann G. 1955

Barleria opaca (Vahl) Nees
F.W.T.A. 2: 421 (1963).

Habit: straggling shrub to 2 m. Guild: pioneer. Distribution: Sierra Leone to Cameroon. Chorology: Guineo-Congolian. Star: green. Alt. range: 800-1000m (1). Flowering: Dec (1). Specimens: - *Eastern slopes:* Maitland T. D. 139

Brachystephanus jaundensis Lindau
Kew Bull. 51: 760 (1996).

Habit: herb. **Habitat:** forest. **Distribution:** Cameroon and Gabon. **Chorology:** Lower Guinea (montane). **Star:** blue. **Alt. range:** 800-1000m (1). **Flowering:** Dec (1). **Specimens:** - *Etinde:* Cable S. 450

Brachystephanus longiflorus Lindau
F.W.T.A. 2: 431 (1963); Kew Bull. 51: 761 (1996).

Habit: herb. **Habitat:** forest. **Distribution:** SE Nigeria, Bioko and SW Cameroon. **Chorology:** Western Cameroon Uplands (montane). **Star:** gold. **Fruiting:** Dec (1). **Specimens:** - *Eastern slopes:* Gregory H. 229; Preuss P. R. 890

Brillantaisia lamium (Nees) Benth.
F.W.T.A. 2: 406 (1963); Fl. Gabon 13: 88 (1966).

Habit: herb to 1.5 m. **Guild:** pioneer. **Habitat:** forest and farmbush. **Distribution:** Guinea to Cameroon, Congo (Kinshasa), Angola, Uganda, W. Kenya and W. Tanzania. **Chorology:** Guineo-Congolian. **Star:** green. **Alt. range:** 1-200m (2), 201-400m (1). **Flowering:** Mar (1), Oct (2). **Fruiting:** Oct (1). **Specimens:** - *Etinde:* Mbani J. M. 71; Tchouto P. 479. *Mabeta-Moliwe:* Sidwell K. 6

Brillantaisia lancifolia Lindau
F.W.T.A. 2: 406 (1963); Fl. Gabon 13: 86 (1966).

Habit: herb. **Habitat:** forest. **Distribution:** SE Nigeria, Cameroon and Gabon. **Chorology:** Lower Guinea. **Star:** blue. **Alt. range:** 600-800m (1). **Flowering:** Nov (1). **Specimens:** - *Etinde:* Cable S. 223

Brillantaisia owariensis P.Beauv.
F.W.T.A. 2: 406 (1963); Fl. Gabon 13: 84 (1966).
    Brillantaisia nitens Lindau, F.W.T.A. 2: 406 (1963).

Habit: herb to 3 m. **Habitat:** forest and forest-grassland transition. **Distribution:** tropical Africa. **Chorology:** Afromontane. **Star:** green. **Alt. range:** 1000-1200m (1), 1401-1600m (2), 1801-2000m (1). **Flowering:** Feb (1), Oct (1). **Specimens:** - *Eastern slopes:* Cable S. 1328; Preuss P. R. 847; Sidwell K. 98. *Etinde:* Thomas D. W. 9222. *No locality:* Mann G. 1959

Brillantaisia cf. owariensis P.Beauv.

Specimens: - *Mokoko River F.R.:* Acworth J. M. 116

Brillantaisia soyauxii Lindau
Fl. Gabon 13: 84 (1966).

Habit: herb. **Habitat:** forest. **Distribution:** Cameroon, Gabon and Congo (Kinshasa). **Chorology:** Guineo-Congolian. **Star:** green. **Alt. range:** 800-1000m (1). **Flowering:** Oct (1). **Specimens:** - *Etinde:* Tchouto P. 446

Brillantaisia vogeliana (Nees) Benth.
F.W.T.A. 2: 406 (1963); Fl. Gabon 13: 92 (1966).

Habit: herb to 1.5 m. **Guild:** pioneer. **Habitat:** forest. **Distribution:** Ghana to Bioko, Cameroon, Gabon, Congo (Kinshasa) CAR, Sudan, Uganda and Kenya. **Chorology:** Guineo-Congolian (montane). **Star:** green. **Alt. range:** 1-200m (1), 801-1000m (1), 1801-2000m (1). **Flowering:** Oct (1), Nov (1), Dec (1). **Fruiting:** Oct (1). **Specimens:** - *Eastern slopes:* Cheek M. 5367. *Etinde:* Maitland T. D. 757; Watts J. 572. *Mabeta-Moliwe:* Cable S. 600

Chlamydocardia buettneri Lindau
F.W.T.A. 2: 423 (1963); Fl. Gabon 13: 185 (1966).

Habit: herb. **Habitat:** forest and farmbush. **Distribution:** Ivory Coast, Nigeria, Cameroon and Gabon. **Chorology:** Guineo-Congolian. **Star:** green. **Alt. range:** 1-200m (2). **Flowering:** Oct (1), Dec (1). **Specimens:** - *Etinde:* Preuss P. R. 1309. *Mabeta-Moliwe:* Cable S. 605. *Onge:* Tchouto P. 1024

Crossandrella dusenii (Lindau) S.Moore
F.W.T.A. 2: 412 (1963); Fl. Gabon 13: 110 (1966).

Habit: shrub to 1 m. **Habitat:** forest. **Distribution:** SE Nigeria, Cameroon, Gabon, Congo (Kinshasa) and Uganda and Tanzania. **Chorology:** Guineo-Congolian. **Star:** green. **Alt. range:** 1-200m (2), 201-400m (2), 401-600m (1). **Flowering:** Mar (1), Oct (2), Nov (1). **Fruiting:** Mar (1), Oct (2), Nov (1). **Specimens:** - *Etinde:* Williams S. 78. *Mokoko River F.R.:* Tchouto P. 625. *Onge:* Akogo M. 98; Ndam N. 795; Tchouto P. 766

Dicliptera alternans Lindau
F.W.T.A. 2: 426 (1963).

Habit: herb. **Distribution:** Cameroon. **Chorology:** Lower Guinea. **Star:** gold. **Fruiting:** Jan (1). **Specimens:** - *Eastern slopes:* Preuss P. R. 604

Dicliptera laxata C.B.Clarke
F.W.T.A. 2: 426 (1963).

Habit: scrambling shrub. **Habitat:** forest. **Distribution:** Cameroon, Bioko, Congo (Kinshasa) and East Africa. **Star:** green. **Flowering:** Jan (1). **Specimens:** - *Eastern slopes:* Deistel H. 489

Dicliptera obanensis S.Moore
F.W.T.A. 2: 426 (1963); Fl. Gabon 13: 195 (1966).

Habit: herb. **Guild:** pioneer. **Habitat:** forest. **Distribution:** Sierra Leone, Ghana and SE Nigeria. **Chorology:** Guineo-Congolian. **Star:** green. **Alt. range:** 200-400m (1), 601-800m (1). **Flowering:** Nov (2). **Specimens:** - *Etinde:* Cable S. 243; Cheek M. 5485

Dicliptera verticillata (Forssk.) C.Chr.
F.W.T.A. 2: 425 (1963); Fl. Gabon 13: 196 (1966).

Habit: herb. **Guild:** pioneer. **Distribution:** tropical Africa. **Star:** excluded. **Flowering:** Feb (1). **Fruiting:** Feb (1). **Specimens:** - *Etinde:* Maitland T. D. 978

Dischistocalyx grandifolius C.B.Clarke
Fl. Gabon 13: 25 (1966).

Habit: herb to 2 m. **Habitat:** forest. **Distribution:** Cameroon and Gabon. **Chorology:** Lower Guinea. **Star:** blue. **Alt. range:** 1-200m (2), 201-400m (1), 801-1000m (1), 1001-1200m (2). **Flowering:** Mar (3), Jul (1), Nov (2). **Fruiting:** Jul (1). **Specimens:** - *Bambuko F.R.:* Watts J. 593. *Etinde:* Cable S. 245, 1548, 1649; Cheek M. 5406; Tchouto P. 258

Dischistocalyx strobilinus C.B.Clarke
Fl. Gabon 13: 17 (1966).

Habit: herb. **Habitat:** forest, sometimes on rocks. **Distribution:** Cameroon, Rio Muni and Gabon. **Chorology:** Lower Guinea. **Star:** blue. **Alt. range:** 1-200m (6), 201-400m (5), 401-600m (1), 601-800m (1). **Flowering:** Mar (2), Aug (2), Oct (8), Nov (1). **Fruiting:** Oct (1), Nov (1). **Specimens:** - *Etinde:* Cheek M. 5433; Tchouto P. 451; Thomas D. W. 9133; Watts J. 476. *Onge:* Akogo M. 58, 59; Baker W. J. 361; Harris D. J. 3876; Ndam N. 729; Tchouto P. 722; Watts J. 693, 745; Wheatley J. I. 700, 846

Elytraria marginata Vahl
F.W.T.A. 2: 418 (1963); Fl. Gabon 13: 155 (1966).

Habit: herb. **Guild:** shade-bearer. **Habitat:** forest. **Distribution:** Guinea to Bioko, Cameroon, Gabon, Congo (Kinshasa), Angola, CAR, Sudan and Uganda. **Chorology:** Guineo-Congolian. **Star:** green. **Alt. range:** 1-200m (5), 201-400m (3), 601-800m (1). **Flowering:** Mar (1), Apr (2), May (2), Jun (1), Oct (1), Nov (2). **Fruiting:** Mar (1), Apr (1), Oct (1), Nov (2). **Specimens:** - *Etinde:* Cable S. 241; Cheek M. 5425. *Mabeta-Moliwe:* Schlechter F. R. R. 12386; Watts J. 356; Wheatley J. I. 106. *Mokoko River F.R.:* Akogo M. 293; Ekema N. 896; Tchouto P. 570; Watts J. 1160. *Onge:* Tchouto P. 968

Eremomastax speciosa (Hochst.) Cufod.
Fl. Gabon 13: 30 (1966); Kew Bull. 44: 69 (1989).
Eremomastax polysperma (Benth.) Dandy, F.W.T.A. 2: 397 (1963).
Clerodendrum eupatorioides Baker, F.W.T.A. 2: 442 (1963).

Habit: herb to 2 m. **Guild:** pioneer. **Habitat:** forest. **Distribution:** tropical Africa. **Star:** excluded. **Alt. range:** 1-200m (1), 601-800m (1). **Flowering:** Dec (1). **Specimens:** - *Mabeta-Moliwe:* Cheek M. 5806. *No locality:* Mann G. 1295

Graptophyllum glandulosum Turrill
F.W.T.A. 2: 423 (1963).

Habit: shrub to 1.5 m. **Habitat:** forest and stream banks. **Distribution:** SE Nigeria and Cameroon. **Chorology:** Lower Guinea. **Star:** blue. **Alt. range:** 1-200m (1), 201-400m (2). **Flowering:** Mar (1), May (1), Oct (1). **Specimens:** - *Bambuko F.R.:* Watts J. 622. *Mokoko River F.R.:* Ekema N. 933. *Onge:* Watts J. 774

Hypoestes aristata (Vahl) Soland. ex Roem. & Schult.
F.W.T.A. 2: 431 (1963); Fl. Gabon 13: 231 (1966).

Habit: herb. **Habitat:** forest and Aframomum thicket. **Distribution:** Widespread in Tropical Africa. **Chorology:** Tropical Africa. **Star:** blue. **Alt. range:** 600-800m (1). **Flowering:** Dec (1). **Specimens:** - *No locality:* Mann G. 1951

Hypoestes consanguinea Lindau
F.W.T.A. 2: 431 (1963).

Habit: herb. **Distribution:** Togo, Nigeria and W. Cameroon. **Chorology:** Guineo-Congolian (montane). **Star:** green. **Flowering:** Jan (1). **Specimens:** - *Eastern slopes:* Preuss P. R. 599

Hypoestes forskalei (Vahl) R.Br.
F.W.T.A. 2: 431 (1963); Fl. Gabon 13: 228 (1966).

Hypoestes verticillaris (L.f.) Sol. ex Roem. & Schult., F.W.T.A. 2: 431 (1963).

Habit: herb. **Distribution:** tropical Africa. **Star:** excluded. **Specimens:** - *No locality:* Preuss P. R. 755

Hypoestes rosea P.Beauv.
F.W.T.A. 2: 431 (1963).

Habit: herb. **Habitat:** forest. **Distribution:** Nigeria and Cameroon. **Chorology:** Lower Guinea. **Star:** blue. **Alt. range:** 400-600m (1), 601-800m (1). **Flowering:** Jan (1), Dec (1). **Specimens:** - *Eastern slopes:* Tekwe C. 15. *Etinde:* Cable S. 401; Dalziel J. M. 8220

Hypoestes triflora (Forssk.) Roem. & Schult.
F.W.T.A. 2: 431 (1963).

Habit: herb. **Habitat:** forest-grassland transition. **Distribution:** tropical Africa. **Chorology:** Afromontane. **Star:** green. **Alt. range:** 800-1000m (1), 1401-1600m (1), 2001-2200m (1), 2201-2400m (1). **Flowering:** Oct (3). **Specimens:** - *Eastern slopes:* Sidwell K. 59; Tchouto P. 375; Watts J. 463. *No locality:* Mann G. 1979

Isoglossa glandulifera Lindau
F.W.T.A. 2: 424 (1963).

Habit: herb to 3 m. **Habitat:** forest-grassland transition. **Distribution:** SE Nigeria, Bioko and W. Cameroon. **Chorology:** Western Cameroon Uplands (montane). **Star:** gold. **Alt. range:** 800-1000m (1), 1001-1200m (3), 1401-1600m (2), 2001-2200m (1), 2201-2400m (1), 2401-2600m (1). **Flowering:** Mar (1), Sep (2), Oct (3), Nov (1). **Fruiting:** Oct (1). **Specimens:** - *Eastern slopes:* Cable S. 1397; Cheek M. 3684, 5355; Preuss P. R. 1062; Tekwe C. 317. *Etinde:* Tchouto P. 429; Tekwe C. 292; Thomas D. W. 9170. *No locality:* Kalbreyer W. 132

Isoglossa nervosa C.B.Clarke
F.W.T.A. 2: 424 (1963).

Habit: herb to 2 m. **Habitat:** forest-grassland transition. **Distribution:** Mount Cameroon. **Chorology:** Endemic (montane). **Star:** black. **Alt. range:** 2200-2400m (2). **Flowering:** Oct (1). **Specimens:** - *Eastern slopes:* Thomas D. W. 9333. *No locality:* Mann G. 2009

Justicia baronii V.A.W.Graham
Kew Bull. 43: 588 (1988).
Adhatoda robusta C.B.Clarke, F.W.T.A. 2: 422 (1963); Fl. Gabon 13: 179 (1966).

Habit: tree or shrub to 5 m. **Guild:** pioneer. **Habitat:** forest. **Distribution:** Ivory Coast to Bioko, Cameroon and Gabon. **Chorology:** Guineo-Congolian (montane). **Star:** green. **Alt. range:** 800-1000m (1), 1001-1200m (1). **Flowering:** Dec (1). **Specimens:** - *Eastern slopes:* Migeod F. 248. *Etinde:* Cheek M. 5610

Justicia biokoensis V.A.W.Graham
Kew Bull. 43: 588 (1988).
Adhatoda maculata C.B.Clarke, F.W.T.A. 2: 422 (1963).

Habit: woody climber. **Guild:** pioneer. **Habitat:** forest and farmbush. **Distribution:** Ivory Coast to Bioko and Cameroon. **Chorology:** Guineo-Congolian. **Star:** green. **Specimens:** - *Southern Bakundu F.R.:* Binuyo A. & Daramola B. O. FHI 35498

3

Justicia extensa T.Anderson
F.W.T.A. 2: 428 (1963); Fl. Gabon 13: 224 (1966); Kew Bull. 43: 589 (1988).

Habit: herb to 3 m. **Guild:** pioneer. **Habitat:** forest. **Distribution:** Guinea to Cameroon, Gabon, Congo (Kinshasa), Angola and CAR. **Chorology:** Guineo-Congolian. **Star:** green. **Alt. range:** 1-200m (2), 401-600m (1), 601-800m (1), 801-1000m (1). **Flowering:** Aug (1), Sep (1), Oct (1), Nov (2). **Specimens:** - *Etinde:* Cheek M. 3763; Kwangue A. 125; Wheatley J. I. 538. *Onge:* Harris D. J. 3695; Watts J. 700

Justicia insularis T.Anderson
F.W.T.A. 2: 427 (1963); Fl. Gabon 13: 216 (1966); Symb. Bot. Ups. 29: 109 (1989); Nord. J. Bot. 10: 384 (1990).

Habit: herb. **Guild:** pioneer. **Habitat:** stream banks, forest clearings. **Distribution:** Nigeria to Congo (Kinshasa). **Chorology:** Guineo-Congolian. **Star:** green. **Alt. range:** 1-200m (2), 201-400m (1). **Flowering:** Oct (2), Nov (1). **Specimens:** - *Limbe Botanic Garden:* Wheatley J. I. 665. *Mabeta-Moliwe:* Sidwell K. 117. *Onge:* Tchouto P. 938. *Southern Bakundu F.R.:* Binuyo A. & Daramola B. O. FHI 35196

Justicia laxa T.Anderson
F.W.T.A. 2: 428 (1963); Fl. Gabon 13: 221 (1966); Kew Bull. 43: 589 (1988).

Habit: herb to 2 m. **Habitat:** forest. **Distribution:** SE Nigeria, Bioko, Cameroon, Gabon, Congo (Brazzaville) and Congo (Kinshasa). **Chorology:** Guineo-Congolian. **Star:** green. **Alt. range:** 200-400m (3). **Flowering:** Oct (2), Nov (1). **Specimens:** - *Etinde:* Cheek M. 5436; Keay R. W. J. FHI 37534. *Onge:* Tchouto P. 850; Watts J. 810

Justicia preussii (Lindau) C.B.Clarke
F.W.T.A. 2: 427 (1963); Kew Bull. 43: 591 (1988).

Habit: shrub to 3 m. **Habitat:** forest. **Distribution:** Cameroon. **Chorology:** Lower Guinea. **Star:** gold. **Alt. range:** 600-800m (2). **Flowering:** Mar (1). **Specimens:** - *Eastern slopes:* Tchouto P. 59. *No locality:* Mann G. 1298

Justicia tenella (Nees) T.Anderson
F.W.T.A. 2: 428 (1963); Fl. Gabon 13: 215 (1966); Kew Bull. 43: 587 (1988).

Habit: herb. **Guild:** pioneer. **Habitat:** forest, on rocks. **Distribution:** Tropical Africa and Madagascar. **Star:** excluded. **Alt. range:** 1-200m (1). **Flowering:** Dec (1). **Specimens:** - *Etinde:* Maitland T. D. 776. *Mabeta-Moliwe:* Cheek M. 5812

Justicia tristis T.Anderson
Bot. J. Linn. Soc. 7: 38 (1864).
    Adhatoda tristis Nees, F.W.T.A. 2: 423 (1963).

Habit: herb to 1.5 m. **Habitat:** forest. **Distribution:** SE Nigeria, Bioko and Cameroon. **Chorology:** Lower Guinea. **Star:** blue. **Alt. range:** 1-200m (2), 201-400m (2). **Flowering:** Oct (2), Nov (1), Dec (1). **Specimens:** - *Mabeta-Moliwe:* Cheek M. 5795. *Onge:* Harris D. J. 3635; Ndam N. 757; Tchouto P. 842

Lankesteria barteri Hook.f.
F.W.T.A. 2: 407 (1963).

Habit: shrub to 1.5 m. **Distribution:** Nigeria and Cameroon. **Chorology:** Lower Guinea. **Star:** blue. **Specimens:** - *Bambuko F.R.:* Keay R. W. J. FHI 37519

Lankesteria brevior C.B.Clarke
F.W.T.A. 2: 407 (1963).

Habit: shrub to 1.5 m. **Guild:** pioneer. **Habitat:** forest. **Distribution:** Sierra Leone to Cameroon. **Chorology:** Guineo-Congolian. **Star:** green. **Alt. range:** 1-200m (2), 201-400m (1). **Flowering:** Mar (3). **Fruiting:** Mar (1). **Specimens:** - *Mokoko River F.R.:* Tchouto P. 522, 616. *Onge:* Wheatley J. I. 843

Lankesteria elegans (P.Beauv.) T.Anderson
F.W.T.A. 2: 407 (1963); Fl. Gabon 13: 96 (1966).

Habit: shrub to 3 m. **Guild:** shade-bearer. **Distribution:** Sierra Leone to Cameroon, Gabon, Congo (Kinshasa), CAR, Uganda and Sudan. **Chorology:** Guineo-Congolian. **Star:** green. **Flowering:** Jan (1). **Specimens:** - *Southern Bakundu F.R.:* Binuyo A. & Daramola B. O. FHI 35061

Mimulopsis solmsii Schweinf.
F.W.T.A. 2: 403 (1963).

Habit: herb to 4 m. **Habitat:** forest-grassland transition. **Distribution:** tropical Africa. **Chorology:** Afromontane. **Star:** green. **Alt. range:** 1400-1600m (1), 1801-2000m (2), 2001-2200m (1). **Flowering:** Oct (2). **Specimens:** - *Eastern slopes:* Cheek M. 3660; Talbot P. A. 814; Thomas D. W. 9421, 9422

Nelsonia canescens (Lam.) Spreng.
F.W.T.A. 2: 418 (1963); Fl. Gabon 13: 158 (1966); Kew Bull. 33: 399 (1979).

Habit: herb. **Guild:** pioneer. **Distribution:** palaeotropical, and introduced in tropical America. **Star:** excluded. **Alt. range:** 800-1000m (1). **Specimens:** - *Eastern slopes:* Maitland T. D. 119

Nelsonia smithii Oersted
Kew Bull. 33: 401 (1979).

Habit: herb. **Habitat:** forest and farmbush, often near streams. **Distribution:** tropical Africa. **Star:** excluded. **Alt. range:** 1-200m (2), 401-600m (1). **Flowering:** Nov (1), Dec (2). **Specimens:** - *Etinde:* Williams S. 75. *Mabeta-Moliwe:* Cable S. 601; Cheek M. 5768

Oreacanthus mannii Benth.
F.W.T.A. 2: 432 (1963).

Habit: shrub to 4 m. **Habitat:** forest. **Distribution:** Bioko and Cameroon. **Chorology:** Lower Guinea (montane). **Star:** gold. **Alt. range:** 1400-1600m (2), 2001-2200m (1). **Flowering:** Oct (1). **Fruiting:** Oct (1). **Specimens:** - *Eastern slopes:* Thomas D. W. 9465. *Etinde:* Thomas D. W. 9238, 9353. *No locality:* Mann G. 1259

Phaulopsis angolana S.Moore
Symb. Bot. Ups. 31: 99 (1995).
    Phaulopsis silvestris (Lindau) Lindau, F.W.T.A. 2: 399 (1963); Fl. Gabon 13: 51 (1966).

Habit: herb. **Guild:** pioneer. **Habitat:** forest. **Distribution:** Sierra Leone to Bioko, Cameroon, Gabon, Congo (Brazzaville), Congo

(Kinshasa), Angola, CAR, Uganda, Ethiopia and Kenya. **Star:** excluded. **Alt. range:** 1-200m (1), 1001-1200m (1). **Flowering:** Dec (1). **Specimens:** - *Eastern slopes:* Maitland T. D. 121. *Etinde:* Cheek M. 5828

Phaulopsis ciliata (Willd.) Hepper
Symb. Bot. Ups. 31: 103 (1995).
Phaulopsis falsisepala C.B.Clarke, F.W.T.A. 2: 399 (1963).

Habit: herb. **Habitat:** forest. **Distribution:** Senegal to Bioko, Cameroon, Gabon, Congo (Brazzaville), Congo (Kinshasa), CAR, Chad and Sudan. **Chorology:** Guineo-Congolian. **Star:** green. **Alt. range:** 1-200m (4), 201-400m (1). **Flowering:** Oct (2), Nov (3). **Specimens:** - *Limbe Botanic Garden:* Wheatley J. I. 648. *Onge:* Harris D. J. 3870; Tchouto P. 937, 952; Thomas D. W. 9904

Phaulopsis imbricata (Forssk.) Sweet subsp. poggei (Lindau) M.Manktelow
Symb. Bot. Ups. 31: 138 (1995).
Phaulopsis poggei Lindau, Fl. Gabon 13: 48 (1966).

Habit: herb. **Guild:** pioneer. **Habitat:** forest. **Distribution:** tropical Africa. **Star:** excluded. **Alt. range:** 1-200m (1). **Flowering:** Dec (1). **Specimens:** - *Mabeta-Moliwe:* Cheek M. 5722

Pseuderanthemum dispermum Milne-Redh.
F.W.T.A. 2: 421 (1963).

Habit: herb. **Distribution:** SE Nigeria and Cameroon. **Chorology:** Lower Guinea. **Star:** blue. **Specimens:** - *No locality:* Schlechter F. R. R. 12681

Pseuderanthemum ludovicianum (Büttner) Lindau
F.W.T.A. 2: 421 (1963); Fl. Gabon 13: 170 (1966).

Habit: shrub to 3 m. **Habitat:** forest. **Distribution:** Liberia to East Africa. **Chorology:** Tropical Africa. **Star:** excluded. **Alt. range:** 200-400m (1), 601-800m (1), 801-1000m (2). **Flowering:** Jan (1), Feb (1), Mar (1), Dec (1). **Specimens:** - *Eastern slopes:* Tekwe C. 14. *Etinde:* Cable S. 1583; Faucher P. 13; Preuss P. R. 598; Wheatley J. I. 22. *Mabeta-Moliwe:* Plot Voucher W 804

Pseuderanthemum tunicatum (Afzel.) Milne-Redh.
F.W.T.A. 2: 421 (1963); Fl. Gabon 13: 168 (1966).

Habit: shrub to 1.5 m. **Guild:** shade-bearer. **Habitat:** forest. **Distribution:** tropical Africa. **Star:** excluded. **Alt. range:** 1-200m (5), 201-400m (1), 401-600m (1), 601-800m (1), 801-1000m (1). **Flowering:** Jan (2), Mar (2), Apr (1), Jun (1), Dec (2). **Fruiting:** Mar (2), Apr (1), Dec (1). **Specimens:** - *Bambuko F.R.:* Watts J. 660. *Eastern slopes:* Tekwe C. 7. *Etinde:* Cable S. 354, 453; Cheek M. 5861; Preuss P. R. 1108; Sunderland T. C. H. 1053. *Mabeta-Moliwe:* Jaff B. 63. *Mokoko River F.R.:* Ekema N. 1189; Tchouto P. 617

Rhinacanthus virens (Nees) Milne-Redh. var. virens
F.W.T.A. 2: 425 (1963); Fl. Gabon 13: 201 (1966).

Habit: herb. **Habitat:** forest. **Distribution:** Guinea to S. Tomé, Gabon, Zaire and East Africa. **Chorology:** Tropical Africa. **Star:**

excluded. **Alt. range:** 1-200m (1), 201-400m (2), 601-800m (1). **Flowering:** Feb (1), Mar (2), Dec (1). **Specimens:** - *Etinde:* Cable S. 1503; Cheek M. 5941; Keay R. W. J. FHI 28673; Tchouto P. 27; Wheatley J. I. 21

Ruellia primuloides (T.Anderson ex Benth.) Heine
Fl. Gabon 13: 14 (1966).
Endosiphon primuloides T.Anderson ex Benth., F.W.T.A. 2: 398 (1963).

Habit: herb. **Habitat:** forest. **Distribution:** Guinea to Bioko, Cameroon and Rio Muni. **Chorology:** Guineo-Congolian. **Star:** green. **Alt. range:** 1-200m (1), 201-400m (1). **Flowering:** Oct (1), Nov (1). **Specimens:** - *Etinde:* Cheek M. 5523. *Onge:* Watts J. 762

Rungia buettneri Lindau
Engl. Bot. Jahrb. 20: 46 (1894).

Habit: shrub to 2 m. **Habitat:** forest and forest-grassland transition. **Distribution:** Cameroon, Congo (Kinshasa), Burundi, Uganda and Sudan. **Chorology:** Guineo-Congolian (montane). **Star:** green. **Alt. range:** 800-1000m (2), 1201-1400m (1), 1401-1600m (3), 1801-2000m (1). **Flowering:** Sep (2), Oct (3), Dec (2). **Specimens:** - *Eastern slopes:* Cheek M. 3714; Tchouto P. 385; Tekwe C. 310. *Etinde:* Cable S. 427; Cheek M. 5614; Wheatley J. I. 583, 610

Rungia congoensis C.B.Clarke
F.W.T.A. 2: 430 (1963); Fl. Gabon 13: 206 (1966).

Habit: herb. **Guild:** pioneer. **Distribution:** Nigeria, Cameroon, Gabon, Congo (Kinshasa) and CAR. **Chorology:** Guineo-Congolian. **Star:** green. **Specimens:** - *Etinde:* Keay R. W. J. FHI 28660

Rungia paxiana (Lindau) C.B.Clarke
F.W.T.A. 2: 430 (1963).

Habit: herb to 2 m. **Guild:** pioneer. **Distribution:** Guinea to Bioko and Cameroon. **Chorology:** Guineo-Congolian (montane). **Star:** green. **Alt. range:** 1000-1200m (1). **Specimens:** - *No locality:* Mann G. 1968

Schaueria populifolia C.B.Clarke
F.W.T.A. 2: 423 (1963).

Habit: herb to 1.5 m. **Habitat:** forest. **Distribution:** SE Nigeria, Bioko and Cameroon. **Chorology:** Lower Guinea. **Star:** blue. **Alt. range:** 1-200m (2). **Flowering:** Mar (1), Dec (1). **Specimens:** - *Mabeta-Moliwe:* Cable S. 585, 1458; Plot Voucher W 23

Sclerochiton preussii (Lindau) C.B.Clarke
F.W.T.A. 2: 408 (1963); Kew Bull. 46: 20 (1991).

Habit: shrub to 4 m. **Habitat:** forest. **Distribution:** SE Nigeria and SW Cameroon. **Chorology:** Western Cameroon Uplands (montane). **Star:** black. **Alt. range:** 1000-1200m (1).**Fruiting:** Oct (1). **Specimens:** - *Eastern slopes:* Preuss P. R. 1073. *Etinde:* Wheatley J. I. 596

Stenandrium guineense (Nees) Vollesen
Kew Bull. 47: 182 (1992).
Crossandra guineensis Nees, F.W.T.A. 2: 409 (1963).

Stenandriopsis guineensis (Nees) Benoist, Fl. Gabon 13: 102 (1966).

Habit: herb. **Habitat:** forest, stream banks and Aframomum thicket. **Distribution:** Guinea to Bioko, Cameroon, Gabon, Congo (Brazzaville), Congo (Kinshasa), Angola (Cabinda), CAR, Uganda and Sudan. **Chorology:** Guineo-Congolian. **Star:** green. **Alt. range:** 1-200m (1), 1001-1200m (3). **Flowering:** Oct (4). **Specimens:** - *Etinde:* Tchouto P. 400; Thomas D. W. 9161, 9539. *Mabeta-Moliwe:* Plot Voucher W 954; Sidwell K. 120

Stenandrium talbotii (S.Moore) Vollesen
Kew Bull. 47: 186 (1992).
Crossandra talbotii S.Moore, F.W.T.A. 2: 409 (1963).
Stenandriopsis talbotii (S.Moore) Heine, Fl. Gabon 13: 105 (1966).

Habit: herb. **Habitat:** forest, stream banks and Aframomum thicket. **Distribution:** Nigeria, Bioko, Cameroon and Gabon. **Chorology:** Lower Guinea. **Star:** blue. **Alt. range:** 1-200m (4), 201-400m (4), 401-600m (2), 601-800m (1). **Flowering:** Oct (7), Nov (4). **Fruiting:** Nov (1). **Specimens:** - *Etinde:* Cable S. 177; Cheek M. 5429; Tchouto P. 468; Thomas D. W. 9125; Watts J. 484, 542. *Onge:* Akogo M. 111, 137; Harris D. J. 3712; Watts J. 833, 921

Stenandrium thomense (Milne-Redh.) Vollesen
Kew Bull. 47: 187 (1992).
Crossandra thomensis Milne-Redh., F.W.T.A. 2: 409 (1963).
Stenandriopsis thomense (Milne-Redh.) Heine, Fl. Gabon 13: 99 (1966).

Habit: herb. **Habitat:** forest. **Distribution:** W. Cameroon and S. Tomé. **Chorology:** Western Cameroon. **Star:** black. **Alt. range:** 400-600m (1). **Flowering:** Nov (1). **Specimens:** - *Etinde:* Williams S. 74

Stenandrium sp.

Specimens: - *Mokoko River F.R.:* Acworth J. M. 172

Thomandersia laurifolia (Benth.) Baill.
F.W.T.A. 2: 413 (1963); Fl. Gabon 13: 144 (1966).

Habit: shrub to 5 m. **Habitat:** forest. **Distribution:** Cameroon, Rio Muni, Gabon and Congo (Kinshasa). **Chorology:** Guineo-Congolian. **Star:** green. **Alt. range:** 1-200m (3), 201-400m (1). **Flowering:** Mar (1), Apr (3), Dec (1). **Fruiting:** Mar (1), Apr (1). **Specimens:** - *Etinde:* Hutchinson J. & Metcalfe C. R. 145. *Mabeta-Moliwe:* Cable S. 519. *Mokoko River F.R.:* Akogo M. 297; Mbani J. M. 344; Tchouto P. 581, 1150

Thunbergia cynanchifolia Benth.
F.W.T.A. 2: 402 (1963).

Habit: herbaceous climber. **Guild:** pioneer. **Habitat:** forest. **Distribution:** Guinea to Cameroon. **Chorology:** Guineo-Congolian. **Star:** green. **Alt. range:** 1-200m (1). **Flowering:** Oct (1). **Specimens:** - *Onge:* Tchouto P. 960

Thunbergia fasciculata Lindau
F.W.T.A. 2: 400 (1963).

Habit: herbaceous climber. **Guild:** pioneer. **Habitat:** forest. **Distribution:** Nigeria to Zaire, Uganda, Sudan and SW Ethiopia. **Chorology:** Guineo-Congolian. **Star:** green. **Alt. range:** 1000-1200m (2). **Flowering:** Oct (2). **Specimens:** - *Etinde:* Sidwell K. 156; Tchouto P. 420

Thunbergia vogeliana Benth.
F.W.T.A. 2: 402 (1963).

Habit: shrub to 5 m. **Guild:** pioneer. **Habitat:** forest and forest gaps. **Distribution:** Ghana to Bioko, SW Cameroon east to Uganda and W. Tanzania, sometimes cultivated. **Chorology:** Guineo-Congolian. **Star:** green. **Alt. range:** 1-200m (1), 601-800m (1), 801-1000m (4), 1001-1200m (1), 1201-1400m (2). **Flowering:** Aug (3), Oct (2), Nov (1), Dec (2). **Fruiting:** Aug (2), Dec (1). **Specimens:** - *Eastern slopes:* Tekwe C. 133. *Etinde:* Cable S. 320, 467, 485; Hunt L. V. 27; Kwangue A. 128; Wheatley J. I. 604, 625. *Mabeta-Moliwe:* von Rege I. 102

AIZOACEAE

Sesuvium portulacastrum L.

Specimens: - *Mabeta-Moliwe* Cheek M. in W1211

ALANGIACEAE

Alangium chinense (Lour.) Harms
F.T.E.A. Alangiaceae: 3 (1958); F.W.T.A. 1: 749 (1958); Fl. Cameroun 10: 6 (1970); Keay R.W.J., Trees of Nigeria: 377 (1989).

Habit: tree to 25 m. **Habitat:** forest and farmbush. **Distribution:** palaeotropical. **Chorology:** Montane. **Star:** excluded. **Alt. range:** 600-800m (1), 1401-1600m (1), 1601-1800m (1). **Fruiting:** Oct (1). **Specimens:** - *Eastern slopes:* Thomas D. W. 9463, 9468. *No locality:* Kalbreyer W. 148

AMARANTHACEAE

C. Townsend (K)

Achyranthes aspera L. var. aspera
F.W.T.A. 1: 152 (1954); Fl. Gabon 7: 37 (1963); Fl. Cameroun 17: 31 (1974); F.T.E.A. Amaranthaceae: 101 (1985); Fl. Zamb. 9(1): 106 (1988).

Habit: herb. **Habitat:** forest and forest-grassland transition. **Distribution:** pantropical. **Chorology:** Montane. **Star:** excluded. **Alt. range:** 1800-2000m (1). **Specimens:** - *No locality:* Mann G. 1307

Achyranthes aspera L. var. sicula L.
Fl. Cameroun 17: 32 (1974); F.T.E.A. Amaranthaceae: 104 (1985); Fl. Zamb. 9(1): 109 (1988).

Habit: herb. **Guild:** pioneer. **Distribution:** pantropical. **Star:** excluded. **Alt. range:** 1800-2000m (1). **Flowering:** Feb (1). **Specimens:** - *No locality:* Breteler F. J. MC 202

Achyranthes talbotii Hutch. & Dalziel

F.W.T.A. 1: 152 (1954); Fl. Cameroun 17: 33 (1974).

Habit: rheophyte. **Habitat:** river banks in forest. **Distribution:** SE Nigeria and SW Cameroon. **Chorology:** Western Cameroon. **Star:** gold. **Alt. range:** 1-200m (3), 201-400m (3). **Flowering:** Oct (5), Nov (1). **Specimens:** - *Onge:* Akogo M. 109; Cheek M. 5013; Ndam N. 742; Tchouto P. 726; Thomas D. W. 9790; Watts J. 806. *Southern Bakundu F.R.:* Brenan J. P. M. 9497

Alternanthera sessilis (L.) R.Br.

F.W.T.A. 1: 154 (1954); Fl. Gabon 7: 42 (1963); Fl. Cameroun 17: 52 (1974); F.T.E.A. Amaranthaceae: 126 (1985); Fl. Zamb. 9(1): 129 (1988).

Habit: herb. **Guild:** pioneer. **Distribution:** pantropical. **Star:** excluded. **Specimens:** - *Eastern slopes:* Maitland T. D. 89

Amaranthus spinosus L.

F.W.T.A. 1: 148 (1954); Fl. Gabon 7: 32 (1963); Fl. Cameroun 17: 17 (1974); F.T.E.A. Amaranthaceae: 26 (1985); Fl. Zamb. 9(1): 51 (1988).

Habit: herb. **Guild:** pioneer. **Distribution:** pantropical. **Star:** excluded. **Specimens:** - *Eastern slopes:* Maitland T. D. 43

Blutaparon vermiculare (L.) Mears

Taxon 31: 113 (1982).
Philoxerus vermicularis (L.) P.Beauv., F.W.T.A. 1: 153 (1954).

Habit: herb. **Distribution:** Atlantic coasts of tropical America and West Africa from Senegal to Angola. **Star:** excluded. **Specimens:** - *Mabeta-Moliwe:* Hutchinson J. & Metcalfe C. R. 136

Celosia globosa Schinz

F.W.T.A. 1: 147 (1954); Fl. Cameroun 17: 12 (1974); F.T.E.A. Amaranthaceae: 17 (1985).

Habit: herb. **Habitat:** forest. **Distribution:** Nigeria, Cameroon, Congo (Kinshasa) and Uganda. **Chorology:** Guineo-Congolian. **Star:** green. **Alt. range:** 200-400m (1). **Flowering:** Mar (1). **Specimens:** - *Mokoko River F.R.:* Tchouto P. 569

Celosia isertii C.C.Towns.

F.T.E.A. Amaranthaceae: 15 (1985); Fl. Zamb. 9(1): 34 (1988).

Habit: herb. **Guild:** pioneer. **Habitat:** forest. **Distribution:** tropical Africa. **Chorology:** Afromontane. **Star:** green. **Alt. range:** 200-400m (2), 601-800m (2), 1001-1200m (1). **Flowering:** Feb (1), Mar (1), Apr (1), Nov (1), Dec (1). **Specimens:** - *Eastern slopes:* Cable S. 1318. *Etinde:* Acworth J. M. 29; Cable S. 1576; Cheek M. 5853; Williams S. 88

Celosia leptostachya Benth.

F.W.T.A. 1: 147 (1954); Fl. Cameroun 17: 8 (1974).

Habit: herb. **Guild:** pioneer. **Habitat:** forest. **Distribution:** Nigeria, Cameroon, Congo (Kinshasa) and Uganda. **Chorology:** Guineo-Congolian. **Star:** green. **Alt. range:** 1-200m (3). **Flowering:** Apr (1), May (1), Oct (1). **Fruiting:** May (1). **Specimens:** - *Mokoko River F.R.:* Pouakouyou D. 7; Watts J. 1147. *Onge:* Cheek M. 5176

Celosia pseudovirgata Schinz

Fl. Cameroun 17: 13 (1974).
Celosia bonnivairii sensu Keay, F.W.T.A. 1: 147 (1954).

Habit: herb. **Guild:** pioneer. **Habitat:** forest. **Distribution:** Nigeria, Bioko, SW Cameroon and Congo (Kinshasa). **Chorology:** Guineo-Congolian. **Star:** green. **Alt. range:** 400-600m (1). **Flowering:** Nov (1). **Specimens:** - *Etinde:* Cable S. 282. *No locality:* Maitland T. D. 1082

Celosia trigyna L.

F.W.T.A. 1: 147 (1954); Fl. Gabon 7: 25 (1963); Fl. Cameroun 17: 7 (1974); F.T.E.A. Amaranthaceae: 12 (1985); Fl. Zamb. 9(1): 31 (1988).
Celosia laxa Schum. & Thonn., F.W.T.A. 1: 147 (1954).

Habit: herb. **Guild:** pioneer. **Habitat:** forest. **Distribution:** tropical and subtropical Africa. **Star:** excluded. **Specimens:** - *Eastern slopes:* Hutchinson J. & Metcalfe C. R. 96; Maitland T. D. 38

Cyathula cylindrica Moq.

Fl. Cameroun 17: 43 (1974); F.T.E.A. Amaranthaceae: 64 (1985); Fl. Zamb. 9(1): 82 (1988).
Cyathula cylindrica Moq. var. *mannii* (Baker) Suesseng., F.W.T.A. 1: 151 (1954).

Habit: herb. **Distribution:** Bioko and Cameroon to East and southern Africa. **Chorology:** Afromontane. **Star:** green. **Alt. range:** 1800-2000m (1). **Specimens:** - *No locality:* Mann G. 2007

Cyathula prostrata (L.) Blume var. pedicellata (C.B.Clarke) Cavaco

Fl. Gabon 7: 35 (1963); Fl. Cameroun 17: 46 (1974); F.T.E.A. Amaranthaceae: 59 (1985); Fl. Zamb. 9(1): 80 (1988).
Cyathula pedicellata C.B.Clarke, F.W.T.A. 1: 149 (1954).

Habit: herb. **Guild:** pioneer. **Habitat:** forest. **Distribution:** Sierra Leone to Congo (Kinshasa), Uganda and Tanzania. **Chorology:** Guineo-Congolian. **Star:** green. **Alt. range:** 1-200m (1), 201-400m (1), 401-600m (1). **Flowering:** Jun (1), Dec (2). **Specimens:** - *Etinde:* Cable S. 351. *Mabeta-Moliwe:* Cable S. 599. *Mokoko River F.R.:* Fraser P. 457; Ndam N. 1095

Cyathula prostrata (L.) Blume var. prostrata

F.W.T.A. 1: 149 (1954); Fl. Gabon 7: 33 (1963); Fl. Cameroun 17: 45 (1974); F.T.E.A. Amaranthaceae: 59 (1985); Fl. Zamb. 9(1): 80 (1988).

Habit: herb. **Guild:** pioneer. **Habitat:** forest. **Distribution:** pantropical. **Star:** excluded. **Alt. range:** 1-200m (3), 601-800m (1), 801-1000m (1). **Flowering:** Nov (1), Dec (3). **Fruiting:** Dec (1). **Specimens:** - *Etinde:* Cable S. 188, 426; Cheek M. 5835. *Mabeta-Moliwe:* Cheek M. 5731; Mildbraed G. W. J. 10789

Sericostachys scandens Gilg & Lopr.

F.W.T.A. 1: 151 (1954); Fl. Cameroun 17: 41 (1974); F.T.E.A. Amaranthaceae: 42 (1985); Fl. Zamb. 9(1): 60 (1988).

Habit: herbaceous climber. **Habitat:** forest. **Distribution:** tropical Africa. **Chorology:** Afromontane. **Star:** green. **Alt. range:** 200-400m (1), 801-1000m (1). **Flowering:** Oct (1). **Fruiting:** Oct (1), Dec (1). **Specimens:** - *Eastern slopes:* Collier F. S. FHI 1757. *Etinde:* Cable S. 455. *Onge:* Ndam N. 617

ANACARDIACEAE

S. Cable (K)

Antrocaryon klaineanum Pierre

F.W.T.A. 1: 728 (1958).

Habit: tree to 30 m. **Habitat:** forest. **Distribution:** SE Nigeria, Bioko, Cameroon, Gabon, Congo (Kinshasa) and Angola (Cabinda). **Chorology:** Guineo-Congolian. **Star:** green. **Specimens:** - *Southern Bakundu F.R.:* Keay R. W. J. s.n.

Lannea nigritana (Scott-Elliot) Keay var. pubescens Keay

F.W.T.A. 1: 733 (1958); Keay R.W.J., Trees of Nigeria: 372 (1989); Hawthorne W., F.G.F.T. Ghana: 182 (1990).

Habit: tree to 15 m. **Habitat:** forest. **Distribution:** Ivory Coast to SW Cameroon. **Chorology:** Guineo-Congolian. **Star:** green. **Alt. range:** 1-200m (1).**Fruiting:** Mar (1). **Specimens:** - *Mokoko River F.R.:* Tchouto P. 650

Lannea welwitschii (Hiern) Engl.

F.W.T.A. 1: 732 (1958); Keay R.W.J., Trees of Nigeria: 372 (1989); Hawthorne W., F.G.F.T. Ghana: 182 (1990).

Habit: tree to 30 m. **Guild:** pioneer. **Habitat:** forest. **Distribution:** Ivory Coast to Cameroon, Gabon, Congo (Kinshasa), Angola and Uganda. **Chorology:** Guineo-Congolian. **Star:** green. **Fruiting:** Feb (1). **Specimens:** - *Etinde:* Maitland T. D. 532; Tchouto P. 3

Pseudospondias microcarpa (A.Rich.) Engl. var. microcarpa

F.W.T.A. 1: 729 (1958); F.T.E.A. Anacardiaceae: 45 (1986); Keay R.W.J., Trees of Nigeria: 367 (1989); Hawthorne W., F.G.F.T. Ghana: 182 (1990).

Habit: tree 15-30 m tall. **Habitat:** forest. **Distribution:** tropical Africa. **Star:** excluded. **Alt. range:** 1-200m (1). **Flowering:** Oct (1), Nov (1). **Fruiting:** May (1), Oct (1), Nov (1). **Specimens:** - *Eastern slopes:* Maitland T. D. 490. *Etinde:* Maitland T. D. 364. *Limbe Botanic Garden:* Cheek M. 4951. *Mabeta-Moliwe:* Nguembock F. 43; Sunderland T. C. H. 1323

Sorindeia grandifolia Engl.

F.W.T.A. 1: 738 (1958); Keay R.W.J., Trees of Nigeria: 374 (1989).

Habit: tree to 15 m. **Habitat:** forest. **Distribution:** Nigeria, W. Cameroon and S. Tomé. **Chorology:** Lower Guinea. **Star:** blue. **Alt. range:** 1-200m (2). **Flowering:** Mar (2). **Specimens:** - *Bambuko F.R.:* Watts J. 683. *Onge:* Wheatley J. I. 690

Sorindeia mildbraedii Engl. & Brehmer

F.W.T.A. 1: 737 (1958); Keay R.W.J., Trees of Nigeria: 374 (1989).

Habit: tree to 13 m. **Habitat:** forest. **Distribution:** SE Nigeria and Cameroon. **Chorology:** Lower Guinea. **Star:** blue. **Alt. range:** 1-200m (2), 201-400m (2). **Flowering:** May (1). **Fruiting:** May (3), Jun (1). **Specimens:** - *Etinde:* Kwangue A. 79, 104. *Mabeta-Moliwe:* Jaff B. 277; Tchouto P. 203. *Southern Bakundu F.R.:* Brenan J. P. M. 9432

Sorindeia nitidula Engl.

F.W.T.A. 1: 737 (1958); Keay R.W.J., Trees of Nigeria: 374 (1989).

Habit: tree to 14 m. **Habitat:** forest. **Distribution:** Nigeria and Cameroon. **Chorology:** Lower Guinea. **Star:** blue. **Alt. range:** 200-400m (1). **Flowering:** Apr (1). **Fruiting:** Apr (1). **Specimens:** - *Etinde:* Maitland T. D. 671. *Mabeta-Moliwe:* Plot Voucher W 276; Wheatley J. I. 103

Sorindeia sp. 2

Specimens: - *Mabeta-Moliwe:* Cheek M. 3514

Sorindeia sp. 3

Specimens: - *Mabeta-Moliwe:* Cheek M. 3533

Sorindeia sp. 4

Specimens: - *Mabeta-Moliwe:* Cheek M. 3553

Sorindeia sp. 5

Specimens: - *Mabeta-Moliwe:* Cheek M. 3595

Spondias mombin L.

F.W.T.A. 1: 728 (1958); Keay R.W.J., Trees of Nigeria: 367 (1989); Hawthorne W., F.G.F.T. Ghana: 182 (1990).

Habit: tree to 18 m. **Habitat:** forest. **Distribution:** tropical Africa and America. **Star:** excluded. **Specimens:** - *Mabeta-Moliwe:* Mann G. 706

Trichoscypha abut Engl. & Brehmer

Engl. Bot. Jahrb. 54: 322 (1917).

Habit: tree to 15 m. **Habitat:** forest. **Distribution:** Cameroon. **Chorology:** Lower Guinea. **Star:** black. **Alt. range:** 1-200m (1). **Flowering:** Apr (1). **Specimens:** - *Mokoko River F.R.:* Acworth J. M. 186

Trichoscypha acuminata Engl.

F.W.T.A. 1: 735 (1958); Keay R.W.J., Trees of Nigeria: 376 (1989).

Habit: tree to 18 m. **Habitat:** forest. **Distribution:** SE Nigeria, Cameroon, Gabon, Congo (Kinshasa) and Angola (Cabinda). **Chorology:** Guineo-Congolian. **Star:** green. **Alt. range:** 1-200m (7). **Flowering:** Apr (2), May (3), Jun (1). **Fruiting:** May (1). **Specimens:** - *Etinde:* Preuss P. R. 1330 (partly). *Mabeta-Moliwe:* Plot Voucher W 555; Sunderland T. C. H. 1367, 1408, 1411. *Mokoko River F.R.:* Acworth J. M. 101, 280; Ekema N. 1126; Ndam N. 1069; Tchouto P. 1093

Trichoscypha arborea (A.Chev.) A.Chev.

F.W.T.A. 1: 736 (1958); Keay R.W.J., Trees of Nigeria: 376 (1989); Hawthorne W., F.G.F.T. Ghana: 181 (1990).

Habit: tree to 20 m. **Guild:** shade-bearer. **Habitat:** forest and farmbush. **Distribution:** Sierra Leone to SW Cameroon. **Chorology:** Guineo-Congolian. **Star:** green. **Alt. range:** 1000-

1200m (2). **Flowering:** Apr (2). **Specimens:** - *Etinde:* Kwangue A. 46; Tchouto P. 134

Trichoscypha bijuga Engl.
F.W.T.A. 1: 736 (1958).

Habit: tree or shrub to 5 m. **Habitat:** forest. **Distribution:** Mount Cameroon and Bioko. **Chorology:** Western Cameroon. **Star:** black. **Alt. range:** 1-200m (16), 201-400m (1), 601-800m (1). **Flowering:** Mar (2), Apr (3), May (2), Jun (2), Nov (1). **Fruiting:** Mar (1), Apr (2), May (6), Jun (4). **Specimens:** - *Mabeta-Moliwe:* Jaff B. 41, 65, 166, 243; Mildbraed G. W. J. 10700; Ndam N. 511; Nguembock F. 1b; Plot Voucher W 77, W 107, W 146, W 177, W 207, W 573, W 855; Sunderland T. C. H. 1175, 1188, 1270, 1462; Watts J. 158, 231, 293, 298, 409, 418; Wheatley J. I. 33, 59

Trichoscypha camerunensis Engl.
F.W.T.A. 1: 736 (1958).

Habit: shrub to 1 m. **Habitat:** forest. **Distribution:** Mount Cameroon. **Chorology:** Endemic. **Star:** black. **Alt. range:** 200-400m (2). **Flowering:** Jun (1). **Fruiting:** Jun (2). **Specimens:** - *Etinde:* Kwangue A. 102; Tchouto P. 234

Trichoscypha mannii Hook.f.
F.W.T.A. 1: 736 (1958); Keay R.W.J., Trees of Nigeria: 376 (1989).

Habit: tree to 12 m. **Habitat:** forest. **Distribution:** SE Nigeria and Cameroon. **Chorology:** Lower Guinea. **Star:** blue. **Alt. range:** 1-200m (3). **Flowering:** Oct (1), Nov (2). **Specimens:** - *Onge:* Thomas D. W. 9841; Watts J. 884, 901. *Southern Bakundu F.R.:* Olorunfemi J. FHI 30528

Trichoscypha patens (Oliv.) Engl.
F.W.T.A. 1: 736 (1958); Keay R.W.J., Trees of Nigeria: 376 (1989).

Habit: tree to 9 m. **Habitat:** forest. **Distribution:** SE Nigeria, Cameroon and Rio Muni. **Chorology:** Lower Guinea. **Star:** blue. **Alt. range:** 1-200m (6). **Flowering:** Apr (3), May (3). **Specimens:** - *Etinde:* Maitland T. D. 182. *Mabeta-Moliwe:* Preuss P. R. 1206; Tchouto P. 186. *Mokoko River F.R.:* Acworth J. M. 138; Akogo M. 212; Ekema N. 1072; Tchouto P. 1087; Watts J. 1182

Trichoscypha preussii Engl.
F.W.T.A. 1: 736 (1958); Keay R.W.J., Trees of Nigeria: 376 (1989).

Habit: tree to 10 m. **Habitat:** forest. **Distribution:** SE Nigeria and Cameroon. **Chorology:** Lower Guinea. **Star:** blue. **Alt. range:** 1-200m (3), 201-400m (2), 401-600m (2), 601-800m (1), 1001-1200m (1). **Flowering:** Oct (4). **Fruiting:** Jan (1), Feb (1), Apr (1), Oct (1), Nov (1). **Specimens:** - *Eastern slopes:* Kwangue A. 48. *Etinde:* Tekwe C. 4; Watts J. 471; Wheatley J. I. 2. *Onge:* Cheek M. 4989, 5185; Ndam N. 753; Tchouto P. 683; Watts J. 911

Trichoscypha sp. 1

Specimens: - *Onge:* Wheatley J. I. 822

Trichoscypha sp. 2

Specimens: - *Bambuko F.R.:* Watts J. 668

Trichoscypha sp. 3

Specimens: - *Etinde:* Tekwe C. 295. *Onge:* Wheatley J. I. 736

Trichoscypha sp. 4

Specimens: - *Mokoko River F.R.:* Tchouto P. 609

Trichoscypha sp. nov.

Habit: shrub to 2 m. **Habitat:** forest. **Distribution:** Mount Cameroon. **Chorology:** Endemic. **Star:** black. **Alt. range:** 200-400m (4), 401-600m (2). **Fruiting:** Oct (6). **Specimens:** - *Etinde:* Thomas D. W. 9147; Watts J. 470. *Onge:* Akogo M. 14; Cheek M. 5003, 5052; Ndam N. 664

ANCISTROCLADACEAE

Ancistrocladus letestui Pellegr.
Bull. Soc. Bot. France 98: 18 (1951).

Habit: woody climber. **Habitat:** forest. **Distribution:** Cameroon and Gabon. **Chorology:** Lower Guinea. **Star:** blue. **Alt. range:** 200-400m (1). **Specimens:** - *Mokoko River F.R.:* Thomas D. W. 10083

Ancistrocladus sp. nov.

Specimens: - *Mabeta-Moliwe* Baker 325; Cheek M. 3436, 3915

ANISOPHYLLEACEAE

Anisophyllea polyneura Floret
Adansonia (ser.4) 4: 377 (1986); Keay R.W.J., Trees of Nigeria: 101 (1989).

Habit: tree to 30 m. **Habitat:** forest. **Distribution:** Nigeria, Cameroon, Gabon and Congo (Kinshasa). **Chorology:** Guineo-Congolian. **Star:** green. **Alt. range:** 1-200m (1). **Specimens:** - *Mokoko River F.R.:* Mbani J. M. 401

Anisophyllea purpurascens Hutch. & Dalziel
F.W.T.A. 1: 282 (1954); Keay R.W.J., Trees of Nigeria: 101 (1989).

Habit: tree to 7 m. **Habitat:** forest. **Distribution:** SE Nigeria, Bioko, Cameroon, Gabon and Congo (Brazzaville). **Chorology:** Guineo-Congolian. **Star:** green. **Alt. range:** 200-400m (1). **Specimens:** - *Southern Bakundu F.R.:* Mbani J. M. 432

Anisophyllea sororia Pierre
Keay R.W.J., Trees of Nigeria: 101 (1989).
Anisophyllea sp. B sensu Keay, F.W.T.A. 1: 282 (1954).

Habit: tree to 35 m. **Habitat:** forest. **Distribution:** Nigeria, Cameroon, S. Tomé, Gabon and Congo (Brazzaville). **Chorology:** Guineo-Congolian. **Star:** green. **Specimens:** - *Mabeta-Moliwe:* Plot Voucher W 122, W 621

Anisophyllea sp.

Specimens: - *Mokoko River F.R.:* Ndam N. 1176

ANNONACEAE
P. Bygrave & S. Cable (K)

Annickia chlorantha (Oliv.) Setten & P.J.Maas
Taxon 39: 676 (1990).
Enantia chlorantha Oliv., F.W.T.A. 1: 51 (1954); Keay R.W.J., Trees of Nigeria: 19 (1989).

Habit: tree to 30 m. **Habitat:** forest. **Distribution:** Nigeria, Cameroon, Bioko, Gabon, Congo (Kinshasa) and Angola. **Chorology:** Guineo-Congolian. **Star:** pink. **Fruiting:** Sep (1). **Specimens:** - *Bambuko F.R.:* Olorunfemi J. FHI 30760

Anonidium mannii (Oliv.) Engl. & Diels var. brieyi (De Wild.) Fries
Hawthorne W., F.G.F.T. Ghana: 78 (1990).
Anonidium friesianum Exell, F.W.T.A. 1: 51 (1954).

Habit: tree to 20 m. **Guild:** shade-bearer. **Habitat:** forest. **Distribution:** SW Cameroon and Angola (Cabinda). **Chorology:** Lower Guinea. **Star:** gold. **Specimens:** - *Southern Bakundu F.R.:* Brenan J. P. M. 9410

Artabotrys cf. rhopalocarpus Le Thomas

Specimens: - *Etinde:* Wheatley J. I. 644. *Mabeta-Moliwe:* Cheek M. 3470

Boutiquea platypetala Le Thomas
Adansonia (ser.2) 5: 531 (1965).

Habit: tree or shrub to 4 m. **Habitat:** forest. **Distribution:** Cameroon. **Chorology:** Lower Guinea. **Star:** gold. **Alt. range:** 1-200m (1). **Fruiting:** Jun (1). **Specimens:** - *Etinde:* Nkeng P. 69

Cleistopholis patens (Benth.) Engl. & Diels
F.W.T.A. 1: 38 (1954); Fl. Gabon 16: 91 (1969); F.T.E.A. Annonaceae: 31 (1971); Keay R.W.J., Trees of Nigeria: 19 (1989); Hawthorne W., F.G.F.T. Ghana: 73 (1990).

Habit: tree to 30 m. **Guild:** pioneer. **Habitat:** forest. **Distribution:** Sierra Leone to Cameroon, Gabon, Congo (Kinshasa), Angola and Uganda. **Chorology:** Guineo-Congolian. **Star:** green. **Specimens:** - *Etinde:* Mildbraed G. W. J. 10640

Dennettia tripetala Baker f.
F.W.T.A. 1: 51 (1954); Keay R.W.J., Trees of Nigeria: 24 (1989); Hawthorne W., F.G.F.T. Ghana: 78 (1990).

Habit: tree to 15 m. **Guild:** shade-bearer. **Habitat:** forest. **Distribution:** Sierra Leone to Cameroon. **Chorology:** Guineo-Congolian. **Star:** blue. **Specimens:** - *Etinde:* Maitland T. D. 626. *No locality:* Box H. E. 3556

Friesodielsia gracilipes (Benth.) Steenis
Blumea 12: 359 (1964); Fl. Gabon 16: 235 (1969).

Friesodielsia longipedicellata (Baker f.) Steenis, Blumea 12: 360 (1964).
Oxymitra longipedicellata (Baker f.) Sprague & Hutch., F.W.T.A. 1: 45 (1954).
Oxymitra gracilipes Benth., F.W.T.A. 1: 45 (1954).

Habit: woody climber. **Habitat:** forest. **Distribution:** SE Nigeria, Bioko, Cameroon and Gabon. **Chorology:** Lower Guinea. **Star:** blue. **Alt. range:** 1-200m (2). **Flowering:** Feb (1). **Specimens:** - *Eastern slopes:* Maitland T. D. 536. *Etinde:* Tekwe C. 21. *Onge:* Thomas D. W. 9767

Friesodielsia cf. gracilipes (Benth.) Steenis

Specimens: - *Etinde:* Tchouto P. 248

Friesodielsia gracilis (Hook.f.) Steenis
Blumea 12: 359 (1964).
Oxymitra gracilis (Hook.f.) Sprague & Hutch., F.W.T.A. 1: 45 (1954).

Habit: woody climber. **Guild:** shade-bearer. **Habitat:** forest. **Distribution:** Sierra Leone to Cameroon. **Chorology:** Guineo-Congolian. **Star:** blue. **Alt. range:** 1-200m (1). **Fruiting:** Mar (1). **Specimens:** - *Mokoko River F.R.:* Tchouto P. 629

Friesodielsia hirsuta (Benth.) Steenis
Blumea 12: 360 (1964).
Oxymitra hirsuta (Benth.) Sprague & Hutch., F.W.T.A. 1: 45 (1954).

Habit: woody climber. **Habitat:** forest. **Distribution:** Guinea to Bioko and SW Cameroon. **Chorology:** Guineo-Congolian. **Star:** green. **Alt. range:** 600-800m (1). **Flowering:** Dec (1). **Specimens:** - *Mabeta-Moliwe:* Mildbraed G. W. J. 10777

Hexalobus crispiflorus A.Rich.
F.W.T.A. 1: 47 (1954); Fl. Gabon 16: 82 (1969); Keay R.W.J., Trees of Nigeria: 24 (1989); Hawthorne W., F.G.F.T. Ghana: 73 (1990).

Habit: tree to 25 m. **Guild:** shade-bearer. **Habitat:** forest. **Distribution:** Guinea to Congo (Kinshasa), Angola and Sudan. **Chorology:** Guineo-Congolian. **Star:** green. **Flowering:** Nov (1). **Specimens:** - *Mabeta-Moliwe:* Mildbraed G. W. J. 10611

Isolona campanulata Engl. & Diels
F.W.T.A. 1: 53 (1954); Fl. Gabon 16: 353 (1969); Keay R.W.J., Trees of Nigeria: 28 (1989); Hawthorne W., F.G.F.T. Ghana: 78 (1990).

Habit: tree to 16 m. **Guild:** shade-bearer. **Habitat:** forest. **Distribution:** Sierra Leone to Cameroon and Gabon. **Chorology:** Guineo-Congolian. **Star:** green. **Alt. range:** 1-200m (1). **Flowering:** Apr (1), May (1). **Specimens:** - *Mokoko River F.R.:* Akogo M. 268; Fraser P. 379

Isolona sp. nov.

Habit: tree to 15 m. **Habitat:** forest. **Distribution:** Mount Cameroon. **Chorology:** Endemic. **Star:** black. **Alt. range:** 1-200m (1), 201-400m (1). **Flowering:** Mar (1). **Fruiting:** Sep (1). **Specimens:** - *Bambuko F.R.:* Watts J. 687. *Etinde:* Wheatley J. I. 501

Monanthotaxis congoensis Baill.

Fl. Gabon 16: 256 (1969); Kew Bull. 25: 20 (1971).

Habit: woody climber. **Habitat:** forest. **Distribution:** Cameroon and Gabon. **Chorology:** Lower Guinea. **Star:** blue. **Alt. range:** 1-200m (1).**Fruiting:** Mar (1). **Specimens:** - *Mokoko River F.R.:* Tchouto P. 632

Monanthotaxis filamentosa (Diels) Verdc.

Kew Bull. 25: 31 (1971).
Popowia filamentosa Diels, F.W.T.A. 1: 44 (1954).

Habit: woody climber. **Habitat:** forest. **Distribution:** Nigeria, Cameroon, Congo (Kinshasa) and CAR. **Chorology:** Guineo-Congolian. **Star:** green. **Specimens:** - *Eastern slopes:* Maitland T. D. 566

Monanthotaxis sp. aff. filamentosa (Diels) Verdc.

Specimens: - *Etinde:* Thomas D. W. 9236

Monanthotaxis foliosa (Engl. & Diels) Verdc.

Kew Bull. 25: 21 (1971); Fl. Cameroun 16: 246 (1973).
Enneastemon foliosus (Engl. & Diels) Robyns & Ghesq., F.W.T.A. 1: 50 (1954).

Habit: woody climber. **Guild:** light demander. **Habitat:** forest. **Distribution:** Ghana, Nigeria, Cameroon and Congo (Kinshasa). **Chorology:** Guineo-Congolian. **Star:** green. **Alt. range:** 600-800m (1). **Specimens:** - *Etinde:* Maitland T. D. 1072

Monanthotaxis oligandra Exell

Fl. Congo B., Rw. & Bur. 2: 374 (1951).

Habit: woody climber. **Habitat:** forest. **Distribution:** Cameroon and Congo (Kinshasa). **Chorology:** Guineo-Congolian. **Star:** green. **Specimens:** - *Mabeta-Moliwe:* Plot Voucher W 316

Monodora brevipes Benth.

F.W.T.A. 1: 54 (1954); Keay R.W.J., Trees of Nigeria: 32 (1989); Hawthorne W., F.G.F.T. Ghana: 78 (1990).

Habit: tree to 13 m. **Habitat:** forest. **Distribution:** Guinea to Bioko and Cameroon. **Chorology:** Guineo-Congolian. **Star:** green. **Specimens:** - *Etinde:* Preuss P. R. 1314

Monodora crispata Engl. & Diels

F.W.T.A. 1: 54 (1954); Fl. Gabon 16: 345 (1969); Keay R.W.J., Trees of Nigeria: 32 (1989).

Habit: tree to 10 m. **Habitat:** forest. **Distribution:** Sierra Leone to Cameroon and Gabon. **Chorology:** Guineo-Congolian. **Star:** green. **Alt. range:** 1-200m (1). **Flowering:** Mar (1). **Specimens:** - *Mokoko River F.R.:* Tchouto P. 627

Monodora myristica (Gaertn.) Dunal

F.W.T.A. 1: 54 (1954); Fl. Gabon 16: 342 (1969); F.T.E.A. Annonaceae: 123 (1971); Keay R.W.J., Trees of Nigeria: 32 (1989); Hawthorne W., F.G.F.T. Ghana: 78 (1990).

Habit: tree to 20 m. **Guild:** shade-bearer. **Habitat:** forest. **Distribution:** Sierra Leone to Congo (Kinshasa), Angola, Uganda and Kenya. **Chorology:** Guineo-Congolian. **Star:** green. **Alt.**

range: 1-200m (2). **Flowering:** Jan (1). **Fruiting:** Jan (1). **Specimens:** - *Etinde:* Sunderland T. C. H. 1051. *Mabeta-Moliwe:* Mann G. 27. *Onge:* Tchouto P. 699

Monodora tenuifolia Benth.

F.W.T.A. 1: 54 (1954); Fl. Gabon 16: 339 (1969); Keay R.W.J., Trees of Nigeria: 32 (1989); Hawthorne W., F.G.F.T. Ghana: 78 (1990).

Habit: tree to 15 m. **Guild:** pioneer. **Habitat:** forest. **Distribution:** Guinea to Bioko, Cameroon, Gabon and Angola. **Chorology:** Guineo-Congolian. **Star:** green. **Alt. range:** 200-400m (2). **Flowering:** Mar (2). **Specimens:** - *Bambuko F.R.:* Watts J. 617. *Mabeta-Moliwe:* Mann G. 111. *Onge:* Wheatley J. I. 671

Pachypodanthium staudtii Engl. & Diels

F.W.T.A. 1: 39 (1954); Fl. Gabon 16: 107 (1969); Keay R.W.J., Trees of Nigeria: 30 (1989); Hawthorne W., F.G.F.T. Ghana: 71 (1990).

Habit: tree to 30 m. **Guild:** light demander. **Habitat:** forest. **Distribution:** Sierra Leone to Cameroon. **Chorology:** Guineo-Congolian. **Star:** green. **Alt. range:** 1-200m (1).**Fruiting:** Nov (1). **Specimens:** - *Onge:* Harris D. J. 3669

Piptostigma sp. aff. glabrescens Oliv.

Habit: tree to 15 m. **Habitat:** forest. **Distribution:** Mount Cameroon. **Chorology:** Endemic. **Star:** black. **Alt. range:** 1-200m (1), 201-400m (3). **Flowering:** Mar (2), Oct (2). **Specimens:** - *Mokoko River F.R.:* Tchouto P. 613. *Onge:* Cheek M. 5071; Watts J. 870; Wheatley J. I. 814

Piptostigma pilosum Oliv.

F.W.T.A. 1: 39 (1954); Fl. Gabon 16: 116 (1969); Keay R.W.J., Trees of Nigeria: 26 (1989).

Habit: tree to 12 m. **Guild:** shade-bearer. **Habitat:** forest. **Distribution:** SE Nigeria, Cameroon and Gabon. **Chorology:** Lower Guinea. **Star:** blue. **Alt. range:** 1-200m (1). **Flowering:** Mar (1). **Specimens:** - *Onge:* Wheatley J. I. 679

Piptostigma sp.

Specimens: - *Onge:* Akogo M. 34; Harris D. J. 3755

Polyceratocarpus parviflorus (Baker f.) Ghesq.

F.W.T.A. 1: 45 (1954); Fl. Gabon 16: 270 (1969); Keay R.W.J., Trees of Nigeria: 28 (1989); Hawthorne W., F.G.F.T. Ghana: 73 (1990).

Habit: tree to 6 m. **Guild:** shade-bearer. **Habitat:** forest. **Distribution:** Guinea to Cameroon and Gabon. **Chorology:** Guineo-Congolian. **Star:** green. **Alt. range:** 200-400m (2).**Fruiting:** Oct (1). **Specimens:** - *Etinde:* Thomas D. W. 9737. *Onge:* Ndam N. 786

Polyceratocarpus sp. nov.

Habit: tree to 12 m. **Habitat:** forest. **Distribution:** Mount Cameroon. **Chorology:** Endemic. **Star:** black. **Alt. range:** 200-400m (1). **Flowering:** Mar (1). **Specimens:** - *Bambuko F.R.:* Watts J. 627

Uvaria anonoides Baker f.
F.W.T.A. 1: 38 (1954).

Habit: woody climber. **Habitat:** forest. **Distribution:** Sierra Leone to Cameroon and Congo (Kinshasa). **Chorology:** Guineo-Congolian. **Star:** green. **Alt. range:** 200-400m (1). **Flowering:** Mar (1). **Specimens:** - *Bambuko F.R.:* Tchouto P. 548

Uvaria bipindensis Engl.
F.W.T.A. 1: 38 (1954).

Habit: woody climber. **Habitat:** forest. **Distribution:** Cameroon. **Chorology:** Lower Guinea. **Star:** gold. **Specimens:** - *Southern Bakundu F.R.:* Brenan J. P. M. 9445

Uvaria obanensis Baker f.
F.W.T.A. 1: 38 (1954).

Habit: woody climber. **Habitat:** forest. **Distribution:** SE Nigeria and Cameroon. **Chorology:** Lower Guinea. **Star:** blue. **Alt. range:** 1-200m (1). **Specimens:** - *Onge:* Thomas D. W. 9802

Uvaria sp. aff. poggei Engl. & Diels

Specimens: - *Etinde:* Tekwe C. 49

Uvaria sp. 1

Specimens: - *Etinde:* Tekwe C. 94. *Mabeta-Moliwe:* Tchouto P. 167

Uvariastrum insculptum (Engl. & Diels) Sprague
F.W.T.A. 1: 47 (1954); Keay R.W.J., Trees of Nigeria: 28 (1989); Hawthorne W., F.G.F.T. Ghana: 77 (1990).

Habit: tree to 15 m. **Guild:** shade-bearer. **Habitat:** forest. **Distribution:** Liberia to Cameroon and Gabon. **Chorology:** Guineo-Congolian. **Star:** green. **Alt. range:** 1-200m (1). **Flowering:** Mar (1). **Specimens:** - *Mabeta-Moliwe:* Wheatley J. I. 46

Uvariastrum zenkeri Engl. & Diels
F.W.T.A. 1: 47 (1954); Keay R.W.J., Trees of Nigeria: 28 (1989).

Habit: tree to 30 m. **Habitat:** forest. **Distribution:** SE Nigeria and Cameroon. **Chorology:** Lower Guinea. **Star:** blue. **Alt. range:** 1-200m (5), 201-400m (3). **Flowering:** Mar (4), Apr (2), Nov (1). **Fruiting:** Oct (1). **Specimens:** - *Bambuko F.R.:* Tchouto P. 533; Watts J. 680. *Mabeta-Moliwe:* Jaff B. 42; Sunderland T. C. H. 1174; Wheatley J. I. 92. *Onge:* Thomas D. W. 9772; Watts J. 770; Wheatley J. I. 857

Uvariodendron connivens (Benth.) R.E.Fr.
F.W.T.A. 1: 46 (1954); Keay R.W.J., Trees of Nigeria: 24 (1989).

Habit: tree to 13 m. **Habitat:** forest. **Distribution:** SE Nigeria, Bioko and Cameroon. **Chorology:** Lower Guinea. **Star:** blue. **Alt. range:** 1-200m (11), 201-400m (3). **Flowering:** Apr (3), Jun (1), Oct (4). **Fruiting:** Mar (1), Apr (1), Jun (4), Oct (2). **Specimens:** - *Eastern slopes:* Maitland T. D. 537. *Etinde:* Nkeng P. 37; Tekwe C. 87. *Mabeta-Moliwe:* Jaff B. 73; Mann G. 763; Nguembock F. 76; Sunderland T. C. H. 1264; Watts J. 245, 336, 394; Wheatley J. I. 194, 326. *Onge:* Cheek M. 5180; Ndam N. 708; Tchouto P. 685; Thomas D. W. 9875; Watts J. 782

Uvariodendron fuscum (Benth.) R.E.Fr.
F.W.T.A. 1: 46 (1954).

Habit: tree to 10 m. **Habitat:** forest. **Distribution:** Mount Cameroon and Bioko. **Chorology:** Western Cameroon Uplands (montane). **Star:** black. **Alt. range:** 1200-1400m (1). **Specimens:** - *Eastern slopes:* Maitland T. D. 453

Uvariodendron cf. fuscum (Benth.) R.E.Fr.

Specimens: - *Onge:* Cheek M. 5145

Uvariodendron giganteum (Engl.) R.E.Fr.
F.W.T.A. 1: 46 (1954); Fl. Gabon 16: 277 (1969).

Habit: tree to 10 m. **Habitat:** forest. **Distribution:** Cameroon and Gabon. **Chorology:** Lower Guinea. **Star:** blue. **Alt. range:** 800-1000m (1). **Specimens:** - *Eastern slopes:* Lehmbach H. 230

Uvariodendron mirabile R.E.Fr.
F.W.T.A. 1: 46 (1954); Keay R.W.J., Trees of Nigeria: 24 (1989).

Habit: tree to 10 m. **Habitat:** forest. **Distribution:** Ivory Coast to SW Cameroon. **Chorology:** Guineo-Congolian. **Star:** green. **Alt. range:** 600-800m (1). **Specimens:** - *Eastern slopes:* Lehmbach H. 178. *Mabeta-Moliwe:* Mildbraed G. W. J. 10720; Preuss P. R. 1378

Uvariodendron sp. 1

Specimens: - *Etinde:* Tchouto P. 136

Uvariodendron sp. 2

Specimens: - *Mokoko River F.R.:* Tchouto P. 611

Uvariopsis bakeriana (Hutch. & Diels) Robyns & Ghesq.
F.W.T.A. 1: 50 (1954); Keay R.W.J., Trees of Nigeria: 26 (1989).

Habit: tree or shrub to 6 m. **Habitat:** forest. **Distribution:** Nigeria and Cameroon. **Chorology:** Lower Guinea. **Star:** blue. **Alt. range:** 1-200m (1). **Fruiting:** Mar (1). **Specimens:** - *Onge:* Wheatley J. I. 802. *Southern Bakundu F.R.:* Brenan J. P. M. 9305

Uvariopsis dioica (Diels) Robyns & Ghesq.
F.W.T.A. 1: 50 (1954); Keay R.W.J., Trees of Nigeria: 26 (1989).

Habit: tree to 8 m. **Habitat:** forest. **Distribution:** Nigeria and Cameroon. **Chorology:** Lower Guinea. **Star:** blue. **Specimens:** - *Etinde:* Mildbraed G. W. J. 10647

Uvariopsis sp.

Specimens: - *Onge:* Tchouto P. 675

Xylopia acutiflora (Dunal) A.Rich.

F.W.T.A. 1: 42 (1954); Fl. Zamb. 1: 137 (1960); Fl. Gabon 16: 169 (1969); Keay R.W.J., Trees of Nigeria: 23 (1989).

Habit: tree to 12 m. **Guild:** light demander. **Habitat:** forest. **Distribution:** Sierra Leone to Congo (Kinshasa), Zambia and CAR. **Chorology:** Guineo-Congolian. **Star:** green. **Alt. range:** 200-400m (3).**Fruiting:** Oct (3). **Specimens:** - *Onge:* Ndam N. 632; Tchouto P. 896, 942

Xylopia africana (Benth.) Oliv.

F.W.T.A. 1: 41 (1954); Keay R.W.J., Trees of Nigeria: 21 (1989); Hawthorne W., F.G.F.T. Ghana: 75 (1990).

Habit: tree to 12 m. **Habitat:** forest. **Distribution:** SE Nigeria, SW Cameroon and S. Tomé. **Chorology:** Western Cameroon Uplands (montane). **Star:** gold. **Alt. range:** 1200-1400m (1), 1401-1600m (1). **Flowering:** Oct (1). **Fruiting:** Sep (1), Oct (1). **Specimens:** - *Eastern slopes:* Tekwe C. 315. *Etinde:* Wheatley J. I. 605

Xylopia sp.

Specimens: - *Onge:* Tchouto P. 737

APOCYNACEAE

A.J.M. Leeuwenberg (WAG)

Alafia barteri Oliv.

F.W.T.A. 2: 73 (1963); Kew Bull. 52: 774 (1997).

Habit: woody climber. **Guild:** light demander. **Habitat:** forest. **Distribution:** Guinea Bissau to Cameroon, Gabon and Congo (Kinshasa). **Chorology:** Guineo-Congolian. **Star:** green. **Alt. range:** 1-200m (12), 201-400m (1), 401-600m (2). **Flowering:** Jan (1), Mar (4), Apr (1), May (8), Oct (1), Nov (1), Dec (2). **Fruiting:** Mar (2), May (1). **Specimens:** - *Etinde:* Khayota B. 560; Mbani J. M. 146. *Mabeta-Moliwe:* Cable S. 607, 682; Cheek M. 5735; Etuge M. 1219; Jaff B. 195; Khayota B. 482; Ndam N. 541; Nguembock F. 11, 49; Sunderland T. C. H. 1123, 1329, 1396; Tchouto P. 179; Wheatley J. I. 195, 245, 297. *Mokoko River F.R.:* Tchouto P. 588. *Southern Bakundu F.R.:* Brenan J. P. M. 9428

Alafia lucida Stapf

F.W.T.A. 2: 73 (1963); Kew Bull. 52: 794 (1997).

Habit: woody climber. **Habitat:** forest. **Distribution:** Guinea to Bioko, Cameroon, Gabon, Congo (Brazzaville), Congo (Kinshasa), Angola, CAR, Uganda and Tanzania. **Chorology:** Guineo-Congolian. **Star:** green. **Specimens:** - *Mabeta-Moliwe:* Nguembock F. 66

Alafia multiflora (Stapf) Stapf

F.W.T.A. 2: 73 (1963); Kew Bull. 52: 799 (1997).

Habit: woody climber. **Guild:** light demander. **Habitat:** lava flows. **Distribution:** Liberia to Bioko, Cameroon, Gabon, Congo (Brazzaville), Congo (Kinshasa) and Sudan. **Chorology:** Guineo-Congolian. **Star:** green. **Specimens:** - *Etinde:* Banks H. 47; Maitland T. D. 754

Alafia schumannii Stapf

F.W.T.A. 2: 74 (1963); Kew Bull. 52: 808 (1997).

Habit: woody climber. **Habitat:** forest. **Distribution:** Sierra Leone to Cameroon, Gabon, Congo (Kinshasa), Angola (Cabinda), Uganda and CAR. **Chorology:** Guineo-Congolian. **Star:** green. **Specimens:** - *Southern Bakundu F.R.:* Binuyo A. & Daramola B. O. FHI 35519

Alafia whytei Stapf

F.W.T.A. 2: 73 (1963); Kew Bull. 52: 813 (1997).

Habit: woody climber to 5 m. **Habitat:** forest. **Distribution:** Liberia, Ivory Coast, Ghana and SW Cameroon. **Chorology:** Guineo-Congolian. **Star:** green. **Specimens:** - *Etinde:* Dusen P. K. H. 251

Alstonia boonei De Wild.

F.W.T.A. 2: 68 (1963); Meded. Land. Wag. 79(13): 5 (1979); Keay R.W.J., Trees of Nigeria: 405 (1989); Hawthorne W., F.G.F.T. Ghana: 53 (1990).

Habit: tree to 40 m. **Guild:** pioneer. **Habitat:** forest. **Distribution:** Senegal to Cameroon, Gabon, Congo (Brazzaville), Congo (Kinshasa), Angola, CAR, Uganda, Sudan and Ethiopia. **Star:** green. **Specimens:** - *Etinde:* Maitland T. D. 765. *Mabeta-Moliwe:* Plot Voucher W 727

Ancylobotrys petersiana (Klotzsch) Pierre

Agric. Univ. Wag. Papers 94(3): 13 (1994).

Habit: woody climber to 5 m. **Habitat:** forest. **Distribution:** Cameroon, Congo (Kinshasa), Burundi and East to South Africa. **Star:** excluded. **Alt. range:** 200-400m (1).**Fruiting:** Oct (1). **Specimens:** - *Onge:* Tchouto P. 709

Ancylobotrys pyriformis Pierre

F.W.T.A. 2: 60 (1963); Agric. Univ. Wag. Papers 94(3): 18 (1994).

Habit: woody climber. **Habitat:** forest. **Distribution:** SE Nigeria, Cameroon and Gabon. **Chorology:** Lower Guinea. **Star:** blue. **Alt. range:** 1-200m (1). **Flowering:** Mar (1). **Specimens:** - *Onge:* Wheatley J. I. 768

Baissea axillaris (Benth.) Hua

Specimens: - *Mokoko:* Thomas 10117, 10126; Watts J. 1132

Baissea baillonii Hua

Bull. Jard. Bot. Nat. Belg. 64: 98 (1995).
Baissea breviloba Stapf, F.W.T.A. 2: 79 (1963).

Habit: woody climber to 5 m. **Habitat:** forest. **Distribution:** Guinea to Cameroon, Gabon, Congo (Kinshasa) and Angola. **Chorology:** Guineo-Congolian. **Star:** green. **Specimens:** - *Southern Bakundu F.R.:* Leeuwenberg A. J. M. 9779

Baissea multiflora A.DC.

F.W.T.A. 2: 79 (1963); Bull. Jard. Bot. Nat. Belg. 64: 130 (1995).
Baissea laxiflora Stapf, F.W.T.A. 2: 79 (1963).

Habit: woody climber. **Habitat:** forest. **Distribution:** Senegal to Cameroon, Gabon, Congo (Brazzaville), Congo (Kinshasa), Angola (Cabinda) and CAR. **Chorology:** Guineo-Congolian. **Star:** green.

Specimens: - *Southern Bakundu F.R.:* Latilo M. G. & Ujor E. U. FHI 31190

Callichilia bequaertii De Wild.
Meded. Land. Wag. 78(7): 12 (1978).
  Callichilia macrocalyx Schellenb., F.W.T.A. 2: 64 (1963).

Habit: tree or shrub to 4 m. **Habitat:** forest and farmbush. **Distribution:** SE Nigeria, Cameroon, Gabon, Congo (Brazzaville), Congo (Kinshasa) and Angola (Cabinda). **Chorology:** Guineo-Congolian. **Star:** green. **Alt. range:** 1-200m (3), 201-400m (1). **Flowering:** Apr (1). **Fruiting:** Apr (3), Jul (1). **Specimens:** - *Mabeta-Moliwe:* Hunt L. V. 19; Jaff B. 110; Sunderland T. C. H. 1258; Wheatley J. I. 114

Callichilia inaequalis Stapf
Meded. Land. Wag. 78(7): 16 (1978).
  Callichilia mannii Stapf, F.W.T.A. 2: 64 (1963).

Habit: woody climber to 10 m. **Habitat:** forest. **Distribution:** SE Nigeria, Cameroon and Gabon. **Chorology:** Lower Guinea. **Star:** blue. **Specimens:** - *Mabeta-Moliwe:* Mann G. 2152

Callichilia sp.

Specimens: - *Mabeta-Moliwe:* Plot Voucher W 41, W 275; *Mokoko:* Ekema 921

Dictyophleba leonensis (Stapf) Pichon
F.W.T.A. 2: 59 (1963); Bull. Jard. Bot. Nat. Belg. 59: 208 (1989).

Habit: woody climber to 15 m. **Habitat:** forest. **Distribution:** Guinea to SW Cameroon. **Chorology:** Guineo-Congolian. **Star:** green. **Fruiting:** Nov (1). **Specimens:** - *Etinde:* Gentry A. FHI 52908

Dictyophleba ochracea (K.Schum. ex Hallier f.) Pichon
F.W.T.A. 2: 59 (1963); Bull. Jard. Bot. Nat. Belg. 59: 218 (1989).

Habit: woody climber. **Habitat:** forest. **Distribution:** SE Nigeria, Cameroon, Gabon, Congo (Kinshasa) and CAR. **Chorology:** Guineo-Congolian. **Star:** green. **Specimens:** - *Etinde:* Gentry A. FHI 52951

Dictyophleba stipulosa (S.Moore ex Wermham) Pichon
F.W.T.A. 2: 59 (1963); Bull. Jard. Bot. Nat. Belg. 59: 223 (1989).

Habit: woody climber. **Habitat:** forest. **Distribution:** Ivory Coast, SE Nigeria, Cameroon, Gabon and Congo (Brazzaville). **Chorology:** Guineo-Congolian. **Star:** green. **Alt. range:** 1-200m (1). **Flowering:** Oct (1). **Specimens:** - *Mokoko:* Ekema 888, 973. *Onge:* Tchouto P. 1026

Funtumia africana (Benth.) Stapf
F.W.T.A. 2: 74 (1963); Meded. Land. Wag. 81(16): 16 (1981); Keay R.W.J., Trees of Nigeria: 409 (1989); Hawthorne W., F.G.F.T. Ghana: 54 (1990).

Habit: tree to 30 m. **Guild:** light demander. **Habitat:** forest. **Distribution:** tropical Africa. **Star:** excluded. **Specimens:** - *Etinde:* Maitland T. D. 319. *Mabeta-Moliwe:* Plot Voucher W 637, W 1181

Funtumia elastica (Preuss) Stapf
F.W.T.A. 2: 74 (1963); Meded. Land. Wag. 81(16): 16 (1981); Keay R.W.J., Trees of Nigeria: 409 (1989).

Habit: tree to 35 m. **Guild:** light demander. **Habitat:** forest and farmbush. **Distribution:** tropical Africa. **Star:** excluded. **Alt. range:** 1-200m (1). **Fruiting:** Jul (1). **Specimens:** - *Mabeta-Moliwe:* Cheek M. 3525

Hunteria camerunensis K.Schum. ex Hallier f.
F.W.T.A. 2: 62 (1963); Agric. Univ. Wag. Papers 96(1): 93 (1996).

Habit: shrub to 3 m. **Habitat:** forest. **Distribution:** Cameroon and Gabon. **Chorology:** Lower Guinea. **Star:** blue. **Specimens:** - *Southern Bakundu F.R.:* Brenan J. P. M. 9401

Landolphia dulcis (Sabine) Pichon
F.W.T.A. 2: 56 (1963); Agric. Univ. Wag. Papers 92(2): 53 (1992).
  Landolphia dulcis (Sabine) Pichon var. *barteri* (Stapf) Pichon, F.W.T.A. 2: 57 (1963).

Habit: woody climber to 10 m. **Guild:** light demander. **Habitat:** forest. **Distribution:** Senegal to Cameroon and Gabon. **Chorology:** Guineo-Congolian. **Star:** green. **Alt. range:** 1-200m (3), 201-400m (2). **Flowering:** May (1), Nov (1). **Fruiting:** Sep (1), Oct (2). **Specimens:** - *Mabeta-Moliwe:* Wheatley J. I. 296. *Onge:* Tchouto P. 821; Watts J. 809, 824, 994

Landolphia flavidiflora (K.Schum.) Persoon
Agric. Univ. Wag. Papers 92(2): 67 (1992).
  Aphanostylis flavidiflora (K.Schum.) Pierre, F.W.T.A. 2: 59 (1963).

Habit: woody climber. **Habitat:** forest. **Distribution:** Cameroon. **Chorology:** Lower Guinea. **Star:** gold. **Specimens:** - *Southern Bakundu F.R.:* Brenan J. P. M. 9322

Landolphia glabra (Pierre ex Stapf) Pichon

Specimens: - Sonké 1230; Tchouto P. 1156

Landolphia hirsuta (Hua) Pichon
F.W.T.A. 2: 57 (1963); Agric. Univ. Wag. Papers 92(2): 90 (1992).

Habit: woody climber. **Guild:** light demander. **Habitat:** forest. **Distribution:** Senegal to Cameroon. **Chorology:** Guineo-Congolian. **Star:** green. **Specimens:** - *Southern Bakundu F.R.:* Brenan J. P. M. 9431

Landolphia incerta (K.Schum.) Persoon
Agric. Univ. Wag. Papers 92(2): 92 (1992).
  Aphanostylis mannii (Stapf) Pierre, F.W.T.A. 2: 59 (1963).

Habit: woody climber. **Habitat:** forest. **Distribution:** Guinea Bissau to Cameroon, Gabon, Congo (Brazzaville), Congo (Kinshasa), Angola, CAR and Zambia. **Chorology:** Guineo-Congolian (montane). **Star:** green. **Alt. range:** 800-1000m (1). **Fruiting:** Dec (1). **Specimens:** - *Etinde:* Cable S. 430

Landolphia landolphioides (Hallier f.) A.Chev.
F.W.T.A. 2: 56 (1963); Agric. Univ. Wag. Papers 92(2): 112 (1992).

Habit: woody climber. Habitat: forest. Distribution: Guinea to Cameroon, Gabon, Congo (Kinshasa), CAR, Uganda and Sudan. Chorology: Guineo-Congolian. Star: green. Alt. range: 1-200m (1), 201-400m (1), 801-1000m (1), 1401-1600m (1). Flowering: Mar (1). Fruiting: Aug (1), Sep (1). Specimens: - *Eastern slopes:* Dalziel J. M. 8242. *Etinde:* Tekwe C. 185. *Mabeta-Moliwe:* Baker W. J. 310. *Onge:* Wheatley J. I. 681

Landolphia leptantha (K.Schum.) Persoon

Agric. Univ. Wag. Papers 92(2): 120 (1992).
  Aphanostylis leptantha (K.Schum.) Pierre, F.W.T.A. 2: 58 (1963).

Habit: woody climber. Habitat: forest. Distribution: SE Nigeria, Cameroon and Gabon. Chorology: Lower Guinea. Star: blue. Alt. range: 1-200m (1). Flowering: Mar (1). Specimens: - *Onge:* Wheatley J. I. 837

Landolphia ligustrifolia (Stapf) Pichon

Agric. Univ. Wag. Papers 92(2): 126 (1992).

Habit: woody climber. Habitat: forest and farmbush. Distribution: Cameroon, Gabon, Congo (Brazzaville) and Congo (Kinshasa). Chorology: Guineo-Congolian. Star: green. Alt. range: 1-200m (2), 201-400m (1). Flowering: Mar (2), Oct (1). Fruiting: Oct (1). Specimens: - *Etinde:* Mbani J. M. 73. *Mokoko River F.R.:* Tchouto P. 607. *Onge:* Tchouto P. 1003

Landolphia owariensis P.Beauv.

F.W.T.A. 2: 55 (1963); Agric. Univ. Wag. Papers 92(2): 153 (1992).

Habit: woody climber or shrub. Guild: light demander. Habitat: forest and Aframomum thicket. Distribution: Senegal to Cameroon, Congo (Kinshasa), Angola, Uganda, Sudan, Tanzania and Zambia. Chorology: Guineo-Congolian. Star: green. Alt. range: 1-200m (1).Fruiting: May (1). Specimens: - *Mabeta-Moliwe:* Sunderland T. C. H. 1370

Landolphia robustior (K.Schum.) Persoon

Agric. Univ. Wag. Papers 92(2): 172 (1992).

Habit: woody climber. Habitat: forest. Distribution: Nigeria, Cameroon, Gabon, Congo (Brazzaville), Congo (Kinshasa) and Angola (Cabinda). Chorology: Guineo-Congolian. Star: green. Alt. range: 1-200m (1). Flowering: Apr (1). Specimens: - *Mabeta-Moliwe:* Cheek M. 3488; Wheatley J. I. 90. *Southern Bakundu F.R.:* Binuyo A. & Daramola B. O. FHI 35095

Landolphia violacea (K.Schum. ex Hallier f.) Pichon

F.W.T.A. 2: 56 (1963); Agric. Univ. Wag. Papers 92(2): 199 (1992).

Habit: woody climber to 10 m. Habitat: forest. Distribution: SE Nigeria, Cameroon, Gabon, Congo (Brazzaville) and Congo (Kinshasa). Chorology: Guineo-Congolian. Star: green. Alt. range: 400-600m (1).Fruiting: May (1). Specimens: - *Etinde:* Mbani J. M. 167. *Mabeta-Moliwe:* Nguembock 39

Landolphia sp.

Specimens: - *Bambuko F.R.:* Watts J. 605. *Mabeta-Moliwe:* Baker W. J. 215

Oncinotis glabrata (Baill.) Stapf ex Hiern

F.W.T.A. 2: 80 (1963); Agric. Univ. Wag. Papers 85(2): 13 (1985).

Habit: woody climber. Guild: light demander. Habitat: forest. Distribution: Guinea to Cameroon, Gabon, Congo (Brazzaville), Congo (Kinshasa), Angola, CAR, Uganda, Burundi and W. Tanzania. Chorology: Guineo-Congolian. Star: green. Alt. range: 1-200m (1), 401-600m (1).Fruiting: Nov (1), Dec (1). Specimens: - *Eastern slopes:* Migeod F. 256. *Etinde:* Cable S. 189, 365

Orthopichonia cirrhosa (Radlk.) H.Huber

Kew Bull. 15: 437 (1962); Agric. Univ. Wag. Papers 89(4): 33 (1989).
  Orthopichonia nigeriana (Pichon) H.Huber, F.W.T.A. 2: 58 (1963).

Habit: woody climber. Habitat: forest. Distribution: SE Nigeria, Cameroon, Gabon, Congo (Brazzaville) and Congo (Kinshasa). Chorology: Guineo-Congolian. Star: green. Alt. range: 1-200m (2).Fruiting: Aug (1), Nov (1). Specimens: - *Etinde:* Cable S. 193. *Mabeta-Moliwe:* Baker W. J. 301. *Southern Bakundu F.R.:* Binuyo A. & Daramola B. O. FHI 35089

Orthopichonia indeniensis (A.Chev.) H.Huber

Kew Bull. 15: 437 (1962); F.W.T.A. 2: 58 (1963); Agric. Univ. Wag. Papers 89(4): 37 (1989).
  Orthopichonia longituba (Wernham) H.Huber, F.W.T.A. 2: 58 (1963).

Habit: woody climber. Habitat: forest. Distribution: Liberia to Cameroon. Chorology: Guineo-Congolian. Star: green. Alt. range: 1-200m (1). Flowering: Apr (1). Specimens: - *Etinde:* Maitland T. D. 405. *Mabeta-Moliwe:* Wheatley J. I. 150

Orthopichonia visciflua (K.Schum. ex Hallier f.) Vonk

Agric. Univ. Wag. Papers 89(4): 46 (1989).
  Orthopichonia staudtii (Stapf) H.Huber, F.W.T.A. 2: 58 (1963).

Habit: woody climber. Habitat: forest. Distribution: Cameroon and Gabon. Chorology: Lower Guinea. Star: blue. Alt. range: 200-400m (1). Flowering: Mar (1). Specimens: - *Etinde:* Tchouto P. 22

Orthopichonia sp.

Specimens:- *Mokoko:* Ekema 1152

Picralima nitida (Stapf) T.Durand & H.Durand

F.W.T.A. 2: 62 (1963); Keay R.W.J., Trees of Nigeria: 406 (1989); Agric. Univ. Wag. Papers 96(1): 128 (1996).

Habit: tree or shrub 4-35 m tall. Guild: shade-bearer. Habitat: forest. Distribution: Ivory Coast to Cameroon, Gabon, Congo (Brazzaville), Congo (Kinshasa), CAR and Uganda. Chorology: Guineo-Congolian. Star: green. Alt. range: 200-400m (1). Flowering: May (1). Fruiting: May (1). Specimens: - *Mabeta-Moliwe:* Tchouto P. 209

Pleiocarpa bicarpellata Stapf

F.W.T.A. 2: 63 (1963); Agric. Univ. Wag. Papers 96(1): 135 (1996).

Habit: tree or shrub to 8 m. Habitat: forest. Distribution: Cameroon, Gabon, Congo (Kinshasa), Angola and Kenya. Chorology: Guineo-Congolian (montane). Star: green. Alt.

15

range: 600-800m (1), 1201-1400m (1). **Flowering:** May (1). **Fruiting:** May (1). **Specimens:** - *Etinde:* Tchouto P. 216. *No locality:* Mann G. 1213

Pleiocarpa mutica Benth.
F.W.T.A. 2: 63 (1963); Hawthorne W., F.G.F.T. Ghana: 54 (1990); Agric. Univ. Wag. Papers 96(1): 142 (1996).

Habit: tree or shrub to 8 m. **Guild:** shade-bearer. **Habitat:** forest. **Distribution:** Sierra Leone to Cameroon and Gabon. **Chorology:** Guineo-Congolian. **Star:** green. **Alt. range:** 1-200m (3), 201-400m (2). **Flowering:** Mar (3). **Fruiting:** Oct (2). **Specimens:** - *Bambuko F.R.:* Watts J. 602, 667. *Mokoko:* Ekema 1114a; Fraser 441; Poukouyouko 57; Sonké 1200; Tchouto P. 1232. *Onge:* Tchouto P. 740; Watts J. 760; Wheatley J. I. 806

Pleiocarpa rostrata Benth.
Agric. Univ. Wag. Papers 96(1): 152 (1996).
Pleiocarpa talbotii Wernham, F.W.T.A. 2: 64 (1963).

Habit: tree or shrub to 5 m. **Habitat:** forest. **Distribution:** SE Nigeria, Cameroon and Gabon. **Chorology:** Lower Guinea. **Star:** blue. **Alt. range:** 1-200m (6), 201-400m (5), 401-600m (2), 801-1000m (1). **Flowering:** Oct (4), Nov (4), Dec (1). **Fruiting:** Mar (3), Nov (2), Dec (2). **Specimens:** - *Bambuko F.R.:* Watts J. 599, 608. *Etinde:* Cable S. 289, 514; Cheek M. 5504; Tchouto P. 449. *Mokoko:* Ekema 1192. *Onge:* Harris D. J. 3797, 3873; Tchouto P. 712, 879; Thomas D. W. 9839; Watts J. 813, 954; Wheatley J. I. 691. *Southern Bakundu F.R.:* Keay R. W. J. FHI 28571

Pleiocarpa sp.

Specimens:- *Mokoko:* Acworth J.M. 104

Pleioceras zenkeri Stapf
F.T.A. 4: 167 (1902); Meded. Land. Wag. 83(7): 41 (1983).
Pleioceras barteri Baill. var. *zenkeri* (Stapf) H.Huber, F.W.T.A. 2: 76 (1963).

Habit: tree or scandent shrub to 4 m. **Habitat:** forest. **Distribution:** SE Nigeria, Cameroon and Gabon. **Chorology:** Lower Guinea. **Star:** blue. **Specimens:** - *Southern Bakundu F.R.:* Olorunfemi J. FHI 30735

Rauvolfia caffra Sond.
F.W.T.A. 2: 69 (1963); Keay R.W.J., Trees of Nigeria: 407 (1989); Bull. Jard. Bot. Nat. Belg. 61: 24 (1991).
Rauvolfia macrophylla Stapf, F.W.T.A. 2: 69 (1963).

Habit: tree to 21 m. **Habitat:** forest and farmbush. **Distribution:** tropical and subtropical Africa. **Star:** excluded. **Alt. range:** 1-200m (1), 601-800m (1). **Flowering:** May (1). **Fruiting:** May (1). **Specimens:** - *Mabeta-Moliwe:* Mildbraed G. W. J. 10783; Ndam N. 522

Rauvolfia mannii Stapf
F.W.T.A. 2: 69 (1963); Bull. Jard. Bot. Nat. Belg. 61: 38 (1991).

Habit: tree to 10 m. **Distribution:** tropical Africa. **Star:** excluded. **Specimens:** - *Mokoko:* Tchouto P. 1148. *Southern Bakundu F.R.:* Brenan J. P. M. 9422

Rauvolfia vomitoria Afzel.
F.W.T.A. 2: 69 (1963); Keay R.W.J., Trees of Nigeria: 407 (1989); Hawthorne W., F.G.F.T. Ghana: 53 (1990); Bull. Jard. Bot. Nat. Belg. 61: 60 (1991).

Habit: tree to 20 m. **Guild:** pioneer. **Habitat:** forest and farmbush. **Distribution:** tropical Africa. **Star:** excluded. **Alt. range:** 1-200m (10), 401-600m (1), 801-1000m (3). **Flowering:** Feb (3), Mar (1), May (2), Aug (1), Oct (1), Nov (1). **Fruiting:** Apr (1), May (4), Jun (1), Jul (1), Aug (1). **Specimens:** - *Eastern slopes:* Etuge M. 1178; Hutchinson J. & Metcalfe C. R. 93; Tekwe C. 113. *Etinde:* Etuge M. 1232; Tchouto P. 19; Tekwe C. 19. *Mabeta-Moliwe:* Baker W. J. 313; Ndam N. 517, 534; Plot Voucher W 753; Sunderland T. C. H. 1191, 1377, 1456; Wheatley J. I. 299. *Mokoko:* Acworth J.M. 91; Akogo 191; Ekema 990. *Onge:* Cheek M. 5102; Tchouto P. 1031

Saba comorensis (Bojer) Pichon
Bull. Jard. Bot. Nat. Belg. 59: 190 (1989).
Saba florida (Benth.) Bullock, F.W.T.A. 2: 61 (1963).

Habit: woody climber. **Habitat:** forest. **Distribution:** tropical Africa. **Star:** excluded. **Specimens:** - *Etinde:* Deistel H. 130

Strophanthus bullenianus Mast.
F.W.T.A. 2: 72 (1963); Meded. Land. Wag. 82(4): 51 (1982).

Habit: woody climber to 12 m. **Habitat:** forest. **Distribution:** SE Nigeria, Cameroon, Gabon, Congo (Brazzaville), Congo (Kinshasa) and Angola (Cabinda). **Chorology:** Guineo-Congolian. **Star:** green. **Specimens:** - *Etinde:* Maitland T. D. 56

Strophanthus gratus (Wall. & Hook.) Baill.
F.W.T.A. 2: 70 (1963); Meded. Land. Wag. 82(4): 81 (1982).

Habit: woody climber, shrub or small tree. **Habitat:** forest. **Distribution:** Senegal to Cameroon, Gabon, Congo (Brazzaville), Congo (Kinshasa) and CAR. **Chorology:** Guineo-Congolian. **Star:** green. **Specimens:** - *Etinde:* Preuss P. R. 1301

Strophanthus hispidus DC.
F.W.T.A. 2: 72 (1963); Meded. Land. Wag. 82(4): 85 (1982).

Habit: woody climber to 5 m. **Guild:** pioneer. **Habitat:** forest. **Distribution:** Senegal to Cameroon, Gabon, Congo (Brazzaville), Congo (Kinshasa), Angola, CAR, Uganda and W. Tanzania. **Chorology:** Guineo-Congolian. **Star:** green. **Specimens:** - *Southern Bakundu F.R.:* Binuyo A. & Daramola B. O. FHI 35617

Strophanthus preussii Engl. & Pax
F.W.T.A. 2: 72 (1963); Meded. Land. Wag. 82(4): 125 (1982).

Habit: woody climber to 5 m. **Guild:** pioneer. **Habitat:** forest and farmbush. **Distribution:** Senegal to Cameroon, Gabon, Congo (Brazzaville), Congo (Kinshasa), Angola, CAR, Uganda and W. Tanzania. **Chorology:** Guineo-Congolian. **Star:** green. **Alt. range:** 1-200m (1). **Flowering:** Apr (1). **Specimens:** - *Etinde:* Preuss P. R. 1114. *Mabeta-Moliwe:* Wheatley J. I. 214

Strophanthus thollonii Franch.
F.W.T.A. 2: 70 (1963); Meded. Land. Wag. 82(4): 147 (1982).

Habit: woody climber. **Habitat:** forest, by streams. **Distribution:** SE Nigeria, Cameroon and Gabon. **Chorology:** Lower Guinea. **Star:** blue. **Alt. range:** 1-200m (5), 201-400m (1). **Flowering:**

Mar (1), Oct (3), Nov (2). **Fruiting:** Mar (1). **Specimens:** - *Onge:* Cheek M. 5170; Harris D. J. 3661; Tchouto P. 954; Thomas D. W. 9892; Watts J. 804; Wheatley J. I. 808

Tabernaemontana brachyantha Stapf
F.W.T.A. 2: 65 (1963); Leeuwenberg A., Rev. of Tabernaemontana I: 15 (1991).

Habit: tree 4-12 m tall. **Habitat:** forest and Aframomum thicket. **Distribution:** SE Nigeria, Bioko, Cameroon, Rio Muni, Gabon and Congo (Kinshasa). **Chorology:** Guineo-Congolian. **Star:** green. **Alt. range:** 1-200m (3), 201-400m (1), 401-600m (1), 1001-1200m (1), 1201-1400m (1), 1401-1600m (2). **Flowering:** Feb (1), Mar (3), Jun (1), Sep (1). **Fruiting:** Mar (2), Jun (1), Aug (1), Sep (1), Dec (1). **Specimens:** - *Eastern slopes:* Cable S. 1381. *Etinde:* Hunt L. V. 30; Khayota B. 584; Sunderland T. C. H. 1080; Tchouto P. 20; Wheatley J. I. 589. *Mabeta-Moliwe:* Cheek M. 5736; Watts J. 414. *No locality:* Keay R. W. J. FHI 37507

Tabernaemontana contorta Stapf
F.W.T.A. 2: 66 (1963); Leeuwenberg A., Rev. of Tabernaemontana I: 18 (1991).

Habit: tree or shrub 3-10 m tall. **Habitat:** forest. **Distribution:** Cameroon. **Chorology:** Lower Guinea. **Star:** gold. **Specimens:** - *Mabeta-Moliwe:* Mann G. 703

Tabernaemontana crassa Benth.
F.W.T.A. 2: 66 (1963); Hawthorne W., F.G.F.T. Ghana: 54 (1990); Leeuwenberg A., Rev. of Tabernaemontana I: 21 (1991).

Habit: tree or shrub 2-15 m tall. **Guild:** shade-bearer. **Habitat:** forest. **Distribution:** Sierra Leone to Cameroon, Gabon, Congo (Brazzaville), Congo (Kinshasa), Angola and CAR. **Chorology:** Guineo-Congolian. **Star:** green. **Alt. range:** 1-200m (4). **Flowering:** Feb (1), Aug (1). **Fruiting:** Feb (1), Apr (1). **Specimens:** - *Etinde:* Mbani J. M. 95. *Mabeta-Moliwe:* Cable S. 1367; Cheek M. 3584; von Rege I. 86; Sunderland T.C.H. 1314. *Mokoko River F.R.:* Acworth J.M. 28. *Southern Bakundu F.R.:* Binuyo A. & Daramola B. O. FHI 35172

Tabernaemontana pachysiphon Stapf
F.W.T.A. 2: 66 (1963); Keay R.W.J., Trees of Nigeria: 411 (1989); Hawthorne W., F.G.F.T. Ghana: 54 (1990); Leeuwenberg A., Rev. of Tabernaemontana I: 51 (1991).

Habit: tree or shrub 2-15 m tall. **Guild:** shade-bearer. **Habitat:** forest. **Distribution:** tropical Africa. **Star:** excluded. **Alt. range:** 1-200m (1). **Fruiting:** Jul (1). **Specimens:** - *Mabeta-Moliwe:* Cheek M. 3590

Tabernaemontana ventricosa Hochst. ex A.DC.
F.W.T.A. 2: 66 (1963); Leeuwenberg A., Rev. of Tabernaemontana I: 71 (1991).

Habit: tree or shrub 3-15 m tall. **Habitat:** forest. **Distribution:** Nigeria to East and southern Africa. **Star:** excluded. **Alt. range:** 600-800m (1), 801-1000m (1), 1001-1200m (1). **Flowering:** Mar (1). **Fruiting:** Dec (2). **Specimens:** - *Etinde:* Cable S. 491, 1628; Faucher P. 21. *No locality:* Dundas J. FHI 15335

Tabernaemontana sp.

Specimens: - *Mabeta-Moliwe:* Plot Voucher W 282, W 297, W 585, W 903, W 1029. *Mokoko:* Ekema 1198; Mbani J.M. 411

Vahadenia laurentii (De Wild.) Stapf
F.W.T.A. 2: 60 (1963); Bull. Jard. Bot. Nat. Belg. 63: 323 (1994).

Habit: woody climber. **Habitat:** forest. **Distribution:** SE Nigeria, Cameroon, Gabon, Congo (Brazzaville), Congo (Kinshasa), Angola and CAR. **Chorology:** Guineo-Congolian. **Star:** green. **Specimens:** - *Mokoko:* Akogo 282. *Southern Bakundu F.R.:* Binuyo A. & Daramola B. O. FHI 35618

Voacanga africana Stapf
F.W.T.A. 2: 67 (1963); Meded. Land. Wag. 82(3): 71 (1982); Agric. Univ. Wag. Papers 85(3): 12 (1985); Keay R.W.J., Trees of Nigeria: 411 (1989); Hawthorne W., F.G.F.T. Ghana: 54 (1990).

Habit: tree or shrub to 10 m. **Guild:** pioneer. **Habitat:** forest and farmbush. **Distribution:** tropical Africa. **Star:** excluded. **Alt. range:** 1-200m (1), 801-1000m (1). **Flowering:** Feb (1), Mar (1), Nov (1). **Specimens:** - *Eastern slopes:* Groves M. 124; Nkeng P. 59. *Etinde:* Cheek M. 5505; Maitland T. D. 422

Voacanga bracteata Stapf
F.W.T.A. 2: 67 (1963); Agric. Univ. Wag. Papers 85(3): 22 (1985). *Voacanga diplochlamys* K.Schum., F.W.T.A. 2: 67 (1963).

Habit: tree or shrub to 6 m. **Guild:** shade-bearer. **Habitat:** forest. **Distribution:** Sierra Leone to Cameroon, Gabon, Congo (Brazzaville) and Congo (Kinshasa). **Chorology:** Guineo-Congolian. **Star:** green. **Specimens:** - *Southern Bakundu F.R.:* Brenan J. P. M. 9300

Voacanga psilocalyx Pierre ex Stapf
F.W.T.A. 2: 67 (1963); Agric. Univ. Wag. Papers 85(3): 36 (1985). *Voacanga bracteata* Stapf var. *zenkeri*, F.W.T.A. 2: 67 (1963).

Habit: tree or shrub to 4 m. **Habitat:** forest. **Distribution:** SE Nigeria, Cameroon, Gabon and Congo (Brazzaville). **Chorology:** Guineo-Congolian. **Star:** green. **Alt. range:** 1-200m (5), 201-400m (3), 401-600m (2), 601-800m (4), 801-1000m (5), 1001-1200m (1). **Flowering:** Feb (4), Mar (10). **Fruiting:** Feb (1), Mar (1), May (1), Oct (4), Nov (1), Dec (1). **Specimens:** - *Bambuko F.R.:* Tchouto P. 559; Watts J. 636. *Eastern slopes:* Cable S. 1356; Etuge M. 1176; Tchouto P. 42, 102. *Etinde:* Cable S. 434, 1589; Groves M. 296; Khayota B. 513; Mbani J. M. 46, 101; Tchouto P. 15; Watts J. 527; Wheatley J. I. 9. *Mokoko:* Ekema 1193; Ndam 1305; Watts J. 1069. *Onge:* Cheek M. 5120; Ndam N. 695; Tchouto P. 770; Watts J. 985; Wheatley J. I. 726, 860

Voacanga sp.

Specimens: - *Mabeta-Moliwe:* Plot Voucher W 289. *Mokoko:* Akogo 209; Batoum 2; Ekema 831, 1093; Ndam 1296; Pouakouyou D. 85. *Onge:* Watts J. 913

AQUIFOLIACEAE

Ilex mitis (L.) Radlk.
F.W.T.A. 1: 623 (1958); Fl. Cameroun 19: 34 (1975); Keay R.W.J., Trees of Nigeria: 305 (1989).

Habit: tree to 20 m. Habitat: forest-grassland transition. Distribution: tropical Africa. Chorology: Afromontane. Star: green. Alt. range: 400-600m (1), 1201-1400m (1), 1801-2000m (1).Fruiting: Sep (2). Specimens: - *Etinde*: Tekwe C. 167, 263. *No locality*: Keay R. W. J. FHI 28644

ARALIACEAE

D. Frodin (K)

Polyscias fulva (Hiern) Harms

F.W.T.A. 1: 750 (1958); F.T.E.A. Araliaceae: 12 (1968); Fl. Cameroun 10: 11 (1970); Keay R.W.J., Trees of Nigeria: 377 (1989).

Habit: tree 6-30 m tall. Habitat: forest and Aframomum thicket. Distribution: tropical Africa. Chorology: Afromontane. Star: green. Alt. range: 1400-1600m (2).Fruiting: Sep (1). Specimens: - *Eastern slopes*: Maitland T. D. FHI 25100. *Etinde*: Tekwe C. 230

Schefflera abyssinica (Hochst. ex. A.Rich.) Harms

F.W.T.A. 1: 751 (1958); F.T.E.A. Araliaceae: 20 (1968); Fl. Cameroun 10: 15 (1970); Keay R.W.J., Trees of Nigeria: 378 (1989).

Habit: tree to 20 m. Habitat: forest-grassland transition. Distribution: SE Nigeria, Cameroon, Uganda, Sudan and Ethiopia. Chorology: Afromontane. Star: green. Alt. range: 1800-2000m (1), 2001-2200m (1).Fruiting: May (1). Specimens: - *Eastern slopes*: Sunderland T. C. H. 1422. *No locality*: Maitland T. D. 529

Schefflera barteri (Seem.) Harms

F.W.T.A. 1: 751 (1958); F.T.E.A. Araliaceae: 17 (1968); Fl. Cameroun 10: 18 (1970); Keay R.W.J., Trees of Nigeria: 378 (1989); Hawthorne W., F.G.F.T. Ghana: 157 (1990).
    Schefflera hierniana Harms, F.W.T.A. 1: 751 (1958).

Habit: tree to 15 m, or epiphytic shrub. Guild: strangler. Habitat: forest. Distribution: Guinea to Cameroon, Gabon, Congo (Kinshasa), Angola, Uganda and Tanzania. Chorology: Guineo-Congolian (montane). Star: green. Alt. range: 1000-1200m (1), 1201-1400m (1). Flowering: Sep (1). Fruiting: Sep (1). Specimens: - *Etinde*: Tekwe C. 155. *No locality*: Maitland T. D. 381

Schefflera mannii (Hook.f.) Harms

F.W.T.A. 1: 751 (1958); Fl. Cameroun 10: 21 (1970); Keay R.W.J., Trees of Nigeria: 378 (1989).

Habit: tree to 15 m. Habitat: forest. Distribution: SE Nigeria, Bioko, W. Cameroon and S. Tomé. Chorology: Western Cameroon Uplands (montane). Star: gold. Alt. range: 400-600m (1), 1601-1800m (1). Flowering: Apr (1). Specimens: - *Etinde*: Tekwe C. 52. *No locality*: Dundas J. FHI 15337

ARISTOLOCHIACEAE

F. Gonzalez (NY) S. Cable & M. Cheek (K)

Pararistolochia ceropegioides (S.Moore) Hutch. & Dalziel

Specimens: - *Etinde*: Groves 127

Pararistolochia goldieana (Hook.f.) Hutch. & Dalziel

Specimens: - *Mabeta-Moliwe*: Sunderland 1318, 1472

Pararistolochia macrocarpa Duchartre

Specimens: - *Mokoko*: Tchouto 1086; Acworth 144

Pararistolochia mannii Hook.f.

Specimens: - No locality: *Watts* s.n.?

Pararistolochia preussii (Engl.) Hutch. & Dalziel

F.W.T.A. 1: 79 (1954); Adansonia (ser.2) 17: 484 (1978).

Habit: woody climber to 10 m. Habitat: forest. Distribution: Cameroon. Chorology: Lower Guinea. Star: black. Alt. range: 1-200m (2). Flowering: Mar (1), Apr (2). Specimens: - *Mabeta-Moliwe*: Sunderland T. C. H. 1197; Wheatley J. I. 137. *Southern Bakundu F.R.*: Brenan J. P. M. 9484

Pararistolochia promissa (Mast.) Keay

F.W.T.A. 1: 79 (1954); Adansonia (ser.2) 17: 491 (1978).
    Pararistolochia talbotii (S.Moore) Keay, F.W.T.A. 1: 81 (1954).
    Pararistolochia tenuicauda (S.Moore) Keay, F.W.T.A. 1: 81 (1954).

Habit: woody climber. Habitat: forest. Distribution: Ivory Coast to Cameroon, Gabon, Congo (Kinshasa) and CAR. Chorology: Guineo-Congolian (montane). Star: green. Specimens: - *Etinde*: Kalbreyer W. 7; *Mokoko*: Tchouto P. 549; Watts J. 653

Pararistolochia zenkeri (Engl.) Hutch. & Dalziel

F.W.T.A. 1: 79 (1954); Adansonia (ser.2) 17: 486 (1978).

Habit: woody climber to 10 m. Habitat: forest. Distribution: Ghana to Cameroon, Congo (Brazzaville) and Congo (Kinshasa). Chorology: Guineo-Congolian. Star: blue. Alt. range: 200-400m (1). Flowering: Nov (1). Specimens: - *Etinde*: Cheek M. 5415

ASCLEPIADACEAE

D. Goyder (K) & U. Meve (Münster)

Anisopus efulensis (N.E.Br.) Goyder

Kew Bull. 49: 743 (1994).

18

Habit: twining shrub. Habitat: forest. Distribution: Ghana, SE Nigeria, Cameroon, Gabon, Congo (Kinshasa) and CAR. Chorology: Guineo-Congolian (montane). Star: green. Alt. range: 1-200m (1), 401-600m (3), 1401-1600m (1), 1601-1800m (1). Flowering: Feb (2), Mar (2), Apr (1). Specimens: - *Eastern slopes:* Meve U. 913, 916a. *Etinde:* Etuge M. 1237; Groves M. 289; Tekwe C. 44. *Mabeta-Moliwe:* Cable S. 1485

Anisopus mannii N.E.Br.
F.W.T.A. 2: 98 (1963); Kew Bull. 49: 740 (1994).

Habit: twining shrub. Habitat: forest. Distribution: Ghana, Nigeria, Cameroon, Rio Muni, Gabon, Congo (Kinshasa) and CAR. Chorology: Guineo-Congolian. Star: green. Alt. range: 1-200m (1). Flowering: Mar (1). Specimens: - *Mabeta-Moliwe:* Wheatley J. I. 65

Batesanthus purpureus N.E.Br.
F.W.T.A. 2: 82 (1963).

Habit: twining shrub. Habitat: forest. Distribution: SE Nigeria and W. Cameroon. Chorology: Western Cameroon. Star: gold. Alt. range: 1-200m (1). Flowering: Apr (1). Specimens: - *Eastern slopes:* Maitland T. D. 666. *Mabeta-Moliwe:* Sunderland T. C. H. 1200

Ceropegia sankuruensis Schltr.
F.W.T.A. 2: 102 (1963).

Habit: herbaceous climber. Habitat: forest. Distribution: tropical Africa. Star: excluded. Alt. range: 600-800m (3). Flowering: Feb (3). Specimens: - *Eastern slopes:* Meve U. 907, 908, 909

Cynanchum adalinae (K.Schum.) K.Schum. subsp. adalinae
F.W.T.A. 2: 89 (1963); Ann. Mo. Bot. Gdn. 83: 294 (1996).

Habit: herbaceous climber. Habitat: forest. Distribution: Mali, Ghana, Nigeria, Bioko and SW Cameroon. Chorology: Guineo-Congolian. Star: green. Alt. range: 400-600m (1), 601-800m (2), 801-1000m (2). Flowering: Feb (3), Mar (2). Specimens: - *Eastern slopes:* Meve U. 902, 903, 904. *Etinde:* Cable S. 1514; Etuge M. 1238. *Mabeta-Moliwe:* Mann G. 765

Dregea schimperi (Decne.) Bullock
F.W.T.A. 2: 97 (1963).

Habit: woody climber. Habitat: forest. Distribution: SE Nigeria, W. Cameroon, Sudan, Somalia, Ethiopia and East Africa. Chorology: Afromontane. Star: green. Specimens: - *Eastern slopes:* Maitland T. D. 654

Gongronema latifolium Benth.
F.W.T.A. 2: 98 (1963).

Habit: herbaceous climber. Guild: pioneer. Habitat: forest. Distribution: tropical Africa. Star: excluded. Alt. range: 600-800m (1). Flowering: Mar (1). Specimens: - *Eastern slopes:* Maitland T. D. 494; Tchouto P. 47

Neoschumannia kamerunensis Schltr.
F.W.T.A. 2: 95 (1963); Pl. Syst. Evol. 197: 233 (1995); Kew Bull. 52: 733 (1997); Bot. Jahrb. Syst. 119: 427 (1997).

Habit: herbaceous climber. Habitat: forest and farmbush. Distribution: Ivory Coast, Cameroon and CAR Chorology: Guineo-Congolian. Star: black. Alt. range: 800-1000m (1). Flowering: Feb (1). Specimens: - *Eastern slopes:* Meve U. 910. *Mabeta-Moliwe:* Schlechter F. R. R. 12384

Pentarrhinum insipidum E.Mey.
F.W.T.A. 2: 90 (1963); Kew Bull. 47: 484 (1992).

Habit: twining herb. Habitat: forest-grassland transition. Distribution: SW Cameroon to East and South Africa. Chorology: Afromontane. Star: green. Alt. range: 2000-2200m (1). Specimens: - *Eastern slopes:* Brenan J. P. M. 9510

Pergularia daemia (Forssk.) Chiov.
F.W.T.A. 2: 90 (1963).

Habit: herbaceous climber. Guild: pioneer. Habitat: Aframomum thicket. Distribution: tropical Africa and Arabia. Star: excluded. Specimens: - *Eastern slopes:* Migeod F. 93

Rhynchostigma racemosum Benth.
F.W.T.A. 2: 88 (1963).

Habit: woody climber. Habitat: forest. Distribution: Mount Cameroon, Bioko, Gabon, Congo (Kinshasa), Burundi, Rwanda and Uganda. Chorology: Guineo-Congolian (montane). Star: black. Alt. range: 1200-1400m (1). Specimens: - *No locality:* Mann G. 1273

Tylophora anomala N.E.Br.

Habit: herbaceous climber. Habitat: forest. Distribution: Mount Cameroon and East to southern Africa. Chorology: Afromontane. Star: black. Alt. range: 1200-1400m (1), 1401-1600m (1), 1601-1800m (1). Flowering: Feb (3). Specimens: - *Eastern slopes:* Meve U. 912, 916b, 917

Tylophora cameroonica N.E.Br.
F.W.T.A. 2: 96 (1963).

Habit: herbaceous climber. Habitat: forest. Distribution: Mount Cameroon and Uganda. Chorology: Guineo-Congolian. Star: black. Alt. range: 600-800m (1). Flowering: Mar (1). Specimens: - *Eastern slopes:* Dalziel J. M. 8236; Tchouto P. 60

Tylophora conspicua N.E.Br.
F.W.T.A. 2: 96 (1963).

Habit: woody climber. Habitat: forest. Distribution: tropical Africa. Star: excluded. Alt. range: 1-200m (1). Flowering: Apr (1). Specimens: - *Mokoko River F.R.:* Tchouto P. 1152

Tylophora oblonga N.E.Br.
F.W.T.A. 2: 96 (1963).

Habit: woody climber. Habitat: forest. Distribution: Liberia, Nigeria and Cameroon. Chorology: Guineo-Congolian (montane). Star: green. Alt. range: 1-200m (3), 1401-1600m (1). Flowering: Feb (1), Apr (1), May (3). Fruiting: Apr (1). Specimens: - *Eastern slopes:* Meve U. 915. *Mabeta-Moliwe:* Wheatley J. I. 216. *Mokoko River F.R.:* Ekema N. 892; Fraser P. 372; Watts J. 1131

Tylophora oculata N.E.Br.
F.W.T.A. 2: 96 (1963).

Habit: herbaceous climber. **Habitat:** forest. **Distribution:** Guinea to SW Cameroon. **Chorology:** Guineo-Congolian. **Star:** green. **Alt. range:** 600-800m (1). **Flowering:** May (1). **Specimens:** - *Etinde:* Tchouto P. 217

Tylophora sylvatica Decne.
F.W.T.A. 2: 96 (1963)

Specimens:- *Etinde:* Rosevear Cam. 45/37

Tylophora urceolata Meve
Kew Bull. 51: 585 (1996).

Habit: twining herb to 4 m. **Habitat:** forest. **Distribution:** Mount Cameroon, Bioko and W. Tanzania. **Chorology:** Guineo-Congolian (montane). **Star:** black. **Alt. range:** 1000-1200m (1). **Flowering:** Feb (1). **Specimens:** - *Eastern slopes:* Meve U. 901

Tylophora sp.

Specimens: - *Eastern slopes:* Meve U. 914

AVICENNIACEAE

Avicennia germinans (L.) L.
Taxon 12: 150 (1963); F.W.T.A. 2: 448 (1963); Fl. Cameroun 19: 60 (1975); Keay R.W.J., Trees of Nigeria: 439 (1989).
Avicennia africana P.Beauv., F.W.T.A. 2: 448 (1963); Hawthorne W., F.G.F.T. Ghana: 43 (1990).
Avicennia nitida Jacq., F.W.T.A. (ed.1) 2: 270 (1931).

Habit: tree to 15 m. **Habitat:** mangrove forest. **Distribution:** coasts of West Africa and eastern tropical America. **Star:** excluded. **Specimens:** - *Etinde:* Maitland T. D. 30; *Mabeta-Moliwe:* Cheek M. 3461.

BALANOPHORACEAE

Thonningia sanguinea Vahl
F.W.T.A. 1: 667 (1958); Fl. Cameroun 33: 48 (1991).

Habit: parasite. **Habitat:** forest. **Distribution:** tropical Africa. **Star:** excluded. **Alt. range:** 1-200m (6), 201-400m (2), 1300m. **Flowering:** May (1), Jun (4), Jul (1). **Fruiting:** Oct (1). **Specimens:** - *Eastern slopes:* Maitland T. D. 376. *Mabeta-Moliwe:* Hurst J. 28; Sunderland T. C. H. 1468; Wheatley J. I. 367; von Rege I. 4. *Mokoko River F.R.:* Acworth J. M. 346; Ekema N. 1201; Fraser P. 411. *Onge:* Akogo M. 65

BALSAMINACEAE

Impatiens burtonii Hook.f.
F.W.T.A. 1: 161 (1954); Grey-Wilson C., Impatiens of Africa: 164 (1980); Fl. Cameroun 22: 16 (1981).

Habit: herb. **Habitat:** forest. **Distribution:** W. Cameroon, E. Congo (Kinshasa), Burundi, Rwanda, Uganda, Kenya and Tanzania. **Chorology:** Guineo-Congolian (montane). **Star:** green. **Alt. range:** 800-1000m (1), 1001-1200m (1), 1201-1400m (1). **Flowering:** Oct (2). **Specimens:** - *Eastern slopes:* Tchouto P. 376. *Etinde:* Thomas D. W. 9168. *No locality:* Mann G. 1247

Impatiens filicornu Hook.f.
F.W.T.A. 1: 162 (1954); Fl. Gabon 4: 42 (1962); Grey-Wilson C., Impatiens of Africa: 95 (1980); Fl. Cameroun 22: 7 (1981).

Habit: herb. **Habitat:** forest. **Distribution:** Bioko, Cameroon, Gabon and Congo (Kinshasa). **Chorology:** Guineo-Congolian (montane). **Star:** green. **Alt. range:** 1-200m (1), 201-400m (1). **Flowering:** Sep (1), Oct (1). **Fruiting:** Sep (1). **Specimens:** - *Etinde:* Tchouto P. 812. *Onge:* Akogo M. 43

Impatiens grandisepala Grey-Wilson
Grey-Wilson C., Impatiens of Africa: 215 (1980); Fl. Cameroun 22: 29 (1981).

Habit: herb. **Habitat:** forest. **Distribution:** Mount Cameroon. **Chorology:** Endemic (montane). **Star:** black. **Alt. range:** 1600-1800m (1). **Flowering:** Oct (1). **Specimens:** - *Efolofo:* Satabié 303.

Impatiens hians Hook.f. var. bipindensis (Gilg) Grey-Wilson
Grey-Wilson C., Impatiens of Africa: 212 (1980); Fl. Cameroun 22: 29 (1981).
Impatiens bipindensis Gilg, Fl. Gabon 4: 19 (1962).

Habit: herb. **Habitat:** forest and stream banks. **Distribution:** SW Cameroon and Gabon. **Chorology:** Lower Guinea. **Star:** gold. **Alt. range:** 1-200m (10), 201-400m (2), 401-600m (2). **Flowering:** Mar (1), Aug (1), Sep (2), Oct (8), Nov (2). **Fruiting:** Sep (1), Oct (2). **Specimens:** - *Etinde:* Tchouto P. 808; Thomas D. W. 9705; Watts J. 517, 727. *Onge:* Akogo M. 41; Cheek M. 4992; Harris D. J. 3647; Ndam N. 659, 774; Tchouto P. 729; Watts J. 692, 876, 926; Wheatley J. I. 848

Impatiens hians Hook.f. var. hians
F.W.T.A. 1: 162 (1954); Fl. Gabon 4: 19 (1962); Grey-Wilson C., Impatiens of Africa: 212 (1980); Fl. Cameroun 22: 28 (1981).

Habit: herb. **Habitat:** forest. **Distribution:** Bioko, Cameroon, Rio Muni, Gabon and Congo (Kinshasa). **Chorology:** Guineo-Congolian (montane). **Star:** green. **Alt. range:** 600-800m (1), 801-1000m (1), 1201-1400m (1), 1401-1600m (1). **Flowering:** May (1), Sep (1), Oct (2). **Fruiting:** Sep (1). **Specimens:** - *Eastern slopes:* Preuss P. R. 881; Sidwell K. 95. *Etinde:* Tekwe C. 152; Watts J. 546; Wheatley J. I. 543. *Mokoko River F.R.:* Ndam N. 1262

Impatiens irvingii Hook.f.
F.W.T.A. 1: 161 (1954); Fl. Gabon 4: 45 (1962); Grey-Wilson C., Impatiens of Africa: 111 (1980); Fl. Cameroun 22: 13 (1981).

Habit: herb. **Distribution:** Guinea to Cameroon, Gabon, Congo (Brazzaville), Congo (Kinshasa), Angola (Cabinda), CAR, Uganda, Sudan, Tanzania, Malawi and Zambia. **Star:** excluded. **Specimens:** - *Eastern slopes:* Dalziel J. M. 8217

Impatiens kamerunensis Warb. subsp. kamerunensis

Grey-Wilson C., Impatiens of Africa: 164 (1980); Fl. Cameroun 22: 9 (1981).

Habit: herb. Habitat: forest. Distribution: Ghana, Togo, Nigeria and W. Cameroon. Chorology: Guineo-Congolian (montane). Star: green. Alt. range: 1000-1200m (1), 1201-1400m (1), 1401-1600m (1), 1601-1800m (2). Flowering: Sep (2). Fruiting: Sep (1). Specimens: - *Eastern slopes:* Brenan J. P. M. 9362; Preuss P. R. 590; Tekwe C. 328; Thomas D. W. 9497. *Etinde:* Wheatley J. I. 570

Impatiens macroptera Hook.f.

F.W.T.A. 1: 162 (1954); Fl. Gabon 4: 30 (1962); Grey-Wilson C., Impatiens of Africa: 217 (1980); Fl. Cameroun 22: 20 (1981).

Habit: herb. Habitat: forest. Distribution: SE Nigeria, Bioko, Cameroon and Gabon. Chorology: Lower Guinea. Star: blue. Alt. range: 200-400m (4), 601-800m (1), 801-1000m (1). Flowering: Sep (1), Oct (3), Nov (1). Fruiting: Sep (1). Specimens: - *Etinde:* Cheek M. 3772; Kwangue A. 139; Preuss P. R. 946, 1352; Wheatley J. I. 500. *Onge:* Ndam N. 600, 637

Impatiens mannii Hook.f.

F.W.T.A. 1: 162 (1954); Fl. Gabon 4: 33 (1962); Grey-Wilson C., Impatiens of Africa: 177 (1980); Fl. Cameroun 22: 18 (1981).
Impatiens deistelii Gilg, F.W.T.A. 1: 162 (1954).

Habit: herb. Habitat: forest and Aframomum thicket. Distribution: Bioko, Cameroon, Gabon, Congo (Kinshasa) and Uganda. Chorology: Guineo-Congolian. Star: green. Alt. range: 600-800m (1), 801-1000m (2), 1001-1200m (2), 1201-1400m (2), 1401-1600m (2). Flowering: Jan (1), Sep (3), Oct (3). Fruiting: Oct (1). Specimens: - *Etinde:* Cheek M. 3765; Sunderland T. C. H. 1075; Tekwe C. 182, 220; Thomas D. W. 9171; Watts J. 539; Wheatley J. I. 541, 600. *No locality:* Deistel H. 659; Mann G. 1258

Impatiens niamniamensis Gilg

F.W.T.A. 1: 161 (1954); Fl. Gabon 4: 18 (1962); Grey-Wilson C., Impatiens of Africa: 164 (1980); Fl. Cameroun 22: 30 (1981).

Habit: herb. Habitat: forest, sometimes on rocks. Distribution: Bioko, Cameroon, Gabon, Congo (Brazzaville), Congo (Kinshasa), Angola, Rwanda, Uganda, Sudan, Kenya and Tanzania. Chorology: Afromontane. Star: green. Alt. range: 1-200m (1), 201-400m (3), 601-800m (2), 1201-1400m (1), 1401-1600m (1), 1601-1800m (1), 1801-2000m (1). Flowering: Jan (1), Mar (1), May (2), Aug (1), Sep (1), Oct (2), Nov (1). Specimens: - *Eastern slopes:* Nkeng P. 80; Sunderland T. C. H. 1440. *Etinde:* Cheek M. 5396b; Mbani J. M. 53; Sunderland T. C. H. 1073; Tchouto P. 222; Thomas D. W. 9116. *No locality:* Mann G. 1248. *Onge:* Akogo M. 87; Tchouto P. 730

Impatiens palpebrata Hook.f.

Grey-Wilson C., Impatiens of Africa: 101 (1980); Fl. Cameroun 22: 10 (1981).
Impatiens preussii Warb., F.W.T.A. 1: 162 (1954).

Habit: herb. Distribution: Cameroon and Gabon. Chorology: Lower Guinea (montane). Star: blue. Alt. range: 1000-1200m (1). Specimens: - *Eastern slopes:* Preuss P. R. 592

Impatiens sakerana Hook.f.

F.W.T.A. 1: 162 (1954); Grey-Wilson C., Impatiens of Africa: 209 (1980); Fl. Cameroun 22: 25 (1981).

Habit: herb. Habitat: forest-grassland transition. Distribution: Bioko and Cameroon. Chorology: Lower Guinea (montane). Star: gold. Alt. range: 1400-1600m (1), 1601-1800m (1), 1801-2000m (2), 2001-2200m (2). Flowering: Sep (2), Oct (3). Fruiting: Oct (1). Specimens: - *Eastern slopes:* Tchouto P. 339; Watts J. 446. *Etinde:* Tekwe C. 219, 262; Thomas D. W. 9261. *No locality:* Mann G. 1256

Impatiens x palpebrata Hook.f.

Grey-Wilson C., Impatiens of Africa: 94 (1980).

Habit: herb. Habitat: forest. Distribution: W. Cameroon. Chorology: Western Cameroon. Star: gold. Alt. range: 200-400m (1), 401-600m (2). Flowering: Oct (3). Fruiting: Oct (1). Specimens: - *Onge:* Cheek M. 4995; Ndam N. 599; Tchouto P. 794

Impatiens sp. nov 1

Specimens: - *Etinde:* Watts 468, 502; Wheatley 544

Impatiens sp. nov. 2 aff grandisepala

Specimens: - *Etinde:* Cheek M. 5539; Cheek M. & Tchouto P. 3600; Thomas et al 9542; Upson 40.

BASELLACEAE

Basella alba L.

F.W.T.A. 1: 155 (1954).

Habit: herbaceous climber. Habitat: forest and forest-grassland transition. Distribution: tropical Africa, Asia and the West Indies. Star: excluded. Alt. range: 200-400m (1). Specimens: - *No locality:* Mann G. 1250

BEGONIACEAE

Hans de Wilde & Marc Sosef (WAG)

Begonia ampla Hook.f.

F.W.T.A. 1: 219 (1954).

Habit: epiphyte. Habitat: forest. Distribution: Cameroon, Bioko, Principe, S. Tomé, Annobon and Uganda. Chorology: Guineo-Congolian. Star: blue. Alt. range: 1-200m (6), 201-400m (3), 401-600m (1), 601-800m (2), 801-1000m (1), 1001-1200m (2), 1201-1400m (1). Flowering: Feb (1), Apr (1), May (1), Sep (3), Oct (6), Nov (2), Dec (2). Fruiting: Sep (1), Oct (1), Dec (1). Specimens: - *Eastern slopes:* Etuge M. 1195; Nkeng P. 92. *Etinde:* Brodie S. 12; Cable S. 185; Cheek M. 5940; Lighava M. 47; Mbani J. M. 97; Tchouto P. 137; Tekwe C. 163; Watts J. 518; Wheatley J. I. 549. *Onge:* Akogo M. 13; Ndam N. 657; Tchouto P. 673; Watts J. 890, 953

Begonia annobonensis A.DC.

F.W.T.A. 1: 220 (1954).

Habit: herb. Habitat: usually on rocks. Distribution: Mount Cameroon, Principe, S. Tomé and Annabon. Chorology: Islands in the Gulf of Guinea, Bioko excepted; naturalized (?) at Limbe. **Star:** black. **Alt. range:** 1-200m (1). **Flowering:** Nov (1). **Specimens:** - *Etinde:* Dundas J. FHI 15319. *Limbe Botanic Garden:* Wheatley J. I. 657

Begonia capillipes Gilg
Engl. Bot. Jahrb. 34: 96 (1904).

Habit: epiphyte. **Habitat:** forest. **Distribution:** Cameroon, Equatorial Guinea and Gabon. **Chorology:** Guineo-Congolian. **Star:** green. **Alt. range:** 800-1000m (1). **Flowering:** Mar (1). **Specimens:** - *Etinde:* Cable S. 1610

Begonia ciliobracteata Warb.
F.W.T.A. 1: 218 (1954); Agric. Univ. Wag. Papers 94(1): 222 (1994).
 Begonia hookeriana Gilg ex Engl., F.W.T.A. 1: 220 (1954).

Habit: herb. **Habitat:** forest. **Distribution:** SE Nigeria and Cameroon. **Chorology:** Lower Guinea. **Star:** blue. **Alt. range:** 800-1000m (1). **Flowering:** Oct (1). **Specimens:** - *Etinde:* Tchouto P. 444. *No locality:* Dusen P. K. H. 18.

Begonia eminii Warb.
F.W.T.A. 1: 220 (1954).

Habit: epiphyte. **Habitat:** forest. **Distribution:** tropical Africa. **Star:** excluded. **Alt. range:** 400-600m (1), 601-800m (1), 801-1000m (2). **Flowering:** Jul (1), Nov (1), Dec (1). **Fruiting:** Jul (1), Nov (1), Dec (1). **Specimens:** - *Eastern slopes:* Preuss P. R. 960; Tekwe C. 116. *Etinde:* Cable S. 246, 452; Kalbreyer W. 96

Begonia fusialata Warb.
F.W.T.A. 1: 220 (1954).

Habit: hemi-epiphyte. **Habitat:** forest, sometimes on rocks. **Distribution:** Liberia to Angola. **Chorology:** Guineo-Congolian. **Star:** green. **Alt. range:** 1-200m (5), 201-400m (7), 401-600m (2), 601-800m (1). **Flowering:** Mar (2), May (1), Sep (1), Oct (3), Nov (5). **Fruiting:** Mar (2), Sep (1), Oct (4), Nov (4), Dec (1). **Specimens:** - *Bambuko F.R.:* Watts J. 638. *Etinde:* Cable S. 267; Cheek M. 5403, 5842; Kwangue A. 131; Mbani J. M. 143; Wheatley J. I. 520. *Onge:* Akogo M. 40; Cheek M. 5040; Harris D. J. 3844; Ndam N. 658; Thomas D. W. 9891; Watts J. 844, 948; Wheatley J. I. 858

Begonia letouzeyi Sosef
Agric. Univ. Wag. Papers 94(1): 162 (1994).

Habit: herb. **Habitat:** forest, sometimes on rocks. **Distribution:** Cameroon, Gabon and Congo (Brazzaville). **Chorology:** Lower Guinea. **Star:** green. **Alt. range:** 200-400m (1), 401-600m (1), 601-800m (1), 801-1000m (3). **Flowering:** Mar (4), Apr (1), Nov (1). **Fruiting:** Mar (1). **Specimens:** - *Etinde:* Cable S. 1523, 1550, 1642; Cheek M. 5548; Wheatley J. I. 136. *Onge:* Wheatley J. I. 750

Begonia longipetiolata Gilg
F.W.T.A. 1: 220 (1954); Agric. Univ. Wag. Papers 91(6): 195 (1991).

Habit: epiphyte. **Habitat:** forest and farmbush. **Distribution:** Nigeria, Bioko, Cameroon, Rio Muni, Gabon, Congo (Brazzaville), Congo (Kinshasa) and Angola (Cabinda). **Chorology:** Guineo-Congolian. **Star:** green. **Alt. range:** 1-200m (9), 201-400m (2), 1001-1200m (1). **Flowering:** Feb (2), Mar (2), Aug (1), Oct (3),

Nov (4). **Fruiting:** Aug (1), Nov (1). **Specimens:** - *Etinde:* Cable S. 149; Cheek M. 5387; Mbani J. M. 23; Nkeng P. 25; Tchouto P. 432. *Onge:* Harris D. J. 3724; Ndam N. 698, 705; Watts J. 703, 976; Wheatley J. I. 704, 722

Begonia loranthoides Hook.f. subsp. rhopalocarpa (Warb.) J.J. De Wilde
F.W.T.A. 1: 219 (1954); Acta Bot. Neerl. 28: 367 (1979)

Habit: epiphyte. **Habitat:** forest. **Distribution:** Cameroon, Gabon and Congo (Kinshasa). **Chorology:** Guineo-Congolian. **Star:** green. **Specimens:** - *Etinde:* Dusen P. K. H. 89

Begonia macrocarpa Warb.
F.W.T.A. 1: 219 (1954).

Habit: herb. **Guild:** shade-bearer. **Habitat:** forest. **Distribution:** Guinea to Angola. **Chorology:** Guineo-Congolian. **Star:** green. **Alt. range:** 1-200m (6), 201-400m (9), 401-600m (4), 601-800m (2). **Flowering:** Aug (1), Sep (1), Oct (9), Nov (7), Dec (1). **Fruiting:** Aug (1), Sep (1), Oct (6), Nov (2), Dec (1). **Specimens:** - *Etinde:* Cable S. 176, 214, 261b, 296; Cheek M. 5423; Nkeng P. 95; Watts J. 490, 491, 545; Wheatley J. I. 508. *Onge:* Akogo M. 108; Cheek M. 5039; Harris D. J. 3725; Ndam N. 630, 631, 764, 769; Tchouto P. 846; Thomas D. W. 9926; Watts J. 704, 764

Begonia mannii Hook.
F.W.T.A. 1: 220 (1954); Begonian 46: 14 (1979).

Habit: epiphyte. **Habitat:** forest, stream banks and Aframomum thicket. **Distribution:** Sierra Leone to Bioko and Gabon. **Chorology:** Guineo-Congolian. **Star:** green. **Alt. range:** 200-400m (5), 401-600m (1), 601-800m (2), 801-1000m (1), 1001-1200m (1), 1201-1400m (2), 1801-2000m (1). **Flowering:** Feb (1), Mar (2), Apr (1), May (1), Jun (1), Oct (5), Nov (1). **Fruiting:** May (1), Oct (1). **Specimens:** - *Bambuko F.R.:* Watts J. 583. *Eastern slopes:* Cable S. 1392; Groves M. 107; Sidwell K. 86; Tekwe C. 97. *Etinde:* Cheek M. 5533; Mbani J. M. 98; Tchouto P. 138; Thomas D. W. 9143. *No locality:* Mann G. 1275. *Onge:* Akogo M. 81; Ndam N. 704, 776

Begonia cf. mannii Hook.

Specimens: - *Etinde:* Cable S. 1538

Begonia oxyanthera Warb.
F.W.T.A. 1: 220 (1954).
 Begonia jussiaeicarpa Warb., F.W.T.A. 1: 220 (1954).

Habit: epiphyte. **Habitat:** forest and forest-grassland transition. **Distribution:** Bioko, Nigeria and SW Cameroon. **Chorology:** Western Cameroon Uplands (montane). **Star:** black. **Alt. range:** 1200-1400m (2), 1601-1800m (1). **Flowering:** Sep (1). **Specimens:** - *Eastern slopes:* Preuss P. R. 867; Thomas D. W. 9478. *Etinde:* Thomas D. W. 9234. *No locality:* Dusen P. K. H. 427

Begonia oxyloba Welw. ex Hook.f.
F.W.T.A. 1: 218 (1954).

Habit: herb. **Habitat:** forest. **Distribution:** Guinea to Bioko, Cameroon, Equatorial Guinea, Gabon, Congo, Zaire, Rwanda, Angola, Uganda, Tanzania and Madagascar. **Chorology:** Guineo-Congolian (montane). **Star:** green. **Alt. range:** 400-600m (1), 601-800m (1), 801-1000m (3), 1001-1200m (3), 1201-1400m (3),

1401-1600m (2). **Flowering:** Feb (2), Aug (2), Sep (3), Oct (3), Dec (2). **Fruiting:** Sep (2), Oct (1), Dec (1). **Specimens:** - *Eastern slopes:* Cable S. 1311, 1323; Kwangue A. 119; Tchouto P. 384. *Etinde:* Cable S. 297, 482; Cheek M. 3759, 3788; Tekwe C. 120, 183, 293; Thomas D. W. 9183. *No locality:* Mann G. 1274

Begonia poculifera Hook.f. var. poculifera
F.W.T.A. 1: 219 (1954).

Habit: epiphytic or terrestrial herb. **Habitat:** forest. **Distribution:** SE Nigeria, Bioko, SW Cameroon, Gabon, Zaire, Rwanda, Burundi, Tanzania and Angola. **Chorology:** Guineo-Congolian (montane). **Star:** green. **Alt. range:** 1000-1200m (3), 1201-1400m (2), 1401-1600m (1), 1601-1800m (3), 1801-2000m (3), 2201-2400m (1). **Flowering:** Jan (1), Mar (2), May (1), Sep (3), Oct (4), Nov (1). **Fruiting:** Oct (1). **Specimens:** - *Eastern slopes:* Cheek M. 5350; Dahl A. 609; Etuge M. 1206; Sidwell K. 73; Sunderland T. C. H. 1441. *Etinde:* Sunderland T. C. H. 1070; Tekwe C. 172, 258; Thomas D. W. 9181; Wheatley J. I. 552, 609, 637. *No locality:* Mann G. 1276

Begonia polygonoides Hook.f.
F.W.T.A. 1: 220 (1954).

Habit: epiphyte. **Habitat:** forest. **Distribution:** Guinea, Liberia, Ivory Coast, Ghana, SE Nigeria, Cameroon, Gabon and Congo (Kinshasa). **Chorology:** Guineo-Congolian. **Star:** green. **Alt. range:** 1-200m (1). **Flowering:** Mar (1). **Fruiting:** Mar (1). **Specimens:** - *Mabeta-Moliwe:* Plot Voucher W908; *Onge:* Wheatley J. I. 737.

Begonia preussii Warb.
Engl. Bot. Jahrb. 22: 36 (1895).
Begonia sessilanthera Warb., F.W.T.A. 1: 220 (1954).

Habit: epiphyte. **Habitat:** forest and farmbush. **Distribution:** SE Nigeria, Bioko and SW Cameroon. **Chorology:** Western Cameroon Uplands (montane). **Star:** gold. **Specimens:** - *Mabeta-Moliwe:* Preuss P. R. 1261

Begonia quadrialata Warb. subsp. dusenii (Warb.) Sosef
Agric. Univ. Wag. Papers 94(1): 194 (1994).
Begonia dusenii Warb., F.W.T.A. 1: 220 (1954).

Habit: terrestrial herb. **Habitat:** Aframomum thicket.
Distribution: Mount Cameroon. **Chorology:** Endemic (montane).
Star: black. **Alt. range:** 400-600m (1). **Flowering:** Oct (1).
Fruiting: Oct (1). **Specimens:** - *Etinde:* Watts J. 501

Begonia quadrialata Warb. subsp. quadrialata
F.W.T.A. 1: 218 (1954); Agric. Univ. Wag. Papers 94(1): 186 (1994).

Habit: epiphytic or terrestrial herb. **Habitat:** forest, sometimes on rocks. **Distribution:** Guinea to Cameroon, Gabon, Congo (Brazzaville), Congo (Kinshasa) and Angola. **Chorology:** Guineo-Congolian. **Star:** green. **Alt. range:** 1-200m (3), 201-400m (9), 401-600m (1). **Flowering:** Jun (1), Aug (1), Oct (8), Nov (3). **Fruiting:** Aug (1), Oct (7), Nov (1). **Specimens:** - *Etinde:* Cable S. 181; Nkeng P. 67; Watts J. 500. *Onge:* Akogo M. 35, 47; Baker W. J. 362; Cheek M. 5047; Harris D. J. 3863; Ndam N. 594, 626; Tchouto P. 858; Watts J. 754, 840, 955

Begonia scapigera Hook.f. subsp. scapigera
F.W.T.A. 1: 218 (1954); Agric. Univ. Wag. Papers 94(1): 199 (1994).

Habit: herb. **Habitat:** forest, sometimes on rocks. **Distribution:** S Nigeria, Cameroon and Gabon. **Chorology:** Lower Guinea (montane). **Star:** blue. **Alt. range:** 200-400m (1), 601-800m (1), 801-1000m (1), 1001-1200m (1), 1201-1400m (2), 1601-1800m (2). **Flowering:** Sep (2), Oct (3), Nov (2), Dec (1). **Fruiting:** Oct (1). **Specimens:** - *Etinde:* Cable S. 226; Cheek M. 5534; Faucher P. 45; Sidwell K. 175; Tekwe C. 144; Thomas D. W. 9535; Wheatley J. I. 553, 602

Begonia scutifolia Hook.f.
Agric. Univ. Wag. Papers 94(1): 206 (1994).

Habit: epiphyte or terrestrial herb. **Habitat:** forest, sometimes on rocks. **Distribution:** Cameroon, Gabon, Congo (Kinshasa) and Angola (Cabinda). **Chorology:** Lower Guinea. **Star:** green. **Alt. range:** 1-200m (2), 201-400m (6), 401-600m (3). **Flowering:** Sep (1), Oct (2), Nov (6), Dec (2). **Fruiting:** Nov (1), Dec (1). **Specimens:** - *Etinde:* Cable S. 138, 139, 258, 356; Cheek M. 5404, 5847; Tchouto P. 457; Thomas D. W. 9128; Wheatley J. I. 522; Williams S. 43, 46

Begonia sessilifolia Hook.f.
F.W.T.A. 1: 219 (1954).

Habit: herb. **Habitat:** forest, stream banks and Aframomum thicket. **Distribution:** Bioko and Cameroon. **Chorology:** Lower Guinea. **Star:** gold. **Alt. range:** 1-200m (2), 201-400m (6), 601-800m (2), 1001-1200m (2). **Flowering:** Sep (2), Oct (5), Nov (3), Dec (1). **Fruiting:** Sep (2), Oct (2), Nov (2), Dec (1). **Specimens:** - *Etinde:* Cable S. 221; Cheek M. 5528, 5908; Sidwell K. 158; Tchouto P. 427; Watts J. 477, 544; Wheatley J. I. 503, 510. *Onge:* Ndam N. 765; Thomas D. W. 9805; Watts J. 779

Begonia squamulosa Hook.f.
F.W.T.A. 1: 220 (1954); Agric. Univ. Wag. Papers 91(6): 210 (1992).

Habit: epiphyte. **Habitat:** forest. **Distribution:** Cameroon, Rio Muni, Gabon and Congo (Kinshasa). **Chorology:** Guineo-Congolian. **Star:** green. **Specimens:** - *No locality:* Kalbreyer W. 155

BIGNONIACEAE

S. Bidgood (K)

Kigelia africana (Lam.) Benth.
F.W.T.A. 2: 385 (1963); Meded. Land. Wag. 82(3): 77 (1982); Fl. Cameroun 27: 32 (1984); Keay R.W.J., Trees of Nigeria: 433 (1989); Hawthorne W., F.G.F.T. Ghana: 163 (1990).

Habit: tree to 21 m. **Guild:** light demander. **Habitat:** forest and Aframomum thicket. **Distribution:** tropical Africa. **Chorology:** Afromontane. **Star:** green. **Alt. range:** 1-200m (1), 1601-1800m (1). **Flowering:** May (1). **Fruiting:** Jul (1). **Specimens:** - *Etinde:* Cheek M. 3618; Wheatley J. I. 232

Markhamia lutea (Benth.) K.Schum.

F.W.T.A. 2: 387 (1963); Fl. Cameroun 27: 36 (1984); Keay
R.W.J., Trees of Nigeria: 435 (1989); Hawthorne W., F.G.F.T.
Ghana: 163 (1990).

Habit: tree or shrub to 10 m. **Habitat:** forest. **Distribution:** Ghana
to Bioko and Congo (Kinshasa). **Chorology:** Guineo-Congolian.
Star: green. **Specimens:** - *Etinde:* Kalbreyer W. 33; *Mabeta-
Moliwe:* Cheek M. 3423

Newbouldia laevis (P.Beauv.) Seeman ex Bureau

F.W.T.A. 2: 388 (1963); Fl. Cameroun 27: 41 (1984); Keay
R.W.J., Trees of Nigeria: 435 (1989); Hawthorne W., F.G.F.T.
Ghana: 163 (1990).

Habit: tree to 18 m. **Guild:** pioneer. **Habitat:** forest. **Distribution:**
Senegal to Bioko and Congo (Kinshasa). **Chorology:** Guineo-
Congolian. **Star:** green. **Specimens:** - *Etinde:* Maitland T. D. 282

Spathodea campanulata P.Beauv. subsp. campanulata

Specimens: - *S. Bakundu:* Thomas, D.W. 5207

Stereospermum acuminatissimum K.Schum.

F.W.T.A. 2: 386 (1963)

Specimens: - *S. Bakundu:* Preuss P.R. 332; Staudt 950

BOMBACACEAE

Bombax brevicuspe Sprague

Specimens: - *S. Bakundu:* Schultze 105, Dundas in FHI 13902

Bombax buonopozense P.Beauv.

F.W.T.A. 1: 334 (1958); Fl. Gabon 22: 47 (1973); Fl. Cameroun
19: 84 (1975); Keay R.W.J., Trees of Nigeria: 137 (1989);
Hawthorne W., F.G.F.T. Ghana: 158 (1990).

Habit: tree to 40 m. **Guild:** pioneer. **Habitat:** forest. **Distribution:**
Sierra Leone to Congo (Kinshasa), Angola and Uganda.
Chorology: Guineo-Congolian. **Star:** green. **Specimens:** - *Etinde:*
Maitland T. D. 175a

Ceiba pentandra (L.) Gaertn.

F.W.T.A. 1: 335 (1958); Fl. Gabon 22: 36 (1973); Fl. Cameroun
19: 76 (1975); Meded. Land. Wag. 82(3): 82 (1982); Keay R.W.J.,
Trees of Nigeria: 138 (1989); Hawthorne W., F.G.F.T. Ghana: 158
(1990).

Habit: tree to 60 m. **Guild:** pioneer. **Habitat:** forest and near
villages. **Distribution:** pantropical. **Star:** pink. **Alt. range:** 1-
200m (1). **Specimens:** - *Etinde:* Rosevear Cam. 80/37. *Onge:*
Harris D. J. 3806

BORAGINACEAE

Cordia aurantiaca Baker

F.W.T.A. 2: 320 (1963); Keay R.W.J., Trees of Nigeria: 432
(1989).

Habit: tree or shrub to 8 m. **Habitat:** forest and farmbush.
Distribution: Nigeria, Bioko and Cameroon. **Chorology:** Lower
Guinea. **Star:** blue. **Alt. range:** 600-800m (1), 801-1000m (3).
Flowering: Mar (1), Apr (1). **Fruiting:** Mar (1), Jul (1), Oct (1).
Specimens: - *Eastern slopes:* Kwangue A. 39; Tchouto P. 55;
Tekwe C. 106. *Etinde:* Watts J. 564

Cordia millenii Baker

F.W.T.A. 2: 320 (1963); Keay R.W.J., Trees of Nigeria: 432
(1989); Hawthorne W., F.G.F.T. Ghana: 141 (1990); F.T.E.A.
Boraginaceae: 13 (1991).

Habit: tree to 12 m. **Guild:** pioneer. **Habitat:** forest. **Distribution:**
tropical Africa. **Star:** excluded. **Specimens:** - *Eastern slopes:*
Maitland T. D. 369

Cordia platythyrsa Baker

F.W.T.A. 2: 321 (1963); Keay R.W.J., Trees of Nigeria: 432
(1989).

Habit: tree to 30 m. **Guild:** pioneer. **Habitat:** forest and
Aframomum thicket. **Distribution:** Sierra Leone to Cameroon.
Chorology: Guineo-Congolian (montane). **Star:** green.
Specimens: - *Mabeta-Moliwe:* Plot Voucher W 1217

Cordia senegalensis Juss.

F.W.T.A. 2: 320 (1963); Keay R.W.J., Trees of Nigeria: 432
(1989); Hawthorne W., F.G.F.T. Ghana: 141 (1990).

Habit: tree to 10 m. **Guild:** pioneer. **Habitat:** forest. **Distribution:**
Senegal to Rwanda. **Chorology:** Guineo-Congolian. **Star:** green.
Specimens: - *Mabeta-Moliwe:* Mann G. 968

Cynoglossum amplifolium A.DC. var. subalpinum (T.C.E.Fr.) Verdc.

F.T.E.A. Boraginaceae: 106 (1991).
   Cynoglossum amplifolium Hochst. ex A.Rich. f. *macrocarpum*
Brand, F.W.T.A. 2: 324 (1963).

Habit: herb. **Habitat:** grassland. **Distribution:** W. Cameroon.
Chorology: Western Cameroon Uplands (montane). **Star:** gold.
Alt. range: 1800-2000m (1), 2201-2400m (1), 2801-3000m (1).
Flowering: Oct (2). **Specimens:** - *Eastern slopes:* Sidwell K. 37;
Thomas D. W. 9263. *No locality:* Mann G. 1266

Cynoglossum coeruleum A.DC. subsp. johnstonii (Baker) Verdc. var. mannii (Baker & Wright) Verdc.

F.W.T.A. 2: 324 (1963); F.T.E.A. Boraginaceae: 110 (1991).
   Cynoglossum lanceolatum Forssk. subsp. *geometricum* (Baker
& Wright) Brand, F.W.T.A. 2: 324 (1963).

Habit: herb. **Distribution:** W. Cameroon and East Africa.
Chorology: Afromontane. **Star:** blue. **Specimens:** - *No locality:*
Deistel H. 657

Ehretia cymosa Thonn. var. **cymosa**
F.W.T.A. 2: 318 (1963); Hawthorne W., F.G.F.T. Ghana: 89
(1990); F.T.E.A. Boraginaceae: 37 (1991).

Habit: tree or shrub to 10 m. **Habitat:** forest. **Distribution:** Sierra
Leone to Cameroon, Gabon, Congo (Kinshasa) and Uganda.
Chorology: Guineo-Congolian. **Star:** green. **Specimens:** - *Eastern
slopes:* Maitland T. D. 694

Myosotis abyssinica Boiss. & Reut.
F.W.T.A. 2: 325 (1963); F.T.E.A. Boraginaceae: 88 (1991).

Habit: herb. **Habitat:** forest-grassland transition. **Distribution:**
Bioko, Mount Cameroon, Ethiopia and East Africa. **Chorology:**
Afromontane. **Star:** blue. **Alt. range:** 2600-2800m (1), 3001-
3200m (1). **Flowering:** Oct (1). **Specimens:** - *Eastern slopes:*
Cheek M. 3668. *No locality:* Mann G. 2033

Myosotis sp. aff. vestergrenii Stroh
F.W.T.A. 2: 325 (1963).

Habit: herb. **Distribution:** Mount Cameroon. **Chorology:**
Endemic (montane). **Star:** black. **Alt. range:** 1-200m (1), 2201-
2400m (1). **Flowering:** Jun (1). **Fruiting:** Jun (1). **Specimens:** -
Eastern slopes: Johnston H. H. 72, Wheatley J. I. 334.

BUDDLEJACEAE

Nuxia congesta R.Br. ex Fresen.
F.W.T.A. 2: 46 (1963); Keay R.W.J., Trees of Nigeria: 401 (1989);
Hawthorne W., F.G.F.T. Ghana: 41 (1990).

Habit: tree to 26 m. **Habitat:** forest-grassland transition.
Distribution: tropical and subtropical Africa. **Chorology:**
Afromontane. **Star:** green. **Alt. range:** 2000-2200m (1).
Specimens: - *No locality:* Mann G. 1206

BURSERACEAE

J.-M. Onana (YA)

Canarium schweinfurthii Engl.
F.W.T.A. 1: 697 (1958); Fl. Gabon 3: 90 (1962); Keay R.W.J.,
Trees of Nigeria: 335 (1989); Hawthorne W., F.G.F.T. Ghana: 171
(1990); F.T.E.A. Burseraceae: 3 (1991).

Habit: tree to 50 m. **Guild:** light demander. **Habitat:** forest.
Distribution: tropical Africa. **Star:** red. **Specimens:** - *Mabeta-
Moliwe:* Plot Voucher W 1038, Cheek M. 3398

Dacryodes edulis (G.Don) H.J.Lam
F.W.T.A. 1: 696 (1958); Fl. Gabon 3: 81 (1962); Keay R.W.J.,
Trees of Nigeria: 337 (1989).

Habit: tree to 20 m. **Habitat:** forest. **Distribution:** Nigeria,
Cameroon, Gabon, Congo (Kinshasa), Angola and Zambia.
Chorology: Guineo-Congolian. **Star:** green. **Alt. range:** 1-200m
(1).**Fruiting:** Apr (1). **Specimens:** - *Etinde:* Maitland T. D. 382.
Mabeta-Moliwe: Wheatley J. I. 201

Dacryodes klaineana (Pierre) H.J.Lam
F.W.T.A. 1: 696 (1958); Fl. Gabon 3: 78 (1962); Keay R.W.J.,
Trees of Nigeria: 339 (1989); Hawthorne W., F.G.F.T. Ghana: 171
(1990).

Habit: tree to 20 m. **Guild:** shade-bearer. **Habitat:** forest.
Distribution: Sierra Leone to Cameroon and Gabon. **Chorology:**
Guineo-Congolian. **Star:** green. **Alt. range:** 1-200m (7).
Flowering: Mar (1), Apr (2). **Fruiting:** Apr (2), May (1).
Specimens: - *Etinde:* Kwangue A. 69. *Mabeta-Moliwe:*
Sunderland T. C. H. 1133; Wheatley J. I. 79, 172, 269. *Mokoko
River F.R.:* Mbani J. M. 330; Tchouto P. 1095. *Southern Bakundu
F.R.:* Olorunfemi J. FHI 30715

Santiria trimera (Oliv.) Aubrév.
F.W.T.A. 1: 696 (1958); Fl. Gabon 3: 94 (1962); Keay R.W.J.,
Trees of Nigeria: 337 (1989).

Habit: tree to 30 m. **Habitat:** forest. **Distribution:** Sierra Leone to
Cameroon, Gabon, Congo (Kinshasa) and Angola (Cabinda).
Chorology: Guineo-Congolian. **Star:** green. **Alt. range:** 1-200m
(1), 201-400m (1).**Fruiting:** Jul (1). **Specimens:** - *Mabeta-
Moliwe:* Cheek M. 3520; Plot Voucher W 506. *Mokoko River
F.R.:* Ndam N. 1222

CACTACEAE

Rhipsalis baccifera (J. Miller) Stearn

Specimens: - *Mabeta-Moliwe:* Cheek M. 3477

CALLITRICHACEAE

Callitriche stagnalis Scop.

Specimens:- *Etinde:* Thomas, D. 2998

CAMPANULACEAE

Lobelia columnaris Hook.f.
F.W.T.A. 2: 313 (1963).

Habit: herb. **Habitat:** grassland. **Distribution:** Bioko and
Cameroon. **Chorology:** Lower Guinea (montane). **Star:** gold. **Alt.
range:** 2400-2600m (1). **Specimens:** - *No locality:* Brenan J. P.
M. 9536

Lobelia molleri Henriq.

Habit: herb. **Distribution:** Cameroon to Tanzania. **Specimens:**
Eastern Slopes: Deistel 39 (652), 91

Lobelia rubescens De Wild.
F.W.T.A. 2: 313 (1963); F.T.E.A. Lobeliaceae: 32 (1984).
Lobelia kamerunensis Engl. ex Hutch. & Dalziel, F.W.T.A. 2:
313 (1963).

Habit: herb. **Habitat**: grassland. **Distribution**: Sierra Leone to Bioko and Cameroon. **Chorology**: Guineo-Congolian (montane). **Star**: green. **Alt. range**: 2200-2400m (1). **Specimens**: - *Eastern slopes*: Maitland 1239, Preuss P. R. 748

Monopsis stellarioides (Presl) Urb. var. schimperiana (Urb.) E.Wimm.
F.W.T.A. 2: 315 (1963).

Habit: herb. **Distribution**: Bioko, Cameroon, Congo (Kinshasa) and East Africa. **Chorology**: Afromontane. **Star**: green. **Alt. range**: 1200-1400m (1). **Specimens**: - *Etinde*: Maitland T. D. 1087

Wahlenbergia krebsii Cham. subsp. arguta (Hook.f.) Thulin
F.T.E.A. Campanulaceae: 13 (1976).
Wahlenbergia arguta Hook.f., F.W.T.A. 2: 309 (1963).

Habit: herb. **Habitat**: grassland and forest-grassland transition. **Distribution**: Bioko and Cameroon. **Chorology**: Lower Guinea (montane). **Star**: gold. **Alt. range**: 2400-2600m (2), 2801-3000m (3). **Flowering**: Oct (4). **Specimens**: - *Eastern slopes*: Sidwell K. 15, 40, 41; Tchouto P. 337. *No locality*: Mann G. 1943

Wahlenbergia ramosissima (Hemsley) Thulin subsp. ramosissima
Symb. Bot. Ups. 21: 189 (1975).
Lightfootia ramosissima (Hemsley) E.Wimm. ex Hepper, F.W.T.A. 2: 311 (1963).

Habit: herb. **Habitat**: grassland. **Distribution**: Mount Cameroon. **Chorology**: Endemic (montane). **Star**: black. **Alt. range**: 1800-2000m (1), 2001-2200m (1), 2401-2600m (1). **Flowering**: Oct (2). **Specimens**: - *Eastern slopes*: Cheek M. 3678; Tchouto P. 269. *No locality*: Mann G. 1992

Wahlenbergia silenoides Hochst. ex A.Rich.
Symb. Bot. Ups. 21: 78 (1975).
Wahlenbergia mannii Vatke, F.W.T.A. 2: 309 (1963).

Habit: herb. **Habitat**: grassland and forest-grassland transition. **Distribution**: Bioko and W. Cameroon. **Chorology**: Western Cameroon Uplands (montane). **Star**: gold. **Alt. range**: 2400-2600m (3). **Flowering**: May (1), Oct (1). **Specimens**: - *Eastern slopes*: Sidwell K. 75; Sunderland T. C. H. 1429. *No locality*: Mann G. 1226

CAPPARACEAE

Capparis tomentosa Lam.
F.W.T.A. 1: 90 (1954); Fl. Cameroun 29: 32 (1986).
Capparis polymorpha A.Rich., F.W.T.A. 1: 90 (1954).

Habit: woody climber to 8 m. **Habitat**: forest. **Distribution**: tropical Africa. **Star**: excluded. **Specimens**: - *Eastern slopes*: Maitland T. D. 241

Cleome gynandra L.
Fl. Cameroun 29: 52 (1986).
Gynandropsis gynandra (L.) Briq., F.W.T.A. 1: 88 (1954).

Habit: herb. **Guild**: pioneer. **Distribution**: pantropical. **Star**: excluded. **Specimens**: - *Etinde*: Ejiofor M. C. FHI 29333

Cleome rutidosperma DC.
Fl. Cameroun 29: 57 (1986).
Cleome ciliata Schum. & Thonn., F.W.T.A. 1: 87 (1954).

Habit: herb. **Guild**: pioneer. **Habitat**: stream banks in forest. **Distribution**: Senegal to Bioko, Cameroon, Gabon, Congo (Brazzaville), Congo (Kinshasa), Angola, Uganda and Tanzania. **Chorology**: Guineo-Congolian. **Star**: green. **Alt. range**: 1-200m (1), 801-1000m (1). **Flowering**: Oct (2). **Fruiting**: Oct (2). **Specimens**: - *Eastern slopes*: Tchouto P. 396. *Onge*: Cheek M. 5205

Cleome spinosa Jacq.
F.W.T.A. 1: 87 (1954); Fl. Cameroun 29: 42 (1986); Fl. Gabon 30: 18 (1987).

Habit: herb to 2 m. **Distribution**: native to tropical America, naturalized in Cameroon. **Star**: excluded. **Specimens**: - *Eastern slopes*: Hutchinson J. & Metcalfe C. R. 87

Euadenia eminens Hook.f.
F.W.T.A. 1: 93 (1954); Fl. Cameroun 29: 74 (1986).
Euadenia alimensis Hua, Fl. Cameroun 29: 74 (1986); Fl. Gabon 30: 39 (1987).
Euadenia pulcherrima Gilg & Benedict, F.W.T.A. 1: 93 (1954).

Habit: shrub to 2 m. **Habitat**: stream banks in forest. **Distribution**: Cameroon, Congo (Brazzaville), Congo (Kinshasa) and Uganda. **Chorology**: Guineo-Congolian. **Star**: green. **Alt. range**: 200-400m (1), 601-800m (1). **Flowering**: Oct (1). **Specimens**: - *Mabeta-Moliwe*: Mildbraed G. W. J. 10626. *Onge*: Cheek M. 5042

Euadenia trifoliolata (Schum. & Thonn.) Oliv.
F.W.T.A. 1: 93 (1954); Fl. Cameroun 29: 72 (1986); Fl. Gabon 30: 36 (1987); Hawthorne W., F.G.F.T. Ghana: 165 (1990).

Habit: shrub to 6 m. **Guild**: shade-bearer. **Habitat**: forest. **Distribution**: Ivory Coast to Cameroon and Gabon. **Chorology**: Guineo-Congolian. **Star**: green. **Alt. range**: 800-1000m (1). **Specimens**: - *No locality*: Mann G. 1178

Ritchiea albersii Gilg & Benedict
F.W.T.A. 1: 92 (1954); Fl. Cameroun 29: 110 (1986); Keay R.W.J., Trees of Nigeria: 42 (1989).

Habit: tree to 11 m. **Habitat**: forest. **Distribution**: SE Nigeria to East Africa. **Chorology**: Afromontane. **Star**: green. **Specimens**: - *Eastern slopes*: Lehmbach H. 180

Ritchiea capparoides (Andr.) Britten var. capparoides
F.W.T.A. 1: 92 (1954); Fl. Cameroun 29: 113 (1986).
Ritchiea fragariodora Gilg, F.W.T.A. 1: 92 (1954).

Habit: woody climber to 10 m. **Habitat**: forest. **Distribution**: tropical Africa. **Star**: excluded. **Specimens**: - *Southern Bakundu F.R.*: Brenan J. P. M. 9429

Ritchiea erecta Hook.f.

F.W.T.A. 1: 91 (1954); Fl. Cameroun 29: 104 (1986); Fl. Gabon
30: 53 (1987).
 Ritchiea polypetala Hook.f., F.W.T.A. 1: 91 (1954).
 Ritchiea brachypoda Gilg, F.W.T.A. 1: 91 (1954).

Habit: shrub to 1 m. **Habitat:** forest. **Distribution:** Nigeria,
Bioko, Cameroon and Gabon. **Chorology:** Lower Guinea. **Star:**
blue. **Alt. range:** 1-200m (3), 201-400m (2), 401-600m (1).
Flowering: Oct (1), Nov (2). **Fruiting:** Jan (1), Mar (2).
Specimens: - *Etinde:* Ludwigs K. 586; Nkeng P. 36; Tekwe C. 1.
Mabeta-Moliwe: Plot Voucher W 101, W 677, W 966. *Onge:*
Harris D. J. 3814; Tchouto P. 872; Watts J. 1033; Wheatley J. I.
756. *Southern Bakundu F.R.:* Keay R. W. J. FHI 28557

Ritchiea cf. erecta Hook.f.

Specimens: - *Mabeta-Moliwe:* Wheatley J. I. 132b, 244. *Mokoko
River F.R.:* Tchouto P. 642

CARICACEAE

Cylicomorpha solmsii (Urb.) Urb.

F.W.T.A. 1: 221 (1954).

Habit: tree to 25 m. **Habitat:** forest and farmbush. **Distribution:**
SW Cameroon. **Chorology:** Western Cameroon Uplands
(montane). **Star:** black. **Flowering:** Jan (1), Mar (1). **Specimens:**
- *Eastern slopes:* Maitland T. D. 1281. *Etinde:* Maitland T. D. 574

CARYOPHYLLACEAE

Cerastium indicum Wright & Arn.

F.W.T.A. 1: 129 (1954).

Habit: herb. **Habitat:** grassland. **Distribution:** Bioko and W.
Cameroon to East and South Africa. **Chorology:** Afromontane.
Star: green. **Alt. range:** 2400-2600m (1). **Specimens:** - *No
locality:* Mann G. 1941

Cerastium octandrum Hochst. ex A.Rich.

F.W.T.A. 1: 129 (1954).

Habit: herb. **Habitat:** grassland and forest-grassland transition.
Distribution: W. Cameroon to East Africa. **Chorology:**
Afromontane. **Star:** green. **Alt. range:** 2200-2400m (2), 2801-
3000m (1). **Flowering:** Oct (2). **Specimens:** - *Eastern slopes:*
Sidwell K. 65; Thomas D. W. 9384. *No locality:* Maitland T. D.
813

Drymaria cordata (L.) Willd.

F.W.T.A. 1: 131 (1954); Fl. Gabon 7: 72 (1963); Kew Bull. 25:
241 (1967).

Habit: herb. **Guild:** pioneer. **Habitat:** forest and forest-grassland
transition. **Distribution:** pantropical. **Chorology:** Montane. **Star:**
excluded. **Alt. range:** 2000-2200m (1), 2201-2400m (1).
Flowering: Oct (1). **Specimens:** - *Eastern slopes:* Tchouto P. 350.
No locality: Mann G. 1319

Drymaria cf. cordata (L.) Willd.

Specimens: - *Etinde:* Thomas D. W. 9226

Drymaria villosa Cham. & Schlecht.

Kew Bull. 25: 241 (1967).

Habit: herb. **Habitat:** forest. **Distribution:** Nigeria and W.
Cameroon, introduced in America. **Chorology:** Montane. **Star:**
blue. **Alt. range:** 1000-1200m (1). **Flowering:** Oct (1).
Specimens: - *Eastern slopes:* Cheek M. 3716

Sagina abyssinica Hochst. ex A.Rich.

F.W.T.A. 1: 130 (1954).

Habit: herb. **Habitat:** grassland. **Distribution:** Bioko, W.
Cameroon, Ethiopia and East Africa. **Chorology:** Afromontane.
Star: green. **Alt. range:** 3000-3200m (1). **Specimens:** - *No
locality:* Mann G. 1288

Silene biafrae Hook.f.

F.W.T.A. 1: 130 (1954).

Habit: herb. **Habitat:** grassland and forest-grassland transition.
Distribution: Mount Cameroon. **Chorology:** Endemic (montane).
Star: black. **Alt. range:** 2000-2200m (1), 2201-2400m (1), 3001-
3200m (1). **Flowering:** Oct (1). **Fruiting:** Oct (1). **Specimens:** -
Eastern slopes: Cheek M. 3661; Thomas D. W. 9326. *No locality:*
Mann G. 2034

Stellaria mannii Hook.f.

F.W.T.A. 1: 129 (1954).

Habit: herb. **Habitat:** forest. **Distribution:** Bioko and W.
Cameroon to East and South Africa. **Chorology:** Afromontane.
Star: green. **Alt. range:** 1400-1600m (1), 1801-2000m (1), 2001-
2200m (1), 2201-2400m (1), 2401-2600m (1). **Flowering:** Oct (2).
Specimens: - *Eastern slopes:* Tchouto P. 338; Thomas D. W.
9351, 9502; Watts J. 462. *No locality:* Mann G. 1940

Stellaria media (L.) Vill.

F.W.T.A. 1: 129 (1954).

Habit: herb. **Habitat:** forest. **Distribution:** worldwide in
temperate and tropical regions. **Chorology:** Montane. **Star:**
excluded. **Alt. range:** 800-1000m (1). **Specimens:** - *Eastern
slopes:* Maitland T. D. 192

Uebelinia abyssinica Hochst.

Bull. Jard. Bot. Nat. Belg. 55: 435 (1985).
 Uebelinia hispida Pax, F.W.T.A. 1: 130 (1954).

Habit: herb. **Habitat:** grassland and forest-grassland transition.
Distribution: W. Cameroon, Rwanda, Uganda and Ethiopia.
Chorology: Afromontane. **Star:** blue. **Alt. range:** 2000-2200m
(1), 2201-2400m (2). **Flowering:** Oct (2). **Fruiting:** Oct (1).
Specimens: - *Eastern slopes:* Cheek M. 3621; Thomas D. W.
9383. *No locality:* Preuss P. R. 971

CECROPIACEAE

Cecropia peltata L.

Specimens: - *Mabeta-Moliwe:* Letouzey 14828

Musanga cecropioides R.Br.

Specimens: - *Etinde:* Struhsaker 17

Myrianthus arboreus P.Beauv.

F.W.T.A. 1: 614 (1958); Bull. Jard. Bot. Nat. Belg. 46: 478 (1976);
Fl. Cameroun 28: 262 (1985); Keay R.W.J., Trees of Nigeria: 303
(1989); Hawthorne W., F.G.F.T. Ghana: 155 (1990).
Habit: tree to 20 m. **Guild:** shade-bearer. **Habitat:** forest and
Aframomum thicket. **Distribution:** tropical Africa. **Star:** excluded.
Alt. range: 1-200m (1).**Fruiting:** Apr (1). **Specimens:** - *Mabeta-
Moliwe:* Mann G. 716; Tchouto P. 170

Myrianthus preussii Engl.

F.W.T.A. 1: 616 (1958); Bull. Jard. Bot. Nat. Belg. 46: 484 (1976).
Habitat: forest. **Alt. range:** 1-200m (3). **Flowering:** Jun (1).
Fruiting: Mar (1), Apr (1), Jun (1). **Specimens:** - *Mabeta-Moliwe:*
Jaff B. 109; Maitland T. D. 391; Nguembock F. 27b; Plot Voucher
W 464, W 546, W 924, W 952, W 1058; Sunderland T. C. H.
1177; Wheatley J. I. 357

Myrianthus preussii Engl. subsp. preussii

F.W.T.A. 1: 616 (1958); Bull. Jard. Bot. Nat. Belg. 46: 486 (1976);
Fl. Cameroun 28: 266 (1985).

Habit: tree or shrub to 6 m. **Habitat:** forest. **Distribution:** SE
Nigeria, Bioko and Cameroon. **Chorology:** Lower Guinea. **Star:**
blue. **Alt. range:** 1-200m (1). **Flowering:** Aug (1). **Specimens:** -
Mabeta-Moliwe: von Rege I. 47

Myrianthus preussii Engl. subsp. seretii (De Wild.) de Ruiter

Bull. Jard. Bot. Nat. Belg. 46: 486 (1976).

Habit: tree or shrub to 10 m. **Habitat:** forest. **Distribution:**
Cameroon, Congo (Kinshasa) and Rwanda. **Chorology:** Guineo-
Congolian. **Star:** green. **Alt. range:** 1-200m (1). **Flowering:** May
(1). **Fruiting:** May (1). **Specimens:** - *Mabeta-Moliwe:* Tchouto P.
211

CELASTRACEAE

Campylostemon angolense Welw. ex Oliv.

F.W.T.A. 1: 626 (1958); Fl. Gabon 29: 266 (1986); Fl. Cameroun
32: 226 (1990).

Habit: woody climber to 10 m. **Habitat:** forest. **Distribution:**
Sierra Leone to Bioko, Cameroon, Congo (Kinshasa), Angola,
CAR and Uganda. **Chorology:** Guineo-Congolian. **Star:** green.
Specimens: - *Mabeta-Moliwe:* Winkler H. J. P. 1467

Cuervea macrophylla (Vahl) R.Wilczek ex N.Hallé

Fl. Gabon 29: 224 (1986); Fl. Cameroun 32: 184 (1990).
Hippocratea macrophylla Vahl, F.W.T.A. 1: 629 (1958).

Habit: woody climber. **Guild:** light demander. **Habitat:** forest.
Distribution: Guinea to Cameroon, Gabon, Congo (Brazzaville),
Congo (Kinshasa), Angola and CAR. **Chorology:** Guineo-
Congolian. **Star:** green. **Specimens:** - *Etinde:* Maitland T. D. 563

Maytenus buchananii (Loes.) Wilczek

Symb. Bot. Ups. 25: 58 (1985).
Maytenus ovatus (Wall. ex Wight & Arn.) L. var. *ovatus* f.
pubescens (Schweinf.) Blakelock, sensu F.W.T.A. 1: 625 (1958)
pro parte.

Habit: shrub to 2 m. **Habitat:** forest and forest-grassland
transition. **Distribution:** Tropical Africa. **Chorology:** Tropical
Africa. **Star:** green. **Alt. range:** 800-1000m (1), 2401-2600m
(1).**Fruiting:** Jul (1). **Specimens:** - *Eastern slopes:* Maitland T. D.
1326; Tekwe C. 118

Maytenus gracilipes (Welw. ex Oliv.) Exell subsp. gracilipes

F.W.T.A. 1: 625 (1958); Symb. Bot. Ups. 25: 50 (1985).
Maytenus ovatus (Wall. ex Wight & Arn.) Loes. var. *argutus*
(Loes.) Blakelock, F.W.T.A. 1: 625 (1958).
Maytenus serrata (Hochst. ex A.Rich.) Wilczek var. *gracilipes*
(Welw. ex Oliv.) Wilczek, Fl. Cameroun 19: 22 (1975).

Habit: spiny shrub to 3 m. **Habitat:** forest. **Distribution:**
Cameroon to Congo (Kinshasa), Angola, Uganda, Sudan and
Ethiopia. **Star:** green. **Specimens:** - *Eastern slopes:* Brenan 9334;
Hutchinson & Metcalfe 106; Maitland T. D. 571, 568, 1309.
Mabeta-Moliwe: Mann G. 2153

Pristimera preussii (Loes.) N.Hallé

Fl. Gabon 29: 214 (1986); Fl. Cameroun 32: 174 (1990).
Hippocratea preussii Loes., F.W.T.A. 1: 627 (1958).

Habit: woody climber. **Habitat:** forest. **Distribution:** Cameroon
and Congo (Kinshasa). **Chorology:** Guineo-Congolian. **Star:**
green. **Specimens:** - *Etinde:* Preuss P. R. 1306

Salacia alata De Wild. var. alata

F.W.T.A. 1: 631 (1958); Fl. Gabon 29: 117 (1986); Fl. Cameroun
32: 94 (1990).

Habit: woody climber. **Guild:** shade-bearer. **Habitat:** forest.
Distribution: Ghana to Cameroon, Gabon, Congo (Brazzaville),
Congo (Kinshasa) and Angola (Cabinda). **Chorology:** Guineo-
Congolian. **Star:** green. **Alt. range:** 1-200m (1), 201-400m (1).
Flowering: Apr (1). **Fruiting:** Aug (1). **Specimens:** - *Mabeta-
Moliwe:* Tchouto P. 154. *Mokoko River F.R.:* Thomas D. W.
10166. *Onge:* Baker W. J. 353

Salacia alata De Wild. var. superba N.Hallé

F.W.T.A. 1: 631 (1958); Fl. Gabon 29: 120 (1986); Fl. Cameroun
32: 95 (1990).

Habit: woody climber or shrub to 5 m. **Habitat:** forest.
Distribution: Nigeria, Cameroon and Gabon. **Chorology:** Lower
Guinea. **Star:** blue. **Alt. range:** 1-200m (1). **Flowering:** Apr (1).
Specimens: - *Mabeta-Moliwe:* Watts J. 179

Salacia debilis (G.Don) Walp.
F.W.T.A. 1: 633 (1958); Fl. Cameroun 32: 66 (1990).

Habit: woody climber. Guild: light demander. Habitat: forest.
Distribution: Guinea Bissau to Bioko, Cameroon, Gabon, Congo
(Brazzaville), Congo (Kinshasa) and CAR. Chorology: Guineo-
Congolian (montane). Star: green. Alt. range: 1000-1200m (1).
Flowering: Mar (1). Specimens: - *Etinde:* Cable S. 1647

Salacia dusenii Loes.
F.W.T.A. 1: 632 (1958); Fl. Gabon 29: 75 (1986); Fl. Cameroun
32: 53 (1990).

Habit: woody climber to 15 m. Habitat: forest. Distribution:
Nigeria, Cameroon, Gabon and Congo (Kinshasa). Chorology:
Guineo-Congolian. Star: green. Alt. range: 1-200m (2), 201-
400m (1). Flowering: Apr (1), May (1), Jun (1). Fruiting: Jun (1).
Specimens: - *Mokoko River F.R.:* Tchouto P. 1153, 1243; Watts J.
1185

Salacia erecta (G.Don) Walp. var. erecta
F.W.T.A. 1: 633 (1958); Fl. Gabon 29: 112 (1986); Fl. Cameroun
32: 88 (1990).

Habit: small tree or shrub, sometimes scandent. Habitat: forest.
Distribution: tropical Africa. Star: excluded. Alt. range: 1400-
1600m (1). Flowering: Oct (1). Specimens: - *Eastern slopes:*
Lehmbach H. 91. *Etinde:* Tchouto P. 418

Salacia fimbrisepala Loes.
F.W.T.A. 1: 632 (1958); Fl. Gabon 29: 54 (1986); Fl. Cameroun
32: 38 (1990).

Habit: woody climber or small tree. Habitat: forest. Distribution:
Mount Cameroon. Chorology: Endemic. Star: black. Specimens:
- *Eastern slopes:* Lehmbach H. 228

Salacia lehmbachii Loes. var. lehmbachii
F.W.T.A. 1: 633 (1958); Fl. Gabon 29: 72 (1986); Fl. Cameroun
32: 52 (1990).

Habit: shrub to 2 m. Habitat: forest. Distribution: Cameroon,
Gabon, Congo (Brazzaville) and Congo (Kinshasa). Chorology:
Guineo-Congolian. Star: green. Alt. range: 1-200m (9), 201-
400m (2), 401-600m (1), 1401-1600m (1). Flowering: Apr (5),
May (6). Fruiting: Jan (1), Mar (1), Apr (2), May (3), Sep (1).
Specimens: - *Eastern slopes:* Lehmbach H. 116. *Etinde:*
Sunderland T. C. H. 1068; Wheatley J. I. 540. *Mabeta-Moliwe:*
Watts J. 222, 237. *Mokoko River F.R.:* Acworth J. M. 303; Akogo
M. 205; Ekema N. 1011; Mbani J. M. 340; Tchouto P. 644, 1079;
Thomas D. W. 10010, 10013; Watts J. 1106, 1198

Salacia lenticellosa Loes. ex Harms
Fl. Gabon 29: 100 (1986); Fl. Cameroun 32: 78 (1990).

Habit: woody climber or shrub. Habitat: forest. Distribution:
Nigeria and Cameroon. Chorology: Lower Guinea. Star: blue.
Alt. range: 1-200m (1).Fruiting: Jun (1). Specimens: - *Mokoko
River F.R.:* Ekema N. 1166

Salacia loloensis Loes. var. loloensis
F.W.T.A. 1: 632 (1958); Fl. Gabon 29: 79 (1986); Fl. Cameroun
32: 58 (1990).

Habit: tree to 8 m. Habitat: forest. Distribution: Nigeria,
Cameroon, Gabon and Congo (Brazzaville). Chorology: Guineo-

Congolian. Star: green. Alt. range: 1-200m (19), 201-400m (6).
Flowering: Mar (2), Apr (6), May (1), Sep (1), Oct (2). Fruiting:
May (2), Jun (5), Aug (1), Sep (2), Oct (5), Nov (1). Specimens: -
Etinde: Kwangue A. 111; Mbani J. M. 79; Nkeng P. 70; Tchouto
P. 80, 462, 807; Tekwe C. 130; Watts J. 729. *Mabeta-Moliwe:* Jaff
B. 158; Watts J. 169; Wheatley J. I. 207. *Mokoko River F.R.:*
Acworth J. M. 336; Akogo M. 163, 216; Ekema N. 1108; Mbani J.
M. 462; Ndam N. 1241; Pouakouyou D. 6; Tchouto P. 1066, 1216.
Onge: Akogo M. 45; Cheek M. 5029; Ndam N. 699, 700, 735;
Tchouto P. 693; Watts J. 939. *Southern Bakundu F.R.:* Olorunfemi
J. FHI 30509

Salacia longipes (Oliv.) N.Hallé var.
camerunensis (Loes.) N.Hallé
Fl. Gabon 29: 66 (1986); Fl. Cameroun 32: 49 (1990).
Salacia camerunensis Loes., F.W.T.A. 1: 633 (1958).

Habit: woody climber. Habitat: forest. Distribution: Liberia to
Cameroon, Gabon, Congo (Brazzaville), Congo (Kinshasa) and
Angola (Cabinda). Chorology: Guineo-Congolian. Star: green.
Alt. range: 1-200m (2), 201-400m (1). Flowering: Apr (1), Jun
(1). Fruiting: Jun (1), Jul (1). Specimens: - *Mabeta-Moliwe:*
Hurst J. 16; Watts J. 220. *Mokoko River F.R.:* Tchouto P. 1251.
Southern Bakundu F.R.: Brenan J. P. M. 9403

Salacia mannii Oliv.
F.W.T.A. 1: 632 (1958); Fl. Gabon 29: 54 (1986); Fl. Cameroun
32: 38 (1990).
Salacia cuspidicoma Loes., F.W.T.A. 1: 632 (1958).

Habit: tree to 6 m. Habitat: forest. Distribution: Nigeria, Bioko,
Cameroon, Rio Muni, Gabon, Congo (Brazzaville) and Congo
(Kinshasa). Chorology: Guineo-Congolian. Star: green. Alt.
range: 1-200m (8), 201-400m (2). Flowering: Mar (2), Apr (3),
May (3), Jun (2). Fruiting: Mar (1), Apr (1), May (2), Jun (1).
Specimens: - *Etinde:* Kwangue A. 66. *Mabeta-Moliwe:*
Sunderland T. C. H. 1129, 1167. *Mokoko River F.R.:* Acworth J.
M. 341; Akogo M. 162, 255; Fraser P. 414; Tchouto P. 1237;
Watts J. 1095, 1156

Salacia cf. mannii Oliv.

Specimens: - *Mabeta-Moliwe:* Sunderland T. C. H. 1184

Salacia nitida (Benth.) N.E.Br.
F.W.T.A. 1: 633 (1958); Fl. Gabon 29: 142 (1986); Fl. Cameroun
32: 110 (1990).

Habit: woody climber. Guild: light demander. Habitat: forest.
Distribution: Sierra Leone to Cameroon, Gabon, Congo
(Brazzaville) and Congo (Kinshasa). Chorology: Guineo-
Congolian. Star: green. Specimens: - *Eastern slopes:* Maitland T.
D. 291. *Etinde:* Brenan J. P. M. 9591

Salacia cf. nitida (Benth.) N.E.Br.

Specimens: - *Etinde:* Kwangue A. 107

Salacia preussii Loes. var. preussii
F.W.T.A. 1: 632 (1958); Fl. Gabon 29: 50 (1986); Fl. Cameroun
32: 35 (1990).

Habit: woody climber. Habitat: forest. Distribution: Cameroon
and Gabon. Chorology: Lower Guinea. Star: blue. Alt. range: 1-
200m (2). Flowering: Apr (2). Specimens: - *Mokoko River F.R.:*
Acworth J. M. 94; Tchouto P. 1067b

Salacia regeliana J.Braun & K.Schum.
Fl. Gabon 29: 45 (1986); Fl. Cameroun 32: 29 (1990).

Habit: woody climber. **Habitat:** forest. **Distribution:** Cameroon, Gabon and Congo (Brazzaville). **Chorology:** Guineo-Congolian. **Star:** green. **Alt. range:** 1-200m (2). **Flowering:** May (2), Jun (1). **Specimens:** - *Mokoko River F.R.:* Ekema N. 1130; Ndam N. 1211; Watts J. 1102

Salacia staudtiana Loes. var. leonensis Loes.
F.W.T.A. 1: 633 (1958); Fl. Gabon 29: 62 (1986); Fl. Cameroun 32: 45 (1990).
Salacia caillei A.Chev., F.W.T.A. 1: 633 (1958).

Habit: woody climber. **Habitat:** forest. **Distribution:** Guinea, Sierra Leone, Cameroon and Gabon. **Chorology:** Guineo-Congolian. **Star:** green. **Alt. range:** 1-200m (2). **Flowering:** Apr (1), Oct (1). **Fruiting:** Oct (1). **Specimens:** - *Mabeta-Moliwe:* Watts J. 146. *Onge:* Watts J. 892

Salacia staudtiana Loes. var. scaphodisca N.Hallé
Fl. Gabon 29: 64 (1986); Fl. Cameroun 32: 45 (1990).

Habit: woody climber. **Habitat:** forest and farmbush. **Distribution:** Gabon and Congo (Brazzaville). **Chorology:** Lower Guinea. **Star:** blue. **Alt. range:** 1-200m (1). **Flowering:** May (1). **Specimens:** - *Mabeta-Moliwe:* Sunderland T. C. H. 1327

Salacia staudtiana Loes. var. staudtiana
F.W.T.A. 1: 633 (1958); Fl. Gabon 29: 60 (1986); Fl. Cameroun 32: 43 (1990).

Habit: woody climber. **Guild:** light demander. **Habitat:** forest. **Distribution:** Cameroon and Gabon. **Chorology:** Lower Guinea. **Star:** blue. **Alt. range:** 1-200m (11), 201-400m (3), 601-800m (1). **Flowering:** Feb (1), Mar (1), Apr (4), May (4), Jun (2). **Fruiting:** Feb (1), Apr (1), Jun (1), Aug (1), Nov (1). **Specimens:** - *Etinde:* Wheatley J. I. 6. *Mabeta-Moliwe:* Baker W. J. 350; Sunderland T. C. H. 1315; Tchouto P. 168; Watts J. 104, 401; Wheatley J. I. 108, 225, 285; von Rege I. 97. *Mokoko River F.R.:* Akogo M. 278; Ekema N. 1182; Tchouto P. 1213, 1229. *Onge:* Tchouto P. 1042

Salacia cf. staudtiana Loes.

Specimens: - *Etinde:* Cheek M. 3609

Salacia talbotii Baker f.
Fl. Gabon 29: 168 (1986); Fl. Cameroun 32: 128 (1990).

Habit: woody climber to 5 m. **Habitat:** forest. **Distribution:** SE Nigeria and Cameroon. **Chorology:** Lower Guinea. **Star:** blue. **Alt. range:** 1-200m (5). **Flowering:** Apr (2), May (2), Jun (1). **Fruiting:** Apr (1), Jun (1). **Specimens:** - *Mokoko River F.R.:* Acworth J. M. 305, 347; Akogo M. 196, 246; Watts J. 1171

Salacia volubilis Loes. & Winkl.
F.W.T.A. 1: 632 (1958); Fl. Gabon 29: 92 (1986); Fl. Cameroun 32: 70 (1990).

Habit: woody climber. **Habitat:** forest. **Distribution:** Cameroon. **Chorology:** Lower Guinea. **Star:** gold. **Specimens:** - *Etinde:* Stossel & Winkler H. J. P. 84

Salacia whytei Loes. var. whytei
Fl. Gabon 29: 95 (1986); Fl. Cameroun 32: 75 (1990).

Habit: woody climber. **Habitat:** forest. **Distribution:** Guinea to Cameroon, Gabon and Angola. **Chorology:** Guineo-Congolian. **Star:** green. **Alt. range:** 1-200m (1).**Fruiting:** Apr (1). **Specimens:** - *Mabeta-Moliwe:* Watts J. 152

Salacia zenkeri Loes.
F.W.T.A. 1: 632 (1958); Fl. Gabon 29: 46 (1986); Fl. Cameroun 32: 32 (1990).

Habit: woody climber or shrub. **Habitat:** forest. **Distribution:** Guinea to Cameroon, Gabon and Congo (Kinshasa). **Chorology:** Guineo-Congolian. **Star:** green. **Alt. range:** 1-200m (4), 201-400m (2). **Flowering:** Mar (3), Apr (2). **Fruiting:** Oct (1). **Specimens:** - *Etinde:* Preuss P. R. 1254. *Mabeta-Moliwe:* Jaff B. 56; Sunderland T. C. H. 1144; Watts J. 114; Wheatley J. I. 53. *Mokoko River F.R.:* Tchouto P. 1067a. *Onge:* Tchouto P. 907

Salacia sp.

Specimens: - *Mabeta-Moliwe:* Cheek M. 3576; Jaff B. 136, 155; Sunderland T. C. H. 1470; Wheatley J. I. 315

Salacia sp. B sensu Hepper & Blakelock
F.W.T.A. 1: 633 (1958).

Habit: shrub to 0.5 m. **Distribution:** Mount Cameroon. **Chorology:** Endemic. **Star:** black. **Alt. range:** 1-200m (1). **Flowering:** Mar (1), May (1). **Specimens:** - *Mokoko River F.R.:* Tchouto P. 1214. *Southern Bakundu F.R.:* Brenan J. P. M. 9483

Salacia sp. nov. 1
Fl. Gabon 29: 82 (1986); Fl. Cameroun 32: 60 (1990).
Salacia sp. A sensu Hepper & Blakelock, F.W.T.A. 1: 632 (1958).

Habit: tree or shrub to 4 m. **Habitat:** forest. **Specimens:** - *Mabeta-Moliwe:* Brenan J. P. M. 9590; Tchouto P. 204. *Mokoko River F.R.:* Acworth J. M. 106, 301; Akogo M. 208; Ekema N. 1170; Fraser P. 409; Mbani J. M. 461; Pouakouyou D. 8, 82; Tchouto P. 1076, 1233; Thomas D. W. 10014; Watts J. 1062

Salacia sp. nov. 2

Specimens: - *Mabeta-Moliwe:* Wheatley J. I. 260

Salacighia letestuana (Pellegr.) Blakelock
F.W.T.A. 1: 634 (1958); Fl. Gabon 29: 18 (1986); Fl. Cameroun 32: 10 (1990).
Salacia letestuana Pellegr., Bull. Mus. Hist. Nat. Paris 28: 312 (1922).

Habit: woody climber. **Habitat:** forest. **Distribution:** Ivory Coast to Cameroon, Gabon, Congo (Kinshasa), Angola (Cabinda) and CAR. **Chorology:** Guineo-Congolian. **Star:** green. **Alt. range:** 1-200m (2). **Flowering:** Apr (2). **Fruiting:** Apr (1). **Specimens:** - *Mabeta-Moliwe:* Watts J. 248. *Mokoko River F.R.:* Pouakouyou D. 25

Simicratea welwitschii (Oliv.) N.Hallé
Fl. Gabon 29: 180 (1986); Fl. Cameroun 32: 140 (1990).
Hippocratea welwitschii Oliv., F.W.T.A. 1: 628 (1958).

Habit: woody climber. **Guild:** light demander. **Habitat:** forest. **Distribution:** Guinea to Cameroon, Gabon, Congo (Kinshasa), Angola, Uganda and Tanzania. **Chorology:** Guineo-Congolian. **Star:** green. **Alt. range:** 800-1000m (1). **Specimens:** - *Etinde:* Preuss P. R. 1304. *Mabeta-Moliwe:* Mildbraed G. W. J. 10709

CHENOPODIACEAE

Chenopodium ambrosioides L.
F.W.T.A. 1: 144 (1954); Fl. Gabon 7: 18 (1963).

Habit: herb. **Distribution:** pantropical. **Star:** excluded. **Specimens:** - *Eastern slopes:* Maitland T. D. 113

CHRYSOBALANACEAE
G. Prance (K)

Chrysobalanus icaco L. subsp. icaco
Bull. Jard. Bot. Brux. 46: 272 (1976); Fl. Cameroun 20: 61 (1978); Keay R.W.J., Trees of Nigeria: 182 (1989).
Chrysobalanus orbicularis Schum., F.W.T.A. 1: 426 (1958).
Chrysobalanus ellipticus Soland. ex Sabine, F.W.T.A. 1: 426 (1958).

Habit: tree to 12 m. **Habitat:** forest and coastal sand dunes. **Distribution:** Senegal to Bioko, Cameroon, Congo (Kinshasa) and Angola, also South America. **Star:** excluded. **Specimens:** - *Etinde:* Kalbreyer W. 30

Dactyladenia cinerea (Engl. ex De Wild.) Prance & F. White
Brittonia 31: 485 (1979).
Acioa cinerea Engl. ex De Wild. Fl. Cameroun 20: 22 (1978)

Specimens: - *Mokoko:* Ndam 1157

Dactyladenia gilletii (De Wild.) Prance & F.White
Brittonia 31: 485 (1979).
Acioa gilletii De Wild.

Habit: tree to 6 m. **Habitat:** forest. **Distribution:** Cameroon, Bioko, CAR, Gabon and Congo (Kinshasa). **Chorology:** Guineo-Congolian. **Star:** green. **Alt. range:** 600-800m (1). **Flowering:** Mar (1). **Specimens:** - *Etinde:* Khayota B. 540

Dactyladenia lehmbachii (Engl.) Prance & F.White
Brittonia 31: 485 (1979); Keay R.W.J., Trees of Nigeria: 189 (1989).
Acioa lehmbachii Engl., Fl. Cameroun 20: 40 (1978).
Acioa rudatisii De Wild., F.W.T.A. 1: 433 (1958).

Habit: tree to 20 m. **Habitat:** forest. **Distribution:** SE Nigeria, Cameroon and Gabon. **Chorology:** Lower Guinea. **Star:** blue. **Specimens:** - *Eastern slopes:* Deistel H. 654

Dactyladenia mannii (Oliv.) Prance & F.White
Brittonia 31: 486 (1979).
Acioa mannii (Oliv.) Engl., F.W.T.A. 1: 433 (1958); Fl. Cameroun 20: 50 (1978).

Habit: tree to 6 m, or scandent shrub. **Habitat:** forest. **Distribution:** Mount Cameroon and Bioko. **Chorology:** Western Cameroon. **Star:** black. **Alt. range:** 1-200m (1). **Flowering:** Mar (1). **Specimens:** - *Eastern slopes:* Maitland T. D. 467. *Onge:* Wheatley J. I. 790

Dactyladenia scabrifolia (Hua) Prance & F.White
Brittonia 31: 486 (1979).
Acioa scabrifolia Hua, F.W.T.A. 1: 431 (1958).

Habit: tree to 10 m. **Habitat:** forest. **Distribution:** Guinea to Cameroon. **Chorology:** Guineo-Congolian. **Star:** green. **Specimens:** - *Eastern slopes:* Lehmbach H. 115

Dactyladenia staudtii (Engl.) Prance & F. White
Acioa staudtii Engl. Fl. Cameroun 20: 58 (1978)

Specimens: - *Mokoko:* Acworth 156, Sonké 1146.

Magnistipula cuneatifolia Hauman
Bull. Jard. Bot. Brux. 21: 175 (1951); Fl. Cameroun 20: 82 (1978).

Specimens: - *Mabeta-Moliwe:* Plot Voucher W 831

Magnistipula glaberrima Engl.
Fl. Gabon 24: 87 (1978); Fl. Cameroun 20: 87 (1978).

Habit: tree 10-15 m tall. **Habitat:** forest. **Distribution:** Cameroon and Gabon. **Chorology:** Lower Guinea. **Star:** gold. **Alt. range:** 200-400m (2). **Specimens:** - *Etinde:* Thomas D. W. 9734. *Mokoko River F.R.:* Ndam N. 1177

Magnistipula tessmannii (Engl.) Prance
Bol. Soc. Brot. (ser.2) 40: 185 (1966); Fl. Cameroun 20: 88 (1978); Keay R.W.J., Trees of Nigeria: 187 (1989).

Habit: tree to 40 m. **Habitat:** forest. **Distribution:** SE Nigeria to Congo (Kinshasa) and Angola (Cabinda). **Chorology:** Guineo-Congolian. **Star:** green. **Alt. range:** 1-200m (1). **Specimens:** - *Southern Bakundu F.R.:* Mbani J. M. 412; *Mokoko:* Thomas, D.W. 10063

Maranthes chrysophylla (Oliv.) Prance
Bull. Jard. Bot. Brux. 46: 295 (1976); Fl. Gabon 24: 100 (1978); Fl. Cameroun 20: 100 (1978); Keay R.W.J., Trees of Nigeria: 187 (1989); Hawthorne W., F.G.F.T. Ghana: 92 (1990).
Parinari chrysophylla Oliv., F.W.T.A. 1: 428 (1958).

Habit: tree to 40 m. **Guild:** shade-bearer. **Habitat:** forest. **Distribution:** Liberia to Cameroon and Gabon. **Chorology:** Guineo-Congolian. **Star:** green. **Alt. range:** 200-400m (1). **Specimens:** - *Onge:* Tchouto P. 760

Maranthes gabunensis (Engl.) Prance

Bull. Jard. Bot. Brux. 46: 299 (1976); Fl. Gabon 24: 105 (1978); Fl. Cameroun 20: 105 (1978); Keay R.W.J., Trees of Nigeria: 185 (1989).
Parinari gabunensis Engl., F.W.T.A. 1: 428 (1958).

Habit: tree to 20 m. **Habitat:** forest. **Distribution:** SE Nigeria to Congo (Kinshasa) and Angola (Cabinda). **Chorology:** Guineo-Congolian. **Star:** green. **Alt. range:** 1-200m (5). **Flowering:** Apr (4), May (1). **Fruiting:** Apr (1). **Specimens:** - *Mabeta-Moliwe:* Cheek M. 3405; Plot Voucher W 629, 898; Sunderland T. C. H. 1216, 1371; Watts J. 192, 223; Wheatley J. I. 224; *Mokoko:* Tchouto 1125.

Maranthes glabra (Oliv.) Prance

Bol. Soc. Brot. (ser.2) 40: 184 (1966); Fl. Cameroun 20: 108 (1978); Keay R.W.J., Trees of Nigeria: 185 (1989).
Parinari glabra Oliv., F.W.T.A. 1: 428 (1958).

Habit: tree to 30 m. **Guild:** shade-bearer. **Habitat:** forest. **Distribution:** Sierra Leone to Congo (Kinshasa), Angola and CAR. **Chorology:** Guineo-Congolian. **Star:** green. **Alt. range:** 1-200m (2). **Specimens:** - *Mokoko River F.R.:* Tchouto P. 1125. *Southern Bakundu F.R.:* Mbani J. M. 419

Parinari hypochrysea Mildbr. ex Letouzey & F.White

Fl. Cameroun 20: 134 (1978); Fl. Gabon 24: 134 (1978); Keay R.W.J., Trees of Nigeria: 184 (1989).
Parinari sp., F.W.T.A. 1: 430 (1958).

Habit: tree to 30 m. **Habitat:** forest. **Distribution:** Nigeria, Cameroon and Gabon. **Chorology:** Lower Guinea. **Star:** blue. **Specimens:** - *Southern Bakundu F.R.:* Ejiofor M. C. FHI 15253; Olorunfemi J. FHI 30502

COMBRETACEAE

C. Jongkind (WAG)

Combretum fuscum Planch. ex Benth.

Combretum batesii Exell F.W.T.A. 1: 272 (1954); Fl. Cameroun 25: 48 (1983).

Habit: woody climber. **Habitat:** forest. **Distribution:** Cameroon. **Chorology:** Lower Guinea. **Star:** gold. **Specimens:** - *No locality:* Dunlap 13

Combretum bipindense Engl. & Diels

F.W.T.A. 1: 272 (1954); Fl. Cameroun 25: 50 (1983).

Habit: woody climber. **Guild:** light demander. **Habitat:** forest. **Distribution:** Ghana and Cameroon. **Chorology:** Guineo-Congolian. **Star:** blue. **Specimens:** - *Etinde:* Maitland T. D. 378

Combretum bracteatum (Laws.) Engl. & Diels

F.W.T.A. 1: 274 (1954); Fl. Cameroun 25: 28 (1983).

Habit: woody climber to 10 m. **Habitat:** forest. **Distribution:** SE Nigeria, Bioko, Cameroon, Gabon, Congo (Kinshasa) and Angola (Cabinda). **Chorology:** Guineo-Congolian. **Star:** green. **Alt. range:** 1-200m (1), 201-400m (1). **Flowering:** Nov (1), Dec (2). **Specimens:** - *Etinde:* Cheek M. 5392c; Preuss P. R. 1320. *Mabeta-Moliwe:* Nguembock F. 52; Sunderland T. C. H. 1031

Combretum cinereipetalum Engl. & Diels

F.W.T.A. 1: 273 (1954); F.T.E.A. Combretaceae: 55 (1973); Fl. Cameroun 25: 39 (1983).

Habit: woody climber. **Habitat:** forest. **Distribution:** Cameroon, Gabon, Congo (Brazzaville), Congo (Kinshasa), Angola, CAR and Uganda. **Chorology:** Guineo-Congolian. **Star:** green. **Specimens:** - *Eastern slopes:* Maitland T. D. 133

Combretum cuspidatum Planch. ex Benth.

F.W.T.A. 1: 272 (1954); Bull. Jard. Bot. Nat. Belg. 49: 59 (1979); Fl. Cameroun 25: 58 (1983).
Combretum insulare Engl. & Diels, F.W.T.A. 1: 272 (1954).

Habit: woody climber. **Habitat:** forest. **Distribution:** Guinea to Bioko and Congo (Kinshasa). **Chorology:** Guineo-Congolian. **Star:** green. **Alt. range:** 1-200m (1). **Flowering:** May (1). **Specimens:** - *Mabeta-Moliwe:* Wheatley J. I. 313

Combretum latialatum Engl. ex Engl. & Diels,

Engl. Monogr. Afr. Pfanzenfam. Combretaceae: 86 (1899); Adansonia 12: 275 (1991); Fl. Cameroun 25: 62 (1983).
Quisqualis latialata (Engl. ex Engl. & Diels) Exell, F.W.T.A. 1: 275 (1954).

Habit: woody climber. **Habitat:** stream banks in forest. **Distribution:** Nigeria, Cameroon, Gabon, Congo (Kinshasa), Angola and CAR. **Chorology:** Guineo-Congolian. **Star:** green. **Alt. range:** 1-200m (5). **Flowering:** May (1), Jun (1), Jul (1), Dec (1). **Fruiting:** Jul (1), Aug (1), Dec (1). **Specimens:** - *Etinde:* Preuss P. R. 1322. *Mabeta-Moliwe:* Baker W. J. 306; Cheek M. 3530, 5734; Ndam N. 539; Wheatley J. I. 371

Combretum paniculatum Vent.

F.W.T.A. 1: 273 (1954); F.T.E.A. Combretaceae: 53 (1973); Meded. Land. Wag. 82(3): 114 (1982); Fl. Cameroun 25: 37 (1983).

Habit: woody climber. **Habitat:** forest and Aframomum thicket. **Distribution:** tropical Africa. **Star:** excluded. **Specimens:** - *Eastern slopes:* Maitland T. D. 887

Combretum racemosum P.Beauv.

F.W.T.A. 1: 272 (1954); F.T.E.A. Combretaceae: 55 (1973); Fl. Cameroun 25: 38 (1983).

Habit: woody climber to 10 m. **Guild:** pioneer. **Habitat:** forest. **Distribution:** Senegal to Bioko, Cameroon, Gabon, Congo (Kinshasa), Angola, CAR, Uganda and Sudan. **Chorology:** Guineo-Congolian. **Star:** green. **Alt. range:** 1-200m (1). **Flowering:** Mar (1). **Specimens:** - *Etinde:* Maitland T. D. 413. *Mabeta-Moliwe:* Sunderland T. C. H. 1132

Conocarpus erectus L.

F.W.T.A. 1: 762 (1958); Fl. Cameroun 25: 84 (1983).

Habit: shrub to 3 m. **Habitat:** forest. **Distribution:** tropical West Africa and America. **Star:** excluded. **Specimens:** - *Etinde:* Maitland T. D. 720

Laguncularia racemosa (L.) Gaertn.f.

F.W.T.A. 1: 281 (1954); Fl. Cameroun 25: 82 (1983); Keay R.W.J., Trees of Nigeria: 95 (1989); Hawthorne W., F.G.F.T. Ghana: 43 (1990).

Habit: tree to 10 m. Habitat: mangrove forest. Distribution: coasts of West Africa and eastern tropical America. Star: excluded. Alt. range: 1-200m (1). Flowering: Nov (1). Fruiting: Nov (1). Specimens: - *Etinde:* Dalziel J. M. 8179; Wheatley J. I. 666; *Mabeta-Moliwe:* Cheek M. 3473

Strephonema mannii Hook.f.
F.W.T.A. 1: 264 (1954); Fl. Cameroun 25: 8 (1983); Keay R.W.J., Trees of Nigeria: 85 (1989); Ann. Missouri Bot. Gard. 82: 536 (1995).

Habit: tree to 30 m. Habitat: forest. Distribution: SE Nigeria, Cameroon and Gabon. Chorology: Lower Guinea. Star: blue. Specimens: - *Eastern slopes:* Maitland T. D. 1145

Strephonema cf. polybotryum Mildbr.

Specimens:- *Mokoko:* Mbani 416

Strephonema sp.

Specimens: - *Southern Bakundu F.R.:* Mbani J. M. 416

Terminalia catappa L.
F.W.T.A. 1: 277 (1954); Fl. Cameroun 25: 79 (1983); Keay R.W.J., Trees of Nigeria: 93 (1989).

Habit: tree to 25 m. Habitat: coastal forest. Distribution: native to India, but introduced throughout the tropics. Star: excluded. Alt. range: 1-200m (2). Flowering: Mar (1), Jul (1). Fruiting: Mar (1), Jul (1). Specimens: - *Mabeta-Moliwe:* Cheek M. 3518; Groves M. 225

Terminalia ivorensis A.Chev.
F.W.T.A. 1: 279 (1954); Fl. Cameroun 25: 68 (1983); Keay R.W.J., Trees of Nigeria: 91 (1989); Hawthorne W., F.G.F.T. Ghana: 139 (1990).

Habit: tree to 50 m. Guild: pioneer. Habitat: forest. Distribution: Guinea to Cameroon. Chorology: Guineo-Congolian. Star: scarlet. Alt. range: 600-800m (1). Specimens: - *Mabeta-Moliwe:* Mildbraed G. W. J. 10740

Terminalia superba Engl. & Diels
F.W.T.A. 1: 277 (1954); Fl. Cameroun 25: 67 (1983); Keay R.W.J., Trees of Nigeria: 89 (1989); Hawthorne W., F.G.F.T. Ghana: 139 (1990).

Habit: tree to 50 m. Guild: pioneer. Habitat: forest. Distribution: Guinea to Cameroon, Gabon, Congo (Kinshasa) and Angola (Cabinda). Chorology: Guineo-Congolian. Star: pink. Specimens: - *Etinde:* Preuss P. R. 1300. *Mabeta-Moliwe:* Plot Voucher W 66, W 917

COMPOSITAE
S. Cable, H. Beentje & N. Hind (K)

Acmella caulirhiza Delile
Fragm. Flor. Geobot. 36(1) Suppl.1: 228 (1991).

Spilanthes filicaulis (Schum. & Thonn.) C.D.Adams, F.W.T.A. 2: 236 (1963).

Habit: herb. Distribution: Guinea to E. Africa and the Mascarenes. Chorology: Tropical Africa. Star: green. Alt. range: 1000-1200m (1). Specimens: - *Eastern slopes:* Boughey A. S. GC 7036

Adenostemma mauritianum D.C.
F.W.T.A. 2: 286 (1963); Fragm. Flor. Geobot. 36(1) Suppl.1: 467 (1991).

Habit: herb. Habitat: forest. Distribution: SE Nigeria, Bioko, Cameroon, Congo (Kinshasa), Zimbabwe and Sudan to East Africa and the Mascarenes. Chorology: Afromontane. Star: green. Alt. range: 1400-1600m (1), 1601-1800m (1). Flowering: Oct (2). Specimens: - *Eastern slopes:* Thomas D. W. 9510. *Etinde:* Thomas D. W. 9259

Adenostemma perrottetii DC.
F.W.T.A. 2: 286 (1963); Fragm. Flor. Geobot. 36(1) Suppl.1: 469 (1991).
Habit: herb. Guild: pioneer. Habitat: forest gaps. Distribution: Senegal to Bioko, Cameroon, Sudan, Uganda and Tanzania. Chorology: Guineo-Congolian (montane). Star: green. Alt. range: 1400-1600m (1). Specimens: - *Eastern slopes:* Maitland T. D. 225

Ageratum conyzoides L.
F.W.T.A. 2: 287 (1963); Fragm. Flor. Geobot. 36(1) Suppl.1: 473 (1991).

Habit: herb. Guild: pioneer. Habitat: lava flows. Distribution: tropical Africa. Chorology: Afromontane. Star: green. Alt. range: 1-200m (1), 801-1000m (1), 1201-1400m (1). Flowering: Oct (2). Specimens: - *Eastern slopes:* Dunlap 74; Tchouto P. 378. *Etinde:* Dawson S. 37

Anisopappus chinensis Hook. & Arn. subsp. buchwaldii (O.Hoffm.) S.Ortiz, Paiva & Rodr.-Oubina var. macrocephala (Humbert) S.Ortiz, Paiva & Rodr.-Oubina
Anales Jard. Bot. Madrid 54: 383 (1996).
Anisopappus suborbicularis Hutch. & B.L.Burtt, F.W.T.A. 2: 258 (1963).
Anisopappus africanus (Hook.f.) Oliv. & Hiern, F.W.T.A. 2: 258 (1963); Kirkia 4: 53 (1964).

Habit: herb. Habitat: grassland and forest-grassland transition. Distribution: tropical Africa. Chorology: Afromontane. Star: green. Alt. range: 1800-2000m (1), 2001-2200m (4). Flowering: Sep (1), Oct (3). Specimens: - *Eastern slopes:* Tchouto P. 374; Thomas D. W. 9279. *Etinde:* Tekwe C. 308; Thomas D. W. 9356. *No locality:* Mann G. 1313

Bidens mannii T.G.J.Rayner
Kew Bull. 48: 459 (1993).
Coreopsis monticola (Hook.f.) Oliv. & Hiern var. *monticola*, F.W.T.A. 2: 232 (1963).
Coreopsis monticola (Hook.f.) Oliv. & Hiern var. *pilosa* Hutch. & Dalziel, F.W.T.A. 2: 232 (1963).

Habit: herb to 2 m. Habitat: grassland and forest-grassland transition. Distribution: W. Cameroon. Chorology: Western Cameroon Uplands (montane). Star: gold. Alt. range: 1800-2000m (2), 2001-2200m (1), 2201-2400m (1), 2401-2600m (1).

Flowering: Sep (1), Oct (2). **Specimens: -** *Eastern slopes:* Tchouto P. 321; Thomas D. W. 9294. *Etinde:* Tekwe C. 271. *No locality:* Mann G. 1219; Migeod F. 211

Bidens pilosa L.
F.W.T.A. 2: 234 (1963); Fragm. Flor. Geobot. 36(1) Suppl.1: 136 (1991); Kew Bull. 48: 500 (1993).

Habit: herb. **Guild:** pioneer. **Distribution:** pantropical. **Chorology:** Montane. **Star:** excluded. **Alt. range:** 1600-1800m (1). **Specimens: -** *No locality:* Dunlap 76

Blumea crispata (Vahl) Merxm. var. **crispata**
Fl. Afr. Cent. Compositae: 13 (1989).
   Laggera alata (D.Don) Sch. Bip. ex Oliv. var. *alata*, F.W.T.A. 2: 262 (1963).
   Laggera pterodonta (DC.) Sch.Bip. ex Oliv., F.W.T.A. 2: 262 (1963).

Habit: herb to 2 m. **Habitat:** forest-grassland transition. **Distribution:** Guinea to Mozambique. **Chorology:** Tropical Africa. **Star:** green. **Alt. range:** 800-1000m (1), 2201-2400m (1). **Specimens: -** *Eastern slopes:* Boughey A. S. GC 10597; Dunlap 160

Blumea crispata (Vahl) Merxm. var. **montana** (Adams) Lebrun & Stork
Lebrun J.-P. & Stork A., E.P.F.A.T. 4: 262 (1997).
   Laggera alata (D.Don) Sch. Bip. ex Oliv. var. *montana* C.D.Adams, F.W.T.A. 2: 262 (1963).

Habit: herb to 2 m. **Habitat:** forest and forest-grassland transition. **Distribution:** W. Cameroon and Bioko. **Chorology:** Western Cameroon Uplands (montane). **Star:** gold. **Alt. range:** 1800-2000m (1), 2001-2200m (1). **Specimens: -** *Etinde:* Thomas D. W. 9355. *No locality:* Dunlap 196

Chromolaena odorata (L.) R.M.King & H.Robinson
Fragm. Flor. Geobot. 36(1) Suppl.1: 450 (1991); Candollea 47: 645 (1992).
   Eupatorium odoratum L., F.W.T.A. 2: 285 (1963).

Habit: shrub to 2.5 m. **Guild:** pioneer. **Habitat:** roadsides and villages. **Distribution:** pantropical. **Star:** excluded. **Alt. range:** 1-200m (1). **Flowering:** Feb (1), Apr (1). **Specimens: -** *Etinde:* Tchouto P. 2. *Mabeta-Moliwe:* Sunderland T. C. H. 1234

Conyza attenuata DC.
Fragm. Flor. Geobot. 36(1) Suppl.1: 89 (1991).
   Conyza persicifolia (Benth.) Oliv. & Hiern, F.W.T.A. 2: 254 (1963).

Habit: herb to 2 m. **Distribution:** tropical Africa. **Chorology:** Afromontane. **Star:** green. **Alt. range:** 1000-1200m (1). **Specimens: -** *Eastern slopes:* Dalziel J. M. 8188

Conyza bonariensis (L.) Cronq.
Fragm. Flor. Geobot. 36(1) Suppl.1: 59 (1991).
   Erigeron bonariensis L., F.W.T.A. 2: 253 (1963).
   Conyza sumatrensis (Retz.) E.Walker, Kirkia 10: 43 (1975).
   Erigeron floribundus (Kunth) Sch.Bip., F.W.T.A. 2: 253 (1963).

Habit: herb. **Habitat:** grassland. **Distribution:** pantropical. **Chorology:** Montane. **Star:** excluded. **Alt. range:** 1800-2000m

(2). **Flowering:** Oct (1). **Specimens: -** *Eastern slopes:* Thomas D. W. 9420. *No locality:* Boughey A. S. GC 7016

Conyza clarenceana (Hook.f.) Oliv. & Hiern
F.W.T.A. 2: 254 (1963).
   Conyza theodori R.E.Fr., Fragm. Flor. Geobot. 36(1) Suppl.1: 88 (1991).

Habit: herb. **Distribution:** Bioko, West Cameroon, Ethiopia, Zaire, Zambia, Uganda & Kenya. **Chorology:** Afromontane. **Star:** black. **Alt. range:** 1200-1400m (1). **Specimens: -** *No locality:* Maitland T. D. 1098a

Conyza gouanii (L.) Willd.
F.W.T.A. 2: 254 (1963); Fragm. Flor. Geobot. 36(1) Suppl.1: 87 (1991).

Habit: herb. **Distribution:** tropical Africa. **Chorology:** Afromontane. **Star:** green. **Alt. range:** 1200-1400m (1). **Specimens: -** *No locality:* Maitland T. D. 1098

Conyza steudelii Sch.Bip. ex A.Rich.
F.W.T.A. 2: 254 (1963); Fragm. Flor. Geobot. 36(1) Suppl.1: 78 (1991).

Habit: herb. **Habitat:** forest-grassland transition. **Distribution:** W. Cameroon, Congo (Kinshasa), Ethiopia, Uganda, Tanzania, Kenya and Zambia. **Chorology:** Afromontane. **Star:** green. **Alt. range:** 2400-2600m (1). **Specimens: -** *No locality:* Keay R. W. J. FHI 28632

Conyza subscaposa O.Hoffm.
F.W.T.A. 2: 254 (1963); Fragm. Flor. Geobot. 36(1) Suppl.1: 81 (1991).

Habit: herb. **Distribution:** W. Cameroon, Congo (Kinshasa), Ethiopia, Uganda, Sudan, Tanzania, Kenya, Zambia, Malawi, Zimbabwe and Mozambique. **Chorology:** Afromontane. **Star:** green. **Specimens: -** *No locality:* Maitland T. D. 675

Crassocephalum bougheyanum C.D.Adams
F.W.T.A. 2: 248 (1963).

Habit: herb. **Habitat:** grassland. **Distribution:** Mount Cameroon and Bioko. **Chorology:** Western Cameroon Uplands (montane). **Star:** black. **Alt. range:** 1800-2000m (1). **Specimens: -** *No locality:* Mann G. 1931

Crassocephalum crepidioides (Benth.) S.Moore
F.W.T.A. 2: 246 (1963); Kew Bull. 41: 908 (1986); Fragm. Flor. Geobot. 36(1) Suppl.1: 354 (1991).

Habit: herb. **Guild:** pioneer. **Habitat:** lava flows. **Distribution:** palaeotropical. **Star:** excluded. **Alt. range:** 1-200m (1), 801-1000m (1). **Specimens: -** *Etinde:* Dawson S. 42. *No locality:* Dunlap 150

Crassocephalum gracile (Hook.f.) Milne-Redh.
F.W.T.A. 2: 248 (1963).

Habit: herb. **Habitat:** grassland and forest-grassland transition. **Distribution:** Guinea, Mount Cameroon and Bioko. **Chorology:** Guineo-Congolian (montane). **Star:** green. **Alt. range:** 1800-

2000m (1), 2001-2200m (2), 2601-2800m (1). **Flowering:** Sep (1), Oct (2). **Specimens:** - *Eastern slopes:* Cheek M. 3677; Thomas D. W. 9303. *Etinde:* Tekwe C. 274. *No locality:* Mann G. 1317

Crassocephalum montuosum (S.Moore) Milne-Redh.
F.W.T.A. 2: 248 (1963); Kew Bull. 41: 906 (1986); Fragm. Flor. Geobot. 36(1) Suppl.1: 331 (1991).

Habit: herb. **Habitat:** forest and forest-grassland transition. **Distribution:** W. Cameroon, Congo (Kinshasa), CAR, Sudan, Uganda, East Africa and Madagascar. **Chorology:** Afromontane. **Star:** green. **Alt. range:** 200-400m (1), 601-800m (1), 801-1000m (1), 1201-1400m (1), 1401-1600m (1), 2001-2200m (1). **Flowering:** Oct (5). **Fruiting:** Oct (1). **Specimens:** - *Eastern slopes:* Tchouto P. 365. *Etinde:* Cheek M. 3752; Tchouto P. 453; Thomas D. W. 9363; Watts J. 568. *No locality:* Maitland T. D. 268

Crassocephalum vitellinum (Benth.) S.Moore
F.W.T.A. 2: 248 (1963); Kew Bull. 41: 907 (1986); Fragm. Flor. Geobot. 36(1) Suppl.1: 339 (1991).

Habit: herb. **Distribution:** SE Nigeria and W. Cameroon to East Africa. **Chorology:** Afromontane. **Star:** green. **Alt. range:** 1000-1200m (1). **Specimens:** - *No locality:* Akpabla G. K. GC 10494

Crassocephalum x picridifolium (DC.) S.Moore
F.W.T.A. 2: 248 (1963).

Habit: herb. **Habitat:** forest. **Distribution:** Gambia to Bioko, Cameroon, Congo (Kinshasa), Angola and Sudan. **Chorology:** Guineo-Congolian. **Star:** green. **Alt. range:** 800-1000m (1). **Flowering:** Oct (1). **Fruiting:** Oct (1). **Specimens:** - *Etinde:* Cheek M. 3760

Crepis hypochoeridea (DC.) Thell.
Fl. Zamb. 6(1): 210 (1992).
Crepis newii Oliv. & Hiern subsp. *kundensis* (Babc.) Babc., F.W.T.A. 2: 294 (1963).
Crepis cameroonica Babc. ex. Hutch. & Dalziel, F.W.T.A. 2: 294 (1963).
Crepis oliveriana (Kuntze) C.Jeffrey, Kew Bull. 18: 458 (1966).

Habit: herb. **Habitat:** grassland and forest-grassland transition. **Distribution:** Nigeria, Mount Cameroon, Zaire, Kenya, Tanzania, Zambia, Zimbabwe, Malawi, Angola, Mozambique, South Africa. **Chorology:** Afromontane. **Star:** black. **Alt. range:** 2000-2200m (2), 2401-2600m (1). **Flowering:** Oct (1), Dec (1). **Specimens:** - *Eastern slopes:* Sunderland T. C. H. 1042; Thomas D. W. 9414. *No locality:* Mann G. 1318

Dichrocephala chrysanthemifolia (Blume) DC. var. chrysanthemifolia
F.W.T.A. 2: 256 (1963); Fragm. Flor. Geobot. 36(1) Suppl.1: 24 (1991).

Habit: herb. **Habitat:** grassland and forest-grassland transition. **Distribution:** palaeotropical. **Chorology:** Montane. **Star:** excluded. **Alt. range:** 1800-2000m (1), 2001-2200m (1), 2201-2400m (1), 2401-2600m (1), 2601-2800m (1). **Flowering:** Oct (4). **Specimens:** - *Eastern slopes:* Cheek M. 3631; Tchouto P. 304, 330; Thomas D. W. 9306. *No locality:* Mann G. 1267

Dichrocephala integrifolia (L.f.) Kuntze subsp. integrifolia
F.W.T.A. 2: 256 (1963); Fragm. Flor. Geobot. 36(1) Suppl.1: 22 (1991).

Habit: herb. **Habitat:** forest and forest-grassland transition. **Distribution:** palaeotropical. **Chorology:** Montane. **Star:** excluded. **Alt. range:** 1400-1600m (2), 2201-2400m (1). **Flowering:** Oct (2). **Specimens:** - *Eastern slopes:* Tchouto P. 358; Thomas D. W. 9331. *No locality:* Mann G. 1268

Elephantopus mollis Kunth
F.W.T.A. 2: 269 (1963); Fl. Zamb. 6(1): 188 (1992).

Habit: herb. **Distribution:** tropical America, introduced in Africa and Asia. **Star:** excluded. **Alt. range:** 800-1000m (1). **Specimens:** - *Eastern slopes:* Dunlap 106

Eleutheranthera ruderalis (Sw.) Sch.Bip.
F.W.T.A. 2: 236 (1963); Fragm. Flor. Geobot. 36(1) Suppl.1: 182 (1991).

Habit: herb. **Guild:** pioneer. **Distribution:** pantropical. **Star:** excluded. **Specimens:** - *Etinde:* Boughey A. S. 1199; *Mabeta-Moliwe:* Cheek M. 3549

Emilia coccinea (Sims) G.Don
F.W.T.A. 2: 244 (1963); Kew Bull. 41: 913 (1986); Fragm. Flor. Geobot. 36(1) Suppl.1: 376 (1991).

Habit: herb. **Guild:** pioneer. **Habitat:** forest and farmbush. **Distribution:** tropical Africa. **Chorology:** Afromontane. **Star:** green. **Alt. range:** 1-200m (2), 201-400m (1), 601-800m (1), 1201-1400m (1), 1401-1600m (1). **Flowering:** Apr (2), Oct (1), Nov (1). **Specimens:** - *Eastern slopes:* Preuss P. R. 1156; Tchouto P. 367. *Etinde:* Acworth J. M. 30; Kwangue A. 141. *Mabeta-Moliwe:* Sunderland T. C. H. 1233. *Mokoko River F.R.:* Acworth J. M. 72

Galinsoga parviflora Cav.
F.W.T.A. 2: 230 (1963); Fragm. Flor. Geobot. 36(1) Suppl.1: 121 (1991).

Habit: herb. **Distribution:** South America, introduced in W. Cameroon. **Chorology:** Montane. **Star:** excluded. **Alt. range:** 1400-1600m (1). **Specimens:** - *No locality:* Dunlap 119

Galinsoga quadriradiata Ruiz. & Pav.
Linneus C., Syst. Veg.: 198 (1797).
Galinsoga ciliata (Raf.) Blake, F.W.T.A. 2: 230 (1963); Fragm. Flor. Geobot. 36(1) Suppl.1: 123 (1991).

Habit: herb. **Habitat:** forest-grassland transition. **Distribution:** South America, introduced in W. Cameroon. **Chorology:** Montane. **Star:** excluded. **Alt. range:** 1400-1600m (1), 2201-2400m (1). **Flowering:** Oct (1). **Specimens:** - *Eastern slopes:* Thomas D. W. 9330. *No locality:* Dunlap 118

Helichrysum biafranum Hook.f.
F.W.T.A. 2: 264 (1963).

Habit: herb. **Habitat:** forest-grassland transition. **Distribution:** Mount Cameroon. **Chorology:** Endemic (montane). **Star:** black. **Alt. range:** 2000-2200m (1). **Specimens:** - *No locality:* Mann G. 1934 (partly)

Helichrysum cameroonense Hutch. & Dalziel

F.W.T.A. 2: 264 (1963).

Habit: herb. Habitat: grassland and forest-grassland transition.
Distribution: W. Cameroon. Chorology: Western Cameroon
Uplands (montane). Star: gold. Alt. range: 2000-2200m (1),
2201-2400m (1), 2601-2800m (1), 3801-4000m (1). Flowering:
Oct (2). Fruiting: Oct (1). Specimens: - *Eastern slopes:* Thomas
D. W. 9317, 9336, 9396. *No locality:* Mann G. 1306

Helichrysum foetidum (L.) Moench

F.W.T.A. 2: 264 (1963).

Habit: herb to 1.5 m. Habitat: grassland. Distribution: SE
Nigeria, W. Cameroon, Bioko, Congo (Kinshasa), S. Tomé, Sudan
and Ethiopia. Chorology: Afromontane. Star: green. Alt. range:
2200-2400m (1), 2601-2800m (1). Flowering: Oct (1).
Specimens: - *Eastern slopes:* Thomas D. W. 9412. *No locality:*
Mann G. 1916

Helichrysum forskahlii (J.F.Gmel.) Hilliard & B.L.Burtt

F.W.T.A. 2: 264 (1963); Fl. Afr. Cent. Compositae: 88 (1989).
Helichrysum cymosum (L.) Less. subsp. *fruticosum* (Forssk.)
Hedberg, Symb. Bot. Ups. 15: 203 (1957).
Helichrysum cymosum sensu Adams, F.W.T.A. 2: 264 (1963).

Habit: herb. Habitat: forest-grassland transition. Distribution:
Nigeria to Zimbabwe and Yemen. Chorology: Afromontane. Star:
green. Alt. range: 2000-2200m (1), 2201-2400m (2), 2601-2800m
(1). Flowering: Oct (4). Specimens: - *Eastern slopes:* Tchouto P.
305, 344; Thomas D. W. 9388, 9415

Helichrysum globosum Sch.Bip. ex A.Rich.

F.W.T.A. 2: 264 (1963); Fl. Afr. Cent. Compositae: 103 (1989).

Habit: herb. Habitat: forest-grassland transition. Distribution:
W. Cameroon, Bioko, Congo (Kinshasa), Ethiopia, Uganda,
Kenya, Tanzania, Angola and Zimbabwe. Chorology:
Afromontane. Star: green. Alt. range: 2000-2200m (2), 2201-
2400m (1), 2401-2600m (1). Flowering: May (1), Oct (2).
Specimens: - *Eastern slopes:* Sunderland T. C. H. 1430; Tchouto
P. 315; Watts J. 441. *No locality:* Mann G. 1912

Helichrysum mannii Hook.f.

F.W.T.A. 2: 264 (1963).

Habit: herb. Habitat: forest-grassland transition. Distribution:
Mount Cameroon and Bioko. Chorology: Western Cameroon
Uplands (montane). Star: black. Alt. range: 2400-2600m (1).
Flowering: Nov (1). Specimens: - *Eastern slopes:* Banks H. 100.
No locality: Mann G. 1937

Helichrysum stenopterum DC

Helichrysum odoratissimum sensu auct non (L.) Sweet F.W.T.A.
2: 264 (1963); Fl. Afr. Cent. Compositae: 76 (1989).

Habit: herb. Habitat: forest-grassland transition. Distribution:
Tropical Africa. Chorology: Afromontane. Star: green. Alt.
range: 2000-2200m (2), 2401-2600m (1). Flowering: May (1),
Oct (1). Specimens: - *Eastern slopes:* Sunderland T. C. H. 1433;
Thomas D. W. 9413. *No locality:* Mann G. 1228

Inula mannii (Hook.f.) Oliv. & Hiern

F.W.T.A. 2: 259 (1963); Fl. Afr. Cent. Compositae: 44 (1989).

Habit: herb to 3 m. Habitat: forest-grassland transition.
Distribution: W. Cameroon, Congo (Kinshasa), Ethiopia, Malawi,
Uganda, Kenya and Tanzania. Chorology: Afromontane. Star:
blue. Alt. range: 2000-2200m (1). Specimens: - *No locality:*
Mann G. 1314

Lactuca glandulifera Hook.f.

F.W.T.A. 2: 293 (1963); Bull. Jard. Bot. Nat. Belg. 52: 370 (1982).
Lactuca glandulifera Hook.f. var. *calva* R.E.Fr., F.W.T.A. 2:
293 (1963).

Habit: herb. Guild: pioneer. Habitat: forest and forest-grassland
transition. Distribution: tropical Africa. Chorology:
Afromontane. Star: green. Alt. range: 1800-2000m (1), 2001-
2200m (1). Specimens: - *No locality:* Dunlap 47; Mann G. 1240

Lactuca inermis Forssk.

Kew Bull. 39: 132 (1984).
Lactuca capensis Thunb., F.W.T.A. 2: 293 (1963).
Habit: herb to 2 m. Guild: pioneer. Habitat: grassland.
Distribution: tropical and subtropical Africa. Chorology:
Afromontane. Star: green. Alt. range: 1400-1600m (1), 1801-
2000m (2), 2401-2600m (2), 2601-2800m (1). Flowering: May
(1), Oct (3), Dec (1). Fruiting: Oct (1). Specimens: - *Eastern
slopes:* Cheek M. 3663; Sunderland T. C. H. 1039, 1428; Thomas
D. W. 9289, 9298. *No locality:* Mann G. 1300

Melanthera scandens (Schumach. & Thonn.) Roberty

F.W.T.A. 2: 240 (1963); Fragm. Flor. Geobot. 36(1) Suppl.1: 205
(1991).

Habit: herbaceous climber. Guild: pioneer. Habitat: forest.
Distribution: tropical Africa. Chorology: Afromontane. Star:
green. Alt. range: 1-200m (3), 401-600m (1), 601-800m (1),
1401-1600m (2), 1601-1800m (2). Flowering: Apr (1), May (1),
Jul (1), Sep (1), Oct (3), Nov (1). Specimens: - *Eastern slopes:*
Tchouto P. 369; Tekwe C. 324; Thomas D. W. 9473. *Etinde:*
Tekwe C. 46; Watts J. 560. *Limbe Botanic Garden:* Wheatley J. I.
650. *Mabeta-Moliwe:* Cheek M. 3540; Sunderland T. C. H. 1358.
No locality: Dunlap 86

Microglossa pyrifolia (Lam.) Kuntze

F.W.T.A. 2: 251 (1963).

Habit: shrub to 3 m. Guild: pioneer. Habitat: forest.
Distribution: palaeotropical. Star: excluded. Alt. range: 800-
1000m (1). Specimens: - *Eastern slopes:* Boughey A. S. GC 7026

Mikania chenopodifolia Willd.

Fragm. Flor. Geobot. 36(1) Suppl.1: 460 (1991).

Habit: scrambling herb. Habitat: forest and Aframomum thicket.
Distribution: palaeotropical. Star: excluded. Alt. range: 1-200m
(1), 601-800m (1). Flowering: Feb (1), Oct (1). Fruiting: Feb (1).
Specimens: - *Etinde:* Nkeng P. 21; Watts J. 553

Mikania cordata (Burm.f.) B.L.Robinson

F.W.T.A. 2: 286 (1963).

Habit: scrambling herb. Guild: pioneer. Habitat: forest and
forest-grassland transition. Distribution: palaeotropical.
Chorology: Montane. Star: excluded. Alt. range: 2000-2200m
(1). Specimens: - *No locality:* Mann G. 1326

Mikania microptera DC.
Fragm. Flor. Geobot. 36(1) Suppl.1: 463 (1991).

Habit: herb. **Guild:** pioneer. **Habitat:** forest-grassland transition. **Distribution:** tropical Africa and South America. **Chorology:** Montane. **Star:** excluded. **Alt. range:** 1600-1800m (1). **Specimens:** - *Eastern slopes:* Thomas D. W. 9495

Mikaniopsis maitlandii C.D.Adams
F.W.T.A. 2: 243 (1963).

Habit: herbaceous climber. **Guild:** pioneer. **Habitat:** forest. **Distribution:** Mount Cameroon. **Chorology:** Endemic (montane). **Star:** black. **Alt. range:** 2200-2400m (2). **Specimens:** - *Eastern slopes:* Brenan J. P. M. 9513. *No locality:* Maitland T. D. 1001

Mikaniopsis tedliei (Oliv. & Hiern) C.D.Adams
F.W.T.A. 2: 243 (1963); Fragm. Flor. Geobot. 36(1) Suppl.1: 441 (1991).

Habit: herbaceous climber. **Guild:** pioneer. **Habitat:** forest. **Distribution:** Ghana and W. Cameroon. **Chorology:** Guineo-Congolian (montane). **Star:** blue. **Alt. range:** 2200-2400m (1). **Specimens:** - *Eastern slopes:* Boughey A. S. GC 10590

Pseudognaphalium luteo-album (L.) Hilliard & Burtt
Fl. Mascar. 109: 74 (1993).
Gnaphalium luteo-album L., F.W.T.A. 2: 266 (1963).

Habit: herb. **Distribution:** tropical and subtropical Africa. **Chorology:** Afromontane. **Star:** green. **Alt. range:** 800-1000m (1). **Specimens:** - *Eastern slopes:* Dalziel J. M. 8229

Senecio burtonii Hook.f.
F.W.T.A. 2: 250 (1963).

Habit: herb. **Habitat:** forest-grassland transition. **Distribution:** W. Cameroon. **Chorology:** Western Cameroon Uplands (montane). **Star:** gold. **Alt. range:** 2000-2200m (1), 2201-2400m (2), 2401-2600m (1), 2801-3000m (1). **Flowering:** Oct (3). **Fruiting:** Oct (1). **Specimens:** - *Eastern slopes:* Tchouto P. 348; Thomas D. W. 9369; Watts J. 452. *Etinde:* Thomas D. W. 9361. *No locality:* Mann G. 1246

Senecio purpureus L.
Kew Bull. 41: 904 (1986).
Senecio clarenceanus Hook.f., F.W.T.A. 2: 250 (1963).

Habit: herb. **Habitat:** grassland and forest-grassland transition. **Distribution:** Mount Cameroon, Bioko, Congo (Kinshasa), Angola, Tanzania and southern Africa. **Chorology:** Afromontane. **Star:** green. **Alt. range:** 2000-2200m (1), 2201-2400m (2). **Flowering:** Oct (2). **Specimens:** - *Eastern slopes:* Thomas D. W. 9316, 9371. *No locality:* Mann G. 1237

Solanecio biafrae (Oliv. & Hiern) C.Jeffrey
Kew Bull. 41: 922 (1986).
Crassocephalum biafrae (Oliv. & Hiern) S.Moore, F.W.T.A. 2: 246 (1963).

Habit: herbaceous climber. **Guild:** pioneer. **Habitat:** forest-grassland transition. **Distribution:** Sierra Leone to Cameroon, Congo (Kinshasa) and Uganda. **Chorology:** Guineo-Congolian

(montane). **Star:** green. **Alt. range:** 2000-2200m (1). **Specimens:** - *No locality:* Mann G. 1325

Solanecio mannii (Hook.f.) C.Jeffrey
Kew Bull. 41: 922 (1986); Fragm. Flor. Geobot. 36(1) Suppl.1: 418 (1991).
Crassocephalum mannii (Hook.f.) Milne-Redh., F.W.T.A. 2: 246 (1963).

Habit: shrub to 3 m. **Habitat:** forest. **Distribution:** SE Nigeria, Bioko, Cameroon, Congo (Kinshasa), Angola, Rwanda, Sudan, Zimbabwe and East Africa. **Chorology:** Afromontane. **Star:** green. **Alt. range:** 800-1000m (1), 1601-1800m (1), 2001-2200m (1). **Flowering:** Feb (1), Oct (1). **Specimens:** - *Eastern slopes:* Tchouto P. 320. *Etinde:* Tchouto P. 10. *No locality:* Mann G. 1935

Sonchus angustissimus Hook.f.
F.W.T.A. 2: 296 (1963); Bot. Not. 127: 407 (1974).

Habit: herb to 3 m. **Guild:** pioneer. **Habitat:** grassland. **Distribution:** Nigeria to East Africa. **Chorology:** Afromontane. **Star:** green. **Alt. range:** 1800-2000m (1), 2001-2200m (1). **Flowering:** Oct (1). **Fruiting:** Oct (1). **Specimens:** - *Eastern slopes:* Thomas D. W. 9304. *No locality:* Mann G. 1311

Sonchus asper (L.) Hill
F.W.T.A. 2: 296 (1963); Fl. Zamb. 6(1): 228 (1992).

Habit: herb. **Guild:** pioneer. **Distribution:** pantropical. **Star:** excluded. **Alt. range:** 800-1000m (1). **Specimens:** - *No locality:* Boughey A. S. GC 7032

Sparganophorus sparganophora (L.) C.Jeffrey
Kew Bull. 43: 271 (1988); Fragm. Flor. Geobot. 37: 379 (1992).
Struchium sparganophora (L.) Kuntze, F.W.T.A. 2: 269 (1963).

Habit: herb. **Guild:** pioneer. **Distribution:** pantropical. **Star:** excluded. **Specimens:** - *Etinde:* Maitland T. D. 358

Synedrella nodiflora Gaertn.
F.W.T.A. 2: 229 (1963).

Habit: herb. **Guild:** pioneer. **Habitat:** forest and farmbush. **Distribution:** pantropical. **Chorology:** Montane. **Star:** excluded. **Alt. range:** 1-200m (1), 801-1000m (1), 1401-1600m (1). **Flowering:** Oct (1), Nov (1). **Specimens:** - *Eastern slopes:* Maitland T. D. 278; Tchouto P. 361. *Limbe Botanic Garden:* Wheatley J. I. 653; *Mabeta-Moliwe:* Cheek M. 3550

Tithonia diversifolia (Hemsley) A. Gray

Specimens: - *Eastern Slopes:* no specimens but a common montane farm weed species.

Tridax procumbens L.

Specimens: - *Eastern slopes:* at c. 100 m. from Mapanja to Buea, but never collected.

Vernonia biafrae Oliv. & Hiern
F.W.T.A. 2: 276 (1963); Fl. Zamb. 6(1): 80 (1992).
Vernonia tufnelliae S.Moore, F.W.T.A. 2: 276 (1963).

Habit: woody climber. Guild: pioneer. Habitat: forest. Distribution: tropical Africa. Chorology: Afromontane. Star: green. Alt. range: 600-800m (1), 801-1000m (2). Flowering: Mar (2). Specimens: - *Eastern slopes:* Tchouto P. 41. *Etinde:* Mbani J. M. 56. *No locality:* Dalziel J. M. 8187; Mildbraed G. W. J. 10800

Vernonia blumeoides Hook.f.
F.W.T.A. 2: 281 (1963).

Habit: shrub to 1.5 m. Habitat: grassland and forest-grassland transition. Distribution: Nigeria and Cameroon. Chorology: Lower Guinea (montane). Star: blue. Alt. range: 1800-2000m (1), 2001-2200m (2), 2201-2400m (1), 2401-2600m (1). Flowering: Oct (4). Specimens: - *Eastern slopes:* Cheek M. 3680; Tchouto P. 284; Thomas D. W. 9291, 9340. *No locality:* Mann G. 1241

Vernonia cinerea (L.) Less.
F.W.T.A. 2: 283 (1963); Fl. Zamb. 6(1): 143 (1992).

Habit: herb to 1.5 m. Guild: pioneer. Distribution: palaeotropical. Star: excluded. Specimens: - *Etinde:* Morton J. K. GC 7130

Vernonia conferta Benth.
F.W.T.A. 2: 277 (1963); Kew Bull. 43: 220 (1988); Keay R.W.J., Trees of Nigeria: 431 (1989); Hawthorne W., F.G.F.T. Ghana: 141 (1990).

Habit: tree to 10 m. Guild: pioneer. Habitat: forest clearings and farmbush. Distribution: Guinea to Bioko, Angola, Uganda and Sudan. Chorology: Guineo-Congolian. Star: green. Alt. range: 800-1000m (1). Specimens: - *Eastern slopes:* Maitland T. D. 190

Vernonia hymenolepis A.Rich.
Kew Bull. 43: 237 (1988).
Vernonia leucocalyx O.Hoffm. var. *acuta* C.D.Adams, F.W.T.A. 2: 276 (1963).
Vernonia leucocalyx O.Hoffm. var. *leucocalyx*, F.W.T.A. 2: 276 (1963).
Vernonia insignis (Hook.f.) Oliv. & Hiern, F.W.T.A. 2: 276 (1963).
Vernonia calvoana (Hook.f.) Hook.f. subsp. *calvoana* var. *acuta* (C.D.Adams) C.Jeffrey, F.W.T.A. 2: 276 (1963); Kew Bull. 43: 237 (1988).
Vernonia calvoana (Hook.f.) Hook.f. subsp. *calvoana* var. *calvoana*, F.W.T.A. 2: 276 (1963); Kew Bull. 43: 237 (1988).
Vernonia calvoana (Hook.f.) Hook.f. subsp. *calvoana* var. *microcephala* C.D.Adams, F.W.T.A. 2: 276 (1963); Kew Bull. 43: 237 (1988).

Habit: herb. Habitat: forest-grassland transition. Distribution: Cameroon, Uganda, Sudan, Ethiopia and Kenya. Chorology: Afromontane. Star: green. Alt. range: 1400-1600m (1), 1601-1800m (2), 2001-2200m (1), 2201-2400m (1). Flowering: Mar (1), Sep (1), Oct (3). Specimens: - *Eastern slopes:* Cheek M. 3644; Nkeng P. 60; Tekwe C. 325; Thomas D. W. 9327; Watts J. 444. *No locality:* Mann G. 1238, 1925

Vernonia ituriensis Muschl.
Kew Bull. 43: 231 (1988).
Vernonia glabra (Steetz) Vatke var. *hillii* (Hutch. & Dalziel) C.D.Adams, F.W.T.A. 2: 281 (1963).

Habit: herb to 1.5 m. Distribution: Cameroon, Rwanda, Burundi, Congo (Kinshasa), Sudan, Ethiopia, Kenya and Tanzania. Chorology: Afromontane. Star: green. Alt. range: 800-1000m

(1). Specimens: - *No locality:* Hill A. W. ?; Mildbraed G. W. J. 10799

Vernonia myriantha Hook.f.
F.W.T.A. 2: 277 (1963); Kew Bull. 43: 219 (1988); Fl. Zamb. 6(1): 82 (1992).
Vernonia subuligera O.Hoffm., F.W.T.A. 2: 279 (1963).

Habit: tree or shrub to 6 m. Habitat: forest. Distribution: Sierra Leone to Ethiopia and South Africa. Chorology: Tropical Africa. Star: black. Alt. range: 1000-1200m (1), 1601-1800m (1). Flowering: Oct (1). Specimens: - *Etinde:* Tchouto P. 422. *No locality:* Mann G. 1913

Vernonia stellulifera (Benth.) C.Jeffrey
Fl. Zamb. 6(1): 142 (1992).
Triplotaxis stellulifera (Benth.) Hutch., F.W.T.A. 2: 269 (1963).
Habit: herb. Habitat: forest and farmbush. Distribution: Guinea to Bioko, Cameroon, Gabon, Congo (Kinshasa), Uganda, Zambia and Angola. Chorology: Guineo-Congolian (montane). Star: green. Alt. range: 1000-1200m (1). Specimens: - *Eastern slopes:* Maitland T. D. 34; *Mabeta-Moliwe:* Cheek 3545

CONNARACEAE

C. Jongkind, F.J. Breteler & R.H.M.J. Lemmens (WAG)

Agelaea pentagyna (Lam.) Baill.
Agric. Univ. Wag. Papers 89(6): 144 (1989); Fl. Gabon 33: 34 (1992).
Agelaea hirsuta De Wild., F.W.T.A. 1: 745 (1958).
Agelaea dewevrei De Wild. & T.Durand, F.W.T.A. 1: 746 (1958).
Agelaea floccosa Schellenb., F.W.T.A. 1: 746 (1958).
Agelaea grisea Schellenb., F.W.T.A. 1: 746 (1958).
Agelaea obliqua (P.Beauv.) Baill., F.W.T.A. 1: 745 (1958).
Agelaea preussii Gilg, F.W.T.A. 1: 746 (1958).
Agelaea pseudobliqua Schellenb., F.W.T.A. 1: 746 (1958).

Habit: woody climber. Habitat: forest. Distribution: tropical Africa. Chorology: Afromontane. Star: green. Alt. range: 200-400m (1), 1401-1600m (2).Fruiting: Sep (2). Specimens: - *Eastern slopes:* Tekwe C. 313. *Etinde:* Ejiofor M. C. FHI 29365; Maitland T. D. 634; Preuss P. R. 1116; Tekwe C. 225. *Mokoko River F.R.:* Thomas D. W. 10110

Cnestis corniculata Lam.
F.W.T.A. 1: 743 (1958); Agric. Univ. Wag. Papers 89(6): 181 (1989); Fl. Gabon 33: 46 (1992).
Cnestis aurantiaca Gilg, F.W.T.A. 1: 743 (1958).
Cnestis congolana De Wild., F.W.T.A. 1: 743 (1958).
Cnestis grisea Baker, F.W.T.A. 1: 743 (1958).
Cnestis longiflora Schellenb., F.W.T.A. 1: 743 (1958).
Cnestis sp. A sensu Hepper, F.W.T.A. 1: 743 (1958).

Habit: woody climber. Guild: light demander. Habitat: forest and forest gaps. Distribution: tropical Africa. Star: excluded. Alt. range: 1-200m (19), 201-400m (1). Flowering: Mar (2), Apr (6), May (2), Nov (1). Fruiting: Apr (3), May (3), Jun (1), Jul (1), Aug (1), Nov (3), Dec (1). Specimens: - *Bambuko F.R.:* Watts J. 682. *Etinde:* Akogo M. 154; Cheek M. 5496; Kwangue A. 58; Maitland T. D. 602; Tekwe C. 57. *Mabeta-Moliwe:* Baker W. J. 303; Cable S. 543; Jaff B. 117; Ndam N. 176; Plot Voucher W 129, W 148, W 396, W 820, W 927, W 981; Sunderland T. C. H. 1118, 1222;

Tchouto P. 195; Wheatley J. I. 308; von Rege I. 10. *Mokoko River F.R.*: Acworth J. M. 338; Akogo M. 294; Ekema N. 1063; Tchouto P. 1138. *Onge:* Harris D. J. 3827; Watts J. 934. *Southern Bakundu F.R.*: Brenan J. P. M. 9419

Cnestis ferruginea Vahl ex DC.

F.W.T.A. 1: 743 (1958); Agric. Univ. Wag. Papers 89(6): 196 (1989); Fl. Gabon 33: 50 (1992).

Habit: woody climber to 5 m. **Guild:** pioneer. **Habitat:** forest. **Distribution:** Gambia to Cameroon, Gabon, Congo (Brazzaville), Congo (Kinshasa), Angola, CAR and Sudan. **Chorology:** Guineo-Congolian. **Star:** green. **Alt. range:** 1-200m (1). **Flowering:** Mar (1). **Fruiting:** Mar (1). **Specimens: -** *Etinde:* Kalbreyer W. 8. *Mokoko River F.R.:* Tchouto P. 618

Cnestis mannii (Baker) Schellenb.

F.W.T.A. 1: 743 (1958); Agric. Univ. Wag. Papers 89(6): 208 (1989); Fl. Gabon 33: 60 (1992).
   Cnestis tomentosa Hepper, F.W.T.A. 1: 743 (1958).

Habit: woody climber. **Habitat:** forest and farmbush. **Distribution:** Nigeria, Cameroon, Gabon, Congo (Kinshasa) and Angola. **Chorology:** Guineo-Congolian. **Star:** green. **Alt. range:** 200-400m (1), 601-800m (1). **Flowering:** Apr (1). **Fruiting:** Feb (1). **Specimens: -** *Etinde:* Wheatley J. I. 3. *Mabeta-Moliwe:* Wheatley J. I. 122. *Southern Bakundu F.R.:* Brenan J. P. M. 9328

Connarus griffonianus Baill.

F.W.T.A. 1: 748 (1958); Agric. Univ. Wag. Papers 89(6): 252 (1989); Fl. Gabon 33: 76 (1992).

Habit: woody climber. **Habitat:** forest. **Distribution:** Nigeria, Cameroon, Bioko, Gabon, Congo (Brazzaville), Congo (Kinshasa), Angola and CAR. **Chorology:** Guineo-Congolian. **Star:** green. **Alt. range:** 1-200m (2). **Flowering:** Apr (1), Dec (1). **Fruiting:** Apr (1), Dec (1). **Specimens: -** *Etinde:* Mildbraed G. W. J. 10584. *Mabeta-Moliwe:* Cheek M. 5759; Wheatley J. I. 91

Hemandradenia mannii Stapf

F.W.T.A. 1: 749 (1958); Agric. Univ. Wag. Papers 89(6): 279 (1989); Fl. Gabon 33: 86 (1992).

Habit: tree or shrub to 15 m. **Guild:** shade-bearer. **Habitat:** forest. **Distribution:** Ivory Coast to Cameroon, Congo (Kinshasa) and CAR. **Chorology:** Guineo-Congolian. **Star:** green. **Alt. range:** 1-200m (2), 201-400m (1).**Fruiting:** Mar (1), Apr (1), May (1). **Specimens: -** *Bambuko F.R.:* Tchouto P. 557. *Mabeta-Moliwe:* Jaff B. 106. *Mokoko River F.R.:* Ekema N. 864

Jollydora duparquetiana (Baill.) Pierre

F.W.T.A. 1: 749 (1958); Agric. Univ. Wag. Papers 89(6): 284 (1989); Fl. Gabon 33: 90 (1992).

Habit: tree to 8 m. **Habitat:** forest. **Distribution:** SE Nigeria, Cameroon, Gabon, Congo (Brazzaville), Congo (Kinshasa) and Angola. **Chorology:** Guineo-Congolian. **Star:** green. **Alt. range:** 1-200m (22), 201-400m (10), 401-600m (1). **Flowering:** Mar (5). **Fruiting:** Jan (2), Mar (3), Apr (4), May (10), Jun (3), Jul (1), Oct (6), Nov (1). **Specimens: -** *Bambuko F.R.:* Watts J. 635, 665. *Etinde:* Kwangue A. 65, 99; Mbani J. M. 1, 84; Nkeng P. 2; Tchouto P. 93; Thomas D. W. 9149; Winkler H. J. P. 406. *Mabeta-Moliwe:* Jaff B. 254; Nguembock F. 60; Plot Voucher W 525, W 916, W 996; Sunderland T. C. H. 1110, 1363; Wheatley J. I. 261; von Rege I. 13. *Mokoko River F.R.:* Acworth J. M. 143; Akogo M. 251; Ekema N. 905, 1013, 1136; Fraser P. 328, 397; Mbani J. M. 319; Ndam N. 1280, 1287; Pouakouyou D. 70; Tchouto P. 598;

Watts J. 1063, 1158. *Onge:* Akogo M. 53, 63; Harris D. J. 3785; Ndam N. 781; Watts J. 758, 846; Wheatley J. I. 739

Jollydora glandulosa Schellenb.

F.W.T.A. 1: 749 (1958); Agric. Univ. Wag. Papers 89(6): 290 (1989).

Habit: tree to 5 m. **Habitat:** forest. **Distribution:** SE Nigeria and SW Cameroon. **Chorology:** Western Cameroon. **Star:** gold. **Specimens: -** *Etinde:* Mildbraed G. W. J. 10607

Manotes griffoniana Baill.

Agric. Univ. Wag. Papers 89(6): 302 (1989); Fl. Gabon 33: 100 (1992).
   Manotes zenkeri Gilg ex Schellenb., F.W.T.A. 1: 747 (1958).

Habit: woody climber to 5 m. **Habitat:** forest. **Distribution:** SE Nigeria, Cameroon, Gabon, Congo (Brazzaville), Congo (Kinshasa), Angola and CAR. **Chorology:** Guineo-Congolian. **Star:** green. **Alt. range:** 1-200m (3). **Flowering:** Apr (1), May (2). **Specimens: -** *Mokoko River F.R.:* Ekema N. 879; Tchouto P. 1155; Watts J. 1194

Rourea myriantha Baill.

Agric. Univ. Wag. Papers 89(6): 342 (1989); Fl. Gabon 33: 124 (1992).
   Paxia liberosepala (Baker f.) Schellenb., F.W.T.A. 1: 741 (1958).
   Paxia cinnabarina Schellenb., F.W.T.A. 1: 741 (1958).

Habit: woody climber. **Habitat:** forest. **Distribution:** SE Nigeria, Cameroon, Gabon, Congo (Brazzaville), Congo (Kinshasa) and Angola. **Chorology:** Guineo-Congolian. **Star:** green. **Specimens: -** *Etinde:* Winkler H. J. P. 526

Rourea solanderi Baker

Agric. Univ. Wag. Papers 89(6): 355 (1989); Fl. Gabon 33: 130 (1992).
   Spiropetalum heterophyllum (Baker) Gilg, F.W.T.A. 1: 748 (1958).
   Spiropetalum solanderi (Baker) Gilg, F.W.T.A. 1: 748 (1958).

Habit: woody climber or shrub. **Guild:** light demander. **Habitat:** forest. **Distribution:** Sierra Leone to Cameroon, Gabon, Congo (Brazzaville), Congo (Kinshasa), Angola (Cabinda) and CAR. **Chorology:** Guineo-Congolian. **Star:** green. **Alt. range:** 200-400m (1).**Fruiting:** Oct (1). **Specimens: -** *Etinde:* Preuss P. R. 1321. *Onge:* Akogo M. 46

Rourea thomsonii (Baker) Jongkind

Agric. Univ. Wag. Papers 89(6): 359 (1989); Fl. Gabon 33: 132 (1992).
   Jaundea pubescens (Baker) Schellenb., F.W.T.A. 1: 742 (1958).
   Jaundea pinnata (P.Beauv.) Schellenb., F.W.T.A. 1: 742 (1958).

Habit: woody climber, shrub or small tree. **Guild:** light demander. **Habitat:** forest. **Distribution:** tropical Africa. **Star:** excluded. **Alt. range:** 1-200m (1). **Flowering:** Apr (1). **Fruiting:** Apr (1). **Specimens: -** *Mabeta-Moliwe:* Sunderland T. C. H. 1218

CONVOLVULACEAE

P. Wilkin (K)

Calycobolus africanus (G.Don) Heine
F.W.T.A. 2: 338 (1963).

Habit: woody climber. **Guild:** light demander. **Habitat:** forest. **Distribution:** Sierra Leone to Cameroon, Gabon, Congo (Kinshasa) and Angola (Cabinda). **Chorology:** Guineo-Congolian. **Star:** green. **Alt. range:** 400-600m (1), 601-800m (2). **Flowering:** May (1). **Fruiting:** Mar (1), Apr (1), May (1). **Specimens:** - *Etinde:* Cable S. 1506; Tchouto P. 220; Tekwe C. 50

Dichondra repens J.R. & G.Forst.
F.T.E.A. Convolvulaceae: 12 (1963); F.W.T.A. 2: 338 (1963); Fl. Zamb. 8(1): 10 (1987).

Habit: herb. **Distribution:** worldwide in tropical and subtropical regions. **Star:** excluded. **Alt. range:** 800-1000m (1). **Specimens:** - *No locality:* Maitland T. D. 1310

Ipomoea aquatica Forssk.
F.T.E.A. Convolvulaceae: 120 (1963); F.W.T.A. 2: 349 (1963); Fl. Zamb. 8(1): 97 (1987).

Habit: herb. **Guild:** pioneer. **Distribution:** pantropical. **Star:** excluded. **Specimens:** - *Etinde:* Maitland T. D. 1346

Ipomoea batatas (L.) Lam.
F.T.E.A. Convolvulaceae: 114 (1963); F.W.T.A. 2: 350 (1963); Fl. Zamb. 8(1): 89 (1987).

Habit: trailing and climbing herb. **Habitat:** forest. **Distribution:** pantropical. **Star:** excluded. **Alt. range:** 1-200m (6), 801-1000m (1). **Flowering:** Feb (1), Jul (1), Oct (1), Nov (3), Dec (1). **Fruiting:** Feb (1), Nov (2). **Specimens:** - *Etinde:* Brodie S. 5; Cheek M. 5520, 5924; Tchouto P. 14. *Limbe Botanic Garden:* Wheatley J. I. 663

Ipomoea cairica (L.) Sweet
F.T.E.A. Convolvulaceae: 125 (1963); F.W.T.A. 2: 351 (1963); Fl. Zamb. 8(1): 105 (1987).

Habit: herbaceous climber. **Guild:** pioneer. **Habitat:** forest. **Distribution:** pantropical. **Star:** excluded. **Alt. range:** 1-200m (2). **Flowering:** Jul (1), Nov (1). **Fruiting:** Jul (1). **Specimens:** - *Etinde:* Cheek M. 5519. *Mabeta-Moliwe:* Cheek M. 3587

Ipomoea indica (Burm.) Merr.
Fl. Zamb. 8(1): 86 (1987).

Habit: herbaceous climber. **Guild:** pioneer. **Habitat:** farmbush. **Distribution:** pantropical. **Star:** excluded. **Alt. range:** 800-1000m (1). **Flowering:** Apr (1). **Specimens:** - *Eastern slopes:* Kwangue A. 38

Ipomoea involucrata P.Beauv.
F.T.E.A. Convolvulaceae: 104 (1963); F.W.T.A. 2: 347 (1963); Fl. Zamb. 8(1): 75 (1987).

Habit: herb. **Guild:** pioneer. **Habitat:** forest and farmbush. **Distribution:** tropical Africa. **Star:** excluded. **Alt. range:** 1-200m (2), 801-1000m (1). **Flowering:** Mar (1), Oct (2). **Specimens:** -

Eastern slopes: Maitland T. D. 220. *Etinde:* Brodie S. 3; Cable S. 1561. *Onge:* Cheek M. 5178

Ipomoea mauritiana Jacq.
F.T.E.A. Convolvulaceae: 135 (1963); F.W.T.A. 2: 351 (1963); Fl. Zamb. 8(1): 117 (1987).

Habit: herbaceous climber. **Guild:** pioneer. **Habitat:** farmbush. **Distribution:** pantropical. **Star:** excluded. **Alt. range:** 1-200m (3). **Flowering:** Oct (1), Nov (1), Dec (1). **Fruiting:** Nov (1), Dec (1). **Specimens:** - *Etinde:* Cheek M. 5503, 5932; Maitland T. D. 470. *Mabeta-Moliwe:* Cheek M. 3564,3937. *Onge:* Cheek M. 5101

Ipomoea pes-caprae (L.) Sweet
F.T.E.A. Convolvulaceae: 121 (1963); F.W.T.A. 2: 347 (1963); Fl. Zamb. 8(1): 98 (1987).

Habit: herb. **Distribution:** pantropical. **Star:** excluded. **Specimens:** - *Etinde:* Nditapah J. K. FHI 50289

Lepistemon owariense (P.Beauv.) Hallier f.
F.T.E.A. Convolvulaceae: 63 (1963); F.W.T.A. 2: 343 (1963); Fl. Zamb. 8(1): 44 (1987).

Habit: herbaceous climber. **Guild:** pioneer. **Habitat:** forest. **Distribution:** tropical Africa. **Star:** excluded. **Alt. range:** 1-200m (2). **Flowering:** Nov (1). **Fruiting:** Dec (1). **Specimens:** - *Etinde:* Cable S. 154; Maitland T. D. 310. *Mabeta-Moliwe:* Cheek M. 5733

Lepistemon parviflorum Pilg. ex Büsgen
F.W.T.A. 2: 343 (1963).

Habit: herbaceous climber. **Guild:** pioneer. **Habitat:** forest. **Distribution:** Sierra Leone to Cameroon. **Chorology:** Guineo-Congolian. **Star:** blue. **Specimens:** - *Etinde:* Maitland T. D. 396

Merremia umbellata (L.) Hallier f. subsp. umbellata
F.T.E.A. Convolvulaceae: 54 (1963); F.W.T.A. 2: 342 (1963).

Habit: herbaceous climber. **Guild:** pioneer. **Habitat:** forest. **Distribution:** pantropical. **Star:** excluded. **Specimens:** - *Etinde:* Hambler D. J. 653

Neuropeltis acuminata (P.Beauv.) Benth.
F.W.T.A. 2: 338 (1963).

Habit: woody climber. **Guild:** light demander. **Habitat:** forest. **Distribution:** Sierra Leone to Cameroon, Gabon, Congo (Kinshasa) and Angola. **Chorology:** Guineo-Congolian. **Star:** green. **Specimens:** - *Etinde:* Kalbreyer W. 11

CRASSULACEAE

Crassula alata (Viv.) Berger subsp. pharnaceoides (Fisch. & Mey.) Wickens & Bywater
F.T.E.A. Crassulaceae: 6 (1987).
Crassula pharnaceoides (Hochst.) Fisch. & Mey., F.W.T.A. 1: 116 (1954).

Habit: herb. Habitat: among rocks in grassland. **Distribution:** Cameroon, Sudan, Ethiopia, Somalia and East Africa. **Chorology:** Afromontane. **Star:** green. **Alt. range:** 2600-2800m (2), 3001-3200m (1). **Flowering:** Oct (1). **Specimens:** - *Eastern slopes:* Cheek M. 3671; Thomas D. W. 9419. *No locality:* Mann G. 1991

Crassula alsinoides (Hook.f.) Engl.
F.W.T.A. 1: 116 (1954); F.T.E.A. Crassulaceae: 12 (1987).

Habit: herb. **Habitat:** wet areas in forest. **Distribution:** W. Cameroon and Bioko to East and South Africa, also in Yemen. **Chorology:** Montane. **Star:** excluded. **Alt. range:** 1800-2000m (1). **Specimens:** - *No locality:* Migeod F. 152

Crassula schimperi Fisch. & Mey. subsp. schimperi
F.T.E.A. Crassulaceae: 7 (1987).
  Crassula pentandra (Edgew.) Schönl., F.W.T.A. 1: 116 (1954).

Habit: terrestrial or epiphytic herb. **Habitat:** forest, sometimes on rocks. **Distribution:** Mount Cameroon, Congo (Kinshasa), Sudan, Ethiopia, Zambia, East Africa, tropical Arabia and India. **Chorology:** Montane. **Star:** blue. **Alt. range:** 2400-2600m (1). **Specimens:** - *No locality:* Mann G. 2037

Crassula vaginata Eckl. & Zeyh.
F.T.E.A. Crassulaceae: 13 (1987).
  Crassula alba sensu Hepper, F.W.T.A. 1: 116 (1954).
  Crassula mannii Hook.f., F.W.T.A. (ed.1) 1: 104 (1927).

Habit: herb. **Habitat:** grassland and forest-grassland transition. **Distribution:** W. Cameroon to East and South Africa. **Chorology:** Afromontane. **Star:** green. **Alt. range:** 1800-2000m (1), 2001-2200m (2), 2401-2600m (2). **Flowering:** May (1), Sep (1), Oct (2), Nov (1). **Specimens:** - *Eastern slopes:* Banks H. 101; Sunderland T. C. H. 1426; Tchouto P. 265; Watts J. 429. *Etinde:* Tekwe C. 245. *No locality:* Mann G. 1304

Kalanchoe crenata (Andrews) Haw.
F.W.T.A. 1: 118 (1954); F.T.E.A. Crassulaceae: 42 (1987).
  Kalanchoe laciniata sensu Hepper, F.W.T.A. 1: 117 (1954).

Habit: herb. **Distribution:** pantropical. **Chorology:** Montane. **Star:** excluded. **Alt. range:** 600-800m (1), 1601-1800m (1). **Specimens:** - *Eastern slopes:* Maitland T. D. 997. *Etinde:* Kalbreyer W. 158. *No locality:* Mann G. 1315

Umbilicus botryoides Hochst. ex A.Rich.
F.W.T.A. 1: 119 (1954); F.T.E.A. Crassulaceae: 60 (1987).

Habit: terrestrial or epiphytic herb. **Habitat:** grassland. **Distribution:** Cameroon, Congo (Kinshasa), Sudan, Ethiopia, Somalia and East Africa. **Chorology:** Afromontane. **Star:** green. **Alt. range:** 2000-2200m (1), 2401-2600m (1). **Flowering:** Oct (1). **Specimens:** - *Eastern slopes:* Tchouto P. 290. *No locality:* Mann G. 1286

CRUCIFERAE

Cardamine africana L.
F.W.T.A. 1: 98 (1954); Fl. Cameroun 21: 14 (1980).

Habit: herb. **Habitat:** forest-grassland transition. **Distribution:** tropical Africa, Brazil, Indonesia and India. **Chorology:** Montane. **Star:** excluded. **Alt. range:** 800-1000m (1), 1801-2000m (1), 2001-2200m (1), 2401-2600m (1). **Flowering:** Oct (2). **Fruiting:** Oct (1). **Specimens:** - *Eastern slopes:* Brenan J. P. M. 9352; Cheek M. 3710; Tchouto P. 391. *No locality:* Maitland T. D. 1345

Cardamine hirsuta L.
F.W.T.A. 1: 98 (1954); Fl. Cameroun 21: 18 (1980).

Habit: herb. **Habitat:** grassland. **Distribution:** temperate regions and tropical Africa. **Chorology:** Montane. **Star:** excluded. **Alt. range:** 2600-2800m (1), 3001-3200m (1). **Fruiting:** Oct (1). **Specimens:** - *Eastern slopes:* Cheek M. 3669. *No locality:* Mann G. 2032

Cardamine trichocarpa Hochst. ex. A.Rich.
F.W.T.A. 1: 98 (1954); Fl. Cameroun 21: 16 (1980).

Habit: herb. **Habitat:** forest. **Distribution:** tropical Africa and India. **Chorology:** Montane. **Star:** excluded. **Alt. range:** 800-1000m (1), 1401-1600m (1), 1601-1800m (1). **Flowering:** Oct (1). **Fruiting:** Oct (1). **Specimens:** - *Eastern slopes:* Migeod F. 228; Tchouto P. 362. *No locality:* Maitland T. D. 1237

Rorippa nasturtium-aquaticum (L.) Hayek
F.W.T.A. 1: 97 (1954); Fl. Cameroun 21: 22 (1980); Fl. Gabon 30: 78 (1987).

Habit: herb. **Guild:** pioneer. **Distribution:** Europe and Asia, naturalized in parts of Africa. **Chorology:** Montane. **Star:** excluded. **Alt. range:** 1600-1800m (1). **Specimens:** - *No locality:* Migeod F. 52

CUCURBITACEAE
C. Jeffrey, S. Cable & M. Cheek (K)

Bambekea racemosa Cogn.
F.W.T.A. 1: 208 (1954); Fl. Cameroun 6: 90 (1967).

Habit: herbaceous climber. **Habitat:** forest. **Distribution:** Nigeria, Cameroon, Gabon and Congo (Kinshasa). **Chorology:** Guineo-Congolian. **Star:** green. **Flowering:** Nov (1). **Fruiting:** Nov (1). **Specimens:** - *Mabeta-Moliwe:* Nguembock F. 26

Cayaponia africana (Hook.f.) Exell
F.W.T.A. 1: 206 (1954); Fl. Cameroun 6: 119 (1967).

Habit: herbaceous climber. **Guild:** pioneer. **Habitat:** forest. **Distribution:** Gambia to Cameroon and S. Tomé. **Chorology:** Guineo-Congolian (montane). **Star:** green. **Alt. range:** 2200-2400m (1). **Specimens:** - *No locality:* Mann G. 1235

Coccinia barteri (Hook.f.) Keay
F.W.T.A. 1: 215 (1954); F.T.E.A. Cucurbitaceae: 60 (1967); Fl. Cameroun 6: 128 (1967).

Habit: herbaceous climber. **Guild:** pioneer. **Habitat:** forest. **Distribution:** tropical Africa. **Star:** excluded. **Alt. range:** 1-200m (7), 201-400m (1), 601-800m (1). **Flowering:** Feb (1), Apr (2), May (1), Jun (2), Oct (2). **Fruiting:** Feb (1), Apr (1), May (2), Jun (1), Oct (2). **Specimens:** - *Etinde:* Dalziel J. M. 8230; Mbani J. M.

137; Nkeng P. 20; Tchouto P. 480; Tekwe C. 81; Watts J. 559. *Mabeta-Moliwe:* Sunderland T. C. H. 1451; Tchouto P. 162; Wheatley J. I. 138, 354

Coccinia cf. barteri (Hook.f.) Keay

Specimens: - *Mabeta-Moliwe:* Tchouto P. 163

Gerrardanthus paniculatus (Mast.) Cogn.
Fl. Cameroun 6: 25 (1967).
Gerrardanthus zenkeri Harms & Gilg ex Cogn., F.W.T.A. 1: 208 (1954).

Habit: herbaceous climber. **Guild:** pioneer. **Habitat:** stream banks in forest. **Distribution:** Ghana, Nigeria, Cameroon, Congo (Kinshasa) and Angola. **Chorology:** Guineo-Congolian. **Star:** green. **Alt. range:** 1-200m (1). **Flowering:** May (1). **Specimens:** - *Mabeta-Moliwe:* Sunderland T. C. H. 1301

Lagenaria breviflora Benth.
F.T.E.A. Cucurbitaceae: 49 (1967).
Adenopus ledermannii Harms, F.W.T.A. 1: 206 (1954).
Adenopus breviflorus Benth., F.W.T.A. 1: 206 (1954).

Habit: herbaceous climber. **Guild:** pioneer. **Habitat:** forest. **Distribution:** tropical Africa. **Star:** excluded. **Alt. range:** 600-800m (1). **Specimens:** - *Mabeta-Moliwe:* Mildbraed G. W. J. 10732

Momordica cf. angustisepala Harms

Specimens: - *Mabeta-Moliwe:* Wheatley J. I. 109

Momordica cf. cabraei (Cogn.) C.Jeffrey

Specimens: - *Mabeta-Moliwe:* Sunderland T. C. H. 1113

Momordica charantia L.
F.W.T.A. 1: 212 (1954); F.T.E.A. Cucurbitaceae: 31 (1967); Fl. Cameroun 6: 172 (1967).

Habit: herbaceous climber. **Guild:** pioneer. **Habitat:** forest. **Distribution:** worldwide in tropical and subtropical regions. **Star:** excluded. **Alt. range:** 1-200m (2). **Flowering:** May (2). **Fruiting:** May (1). **Specimens:** - *Eastern slopes:* Hutchinson J. & Metcalfe C. R. 99. *Mabeta-Moliwe:* Sunderland T. C. H. 1387; Wheatley J. I. 263

Momordica cissoides Planch. ex Benth.
F.W.T.A. 1: 211 (1954); F.T.E.A. Cucurbitaceae: 26 (1967); Fl. Cameroun 6: 160 (1967).

Habit: herbaceous climber. **Guild:** pioneer. **Habitat:** forest and stream banks. **Distribution:** tropical Africa. **Star:** excluded. **Alt. range:** 1-200m (3), 201-400m (1), 401-600m (1), 601-800m (1). **Flowering:** Feb (1), Apr (3), May (1). **Fruiting:** May (1), Aug (1). **Specimens:** - *Etinde:* Maitland T. D. 421; Mbani J. M. 144; Tekwe C. 122. *Mabeta-Moliwe:* Plot Voucher W 682, W 728; Sunderland T. C. H. 1198, 1236, 1322; Wheatley J. I. 101. *No locality:* Nkeng P. 15

Momordica foetida Schum. & Thonn.
F.W.T.A. 1: 212 (1954); F.T.E.A. Cucurbitaceae: 29 (1967); Fl. Cameroun 6: 180 (1967).

Momordica cordata Cogn., F.W.T.A. 1: 212 (1954).

Habit: herbaceous climber. **Guild:** pioneer. **Habitat:** forest and forest-grassland transition. **Distribution:** tropical and subtropical Africa. **Chorology:** Afromontane. **Star:** green. **Alt. range:** 1-200m (2), 201-400m (1), 601-800m (2), 1001-1200m (1), 1601-1800m (1). **Flowering:** Apr (1), May (1), Sep (1), Oct (2). **Fruiting:** Feb (1), Apr (1), May (1), Oct (1), Nov (1). **Specimens:** - *Etinde:* Kwangue A. 145; Preuss P. R. 1203; Sidwell K. 181; Thomas D. W. 9218; Watts J. 555; Wheatley J. I. 20. *Mabeta-Moliwe:* Mann G. 766; Sunderland T. C. H. 1307; Tchouto P. 161

Momordica gilgiana Cogn.
F.W.T.A. 1: 212 (1954); Fl. Cameroun 6: 164 (1967).

Habit: herbaceous climber. **Habitat:** forest. **Distribution:** Cameroon and Gabon. **Chorology:** Lower Guinea. **Star:** blue. **Specimens:** - *Eastern slopes:* Deistel H. 486

Momordica multiflora Hook.f.
F.W.T.A. 1: 212 (1954); F.T.E.A. Cucurbitaceae: 27 (1967); Fl. Cameroun 6: 175 (1967).

Habit: herbaceous climber. **Guild:** pioneer. **Habitat:** forest and farmbush. **Distribution:** Ghana to Bioko, Cameroon, Gabon, Congo (Brazzaville), Congo (Kinshasa), Angola, Sudan, Uganda and Tanzania. **Star:** green. **Alt. range:** 1-200m (2). **Flowering:** Jun (2). **Specimens:** - *Mabeta-Moliwe:* Plot Voucher W 921; Sunderland T. C. H. 1471; Wheatley J. I. 339

Oreosyce africana Hook.f.
F.W.T.A. 1: 210 (1954); F.T.E.A. Cucurbitaceae: 110 (1967); Fl. Cameroun 6: 65 (1967).

Habit: herbaceous climber. **Habitat:** forest. **Distribution:** Bioko and SW Cameroon to East and South Africa. **Chorology:** Afromontane. **Star:** green. **Specimens:** - *No locality:* Mann G. 1285

Raphidiocystis mannii Hook.f.
F.W.T.A. 1: 215 (1954); Fl. Cameroun 6: 106 (1967).

Habit: herbaceous climber. **Habitat:** stream banks in forest. **Distribution:** Nigeria, Bioko and Cameroon. **Chorology:** Lower Guinea. **Star:** blue. **Alt. range:** 1-200m (1).**Fruiting:** May (1). **Specimens:** - *Etinde:* Maitland T. D. 373; Mbani J. M. 132

Ruthalicia longipes (Hook.f.) C.Jeffrey
Fl. Cameroun 6: 147 (1967).
Physedra longipes Hook.f., F.W.T.A. 1: 214 (1954).

Habit: herbaceous climber. **Guild:** pioneer. **Habitat:** stream banks in forest. **Distribution:** Liberia to Cameroon, Congo (Kinshasa) and Angola. **Chorology:** Guineo-Congolian. **Star:** green. **Alt. range:** 1-200m (1). **Flowering:** Apr (1). **Specimens:** - *Mabeta-Moliwe:* Cheek M. 3249; Nguembock F. 2; Tchouto P. 160

Zehneria capillacea (Schumach.) C.Jeffrey
F.T.E.A. Cucurbitaceae: 129 (1967); Fl. Cameroun 6: 40 (1967).
Melothria capillacea (Schum. & Thonn.) Cogn., F.W.T.A. 1: 209 (1954).

Habit: herbaceous climber. **Guild:** pioneer. **Habitat:** forest. **Distribution:** Sierra Leone to Bioko, Cameroon, Gabon, Congo (Kinshasa), Angola and Uganda. **Chorology:** Guineo-Congolian.

42

Star: green. **Alt. range:** 1-200m (1). **Flowering:** May (1).
Fruiting: May (1). **Specimens:** - *Mabeta-Moliwe:* Tchouto P. 200

Zehneria minutiflora (Cogn.) C.Jeffrey
F.T.E.A. Cucurbitaceae: 126 (1967); Fl. Cameroun 6: 37 (1967).
Melothria minutiflora Cogn. var. *hirtella* Cogn., F.W.T.A. 1:
209 (1954).
Melothria minutiflora Cogn. var. *minutiflora*, F.W.T.A. 1: 209
(1954).
Melothria minutiflora Cogn. var. *parviflora* Cogn., F.W.T.A. 1:
209 (1954).

Habit: herbaceous climber. **Habitat:** forest-grassland transition.
Distribution: Cameroon, Congo (Kinshasa), Zimbabwe, Ethiopia
and East Africa. **Chorology:** Afromontane. **Star:** green. **Alt.
range:** 800-1000m (1), 1201-1400m (1), 1401-1600m (1), 2001-
2200m (1), 2201-2400m (1), 2401-2600m (1). **Flowering:** Sep (1),
Oct (1). **Specimens:** - *Eastern slopes:* Thomas D. W. 9342, 9512.
Etinde: Thomas D. W. 9189. *No locality:* Mann G. 2010; Preuss P.
R. 929, 958

Zehneria scabra (L.F.) Sond.
F.T.E.A. Cucurbitaceae: 122 (1967); Fl. Cameroun 6: 44 (1967).
Melothria punctata (Thunb.) Cogn., F.W.T.A. 1: 209 (1954).
Melothria mannii Cogn., F.W.T.A. 1: 209 (1954).

Habit: herbaceous climber. **Habitat:** forest. **Distribution:**
palaeotropical. **Chorology:** Montane. **Star:** excluded. **Alt. range:**
1200-1400m (1). **Specimens:** - *No locality:* Mann G. 1277

DICHAPETALACEAE

F.J. Breteler (WAG)

Dichapetalum affine (Planch. ex Benth.) Breteler
Meded. Land. Wag. 73(13): 45 (1973); Fl. Gabon 32: 38 (1991).
Dichapetalum affinis Planch. ex Benth., F.W.T.A. 1: 438
(1958).

Habit: woody climber or shrub. **Habitat:** forest. **Distribution:**
Bioko, Cameroon and Gabon. **Chorology:** Lower Guinea. **Star:**
blue. **Alt. range:** 1-200m (1). **Flowering:** Apr (1). **Specimens:** -
Mabeta-Moliwe: Wheatley J. I. 169

Dichapetalum angolense Chodat
F.W.T.A. 1: 436 (1958); Meded. Land. Wag. 73(13): 55 (1973); Fl.
Gabon 32: 44 (1991).

Habit: woody climber. **Guild:** light demander. **Habitat:** forest.
Distribution: Liberia to Cameroon, Gabon, Congo (Kinshasa),
Angola, CAR, Uganda and Sudan. **Chorology:** Guineo-Congolian.
Star: green. **Specimens:** - *Eastern slopes:* Dunlap 157

Dichapetalum choristilum Engl.
Meded. Land. Wag. 78(10): 11 (1978); Agric. Univ. Wag. Papers
86(3): 32 (1986); Fl. Gabon 32: 69 (1991).

Habit: woody climber or shrub. **Habitat:** forest. **Distribution:**
Liberia to Cameroon, Gabon, Congo (Brazzaville) and Congo
(Kinshasa). **Chorology:** Guineo-Congolian. **Star:** green. **Alt.
range:** 1-200m (1).**Fruiting:** Mar (1). **Specimens:** - *Mokoko River
F.R.:* Tchouto P. 648

Dichapetalum gabonense Engl.
Meded. Land. Wag. 79(16): 3 (1979); Fl. Gabon 32: 88 (1991).
Dichapetalum nitidulum Engl. & Ruhl., F.W.T.A. 1: 438 (1958).

Habit: woody climber or shrub. **Habitat:** forest. **Distribution:** SE
Nigeria, Cameroon, Angola (Cabinda), Gabon and Congo
(Brazzaville). **Chorology:** Guineo-Congolian. **Star:** green. **Alt.
range:** 1-200m (1), 201-400m (1). **Flowering:** Jun (1). **Fruiting:**
Mar (1), Jun (1). **Specimens:** - *Mabeta-Moliwe:* Wheatley J. I.
347. *Mokoko River F.R.:* Tchouto P. 584. *No locality:* Box H. E.
3552

Dichapetalum heudelotii (Planch. ex Oliv.) Baill. var. heudelotii
F.W.T.A. 1: 438 (1958); Meded. Land. Wag. 79(16): 27 (1979); Fl.
Gabon 32: 101 (1991).
Dichapetalum subauriculatum (Oliv.) Engl., F.W.T.A. 1: 438
(1958).
Dichapetalum ferrugineum Engl., F.W.T.A. 1: 438 (1958).
Dichapetalum johnstonii Engl., F.W.T.A. 1: 438 (1958).
Habit: woody climber or shrub. **Habitat:** forest. **Distribution:**
Guinea Bissau to Cameroon, Gabon, Congo (Brazzaville), Congo
(Kinshasa), Angola, CAR, Sudan and Zambia. **Chorology:**
Guineo-Congolian. **Star:** green. **Alt. range:** 1-200m (3).
Flowering: Apr (1). **Fruiting:** Apr (1), May (1). **Specimens:** -
Etinde: Preuss P. R. 1275. *Mabeta-Moliwe:* Maitland T. D. 394;
Plot Voucher W 944. *Mokoko River F.R.:* Ekema N. 976; Tchouto
P. 1077. *Onge:* Harris D. J. 3750

Dichapetalum heudelotii (Planch. ex Oliv.) Baill. var. longitubulosum (Engl.) Breteler
Meded. Land. Wag. 79(16): 35 (1979).
Dichapetalum scabrum Engl., F.W.T.A. 1: 438 (1958).
Dichapetalum longitubulosum Engl., F.W.T.A. 1: 438 (1958).

Habit: woody climber or shrub. **Guild:** shade-bearer. **Habitat:**
forest. **Distribution:** SE Nigeria and Cameroon. **Chorology:**
Lower Guinea. **Star:** blue. **Alt. range:** 1-200m (4), 201-400m (2).
Flowering: Mar (1), Apr (1). **Fruiting:** Mar (2), Apr (2), Dec (1).
Specimens: - *Etinde:* Cheek M. 5867. *Mabeta-Moliwe:* Jaff B.
113; Sunderland T. C. H. 1116; Watts J. 239; Wheatley J. I. 40.
Mokoko River F.R.: Tchouto P. 593

Dichapetalum cf. heudelotii (Planch. ex Oliv.) Baill.

Specimens: - *Mabeta-Moliwe:* Baker W. J. 302

Dichapetalum insigne Engl.
Meded. Land. Wag. 79(16): 44 (1979); Fl. Gabon 32: 109 (1991).

Habit: woody climber, shrub or small tree. **Habitat:** forest.
Distribution: Cameroon, Angola, Gabon, Congo (Brazzaville) and
Congo (Kinshasa). **Chorology:** Guineo-Congolian. **Star:** green.
Alt. range: 1-200m (4). **Flowering:** Apr (2), May (1), Nov (1).
Specimens: - *Mabeta-Moliwe:* Plot Voucher W 51, W 345, W 610,
W 649, W 769; Wheatley J. I. 230. *Mokoko River F.R.:* Tchouto P.
1065; Watts J. 1057. *Onge:* Thomas D. W. 9874

Dichapetalum madagascariense Poir. var. madagascariense
Agric. Univ. Wag. Papers 86(3): 39 (1986); Keay R.W.J., Trees of
Nigeria: 189 (1989).
Dichapetalum floribundum (Planch.) Engl., F.W.T.A. 1: 437
(1958).

Dichapetalum guineense (DC.) Keay, F.W.T.A. 1: 436 (1958).
Dichapetalum thomsonii (Oliv.) Engl., F.W.T.A. 1: 438 (1958).

Habit: tree, scandent shrub or shrub to 10 m. **Habitat:** forest.
Distribution: tropical Africa. **Star:** excluded. **Specimens:** -
Eastern slopes: Preuss P. R. 904

Dichapetalum mundense Engl.
Meded. Land. Wag. 81(10): 46 (1981); Fl. Gabon 32: 136 (1991).

Habit: woody climber. **Habitat:** forest. **Distribution:** SE Nigeria,
Cameroon, Angola, Gabon and Congo (Kinshasa). **Chorology:**
Guineo-Congolian. **Star:** green. **Alt. range:** 200-400m (4), 401-
600m (1). **Flowering:** Oct (1). **Fruiting:** Oct (5). **Specimens:** -
Etinde: Tchouto P. 465. *Onge:* Akogo M. 52; Ndam N. 726, 751;
Tchouto P. 780

Dichapetalum oblongum (Hook.f. ex Benth.) Engl.
F.W.T.A. 1: 437 (1958); Meded. Land. Wag. 81(10): 56 (1981); Fl.
Gabon 32: 143 (1991).

Habit: woody climber. **Guild:** shade-bearer. **Habitat:** forest.
Distribution: Sierra Leone to Bioko, Cameroon and Gabon.
Chorology: Guineo-Congolian. **Star:** green. **Alt. range:** 1-200m
(3), 801-1000m (1). **Flowering:** Aug (1), Dec (1). **Fruiting:** May
(1), Aug (1), Dec (1). **Specimens:** - *Mabeta-Moliwe:* Baker W. J.
298; Cheek M. 5685; Mildbraed G. W. J. 10686; Sunderland T. C.
H. 1342

Dichapetalum pallidum (Oliv.) Engl.
F.W.T.A. 1: 436 (1958); Meded. Land. Wag. 81(10): 65 (1981); Fl.
Gabon 32: 146 (1991).

Habit: woody climber. **Guild:** shade-bearer. **Habitat:** forest.
Distribution: Guinea to Cameroon, Gabon, Congo (Kinshasa) and
Angola. **Chorology:** Guineo-Congolian. **Star:** green. **Alt. range:**
1-200m (6). **Flowering:** Mar (2), May (1), Dec (1). **Fruiting:** Apr
(1), May (1). **Specimens:** - *Etinde:* Akogo M. 153; Tekwe C. 74.
Mabeta-Moliwe: Cheek M. 5757; Plot Voucher W 507, W 581;
Tchouto P. 196; Wheatley J. I. 38. *Onge:* Wheatley J. I. 680

Dichapetalum rudatisii Engl.
F.W.T.A. 1: 438 (1958); Meded. Land. Wag. 82(8): 8 (1982);
Agric. Univ. Wag. Papers 86(3): 42 (1986); Fl. Gabon 32: 160
(1991).

Habit: woody climber or shrub. **Habitat:** forest. **Distribution:** SE
Nigeria and Cameroon. **Chorology:** Lower Guinea. **Star:** blue.
Alt. range: 1-200m (5), 201-400m (1). **Flowering:** Mar (1), Dec
(1). **Fruiting:** May (1), Sep (1), Oct (1), Nov (1), Dec (1).
Specimens: - *Bambuko F.R.:* Watts J. 676. *Mabeta-Moliwe:*
Cheek M. 5758; Sunderland T. C. H. 1275. *Onge:* Harris D. J.
3839; Tchouto P. 823; Watts J. 766

Dichapetalum tomentosum Engl.
F.W.T.A. 1: 438 (1958); Meded. Land. Wag. 82(8): 51 (1982); Fl.
Gabon 32: 177 (1991).

Habit: woody climber. **Habitat:** forest. **Distribution:** SE Nigeria,
Cameroon, Gabon, Congo (Brazzaville) and Congo (Kinshasa).
Chorology: Guineo-Congolian. **Star:** green. **Alt. range:** 1-200m
(6). **Flowering:** May (2), Jun (1). **Fruiting:** Aug (1), Oct (1), Nov
(1). **Specimens:** - *Etinde:* Preuss P. R. 1103. *Mabeta-Moliwe:*
Baker W. J. 238; Sunderland T. C. H. 1319; Wheatley J. I. 234,
351. *Onge:* Tchouto P. 1008; Watts J. 1013

Dichapetalum sp. nov.?

Specimens: - *Mabeta-Moliwe:* von Rege 62a

Dichapetalum sp.

Specimens: - *Mokoko River F.R.:* Tchouto P. 1151

Tapura africana Oliv.
F.W.T.A. 1: 438 (1958); Agric. Univ. Wag. Papers 86(3): 46
(1986); Keay R.W.J., Trees of Nigeria: 191 (1989); Fl. Gabon 32:
196 (1991).

Habit: tree to 25 m. **Habitat:** forest and farmbush. **Distribution:**
SE Nigeria, Bioko, Cameroon and Gabon. **Chorology:** Lower
Guinea. **Star:** blue. **Alt. range:** 1-200m (11), 201-400m (1).
Flowering: Apr (4), May (4). **Fruiting:** May (3). **Specimens:** -
Mabeta-Moliwe: Jaff B. 70, 79, 139; Maitland T. D. 611; Ndam N.
509; Plot Voucher W 135, W 382, W 457; Sunderland T. C. H.
1186, 1304, 1320; Tchouto P. 174; Watts J. 291; Wheatley J. I.
115. *Mokoko River F.R.:* Mbani J. M. 328; Tchouto P. 1175

DILLENIACEAE

Tetracera alnifolia Willd.
F.W.T.A. 1: 180 (1954); Meded. Land. Wag. 82(3): 128 (1982).
Tetracera alnifolia Willd. var. *podotricha* (Gilg) Staner, Fl. Afr.
Cent. Dilleniaceae: 10 (1967).
Tetracera podotricha Gilg, F.W.T.A. 1: 181 (1954).

Habit: woody climber. **Habitat:** lava flows. **Distribution:** Senegal
to Bioko, Cameroon, CAR, Gabon, Congo (Kinshasa) and Angola.
Chorology: Guineo-Congolian. **Star:** green. **Alt. range:** 1-200m
(5), 201-400m (1). **Flowering:** Jan (1), Feb (1), May (1). **Fruiting:**
Mar (1), May (2), Jul (1). **Specimens:** - *Etinde:* Keay R. W. J. FHI
28659; Tchouto P. 1, 240, 1050; Tekwe C. 26. *Mabeta-Moliwe:*
Sunderland T. C. H. 1321. *Mokoko River F.R.:* Thomas D. W.
10088. *Onge:* Wheatley J. I. 769

Tetracera masuiana De Wild. & T.Durand
Fl. Zamb. 1: 105 (1960).

Habit: woody climber. **Habitat:** forest. **Distribution:** Cameroon,
Congo (Kinshasa) and Burundi. **Chorology:** Guineo-Congolian
(montane). **Star:** green. **Specimens:** - *Mabeta-Moliwe:* Plot
Voucher W 740

Tetracera potatoria Afzel. ex G.Don
F.W.T.A. 1: 180 (1954).

Habit: woody climber. **Guild:** light demander. **Habitat:** forest.
Distribution: Senegal to Cameroon, CAR, Congo (Kinshasa),
Uganda and Sudan. **Chorology:** Guineo-Congolian. **Star:** green.
Alt. range: 1-200m (1).**Fruiting:** Jul (1). **Specimens:** - *Mabeta-
Moliwe:* Cheek M. 3532

Tetracera sp.

Specimens: - *Onge:* Harris D. J. 3667, 3674

DIPSACACEAE

Succisa trichotocephala Baksay
F.W.T.A. 2: 223 (1963); Fl. Cameroun 21: 34 (1980).

Habit: herb. **Habitat:** grassland and forest-grassland transition. **Distribution:** W. Cameroon. **Chorology:** Western Cameroon Uplands (montane). **Star:** gold. **Alt. range:** 2200-2400m (1), 2401-2600m (2). **Flowering:** Oct (2). **Specimens:** - *Eastern slopes:* Sidwell K. 57; Thomas D. W. 9325. *No locality:* Mann G. 1309

EBENACEAE
G. Gosline (K)

Diospyros bipindensis Gürke
F.W.T.A. 2: 11 (1963); Fl. Gabon 18: 35 (1970); Fl. Cameroun 11: 35 (1970); Bull. Jard. Bot. Nat. Belg. 48: 303 (1978).

Habit: tree 8-20 m tall, or shrub to 3 m. **Guild:** shade-bearer. **Habitat:** forest. **Distribution:** Cameroon, Gabon, Congo (Brazzaville), Congo (Kinshasa), Angola and Uganda. **Chorology:** Guineo-Congolian. **Star:** green. **Alt. range:** 1-200m (18). **Flowering:** Mar (6), Apr (5), May (1). **Fruiting:** Mar (1), Apr (2), May (3), Jun (3), Aug (2). **Specimens:** - *Etinde:* Tchouto P. 91; Tekwe C. 64. *Mabeta-Moliwe:* Baker W. J. 221, 326; Cable S. 1404; Etuge M. 1218; Groves M. 221; Jaff B. 101, 182; Ndam N. 498; Sunderland T. C. H. 1122, 1364; Tchouto P. 147; Watts J. 107, 381, 407; Wheatley J. I. 206, 332. *Mokoko River F.R.:* Acworth J.M. 60, 92; Akogo 277; Ekema 1116b, 1184; Fraser 348; Pouakouyou D. 55; Tchouto P. 1092. *Southern Bakundu F.R.:* Brenan J. P. M. 9418

Diospyros cf. canaliculata De Wilde
Specimens: - *Mokoko:* Mbani 386, 414

Diospyros cinnabarina (Gürke) F.White
F.W.T.A. 2: 15 (1963); Fl. Gabon 18: 48 (1970); Fl. Cameroun 11: 48 (1970); Bull. Jard. Bot. Nat. Belg. 48: 329 (1978); Keay R.W.J., Trees of Nigeria: 387 (1989).
    Diospyros simulans F.White, F.W.T.A. 2: 15 (1963); Fl. Cameroun 11: 147 (1970).

Habit: tree to 15 m. **Guild:** shade-bearer. **Habitat:** forest. **Distribution:** Cameroon, Gabon, Congo (Kinshasa) and Angola (Cabinda). **Chorology:** Guineo-Congolian. **Star:** green. **Alt. range:** 1-200m (1), 201-400m (3). **Flowering:** Jun (1). **Fruiting:** Apr (1), Jun (2), Oct (1). **Specimens:** - *Etinde:* Kwangue A. 62, 103; Tchouto P. 235. *Mokoko:* Mbani 427; Thomas 10039. *Onge:* Akogo M. 73. *Southern Bakundu F.R.:* Binuyo A. & Daramola B. O. FHI 35602; Brenan J. P. M. 9320

Diospyros conocarpa Gürke & K.Schum.
F.W.T.A. 2: 11 (1963); Fl. Gabon 18: 52 (1970); Fl. Cameroun 11: 52 (1970); Bull. Jard. Bot. Nat. Belg. 48: 297 (1978); Keay R.W.J., Trees of Nigeria: 382 (1989).

Habit: tree to 9 m. **Guild:** shade-bearer. **Habitat:** forest. **Distribution:** Nigeria, Cameroon, Rio Muni, Gabon, Congo (Brazzaville), Congo (Kinshasa) and Angola (Cabinda). **Chorology:** Guineo-Congolian. **Star:** green. **Alt. range:** 1-200m (2), 201-400m (2). **Flowering:** Apr (1), May (2). **Fruiting:** May (1), Oct (1). **Specimens:** - *Mokoko River F.R.:* Acworth J.M. 103; Akogo 158; Ekema 840; Mbani 428; Ndam N. 1165, 1360; Pouakouyou D. 50; Tchouto P. 1062; Watts J. 1125. *Onge:* Tchouto P. 905. *Southern Bakundu F.R.:* Brenan J. P. M. 9424

Diospyros crassiflora Hiern
F.W.T.A. 2: 12 (1963); Fl. Gabon 18: 57 (1970); Fl. Cameroun 11: 57 (1970); Bull. Jard. Bot. Nat. Belg. 48: 298 (1978); Keay R.W.J., Trees of Nigeria: 384 (1989).

Habit: tree to 25 m. **Guild:** shade-bearer. **Habitat:** forest. **Distribution:** Nigeria, Cameroon, Gabon, Congo (Brazzaville), Congo (Kinshasa), Angola and CAR. **Chorology:** Guineo-Congolian. **Star:** green. **Specimens:** - *Etinde:* Maitland T. D. 407. *Mokoko:* Thomas 10142

Diospyros dendo Welw. ex Hiern
F.W.T.A. 2: 10 (1963); Fl. Gabon 18: 63 (1970); Fl. Cameroun 11: 63 (1970); Bull. Jard. Bot. Nat. Belg. 48: 290 (1978); Keay R.W.J., Trees of Nigeria: 382 (1989).

Habit: tree to 15 m. **Guild:** shade-bearer. **Habitat:** forest. **Distribution:** Nigeria, Cameroon, Gabon, Congo (Brazzaville), Congo (Kinshasa) and Angola. **Chorology:** Guineo-Congolian. **Star:** green. **Specimens:** - *Bambuko F.R.:* Keay R. W. J. FHI 37474. *Mokoko:* Akogo 298; Tchouto P. 1149

Diospyros fragrans Gürke
F.W.T.A. 2: 11 (1963); Fl. Gabon 18: 72 (1970); Fl. Cameroun 11: 72 (1970); Bull. Jard. Bot. Nat. Belg. 48: 309 (1978).

Habit: tree to 15 m. **Guild:** shade-bearer. **Habitat:** forest. **Distribution:** Cameroon, Rio Muni and Gabon. **Chorology:** Lower Guinea. **Star:** blue. **Specimens:** - *Southern Bakundu F.R.:* Binuyo A. & Daramola B. O. FHI 35629

Diospyros gabunensis Gürke
F.W.T.A. 2: 12 (1963); Fl. Gabon 18: 77 (1970); Fl. Cameroun 11: 77 (1970); Bull. Jard. Bot. Nat. Belg. 48: 317 (1978); Keay R.W.J., Trees of Nigeria: 383 (1989); Hawthorne W., F.G.F.T. Ghana: 68 (1990).

Habit: tree to 20 m. **Guild:** shade-bearer. **Habitat:** forest. **Distribution:** Sierra Leone to Cameroon, Gabon, Congo (Brazzaville), Congo (Kinshasa), Rwanda and Zambia. **Chorology:** Guineo-Congolian. **Star:** green. **Specimens:** - *Southern Bakundu F.R.:* Brenan J. P. M. 9408

Diospyros gracilescens Gürke
Fl. Cameroun 11: 86 (1970); Bull. Jard. Bot. Nat. Belg. 48: 321 (1978); Keay R.W.J., Trees of Nigeria: 385 (1989).
    Diospyros nigerica F.White, F.W.T.A. 2: 12 (1963).

Habit: tree to 25 m. **Guild:** shade-bearer. **Habitat:** forest. **Distribution:** Nigeria, Cameroon, Gabon, Congo (Brazzaville) and Congo (Kinshasa). **Chorology:** Guineo-Congolian. **Star:** green. **Alt. range:** 600-800m (1). **Specimens:** - *Mabeta-Moliwe:* Mildbraed G. W. J. 10595. *Mokoko:* Batoum 7; Watts J. 1179

Diospyros hoyleana F.White subsp. hoyleana
F.W.T.A. 2: 15 (1963); Fl. Gabon 18: 88 (1970); Fl. Cameroun 11: 88 (1970); Bull. Jard. Bot. Nat. Belg. 48: 344 (1978); Keay R.W.J., Trees of Nigeria: 384 (1989).

Habit: tree 8-15 m tall, or shrub to 3 m. **Habitat:** forest.
Distribution: SE Nigeria, Cameroon, Gabon, Congo (Brazzaville),
Congo (Kinshasa), Angola and Zambia. **Chorology:** Guineo-
Congolian. **Star:** green. **Alt. range:** 1-200m (1). **Specimens:** -
Mokoko River F.R.: Mbani 309; Tchouto P. 1109. *Southern
Bakundu F.R.:* Brenan J. P. M. 9288

Diospyros iturensis (Gürke) Letouzey & F.White

Fl. Gabon 18: 32 (1970); Fl. Cameroun 11: 93 (1970); Bull. Jard.
Bot. Nat. Belg. 48: 337 (1978); Keay R.W.J., Trees of Nigeria: 387
(1989).

Habit: tree to 20 m. **Habitat:** forest. **Distribution:** Nigeria,
Cameroon, Gabon, Congo (Brazzaville), Congo (Kinshasa), Angola
and CAR. **Chorology:** Guineo-Congolian. **Star:** green. **Alt. range:**
1-200m (5), 201-400m (1). **Flowering:** Mar (1), Apr (1). **Fruiting:**
May (2), Jun (1). **Specimens:** - *Mabeta-Moliwe:* Sunderland T. C.
H. 1362; Watts J. 273; Wheatley J. I. 49, 142, 342. *Mokoko River
F.R.:* Mbani 337; Ndam N. 1161, 1065; Thomas 10040, 10147

Diospyros cf. iturensis (Gürke) Letouzey & F.White

Specimens: - *Etinde:* Kwangue A. 70

Diospyros kamerunensis Gürke

F.W.T.A. 2: 11 (1963); Fl. Cameroun 11: 97 (1970); Bull. Jard.
Bot. Nat. Belg. 48: 310 (1978).

Habit: tree to 15 m. **Guild:** shade-bearer. **Habitat:** forest.
Distribution: Liberia, Ivory Coast, Ghana, Cameroon and Gabon.
Chorology: Guineo-Congolian. **Star:** green. **Specimens:** - *Mokoko
River F.R.:* Mbani J. M. 359; Thomas 10095

Diospyros mannii Hiern

F.W.T.A. 2: 11 (1963); Fl. Cameroun 11: 104 (1970); Bull. Jard.
Bot. Nat. Belg. 48: 281 (1978); Keay R.W.J., Trees of Nigeria: 383
(1989).

Habit: tree to 30 m. **Guild:** shade-bearer. **Habitat:** forest.
Distribution: Sierra Leone to Cameroon, Gabon, Congo
(Brazzaville), Congo (Kinshasa), Angola (Cabinda) and CAR.
Chorology: Guineo-Congolian. **Star:** green. **Alt. range:** 1-200m
(1). **Specimens:** - *Mokoko River F.R.:* Mbani J. M. 336, 377, 428;
Thomas 10032, 10148. *Southern Bakundu F.R.:* Mbani J. M. 425

Diospyros melocarpa F.White

Specimens:- *Mokoko:* Thomas 10031

Diospyros obliquifolia (Hiern ex Gürke) F.White

F.W.T.A. 2: 15 (1963); Fl. Gabon 18: 119 (1970); Fl. Cameroun
11: 119 (1970); Bull. Jard. Bot. Nat. Belg. 48: 321 (1978); Keay
R.W.J., Trees of Nigeria: 384 (1989).

Habit: tree to 6 m. **Habitat:** forest. **Distribution:** Nigeria,
Cameroon, Rio Muni, Gabon and Congo (Brazzaville).
Chorology: Guineo-Congolian. **Star:** green. **Alt. range:** 1-200m
(2), 201-400m (1). **Flowering:** Apr (2). **Fruiting:** Oct (1).
Specimens: - *Mabeta-Moliwe:* Sunderland T. C. H. 1223;
Wheatley J. I. 157. *Onge:* Cheek M. 5072. *Southern Bakundu
F.R.:* Brenan J. P. M. 9448

Diospyros physocalycina Gürke

F.W.T.A. 2: 11 (1963); Fl. Gabon 18: 122 (1970); Fl. Cameroun
11: 122 (1970); Bull. Jard. Bot. Nat. Belg. 48: 302 (1978); Keay
R.W.J., Trees of Nigeria: 383 (1989).

Habit: tree to 15 m. **Habitat:** forest. **Distribution:** Nigeria,
Cameroon and Gabon. **Chorology:** Lower Guinea. **Star:** blue. **Alt.
range:** 1-200m (1). **Flowering:** May (1). **Fruiting:** May (1).
Specimens: - *Etinde:* Tchouto P. 226

Diospyros piscatoria Gürke

F.W.T.A. 2: 10 (1963); Fl. Gabon 18: 29 (1970); Fl. Cameroun 11:
126 (1970); Bull. Jard. Bot. Nat. Belg. 48: 295 (1978); Keay
R.W.J., Trees of Nigeria: 385 (1989); Hawthorne W., F.G.F.T.
Ghana: 68 (1990).

Habit: tree to 30 m. **Guild:** shade-bearer. **Habitat:** forest.
Distribution: Guinea to Bioko, Cameroon, Gabon, Congo
(Brazzaville), Congo (Kinshasa) and Angola (Cabinda).
Chorology: Guineo-Congolian. **Star:** green. **Specimens:** -
Bambuko F.R.: Keay R. W. J. FHI 37472

Diospyros polystemon Gürke

F.W.T.A. 2: 10 (1963); Fl. Gabon 18: 132 (1970); Fl. Cameroun
11: 132 (1970); Bull. Jard. Bot. Nat. Belg. 48: 294 (1978).

Habit: tree 10-30 m tall, or shrub to 5 m. **Habitat:** forest and
farmbush. **Distribution:** Cameroon, Congo (Brazzaville), Congo
(Kinshasa) and Angola. **Chorology:** Guineo-Congolian. **Star:**
green. **Alt. range:** 1-200m (1). **Flowering:** May (1). **Specimens:** -
Mabeta-Moliwe: Mildbraed G. W. J. 10754; Tchouto P. 173

Diospyros preussii Gürke

F.W.T.A. 2: 14 (1963); Fl. Gabon 18: 135 (1970); Fl. Cameroun
11: 135 (1970); Bull. Jard. Bot. Nat. Belg. 48: 304 (1978); Keay
R.W.J., Trees of Nigeria: 382 (1989).

Habit: tree to 9 m. **Habitat:** forest and stream banks.
Distribution: SE Nigeria, Cameroon and Gabon. **Chorology:**
Lower Guinea. **Star:** blue. **Alt. range:** 1-200m (22), 201-400m
(2), 401-600m (2). **Flowering:** Feb (1), Mar (6), Apr (2), May (3).
Fruiting: Mar (2), Apr (1), May (10), Jun (1), Jul (1), Aug (1), Sep
(1), Oct (2), Dec (1). **Specimens:** - *Etinde:* Khayota B. 588;
Kwangue A. 113; Mbani J. M. 38; Tchouto P. 90, 125; Wheatley J.
I. 524. *Mabeta-Moliwe:* Cable S. 594, 1371, 1426; Groves M. 247;
Hunt L. V. 1; Jaff B. 219; Ndam N. 483, 485; Nguembock F. 58;
Sunderland T. C. H. 1101, 1292, 1382; Watts J. 253, 275, 292;
Wheatley J. I. 254. *Mokoko River F.R.:* Acworth J. M. 52; 279;
Akogo 200, 262; Ekema 1053; Fraser 377, 423; Ndam 1105;
Sonké 1094; Watts J. 1113. *Onge:* Akogo M. 117; Cheek M. 5022;
Watts J. 701. *Southern Bakundu F.R.:* Binuyo A. & Daramola B.
O. FHI 35548

Diospyros suaveolens Gürke

F.W.T.A. 2: 11 (1963); Fl. Gabon 18: 154 (1970); Fl. Cameroun
11: 154 (1970); Bull. Jard. Bot. Nat. Belg. 48: 282 (1978); Keay
R.W.J., Trees of Nigeria: 383 (1989).

Habit: tree 8-30 m tall. **Habitat:** forest. **Distribution:** Nigeria,
Cameroon and Gabon. **Chorology:** Lower Guinea. **Star:** blue. **Alt.
range:** 1-200m (1). **Flowering:** Apr (1). **Fruiting:** Apr (1).
Specimens: - *Bambuko F.R.:* Keay R. W. J. FHI 37438. *Mabeta-
Moliwe:* Wheatley J. I. 221. *Mokoko:* Thomas 10012, 10028

Diospyros viridicans Hiern

F.W.T.A. 2: 10 (1963); Fl. Gabon 18: 163 (1970); Fl. Cameroun 11: 163 (1970); Bull. Jard. Bot. Nat. Belg. 48: 296 (1978); Keay R.W.J., Trees of Nigeria: 385 (1989); Hawthorne W., F.G.F.T. Ghana: 68 (1990).

Habit: tree to 20 m. **Guild:** shade-bearer. **Habitat:** forest. **Distribution:** Sierra Leone to Cameroon, Gabon, Congo (Brazzaville), Congo (Kinshasa) and Angola (Cabinda). **Chorology:** Guineo-Congolian. **Star:** green. **Alt. range:** 1-200m (1), 201-400m (1). **Flowering:** Apr (1). **Specimens:** - *Etinde:* Maitland T. D. 768. *Mokoko River F.R.:* Tchouto P. 1149

Diospyros zenkeri (Gürke) F.White

F.W.T.A. 2: 14 (1963); Fl. Gabon 18: 165 (1970); Fl. Cameroun 11: 165 (1970); Bull. Jard. Bot. Nat. Belg. 48: 336 (1978); Keay R.W.J., Trees of Nigeria: 387 (1989).

Habit: tree to 20 m. **Habitat:** forest. **Distribution:** Nigeria, Cameroon, Gabon, Congo (Brazzaville) and Congo (Kinshasa). **Chorology:** Guineo-Congolian. **Star:** green. **Alt. range:** 1-200m (5), 201-400m (1). **Flowering:** Apr (1). **Fruiting:** Sep (1), Oct (1). **Specimens:** - *Etinde:* Maitland T. D. 619. *Mokoko River F.R.:* Mbani J. M. 351, 381; Ndam 1164; Sonké 1224; Tchouto P. 1072, 1106; Thomas 10150. *Onge:* Akogo M. 118; Tchouto P. 705; Watts J. 712

Diospyros sp. nov.

Specimens: - *Mokoko River F.R.:* Ndam N. 1223

ERICACEAE

Agauria salicifolia (Comm. ex Lam.) Hook.f. ex Oliv.

F.W.T.A. 2: 2 (1963); Fl. Cameroun 11: 187 (1970); Keay R.W.J., Trees of Nigeria: 380 (1989).

Habit: tree to 12 m. **Habitat:** forest-grassland transition. **Distribution:** SE Nigeria, W. Cameroon and Bioko. **Chorology:** Western Cameroon Uplands (montane). **Star:** gold. **Alt. range:** 1200-1400m (1), 1801-2000m (1), 2001-2200m (2). **Flowering:** Oct (1). **Fruiting:** Sep (2). **Specimens:** - *Eastern slopes:* Watts J. 427. *Etinde:* Tekwe C. 169, 264. *No locality:* Mann G. 1209

Erica mannii (Hook.f.) Beentje

Fl. Cameroun 11: 201 (1970); Utafiti 3: 13 (1990).
    Philippia mannii (Hook.f.) Alm & Fries, F.W.T.A. 2: 2 (1963).

Habit: shrub to 4 m. **Habitat:** forest-grassland transition. **Distribution:** Bioko and W. Cameroon. **Chorology:** Western Cameroon Uplands (montane). **Star:** gold. **Alt. range:** 1800-2000m (1), 2001-2200m (1), 2201-2400m (2), 2801-3000m (1). **Flowering:** Sep (1), Oct (2), Nov (1). **Specimens:** - *Eastern slopes:* Banks H. 90; Tchouto P. 282; Thomas D. W. 9328. *Etinde:* Tekwe C. 269. *No locality:* Mann G. 1289

Erica tenuipilosa (Engl. ex Alm & T.C.E.Fr.) Cheek subsp. tenuipilosa

Fl. Cameroun 11: 195 (1970); Kew Bull. 52: 753 (1997).
    Blaeria spicata Hochst ex A.Rich subsp. *mannii* (Engl.) Wickens, Kew Bull. 27: 513 (1972); Nord. J. Bot. 5: 463 (1985).
    Blaeria mannii (Engl.) Engl., F.W.T.A. 2: 2 (1963).

Habit: shrub to 0.5 m. **Habitat:** forest-grassland transition. **Distribution:** Ivory Coast to Bioko and W. Cameroon. **Chorology:** Guineo-Congolian (montane). **Star:** green. **Alt. range:** 1800-2000m (2), 2001-2200m (1), 2401-2600m (1). **Flowering:** Sep (1), Oct (2). **Specimens:** - *Eastern slopes:* Thomas D. W. 9288; Watts J. 447. *Etinde:* Tekwe C. 275. *No locality:* Mann G. 1280

EUPHORBIACEAE

A. Radcliffe-Smith, S. Cable & M. Cheek (K)

Acalypha brachystachya Hornem.

F.W.T.A. 1: 409 (1958); F.T.E.A. Euphorbiaceae (1): 203 (1987).

Habit: herb. **Distribution:** palaeotropical. **Star:** excluded. **Specimens:** - *Eastern slopes:* Lehmbach H. 190

Acalypha manniana Müll.Arg.

F.W.T.A. 1: 409 (1958).

Habit: shrub to 2.5 m. **Habitat:** forest. **Distribution:** Ghana, Cameroon, Bioko, Congo (Kinshasa), Rwanda, Burundi and Uganda. **Chorology:** Guineo-Congolian (montane). **Star:** green. **Alt. range:** 1-200m (2), 601-800m (1), 1201-1400m (2). **Flowering:** Mar (1), May (1), Jul (1), Oct (1). **Fruiting:** May (1), Jul (1). **Specimens:** - *Eastern slopes:* Tchouto P. 52; Thomas D. W. 9513. *Mabeta-Moliwe:* Cheek M. 3542; Tchouto P. 197. *No locality:* Mann G. 1270

Acalypha racemosa Wall. ex Baill.

F.W.T.A. 1: 409 (1958); F.T.E.A. Euphorbiaceae (1): 187 (1987).

Habit: shrub to 3 m. **Guild:** pioneer. **Distribution:** tropical Africa, India, Sri Lanka and Java. **Star:** excluded. **Alt. range:** 200-400m (1). **Specimens:** - *Mabeta-Moliwe:* Mann G. 767. *No locality:* Mann G. 1254

Acalypha sp.

Specimens: - *Etinde:* Tekwe C. 79

Alchornea cordifolia (Schum. & Thonn.) Müll.Arg.

F.W.T.A. 1: 403 (1958); Meded. Land. Wag. 82(3): 131 (1982); F.T.E.A. Euphorbiaceae (1): 252 (1987); Hawthorne W., F.G.F.T. Ghana: 131 (1990).

Habit: tree or shrub to 8 m. **Guild:** pioneer. **Habitat:** forest. **Distribution:** tropical Africa. **Star:** excluded. **Alt. range:** 1-200m (6). **Flowering:** Feb (1), Apr (1), May (2). **Fruiting:** Feb (1), Mar (1), Apr (3), May (2). **Specimens:** - *Etinde:* Maitland T. D. 181; Tchouto P. 5; Tekwe C. 60. *Mabeta-Moliwe:* Ndam N. 493, 536; Sunderland T. C. H. 1107; Watts J. 229; Wheatley J. I. 187

Alchornea floribunda Müll.Arg.

F.W.T.A. 1: 403 (1958); F.T.E.A. Euphorbiaceae (1): 253 (1987); Hawthorne W., F.G.F.T. Ghana: 102 (1990).

Habit: tree or shrub to 4.5 m. **Guild:** shade-bearer. **Habitat:** forest. **Distribution:** Sierra Leone to Bioko, Cameroon, Gabon, Congo (Kinshasa), Uganda and Sudan. **Chorology:** Guineo-Congolian. **Star:** green. **Alt. range:** 200-400m (1), 801-1000m (2). **Flowering:** Feb (1). **Fruiting:** Jul (1). **Specimens:** - *Eastern*

47

slopes: Maitland T. D. 1103; Tekwe C. 105. *Etinde:* Maitland T. D. 564; Wheatley J. I. 17

Alchornea hirtella Benth.

Specimens: - *Mabeta-Moliwe:* Plot Voucher W238

Alchornea laxiflora (Benth.) Pax & K.Hoffm.
F.W.T.A. 1: 403 (1958); Kew Bull. 29: 435 (1974); Meded. Land. Wag. 82(3): 133 (1982); F.T.E.A. Euphorbiaceae (1): 257 (1987).

Habit: tree or shrub to 8 m. Habitat: forest and farmbush. Distribution: Nigeria to East and South Africa. Star: excluded. Alt. range: 800-1000m (1).Fruiting: Apr (1). Specimens: - *Eastern slopes:* Kwangue A. 37; Maitland T. D. 454

Amanoa strobilacea Müll.Arg.
F.W.T.A. 1: 371 (1958); Hawthorne W., F.G.F.T. Ghana: 86 (1990).

Habit: tree to 10 m. Habitat: forest. Distribution: Liberia, Ghana, Cameroon and Angola (Cabinda). Chorology: Guineo-Congolian. Star: green. Specimens: - *Mabeta-Moliwe:* Plot Voucher W 373

Antidesma laciniatum Müll.Arg. var. laciniatum
F.W.T.A. 1: 374 (1958); F.T.E.A. Euphorbiaceae (2): 572 (1988); Keay R.W.J., Trees of Nigeria: 177 (1989).

Habit: tree or shrub to 14 m. Habitat: forest, gaps, stream banks and farmbush. Distribution: Ivory Coast to Bioko, Cameroon, Rio Muni and Congo (Kinshasa). Chorology: Guineo-Congolian. Star: green. Alt. range: 1-200m (24), 801-1000m (1). Flowering: Mar (4), Apr (2), May (3), Aug (1), Dec (1). Fruiting: Mar (1), Apr (1), May (4), Jun (6), Jul (1), Aug (2), Dec (3). Specimens: - *Eastern slopes:* Maitland T. D. 561. *Etinde:* Hunt L. V. 23; Kwangue A. 89; Preuss P. R. 1104; Tchouto P. 76; Tekwe C. 63, 75. *Mabeta-Moliwe:* Baker W. J. 201; Cable S. 608; Jaff B. 135; Ndam N. 501; Plot Voucher W 97, W 521, W 564, W 867, W 928; Sunderland T. C. H. 1013, 1014, 1102, 1136, 1288, 1297, 1454, 1467; Watts J. 211, 371; Wheatley J. I. 256, 359; von Rege I. 50. *Mokoko River F.R.:* Acworth J. M. 55; Batoum A. 30; Tchouto P. 622

Antidesma laciniatum Müll.Arg. var. membranaceum Müll.Arg.
F.W.T.A. 1: 375 (1958); F.T.E.A. Euphorbiaceae (2): 573 (1988); Keay R.W.J., Trees of Nigeria: 177 (1989); Hawthorne W., F.G.F.T. Ghana: 89 (1990).

Habit: tree or shrub to 14 m. Guild: shade-bearer. Habitat: forest and farmbush. Distribution: Guinea to Cameroon, Rio Muni and Congo (Kinshasa). Chorology: Guineo-Congolian. Star: green. Alt. range: 1-200m (2). Flowering: May (1). Fruiting: May (1). Specimens: - *Mabeta-Moliwe:* Sunderland T. C. H. 1309; Wheatley J. I. 247

Antidesma membranaceum Müll.Arg.
F.W.T.A. 1: 375 (1958); F.T.E.A. Euphorbiaceae (2): 574 (1988); Keay R.W.J., Trees of Nigeria: 177 (1989); Hawthorne W., F.G.F.T. Ghana: 89 (1990).

Habit: tree to 18 m. Guild: pioneer. Habitat: forest. Distribution: tropical Africa. Star: excluded. Alt. range: 1-200m (4), 201-400m (2). Flowering: Apr (1), Jun (1). Fruiting: Jun (2). Specimens: -

Mabeta-Moliwe: Jaff B. 75. *Mokoko River F.R.:* Acworth J. M. 265; Fraser P. 454; Mbani J. M. 378; Pouakouyou D. 89; Tchouto P. 1132, 1244

Antidesma vogelianum Müll.Arg.
F.W.T.A. 1: 375 (1958); F.T.E.A. Euphorbiaceae (2): 576 (1988); Keay R.W.J., Trees of Nigeria: 179 (1989).

Habit: tree to 10 m. Habitat: forest and stream banks. Distribution: SE Nigeria, Cameroon, Gabon, Congo (Kinshasa) and East Africa. Star: excluded. Alt. range: 1-200m (13), 201-400m (1). Flowering: Mar (1), Apr (2), May (3), Aug (1), Oct (1), Nov (1). Fruiting: Mar (1), Apr (1), May (4), Jun (2), Aug (1), Oct (1), Nov (1). Specimens: - *Etinde:* Mbani J. M. 87; Preuss P. R. 1321. *Mabeta-Moliwe:* Watts J. 138, 183, 252. *Mokoko River F.R.:* Batoum A. 31; Ekema N. 1057, 1070, 1206; Pouakouyou D. 53. *Onge:* Harris D. J. 3662; Tchouto P. 501, 931; Watts J. 695, 1010, 1010a

Antidesma sp.

Specimens: - *Eastern slopes:* Maitland T. D. 191

Bridelia micrantha (Hochst.) Baill.
F.W.T.A. 1: 370 (1958); F.T.E.A. Euphorbiaceae (1): 127 (1987); Keay R.W.J., Trees of Nigeria: 158 (1989); Hawthorne W., F.G.F.T. Ghana: 95 (1990).

Habit: tree to 18 m. Guild: pioneer. Habitat: forest regrowth. Distribution: tropical and subtropical Africa. Star: excluded. Alt. range: 1-200m (2), 201-400m (1), 401-600m (2). Flowering: Feb (1), Apr (1), Oct (1). Fruiting: Apr (1), Oct (2). Specimens: - *Eastern slopes:* Deistel H. 615. *Etinde:* Cheek M. 5891; Nkeng P. 24; Tchouto P. 121, 452. *Mabeta-Moliwe:* Plot Voucher W 725, W 1150. *Onge:* Tchouto P. 814

Bridelia scleroneura Müll.Arg.
F.W.T.A. 1: 370 (1958); Meded. Land. Wag. 82(3): 134 (1982); F.T.E.A. Euphorbiaceae (1): 122 (1987); Keay R.W.J., Trees of Nigeria: 159 (1989).

Habit: tree or shrub to 5 m. Habitat: forest gaps. Distribution: Guinea to Congo (Kinshasa), Zimbabwe, Ethiopia and Yemen. Star: excluded. Alt. range: 1-200m (1). Flowering: Apr (1). Specimens: - *Mokoko River F.R.:* Acworth J. M. 236

Bridelia speciosa Müll.Arg.
F.W.T.A. 1: 370 (1958); Keay R.W.J., Trees of Nigeria: 158 (1989).

Habit: tree to 10 m. Habitat: forest. Distribution: SE Nigeria, Cameroon and CAR. Chorology: Guineo-Congolian (montane). Star: green. Alt. range: 1400-1600m (1). Specimens: - *No locality:* Mann G. 1215

Caperonia latifolia Pax
F.W.T.A. 1: 397 (1958).

Habit: herb. Distribution: Nigeria, SW Cameroon and S. Tomé, also in tropical America. Star: blue. Specimens: - *Etinde:* Dusen P. K. H. 281

Cleistanthus letouzeyi J.Léonard
Adansonia (ser.2) 3: 66 (1963).

Habit: tree or shrub to 8 m. **Habitat:** forest. **Distribution:** Cameroon. **Chorology:** Lower Guinea. **Star:** gold. **Alt. range:** 1-200m (1).**Fruiting:** Aug (1). **Specimens:** - *Onge:* Watts J. 708

Cleistanthus sp.

Specimens: - *Onge:* Tchouto P. 818, 853

Croton lobatus L.

F.W.T.A. 1: 394 (1958).

Habit: herb. **Guild:** pioneer. **Habitat:** roadsides and villages. **Distribution:** Senegal to Bioko, Cameroon, Sudan and Ethiopia. **Star:** green. **Alt. range:** 1-200m (1). **Flowering:** Jul (1). **Specimens:** - *Mabeta-Moliwe:* Cheek M. 3526

Croton longiracemosus Hutch.

F.W.T.A. 1: 396 (1958); Keay R.W.J., Trees of Nigeria: 156 (1989).

Habit: tree to 25 m. **Habitat:** forest. **Distribution:** Nigeria, Cameroon and Congo (Kinshasa). **Chorology:** Guineo-Congolian. **Star:** green. **Alt. range:** 200-400m (1), 801-1000m (1). **Flowering:** Mar (1). **Fruiting:** Mar (1). **Specimens:** - *Eastern slopes:* Maitland T. D. 1106; Tchouto P. 94b. *Etinde:* Thomas D. W. 9731

Crotonogyne preussii Pax

F.W.T.A. 1: 400 (1958).

Habit: tree or shrub to 8 m. **Habitat:** forest and forest gaps. **Distribution:** Nigeria and Cameroon. **Chorology:** Lower Guinea. **Star:** blue. **Alt. range:** 1-200m (13), 201-400m (5). **Flowering:** Mar (3), Apr (2), May (2), Jun (2), Aug (1), Sep (1), Oct (3), Nov (1). **Fruiting:** Mar (2), Apr (2), May (1), Jun (1), Aug (1), Sep (1), Oct (2), Nov (1). **Specimens:** - *Bambuko F.R.:* Watts J. 669. *Etinde:* Kwangue A. 73; Preuss P. R. 1220. *Mabeta-Moliwe:* Baker W. J. 327; Sunderland T. C. H. 1117, 1243; Watts J. 182, 186, 380, 383; Wheatley J. I. 36, 252. *Mokoko River F.R.:* Ndam N. 1074, 1125, 1309. *Onge:* Harris D. J. 3729; Tchouto P. 711; Watts J. 710, 848, 850, 966

Crotonogynopsis usambarica Pax

F.W.T.A. 1: 404 (1958); F.T.E.A. Euphorbiaceae (1): 215 (1987); Bull. Jard. Bot. Nat. Belg. 65: 138 (1996).

Habit: tree or shrub 2-5 m tall. **Distribution:** Cameroon to Congo (Kinshasa), Uganda and Tanzania. **Chorology:** Guineo-Congolian. **Star:** green. **Specimens:** - *Eastern slopes:* Deistel H. 218

Cyathogyne viridis Müll.Arg.

F.W.T.A. 1: 375 (1958).

Habit: herb. **Habitat:** forest. **Distribution:** Nigeria, Cameroon, Rio Muni, Gabon and Congo (Brazzaville). **Chorology:** Guineo-Congolian. **Star:** green. **Alt. range:** 1-200m (7). **Flowering:** Mar (1), May (2), Jun (2), Nov (2), Dec (1). **Specimens:** - *Etinde:* Cheek M. 5936; Nkeng P. 46; Preuss P. R. 1210. *Mabeta-Moliwe:* Cheek M. 3498; Ndam N. 481; Plot Voucher W 681; Sunderland T. C. H. 1376; Wheatley J. I. 324. *Onge:* Thomas D. W. 9757, 9925

Cyrtogonone argentea (Pax) Prain

F.W.T.A. 1: 399 (1958); Keay R.W.J., Trees of Nigeria: 158 (1989).

Habit: tree to 30 m. **Habitat:** forest. **Distribution:** Nigeria, Cameroon and Rio Muni. **Chorology:** Lower Guinea. **Star:** blue. **Alt. range:** 1-200m (3), 801-1000m (1). **Flowering:** Apr (3). **Specimens:** - *Mabeta-Moliwe:* Mildbraed G. W. J. 10523; Sunderland T. C. H. 1210; Tchouto P. 146; Watts J. 195; Wheatley 204

Dichostemma glaucescens Pierre

F.W.T.A. 1: 416 (1958); Keay R.W.J., Trees of Nigeria: 151 (1989).

Habit: tree to 12 m. **Habitat:** forest. **Distribution:** SE Nigeria, Cameroon, Gabon, Congo (Kinshasa) and Angola (Cabinda). **Chorology:** Guineo-Congolian. **Star:** green. **Specimens:** - *Etinde:* Dundas J. FHI 8403

Discoclaoxylon hexandrum (Müll.Arg.) Pax & K.Hoffm.

F.T.E.A. Euphorbiaceae (1): 280 (1987); Keay R.W.J., Trees of Nigeria: 176 (1989); Hawthorne W., F.G.F.T. Ghana: 132 (1990). *Claoxylon hexandrum* Müll.Arg., F.W.T.A. 1: 401 (1958).

Habit: tree to 20 m. **Guild:** shade-bearer. **Habitat:** forest and stream banks. **Distribution:** Sierra Leone to Congo (Kinshasa) and Uganda. **Chorology:** Guineo-Congolian. **Star:** green. **Alt. range:** 1-200m (8), 1001-1200m (1). **Flowering:** Feb (1), Mar (1), Apr (3), Oct (1). **Fruiting:** Apr (4). **Specimens:** - *Eastern slopes:* Maitland T. D. 234. *Etinde:* Nkeng P. 28; Wheatley J. I. 606. *Mabeta-Moliwe:* Jaff B. 120; Sunderland T. C. H. 1189; Watts J. 120, 155; Wheatley J. I. 131, 204. *Mokoko River F.R.:* Akogo M. 284

Discoglypremna caloneura (Pax) Prain

F.W.T.A. 1: 403 (1958); F.T.E.A. Euphorbiaceae (1): 223 (1987); Keay R.W.J., Trees of Nigeria: 169 (1989); Hawthorne W., F.G.F.T. Ghana: 131 (1990); Bull. Jard. Bot. Nat. Belg. 64: 202 (1995).

Habit: tree 9-35 m tall. **Guild:** pioneer. **Habitat:** forest. **Distribution:** Guinea to Bioko, Cameroon, Gabon, Congo (Brazzaville), Congo (Kinshasa), Angola (Cabinda), CAR and Uganda. **Chorology:** Guineo-Congolian. **Star:** green. **Specimens:** - *Etinde:* Maitland T. D. 590

Drypetes gilgiana (Pax) Pax & K.Hoffm.

F.W.T.A. 1: 382 (1958); Keay R.W.J., Trees of Nigeria: 164 (1989); Hawthorne W., F.G.F.T. Ghana: 101 (1990).

Habit: tree to 12 m. **Guild:** shade-bearer. **Habitat:** forest. **Distribution:** Guinea Bissau to Cameroon. **Chorology:** Guineo-Congolian. **Star:** green. **Alt. range:** 800-1000m (1). **Specimens:** - *Mabeta-Moliwe:* Mildbraed G. W. J. 10758

Drypetes leonensis (Pax) Pax & K. Hoffm.

F.W.T.A. 1: 381 (1958); Keay R.W.J., Trees of Nigeria: 164 (1989); Hawthorne W., F.G.F.T. Ghana: 101 (1990).

Habit: tree to 23 m. **Guild:** shade-bearer. **Habitat:** forest. **Distribution:** Guinea to Cameroon. **Chorology:** Guineo-Congolian. **Star:** green. **Alt. range:** 1-200m (1). **Specimens:** - *Mokoko River F.R.:* Mbani J. M. 387

Drypetes molunduana Pax & K.Hoffm.

F.W.T.A. 1: 381 (1958); Keay R.W.J., Trees of Nigeria: 164 (1989).

Habit: tree or shrub to 4 m. **Habitat:** forest. **Distribution:** Nigeria and Cameroon. **Chorology:** Lower Guinea. **Star:** blue. **Alt. range:** 1-200m (3). **Flowering:** Jun (1). **Fruiting:** May (1), Jun (2). **Specimens:** - *Etinde:* Kwangue A. 95; Mbani J. M. 123; Tchouto P. 231. *Southern Bakundu F.R.:* Brenan J. P. M. 9296

Drypetes paxii Hutch.

F.W.T.A. 1: 381 (1958); Keay R.W.J., Trees of Nigeria: 164 (1989).

Habit: tree to 30 m. **Habitat:** forest. **Distribution:** Nigeria, Cameroon, Gabon and Congo (Kinshasa). **Chorology:** Guineo-Congolian. **Star:** green. **Specimens:** - *Mabeta-Moliwe:* Maitland T. D. 423

Drypetes preussii (Pax) Hutch.

F.W.T.A. 1: 382 (1958); Keay R.W.J., Trees of Nigeria: 165 (1989).

Habit: tree to 21 m. **Habitat:** forest. **Distribution:** SE Nigeria and SW Cameroon. **Chorology:** Western Cameroon. **Star:** gold. **Alt. range:** 1-200m (4).**Fruiting:** Apr (2), May (1), Jun (1), Aug (1). **Specimens:** - *Mokoko River F.R.:* Acworth J. M. 58; Akogo M. 238; Batoum A. 18; Pouakouyou D. 99. *Southern Bakundu F.R.:* Olorunfemi J. FHI 30739

Drypetes principum (Müll.Arg.) Hutch.

F.W.T.A. 1: 381 (1958); Keay R.W.J., Trees of Nigeria: 165 (1989); Hawthorne W., F.G.F.T. Ghana: 101 (1990).

Habit: tree to 10 m. **Guild:** shade-bearer. **Habitat:** forest. **Distribution:** Guinea to Bioko and Cameroon. **Chorology:** Guineo-Congolian. **Star:** green. **Alt. range:** 1-200m (2), 1201-1400m (1).**Fruiting:** May (1), Jun (1), Sep (1). **Specimens:** - *Etinde:* Wheatley J. I. 557; Winkler H. J. P. 551. *Mabeta-Moliwe:* Cheek M. 3568. *Mokoko River F.R.:* Ndam N. 1302; Pouakouyou D. 106; Tchouto P. 1102

Drypetes cf. principum (Müll.Arg.) Hutch.

Specimens: - *Etinde:* Mbani J. M. 15. *Mokoko River F.R.:* Acworth J. M. 102

Drypetes sp. aff. principum (Müll.Arg.) Hutch.

Specimens: - *Mokoko River F.R.:* Akogo M. 160; Tchouto P. 1069; Watts J. 1111

Drypetes staudtii (Pax) Hutch.

F.W.T.A. 1: 382 (1958); Keay R.W.J., Trees of Nigeria: 164 (1989).

Habit: tree to 13 m. **Habitat:** forest. **Distribution:** Nigeria and Cameroon. **Chorology:** Lower Guinea. **Star:** blue. **Alt. range:** 1-200m (7), 201-400m (1).**Fruiting:** Apr (4), May (1), Jun (1). **Specimens:** - *Etinde:* Kwangue A. 68. *Mokoko River F.R.:* Acworth J. M. 107; Akogo M. 222; Ndam N. 1061; Pouakouyou D. 46; Tchouto P. 1085, 1131; Watts J. 1056

Drypetes tessmanniana (Pax) Pax & K.Hoffm.

F.W.T.A. 1: 382 (1958).

Habit: shrub to 2 m. **Habitat:** forest. **Distribution:** Mount Cameroon and Rio Muni. **Chorology:** Lower Guinea. **Star:** black.

Flowering: Nov (1). **Specimens:** - *Mabeta-Moliwe:* Mildbraed G. W. J. 10600

Drypetes sp. nov. aff. pellegrini Leandri

Specimens: - *Mabeta-Moliwe:* Cheek M. 3571

Erythrococca africana (Baill.) Prain

F.W.T.A. 1: 401 (1958).

Habit: shrub to 3 m. **Guild:** shade-bearer. **Habitat:** forest. **Distribution:** Senegal to Cameroon. **Chorology:** Guineo-Congolian (montane). **Star:** green. **Alt. range:** 1400-1600m (1). **Flowering:** Sep (1). **Specimens:** - *Eastern slopes:* Tekwe C. 321

Erythrococca cf. africana (Baill.) Prain

Specimens: - *Etinde:* Wheatley J. I. 491

Erythrococca anomala (Juss. ex Poir.) Prain

F.W.T.A. 1: 401 (1958).

Habit: shrub to 3 m. **Guild:** shade-bearer. **Habitat:** forest. **Distribution:** Guinea to Bioko and Cameroon. **Chorology:** Guineo-Congolian. **Star:** green. **Alt. range:** 1-200m (8), 201-400m (5). **Flowering:** Jul (2), Aug (3), Nov (2), Dec (1). **Fruiting:** Mar (1), Jul (1), Aug (3), Oct (2), Nov (1). **Specimens:** - *Etinde:* Cheek M. 5418, 5457, 5875; Maitland T. D. 624. *Mabeta-Moliwe:* Baker W. J. 229; Cheek M. 3534; Hurst J. 6; Plot Voucher W 310, W 399, W 1046; Wheatley J. I. 55; von Rege I. 7, 45, 101. *Onge:* Cheek M. 5139; Thomas D. W. 9882; Watts J. 854

Erythrococca hispida (Pax) Prain

F.W.T.A. 1: 401 (1958).

Habit: tree or shrub to 5 m. **Habitat:** forest. **Distribution:** Cameroon. **Chorology:** Lower Guinea (montane). **Star:** gold. **Alt. range:** 1200-1400m (2), 1401-1600m (1). **Flowering:** May (2). **Specimens:** - *Eastern slopes:* Mbani J. M. 104, 117. *No locality:* Preuss P. R. 908

Erythrococca membranacea (Müll.Arg.) Prain

F.W.T.A. 1: 401 (1958).

Habit: shrub to 3 m. **Habitat:** forest. **Distribution:** Nigeria and SW Cameroon. **Chorology:** Lower Guinea. **Star:** gold. **Alt. range:** 200-400m (3), 801-1000m (1), 1001-1200m (1), 1201-1400m (1). **Flowering:** Mar (1), Oct (2), Dec (1). **Fruiting:** Mar (1). **Specimens:** - *Bambuko F.R.:* Watts J. 592. *Eastern slopes:* Lehmbach H. 212. *Etinde:* Cable S. 340; Tchouto P. 424. *Mokoko River F.R.:* Tchouto P. 527. *No locality:* Mann G. 1197. *Onge:* Tchouto P. 869

Euphorbia heterophylla L.

F.W.T.A. 1: 421 (1958); F.T.E.A. Euphorbiaceae (2): 431 (1988).

Habit: herb. **Guild:** pioneer. **Habitat:** farmbush. **Distribution:** tropical Africa and Central America. **Star:** excluded. **Alt. range:** 1-200m (1). **Flowering:** Apr (1). **Fruiting:** Apr (1). **Specimens:** - *Mabeta-Moliwe:* Sunderland T. C. H. 1219

Euphorbia hirta L.

F.W.T.A. 1: 419 (1958); F.T.E.A. Euphorbiaceae (2): 415 (1988).

Habit: herb. **Guild:** pioneer. **Distribution:** pantropical. **Star:** excluded. **Specimens:** - *Mabeta-Moliwe:* Hutchinson J. & Metcalfe C. R. 135

Euphorbia schimperiana Scheele
F.W.T.A. 1: 421 (1958); F.T.E.A. Euphorbiaceae (2): 433 (1988).

Habit: herb. **Habitat:** forest-grassland transition. **Distribution:** tropical Africa. **Chorology:** Afromontane. **Star:** green. **Alt. range:** 2000-2200m (1), 2201-2400m (1). **Specimens:** - *Eastern slopes:* Thomas D. W. 9332. *No locality:* Mann G. 1265

Excoecaria guineensis (Benth.) Müll.Arg.
Bull. Jard. Bot. Nat. Belg. 29: 133 (1959); Keay R.W.J., Trees of Nigeria: 175 (1989).
Sapium guineense (Benth.) Kuntze, F.W.T.A. 1: 415 (1958).

Habit: shrub to 5 m. **Distribution:** Sierra Leone to Cameroon, Congo (Brazzaville) and Congo (Kinshasa). **Chorology:** Guineo-Congolian. **Star:** green. **Alt. range:** 800-1000m (1). **Specimens:** - *Mabeta-Moliwe:* Mildbraed G. W. J. 10548

Grossera paniculata Pax
Adansonia (ser.2) 3: 70 (1963).

Habit: tree to 6 m. **Habitat:** forest. **Distribution:** Cameroon and Gabon. **Chorology:** Lower Guinea. **Star:** blue. **Alt. range:** 1-200m (5). **Flowering:** Apr (3), May (1). **Fruiting:** Jun (1). **Specimens:** - *Mabeta-Moliwe:* Jaff B. 124; Sunderland T. C. H. 1238, 1369; Watts J. 337; Wheatley J. I. 130

Grossera vignei Hoyle
F.W.T.A. 1: 398 (1958); Keay R.W.J., Trees of Nigeria: 172 (1989); Hawthorne W., F.G.F.T. Ghana: 132 (1990).

Habit: tree to 15 m. **Guild:** shade-bearer. **Habitat:** forest. **Distribution:** Ivory Coast to Cameroon. **Chorology:** Guineo-Congolian. **Star:** green. **Specimens:** - *No locality:* Box H. E. 3565

Gymnanthes inopinata (Prain) Esser, adhoc ined.
Duvigneaudia inopinata (Prain) J.Léonard Adansonia (ser.2) 3: 69 (1963).
Sebastiania inopinata Prain, F.W.T.A. 1: 415 (1958).

Habit: shrub to 5 m. **Distribution:** Cameroon. **Chorology:** Lower Guinea. **Star:** gold. **Specimens:** - *No locality:* Mann G. 755

Hamilcoa zenkeri (Pax) Prain
F.W.T.A. 1: 414 (1958).

Habit: woody climber or tree to 15 m. **Habitat:** forest. **Distribution:** Cameroon. **Chorology:** Lower Guinea. **Star:** gold. **Specimens:** - *Southern Bakundu F.R.:* Brenan J. P. M. 9430; Olorunfemi J. FHI 30705

Klaineanthus gaboniae Pierre ex Prain
F.W.T.A. 1: 413 (1958); Keay R.W.J., Trees of Nigeria: 172 (1989).

Habit: tree to 30 m. **Habitat:** forest. **Distribution:** SE Nigeria, Cameroon, Gabon and Congo (Kinshasa). **Chorology:** Guineo-Congolian. **Star:** green. **Alt. range:** 1-200m (5), 201-400m (1). **Flowering:** Apr (3). **Fruiting:** Apr (1), May (1), Aug (1). **Specimens:** - *Etinde:* Thomas D. W. 9736. *Mabeta-Moliwe:* Baker W. J. 332; Jaff B. 78; Ndam N. 499; Sunderland T. C. H. 1253; Wheatley J. I. 159. *Southern Bakundu F.R.:* Olorunfemi J. FHI 30711

Leeuwenbergia africana Letouzey & N.Hallé
Adansonia (ser.2) 14: 387 (1974).

Habit: tree to 40 m. **Habitat:** forest. **Distribution:** Cameroon, Gabon and Congo (Kinshasa). **Chorology:** Guineo-Congolian. **Star:** green. **Alt. range:** 1-200m (1), 201-400m (2). **Flowering:** Jun (1). **Fruiting:** Oct (1). **Specimens:** - *Etinde:* Thomas D. W. 9728. *Mokoko River F.R.:* Ekema N. 1117b. *Onge:* Tchouto P. 835

Macaranga monandra Müll.Arg.
F.W.T.A. 1: 407 (1958); F.T.E.A. Euphorbiaceae (1): 248 (1987); Keay R.W.J., Trees of Nigeria: 157 (1989).

Habit: tree to 16 m. **Habitat:** forest. **Distribution:** SE Nigeria, Cameroon, Gabon, Congo (Kinshasa), Angola, Uganda and Tanzania. **Chorology:** Guineo-Congolian. **Star:** green. **Alt. range:** 1-200m (5), 601-800m (1). **Flowering:** Mar (2), Apr (4). **Fruiting:** Jun (1). **Specimens:** - *Eastern slopes:* Tekwe C. 96. *Etinde:* Maitland T. D. 522; Tchouto P. 74. *Mabeta-Moliwe:* Sunderland T. C. H. 1204; Watts J. 100. *Mokoko River F.R.:* Acworth J. M. 226; Akogo M. 233; Mbani J. M. 342

Macaranga occidentalis (Müll.Arg.) Müll.Arg.
F.W.T.A. 1: 407 (1958); Keay R.W.J., Trees of Nigeria: 157 (1989).

Habit: tree to 25 m. **Habitat:** forest. **Distribution:** SE Nigeria, Bioko and Cameroon. **Chorology:** Lower Guinea (montane). **Star:** blue. **Alt. range:** 200-400m (1), 801-1000m (1), 1201-1400m (2). **Flowering:** Feb (1), Mar (1). **Fruiting:** Jul (1), Sep (1). **Specimens:** - *Eastern slopes:* Nkeng P. 65; Tekwe C. 112. *Etinde:* Tekwe C. 179; Wheatley J. I. 24. *No locality:* Mann G. 771

Macaranga schweinfurthii Pax
F.W.T.A. 1: 407 (1958); F.T.E.A. Euphorbiaceae (1): 241 (1987); Keay R.W.J., Trees of Nigeria: 157 (1989).

Habit: tree to 25 m. **Habitat:** forest. **Distribution:** SE Nigeria, Cameroon, Gabon, Congo (Kinshasa), Angola, Tanzania and Sudan **Chorology:** Guineo-Congolian. **Star:** green. **Alt. range:** 200-400m (1). **Flowering:** Apr (1). **Specimens:** - *Mabeta-Moliwe:* Plot Voucher W 445; Wheatley J. I. 105

Macaranga spinosa Müll.Arg.
F.W.T.A. 1: 408 (1958); F.T.E.A. Euphorbiaceae (1): 249 (1987); Keay R.W.J., Trees of Nigeria: 158 (1989); Hawthorne W., F.G.F.T. Ghana: 135 (1990).

Habit: tree or shrub to 12 m. **Guild:** pioneer. **Habitat:** forest and farmbush. **Distribution:** Liberia and Ivory Coast, then SE Nigeria, Cameroon, Bioko, Gabon, Angola, Uganda and Tanzania. **Chorology:** Guineo-Congolian. **Star:** green. **Alt. range:** 1-200m (3). **Flowering:** Apr (1), May (2). **Specimens:** - *Mabeta-Moliwe:* Ndam N. 523, 535; Wheatley J. I. 153

Maesobotrya barteri (Baill.) Hutch.
F.W.T.A. 1: 374 (1958); Keay R.W.J., Trees of Nigeria: 176 (1989); Hawthorne W., F.G.F.T. Ghana: 132 (1990).

Habit: tree to 10 m. **Guild:** shade-bearer. **Habitat:** forest.
Distribution: Sierra Leone to Cameroon and Rio Muni.
Chorology: Guineo-Congolian. **Star:** green. **Alt. range:** 1-200m
(10), 401-600m (2), 601-800m (1). **Flowering:** Apr (3), May (3),
Jun (2). **Fruiting:** Apr (3), May (4), Jun (1). **Specimens:** - *Eastern
slopes:* Tchouto P. 142. *Etinde:* Mbani J. M. 157; Tchouto P. 117.
Mabeta-Moliwe: Jaff B. 172; Sunderland T. C. H. 1213; Tchouto
P. 194; Watts J. 339, 384. *Mokoko River F.R.:* Acworth J. M. 320;
Akogo M. 281, 306; Batoum A. 24; Watts J. 1067

Maesobotrya cf. barteri (Baill.) Hutch.

Specimens: - *Etinde:* Lighava M. 58

Maesobotrya dusenii (Pax) Pax
F.W.T.A. 1: 374 (1958); Keay R.W.J., Trees of Nigeria: 176
(1989).

Habit: tree to 14 m. **Habitat:** forest. **Distribution:** Nigeria,
Cameroon, Bioko and Rio Muni. **Chorology:** Lower Guinea. **Star:**
blue. **Alt. range:** 1-200m (6), 201-400m (1), 401-600m (1).
Flowering: Apr (1), May (2). **Fruiting:** Apr (1), May (4).
Specimens: - *Etinde:* Kwangue A. 60; Mbani J. M. 160. *Mokoko
River F.R.:* Akogo M. 286; Ekema N. 1022; Fraser P. 432;
Tchouto P. 1174; Watts J. 1085, 1193. *Southern Bakundu F.R.:*
Olorunfemi J. FHI 30544

Maesobotrya staudtii (Pax) Hutch.
F.W.T.A. 1: 374 (1958); Keay R.W.J., Trees of Nigeria: 176
(1989).

Habit: tree to 8 m. **Habitat:** forest. **Distribution:** SE Nigeria,
Cameroon, Gabon and Congo (Kinshasa). **Chorology:** Guineo-
Congolian. **Star:** green. **Alt. range:** 1-200m (12), 201-400m (5),
801-1000m (1). **Flowering:** May (5), Jun (3), Oct (1). **Fruiting:**
Jun (2), Aug (2), Oct (5). **Specimens:** - *Etinde:* Preuss P. R. 1366.
Mabeta-Moliwe: Baker W. J. 250; Mildbraed G. W. J. 10564; Plot
Voucher W 1172. *Mokoko River F.R.:* Ekema N. 1045, 1073,
1086, 1187; Fraser P. 440; Pouakouyou D. 34, 78, 97; Tchouto P.
1099; Watts J. 1089. *Onge:* Akogo M. 62; Ndam N. 744; Tchouto
P. 689; Watts J. 696, 789, 792

Mallotus oppositifolius (Geiseler) Müll.Arg.
F.W.T.A. 1: 402 (1958); F.T.E.A. Euphorbiaceae (1): 236 (1987);
Keay R.W.J., Trees of Nigeria: 152 (1989); Hawthorne W.,
F.G.F.T. Ghana: 131 (1990).

Habit: tree or shrub to 12 m. **Guild:** shade-bearer. **Habitat:** forest
and farmbush. **Distribution:** tropical Africa. **Star:** excluded. **Alt.
range:** 1-200m (8). **Flowering:** Mar (1), Apr (1), May (1), Jun (2),
Nov (1). **Fruiting:** Apr (2). **Specimens:** - *Limbe Botanic Garden:*
Wheatley J. I. 647. *Mabeta-Moliwe:* Cable S. 1430; Dunlap 251;
Plot Voucher W 463, W 898, W 1055; Sunderland T. C. H. 1266,
1303; Watts J. 364; Wheatley J. I. 213, 229. *Mokoko River F.R.:*
Acworth J. M. 266b

Mallotus subulatus Müll.Arg.
F.W.T.A. 1: 402 (1958); Hawthorne W., F.G.F.T. Ghana: 131
(1990).

Habit: tree or shrub to 5 m. **Habitat:** swamp forest. **Distribution:**
Sierra Leone to Bioko, Cameroon, Gabon and Congo (Kinshasa).
Chorology: Guineo-Congolian. **Star:** green. **Alt. range:** 1-200m
(1), 201-400m (1). **Flowering:** Feb (1), Apr (1). **Fruiting:** Feb (1).
Specimens: - *Etinde:* Maitland T. D. 368; Wheatley J. I. 16.
Mabeta-Moliwe: Dunlap 252. *Mokoko River F.R.:* Acworth J. M.
57

Mallotus cf. subulatus Müll.Arg.

Specimens: - *Limbe Botanic Garden:* Wheatley J. I. 660

Manniophyton africanum Müll.Arg.
Manniophyton fulvum Müll. Arg. F.W.T.A. 1: 400 (1958).

Habit: shrub or climber. **Guild:** light demander. **Habitat:** forest.
Distribution: Sierra Leone to Cameroon, Gabon, Congo
(Kinshasa), Angola and Sudan. **Chorology:** Guineo-Congolian.
Star: green. **Specimens:** - *Southern Bakundu F.R.:* Olorunfemi J.
FHI 30540

Maprounea membranacea Pax & K.Hoffm.
F.W.T.A. 1: 416 (1958); Keay R.W.J., Trees of Nigeria: 174
(1989).

Habit: tree to 15 m. **Habitat:** forest. **Distribution:** Nigeria,
Cameroon, Gabon and Congo (Kinshasa). **Chorology:** Guineo-
Congolian. **Star:** green. **Alt. range:** 1-200m (1). **Flowering:** Apr
(1). **Specimens:** - *Mokoko River F.R.:* Akogo M. 168

Mareya micrantha (Benth.) Müll.Arg.
F.W.T.A. 1: 404 (1958); Keay R.W.J., Trees of Nigeria: 175
(1989); Hawthorne W., F.G.F.T. Ghana: 132 (1990); Bull. Jard.
Bot. Nat. Belg. 65: 9 (1996).

Habit: tree or shrub to 8 m. **Guild:** shade-bearer. **Habitat:** forest.
Distribution: Guinea Bissau to Bioko, Cameroon, Gabon, Congo
(Brazzaville) and Angola (Cabinda). **Chorology:** Guineo-
Congolian. **Star:** green. **Alt. range:** 1-200m (1). **Flowering:** May
(1). **Specimens:** - *Etinde:* Maitland T. D. 582; Winkler H. J. P.
664. *Mabeta-Moliwe:* Tchouto P. 190

Mareyopsis longifolia (Pax) Pax & K.Hoffm.
F.W.T.A. 1: 403 (1958); Bull. Jard. Bot. Nat. Belg. 65: 16 (1996).

Habit: tree or shrub to 10 m. **Habitat:** forest. **Distribution:**
Nigeria, Cameroon, Gabon and Congo (Kinshasa). **Chorology:**
Guineo-Congolian. **Star:** green. **Alt. range:** 1-200m (3).
Flowering: May (1). **Fruiting:** May (1), Jun (1). **Specimens:** -
Mokoko River F.R.: Ekema N. 1205; Mbani J. M. 405; Ndam N.
1301; Tchouto P. 1096

Margaritaria discoidea (Baill.) Webster
J. Arnold. Arb. 48: 311 (1967); Meded. Land. Wag. 82(3): 145
(1982); F.T.E.A. Euphorbiaceae (2): 63 (1988); Keay R.W.J., Trees
of Nigeria: 167 (1989); Ann. Missouri Bot. Gard. 77: 217 (1990);
Hawthorne W., F.G.F.T. Ghana: 85 (1990).
Phyllanthus discoideus (Baill.) Müll.Arg., F.W.T.A. 1: 387
(1958).

Habit: tree 15-30 m tall. **Habitat:** forest. **Distribution:** tropical
and subtropical Africa. **Star:** excluded. **Alt. range:** 600-800m (1).
Specimens: - *Eastern slopes:* Maitland T. D. 443. *Etinde:*
Maitland T. D. 1138

Necepsia afzelii Prain
F.W.T.A. 1: 405 (1958); Hawthorne W., F.G.F.T. Ghana: 132
(1990).

Habit: tree to 13 m. **Habitat:** forest. **Distribution:** Sierra Leone to
Cameroon. **Chorology:** Guineo-Congolian. **Star:** blue.
Specimens: - *No locality:* Winkler H. J. P. 1078

Neoboutonia mannii Benth. & Hook.f.
F.W.T.A. 1: 404 (1958); Kew Bull. 29: 438 (1974); Keay R.W.J., Trees of Nigeria: 154 (1989).
Neoboutonia glabrescens Prain, F.W.T.A. 1: 404 (1958).

Habit: tree to 20 m. **Habitat:** forest and farmbush. **Distribution:** Liberia to Bioko, Cameroon and Rio Muni. **Chorology:** Guineo-Congolian. **Star:** green. **Alt. range:** 200-400m (2). **Specimens:** - *Eastern slopes:* Deistel H. 643. *Mabeta-Moliwe:* Preuss P. R. 1288. *Onge:* Rosevear D. R. 89/37, 90/37

Phyllanthus amarus Schumach. & Thonn.
F.W.T.A. 1: 387 (1958); F.T.E.A. Euphorbiaceae (1): 58 (1988).
Habit: herb. **Guild:** pioneer. **Habitat:** lava flows. **Distribution:** pantropical. **Star:** excluded. **Alt. range:** 1-200m (1). **Flowering:** Oct (1). **Fruiting:** Oct (1). **Specimens:** - *Etinde:* Brodie S. 4

Phyllanthus mannianus Müll.Arg.
F.W.T.A. 1: 388 (1958).

Habit: shrub to 1.5 m. **Habitat:** forest-grassland transition. **Distribution:** Guinea, Ivory Coast and W. Cameroon. **Chorology:** Guineo-Congolian (montane). **Star:** green. **Alt. range:** 1400-1600m (1), 2001-2200m (2). **Flowering:** Oct (2), Nov (1). **Specimens:** - *Eastern slopes:* Banks H. 84; Tchouto P. 279. *Etinde:* Thomas D. W. 9248. *No locality:* Mann G. 1231

Phyllanthus muellerianus (Kuntze) Exell
F.W.T.A. 1: 385 (1958); Meded. Land. Wag. 82(3): 146 (1982); F.T.E.A. Euphorbiaceae (1): 24 (1987).

Habit: tree or scandent shrub to 12 m. **Guild:** pioneer. **Habitat:** forest. **Distribution:** tropical Africa. **Star:** excluded. **Alt. range:** 1-200m (1), 601-800m (1). **Flowering:** Mar (1). **Fruiting:** Mar (1), Apr (1). **Specimens:** - *Eastern slopes:* Maitland T. D. 499; Tchouto P. 48. *Mabeta-Moliwe:* Plot Voucher W 734; Wheatley J. I. 128

Phyllanthus nummulariifolius Poir.
F.T.E.A. Euphorbiaceae (1): 28 (1987).
Phyllanthus capillaris Schumach. & Thonn., F.W.T.A. 1: 387 (1958).

Habit: shrub to 4.5 m. **Guild:** pioneer. **Habitat:** forest. **Distribution:** tropical and subtropical Africa. **Chorology:** Afromontane. **Star:** green. **Alt. range:** 1400-1600m (1). **Flowering:** Oct (1). **Specimens:** - *Etinde:* Wheatley J. I. 611

Phyllanthus odontadenius Müll.Arg.
F.W.T.A. 1: 388 (1958); F.T.E.A. Euphorbiaceae (2): 47 (1988).

Habit: herb. **Guild:** pioneer. **Habitat:** grassland. **Distribution:** tropical Africa. **Chorology:** Afromontane. **Star:** green. **Alt. range:** 1-200m (2). **Flowering:** Apr (1), Jul (1). **Fruiting:** Jul (1). **Specimens:** - *Eastern slopes:* Dundas J. FHI 15237. *Mabeta-Moliwe:* Cheek M. 3536. *Mokoko River F.R.:* Acworth J. M. 77

Phyllanthus cf. odontadenius Müll.Arg.

Specimens: - *Etinde:* Watts J. 575

Phyllanthus reticulatus Poir.
F.W.T.A. 1: 387 (1958); Meded. Land. Wag. 82(3): 147 (1982); F.T.E.A. Euphorbiaceae (1): 34 (1987).

Habit: shrub to 5 m. **Guild:** pioneer. **Habitat:** stream banks in forest. **Distribution:** palaeotropical. **Star:** excluded. **Alt. range:** 1-200m (1).**Fruiting:** May (1). **Specimens:** - *Mokoko River F.R.:* Batoum A. 36

Plagiostyles africana (Müll.Arg.) Prain
F.W.T.A. 1: 414 (1958).

Habit: tree to 15 m. **Habitat:** forest. **Distribution:** SE Nigeria, Cameroon, Gabon, Congo (Brazzaville) and Congo (Kinshasa). **Chorology:** Guineo-Congolian. **Star:** green. **Specimens:** - *Southern Bakundu F.R.:* Brenan J. P. M. 9330

Protomegabaria stapfiana (Beille) Hutch.
F.W.T.A. 1: 373 (1958); Keay R.W.J., Trees of Nigeria: 177 (1989); Hawthorne W., F.G.F.T. Ghana: 137 (1990).

Habit: tree to 25 m. **Guild:** shade-bearer. **Habitat:** forest. **Distribution:** Sierra Leone to S. Tomé, Cameroon and Gabon. **Chorology:** Guineo-Congolian. **Star:** green. **Alt. range:** 1-200m (5), 201-400m (2).**Fruiting:** Mar (1), Jun (1), Oct (4), Nov (1). **Specimens:** - *Mokoko River F.R.:* Ndam N. 1097; Pouakouyou D. 87. *Onge:* Akogo M. 144; Ndam N. 796; Tchouto P. 708; Watts J. 875, 1028; Wheatley J. I. 730

Pseudagrostistachys africana (Müll.Arg.) Pax & K.Hoffm.
F.W.T.A. 1: 399 (1958); Keay R.W.J., Trees of Nigeria: 175 (1989); Hawthorne W., F.G.F.T. Ghana: 132 (1990).

Habit: tree to 8 m. **Habitat:** forest. **Distribution:** SE Nigeria, Bioko, Cameroon and S. Tomé. **Chorology:** Lower Guinea (montane). **Star:** gold. **Alt. range:** 1000-1200m (1), 1401-1600m (1). **Flowering:** Oct (2). **Specimens:** - *Etinde:* Cheek M. 3800; Thomas D. W. 9520

Pycnocoma macrophylla Benth.
F.W.T.A. 1: 405 (1958); Bull. Jard. Bot. Nat. Belg. 65: 54 (1996).
Pycnocoma brachystachya Pax, F.W.T.A. 1: 405 (1958).

Habit: shrub to 3 m. **Guild:** shade-bearer. **Habitat:** forest. **Distribution:** Ivory Coast to Bioko, Cameroon, Congo (Brazzaville) and Congo (Kinshasa). **Chorology:** Guineo-Congolian. **Star:** green. **Alt. range:** 1-200m (7), 201-400m (4), 401-600m (1), 801-1000m (1). **Flowering:** Mar (1), Oct (4), Nov (2), Dec (1). **Fruiting:** Nov (2), Dec (3). **Specimens:** - *Etinde:* Cheek M. 5898, 5931; Faucher P. 24; Maitland T. D. 789; Winkler H. J. P. 367. *Mabeta-Moliwe:* Cheek M. 5796; Mildbraed G. W. J. 10543; Plot Voucher 1003. *Onge:* Ndam N. 782; Tchouto P. 776; Thomas D. W. 9822; Watts J. 750, 825, 938, 1021; Wheatley J. I. 811

Sclerocroton ellipticus Hochst. ex Krauss
Sapium ellipticum (Krauss) Pax F.W.T.A. 1: 415 (1958); Meded. Land. Wag. 82(3): 148 (1982); F.T.E.A. Euphorbiaceae (1): 390 (1987); Keay R.W.J., Trees of Nigeria: 175 (1989); Hawthorne W., F.G.F.T. Ghana: 117 (1990).

Habit: tree 15-35 m tall. **Guild:** pioneer. **Habitat:** forest. **Distribution:** tropical Africa and Natal. **Star:** excluded. **Specimens:** - *Eastern slopes:* Maitland T. D. 116

Sibangea similis (Hutch.) Radcl.-Sm.
Kew Bull. 32: 481 (1978).
Drypetes similis Hutch., F.W.T.A. 1: 381 (1958).

Habit: tree or shrub to 5 m. **Habitat:** forest. **Distribution:** SE Nigeria and Cameroon. **Chorology:** Lower Guinea. **Star:** blue. **Alt. range:** 200-400m (1).**Fruiting:** Jun (1). **Specimens:** - *Etinde:* Kwangue A. 106

Sibangea sp. nov. aff. similis (Hutch.) Radcl.-Sm.

Specimens: - *Mabeta-Moliwe:* Watts 171; Wheatley 208, 218

Spondianthus preussii Engl. var. preussii

F.W.T.A. 1: 372 (1958); F.T.E.A. Euphorbiaceae (1): 107 (1987); Keay R.W.J., Trees of Nigeria: 172 (1989); Hawthorne W., F.G.F.T. Ghana: 137 (1990).

Habit: tree to 30 m. **Habitat:** swamp forest. **Distribution:** Liberia to Bioko, Cameroon and Gabon. **Chorology:** Guineo-Congolian. **Star:** green. **Alt. range:** 800-1000m (1). **Specimens:** - *Etinde:* Maitland T. D. 600. *Mabeta-Moliwe:* Mildbraed G. W. J. 10587; Plot Voucher W 794, W 1151

Suregada occidentalis (Hoyle) Croizat

Bull. Jard. Bot. Brux. 28: 449 (1958).
Gelonium occidentale Hoyle, F.W.T.A. 1: 413 (1958).

Habit: shrub to 3 m. **Habitat:** forest. **Distribution:** Ghana, Nigeria and Cameroon. **Chorology:** Guineo-Congolian. **Star:** green. **Alt. range:** 1-200m (1). **Specimens:** - *Mokoko River F.R.:* Tchouto P. 1126

Plukenetia conophora Müll. Arg.

Tetracarpidium conophorum (Müll.Arg.) Hutch. & Dalziel F.W.T.A. 1: 410 (1958).

Habit: woody climber. **Habitat:** forest. **Distribution:** Sierra Leone to Bioko, Cameroon, Gabon and Congo (Kinshasa). **Chorology:** Guineo-Congolian. **Star:** pink. **Specimens:** - *Etinde:* Maitland T. D. 184

Ricinodendron heudelotii (Baill.) Pierre ex Pax subsp. africanum (Müll. Arg.) J.Léonard

Specimens:- *S. Bakundu:* Thomas 3477

Tetrorchidium didymostemon (Baill.) Pax & K.Hoffm.

F.W.T.A. 1: 414 (1958); Bull. Soc. Roy. Bot. Belg. 94: 29 (1962); F.T.E.A. Euphorbiaceae (1): 374 (1987); Keay R.W.J., Trees of Nigeria: 167 (1989); Hawthorne W., F.G.F.T. Ghana: 83 (1990).
Tetrorchidium minus (Prain) Pax & K.Hoffm., F.W.T.A. 1: 414 (1958).

Habit: tree to 12 m. **Guild:** pioneer. **Habitat:** forest regrowth. **Distribution:** tropical Africa. **Star:** excluded. **Alt. range:** 1-200m (9), 201-400m (1). **Flowering:** Mar (1), Apr (3). **Fruiting:** Apr (1), May (4), Dec (1). **Specimens:** - *Etinde:* Maitland T. D. 594. *Mabeta-Moliwe:* Cable S. 517; Jaff B. 224; Plot Voucher W 303, W 742; Sunderland T. C. H. 1192, 1390; Watts J. 185, 282; Wheatley J. I. 145, 176, 301. *Mokoko River F.R.:* Tchouto P. 579

Thecacoris cf. annobonae Pax & K.Hoffm.

F.W.T.A. 1: 372 (1958).

Habit: tree to 10 m. **Distribution:** SW Cameroon and Annobon Island. **Chorology:** Western Cameroon. **Star:** black. **Alt. range:** 1-200m (1).**Fruiting:** Jun (1). **Specimens:** - *Mokoko River F.R.:* Ekema N. 1167. *Southern Bakundu F.R.:* Keay R. W. J. & Russell T. A. FHI 28679

Thecacoris leptobotrya (Müll.Arg.) Brenan

F.W.T.A. 1: 372 (1958); Keay R.W.J., Trees of Nigeria: 177 (1989).

Habit: tree or shrub 2-10 m tall. **Habitat:** forest. **Distribution:** SE Nigeria, Cameroon, Rio Muni, Gabon and Congo (Kinshasa). **Chorology:** Guineo-Congolian. **Star:** green. **Alt. range:** 1-200m (3), 201-400m (1). **Flowering:** Apr (1), May (1). **Fruiting:** May (1), Sep (1). **Specimens:** - *Mabeta-Moliwe:* Plot Voucher W 12. *Mokoko River F.R.:* Ndam N. 1071; Watts J. 1088, 1126b. *Onge:* Watts J. 713

Thecacoris stenopetala (Müll.Arg.) Müll.Arg.

F.W.T.A. 1: 372 (1958).

Habit: shrub to 4 m. **Guild:** shade-bearer. **Habitat:** forest. **Distribution:** Sierra Leone to Bioko, SW Cameroon and S. Tomé. **Chorology:** Guineo-Congolian. **Star:** green. **Alt. range:** 1-200m (1), 201-400m (3). **Flowering:** Oct (1). **Fruiting:** Mar (1), Nov (1). **Specimens:** - *Bambuko F.R.:* Tchouto P. 545. *Mokoko River F.R.:* Ndam N. 1276a. *Onge:* Tchouto P. 761, 1041

Tragia benthami Baker

F.W.T.A. 1: 412 (1958); F.T.E.A. Euphorbiaceae (1): 303 (1987).

Habit: herbaceous climber. **Habitat:** forest. **Distribution:** tropical and subtropical Africa. **Chorology:** Afromontane. **Star:** green. **Alt. range:** 200-400m (1). **Specimens:** - *Etinde:* Mann G. 1255

Tragia preussii Pax

F.W.T.A. 1: 412 (1958).

Habit: herbaceous climber. **Habitat:** forest. **Distribution:** SE Nigeria, SW Cameroon and Congo (Kinshasa). **Chorology:** Guineo-Congolian. **Star:** green. **Alt. range:** 1-200m (3), 201-400m (2), 401-600m (1), 601-800m (1). **Flowering:** Oct (3), Nov (1). **Fruiting:** Oct (4), Nov (1), Dec (1). **Specimens:** - *Etinde:* Cheek M. 5603; Faucher P. 34. *Onge:* Tchouto P. 778, 843, 867; Watts J. 881, 940

Uapaca acuminata (Hutch.) Pax & K.Hoffm.

F.W.T.A. 1: 390 (1958); Keay R.W.J., Trees of Nigeria: 161 (1989).

Habit: tree to 23 m. **Habitat:** forest. **Distribution:** SE Nigeria, Cameroon and Angola (Cabinda). **Chorology:** Lower Guinea. **Star:** blue. **Alt. range:** 1-200m (1). **Specimens:** - *Southern Bakundu F.R.:* Mbani J. M. 426

Uapaca guineensis Müll.Arg.

F.W.T.A. 1: 390 (1958); F.T.E.A. Euphorbiaceae (2): 570 (1988); Keay R.W.J., Trees of Nigeria: 163 (1989); Hawthorne W., F.G.F.T. Ghana: 137 (1990).

Habit: tree to 30 m. **Guild:** light demander. **Habitat:** forest. **Distribution:** Sierra Leone to Cameroon, Gabon and Congo (Kinshasa). **Chorology:** Guineo-Congolian. **Star:** green. **Alt. range:** 1-200m (1), 201-400m (1). **Flowering:** Jul (1). **Specimens:** - *Etinde:* Maitland T. D. 596. *Mabeta-Moliwe:* Cheek M. 3435;

3517; Plot Voucher W 226, W 849, W 1173. *Mokoko River F.R.:* Ndam N. 1221

Uapaca staudtii Pax
F.W.T.A. 1: 390 (1958); Keay R.W.J., Trees of Nigeria: 161 (1989).

Habit: tree to 30 m. **Habitat:** swamp forest. **Distribution:** Nigeria, Bioko and Cameroon. **Chorology:** Lower Guinea. **Star:** blue. **Alt. range:** 1-200m (3). **Flowering:** May (2). **Specimens:** - *Bambuko F.R.:* Olorunfemi J. FHI 30773. *Etinde:* Preuss P. R. 1171. *Mabeta-Moliwe:* Plot Voucher W 526, W 735; Sunderland T. C. H. 1282; Wheatley J. I. 251. *Mokoko River F.R.:* Acworth J. M. 174; Fraser P. 403

Uapaca vanhouttei De Wild.
F.W.T.A. 1: 390 (1958); Keay R.W.J., Trees of Nigeria: 161 (1989).

Habit: tree to 10 m. **Habitat:** forest. **Distribution:** SE Nigeria, Cameroon and Congo (Kinshasa). **Chorology:** Guineo-Congolian. **Star:** gold. **Alt. range:** 1-200m (1).**Fruiting:** Jun (1). **Specimens:** - *Mabeta-Moliwe:* Plot Voucher W 290. *Mokoko River F.R.:* Ekema N. 1209

FLACOURTIACEAE

Casearia barteri Mast.
F.W.T.A. 1: 198 (1954); Keay R.W.J., Trees of Nigeria: 64 (1989); Hawthorne W., F.G.F.T. Ghana: 81 (1990); Fl. Cameroun 34: 20 (1995).

Habit: tree to 16 m. **Habitat:** swamp forest. **Distribution:** Nigeria, Cameroon and Gabon. **Chorology:** Lower Guinea. **Star:** blue. **Alt. range:** 1-200m (1), 601-800m (1).**Fruiting:** Jul (1), Aug (1). **Specimens:** - *Etinde:* Tekwe C. 121. *Mabeta-Moliwe:* Cheek M. 3531; Mann G. 714

Casearia stipitata Mast.
F.W.T.A. 1: 198 (1954); Keay R.W.J., Trees of Nigeria: 64 (1989); Fl. Cameroun 34: 19 (1995).
  Casearia zenkeri Gilg, Engl. Bot. Jahrb. 40: 512 (1908).

Habit: tree to 6 m. **Habitat:** stream banks in forest. **Distribution:** Ivory Coast to Cameroon, Gabon and Congo (Kinshasa). **Chorology:** Guineo-Congolian. **Star:** green. **Alt. range:** 200-400m (1).**Fruiting:** Oct (1). **Specimens:** - *Etinde:* Thomas D. W. 9725

Dasylepis blackii (Oliv.) Chipp
F.W.T.A. 1: 186 (1954); Keay R.W.J., Trees of Nigeria: 56 (1989); Fl. Cameroun 34: 30 (1995).

Habit: tree to 8 m. **Habitat:** forest. **Distribution:** SE Nigeria, Cameroon and Rio Muni. **Chorology:** Lower Guinea. **Star:** blue. **Alt. range:** 1-200m (4), 201-400m (5). **Flowering:** Apr (2). **Specimens:** - *Mabeta-Moliwe:* Jaff B. 97. *Mokoko River F.R.:* Mbani J. M. 311, 354, 356, 406; Ndam N. 1123, 1156, 1253. *Onge:* Tchouto P. 758. *Southern Bakundu F.R.:* Ejiofor M. C. FHI 29317

Dasylepis racemosa Oliv.
F.W.T.A. 1: 186 (1954); Keay R.W.J., Trees of Nigeria: 56 (1989).

Habit: tree to 15 m. **Habitat:** forest and Aframomum thicket. **Distribution:** SE Nigeria, Cameroon, Congo (Kinshasa) and Uganda. **Chorology:** Guineo-Congolian (montane). **Star:** green. **Alt. range:** 1000-1200m (2). **Flowering:** Jan (1). **Specimens:** - *Etinde:* Sunderland T. C. H. 1066. *No locality:* Mann G. 2149

Dovyalis zenkeri Gilg
F.W.T.A. 1: 190 (1954); Hawthorne W., F.G.F.T. Ghana: 95 (1990); Fl. Cameroun 34: 10 (1995).

Habit: shrub to 4 m. **Guild:** pioneer. **Habitat:** forest. **Distribution:** SE Nigeria and Cameroon. **Chorology:** Lower Guinea. **Star:** blue. **Alt. range:** 1-200m (1).**Fruiting:** Apr (1). **Specimens:** - *Mokoko River F.R.:* Tchouto P. 1144

Dovyalis sp. nov.

Specimens: - *Bambuko F.R.:* Watts J. 625. *Mokoko River F.R.:* Tchouto P. 620. *Onge:* Harris D. J. 3665, 3666

Flacourtia vogelii Hook.f.
F.W.T.A. 1: 189 (1954); Keay R.W.J., Trees of Nigeria: 59 (1989); Fl. Cameroun 34: 8 (1995).

Habit: tree to 10 m. **Distribution:** Sierra Leone to Cameroon and CAR. **Chorology:** Guineo-Congolian. **Star:** green. **Alt. range:** 600-800m (1). **Specimens:** - *Mabeta-Moliwe:* Cheek in W1216; Mildbraed G. W. J. 10762

Flacourtia sp.

Specimens: - *Limbe Botanic Garden:* Cheek M. 4953

Homalium cf. africanum (Hook.f.) Benth.

Specimens: - *Mabeta-Moliwe:* Plot Voucher W178, W344, W781, W830 W943

Homalium letestui Pellegr.
F.W.T.A. 1: 196 (1954); Keay R.W.J., Trees of Nigeria: 62 (1989); Hawthorne W., F.G.F.T. Ghana: 109 (1990); Fl. Cameroun 34: 74 (1995).
  Homalium skirlii Gilg ex Engl., F.W.T.A. 1: 197 (1954).

Habit: tree 8-30 m tall. **Guild:** light demander. **Habitat:** forest. **Distribution:** Guinea to Bioko, Cameroon and Gabon. **Chorology:** Guineo-Congolian. **Star:** green. **Specimens:** - *Etinde:* Maitland T. D. 176; Preuss P. R. 1379. *Mabeta-Moliwe:* Plot Voucher W380, W420, W520, W1176. *Southern Bakundu F.R.:* Ejiofor M. C. FHI 14078

Homalium longistylum Mast.
Bull. Jard. Bot. Nat. Belg. 43: 270 (1973); Keay R.W.J., Trees of Nigeria: 64 (1989); Hawthorne W., F.G.F.T. Ghana: 109 (1990); Fl. Cameroun 34: 63 (1995).
  Homalium macropterum Gilg, F.W.T.A. 1: 196 (1954).
  Homalium aylmeri Hutch., Dalziel & Chipp, F.W.T.A. 1: 196 (1954).

Habit: tree to 30 m. **Guild:** light demander. **Habitat:** forest. **Distribution:** tropical Africa. **Star:** excluded. **Alt. range:** 1-200m (5), 201-400m (1). **Flowering:** Jul (1), Sep (1), Oct (2), Nov (2). **Specimens:** - *Mabeta-Moliwe:* Cheek M. 3559. *Onge:* Harris D. J. 3782; Tchouto P. 824, 1012; Watts J. 845, 993

Oncoba dentata Oliv.

Keay R.W.J., Trees of Nigeria: 62 (1989); Fl. Cameroun 34: 52 (1995); Adansonia (ser.3) 19: 257 (1997).
 Lindackeria dentata (Oliv.) Gilg, Hawthorne W., F.G.F.T. Ghana: 131 (1990).

Habit: tree or shrub to 15 m. **Guild:** pioneer. **Habitat:** forest. **Distribution:** Guinea to Congo (Brazzaville), Angola and Sudan. **Chorology:** Guineo-Congolian. **Star:** green. **Alt. range:** 1-200m (5), 201-400m (1), 801-1000m (2). **Flowering:** Feb (1), Mar (1), Apr (2), May (1). **Fruiting:** Mar (1), Apr (1), May (2), Aug (1), Oct (1). **Specimens:** - *Eastern slopes:* Mann G. 1177; Tekwe C. 134. *Etinde:* Tchouto P. 18, 94a; Tekwe C. 77. *Mabeta-Moliwe:* Wheatley J. I. 242. *Mokoko River F.R.:* Akogo M. 157; Tchouto P. 1063. *Onge:* Tchouto P. 939

Oncoba glauca (P.Beauv.) Planch.

Keay R.W.J., Trees of Nigeria: 61 (1989); Fl. Cameroun 34: 44 (1995); Adansonia (ser.3) 19: 257 (1997).
 Caloncoba glauca (P.Beauv.) Gilg, Hawthorne W., F.G.F.T. Ghana: 146 (1990).

Habit: tree to 20 m. **Habitat:** forest. **Distribution:** Nigeria, Cameroon, Bioko, Gabon and Congo (Kinshasa). **Chorology:** Guineo-Congolian (montane). **Star:** green. **Alt. range:** 1-200m (4), 201-400m (3), 1401-1600m (1). **Flowering:** Jan (1), Mar (1), Oct (3), Nov (1). **Fruiting:** Jan (1), Feb (1), Mar (2). **Specimens:** - *Etinde:* Sunderland T. C. H. 1069; Tekwe C. 23. *Mabeta-Moliwe:* Mann G. 18. *Mokoko River F.R.:* Tchouto P. 531. *Onge:* Cheek M. 5068; Ndam N. 622; Tchouto P. 728; Thomas D. W. 9877; Wheatley J. I. 827

Oncoba lophocarpa Oliv.

F.T.A. 1: 117 (1868).
 Caloncoba lophocarpa (Oliv.) Gilg, F.W.T.A. 1: 188 (1954).

Habit: tree to 12 m. **Habitat:** forest. **Distribution:** Mount Cameroon. **Chorology:** Endemic (montane). **Star:** black. **Alt. range:** 1200-1400m (1), 1401-1600m (1). **Flowering:** Oct (1). **Fruiting:** Sep (1), Oct (1). **Specimens:** - *Eastern slopes:* Mann G. 1195; Tchouto P. 355. *Etinde:* Tekwe C. 165

Oncoba mannii Oliv.

Keay R.W.J., Trees of Nigeria: 61 (1989); Fl. Cameroun 34: 56 (1995).
 Camptostylus mannii (Oliv.) Gilg, F.W.T.A. 1: 187 (1954).

Habit: tree to 16 m. **Habitat:** forest. **Distribution:** SE Nigeria, Bioko, Cameroon, Gabon, Congo (Kinshasa) and Angola. **Chorology:** Guineo-Congolian. **Star:** green. **Alt. range:** 1-200m (7), 201-400m (1). **Flowering:** Feb (1), Mar (1), Apr (1). **Fruiting:** Mar (1), Apr (1), May (1), Jun (3), Jul (1). **Specimens:** - *Etinde:* Maitland T. D. 317; Nkeng P. 29; Tekwe C. 93. *Mabeta-Moliwe:* Cheek M. 3594; Mann G. 11; Sunderland T. C. H. 1163. *Mokoko River F.R.:* Acworth J. M. 71; Ekema N. 1071, 1129; Tchouto P. 1242. *Southern Bakundu F.R.:* Brenan J. P. M. 9478

Oncoba ovalis Oliv.

F.T.A. 1: 118 (1868); Adansonia (ser.3) 19: 259 (1997).
 Camptostylus ovalis (Oliv.) Chipp, F.W.T.A. 1: 187 (1954).

Habit: tree to 10 m. **Habitat:** forest. **Distribution:** Mount Cameroon. **Chorology:** Endemic (montane). **Star:** black. **Alt. range:** 1000-1200m (1), 1201-1400m (2). **Flowering:** Oct (1). **Fruiting:** Aug (1). **Specimens:** - *Eastern slopes:* Kwangue A. 122. *Etinde:* Thomas D. W. 9538. *No locality:* Mann G. 1196

Oncoba spinosa Forssk.

F.W.T.A. 1: 188 (1954); F.T.E.A. Flacourtiaceae: 16 (1975); Meded. Land. Wag. 82(3): 151 (1982); Keay R.W.J., Trees of Nigeria: 59 (1989); Hawthorne W., F.G.F.T. Ghana: 95 (1990).

Habit: tree or shrub to 12 m. **Habitat:** forest. **Distribution:** tropical Africa. **Star:** excluded. **Alt. range:** 600-800m (1). **Specimens:** - *Etinde:* Maitland T. D. 615. *No locality:* Maitland T. D. 1109

Oncoba welwitschii Oliv.

Keay R.W.J., Trees of Nigeria: 61 (1989); Fl. Cameroun 34: 46 (1995); Adansonia (ser.3) 19: 261 (1997).
 Caloncoba welwitschii (Oliv.) Gilg, F.W.T.A. 1: 188 (1954).

Habit: tree or shrub to 10 m. **Habitat:** forest and forest gaps. **Distribution:** SE Nigeria, Cameroon, Gabon, Congo (Brazzaville), Congo (Kinshasa) and Angola. **Chorology:** Guineo-Congolian. **Star:** green. **Alt. range:** 1-200m (5). **Flowering:** Apr (3). **Fruiting:** Apr (3), May (2). **Specimens:** - *Mabeta-Moliwe:* Sunderland T. C. H. 1226, 1250, 1317; Wheatley J. I. 167, 309

Oncoba sp.

Specimens: - *Mokoko River F.R.:* Fraser P. 360

Ophiobotrys zenkeri Gilg

F.W.T.A. 1: 189 (1954); Keay R.W.J., Trees of Nigeria: 56 (1989); Hawthorne W., F.G.F.T. Ghana: 111 (1990); Fl. Cameroun 34: 18 (1995).

Habit: tree to 30 m. **Guild:** light demander. **Habitat:** forest. **Distribution:** Ivory Coast to Cameroon and Gabon. **Chorology:** Guineo-Congolian. **Star:** green. **Specimens:** - *Etinde:* Maitland T. D. 425

Phyllobotryon spathulatum Müll.Arg.

Fl. Cameroun 34: 24 (1995).
 Phyllobotryon soyauxianum Baill., F.W.T.A. 1: 191 (1954).

Habit: tree to 5 m. **Habitat:** forest. **Distribution:** Nigeria, Cameroon and Gabon. **Chorology:** Lower Guinea. **Star:** blue. **Specimens:** - *Southern Bakundu F.R.:* Brenan J. P. M. 9306

Scottellia klaineana Pierre

F.W.T.A. 1: 187 (1954); Keay R.W.J., Trees of Nigeria: 56 (1989); Fl. Cameroun 34: 33 (1995).
 Scottellia mimfiensis Gilg, F.W.T.A. 1: 186 (1954).
 Scottellia klaineana Pierre var. *mimfiensis* (Gilg) Pellegrin, Blumea 20: 281 (1972).
 Scottellia coriacea A.Chev. ex Hutch. & Dalziel, F.W.T.A. 1: 187 (1954).

Habit: tree to 30 m. **Habitat:** forest. **Distribution:** Sierra Leone to Cameroon, Gabon, Congo (Kinshasa) and Angola. **Chorology:** Guineo-Congolian. **Star:** green. **Alt. range:** 1-200m (4). **Flowering:** Jun (1). **Fruiting:** Jun (1). **Specimens:** - *Mabeta-Moliwe:* Wheatley J. I. 349; Plot Vouchers W98, W161, W634, W865, W871. *Mokoko River F.R.:* Mbani J. M. 318, 327; Tchouto P. 1186

GENTIANACEAE

N. Silesh (ETH)

Sebaea brachyphylla Griseb.
F.W.T.A. 2: 298 (1963).

Habit: herb. **Habitat:** grassland. **Distribution:** tropical Africa. **Chorology:** Afromontane. **Star:** green. **Alt. range:** 1800-2000m (1), 2401-2600m (1), 2601-2800m (1). **Flowering:** Oct (2). **Specimens:** - *Eastern slopes:* Sidwell K. 38; Thomas D. W. 9280. *No locality:* Brenan J. P. M. 9522a

Sebaea oligantha (Gilg) Schinz

Specimens: - *Mabeta-Moliwe:* Cheek M. 3220

Swertia abyssinica Hochst.
F.W.T.A. 2: 299 (1963).

Habit: herb. **Habitat:** grassland and forest-grassland transition. **Distribution:** Bioko, W. Cameroon, Ethiopia and East Africa. **Chorology:** Afromontane. **Star:** green. **Alt. range:** 1800-2000m (2), 2201-2400m (1), 2401-2600m (1). **Flowering:** Sep (1), Oct (2). **Specimens:** - *Eastern slopes:* Sidwell K. 84; Thomas D. W. 9313. *Etinde:* Tekwe C. 267. *No locality:* Mann G. 1216

Swertia mannii Hook.f.
F.W.T.A. 2: 299 (1963).

Habit: herb. **Habitat:** grassland. **Distribution:** Guinea to W. Cameroon. **Chorology:** Guineo-Congolian (montane). **Star:** green. **Alt. range:** 2000-2200m (2), 2201-2400m (1), 2801-3000m (1). **Flowering:** Sep (1), Oct (1), Nov (1). **Specimens:** - *Eastern slopes:* Banks H. 91; Tchouto P. 270. *Etinde:* Tekwe C. 300. *No locality:* Mann G. 2000

GERANIACEAE

Geranium arabicum Forssk. subsp. arabicum
Notes Roy. Bot. Gard. Edinburgh 42: 171 (1985).
Geranium simense Hochst. ex. A.Rich., F.W.T.A. 1: 157 (1954).

Habit: scrambling herb to 2.5 m. **Habitat:** grassland and forest-grassland transition. **Distribution:** tropical Africa and Arabia. **Chorology:** Montane. **Star:** excluded. **Alt. range:** 1800-2000m (1), 2001-2200m (4). **Flowering:** May (1), Oct (2). **Specimens:** - *Eastern slopes:* Banks H. 70; Sunderland T. C. H. 1416; Tchouto P. 318; Watts J. 428. *No locality:* Mann G. 1323

Geranium mascatense Boiss.
Webbia 25: 644 (1971).
Geranium ocellatum Cambess., F.W.T.A. 1: 157 (1954).

Habit: herb. **Habitat:** forest-grassland transition. **Distribution:** tropical Africa and Asia. **Chorology:** Montane. **Star:** excluded. **Alt. range:** 2000-2200m (2), 2201-2400m (1). **Flowering:** Oct (2). **Specimens:** - *Eastern slopes:* Tchouto P. 317; Watts J. 443. *No locality:* Mann G. 1261

GESNERIACEAE

B.L. Burtt (E)

Acanthonema strigosum Hook.f.
F.W.T.A. 2: 383 (1963); Fl. Cameroun 27: 18 (1984).

Habit: herb. **Habitat:** on rocks in forest. **Distribution:** SE Nigeria, Bioko and Cameroon. **Chorology:** Lower Guinea (montane). **Star:** blue. **Alt. range:** 200-400m (1), 401-600m (3), 601-800m (1), 801-1000m (1), 1001-1200m (2), 1201-1400m (1). **Flowering:** Sep (1), Oct (4), Nov (2), Dec (1). **Fruiting:** Oct (2). **Specimens:** - *Etinde:* Cable S. 278, 439; Sidwell K. 163; Tchouto P. 434, 435; Watts J. 504; Wheatley J. I. 534; Williams S. 45. *No locality:* Mann G. 1948

Epithema tenue C.B.Clarke
F.W.T.A. 2: 383 (1963); Fl. Cameroun 27: 6 (1984).

Habit: herb. **Habitat:** on rocks in forest. **Distribution:** Guinea to Bioko, Cameroon and Uganda. **Chorology:** Guineo-Congolian (montane). **Star:** green. **Alt. range:** 400-600m (1), 601-800m (1), 801-1000m (1), 1001-1200m (1), 1601-1800m (1). **Flowering:** Jul (1), Aug (1), Sep (1), Oct (1), Dec (1). **Specimens:** - *Eastern slopes:* Nkeng P. 90. *Etinde:* Cable S. 283; Cheek M. 3617, 3761; Tekwe C. 119

Schizoboea kamerunensis (Engl.) B.L.Burtt
Fl. Cameroun 27: 15 (1984).
Didymocarpus kamerunensis Engl., F.W.T.A. 2: 382 (1963).

Habit: herb. **Habitat:** forest. **Distribution:** SW Cameroon and Bioko. **Chorology:** Western Cameroon Uplands to Tanzania (montane). **Star:** blue. **Alt. range:** 1600-1800m (1). **Flowering:** Oct (1). **Fruiting:** Oct (1). **Specimens:** - *Etinde:* Thomas D. W. 9532

Streptocarpus elongatus Engl.
F.W.T.A. 2: 382 (1963); Fl. Cameroun 27: 10 (1984).

Habit: herb. **Habitat:** on rocks in forest. **Distribution:** SW Cameroon and Sudan. **Chorology:** Guineo-Congolian. **Star:** green. **Alt. range:** 1-200m (1); 1200-1400m (1). **Specimens:** - *Eastern slopes:* Maitland T. D. s.n.; Preuss P. R. 1010. *Mabeta-Moliwe:* Cheek M. 3570

Streptocarpus nobilis C.B.Clarke
F.W.T.A. 2: 382 (1963); Fl. Cameroun 27: 12 (1984).

Habit: herb. **Habitat:** forest and farmbush. **Distribution:** Guinea to Cameroon, S. Tomé and CAR. **Chorology:** Guineo-Congolian. **Star:** green. **Alt. range:** 1-200m (1), 201-400m (1). **Flowering:** Oct (1). **Fruiting:** Jul (1), Oct (1). **Specimens:** - *Etinde:* Maitland T. D. 743. *Mabeta-Moliwe:* Cheek M. s.n. *Onge:* Tchouto P. 950

GUTTIFERAE

Allanblackia floribunda Oliv.
F.W.T.A. 1: 291 (1954); Keay R.W.J., Trees of Nigeria: 104 (1989); Hawthorne W., F.G.F.T. Ghana: 50 (1990).

Habit: tree to 30 m. **Guild:** shade-bearer. **Habitat:** forest.
Distribution: Sierra Leone to Congo (Kinshasa) and Angola.
Chorology: Guineo-Congolian. **Star:** green. **Alt. range:** 1-200m
(1), 201-400m (1), 401-600m (1). **Flowering:** May (1), Oct (1).
Fruiting: May (1). **Specimens:** - *Mokoko River F.R.:* Thomas D.
W. 10089. *Onge:* Cheek M. 5011. *Southern Bakundu F.R.:*
Leeuwenberg A. J. M. 9801

Calophyllum inophyllum L.

Specimens: - *Mabeta-Moliwe:* Cheek M. 3574

Endodesmia calophylloides Benth.
F.W.T.A. 1: 287 (1954); Fl. Afr. Cent. Guttiferae: 43 (1970); Keay
R.W.J., Trees of Nigeria: 110 (1989).

Habit: tree to 15 m. **Habitat:** swamp forest or by streams in forest.
Distribution: Nigeria, Cameroon, Gabon, Congo (Kinshasa) and
Angola (Cabinda). **Chorology:** Guineo-Congolian. **Star:** green.
Alt. range: 1-200m (3), 201-400m (2). **Specimens:** - *Mokoko
River F.R.:* Mbani J. M. 316; Ndam N. 1251; Tchouto P. 1114;
Thomas D. W. 10064, 10151

Garcinia chromocarpa Engl.
Fl. Congo B., Rw. & Bur. Guttiferae: 56 (1970).

Habit: tree to 15 m. **Habitat:** forest. **Distribution:** Cameroon and
Congo (Kinshasa). **Chorology:** Guineo-Congolian. **Star:** green.
Alt. range: 1-200m (1). **Flowering:** Nov (1). **Specimens:** - *Onge:*
Thomas D. W. 9878

Garcinia conrauana Engl.
F.W.T.A. 1: 295 (1954).

Habit: tree to 20 m. **Habitat:** forest. **Distribution:** W. Cameroon.
Chorology: Western Cameroon. **Star:** gold. **Alt. range:** 1-200m
(2), 201-400m (3). **Flowering:** Oct (1), Nov (1). **Fruiting:** May
(1). **Specimens:** - *Mokoko River F.R.:* Mbani J. M. 433; Thomas
D. W. 10104; Watts J. 1174. *Onge:* Ndam N. 779; Watts J. 1039

Garcinia densivenia Engl.
Engl. Bot. Jahrb. 40: 564 (1908).

Habit: tree to 15 m. **Habitat:** forest. **Distribution:** Cameroon.
Chorology: Lower Guinea. **Star:** gold. **Specimens:** - *Mabeta-
Moliwe:* Plot Voucher W 236, W 240

Garcinia kola Heckel
F.W.T.A. 1: 294 (1954); Keay R.W.J., Trees of Nigeria: 108
(1989); Hawthorne W., F.G.F.T. Ghana: 50 (1990).

Habit: tree 10-30 m tall. **Guild:** shade-bearer. **Habitat:** forest and
farmbush. **Distribution:** Sierra Leone to Congo (Kinshasa) and
Angola. **Chorology:** Guineo-Congolian. **Star:** scarlet. **Alt. range:**
200-400m (1).**Fruiting:** Oct (1). **Specimens:** - *Onge:* Ndam N.
760

Garcinia mannii Oliv.
F.W.T.A. 1: 295 (1954); Keay R.W.J., Trees of Nigeria: 110
(1989).

Habit: tree to 30 m. **Habitat:** forest. **Distribution:** Nigeria,
Cameroon, Rio Muni and Gabon. **Chorology:** Lower Guinea.
Star: scarlet. **Alt. range:** 1-200m (10), 201-400m (5). **Flowering:**

Jan (1), Mar (3), Apr (2), May (2), Jul (1), Nov (3). **Fruiting:** Apr
(1), Jun (2), Nov (1). **Specimens:** - *Etinde:* Tchouto P. 263.
Mabeta-Moliwe: Mann G. 711; Plot Voucher W 590, W 1170;
Sunderland T. C. H. 1062; Tchouto P. 187; Wheatley J. I. 42, 81.
Mokoko River F.R.: Acworth J. M. 66; Ekema N. 1096, 1176;
Mbani J. M. 458; Ndam N. 1121; Tchouto P. 576, 608; Thomas D.
W. 10118. *Onge:* Thomas D. W. 9770; Watts J. 899, 984

Garcinia ovalifolia Oliv.
F.W.T.A. 1: 295 (1954); Meded. Land. Wag. 82(3): 152 (1982);
Keay R.W.J., Trees of Nigeria: 108 (1989).
Habit: tree or shrub to 10 m. **Habitat:** forest. **Distribution:**
Guinea to Congo (Kinshasa), Angola and Ethiopia. **Chorology:**
Guineo-Congolian. **Star:** green. **Alt. range:** 200-400m (1).
Specimens: - *Mokoko River F.R.:* Thomas D. W. 10157

Garcinia cf. ovalifolia Oliv.

Specimens: - *Mokoko River F.R.:* Ndam N. 1163

Garcinia smeathmannii (Planch. & Triana) Oliv.
F.W.T.A. 1: 295 (1954); Fl. Zamb. 1(2): 399 (1961); F.T.E.A.
Guttiferae: 21 (1978); Keay R.W.J., Trees of Nigeria: 108 (1989);
Hawthorne W., F.G.F.T. Ghana: 50 (1990).
  Garcinia polyantha Oliv., F.W.T.A. 1: 294 (1954).

Habit: tree to 25 m. **Guild:** shade-bearer. **Habitat:** forest.
Distribution: Guinea to Cameroon, Congo (Kinshasa), Angola,
Zambia and Tanzania. **Chorology:** Guineo-Congolian (montane).
Star: green. **Alt. range:** 1-200m (1), 601-800m (1), 1401-1600m
(1), 1601-1800m (2). **Flowering:** Jul (1), Nov (1). **Fruiting:** Jul
(1), Sep (1), Oct (1). **Specimens:** - *Etinde:* Tchouto P. 252; Tekwe
C. 227; Thomas D. W. 9531. *Mabeta-Moliwe:* Mildbraed G. W. J.
10685; Plot Voucher W 853. *Onge:* Thomas D. W. 9921

Garcinia cf. smeathmannii (Planch. & Triana) Oliv.

Specimens: - *Mabeta-Moliwe:* Watts J. 124, 272, 295

Garcinia staudtii Engl.
F.W.T.A. 1: 294 (1954); Keay R.W.J., Trees of Nigeria: 108
(1989).

Habit: tree or shrub to 12 m. **Habitat:** forest. **Distribution:** SE
Nigeria and Cameroon. **Chorology:** Lower Guinea. **Star:** blue.
Alt. range: 1-200m (1), 201-400m (1). **Flowering:** May (1).
Specimens: - *Mokoko River F.R.:* Thomas D. W. 10090; Watts J.
1086

Garcinia sp.

Specimens: - *Limbe Botanic Garden:* Cable S. 1306; Cheek M.
4954. *Mokoko River F.R.:* Thomas D. W. 10156. *Onge:* Akogo M.
97; Ndam N. 697

Harungana madagascariensis Lam. ex Poir.
F.W.T.A. 1: 290 (1954); Fl. Zamb. 1: 391 (1961); Meded. Land.
Wag. 82(3): 153 (1982); Keay R.W.J., Trees of Nigeria: 110
(1989); Hawthorne W., F.G.F.T. Ghana: 49 (1990).

Habit: tree or shrub to 6 m. **Guild:** pioneer. **Habitat:** forest gaps
and farmbush. **Distribution:** tropical Africa. **Chorology:**

Afromontane. **Star:** green. **Alt. range:** 1-200m (1), 801-1000m (1). **Specimens:** - *Etinde:* Rosevear Cam. 110/36; Rumball s.n. *Mabeta-Moliwe:* Ndam N. 183. *No locality:* Mann G. 1183

Hypericum peplidifolium A.Rich.
F.W.T.A. 1: 287 (1954); Webbia 22: 265 (1967); Bull. Jard. Bot. Nat. Belg. 41: 433 (1971).

Habit: herb. **Habitat:** grassland. **Distribution:** tropical Africa. **Chorology:** Afromontane. **Star:** green. **Alt. range:** 1200-1400m (1), 2001-2200m (1). **Specimens:** - *Eastern slopes:* Richards P. W. 9521. *Etinde:* Maitland T. D. 1088

Hypericum revolutum Vahl subsp. revolutum
Webbia 22: 239 (1967); Bull. Jard. Bot. Nat. Belg. 41: 438 (1971); Kew Bull. 33: 581 (1979).
Hypericum lanceolatum Lam., F.W.T.A. 1: 287 (1954).

Habit: tree or shrub to 10 m. **Habitat:** grassland and forest-grassland transition. **Distribution:** tropical Africa. **Chorology:** Afromontane. **Star:** green. **Alt. range:** 1800-2000m (1), 2001-2200m (2), 2401-2600m (2). **Flowering:** Sep (1), Oct (2). **Fruiting:** May (1). **Specimens:** - *Eastern slopes:* Sidwell K. 19; Sunderland T. C. H. 1425; Tchouto P. 287. *Etinde:* Tekwe C. 244. *No locality:* Mann G. 1199

Hypericum roeperanum Schimp. ex A.Rich.
Kew Bull. 3: 444 (1957
Hypericum riparium A.Chev. F.W.T.A. 1: 287 (1954).

Specimens:- none located, though common in montane grassland - forest transition.

Mammea africana Sabine
F.W.T.A. 1: 293 (1954); Keay R.W.J., Trees of Nigeria: 107 (1989); Hawthorne W., F.G.F.T. Ghana: 50 (1990).

Habit: tree to 30 m. **Guild:** shade-bearer. **Habitat:** forest. **Distribution:** Sierra Leone to Congo (Kinshasa), Angola and Uganda. **Chorology:** Guineo-Congolian. **Star:** pink. **Alt. range:** 200-400m (1), 601-800m (1). **Flowering:** Mar (1). **Specimens:** - *Mabeta-Moliwe:* Mildbraed G. W. J. 10610. *Mokoko River F.R.:* Tchouto P. 562

Pentadesma butyracea Sabine
F.W.T.A. 1: 291 (1954); Bull. Jard. Bot. Brux. 35: 414 (1965); Keay R.W.J., Trees of Nigeria: 104 (1989); Hawthorne W., F.G.F.T. Ghana: 50 (1990).

Habit: tree to 25 m. **Guild:** shade-bearer. **Habitat:** forest. **Distribution:** Guinea to Congo (Kinshasa). **Chorology:** Guineo-Congolian. **Star:** green. **Alt. range:** 1-200m (5), 201-400m (2). **Flowering:** Mar (3), Apr (1), May (1), Jun (2). **Specimens:** - *Etinde:* Mbani J. M. 34. *Mabeta-Moliwe:* Plot Voucher W 102; Tchouto P. 215; Watts J. 121, 147; Wheatley J. I. 54. *Mokoko River F.R.:* Ekema N. 1120b; Pouakouyou D. 69. *Southern Bakundu F.R.:* Ejiofor M. C. FHI 29364

Psorospermum staudtii Engl.
Bull. Jard. Bot. Nat. Belg. 39: 347 (1969); Fl. Afr. Cent. Guttiferae: 19 (1970).

Habit: tree or shrub to 6 m. **Habitat:** forest. **Distribution:** Cameroon. **Chorology:** Lower Guinea. **Star:** gold. **Alt. range:** 1-200m (3). **Flowering:** May (1). **Fruiting:** May (3). **Specimens:** -

Etinde: Brodie 26. *Mokoko River F.R.:* Ekema N. 936; Tchouto P. 1194; Watts J. 1151

Psorospermum tenuifolium Hook.f.
F.W.T.A. 1: 290 (1954).

Habit: shrub to 3 m. **Habitat:** swamp and coastal forest. **Distribution:** Nigeria, Cameroon and Congo (Kinshasa). **Chorology:** Guineo-Congolian. **Star:** green. **Alt. range:** 200-400m (2). **Flowering:** Mar (1). **Fruiting:** Oct (1). **Specimens:** - *Bambuko F.R.:* Tchouto P. 556. *Etinde:* Fraser 1005. *Onge:* Watts J. 868

Symphonia globulifera L.f.
F.W.T.A. 1: 293 (1954); Keay R.W.J., Trees of Nigeria: 102 (1989); Hawthorne W., F.G.F.T. Ghana: 50 (1990).

Habit: tree to 30 m. **Habitat:** forest. **Distribution:** West and Central Africa and tropical America. **Star:** pink. **Alt. range:** 200-400m (2), 601-800m (1). **Specimens:** - *Mabeta-Moliwe:* Mann G. 761; Mildbraed G. W. J. 10542; Plot Voucher W 697. *Mokoko River F.R.:* Thomas D. W. 10030, 10066

Vismia guineensis (L.) Choisy
F.W.T.A. 1: 288 (1954); Bull. Jard. Bot. Nat. Belg. 36: 438 (1966); Keay R.W.J., Trees of Nigeria: 110 (1989); Hawthorne W., F.G.F.T. Ghana: 49 (1990).

Habit: shrub to 3 m. **Guild:** pioneer. **Habitat:** lava flows. **Distribution:** Guinea Bissau to Cameroon, Gabon and Congo (Kinshasa). **Chorology:** Guineo-Congolian. **Star:** green. **Alt. range:** 1-200m (1). **Flowering:** Feb (1), Jul (1). **Fruiting:** Feb (1), Jul (1). **Specimens:** - *Etinde:* Tchouto P. 6, 239

HOPLESTIGMATACEAE

Hoplestigma pierreanum Gilg
F.W.T.A. 2: 16 (1963).

Habit: tree to 25 m. **Habitat:** forest. **Distribution:** Ivory Coast, Cameroon and Gabon. **Chorology:** Guineo-Congolian. **Star:** gold. **Alt. range:** 200-400m (1). **Specimens:** - *Mabeta-Moliwe:* Mildbraed G. W. J. 10774. *Mokoko River F.R.:* Thomas D. W. 10022

HUACEAE

Afrostyrax kamerunensis Perkins & Gilg
Bull. Soc. Roy. Bot. Belg. 91: 94 (1958); F.W.T.A. 2: 34 (1963).

Habit: tree to 12 m. **Habitat:** forest. **Distribution:** Cameroon. **Chorology:** Lower Guinea. **Star:** gold. **Alt. range:** 1-200m (5), 601-800m (1). **Flowering:** Mar (1), Jun (1), Aug (1). **Fruiting:** Mar (1), Jun (1), Aug (3). **Specimens:** - *Mabeta-Moliwe:* Baker W. J. 282; Cheek M. 3484; Mildbraed G. W. J. 10888; Nguembock F. 65; Plot Voucher W 462, W 623; Wheatley J. I. 35, 344; von Rege I. 89, 96

Alsodeiopsis mannii Oliv.

F.W.T.A. 1: 638 (1958); Fl. Gabon 20: 23 (1973); Fl. Cameroun 15: 23 (1973).

Habit: shrub to 1 m. **Habitat:** forest. **Distribution:** Cameroon and Gabon. **Chorology:** Lower Guinea. **Star:** blue. **Specimens:** - *Southern Bakundu F.R.:* Brenan J. P. M. 9267

Alsodeiopsis weissenborniana J.Braun & K.Schum.

F.W.T.A. 1: 638 (1958); Fl. Gabon 20: 20 (1973); Fl. Cameroun 15: 20 (1973).

Habit: shrub to 1.5 m. **Habitat:** forest. **Distribution:** Cameroon, Gabon and Angola. **Chorology:** Lower Guinea. **Star:** blue. **Alt. range:** 1-200m (8), 201-400m (3), 601-800m (1). **Flowering:** Mar (2), Jun (1), Nov (1), Dec (2). **Fruiting:** Mar (2), Apr (1), May (1), Jun (2), Aug (1), Sep (1), Nov (2), Dec (2). **Specimens:** - *Bambuko F.R.:* Watts J. 663. *Etinde:* Cheek M. 5569, 5913; Faucher P. 20; Kwangue A. 77, 90; Tchouto P. 229; Watts J. 728. *Mokoko River F.R.:* Akogo M. 229; Tchouto P. 641. *Onge:* Thomas D. W. 9800; Watts J. 699

Chlamydocarya thomsoniana Baill.

F.W.T.A. 1: 643 (1958); Fl. Gabon 20: 95 (1973); Fl. Cameroun 15: 95 (1973).

Habit: woody climber. **Guild:** shade-bearer. **Habitat:** forest. **Distribution:** Sierra Leone to Cameroon, Gabon and Congo (Kinshasa). **Chorology:** Guineo-Congolian. **Star:** green. **Alt. range:** 1-200m (3). **Fruiting:** Nov (1). **Fruiting:** Apr (1), May (1). **Specimens:** - *Mabeta-Moliwe:* Tchouto P. 152. *Mokoko River F.R.:* Watts J. 1195. *Onge:* Tchouto P. 1043

Desmostachys tenuifolius Oliv. var. tenuifolius

F.W.T.A. 1: 639 (1958); Fl. Gabon 20: 44 (1973); Fl. Cameroun 15: 44 (1973); Keay R.W.J., Trees of Nigeria: 310 (1989).

Habit: tree or shrub to 10 m. **Habitat:** forest and forest gaps. **Distribution:** SE Nigeria, Bioko, Cameroon and Gabon. **Chorology:** Lower Guinea. **Star:** blue. **Alt. range:** 1-200m (19), 201-400m (2), 401-600m (1). **Flowering:** Apr (3), May (8), Jun (2), Jul (1). **Fruiting:** Jan (1), Apr (1), May (4), Jun (3), Oct (1). **Specimens:** - *Etinde:* Akogo M. 156; Kwangue A. 57, 78; Mbani J. M. 152; Nkeng P. 3; Tchouto P. 128; Tekwe C. 76. *Mabeta-Moliwe:* Baker W. J. 203; Plot Voucher W 1022; Sunderland T. C. H. 1293, 1296, 1404; Watts J. 346, 362, 402; Wheatley J. I. 127, 173, 305. *Mokoko River F.R.:* Batoum A. 13; Ekema N. 1149; Tchouto P. 1207; Thomas D. W. 10100. *Onge:* Cheek M. 5143. *Southern Bakundu F.R.:* Brenan J. P. M. 9276

Iodes africana Welw. ex Oliv.

F.W.T.A. 1: 643 (1958); Fl. Gabon 20: 6 (1973); Fl. Cameroun 15: 6 (1973).

Habit: woody climber. **Guild:** light demander. **Habitat:** forest. **Distribution:** Nigeria, Cameroon, Gabon, Congo (Kinshasa) and Angola. **Chorology:** Guineo-Congolian. **Star:** green. **Alt. range:** 400-600m (1), 601-800m (2), 801-1000m (1). **Flowering:** Feb (1). **Fruiting:** Mar (1), Apr (2). **Specimens:** - *Eastern slopes:* Tchouto P. 144. *Etinde:* Khayota B. 514; Maitland T. D. 436; Tekwe C. 51; Wheatley J. I. 29

Iodes kamerunensis Engl.

F.W.T.A. 1: 643 (1958); Fl. Gabon 20: 8 (1973); Fl. Cameroun 15: 8 (1973).

Habit: woody climber. **Habitat:** forest. **Distribution:** Cameroon. **Chorology:** Lower Guinea. **Star:** gold. **Specimens:** - *Eastern slopes:* Maitland T. D. 491

Iodes klaineana Pierre

F.W.T.A. 1: 643 (1958); Fl. Gabon 20: 10 (1973); Fl. Cameroun 15: 10 (1973).

Habit: woody climber. **Habitat:** forest. **Distribution:** Nigeria, Cameroon, Gabon, Congo (Kinshasa) and Angola. **Chorology:** Guineo-Congolian. **Star:** green. **Alt. range:** 1-200m (1).**Fruiting:** Oct (1). **Specimens:** - *Onge:* Tchouto P. 958

Iodes seretii (De Wild.) Boutique

Fl. Cameroun 15: 13 (1973).

Habit: woody climber. **Habitat:** forest. **Distribution:** Cameroon, Gabon and Congo (Kinshasa). **Chorology:** Guineo-Congolian. **Star:** green. **Alt. range:** 1-200m (1).**Fruiting:** Apr (2). **Specimens:** - *Mokoko River F.R.:* Tchouto P. 1137. *Southern Bakundu F.R.:* Daramola B. O. FHI 29813

Lasianthera africana P.Beauv.

F.W.T.A. 1: 638 (1958); Fl. Gabon 20: 14 (1973); Fl. Cameroun 15: 14 (1973).

Habit: tree or shrub to 6 m. **Habitat:** forest and forest gaps. **Distribution:** Nigeria, Cameroon, Bioko, Gabon, Congo (Kinshasa) and Angola (Cabinda). **Chorology:** Guineo-Congolian. **Star:** green. **Alt. range:** 1-200m (23), 201-400m (4), 401-600m (2), 601-800m (1). **Flowering:** Mar (2), Apr (7), May (13), Jun (2), Oct (3), Nov (2), Dec (1). **Fruiting:** Apr (6), May (5), Jun (1), Jul (1). **Specimens:** - *Eastern slopes:* Tekwe C. 100. *Etinde:* Lighava M. 38; Maitland T. D. 93; Mbani J. M. 140; Tchouto P. 261. *Mabeta-Moliwe:* Jaff B. 131, 184, 228; Ndam N. 527; Plot Voucher W 100, W 212, W 241, W 330; Sunderland T. C. H. 1183, 1248, 1345, 1365; Watts J. 166, 227; Wheatley J. I. 215. *Mokoko River F.R.:* Acworth J. M. 56; Ekema N. 838, 881, 963, 1049; Fraser P. 438; Ndam N. 1090; Pouakouyou D. 90; Watts J. 1051, 1054. *Onge:* Akogo M. 142; Cheek M. 5117; Harris D. J. 3640, 3795; Watts J. 814; Wheatley J. I. 833

Lavigeria macrocarpa (Oliv.) Pierre

F.W.T.A. 1: 641 (1958); Fl. Gabon 20: 65 (1973); Fl. Cameroun 15: 65 (1973).

Habit: woody climber. **Habitat:** forest. **Distribution:** SE Nigeria, Bioko, Cameroon, Gabon and Congo (Kinshasa). **Chorology:** Guineo-Congolian. **Star:** green. **Alt. range:** 1-200m (8), 201-400m (1), 601-800m (2), 801-1000m (1). **Flowering:** Jan (1), Apr (1), May (1), Oct (2). **Fruiting:** Jan (1), Mar (2), Apr (1), Jun (1), Aug (2). **Specimens:** - *Eastern slopes:* Groves M. 193; Tekwe C. 12. *Mabeta-Moliwe:* Baker W. J. 243; Mildbraed G. W. J. 10536; Plot Voucher W 450, W 814; Sunderland T. C. H. 1178; Wheatley J. I. 87, 287; von Rege I. 49. *Mokoko River F.R.:* Ekema N. 1150; Pouakouyou D. 2. *Onge:* Tchouto P. 887; Watts J. 882

Leptaulus daphnoides Benth.

F.W.T.A. 1: 637 (1958); Fl. Gabon 20: 59 (1973); Fl. Cameroun 15: 59 (1973); Keay R.W.J., Trees of Nigeria: 310 (1989); Hawthorne W., F.G.F.T. Ghana: 85 (1990).

Habit: tree or shrub to 15 m. **Guild:** shade-bearer. **Habitat:** forest. **Distribution:** Sierra Leone to Cameroon, Gabon, Congo (Kinshasa), Angola (Cabinda), Uganda, Sudan and Tanzania. **Chorology:** Guineo-Congolian. **Star:** green. **Alt. range:** 1-200m (7), 201-400m (2), 1201-1400m (1). **Flowering:** Mar (1), Apr (4). **Fruiting:** Mar (2), Apr (2). **Specimens:** - *Bambuko F.R.:* Watts J. 651. *Eastern slopes:* Maitland T. D. 686. *Mabeta-Moliwe:* Plot Voucher W 337, W 545, W 606; Sunderland 1209; Watts J. 149; Wheatley J. I. 44. 143. *Mokoko River F.R.:* Akogo M. 299; Tchouto P. 1100, 1107, 1117; Thomas D. W. 10106

Leptaulus grandifolius Engl.
F.W.T.A. 1: 637 (1958); Fl. Gabon 20: 57 (1973); Fl. Cameroun 15: 57 (1973).

Habit: tree to 5 m. **Habitat:** forest. **Distribution:** Cameroon, Rio Muni, Gabon and Congo (Brazzaville). **Chorology:** Lower Guinea. **Star:** blue. **Specimens:** - *Southern Bakundu F.R.:* Brenan J. P. M. 9274

Pyrenacantha acuminata Engl.
F.W.T.A. 1: 642 (1958); Fl. Cameroun 15: 78 (1973).

Habit: woody climber. **Habitat:** forest. **Distribution:** Sierra Leone to Cameroon, Gabon and Congo (Kinshasa). **Chorology:** Guineo-Congolian. **Star:** green. **Alt. range:** 1-200m (3). **Flowering:** May (1). **Fruiting:** May (2). **Specimens:** - *Mokoko River F.R.:* Ekema N. 923; Tchouto P. 1193, 1217

Pyrenacantha glabrescens (Engl.) Engl.
Fl. Cameroun 15: 72 (1973).
Pyrenacantha mangenotiana J.Miege, F.W.T.A. 1: 641 (1958).

Habit: woody climber. **Guild:** shade-bearer. **Habitat:** forest. **Distribution:** Guinea, Liberia, Ivory Coast, Cameroon and Gabon. **Chorology:** Guineo-Congolian. **Star:** gold. **Alt. range:** 1-200m (1).**Fruiting:** Apr (1). **Specimens:** - *Mabeta-Moliwe:* Jaff B. 126

Pyrenacantha longirostrata Villiers
Fl. Gabon 20: 85 (1973); Fl. Cameroun 15: 85 (1973).
Pyrenacantha sp. C sensu Keay, F.W.T.A. 1: 642 (1958).

Habit: woody climber. **Habitat:** forest. **Distribution:** Cameroon and Gabon. **Chorology:** Lower Guinea. **Star:** blue. **Alt. range:** 200-400m (1).**Fruiting:** Mar (1). **Specimens:** - *Mokoko River F.R.:* Tchouto P. 602

Pyrenacantha staudtii (Engl.) Engl.
F.W.T.A. 1: 642 (1958); Fl. Gabon 20: 79 (1973); Fl. Cameroun 15: 79 (1973).

Habit: woody climber. **Habitat:** forest. **Distribution:** Nigeria, Cameroon, Gabon, Congo (Kinshasa), Angola and Uganda. **Chorology:** Guineo-Congolian. **Star:** green. **Fruiting:** Apr (1). **Specimens:** - *Mabeta-Moliwe:* Nguembock 64. *Southern Bakundu F.R.:* Ejiofor M. C. FHI 29338

Pyrenacantha vogeliana Baill.
F.W.T.A. 1: 642 (1958); Fl. Gabon 20: 76 (1973); Fl. Cameroun 15: 76 (1973).

Habit: woody climber. **Guild:** shade-bearer. **Habitat:** forest. **Distribution:** Sierra Leone to Cameroon, Gabon, Congo (Kinshasa) and Tanzania. **Chorology:** Guineo-Congolian. **Star:** green. **Alt. range:** 1-200m (1), 201-400m (1).**Fruiting:** Mar (1), Sep (1). **Specimens:** - *Bambuko F.R.:* Watts J. 624. *Onge:* Tchouto P. 829

Pyrenacantha sp. nov. aff. grandifolia Engl.

Specimens: - *Mabeta-Moliwe:* - Jaff 126.

Pyrenacantha sp.

Specimens: - *Etinde:* Banks H. 35

Rhaphiostylis beninensis (Hook.f. ex Planch.) Planch. ex Benth.
F.W.T.A. 1: 638 (1958); Fl. Gabon 20: 32 (1973); Fl. Cameroun 15: 32 (1973).

Habit: woody climber. **Guild:** shade-bearer. **Habitat:** forest. **Distribution:** tropical Africa. **Star:** excluded. **Alt. range:** 200-400m (1). **Flowering:** Oct (1). **Specimens:** - *Onge:* Tchouto P. 852

Rhaphiostylis ferruginea Engl.
F.W.T.A. 1: 639 (1958); Fl. Gabon 20: 37 (1973); Fl. Cameroun 15: 37 (1973).

Habit: woody climber. **Guild:** shade-bearer. **Habitat:** forest. **Distribution:** Ivory Coast to Cameroon, Gabon and Congo (Kinshasa). **Chorology:** Guineo-Congolian. **Star:** green. **Specimens:** - *Etinde:* Maitland T. D. 1062

Rhaphiostylis preussii Engl.
F.W.T.A. 1: 639 (1958); Fl. Gabon 20: 34 (1973); Fl. Cameroun 15: 34 (1973).

Habit: woody climber. **Guild:** shade-bearer. **Habitat:** forest. **Distribution:** Sierra Leone to Cameroon and Gabon. **Chorology:** Guineo-Congolian. **Star:** green. **Alt. range:** 200-400m (1). **Flowering:** Oct (1). **Specimens:** - *Onge:* Watts J. 862

Stachyanthus zenkeri Engl.
Fl. Gabon 20: 68 (1973); Fl. Cameroun 15: 68 (1973).
Neostachyanthus zenkeri (Engl.) Exell & Mendonça, F.W.T.A. 1: 643 (1958).

Habit: woody climber. **Habitat:** forest. **Distribution:** Nigeria, Bioko, Cameroon, Congo (Kinshasa) and Angola (Cabinda). **Chorology:** Guineo-Congolian. **Star:** green. **Alt. range:** 1-200m (5), 201-400m (7), 401-600m (2). **Flowering:** Mar (1), Jun (1), Oct (6), Nov (1). **Fruiting:** Mar (1), Jun (2), Oct (5), Nov (1), Dec (1). **Specimens:** - *Bambuko F.R.:* Watts J. 598. *Etinde:* Cheek M. 5899; Kwangue A. 93. *Mokoko River F.R.:* Pouakouyou D. 95. *Onge:* Akogo M. 96; Cheek M. 5009, 5149; Ndam N. 665, 716; Tchouto P. 817; Thomas D. W. 9867; Watts J. 791, 957; Wheatley J. I. 687

IRVINGIACEAE

D.J. Harris (E)

Desbordesia glaucescens (Engl.) Tiegh.
F.W.T.A. 1: 694 (1958); Fl. Gabon 3: 29 (1962); Keay R.W.J., Trees of Nigeria: 333 (1989); Bull. Jard. Bot. Nat. Belg. 65: 147 (1996).

Habit: tree to 50 m. **Habitat:** forest. **Distribution:** Nigeria, Cameroon, Gabon, Congo (Brazzaville) and Congo (Kinshasa).

Chorology: Guineo-Congolian. **Star:** green. **Alt. range:** 1-200m (1), 601-800m (1). **Flowering:** Apr (1). **Fruiting:** May (1). **Specimens:** - *Etinde:* Maitland T. D. 625. *Mabeta-Moliwe:* Jaff B. 111; Mildbraed G. W. J. 10741; Plot Voucher W 1177. *Mokoko River F.R.:* Ndam N. 1195

Irvingia gabonensis (Aubry-Lecomte ex O'Rorke) Baill.

F.W.T.A. 1: 693 (1958); Fl. Gabon 3: 22 (1962); Keay R.W.J., Trees of Nigeria: 332 (1989); Hawthorne W., F.G.F.T. Ghana: 86 (1990); Bull. Jard. Bot. Nat. Belg. 65: 172 (1996).

Habit: tree 10-40 m tall. **Guild:** light demander. **Habitat:** forest. **Distribution:** Nigeria, Cameroon, Gabon, Congo (Brazzaville), Congo (Kinshasa) and CAR. **Chorology:** Guineo-Congolian. **Star:** pink. **Alt. range:** 200-400m (2). **Specimens:** - *Mokoko River F.R.:* Ndam N. 1149, 1325

Klainedoxa gabonensis Pierre ex Engl.

Bull. Jard. Bot. Nat. Belg. 65: 153 (1996).
   Klainedoxa gabonensis Pierre ex Engl. var. *oblongifolia* Engl., F.W.T.A. 1: 698 (1958); Hawthorne W., F.G.F.T. Ghana: 86 (1990).

Habit: tree to 50 m. **Guild:** light demander. **Habitat:** forest. **Distribution:** Guinea Bissau to Cameroon, Gabon, Congo (Brazzaville), Congo (Kinshasa), CAR, Uganda, Tanzania and Zambia. **Chorology:** Guineo-Congolian. **Star:** green. **Alt. range:** 200-400m (1). **Specimens:** - *Etinde:* Kalbreyer W. 10. *Mokoko River F.R.:* Thomas D. W. 10116

Klainedoxa trillesii Pierre ex Tiegh.

Bull. Jard. Bot. Nat. Belg. 65: 161 (1996).

Habit: tree to 30 m. **Habitat:** forest. **Distribution:** Sierra Leone to Cameroon, Gabon, Congo (Brazzaville) and Congo (Kinshasa). **Chorology:** Guineo-Congolian. **Star:** green. **Alt. range:** 1-200m (1). **Specimens:** - *Onge:* Harris D. J. 3698

IXONANTHACEAE

Phyllocosmus africanus (Hook.f.) Klotzsch

Kew Bull. 19: 517 (1965); Fl. Cameroun 14: 62 (1972); Keay R.W.J., Trees of Nigeria: 141 (1989).
   Ochthocosmus africanus Hook.f., F.W.T.A. 1: 355 (1958).

Habit: tree to 20 m. **Guild:** light demander. **Habitat:** forest. **Distribution:** Guinea Bissau to Congo (Brazzaville) and Congo (Kinshasa). **Chorology:** Guineo-Congolian. **Star:** green. **Alt. range:** 1-200m (1). **Specimens:** - *Mokoko River F.R.:* Tchouto P. 1133

LABIATAE

A. Paton (K)

Achyrospermum africanum Hook.f. ex Baker

F.W.T.A. 2: 469 (1963).

Habit: shrub to 2.5 m. **Guild:** pioneer. **Habitat:** forest. **Distribution:** Guinea to Cameroon. **Chorology:** Guineo-

62

Congolian (montane). **Star:** green. **Alt. range:** 1000-1200m (1). **Specimens:** - *No locality:* Morton J. K. K 700

Achyrospermum oblongifolium Baker

F.W.T.A. 2: 469 (1963).

Habit: shrub to 1 m. **Guild:** pioneer. **Habitat:** forest. **Distribution:** Guinea to Bioko, Cameroon and Angola (Cabinda) **Chorology:** Guineo-Congolian (montane). **Star:** green. **Alt. range:** 1-200m (1), 201-400m (3), 601-800m (1). **Flowering:** Oct (1), Nov (3), Dec (1). **Fruiting:** Dec (1). **Specimens:** - *Etinde:* Cheek M. 5411, 5440, 5604, 5887. *Onge:* Cheek M. 5037

Achyrospermum schlechteri Gürke

F.W.T.A. 2: 468 (1963).

Habit: shrub to 1 m. **Guild:** pioneer. **Distribution:** SE Nigeria and SW Cameroon. **Chorology:** Western Cameroon Uplands (montane). **Star:** gold. **Alt. range:** 600-800m (1). **Specimens:** - *Eastern slopes:* Schlechter F. R. R. 12850

Hoslundia opposita Vahl

F.W.T.A. 2: 456 (1963); Meded. Land. Wag. 82(3): 157 (1982).

Habit: woody climber or shrub. **Guild:** pioneer. **Habitat:** forest. **Distribution:** tropical and subtropical Africa. **Chorology:** Afromontane. **Star:** green. **Alt. range:** 1400-1600m (1). **Flowering:** Mar (1), Oct (1). **Fruiting:** Oct (1). **Specimens:** - *Eastern slopes:* Morton J. K. K 919; Nkeng P. 56; Tchouto P. 356

Hyptis lanceolata Poir.

F.W.T.A. 2: 466 (1963).

Habit: herb. **Distribution:** tropical Africa and America. **Star:** excluded. **Specimens:** - *Etinde:* Boughey A. S. 1175

Isodon ramosissimus (Hook.f.) Codd

Bothalia 15: 8 (1984); Fl. Rwanda 3: 311 (1985).
   Homalocheilos ramosissimus (Hook.f.) J.K.Morton, F.W.T.A. 2: 460 (1963).

Habit: herb to 4 m. **Habitat:** forest-grassland transition. **Distribution:** Sierra Leone to Bioko, Cameroon, Uganda and Zimbabwe. **Chorology:** Afromontane. **Star:** green. **Alt. range:** 800-1000m (1), 1801-2000m (1), 2001-2200m (3), 2201-2400m (1). **Flowering:** Oct (4), Nov (1), Dec (1). **Specimens:** - *Eastern slopes:* Banks H. 86; Cheek M. 3662; Tchouto P. 308, 316; Thomas D. W. 9431. *Etinde:* Cheek M. 5620. *No locality:* Dunlap 213

Leocus africanus (Baker ex Sc.Elliot) J.K.Morton

F.W.T.A. 2: 461 (1963).

Habit: herb. **Distribution:** Guinea to Cameroon, Congo (Kinshasa), CAR, Chad and Uganda. **Chorology:** Guineo-Congolian. **Star:** green. **Specimens:** - *Eastern slopes:* Mildbraed G. W. J. 9483

Leonotis nepetifolia (L.) Ait. var. africana (P.Beauv.) J.K.Morton

F.W.T.A. 2: 470 (1963).

Habit: herb. **Distribution:** tropical Africa. **Star:** excluded. **Specimens:** - *No locality:* Boughey A. S. ?

Leonotis nepetifolia (L.) Ait. var. nepetifolia
F.W.T.A. 2: 470 (1963).

Habit: herb to 2.5 m. **Distribution:** pantropical. **Star:** excluded. **Specimens:** - *Eastern slopes:* Morton J. K. K 661. *Mabeta-Moliwe:* Maitland T. D. s.n.

Leucas deflexa Hook.f.
F.W.T.A. 2: 470 (1963).

Habit: herb. **Guild:** pioneer. **Distribution:** Ghana, Bioko, W. Cameroon and Angola. **Chorology:** Guineo-Congolian (montane). **Star:** green. **Alt. range:** 1600-1800m (1). **Specimens:** - *No locality:* Mann G. 1976

Leucas oligocephala Hook.f. var. oligocephala
F.W.T.A. 2: 470 (1963).

Habit: herb. **Habitat:** grassland and forest-grassland transition. **Distribution:** W. Cameroon to East and South Africa. **Chorology:** Afromontane. **Star:** green. **Alt. range:** 1800-2000m (1), 2001-2200m (1), 2201-2400m (2). **Flowering:** Oct (3). **Specimens:** - *Eastern slopes:* Tchouto P. 347; Thomas D. W. 9287, 9386. *No locality:* Mann G. 1220

Ocimum gratissimum L.
F.W.T.A. 2: 452 (1963); Kew Bull. 43: 411 (1992).

Habit: herb. **Distribution:** tropical Africa, America and India. **Star:** excluded. **Specimens:** - *Etinde:* Schlechter F. R. R. 12570. *Mabeta-Moliwe:* Cheek M. 3437

Platostoma africanum P.Beauv.
F.W.T.A. 2: 453 (1963).

Habit: herb. **Guild:** pioneer. **Habitat:** forest and farmbush. **Distribution:** tropical Africa and India. **Star:** excluded. **Alt. range:** 1-200m (3), 401-600m (1), 601-800m (1), 801-1000m (1). **Flowering:** May (1), Oct (4), Nov (1). **Specimens:** - *Eastern slopes:* Morton J. K. GC 6742. *Etinde:* Kwangue A. 127; Watts J. 574. *Mabeta-Moliwe:* Cheek M. 3544; Sidwell K. 3; Sunderland T. C. H. 1280. *Onge:* Cheek M. 4973, 5156

Plectranthus decurrens (Gürke) J.K.Morton
F.W.T.A. 2: 460 (1963).

Habit: herb to 2 m. **Habitat:** forest and Aframomum thicket. **Distribution:** SE Nigeria, Bioko, Cameroon and Gabon. **Chorology:** Lower Guinea (montane). **Star:** blue. **Alt. range:** 400-600m (2), 801-1000m (2), 1001-1200m (1), 1201-1400m (1). **Flowering:** Sep (1), Oct (4). **Fruiting:** Oct (1). **Specimens:** - *Eastern slopes:* Preuss P. R. 948. *Etinde:* Cheek M. 3773; Tchouto P. 430; Tekwe C. 150; Thomas D. W. 9139; Watts J. 499

Plectranthus dissitiflorus (Gürke) J.K.Morton
F.W.T.A. 2: 460 (1963).

Habit: herb. **Distribution:** Mount Cameroon. **Chorology:** Endemic. **Star:** black. **Flowering:** Oct (1). **Fruiting:** Oct (1). **Specimens:** - *Eastern slopes:* Preuss P. R. 1055

Plectranthus glandulosus Hook.f.
F.W.T.A. 2: 460 (1963).

Habit: herb to 3 m. **Habitat:** forest. **Distribution:** Mali to Bioko and Cameroon. **Chorology:** Guineo-Congolian (montane). **Star:** green. **Alt. range:** 800-1000m (1), 1401-1600m (1), 2001-2200m (1), 2201-2400m (1), 2401-2600m (1). **Flowering:** Oct (4). **Specimens:** - *Eastern slopes:* Tchouto P. 354; Thomas D. W. 9277; Watts J. 461. *Etinde:* Watts J. 578. *No locality:* Morton J. K. GC 6908

Plectranthus insignis Hook.f.
F.W.T.A. 2: 460 (1963).

Habit: shrub to 5 m. **Habitat:** forest. **Distribution:** W. Cameroon. **Chorology:** Western Cameroon Uplands (montane). **Star:** gold. **Alt. range:** 1600-1800m (1), 1801-2000m (1), 2001-2200m (1). **Flowering:** Oct (1). **Specimens:** - *Eastern slopes:* Cheek M. 3659. *Etinde:* Thomas D. W. 9254. *No locality:* Preuss P. R. 991

Plectranthus kamerunensis Gürke
F.W.T.A. 2: 460 (1963).

Habit: herb. **Habitat:** forest-grassland transition. **Distribution:** SE Nigeria and W. Cameroon. **Chorology:** Western Cameroon Uplands (montane). **Star:** gold. **Alt. range:** 1000-1200m (3), 1201-1400m (1). **Flowering:** Oct (3). **Specimens:** - *Eastern slopes:* Thomas D. W. 9515. *Etinde:* Sidwell K. 157; Tchouto P. 421. *No locality:* Mann G. 1947

Plectranthus luteus Gürke
F.W.T.A. 2: 460 (1963).

Habit: herb. **Habitat:** forest. **Distribution:** Liberia, Cameroon, Congo (Kinshasa) and East Africa. **Chorology:** Afromontane. **Star:** green. **Alt. range:** 1600-1800m (1). **Flowering:** Oct (1). **Specimens:** - *Etinde:* Tchouto P. 412

Plectranthus punctatus L'Hér. subsp. punctatus
F.W.T.A. 2: 460 (1963).

Habit: herb. **Habitat:** forest-grassland transition. **Distribution:** Mount Cameroon and Bioko. **Chorology:** Western Cameroon Uplands (montane). **Star:** black. **Alt. range:** 1600-1800m (1), 2201-2400m (3). **Flowering:** Oct (2), Nov (1). **Specimens:** - *Eastern slopes:* Cheek M. 5351; Tchouto P. 341; Thomas D. W. 9378. *No locality:* Mann G. 1301

Plectranthus tenuicaulis (Hook.f.) J.K.Morton
F.W.T.A. 2: 460 (1963).

Habit: herb. **Habitat:** forest-grassland transition. **Distribution:** W. Cameroon. **Chorology:** Western Cameroon Uplands (montane). **Star:** gold. **Alt. range:** 1800-2000m (1), 2001-2200m (1), 2401-2600m (1). **Flowering:** Oct (2). **Specimens:** - *Eastern slopes:* Cheek M. 3683; Watts J. 432. *No locality:* Mann G. 1939

Pycnostachys meyeri Gürke
F.W.T.A. 2: 458 (1963).

Habit: shrub to 1 m. **Habitat:** grassland and forest-grassland transition. **Distribution:** tropical Africa. **Chorology:** Afromontane. **Star:** green. **Alt. range:** 1800-2000m (1), 2401-2600m (2). **Flowering:** Sep (1), Oct (1). **Fruiting:** Sep (1). **Specimens:** - *Eastern slopes:* Thomas D. W. 9429. *Etinde:* Tekwe C. 270. *No locality:* Keay R. W. J. FHI 28598

Satureja pseudosimensis Brenan
F.W.T.A. 2: 467 (1963).

Habit: herb. Habitat: grassland. Distribution: W. Cameroon, Bioko, Congo (Kinshasa), Sudan and East Africa. Chorology: Afromontane. Star: green. Alt. range: 2200-2400m (1), 2601-2800m (1), 2801-3000m (1). Flowering: Oct (1), Nov (1). Specimens: - *Eastern slopes:* Cheek M. 3676, 5358. *No locality:* Dunlap 241

Satureja punctata (Benth.) Brig.
F.W.T.A. 2: 467 (1963).

Habit: herb. Habitat: grassland and forest-grassland transition. Distribution: Cameroon to East and South Africa. Chorology: Afromontane. Star: green. Alt. range: 1800-2000m (1), 2201-2400m (1), 2801-3000m (1). Flowering: Oct (1), Nov (1). Specimens: - *Eastern slopes:* Cheek M. 5303; Thomas D. W. 9272. *No locality:* Dundas J. FHI 20358

Satureja robusta (Hook.f.) Brenan
F.W.T.A. 2: 467 (1963).

Habit: herb. Habitat: forest-grassland transition. Distribution: W. Cameroon. Chorology: Western Cameroon Uplands (montane). Star: gold. Alt. range: 1800-2000m (1), 2001-2200m (3), 2401-2600m (1). Flowering: May (1), Oct (3), Nov (1). Specimens: - *Eastern slopes:* Banks H. 94; Sidwell K. 18; Sunderland T. C. H. 1432; Tchouto P. 307; Thomas D. W. 9302. *No locality:* Dundas J. FHI 20357

Solenostemon decumbens (Hook.f.) Baker
F.W.T.A. 2: 464 (1963).

Habit: herb. Habitat: grassland. Distribution: W. Cameroon. Chorology: Western Cameroon Uplands (montane). Star: gold. Alt. range: 1800-2000m (1), 2001-2200m (1), 2201-2400m (2). Flowering: Sep (1), Oct (2). Specimens: - *Eastern slopes:* Thomas D. W. 9285, 9314. *Etinde:* Tekwe C. 272. *No locality:* Johnston H. H. 79

Solenostemon mannii (Hook.f.) Baker
F.W.T.A. 2: 464 (1963).

Habit: herb. Guild: pioneer. Habitat: forest and scrub. Distribution: Sierra Leone, Ghana, Nigeria, Bioko and W. Cameroon. Chorology: Guineo-Congolian (montane). Star: green. Alt. range: 200-400m (2), 401-600m (1), 601-800m (1). Flowering: Nov (2), Dec (1). Specimens: - *Etinde:* Cheek M. 5535, 5584, 5854. *No locality:* Mann G. 1251

Solenostemon monostachyus (P.Beauv.) Briq. subsp. **monostachyus**
F.W.T.A. 2: 464 (1963).

Habit: herb. Guild: pioneer. Habitat: roadsides and villages. Distribution: tropical Africa. Star: excluded. Alt. range: 1-200m (1). Flowering: Oct (1). Specimens: - *Etinde:* Morton J. K. K 690. *Mabeta-Moliwe:* Sidwell K. 4

Solenostemon repens (Gürke) J.K.Morton
F.W.T.A. 2: 463 (1963).

Habit: herb. Guild: pioneer. Habitat: forest and Aframomum thicket. Distribution: Guinea to Bioko and SW Cameroon.

Chorology: Guineo-Congolian (montane). Star: green. Alt. range: 400-600m (1), 801-1000m (3), 1001-1200m (1), 1201-1400m (1), 1401-1600m (1). Flowering: Sep (3), Oct (3). Fruiting: Oct (1). Specimens: - *Eastern slopes:* Nkeng P. 88; Preuss P. R. 949. *Etinde:* Cheek M. 3764; Tchouto P. 448; Tekwe C. 213; Watts J. 535; Wheatley J. I. 542

Solenostemon sp. nov.

Habit: herb. Distribution: Mount Cameroon. Chorology: Endemic. Star: black. Alt. range: 200-400m (2), 601-800m (2). Flowering: Nov (4). Specimens: - *Etinde:* Cable S. 216, 227; Cheek M. 5563; Williams S. 52

Stachys aculeolata Hook.f.
F.W.T.A. 2: 469 (1963).

Habit: herb. Habitat: forest-grassland transition. Distribution: Cameroon, Bioko, Congo (Kinshasa), Ethiopia, Uganda and Kenya. Chorology: Afromontane. Star: green. Alt. range: 1000-1200m (1), 1201-1400m (1), 2201-2400m (1), 2401-2600m (1). Flowering: Oct (4). Specimens: - *Eastern slopes:* Thomas D. W. 9423, 9516; Watts J. 453. *Etinde:* Thomas D. W. 9172. *No locality:* Maitland T. D. 926

LAURACEAE

Beilschmiedia gaboonensis (Meisn.) Benth. & Hook.f.
F.W.T.A. 1: 57 (1954); Fl. Gabon 10: 60 (1965); Fl. Cameroun 18: 52 (1974); Keay R.W.J., Trees of Nigeria: 34 (1989).

Habit: tree to 30 m. Habitat: forest. Distribution: SE Nigeria, Cameroon, Gabon and Congo (Kinshasa). Chorology: Guineo-Congolian. Star: green. Specimens: - *Etinde:* Maitland T. D. 535

Beilschmiedia hutchinsoniana Robyns & R.Wilczek
F.W.T.A. 1: 57 (1954); Fl. Gabon 10: 26 (1965); Fl. Cameroun 18: 36 (1974).

Habit: small tree. Distribution: SE Nigeria and Cameroon. Chorology: Lower Guinea (montane). Star: black. Specimens: - *Etinde:* Maitland T. D. 559

Beilschmiedia mannii (Meisn.) Benth. & Hook.f.
F.W.T.A. 1: 57 (1954); Fl. Gabon 10: 50 (1965); Fl. Cameroun 18: 66 (1974); Keay R.W.J., Trees of Nigeria: 34 (1989); Hawthorne W., F.G.F.T. Ghana: 71 (1990).

Habit: tree to 10 m. Habitat: forest. Distribution: Sierra Leone to Cameroon, Gabon and Congo (Kinshasa). Chorology: Guineo-Congolian. Star: green. Alt. range: 1-200m (1). Flowering: Mar (1). Specimens: - *Onge:* Wheatley J. I. 729

Beilschmiedia preussii Engl.
F.W.T.A. 1: 57 (1954); Fl. Gabon 10: 24 (1965); Fl. Cameroun 18: 50 (1974).

Habit: small tree or shrub. Habitat: forest. Distribution: Mount Cameroon. Chorology: Endemic. Star: black. Flowering: May (1). Specimens: - *Mabeta-Moliwe:* Preuss P. R. 1272

Beilschmiedia talbotiae (S.Moore) Robyns & R.Wilczek

F.W.T.A. 1: 57 (1954); Fl. Gabon 10: 23 (1965); Fl. Cameroun 18: 33 (1974); Keay R.W.J., Trees of Nigeria: 34 (1989).

Habit: tree or shrub to 8 m. Habitat: forest. Distribution: SE Nigeria and SW Cameroon. Chorology: Western Cameroon. Star: gold. Specimens: - *Etinde:* Maitland T. D. 558

Beilschmiedia sp.

Specimens: - *Mabeta-Moliwe:* Baker W. J. 226; Hunt L. V. 14. *Onge:* Cheek M. 5183; Tchouto P. 757; Wheatley J. I. 720

Hypodaphnis zenkeri (Engl.) Stapf

Specimens: - *S. Bakundu:* Staudt. 961

LECYTHIDACEAE

Napoleonaea talbotii Baker f.

F.W.T.A. 1: 244 (1954); Bull. Jard. Bot. Nat. Belg. 41: 369 (1971); Keay R.W.J., Trees of Nigeria: 81 (1989).
 Napoleonaea megacarpa Baker f., F.W.T.A. 1: 244 (1954).
 Napoleonaea gascoignei Baker f., F.W.T.A. 1: 244 (1954).

Habit: tree or shrub to 5 m. Habitat: forest. Distribution: SE Nigeria, Cameroon and Gabon. Chorology: Lower Guinea. Star: blue. Alt. range: 400-600m (1).Fruiting: May (1). Specimens: - *Etinde:* Mbani J. M. 162

Napoleonaea vogelii Hook. & Planch.

F.W.T.A. 1: 244 (1954); Bull. Jard. Bot. Nat. Belg. 41: 371 (1971); Keay R.W.J., Trees of Nigeria: 81 (1989); Hawthorne W., F.G.F.T. Ghana: 107 (1990).
 Napoleonaea parviflora Baker f., F.W.T.A. 1: 244 (1954).

Habit: tree to 15 m. Guild: shade-bearer. Habitat: forest. Distribution: Guinea to Angola. Chorology: Guineo-Congolian. Star: green. Alt. range: 1-200m (5), 201-400m (1). Flowering: Apr (1). Fruiting: Apr (4), May (1). Specimens: - *Mabeta-Moliwe:* Jaff B. 47, 85; Ndam N. 474; Plot Voucher W 174, W 482, W 541, W 570; Watts J. 221; Wheatley J. I. 94. *Mokoko River F.R.:* Ndam N. 1324

Petersianthus macrocarpus (P.Beauv.) Liben

Bull. Jard. Bot. Nat. Belg. 38: 207 (1968); Keay R.W.J., Trees of Nigeria: 79 (1989).
 Combretodendron macrocarpum (P.Beauv.) Keay, F.W.T.A. 1: 761 (1958).
 Combretodendron africanum (Welw. ex Benth. & Hook.f.) Exell, F.W.T.A. 1: 242 (1954).

Habit: tree to 40 m. Guild: pioneer. Habitat: forest. Distribution: Guinea to Cameroon, Gabon, Congo (Kinshasa) and Angola. Chorology: Guineo-Congolian. Star: green. Alt. range: 1-200m (1).Fruiting: Jul (1). Specimens: - *Mabeta-Moliwe:* Cheek M. 3591

LEEACEAE

Leea guineensis G.Don

F.W.T.A. 1: 683 (1958); Fl. Gabon 14: 116 (1968); Fl. Cameroun 13: 134 (1972).

Habit: tree or shrub to 10 m. Guild: pioneer. Habitat: forest and forest gaps. Distribution: tropical Africa. Star: excluded. Alt. range: 1-200m (16), 201-400m (2), 401-600m (1), 601-800m (3), 1201-1400m (1). Flowering: Mar (3), Apr (1), May (5), Jun (1), Jul (2), Aug (1), Sep (1). Fruiting: Mar (2), Apr (1), May (1), Jun (1), Jul (1), Aug (1), Sep (1), Oct (1), Nov (3), Dec (2). Specimens: - *Eastern slopes:* Migeod F. 64; Tchouto P. 43. *Etinde:* Cable S. 198, 264; Lighava M. 25, 26; Tchouto P. 75; Tekwe C. 61, 78, 154. *Mabeta-Moliwe:* Cable S. 1479; Hurst J. 5; Jaff B. 191; Ndam N. 524; Nguembock F. 28b; Tchouto P. 172; Watts J. 112; Wheatley J. I. 243; von Rege I. 11, 63. *Mokoko River F.R.:* Acworth J. M. 264; Fraser P. 458; Ndam N. 1259. *Onge:* Ndam N. 638; Thomas D. W. 9836

LEGUMINOSAE

B. Mackinder, R.M. Polhill, J.M. Lock, L. Rico-Arce, J. Barham, CS. Stirton & B.D. Schrire (K)

LEGUMINOSAE-CAESALPINIOIDEAE

Afzelia bella Harms

Fl. Congo B., Rw. & Bur. 3: 358 (1952); F.W.T.A. 1: 461 (1958); Keay R.W.J., Trees of Nigeria: 218 (1989); Hawthorne W., F.G.F.T. Ghana: 202 (1990).

Habit: shrub or tree to 12 m. Guild: light demander. Habitat: forest. Distribution: Nigeria, Cameroon, Gabon and Congo (Kinshasa). Chorology: Guineo-Congolian. Star: pink. Alt. range: 1-200m (1).Fruiting: Jul (1). Specimens: - *Etinde:* Maitland T. D. 742. *Mabeta-Moliwe:* Cheek M. 3554; Plot Voucher W 394

Anthonotha fragrans (Baker f.) Exell & Hillc.

F.W.T.A. 1: 473 (1958); Keay R.W.J., Trees of Nigeria: 224 (1989); Hawthorne W., F.G.F.T. Ghana: 205 (1990).

Habit: tree to 40 m. Guild: light demander. Habitat: forest. Distribution: Sierra Leone to Congo (Kinshasa) and Angola (Cabinda). Chorology: Guineo-Congolian. Star: green. Alt. range: 1-200m (1). Specimens: - *Mokoko River F.R.:* Mbani J. M. 393

Anthonotha leptorrhachis (Harms) J.Léonard

F.W.T.A. 1: 473 (1958).
 Isomacrolobium leptorhachis (Harms) Aubrév. & Pellegr., Fl. Cameroun 9: 182 (1970).

Habit: tree to 16 m. Habitat: forest, often by streams. Distribution: Cameroon. Chorology: Lower Guinea. Star: gold. Alt. range: 1-200m (1). Specimens: - *Mabeta-Moliwe:* Plot Voucher W 1161. *Onge:* Thomas D. W. 9779. *Southern Bakundu F.R.:* Olorunfemi J. FHI 30737

Anthonotha macrophylla P.Beauv.

F.W.T.A. 1: 473 (1958); Keay R.W.J., Trees of Nigeria: 223 (1989); Hawthorne W., F.G.F.T. Ghana: 205 (1990).

Habit: shrub or tree to 12 m. **Guild:** shade-bearer. **Habitat:** forest. **Distribution:** Guinea to Bioko, Congo (Kinshasa) and Angola. **Chorology:** Guineo-Congolian. **Star:** green. **Alt. range:** 1-200m (11), 201-400m (1). **Flowering:** Feb (1), Mar (3), Apr (6), May (1). **Fruiting:** Mar (2). **Specimens:** - *Etinde:* Maitland T. D. 623. *Mabeta-Moliwe:* Groves M. 260, 264; Nguembock F. 86; Plot Voucher W 256, W 359, W 510, W 651, W 919, W 1006; Sunderland T. C. H. 1105, 1128; Wheatley J. I. 93, 140. *Mokoko River F.R.:* Acworth J. M. 65, 135, 306; Mbani J. M. 314; Pouakouyou D. 26; Tchouto P. 572

Anthonotha sp.

Specimens: - *Etinde:* Maitland T. D. 1076. *Mabeta-Moliwe:* Etuge M. 1227

Baikiaea insignis Benth.

Fl. Congo B., Rw. & Bur. 3: 298 (1952); F.W.T.A. 1: 456 (1958); F.T.E.A. Caesalpinioideae: 109 (1967); Keay R.W.J., Trees of Nigeria: 209 (1989).

Habit: tree to 26 m. **Habitat:** forest. **Distribution:** Sierra Leone to Bioko, Cameroon, Gabon, Congo (Brazzaville), Congo (Kinshasa), Angola, Uganda and Tanzania. **Chorology:** Guineo-Congolian. **Star:** green. **Alt. range:** 1-200m (5). **Specimens:** - *Etinde:* Maitland T. D. 1307. *Mokoko River F.R.:* Mbani J. M. 310, 323, 325; Tchouto P. 1118. *Onge:* Thomas D. W. 9859

Berlinia bracteosa Benth.

F.W.T.A. 1: 470 (1958); Keay R.W.J., Trees of Nigeria: 221 (1989).

Habit: tree to 26 m. **Habitat:** forest. **Distribution:** SE Nigeria, Bioko, Cameroon, Gabon, Congo (Kinshasa) and Angola (Cabinda). **Chorology:** Guineo-Congolian. **Star:** green. **Specimens:** - *Etinde:* Winkler H. J. P. 1087

Berlinia congolensis vel. aff. (Baker f.) Keay

Specimens: - *Mabeta-Moliwe:* von Rege I. 77

Berlinia craibiana Baker f.

F.W.T.A. 1: 470 (1958); Keay R.W.J., Trees of Nigeria: 221 (1989).

Habit: tree to 30 m. **Habitat:** forest. **Distribution:** Nigeria, Cameroon and Gabon. **Chorology:** Lower Guinea. **Star:** blue. **Alt. range:** 1-200m (2). **Flowering:** Mar (1). **Fruiting:** May (1). **Specimens:** - *Mabeta-Moliwe:* Nguembock F. 93; Plot Voucher W 589; Sunderland T. C. H. 1106, 1407

Berlinia sp.

Specimens: - *Mabeta-Moliwe:* Baker W. J. 300

Brachystegia cynometroides Harms

Fl. Cameroun 9: 260 (1970).

Habit: tree to 50 m. **Habitat:** forest. **Distribution:** Cameroon. **Chorology:** Lower Guinea. **Star:** gold. **Alt. range:** 1-200m (3).

Flowering: Apr (1). **Specimens:** - *Mokoko River F.R.:* Acworth J. M. 136; Tchouto P. 1116. *Southern Bakundu F.R.:* Mbani J. M. 415

Brachystegia laurentii (De Wild.) Louis ex Hoyle

Fl. Congo B., Rw. & Bur. 3: 461 (1952); F.W.T.A. 1: 479 (1958).

Habit: tree to 50 m. **Habitat:** forest. **Distribution:** Cameroon, Gabon and Congo (Kinshasa). **Chorology:** Guineo-Congolian. **Star:** green. **Specimens:** - *Southern Bakundu F.R.:* Ngalame M. FHI 24805

Caesalpinia bonduc (L.) Roxb.

Fl. Congo B., Rw. & Bur. 3: 250 (1952); F.W.T.A. 1: 481 (1958); F.T.E.A. Caesalpinioideae: 37 (1967); Opera Botanica 68: 18 (1983).

Habit: scrambling shrub. **Habitat:** forest. **Distribution:** pantropical. **Star:** excluded. **Specimens:** - *Etinde:* Maitland T. D. 777. *Mabeta-Moliwe:* Plot Voucher W 1196

Cassia mannii Oliv.

Fl. Congo B., Rw. & Bur. 3: 499 (1952); F.W.T.A. 1: 452 (1958); F.T.E.A. Caesalpinioideae: 58 (1967); Keay R.W.J., Trees of Nigeria: 227 (1989).

Habit: tree to 26 m. **Habitat:** forest. **Distribution:** Nigeria, Cameroon, Gabon, Congo (Kinshasa), Uganda and Sudan. **Chorology:** Guineo-Congolian. **Star:** green. **Specimens:** - *Etinde:* Winkler H. J. P. 602

Chamaecrista kirkii (Oliv.) Standl.

Lock J.M., Legumes of Africa: 31 (1989).
 Cassia kirkii Oliv., F.W.T.A. 1: 452 (1958).

Habit: shrub to 2 m. **Distribution:** tropical Africa. **Star:** excluded. **Specimens:** - *Eastern slopes:* Maitland T. D. 810

Chamaecrista mimosoides (L.) Greene

Kew Bull. 43: 333-342; Lock J.M., Legumes of Africa: 32 (1989)
 Cassia mimosoides L. Fl. Congo B., Rw. & Bur. 3: 513 (1952); F.W.T.A. 1: 452 (1958); F.T.E.A. Caesalpinioideae: 100 (1967); Opera Botanica 68: 30 (1983).

Habit: herb. **Distribution:** palaeotropical. **Star:** excluded. **Specimens:** - *Bambuko F.R.:* Olorunfemi J. FHI 30775

Copaifera mildbraedii Harms

Fl. Congo B., Rw. & Bur. 3: 307 (1952); F.W.T.A. 1: 457 (1958); Keay R.W.J., Trees of Nigeria: 203 (1989).

Habit: tree to 50 m. **Habitat:** forest. **Distribution:** Nigeria, Cameroon, Gabon, Congo (Kinshasa) and CAR. **Chorology:** Guineo-Congolian. **Star:** green. **Alt. range:** 800-1000m (1). **Specimens:** - *Mabeta-Moliwe:* Mildbraed G. W. J. 10572. *Southern Bakundu F.R.:* Ngalame M. FHI 24807

Crudia bibundina Harms

Lock M., Legumes of Africa: 80 (1989).
 Crudia sp. A sensu Keay pro parte, F.W.T.A. 1: 467 (1958).

Habit: large tree. **Distribution:** Mount Cameroon. **Chorology:** Endemic. **Star:** black. **Flowering:** Nov (1). **Fruiting:** Feb (1).

Specimens: - *Etinde:* Dalziel J. M. 8246; Mildbraed G. W. J. 10645

Cynometra hankei Harms
Fl. Congo B., Rw. & Bur. 3: 318 (1952); F.W.T.A. 1: 458 (1958); Keay R.W.J., Trees of Nigeria: 203 (1989).

Habit: tree to 45 m. **Habitat:** forest. **Distribution:** SE Nigeria, Cameroon and Congo (Kinshasa). **Chorology:** Guineo-Congolian. **Star:** green. **Specimens:** - *Southern Bakundu F.R.:* Ejiofor M. C. FHI 29361

Cynometra mannii Oliv.
Fl. Congo B., Rw. & Bur. 3: 313 (1952); F.W.T.A. 1: 458 (1958); Keay R.W.J., Trees of Nigeria: 201 (1989).

Habit: tree to 20 m. **Habitat:** forest. **Distribution:** SE Nigeria, Cameroon, Gabon, Congo (Kinshasa) and Angola. **Chorology:** Guineo-Congolian. **Star:** green. **Specimens:** - *Etinde:* Kalbreyer W. 83. *Mabeta-Moliwe:* Mann G. 707; Plot Voucher W 1186 a

Daniellia oblonga Oliv.
F.W.T.A. 1: 463 (1958); Keay R.W.J., Trees of Nigeria: 226 (1989).

Habit: tree to ?! 60 m. **Habitat:** forest. **Distribution:** SE Nigeria, Bioko and SW Cameroon. **Chorology:** Western Cameroon. **Star:** gold. **Alt. range:** 600-800m (1). **Specimens:** - *Mabeta-Moliwe:* Mildbraed G. W. J. 10759

Didelotia letouzeyi Pellegr.
Fl. Cameroun 9: 228 (1970).

Habit: tree to 40 m. **Habitat:** forest. **Distribution:** Cameroon, Gabon and Congo (Kinshasa). **Chorology:** Guineo-Congolian. **Star:** green. **Alt. range:** 1-200m (1). **Specimens:** - *Onge:* Thomas D. W. 9856

Distemonanthus benthamianus Baill.
F.W.T.A. 1: 449 (1958); Keay R.W.J., Trees of Nigeria: 208 (1989); Hawthorne W., F.G.F.T. Ghana: 209 (1990).

Habit: tree to 35 m. **Guild:** light demander. **Habitat:** forest. **Distribution:** Sierra Leone to Cameroon, Rio Muni and Gabon. **Chorology:** Guineo-Congolian. **Star:** pink. **Alt. range:** 200-400m (1). **Specimens:** - *Mabeta-Moliwe:* Mildbraed G. W. J. 10738

Duparquetia orchidacea Baill.
Fl. Congo B., Rw. & Bur. 3: 545 (1952); F.W.T.A. 1: 448 (1958).

Habit: tree or scandent shrub to 6 m. **Guild:** light demander. **Habitat:** forest and stream banks. **Distribution:** Liberia to Cameroon, Gabon and Congo (Kinshasa). **Chorology:** Guineo-Congolian. **Star:** green. **Alt. range:** 1-200m (4). **Flowering:** Apr (2), Nov (2). **Fruiting:** Apr (1), Nov (1). **Specimens:** - *Etinde:* Olorunfemi J. FHI 30533. *Mokoko River F.R.:* Tchouto P. 1068. *Onge:* Harris D. J. 3688; Watts J. 999; Wheatley J. I. 189

Erythrophleum ivorense A.Chev.
F.W.T.A. 1: 484 (1958); Keay R.W.J., Trees of Nigeria: 231 (1989); Hawthorne W., F.G.F.T. Ghana: 217 (1990).

Habit: tree to 40 m. **Guild:** light demander. **Habitat:** forest. **Distribution:** Sierra Leone to Cameroon, Rio Muni and Gabon. **Chorology:** Guineo-Congolian. **Star:** red. **Alt. range:** 800-1000m

(1). **Fruiting:** Jun (1). **Specimens:** - *Mabeta-Moliwe:* Mildbraed G. W. J. 10696; Plot Voucher W 1156

Eurypetalum unijugum Harms
F.W.T.A. 1: 464 (1958); Keay R.W.J., Trees of Nigeria: 200 (1989).

Habit: tree to 30 m. **Habitat:** forest. **Distribution:** SE Nigeria, Cameroon and Gabon. **Chorology:** Lower Guinea. **Star:** blue. **Specimens:** - *Southern Bakundu F.R.:* Ejiofor M. C. FHI 29308

Gilbertiodendron brachystegioides (Harms) J.Léonard
F.W.T.A. 1: 477 (1958).

Habit: tree to 40 m. **Guild:** light demander. **Habitat:** forest. **Distribution:** Cameroon and Gabon. **Chorology:** Lower Guinea. **Star:** gold. **Alt. range:** 1-200m (1), 201-400m (1). **Specimens:** - *Mokoko River F.R.:* Mbani J. M. 349. *Onge:* Thomas D. W. 9773. *Southern Bakundu F.R.:* Lobe Babute F. G. 51/36

Gilbertiodendron dewevrei (De Wild.) J.Léonard
Fl. Congo B., Rw. & Bur. 3: 429 (1952); F.W.T.A. 1: 477 (1958); Keay R.W.J., Trees of Nigeria: 212 (1989).

Habit: tree to 30 m. **Habitat:** swamp forest and by streams. **Distribution:** Nigeria, Cameroon, Gabon, Congo (Brazzaville), Congo (Kinshasa) and Angola. **Chorology:** Guineo-Congolian. **Star:** pink. **Alt. range:** 1-200m (1). **Flowering:** Apr (1). **Specimens:** - *Mokoko River F.R.:* Acworth J. M. 167

Gilbertiodendron grandiflorum (De Wild.) J.Léonard
Fl. Congo B., Rw. & Bur. 3: 433 (1952); F.W.T.A. 1: 475 (1958); Keay R.W.J., Trees of Nigeria: 212 (1989).

Habit: tree to 12 m. **Habitat:** forest. **Distribution:** SE Nigeria, Cameroon, Gabon and Congo (Kinshasa). **Chorology:** Guineo-Congolian. **Star:** green. **Alt. range:** 1-200m (2). **Flowering:** May (3). **Specimens:** - *Mabeta-Moliwe:* Sunderland T. C. H. 1269; Tchouto P. 192; Wheatley J. I. 267

Griffonia physocarpa Baill.
F.W.T.A. 1: 446 (1958).

Habit: woody climber. **Habitat:** forest. **Distribution:** Nigeria, Cameroon, Bioko, Gabon and Congo (Kinshasa). **Chorology:** Guineo-Congolian. **Star:** green. **Alt. range:** 1-200m (1). **Flowering:** May (1). **Specimens:** - *Mokoko River F.R.:* Tchouto P. 1206

Guibourtia demeusei (Harms) J.Léonard
Fl. Congo B., Rw. & Bur. 3: 361 (1952); F.W.T.A. 1: 465 (1958).

Habit: tree to 40 m. **Habitat:** forest. **Distribution:** Cameroon, Gabon, Congo (Brazzaville), Congo (Kinshasa) and CAR. **Chorology:** Guineo-Congolian. **Star:** green. **Alt. range:** 1-200m (1), 801-1000m (1). **Specimens:** - *Etinde:* Maitland T. D. 755. *Mabeta-Moliwe:* Mildbraed G. W. J. 10753, 10768

Guibourtia pellegriniana J.Léonard
F.W.T.A. 1: 465 (1958); Keay R.W.J., Trees of Nigeria: 200 (1989).

Habit: tree to 35 m. **Habitat:** forest. **Distribution:** SE Nigeria, Cameroon, Gabon, Congo (Brazzaville), Congo (Kinshasa) and Angola (Cabinda). **Chorology:** Guineo-Congolian. **Star:** green. **Specimens:** - *Southern Bakundu F.R.:* Ejiofor M. C. FHI 15252

Guibourtia tessmannii (Harms) J.Léonard
F.W.T.A. 1: 466 (1958).

Habit: tree to 40 m. **Habitat:** forest. **Distribution:** Cameroon, Rio Muni and Gabon. **Chorology:** Lower Guinea. **Star:** pink. **Flowering:** Dec (1). **Specimens:** - *Mabeta-Moliwe:* Mildbraed G. W. J. 10771

Hylodendron gabunense Taub.
Fl. Congo B., Rw. & Bur. 3: 303 (1952); F.W.T.A. 1: 456 (1958); Keay R.W.J., Trees of Nigeria: 203 (1989).

Habit: tree to 30 m. **Habitat:** forest. **Distribution:** Nigeria, Cameroon, Gabon and Congo (Kinshasa). **Chorology:** Guineo-Congolian. **Star:** pink. **Alt. range:** 1-200m (4). **Flowering:** Apr (1). **Fruiting:** Jun (1). **Specimens:** - *Etinde:* Winkler H. J. P. 1164. *Mabeta-Moliwe:* Plot Voucher W 201, W 202, W 292, W 704, W 762, W 839; Wheatley J. I. 348. *Mokoko River F.R.:* Acworth J. M. 178; Tchouto P. 1183. *Onge:* Harris D. J. 3720

Hymenostegia afzelii (Oliv.) Harms
F.W.T.A. 1: 464 (1958); Keay R.W.J., Trees of Nigeria: 224 (1989); Hawthorne W., F.G.F.T. Ghana: 201 (1990).

Habit: tree to 15 m. **Guild:** shade-bearer. **Habitat:** forest. **Distribution:** Guinea to Cameroon. **Chorology:** Guineo-Congolian. **Star:** green. **Alt. range:** 200-400m (1), 801-1000m (1). **Specimens:** - *Mabeta-Moliwe:* Mildbraed G. W. J. 10537; Plot Voucher W 771. *Mokoko River F.R.:* Ndam N. 1062

Hymenostegia sp.

Specimens: - *Bambuko F.R.:* Tchouto P. 535. *Etinde:* Thomas D. W. 9729, 9730

Isomacrolobium sp.

Specimens: - *Onge:* Thomas D. W. 9860

Julbernardia seretii (De Wild.) Troupin
F.W.T.A. 1: 471 (1958); Keay R.W.J., Trees of Nigeria: 200 (1989).

Habit: tree to 40 m. **Habitat:** forest. **Distribution:** SE Nigeria, Cameroon, Gabon, Congo (Brazzaville), Congo (Kinshasa) and Angola (Cabinda). **Chorology:** Guineo-Congolian. **Star:** green. **Specimens:** - *Etinde:* Onochie C. F. A. FHI 32084. *Southern Bakundu F.R.:* Ejiofor M. C. FHI 29310

Leonardoxa africana (Baill.) Aubrév.
Adansonia (ser.2) 8: 178 (1968); Keay R.W.J., Trees of Nigeria: 214 (1989).
    Schotia africana (Baill.) Keay, F.W.T.A. 1: 459 (1958).

Habit: tree to 10 m. **Habitat:** forest. **Distribution:** SE Nigeria, Cameroon, Rio Muni and Gabon. **Chorology:** Lower Guinea. **Star:** blue. **Specimens:** - *Mabeta-Moliwe:* Plot Voucher W 277

Loesenera talbotii Baker f.
F.W.T.A. 1: 461 (1958); Keay R.W.J., Trees of Nigeria: 219 (1989).

Habit: tree to 15 m. **Habitat:** forest. **Distribution:** SE Nigeria and SW Cameroon. **Chorology:** Western Cameroon. **Star:** black. **Alt. range:** 200-400m (1). **Specimens:** - *Mokoko River F.R.:* Ndam N. 1067, 1276b

Microberlinia bisulcata A.Chev.
F.W.T.A. 1: 471 (1958).

Habit: tree to 40 m. **Habitat:** forest. **Distribution:** S.W. Cameroon. **Chorology:** Lower Guinea. **Star:** gold. **Alt. range:** 1-200m (1), 201-400m (1). **Specimens:** - *Mokoko River F.R.:* Mbani J. M. 334; Ndam N. 1175. *Southern Bakundu F.R.:* Brenan J. P. M. 9319; Maitland T. D. 1067

Monopetalanthus letestui Pellegr.
Fl. Cameroun 9: 287 (1970).

Habit: tree to 40 m. **Habitat:** forest. **Distribution:** Cameroon. **Chorology:** Lower Guinea. **Star:** gold. **Alt. range:** 1-200m (1). **Specimens:** - *Onge:* Thomas D. W. 9816

Monopetalanthus sp.

Specimens: - *Mokoko River F.R.:* Tchouto P. 614; Thomas D. W. 10057

Oxystigma mannii (Baill.) Harms
F.W.T.A. 1: 466 (1958); Keay R.W.J., Trees of Nigeria: 200 (1989).

Habit: tree to 26 m. **Habitat:** forest. **Distribution:** SE Nigeria, Cameroon, Rio Muni and Gabon. **Chorology:** Lower Guinea. **Star:** blue. **Alt. range:** 1-200m (3). **Flowering:** Apr (1). **Fruiting:** Apr (1), May (1). **Specimens:** - *Mabeta-Moliwe:* Ndam N. 487; Wheatley J. I. 188. *Onge:* Thomas D. W. 9775

Plagiosiphon emarginatus (Hutch. & Dalziel) J.Léonard
F.W.T.A. 1: 464 (1958).

Habit: tree to 30 m. **Habitat:** forest, by streams. **Distribution:** Sierra Leone to Cameroon and Gabon. **Chorology:** Guineo-Congolian. **Star:** green. **Alt. range:** 1-200m (3). **Specimens:** - *Mokoko River F.R.:* Tchouto P. 621. *Onge:* Tchouto P. 773; Thomas D. W. 9817

Plagiosiphon longitubus (Harms) J.Léonard
Fl. Cameroun 9: 93 (1970).

Habit: tree to 8 m. **Habitat:** forest. **Distribution:** Cameroon. **Chorology:** Lower Guinea. **Star:** gold. **Alt. range:** 1-200m (1), 201-400m (2).**Fruiting:** Oct (1). **Specimens:** - *Mokoko River F.R.:* Ndam N. 1249; Thomas D. W. 10048. *Onge:* Ndam N. 749

Stemonocoleus micranthus Harms
F.W.T.A. 1: 466 (1958); Hawthorne W., F.G.F.T. Ghana: 202 (1990).

Habit: tree to 50 m. **Guild:** light demander. **Habitat:** forest. **Distribution:** Ivory Coast to Cameroon and Gabon. **Chorology:**

Guineo-Congolian. **Star:** green. **Specimens:** - *Southern Bakundu F.R.:* Keay R. W. J. s.n.

Tessmannia africana Harms

Fl. Congo B., Rw. & Bur. 3: 288 (1952); F.W.T.A. 1: 455 (1958).

Habit: tree to 50 m. **Distribution:** Cameroon, Gabon, Congo (Brazzaville), Congo (Kinshasa) and CAR. **Chorology:** Guineo-Congolian. **Star:** green. **Specimens:** - *Southern Bakundu F.R.:* Ejiofor M. C. FHI 29316

Tetraberlinia bifoliolata (Harms) Hauman

Fl. Congo B., Rw. & Bur. 3: 218 (1952).
Julbernardia bifoliolata (Harms.) Troupin, Bull. Jard. Bot. Brux. 20: 319 (1950).

Habit: tree to 30 m. **Habitat:** forest. **Distribution:** Cameroon, Rio Muni, Gabon, Congo (Kinshasa) and Angola. **Chorology:** Guineo-Congolian. **Star:** green. **Alt. range:** 1-200m (2). **Specimens:** - *Onge:* Tchouto P. 772; Thomas D. W. 9857. *Southern Bakundu F.R.:* Brenan J. P. M. 9312

LEGUMINOSAE-MIMOSOIDEAE

Acacia kamerunensis Gand.

Lock J.M., Legumes of Africa: 69 (1989).

Habit: woody climber. **Habitat:** forest. **Distribution:** Liberia to S. Tomé, Cameroon, Gabon and Congo (Kinshasa). **Chorology:** Guineo-Congolian. **Star:** green. **Specimens:** - *Mabeta-Moliwe:* Plot Voucher W 447, W 937

Acacia pentagona (Schum.) Hook.f.

F.T.E.A. Mimosoideae: 100 (1959); Opera Botanica 68: 42 (1983).
Acacia pennata sensu Keay, F.W.T.A. 1: 500 (1958).

Habit: woody climber. **Habitat:** forest. **Distribution:** tropical Africa. **Star:** excluded. **Specimens:** - *Eastern slopes:* Maitland T. D. 658

Albizia zygia (DC.) J.F.Macbr.

F.W.T.A. 1: 502 (1958); F.T.E.A. Mimosoideae: 161 (1959); Meded. Land. Wag. 82(3): 191 (1982); Keay R.W.J., Trees of Nigeria: 244 (1989); Hawthorne W., F.G.F.T. Ghana: 219 (1990).

Habit: tree to 25 m. **Guild:** light demander. **Habitat:** forest and farmbush. **Distribution:** tropical Africa. **Star:** pink. **Alt. range:** 200-600m (1). **Flowering:** Mar (1). **Specimens:** - *Etinde:* Khayota B. 600. *Mabeta-Moliwe:* Plot Voucher W 807, W 897, W 1182, W 1200

Calpocalyx dinklagei Harms

F.W.T.A. 1: 488 (1958); Keay R.W.J., Trees of Nigeria: 243 (1989).

Habit: tree to 15 m. **Habitat:** forest. **Distribution:** SE Nigeria, Cameroon and Gabon. **Chorology:** Lower Guinea. **Star:** blue. **Alt. range:** 1-200m (1). **Specimens:** - *Mokoko River F.R.:* Tchouto P. 1090. *Southern Bakundu F.R.:* Brenan J. P. M. 9298

Calpocalyx winkleri (Harms) Harms

F.W.T.A. 1: 488 (1958); Keay R.W.J., Trees of Nigeria: 243 (1989).

Habit: tree to 21 m. **Habitat:** forest. **Distribution:** Nigeria and Cameroon. **Chorology:** Lower Guinea. **Star:** blue. **Alt. range:** 800-1000m (1). **Specimens:** - *Mabeta-Moliwe:* Mildbraed G. W. J. 10721

Entada mannii (Oliv.) Tisserant

F.W.T.A. 1: 491 (1958).

Habit: woody climber. **Habitat:** forest. **Distribution:** Senegal to Congo (Kinshasa) and Angola. **Chorology:** Guineo-Congolian. **Star:** green. **Alt. range:** 1-200m (1). **Flowering:** Aug (1). **Specimens:** - *Mabeta-Moliwe:* von Rege I. 81

Leucaena leucocephala (Lam.) De Wit

Opera Botanica 68: 38 (1983); Lock M., Legumes of Africa: 93 (1989).
Leucaena glauca sensu Keay, F.W.T.A. 1: 495 (1958).

Habit: tree to 15 m. **Guild:** pioneer. **Habitat:** farmbush. **Distribution:** Mexico, widely introduced throughout tropics. **Star:** excluded. **Specimens:** - *Limbe Botanic Garden:* Banks H. 25

Mimosa pudica L.

Specimens:- *Etinde:* Brodie 7, Fraser 1009

Newtonia duparquetiana (Baill.) Keay

F.W.T.A. 1: 489 (1958); Keay R.W.J., Trees of Nigeria: 242 (1989).

Habit: tree to 26 m. **Guild:** shade-bearer. **Habitat:** forest. **Distribution:** Sierra Leone to Cameroon and Gabon. **Chorology:** Guineo-Congolian. **Star:** green. **Alt. range:** 200-400m (2). **Specimens:** - *Mokoko River F.R.:* Ndam N. 1147; Thomas D. W. 10172

Newtonia griffoniana (Baill.) Baker f.

Fl. Gabon 31: 49 (1989).
Newtonia zenkeri Harms, F.W.T.A. 1: 489 (1958).

Habit: tree to 35 m. **Habitat:** forest. **Distribution:** Cameroon, Gabon, Congo (Brazzaville) and Angola. **Chorology:** Guineo-Congolian. **Star:** green. **Specimens:** - *Etinde:* Maitland T. D. 575

Parkia bicolor A.Chev.

F.W.T.A. 1: 487 (1958); Bull. Jard. Bot. Nat. Belg. 54: 241 (1984); Keay R.W.J., Trees of Nigeria: 251 (1989); Hawthorne W., F.G.F.T. Ghana: 220 (1990).

Habit: tree to 30 m. **Guild:** light demander. **Habitat:** forest. **Distribution:** Guinea to Cameroon, Gabon, Congo (Brazzaville), Congo (Kinshasa) and Angola (Cabinda). **Chorology:** Guineo-Congolian. **Star:** green. **Specimens:** - *Bambuko F.R.:* Olorunfemi J. FHI 30774

Pentaclethra macrophylla Benth.

Specimens:- *Etinde:* Maitland 762.

Piptadeniastrum africanum (Hook.f.) Brenan

F.W.T.A. 1: 489 (1958); F.T.E.A. Mimosoideae: 21 (1959); Keay R.W.J., Trees of Nigeria: 257 (1989); Hawthorne W., F.G.F.T. Ghana: 220 (1990).

Habit: tree to 45 m. **Guild:** light demander. **Habitat:** forest. **Distribution:** Senegal to Cameroon, Gabon, Congo (Brazzaville), Congo (Kinshasa), Angola, CAR, Uganda and Sudan. **Chorology:** Guineo-Congolian. **Star:** red. **Specimens:** - *Etinde:* Maitland T. D. 427. *Mabeta-Moliwe:* Plot Voucher W 302, W 443

Tetrapleura tetraptera (Schum. & Thonn.) Taub.

F.W.T.A. 1: 493 (1958); F.T.E.A. Mimosoideae: 32 (1959); Keay R.W.J., Trees of Nigeria: 249 (1989); Hawthorne W., F.G.F.T. Ghana: 219 (1990).

Habit: tree to 24 m. **Guild:** pioneer. **Habitat:** forest. **Distribution:** tropical Africa. **Star:** pink. **Specimens:** - *Mabeta-Moliwe:* Dunlap 260

LEGUMINOSAE-PAPILIONOIDEAE

Abrus precatorius L.

F.W.T.A. 1: 574 (1958); F.T.E.A. Papilionoideae: 114 (1971); Opera Botanica 68: 70 (1983).

Habit: twining shrub. **Habitat:** grassland. **Distribution:** pantropical. **Chorology:** Montane. **Star:** excluded. **Alt. range:** 2000-2200m (1). **Flowering:** Sep (1). **Specimens:** - *Etinde:* Tekwe C. 282

Adenocarpus mannii (Hook.f.) Hook.f.

F.W.T.A. 1: 552 (1958); F.T.E.A. Papilionoideae: 1009 (1971); Opera Botanica 68: 181 (1983).

Habit: shrub to 2 m. **Habitat:** grassland and forest-grassland transition. **Distribution:** Bioko, Cameroon, East Africa, Zambia, Malawi, Angola, Rwanda, Sudan and Ethiopia. **Chorology:** Afromontane. **Star:** blue. **Alt. range:** 3000-3200m (1). **Flowering:** Nov (1). **Fruiting:** Nov (1). **Specimens:** - *Eastern slopes:* Banks H. 97. *No locality:* Mann G. 1308

Airyantha schweinfurthii (Taub.) Brummitt subsp. schweinfurthii

Lock J.M., Legumes of Africa: 465 (1989); Agric. Univ. Wag. Papers 94(4): 35 (1994).

Habit: woody climber. **Habitat:** forest. **Distribution:** Ivory Coast, Ghana, Nigeria, Cameroon, Rio Muni, Congo (Kinshasa) and CAR. **Chorology:** Guineo-Congolian. **Star:** green. **Alt. range:** 1-200m (1). **Flowering:** Nov (1). **Specimens:** - *Etinde:* Cheek M. 5506. *Mabeta-Moliwe:* Plot Voucher W 52

Amphicarpaea africana (Hook.f.) Harms

F.W.T.A. 1: 560 (1958); F.T.E.A. Papilionoideae: 511 (1971); Opera Botanica 68: 116 (1983).

Habit: herbaceous climber. **Habitat:** forest and forest-grassland transition. **Distribution:** tropical Africa. **Chorology:** Afromontane. **Star:** green. **Alt. range:** 2000-2200m (2). **Flowering:** Oct (1). **Specimens:** - *Eastern slopes:* Tchouto P. 311. *No locality:* Mann G. 1974

Andira inermis (Wright) DC. subsp. inermis

F.W.T.A. 1: 518 (1958); Kew Bull. 23: 488 (1969); F.T.E.A. Papilionoideae: 63 (1971); F.T.E.A. Papilionoideae: 63 (1971); Meded. Land. Wag. 82(3): 201 (1982); Keay R.W.J., Trees of Nigeria: 264 (1989).

Habit: tree to 12 m. **Habitat:** forest. **Distribution:** SE Nigeria and Cameroon. **Chorology:** Lower Guinea. **Star:** blue. **Alt. range:** 1-200m (2). **Flowering:** May (1), Jul (1). **Fruiting:** Jul (1). **Specimens:** - *Mabeta-Moliwe:* Cheek M. 3579; Wheatley J. I. 237

Angylocalyx oligophyllus (Baker) Baker f.

F.W.T.A. 1: 510 (1958).

Habit: shrub to 2 m. **Guild:** shade-bearer. **Habitat:** forest and forest gaps. **Distribution:** Liberia to Cameroon, Gabon, Congo (Kinshasa) and Angola (Cabinda). **Chorology:** Guineo-Congolian. **Star:** green. **Alt. range:** 1-200m (9), 401-600m (1). **Flowering:** Mar (4). **Fruiting:** Feb (1), Mar (3), May (3), Aug (1). **Specimens:** - *Etinde:* Khayota B. 580; Preuss P. R. 1374. *Mabeta-Moliwe:* Baker W. J. 241; Cable S. 1486; Sunderland T. C. H. 1085, 1124, 1181, 1274b; Wheatley J. I. 66. *Mokoko River F.R.:* Batoum A. 9; Ekema N. 1016; Mbani J. M. 498

Angylocalyx talbotii Baker f.

F.W.T.A. 1: 510 (1958).

Habit: shrub to 2 m. **Habitat:** forest. **Distribution:** SE Nigeria and SW Cameroon. **Chorology:** Western Cameroon. **Star:** gold. **Alt. range:** 1-200m (1), 201-400m (1). **Flowering:** Aug (1). **Fruiting:** Oct (1). **Specimens:** - *Mabeta-Moliwe:* Baker W. J. 305. *Onge:* Tchouto P. 889

Angylocalyx sp.

Specimens: - *Mabeta-Moliwe:* Khayota B. 473

Antopetitia abyssinica A.Rich.

F.W.T.A. 1: 577 (1958); F.T.E.A. Papilionoideae: 1049 (1971); Opera Botanica 68: 186 (1983).

Habit: herb. **Distribution:** Cameroon to East Africa, Zambia, Zimbabwe, Mozambique and Malawi. **Chorology:** Afromontane. **Star:** green. **Specimens:** - *No locality:* Migeod F. 202

Baphia laurifolia Baill.

F.W.T.A. 1: 512 (1958); Kew Bull. 40 (2): 363 (1985); Keay R.W.J., Trees of Nigeria: 261 (1989).

Habit: tree or shrub to 15 m. **Habitat:** forest. **Distribution:** SE Nigeria, Cameroon, Gabon, C.A.R., Congo (Kinshasa), Rio Muni. **Chorology:** Guineo-Congolian. **Star:** green. **Alt. range:** 1-200m (8). **Flowering:** Apr (3), Oct (1), Dec (1). **Fruiting:** Apr (2), Nov (1), Dec (1). **Specimens:** - *Mabeta-Moliwe:* Cheek M. 5767. *Mokoko River F.R.:* Acworth J. M. 117; Akogo M. 199; Pouakouyou D. 30; Tchouto P. 1071. *Onge:* Cheek M. 5198; Harris D. J. 3675, 3683

Baphia leptobotrys Harms

F.W.T.A. 1: 512 (1958); Kew Bull. 40: 343 (1985)

Habit: tree or shrub to 6 m. **Habitat:** forest and farmbush. **Distribution:** SE Nigeria, Cameroon and Gabon. **Chorology:** Lower Guinea. **Star:** blue. **Alt. range:** 1-200m (2). **Flowering:** May (2). **Fruiting:** May (1). **Specimens:** - *Mabeta-Moliwe:*

Dunlap 166; Nguembock F. 82; Sunderland T. C. H. 1289;
Tchouto P. 175

Baphia leptostemma Baill. var. gracilipes (Harms) Soladoye

Kew Bull. 40: 371 (1985); Keay R.W.J., Trees of Nigeria: 263
(1989).
Baphia gracilipes Harms., F.W.T.A. 1: 512 (1958).

Habit: tree or shrub to 12 m. **Habitat:** forest. **Distribution:** SE
Nigeria and Cameroon. **Chorology:** Lower Guinea. **Star:** blue.
Alt. range: 1-200m (2). **Flowering:** Apr (2). **Specimens:** -
Mabeta-Moliwe: Jaff B. 74; Wheatley J. I. 161

Baphia nitida Lodd.

F.W.T.A. 1: 512 (1958); Kew Bull. 40: 352 (1985); Keay R.W.J.,
Trees of Nigeria: 261 (1989); Hawthorne W., F.G.F.T. Ghana: 199
(1990).

Habit: tree or shrub to 4 m. **Guild:** shade-bearer. **Habitat:** forest.
Distribution: Senegal to Cameroon and Gabon. **Chorology:**
Guineo-Congolian. **Star:** green. **Alt. range:** 1-200m (3), 401-
600m (1). **Flowering:** Mar (1), Apr (1), May (1). **Fruiting:** Apr
(1). **Specimens:** - *Etinde:* Khayota B. 590; Mildbraed G. W. J.
10533. *Mabeta-Moliwe:* Tchouto P. 214. *Mokoko River F.R.:*
Pouakouyou D. 23, 28

Baphiopsis parviflora Benth. ex Baker

F.W.T.A. 1: 446 (1958).

Habit: tree to 15 m. **Habitat:** forest. **Distribution:** Cameroon,
Gabon, Congo (Kinshasa), Angola, Uganda and Tanzania.
Chorology: Guineo-Congolian (montane). **Star:** green. **Alt.
range:** 1-200m (4), 801-1000m (1), 1201-1400m (1). **Flowering:**
Feb (1), Mar (2), Nov (1). **Fruiting:** Nov (1). **Specimens:** -
Eastern slopes: Etuge M. 1162. *Etinde:* Tchouto P. 89. *Mabeta-
Moliwe:* Mann G. 715; Mildbraed G. W. J. 10561. *Mokoko River
F.R.:* Tchouto P. 647. *Onge:* Harris D. J. 3693; Thomas D. W.
9887

Calopogonium muconoides Desv.

F.W.T.A. 1: 563 (1958); F.T.E.A. Papilionoideae: 577 (1971).

Habit: herb. **Habitat:** forest. **Distribution:** tropical America,
cultivated in Africa. **Star:** excluded. **Alt. range:** 1-200m (1).
Flowering: Dec (1). **Specimens:** - *Mabeta-Moliwe:* Cable S. 556

Centrosema virginiana (L.) Benth.

F.W.T.A. 1: 560 (1958).

Habit: creeping herb. **Habitat:** lava flows. **Distribution:**
introduced to West Africa from America. **Star:** excluded.
Flowering: Oct (1). **Specimens:** - *Etinde:* Banks H. 12

Clitoria rubiginosa Juss. ex Pers.

F.W.T.A. 1: 560 (1958).

Habit: herbaceous climber. **Habitat:** disturbed ground.
Distribution: tropical Africa (widespread and introduced) and
America (native of C. America and W. Indies). **Star:** excluded.
Flowering: Oct (1). **Specimens:** - *Etinde:* Banks H. 6

Craibia atlantica Dunn

F.W.T.A. 1: 527 (1958); Meded. Land. Wag. 82(3): 205 (1982);
Keay R.W.J., Trees of Nigeria: 264 (1989).

Habit: tree to 10 m. **Guild:** shade-bearer. **Habitat:** forest.
Distribution: Ghana, Ivory Coast, Nigeria and Cameroon.
Chorology: Guineo-Congolian. **Star:** green. **Alt. range:** 1-200m
(1). **Flowering:** Jul (1). **Fruiting:** Jul (1). **Specimens:** - *Mabeta-
Moliwe:* Cheek M. 3562

Crotalaria micans Link

Lock M., Legumes of Africa: 191 (1989).
Crotalaria sp. B sensu Hepper, F.W.T.A. 1: 552 (1958).

Habit: shrub to 2.5 m. **Distribution:** introduced, native of South
America and the Indian Ocean. **Star:** excluded. **Specimens:** -
Etinde: Maitland T. D. 834

Crotalaria pallida Aiton

F.T.E.A. Papilionoideae: 905 (1971); Polhill R., Crotalaria of
Africa: 184 (1982); Opera Botanica 68: 166 (1983).
Crotalaria mucronata Desv., F.W.T.A. 1: 550 (1958).

Habit: herb. **Distribution:** pantropical. **Star:** excluded. **Alt.
range:** 800-1000m (1). **Specimens:** - *Eastern slopes:* Maitland T.
D. 115

Crotalaria retusa L.

F.W.T.A. 1: 548 (1958); F.T.E.A. Papilionoideae: 958 (1971);
Polhill R., Crotalaria of Africa: 272 (1982); Opera Botanica 68:
176 (1983).

Habit: herb. **Guild:** pioneer. **Habitat:** forest. **Distribution:**
pantropical. **Star:** excluded. **Alt. range:** 1-200m (3). **Flowering:**
May (1), Aug (1), Oct (1), Dec (1). **Specimens:** - *Etinde:* Banks H.
27. *Mabeta-Moliwe:* Baker W. J. 308; Cheek M. 5811; Ndam N.
528

Crotalaria subcapitata De Wild. subsp. oreadum (Baker f.) Polhill

Polhill R., Crotalaria of Africa: 197 (1982).
Crotalaria acervata sensu Hepper, F.W.T.A. 1: 550 (1958).

Habit: herb. **Habitat:** grassland. **Distribution:** tropical Africa.
Chorology: Afromontane. **Star:** green. **Alt. range:** 1800-2000m
(1), 2001-2200m (1). **Flowering:** Oct (2). **Specimens:** - *Eastern
slopes:* Sidwell K. 91; Tchouto P. 285

Dalbergia ecastaphyllum (L.) Taub.

F.W.T.A. 1: 515 (1958).

Habit: shrub to 6 m. **Habitat:** forest. **Distribution:** Senegal to
Bioko and Angola. **Chorology:** Guineo-Congolian. **Star:** green.
Alt. range: 1-200m (2). **Flowering:** Mar (1), Jul (1). **Fruiting:**
Mar (1), Jul (1). **Specimens:** - *Etinde:* Preuss P. R. 1303. *Mabeta-
Moliwe:* Cheek M. 3585; Groves M. 235

Dalbergia lactea Vatke

F.W.T.A. 1: 516 (1958); F.T.E.A. Papilionoideae: 111 (1971);
Opera Botanica 68: 70 (1983).

Habit: woody climber. **Guild:** light demander. **Habitat:** forest.
Distribution: SE Nigeria, Cameroon, Gabon, Congo (Kinshasa)
and East Africa. **Chorology:** Tropical Africa. **Star:** green.
Specimens: - *Eastern slopes:* Preuss P. R. 897

71

Dalbergia oligophylla Baker ex Hutch. & Dalziel
F.W.T.A. 1: 516 (1958).

Habit: shrub to 1 m. **Habitat:** forest-grassland transition. **Distribution:** Sierra Leone, Nigeria and Cameroon. **Chorology:** Guineo-Congolian (montane). **Star:** green. **Alt. range:** 1600-1800m (1). **Specimens:** - *No locality:* Mann G. 2172

Dalbergia saxatilis Hook.f.
F.W.T.A. 1: 516 (1958).

Habit: woody climber. **Guild:** light demander. **Habitat:** forest. **Distribution:** Guinea Bissau to Cameroon, Gabon, Congo (Kinshasa) and Angola. **Chorology:** Guineo-Congolian. **Star:** green. **Specimens:** - *Etinde:* Maitland T. D. 285

Desmodium adscendens (Sw.) DC. var. adscendens
F.W.T.A. 1: 585 (1958); F.T.E.A. Papilionoideae: 461 (1971); Opera Botanica 68: 111 (1983).

Habit: herb. **Habitat:** forest. **Distribution:** tropical and subtropical Africa. **Star:** excluded. **Alt. range:** 1-200m (3), 201-400m (1), 601-800m (1). **Flowering:** Apr (1), May (1), Oct (1), Nov (1). **Fruiting:** Nov (1). **Specimens:** - *Etinde:* Banks H. 10; Cable S. 209; Rosevear D. R. Cam 46/37. *Mokoko River F.R.:* Acworth J. M. 197; Watts J. 1045. *Onge:* Ndam N. 803; Thomas D. W. 9915

Desmodium adscendens (Sw.) DC. var. robustum Schubert
F.W.T.A. 1: 585 (1958); F.T.E.A. Papilionoideae: 462 (1971); Opera Botanica 68: 111 (1983).

Habit: herb. **Distribution:** Guinea Bissau to Bioko, Congo (Kinshasa), Sudan and Zimbabwe. **Chorology:** Tropical Africa. **Star:** green. **Alt. range:** 800-1000m (1). **Specimens:** - *Eastern slopes:* Migeod F. 112

Desmodium ospriostreblum Chiov.
Opera Botanica 68: 112 (1983).
Desmodium tortuosum sensu Hepper, F.W.T.A. 1: 585 (1985).

Habit: herb. **Distribution:** pantropical. **Star:** excluded. **Specimens:** - *Etinde:* Maitland T. D. FHI 12111

Desmodium repandum (Vahl) DC.
F.W.T.A. 1: 584 (1958); F.T.E.A. Papilionoideae: 465 (1971); Opera Botanica 68: 112 (1983).

Habit: herb. **Habitat:** forest and forest-grassland transition. **Distribution:** Tropical Africa and Asia. **Chorology:** Montane. **Star:** excluded. **Alt. range:** 400-600m (1), 801-1000m (3), 1201-1400m (1), 2001-2200m (1). **Flowering:** Mar (1), May (1), Dec (3). **Fruiting:** May (1). **Specimens:** - *Eastern slopes:* Sunderland T. C. H. 1420. *Etinde:* Cable S. 343, 509, 1641; Lighava M. 12. *No locality:* Mann G. 1229

Desmodium setigerum (E.Mey.) Harv.

Specimens: - *Etinde:* Nning 20

Desmodium velutinum (Willd.) DC.
F.W.T.A. 1: 584 (1958); Opera Botanica 68: 112 (1983).

Habit: herb. **Guild:** pioneer. **Habitat:** roadsides and villages. **Distribution:** tropical Africa and Asia. **Star:** excluded. **Flowering:** Oct (1). **Specimens:** - *Etinde:* Banks H. 4

Dolichos sericeus E.Mey. subsp. formosus (A.Rich.) Verdc.
Opera Botanica 68: 139 (1983).
Dolichos formosus A.Rich., F.W.T.A. 1: 571 (1958); F.T.E.A. Pap: 682 (1971).

Habit: herbaceous climber. **Habitat:** forest and roadsides. **Distribution:** Cameroon, Uganda, Sudan, Ethiopia and Kenya. **Chorology:** Afromontane. **Star:** green. **Alt. range:** 1600-1800m (1). **Flowering:** Oct (1). **Specimens:** - *Eastern slopes:* Banks H. 24

Eriosema psoraleoides (Lam.) G.Don
F.W.T.A. 1: 557 (1958); F.T.E.A. Papilionoideae: 772 (1971); Opera Botanica 68: 151 (1983).

Habit: shrub to 1.5 m. **Habitat:** forest. **Distribution:** tropical and subtropical Africa. **Star:** excluded. **Flowering:** Sep (1). **Specimens:** - *Bambuko F.R.:* Olorunfemi J. FHI 30763

Erythrina excelsa Baker
F.W.T.A. 1: 562 (1958); F.T.E.A. Papilionoideae: 559 (1971).

Habit: tree to 25 m. **Guild:** pioneer. **Habitat:** forest. **Distribution:** SW Cameroon, Congo (Kinshasa), Sudan, Uganda, Tanzania and Zambia. **Chorology:** Guineo-Congolian. **Star:** green. **Specimens:** - *Mabeta-Moliwe:* Mann G. 704

Indigofera atriceps Hook.f. subsp. atriceps
F.W.T.A. 1: 541 (1958); F.T.E.A. Papilionoideae: 282 (1971); Opera Botanica 68: 192 (1983).
Indigofera atriceps Hook.f. subsp. *alboglandulosa* (Engl.) J.B.Gillett, F.W.T.A. 1: 541 (1958).

Habit: herb. **Distribution:** tropical Africa. **Chorology:** Afromontane. **Star:** green. **Alt. range:** 1600-1800m (1), 2201-2400m (1). **Specimens:** - *No locality:* Mann G. 1303; Mildbraed G. W. J. 10831

Indigofera spicata Forrsk.

Specimens: - *Etinde:* Nning 14

Lablab purpureus Sweet
F.T.E.A. Papilionoideae: 696 (1971); Opera Botanica 68: 140 (1983).
Lablab niger L.W.Medicus, F.W.T.A. 1: 571 (1958).

Habit: herbaceous climber. **Guild:** pioneer. **Habitat:** forest. **Distribution:** tropical and subtropical Africa, often cultivated. **Star:** excluded. **Alt. range:** 200-400m (1). **Flowering:** Oct (1). **Fruiting:** Oct (1). **Specimens:** - *Etinde:* Maitland T. D. 778. *Onge:* Tchouto P. 935

Leptoderris aurantiaca Dunn
F.W.T.A. 1: 521 (1958).

Habit: woody climber. **Habitat:** forest. **Distribution:** Nigeria, Cameroon and Gabon. **Chorology:** Lower Guinea. **Star:** blue. **Alt. range:** 1-200m (1), 201-400m (1). **Specimens:** - *Etinde:* Thomas D. W. 9727. *Onge:* Thomas D. W. 9778

Leptoderris brachyptera (Benth.) Dunn
F.W.T.A. 1: 521 (1958).

Habit: woody climber. **Guild:** light demander. **Habitat:** forest. **Distribution:** Senegal to Bioko, Cameroon and Angola. **Chorology:** Guineo-Congolian. **Star:** green. **Specimens:** - *Etinde:* Maitland T. D. 385

Leptoderris sp.

Specimens: - *Onge:* Harris D. J. 3715

Leucomphalos capparideus Benth. ex Planch.
F.W.T.A. 1: 511 (1958); Agric. Univ. Wag. Papers 94(4): 21 (1994).

Habit: woody climber. **Habitat:** forest. **Distribution:** SE Nigeria, Bioko, Cameroon, Rio Muni and Gabon. **Chorology:** Lower Guinea. **Star:** blue. **Alt. range:** 1-200m (8), 201-400m (1). **Flowering:** Oct (1), Nov (2). **Fruiting:** Mar (1), Apr (1), May (3), Oct (2), Nov (1). **Specimens:** - *Bambuko F.R.:* Watts J. 675. *Mabeta-Moliwe:* Wheatley J. I. 259. *Mokoko River F.R.:* Fraser P. 387; Mbani J. M. 339; Watts J. 1042. *Onge:* Harris D. J. 3731; Tchouto P. 966, 1002; Watts J. 826, 903, 935

Lonchocarpus sericeus (Poir.) Kunth
F.W.T.A. 1: 522 (1958); Meded. Land. Wag. 82(3): 211 (1982); Keay R.W.J., Trees of Nigeria: 277 (1989); Hawthorne W., F.G.F.T. Ghana: 213 (1990).

Habit: tree to 15 m. **Guild:** light demander. **Habitat:** forest. **Distribution:** Senegal to Bioko and Angola, also tropical America. **Star:** excluded. **Alt. range:** 1-200m (2). **Flowering:** Dec (1). **Specimens:** - *Etinde:* Maitland T. D. 288. *Mabeta-Moliwe:* Cable S. 579; Cheek M. 3420, 5737

Machaerium lunatum (L.f.) Ducke
Lock J.M., Legumes of Africa: 241 (1989).
    Drepanocarpus lunatus (L.f.) G.Mey., F.W.T.A. 1: 519 (1958).

Habit: straggling shrub. **Habitat:** forest. **Distribution:** Senegal to Bioko, Cameroon, Congo (Kinshasa) and Angola, also South America. **Star:** excluded. **Specimens:** - *Etinde:* Maitland T. D. 399; *Mabeta-Moliwe:* Cheek 3503

Mildbraediodendron excelsum Harms
F.W.T.A. 1: 448 (1958); Keay R.W.J., Trees of Nigeria: 266 (1989); Hawthorne W., F.G.F.T. Ghana: 213 (1990).

Habit: tree to 40 m. **Habitat:** forest. **Distribution:** Ghana to Congo (Kinshasa), Uganda and Sudan. **Chorology:** Guineo-Congolian. **Star:** green. **Specimens:** - *Etinde:* Mildbraed G. W. J. 10643

Millettia barteri (Benth.) Dunn
F.W.T.A. 1: 526 (1958).

Habit: woody climber. **Habitat:** forest and stream banks. **Distribution:** Senegal to Cameroon, Rio Muni, Gabon, Congo (Brazzaville), S. Tomé, Congo (Kinshasa) and Sudan. **Chorology:**

Guineo-Congolian. **Star:** green. **Alt. range:** 1-200m (7). **Flowering:** Feb (1), Apr (1), May (2), Aug (1), Nov (1). **Fruiting:** May (1). **Specimens:** - *Mabeta-Moliwe:* Cable S. 1366; Ndam N. 491, 495; Sunderland T. C. H. 1375; Wheatley J. I. 181; von Rege I. 80. *Onge:* Harris D. J. 3682

Millettia bipindensis Harms
Lock J.M., Legumes of Africa: 356 (1989).

Habit: woody climber. **Habitat:** forest. **Distribution:** Cameroon, Gabon and Congo (Kinshasa). **Chorology:** Guineo-Congolian. **Star:** green. **Specimens:** - *Mabeta-Moliwe:* Plot Voucher W 293

Millettia dinklagei Harms
F.W.T.A. 1: 525 (1958).

Habit: tree or shrub to 5 m. **Habitat:** stream banks in forest. **Distribution:** Liberia, Sierra Leone, Nigeria and Cameroon. **Chorology:** Guineo-Congolian. **Star:** green. **Alt. range:** 1-200m (2).**Fruiting:** Oct (1), Nov (1). **Specimens:** - *Onge:* Tchouto P. 910; Thomas D. W. 9811

Millettia drastica Welw.
F.W.T.A. 1: 527 (1958); Keay R.W.J., Trees of Nigeria: 276 (1989).

Habit: tree to 20 m. **Habitat:** forest. **Distribution:** SE Nigeria, Cameroon, Rio Muni, Gabon, Congo (Kinshasa), Angola and Sudan. **Chorology:** Guineo-Congolian (montane). **Star:** green. **Alt. range:** 1-200m (1). **Specimens:** - *Onge:* Harris D. J. 3855

Millettia griffoniana Baill.
Kew Bull. 25: 260 (1971); Keay R.W.J., Trees of Nigeria: 276 (1989); Hawthorne W., F.G.F.T. Ghana: 213 (1990).
    Lonchocarpus griffonianus (Baill.) Dunn, F.W.T.A. 1: 523 (1958).

Habit: tree to 10 m. **Guild:** shade-bearer. **Habitat:** forest. **Distribution:** Ivory Coast to Bioko and Angola. **Chorology:** Guineo-Congolian. **Star:** green. **Alt. range:** 1-200m (3). **Flowering:** Apr (1), Jul (1). **Specimens:** - *Mabeta-Moliwe:* Cheek M. 3513, 3519. *Onge:* Wheatley J. I. 190

Millettia macrophylla Benth.
F.W.T.A. 1: 526 (1958); Keay R.W.J., Trees of Nigeria: 277 (1989).

Habit: tree to 10 m. **Habitat:** forest and stream banks. **Distribution:** SE Nigeria, Bioko, Cameroon and Congo (Brazzaville). **Chorology:** Lower Guinea. **Star:** blue. **Alt. range:** 1-200m (7), 201-400m (6). **Flowering:** Mar (1), Oct (3), Nov (4), Dec (1). **Fruiting:** Mar (1), Oct (3), Nov (4). **Specimens:** - *Bambuko F.R.:* Tchouto P. 560. *Etinde:* Banks H. 29; Cheek M. 5405, 5494, 5515, 5945; Maitland T. D. 782. *Onge:* Akogo M. 27; Cheek M. 5053, 5174; Harris D. J. 3684, 3800; Ndam N. 650; Watts J. 1008

Millettia pilosa Hutch. & Dalziel
F.W.T.A. 1: 526 (1958).

Habit: woody climber. **Habitat:** forest. **Distribution:** SE Nigeria and SW Cameroon. **Chorology:** Western Cameroon. **Star:** gold. **Specimens:** - *Mabeta-Moliwe:* Plot Voucher W 28, W 46, W 315, W 446, W 630, W 709, W 825

Millettia sanagana Harms
F.W.T.A. 1: 526 (1958).

Habit: tree or shrub to 5 m. Habitat: forest. Distribution: Guinea to Bioko and Cameroon. Chorology: Guineo-Congolian. Star: green. Specimens: - *Etinde:* Maitland T. D. 521

Millettia sp.

Specimens: - *Etinde:* Williams S. 79. *Onge:* Baker W. J. 356

Mucuna flagellipes Hook.f.
F.W.T.A. 1: 561 (1958); F.T.E.A. Papilionoideae: 562 (1971).

Habit: woody climber. Habitat: forest. Distribution: Sierra Leone to Bioko, Cameroon, Congo (Kinshasa), Angola and Uganda. Chorology: Guineo-Congolian. Star: green. Alt. range: 1-200m (5), 401-600m (1). Flowering: May (1), Jun (1), Jul (1), Oct (1). Fruiting: May (1), Jul (2), Nov (1). Specimens: - *Mabeta-Moliwe:* Cheek M. 3566; Plot Voucher W 530; Watts J. 421; Wheatley J. I. 265. 378. *Onge:* Tchouto P. 796; Thomas D. W. 9912

Mucuna pruriens (L.) DC. var. pruriens
F.W.T.A. 1: 561 (1958); F.T.E.A. Papilionoideae: 567 (1971); Opera Botanica 68: 124 (1983).

Habit: woody climber. Habitat: forest. Distribution: pantropical. Star: excluded. Alt. range: 1-200m (1). Flowering: Aug (1). Fruiting: Aug (1). Specimens: - *Mabeta-Moliwe:* von Rege I. 43

Mucuna sloanei Fawc. & Rendle
F.W.T.A. 1: 561 (1958).

Habit: woody climber. Habitat: forest. Distribution: pantropical. Star: excluded. Alt. range: 1-200m (2). Flowering: Nov (1), Dec (1). Fruiting: Dec (1). Specimens: - *Etinde:* Cheek M. 5510, 5928

Myroxylon balsamum (L.) Harms
F.W.T.A. 1: 509 (1958); F.T.E.A. Pap.: 31.

Habit: tree to 30 m. Distribution: introduced in West and East Africa from tropical America. Star: excluded. Fruiting: Nov (1). Specimens: - *Limbe Botanic Garden:* Cheek M. 5379

Ormocarpum klainei Tisser.
Lock J.M., Legumes of Africa: 119 (1989).
 Ormocarpum sp. A sensu Hepper, F.W.T.A. 1: 577 (1958).

Habit: shrub to 1 m. Habitat: forest. Distribution: Cameroon and Gabon. Chorology: Lower Guinea. Star: black. Specimens: - *Eastern slopes:* Maitland T. D. 1265

Ormocarpum megalophyllum Harms
F.W.T.A. 1: 577 (1958).

Habit: shrub to 2 m. Guild: pioneer. Habitat: forest. Distribution: Guinea to Cameroon and Gabon. Chorology: Guineo-Congolian. Star: green. Specimens: - *Etinde:* Preuss P. R. 1193

Ormocarpum sennoides (Willd.) DC. subsp. hispidum (Willd.) Brenan & J.Léonard
F.W.T.A. 1: 576 (1958).

Habit: shrub to 4 m. Habitat: forest. Distribution: tropical Africa. Star: excluded. Alt. range: 1-200m (2).Fruiting: Apr (1), May (1). Specimens: - *Mabeta-Moliwe:* Schlechter F. R. R. 12385. *Mokoko River F.R.:* Akogo M. 272; Watts J. 1155

Ostryocarpus riparius Hook.f.
F.W.T.A. 1: 519 (1958).

Habit: woody climber. Guild: light demander. Habitat: forest. Distribution: Guinea Bissau to Bioko, Cameroon, Gabon and Congo (Kinshasa). Chorology: Guineo-Congolian. Star: green. Alt. range: 1-200m (1). Flowering: May (1). Specimens: - *Etinde:* Maitland T. D. 604. *Mabeta-Moliwe:* Sunderland T. C. H. 1374

Physostigma venenosum Balf.
F.W.T.A. 1: 564 (1958).

Habit: scrambling shrub. Guild: pioneer. Habitat: forest. Distribution: Sierra Leone to Bioko, Cameroon, Gabon and Congo (Kinshasa). Chorology: Guineo-Congolian. Star: green. Specimens: - *Etinde:* Kalbreyer W. 90

Platysepalum violaceum Welw. ex Baker var. vanhouttei (De Wild.) Hauman
F.W.T.A. 1: 524 (1958); Keay R.W.J., Trees of Nigeria: 272 (1989).

Habit: tree or shrub to 13 m. Habitat: forest. Distribution: SE Nigeria, Cameroon, Rio Muni, Gabon, Congo (Kinshasa) and Angola (Cabinda). Chorology: Guineo-Congolian. Star: green. Specimens: - *Eastern slopes:* Maitland T. D. 897

Psophocarpus palustris Desv.
F.W.T.A. 1: 572 (1958).

Habit: twining herb. Guild: pioneer. Habitat: forest. Distribution: tropical Africa. Star: excluded. Specimens: - *Etinde:* Maitland T. D. 515

Pterocarpus mildbraedii Harms subsp. mildbraedii
F.W.T.A. 1: 517 (1958); Keay R.W.J., Trees of Nigeria: 271 (1989); Hawthorne W., F.G.F.T. Ghana: 210 (1990).

Habit: tree to 15 m. Habitat: forest. Distribution: Ivory Coast to Bioko and Gabon. Chorology: Guineo-Congolian. Star: green. Specimens: - *Mabeta-Moliwe:* Plot Voucher W 480

Pterocarpus santalinoides L'Hér. ex DC.
F.W.T.A. 1: 517 (1958); Meded. Land. Wag. 82(3): 215 (1982); Keay R.W.J., Trees of Nigeria: 269 (1989); Hawthorne W., F.G.F.T. Ghana: 210 (1990).

Habit: tree to 12 m. Habitat: forest. Distribution: Senegal to Cameroon, also tropical America. Star: excluded. Alt. range: 1-200m (1). Flowering: Dec (1). Specimens: - *Etinde:* Cheek M. 5923. *Mabeta-Moliwe:* Mann G. 713

Pterocarpus soyauxii Taub.
F.W.T.A. 1: 517 (1958); Keay R.W.J., Trees of Nigeria: 269 (1989).

Habit: tree to 30 m. **Habitat:** forest. **Distribution:** Nigeria, Cameroon, Rio Muni, Gabon, Congo (Kinshasa), Angola (Cabinda) and CAR. **Chorology:** Guineo-Congolian. **Star:** red. **Alt. range:** 1-200m (3). **Flowering:** Apr (1), May (1). **Fruiting:** Jul (1). **Specimens:** - *Etinde:* Maitland T. D. 100. *Mabeta-Moliwe:* Cheek M. 3552; Plot Voucher W 550; Sunderland T. C. H. 1380; Wheatley J. I. 219

Pueraria phaseoloides (Roxb.) Benth.

F.W.T.A. 1: 573 (1958); F.T.E.A. Papilionoideae: 596 (1971).

Habit: climbing or straggling herb. **Habitat:** forest. **Distribution:** tropical Asia, introduced to Africa. **Star:** excluded. **Alt. range:** 1-200m (2). **Flowering:** Dec (2). **Specimens:** - *Mabeta-Moliwe:* Cheek M. 5805, 5810

Rhynchosia densiflora (Roth.) DC.

F.W.T.A. 1: 555 (1958); F.T.E.A. Papilionoideae: 723 (1971); Opera Botanica 68: 143 (1983).

Habit: twining herb. **Guild:** pioneer. **Habitat:** forest. **Distribution:** tropical Africa and India. **Star:** excluded. **Specimens:** - *Etinde:* Mann G. 1249

Rhynchosia mannii Baker

F.W.T.A. 1: 554 (1958); F.T.E.A. Papilionoideae: 722 (1971).

Habit: herb. **Guild:** pioneer. **Habitat:** forest gaps. **Distribution:** Nigeria, Bioko, Cameroon, Gabon, Congo (Kinshasa), Angola, CAR and Uganda. **Chorology:** Guineo-Congolian. **Star:** green. **Alt. range:** 200-400m (1), 401-600m (1), 1201-1400m (1). **Flowering:** Mar (1). **Fruiting:** Mar (2). **Specimens:** - *Eastern slopes:* Maitland T. D. 195. *Etinde:* Acworth J. M. 26; Khayota B. 556. *Mokoko River F.R.:* Tchouto P. 590

Rhynchosia pycnostachya (DC.) Meikle

F.W.T.A. 1: 554 (1958).

Habit: herb. **Guild:** pioneer. **Habitat:** forest. **Distribution:** Senegal to Bioko and SW Cameroon. **Chorology:** Guineo-Congolian. **Star:** green. **Specimens:** - *Mabeta-Moliwe:* Mann G. 730; Nguembock F. 75

Tephrosia noctiflora Bojer ex Baker

F.W.T.A. 1: 530 (1958); F.T.E.A. Papilionoideae: 182 (1971).

Habit: shrub to 1 m. **Habitat:** forest. **Distribution:** Introduced in W. Africa. Native of E. and S. Africa, Asia and Indian Ocean. **Star:** excluded. **Specimens:** - *Etinde:* Maitland T. D. FHI 8768

Tephrosia paniculata Baker

Lock J.M., Legumes of Africa: 380 (1989).
Tephrosia preussii Taub., F.W.T.A. 1: 531 (1958).

Habit: shrub to 2 m. **Habitat:** forest. **Distribution:** Sierra Leone to Cameroon, East Africa, Angola and Mozambique. **Chorology:** Tropical Africa. **Star:** green. **Specimens:** - *Eastern slopes:* Lehmbach H. 63

Tephrosia vogelii Hook.f.

F.W.T.A. 1: 530 (1958); F.T.E.A. Papilionoideae: 210 (1971); Meded. Land. Wag. 82(3): 217 (1982); Opera Botanica 68: 18 (1983).

Habit: shrub to 3 m. **Habitat:** forest. **Distribution:** tropical Africa. **Star:** excluded. **Specimens:** - *Eastern slopes:* Hutchinson J. & Metcalfe C. R. 109

Trifolium rueppellianum Fresen. var. rueppellianum

Opera Botanica 68: 198 (1983); F.T.E.A. Pap: 1031 (1971).
Trifolium rueppellianum Fresen. var. *preussii* (Baker f.) J.B.Gillet, F.W.T.A. 1: 553 (1958).

Habit: herb. **Habitat:** grassland. **Distribution:** Bioko, Cameroon, E. Congo (Kinshasa), Sudan, Ethiopia and East Africa. Introduced to Malawi and Zimbabwe. **Chorology:** Afromontane. **Star:** green. **Alt. range:** 2400-2600m (1). **Specimens:** - *No locality:* Preuss P. R. 972

Trifolium simense Fresen.

F.W.T.A. 1: 553 (1958); F.T.E.A. Papilionoideae: 1022 (1971); Opera Botanica 68: 193 (1983).

Habit: herb. **Habitat:** grassland. **Distribution:** Bioko, Cameroon, E. Congo (Kinshasa), Rwanda, Burundi, Sudan, Ethiopia, Malawi, Zambia and East Africa. **Chorology:** Afromontane. **Star:** green. **Alt. range:** 2400-2600m (1). **Specimens:** - *No locality:* Johnston H. H. 4

Vigna adenantha (G.Mey.) Maréchal, Mascherpa & Stainier

Taxon 27: 199 (1978).
Phaseolus adenanthus G.Mey., F.W.T.A. 1: 565 (1958); F.T.E.A. Pap. 615 (1971).

Habit: herbaceous climber. **Habitat:** forest. **Distribution:** Senegal to Cameroon, Gabon, Congo (Kinshasa) and Tanzania. **Chorology:** Guineo-Congolian. **Star:** green. **Specimens:** - *Etinde:* Maitland T. D. 760

Vigna gracilis (Guill. & Perr.) Hook.f. var. multiflora (Hook.f.) Maréchal, Mascherpa & Stainer

Taxon 27: 199 (1978).
Vigna multiflora Hook.f., F.W.T.A. 1: 569 (1958).

Habit: twining herb. **Guild:** pioneer. **Habitat:** forest. **Distribution:** Sierra Leone to Bioko, Cameroon and Congo (Kinshasa). **Chorology:** Guineo-Congolian. **Star:** green. **Specimens:** - *Eastern slopes:* Migeod F. 11

Vigna marina (Burm.) Merr.

F.W.T.A. 1: 569 (1958); F.T.E.A. Papilionoideae: 626 (1971).

Habit: herb. **Guild:** pioneer. **Distribution:** pantropical. **Star:** excluded. **Specimens:** - *Etinde:* Preuss P. R. 1126

LENTIBULARIACEAE

Genlisea hispidula Stapf

F.W.T.A. 2: 375 (1963).

Habit: herb. **Distribution:** Nigeria, W. Cameroon and East and southern Africa. **Chorology:** Afromontane. **Star:** green. **Specimens:** - *Eastern slopes:* Mildbraed G. W. J. 9468

Utricularia livida E.Mey.
Taylor P., Utricularia: 225 (1989).

Habit: herb. **Habitat:** grassland. **Distribution:** tropical and southern Africa and Central America. **Chorology:** Montane. **Star:** excluded. **Alt. range:** 1800-2000m (1). **Flowering:** Oct (1). **Specimens:** - *Eastern slopes:* Cheek M. 3688

Utricularia mannii Oliv.
F.W.T.A. 2: 377 (1963); Taylor P., Utricularia: 405 (1989).

Habit: epiphyte. **Habitat:** forest-grassland transition. **Distribution:** Bioko, W. Cameroon and S. Tomé. **Chorology:** Western Cameroon Uplands (montane). **Star:** gold. **Alt. range:** 1400-1600m (2), 1601-1800m (1), 1801-2000m (1). **Flowering:** Sep (1), Oct (2). **Specimens:** - *Eastern slopes:* Cheek M. 3690. *Etinde:* Tekwe C. 241; Thomas D. W. 9176. *No locality:* Mann G. 2112

Utricularia striatula Sm.
F.W.T.A. 2: 378 (1963); Taylor P., Utricularia: 481 (1989).

Habit: epiphyte. **Habitat:** forest and stream banks. **Distribution:** palaeotropical. **Star:** excluded. **Alt. range:** 1-200m (1), 401-600m (1), 1201-1400m (1). **Flowering:** Sep (1), Oct (2). **Fruiting:** Oct (1). **Specimens:** - *Eastern slopes:* Cheek M. 3715. *Etinde:* Thomas D. W. 9154. *No locality:* Mann G. 1964. *Onge:* Tchouto P. 820

LEPIDOBOTRYACEAE

Lepidobotrys staudtii Engl.
F.W.T.A. 1: 357 (1958); Fl. Cameroun 14: 44 (1972); Fl. Gabon 21: 42 (1973); Keay R.W.J., Trees of Nigeria: 144 (1989).

Habit: tree to 15 m. **Habitat:** forest. **Distribution:** Nigeria, Cameroon, Gabon, Congo (Kinshasa) and CAR. **Chorology:** Guineo-Congolian. **Star:** green. **Alt. range:** 1-200m (2), 201-400m (2). **Flowering:** Oct (1). **Specimens:** - *Mokoko River F.R.:* Tchouto P. 1112, 1129; Thomas D. W. 10096. *Onge:* Tchouto P. 752

LINACEAE

Hugonia macrophylla Oliv.
F.W.T.A. 1: 359 (1958); Fl. Cameroun 14: 30 (1972); Fl. Gabon 21: 28 (1973).

Habit: woody climber. **Habitat:** forest. **Distribution:** SE Nigeria, Cameroon and Gabon. **Chorology:** Lower Guinea. **Star:** blue. **Alt. range:** 1-200m (2). **Specimens:** - *Mokoko River F.R.:* Tchouto P. 1078, 1218

Hugonia obtusifolia C.H.Wright
F.W.T.A. 1: 359 (1958); Fl. Cameroun 14: 28 (1972); Fl. Gabon 21: 26 (1973).

Habit: woody climber. **Habitat:** forest. **Distribution:** Nigeria, Cameroon, Gabon and Congo (Kinshasa). **Chorology:** Guineo-Congolian. **Star:** green. **Flowering:** Jan (1). **Specimens:** - *Southern Bakundu F.R.:* Binuyo A. & Daramola B. O. FHI 35480

Hugonia platysepala Welw. ex Oliv.
F.W.T.A. 1: 359 (1958); Fl. Cameroun 14: 26 (1972); Fl. Gabon 21: 25 (1973).

Habit: woody climber. **Guild:** light demander. **Habitat:** forest. **Distribution:** Guinea to Bioko, Congo (Kinshasa), Angola and Uganda. **Chorology:** Guineo-Congolian. **Star:** green. **Alt. range:** 1-200m (2), 201-400m (1). **Flowering:** Mar (1), Apr (2). **Fruiting:** Apr (1). **Specimens:** - *Etinde:* Tekwe C. 73. *Mabeta-Moliwe:* Groves M. 276; Plot Voucher W 505; Wheatley J. I. 118

Hugonia sp.

Specimens: - *Etinde:* Cable S. 1588. *Mokoko River F.R.:* Tchouto P. 1211, 1226

Radiola linoides Roth
F.W.T.A. 1: 361 (1958); Fl. Cameroun 14: 41 (1972).

Habit: herb. **Habitat:** grassland. **Distribution:** Europe and Africa in temperate areas. **Chorology:** Montane. **Star:** excluded. **Alt. range:** 1800-2000m (1), 2001-2200m (1), 2401-2600m (1), 2801-3000m (3). **Flowering:** Oct (3). **Specimens:** - *Eastern slopes:* Cheek M. 3679; Tchouto P. 272; Thomas D. W. 9449. *No locality:* Maitland T. D. 816; Mann G. 1334, 2021

LOGANIACEAE

A.J.M. Leeuwenberg (WAG)

Anthocleista obanensis Wernham
Acta Bot. Neerl. 10: 34 (1961); F.W.T.A. 2: 37 (1963); Fl. Gabon 19: 10 (1972); Fl. Cameroun 12: 10 (1972).

Habit: woody climber. **Habitat:** forest. **Distribution:** Nigeria, Cameroon, Gabon and Congo (Kinshasa). **Chorology:** Guineo-Congolian (montane). **Star:** green. **Alt. range:** 1-200m (2). **Flowering:** Oct (1). **Specimens:** - *Onge:* Cheek M. 5202; Harris D. J. 3643

Anthocleista scandens Hook.f.
Acta Bot. Neerl. 10: 34 (1961); F.W.T.A. 2: 37 (1963); Fl. Gabon 19: 11 (1972); Fl. Cameroun 12: 11 (1972).

Habit: woody climber or tree to 15 m. **Habitat:** forest and forest-grassland transition. **Distribution:** Bioko, W. Cameroon and S. Tomé. **Chorology:** Western Cameroon Uplands (montane). **Star:** gold. **Alt. range:** 1600-1800m (1). **Flowering:** Aug (1). **Specimens:** - *Etinde:* Hunt L. V. 34

Anthocleista schweinfurthii Gilg
Acta Bot. Neerl. 10: 24 (1961); F.W.T.A. 2: 35 (1963); Fl. Gabon 19: 12 (1972).

Habit: tree to 20 m. **Habitat:** forest. **Distribution:** tropical Africa. **Star:** excluded. **Alt. range:** 1-200m (1). **Fruiting:** Dec (1). **Specimens:** - *Mabeta-Moliwe:* Cheek M. 5692

Anthocleista vogelii Planch.

Acta Bot. Neerl. 10: 16 (1961); F.W.T.A. 2: 35 (1963); Fl. Gabon 19: 14 (1972); Fl. Cameroun 12: 14 (1972); Keay R.W.J., Trees of Nigeria: 400 (1989); Hawthorne W., F.G.F.T. Ghana: 43 (1990).

Habit: tree to 20 m. **Guild:** shade-bearer. **Habitat:** forest. **Distribution:** tropical Africa. **Star:** excluded. **Alt. range:** 1-200m (2).**Fruiting:** Apr (1), May (1). **Specimens:** - *Mabeta-Moliwe:* Mann G. s.n.; Ndam N. 489; Wheatley J. I. 185

Mostuea batesii Bak.

Specimens: - Akogo 194; Ekema 867; Tchouto P. 1061

Mostuea brunonis Didr. var. brunonis

Meded. Land. Wag. 61(4): 14 (1961); F.W.T.A. 2: 45 (1963); Fl. Gabon 19: 28 (1972); Fl. Cameroun 12: 26 (1972).

Habit: tree or shrub to 7 m. **Guild:** pioneer. **Habitat:** forest. **Distribution:** tropical Africa. **Star:** excluded. **Alt. range:** 1-200m (5). **Flowering:** Mar (1), Aug (1), Nov (2). **Fruiting:** Mar (1), Aug (1), Nov (2). **Specimens:** - *Onge:* Harris D. J. 3786; Thomas D. W. 9799; Watts J. 705, 924; Wheatley J. I. 859

Mostuea hirsuta (T.Anderson ex Benth.) Baill. ex Baker

Meded. Land. Wag. 61(4): 3 (1961); F.W.T.A. 2: 45 (1963); Fl. Gabon 19: 31 (1972); Fl. Cameroun 12: 31 (1972).

Habit: shrub to 3 m. **Guild:** pioneer. **Distribution:** tropical Africa. **Star:** excluded. **Specimens:** - *Southern Bakundu F.R.:* Brenan J. P. M. 9453

Strychnos aculeata Soler.

F.W.T.A. 2: 43 (1963); Meded. Land. Wag. 69(1): 49 (1969); Fl. Gabon 19: 55 (1972); Fl. Cameroun 12: 55 (1972).

Habit: woody climber. **Guild:** pioneer. **Habitat:** forest. **Distribution:** Sierra Leone to Bioko, Cameroon, Gabon, Congo (Brazzaville), Congo (Kinshasa), Angola (Cabinda) and CAR. **Chorology:** Guineo-Congolian. **Star:** green. **Alt. range:** 1-200m (2).**Fruiting:** Jun (1). **Specimens:** - *Mabeta-Moliwe:* Watts J. 345. *Onge:* Harris D. J. 3685

Strychnos barteri Soler.

F.W.T.A. 2: 44 (1963).

Habit: woody climber. **Habitat:** forest. **Distribution:** Liberia to Cameroon. **Chorology:** Guineo-Congolian (montane). **Star:** green. **Alt. range:** 1200-1400m (1).**Fruiting:** Feb (1). **Specimens:** - *Eastern slopes:* Etuge M. 1150

Strychnos campicola Gilg ex Leeuwenb.

Specimens: - *Mokoko:* Akogo 296; Tchouto P. 1154

Strychnos camptoneura Gilg & Busse

F.W.T.A. 2: 44 (1963); Meded. Land. Wag. 69(1): 75 (1969); Fl. Gabon 19: 66 (1972); Fl. Cameroun 12: 66 (1972).

Habit: woody climber. **Guild:** light demander. **Habitat:** forest. **Distribution:** Liberia to Cameroon, Gabon, Congo (Brazzaville), Congo (Kinshasa) and CAR. **Chorology:** Guineo-Congolian. **Star:** green. **Fruiting:** Jul (1). **Specimens:** - *Etinde:* Buchholz R. W. s.n.

Strychnos chrysophylla Gilg

F.W.T.A. 2: 43 (1963); Meded. Land. Wag. 69(1): 82 (1969); Fl. Gabon 19: 70 (1972).

Habit: woody climber. **Habitat:** forest. **Distribution:** SE Nigeria, Cameroon and Gabon. **Chorology:** Lower Guinea. **Star:** blue. **Alt. range:** 1-200m (1).**Fruiting:** Oct (1). **Specimens:** - *Mokoko:* Ekema 1194b. *Onge:* Tchouto P. 927

Strychnos dale De Wild.

Fl. Cameroun 12: 75 (1972).

Habit: woody climber. **Habitat:** forest. **Distribution:** Cameroon, Gabon, Congo (Brazzaville) and Congo (Kinshasa). **Chorology:** Guineo-Congolian. **Star:** green. **Alt. range:** 200-400m (1).**Fruiting:** Mar (1). **Specimens:** - *Mokoko River F.R.:* Tchouto P. 603

Strychnos densiflora Baill.

F.W.T.A. 2: 43 (1963); Meded. Land. Wag. 69(1): 104 (1969); Fl. Gabon 19: 76 (1972); Fl. Cameroun 12: 76 (1972).

Habit: woody climber. **Guild:** light demander. **Habitat:** forest. **Distribution:** Guinea to Cameroon, Gabon, Congo (Kinshasa) and CAR. **Chorology:** Guineo-Congolian. **Star:** green. **Alt. range:** 1-200m (1), 201-400m (2).**Fruiting:** Oct (2), Nov (1). **Specimens:** - *Onge:* Akogo M. 75; Thomas D. W. 9883; Watts J. 835

Strychnos elaeocarpa Gilg ex Leeuwenb.

Meded. Land. Wag. 69(1): 114 (1969); Fl. Gabon 19: 79 (1972); Fl. Cameroun 12: 79 (1972).

Habit: tree to 10 m. **Habitat:** forest. **Distribution:** Cameroon. **Chorology:** Lower Guinea. **Star:** gold. **Alt. range:** 1-200m (3), 201-400m (1).**Fruiting:** Apr (1), Jun (1), Oct (1), Nov (1). **Specimens:** - *Mabeta-Moliwe:* Jaff B. 324; Wheatley J. I. 148. *Onge:* Tchouto P. 1034; Watts J. 869

Strychnos fallax Leeuwenb.

Fl. Cameroun 12: 81 (1972).

Habit: woody climber. **Habitat:** forest. **Distribution:** Cameroon and Congo (Kinshasa). **Chorology:** Guineo-Congolian. **Star:** green. **Alt. range:** 1-200m (1). **Flowering:** Dec (1). **Fruiting:** Dec (1). **Specimens:** - *Mabeta-Moliwe:* Cheek M. 5748

Strychnos floribunda Gilg

F.W.T.A. 2: 44 (1963); Meded. Land. Wag. 69(1): 120 (1969); Fl. Gabon 19: 82 (1972); Fl. Cameroun 12: 82 (1972).

Habit: woody climber. **Guild:** light demander. **Habitat:** forest. **Distribution:** Sierra Leone to Cameroon, Gabon, Congo (Kinshasa), Angola and CAR. **Chorology:** Guineo-Congolian. **Star:** green. **Specimens:** - *Mabeta-Moliwe:* Dunlap 168

Strychnos gnetifolia Gilg ex Onochie & Hepper

F.W.T.A. 2: 44 (1963); Meded. Land. Wag. 69(1): 124 (1969); Fl. Gabon 19: 84 (1972); Fl. Cameroun 12: 84 (1972).

Habit: tree to 20 m. **Distribution:** SE Nigeria and Cameroon. **Chorology:** Lower Guinea. **Star:** gold. **Specimens:** - *Southern Bakundu F.R.:* Lobe Babute F. G. Cam 37/36

Strychnos icaja Baill.

F.W.T.A. 2: 44 (1963); Meded. Land. Wag. 69(1): 133 (1969); Fl. Gabon 19: 86 (1972); Fl. Cameroun 12: 86 (1972).

Habit: woody climber. **Guild:** light demander. **Habitat:** forest. **Distribution:** Guinea to Cameroon, Gabon, Congo (Brazzaville), Congo (Kinshasa), Angola and CAR. **Chorology:** Guineo-Congolian. **Star:** green. **Alt. range:** 1-200m (2), 201-400m (1).**Fruiting:** Apr (2), Oct (1). **Specimens:** - *Mabeta-Moliwe:* Plot Voucher W 476, W 563, W 620, W 810, W 1024; Watts J. 151; Wheatley J. I. 80. *Onge:* Akogo M. 80

Strychnos johnsonii Hutch. & M.B.Moss

F.W.T.A. 2: 44 (1963); Meded. Land. Wag. 69(1): 147 (1969); Fl. Cameroun 12: 90 (1972).

Habit: woody climber. **Guild:** light demander. **Habitat:** forest. **Distribution:** Guinea to Congo (Brazzaville), Congo (Kinshasa), Angola and Uganda. **Chorology:** Guineo-Congolian. **Star:** green. **Specimens:** - *Mabeta-Moliwe:* Cheek 3463; Plot Voucher W 80

Strychnos malacoclados C.H.Wright

F.W.T.A. 2: 43 (1963); Meded. Land. Wag. 69(1): 171 (1969); Fl. Gabon 19: 94 (1972).

Habit: woody climber. **Guild:** light demander. **Habitat:** forest. **Distribution:** Sierra Leone to Cameroon. **Chorology:** Guineo-Congolian. **Star:** green. **Alt. range:** 1-200m (1). **Flowering:** Mar (1). **Specimens:** - *Bambuko F.R.:* Watts J. 684

Strychnos memecyloides S.Moore

F.W.T.A. 2: 41 (1963); Meded. Land. Wag. 69(1): 182 (1969); Fl. Gabon 19: 97 (1972).

Habit: woody climber. **Habitat:** forest. **Distribution:** Nigeria, Cameroon, Gabon and Congo (Kinshasa). **Chorology:** Guineo-Congolian. **Star:** green. **Alt. range:** 1-200m (1).**Fruiting:** Nov (1). **Specimens:** - *Onge:* Thomas D. W. 9888

Strychnos ndengensis Pellegr.

F.W.T.A. 2: 44 (1963); Meded. Land. Wag. 69(1): 200 (1969); Fl. Gabon 19: 100 (1972); Fl. Cameroun 12: 100 (1972).

Habit: woody climber. **Habitat:** forest. **Distribution:** Gabon and Congo (Kinshasa). **Chorology:** Guineo-Congolian. **Star:** green. **Alt. range:** 600-800m (1). **Specimens:** - *Mabeta-Moliwe:* Mildbraed G. W. J. 10576

Strychnos ngouniensis Pellegr.

F.W.T.A. 2: 43 (1963).

Habit: woody climber. **Habitat:** forest. **Distribution:** Cameroon and Gabon. **Chorology:** Lower Guinea. **Star:** blue. **Alt. range:** 1-200m (1).**Fruiting:** Nov (1). **Specimens:** - *Mabeta-Moliwe:* Plot Voucher W 81, W 159. *Onge:* Harris D. J. 3686

Strychnos phaeotricha Gilg

F.W.T.A. 2: 41 (1963); Meded. Land. Wag. 69(1): 215 (1969); Fl. Cameroun 12: 104 (1972).

Habit: woody climber. **Guild:** light demander. **Habitat:** forest. **Distribution:** Ghana to Cameroon, Gabon, Congo (Brazzaville) and Congo (Kinshasa). **Chorology:** Guineo-Congolian. **Star:** green. **Alt. range:** 1-200m (1). **Flowering:** Apr (1). **Specimens:** - *Mabeta-Moliwe:* Cheek 3508, Plot Voucher W 209, W 593, W

774, W 857, W 920. *Mokoko River F.R.:* Acworth J. M. 131; Pouakouyou 21

Strychnos staudtii Gilg

F.W.T.A. 2: 43 (1963); Meded. Land. Wag. 69(1): 251 (1969); Fl. Gabon 19: 113 (1972); Fl. Cameroun 12: 113 (1972).

Habit: tree to 20 m. **Habitat:** forest. **Distribution:** Cameroon and Gabon. **Chorology:** Lower Guinea (montane). **Star:** blue. **Alt. range:** 1-200m (3).**Fruiting:** Apr (1), May (1), Sep (1). **Specimens:** - *Etinde:* Tchouto P. 802. *Mabeta-Moliwe:* Jaff B. 107; Plot Voucher W 1158; Sunderland T. C. H. 1324. *Mokoko:* Mbani 332, 422; Tchouto P. 1185, 1230

Strychnos tricalysioides Hutch. & M.B.Moss

F.W.T.A. 2: 41 (1963); Meded. Land. Wag. 69(1): 262 (1969); Fl. Cameroun 12: 119 (1972).

Habit: woody climber. **Habitat:** forest. **Distribution:** SE Nigeria, Bioko, Cameroon, Gabon and Congo (Brazzaville). **Chorology:** Guineo-Congolian. **Star:** green. **Alt. range:** 1-200m (1).**Fruiting:** Apr (1). **Specimens:** - *Mabeta-Moliwe:* Plot Voucher W 217; Wheatley J. I. 226

LORANTHACEAE

R.M. Polhill (K)

Agelanthus brunneus (Engl.) Balle & N.Hallé

F.W.T.A. 1: 660 (1958); Lebrun J.-P. & Stork A., E.P.F.A.T. 2: 162 (1992).
Tapinanthus brunneus (Engl.) Danser, Fl. Cameroun 23: 64 (1982).

Habit: parasite. **Habitat:** forest and farmbush. **Distribution:** Senegal to W. Kenya. **Chorology:** Guineo-Conglian (montane). **Star:** green. **Alt. range:** 800-1000m (1). **Flowering:** Feb (1). **Fruiting:** Feb (1). **Specimens:** - *Eastern slopes:* Groves M. 152

Agelanthus dodoneifolius (DC.) Polhill & Wiens

Lebrun J.-P. & Stork A., E.P.F.A.T. 2: 163 (1992).
Tapinanthus dodoneifolius (DC.) Danser, F.W.T.A. 1: 662 (1958).

Habit: parasite. **Habitat:** savanna. **Distribution:** Senegal to CAR, Sudan, Uganda and Ethiopia. **Star:** green. **Specimens:** - *Eastern slopes:* Mildbraed G. W. J. 9426

Agelanthus glaucoviridis (Engl.) Polhill & Wiens

Lebrun J.-P. & Stork A., E.P.F.A.T. 2: 163 (1992).
Tapinanthus dodoneifolius (DC.) Danser subsp. *glaucoviridis* (Engl.) Balle, Fl. Cameroun 23: 54 (1982).

Habit: parasite. **Habitat:** forest. **Distribution:** Cameroon and Gabon. **Chorology:** Lower Guinea. **Star:** blue. **Alt. range:** 200-400m (1). **Flowering:** Mar (1). **Specimens:** - *Bambuko F.R.:* Watts J. 604

Globimetula dinklagei (Engl.) Danser

F.W.T.A. 1: 660 (1958); Fl. Cameroun 23: 20 (1982).

Habit: parasite. **Habitat:** forest and plantations. **Distribution:** Cameroon, Gabon and Congo (Brazzaville). **Chorology:** Lower Guinea. **Star:** blue. **Alt. range:** 1000-1200m (1). **Flowering:** Oct (1). **Specimens:** - *Etinde:* Hanke F. 622; Wheatley J. I. 603

Globimetula oreophila (Oliv.) Tiegh.
F.W.T.A. 1: 660 (1958); Fl. Cameroun 23: 12 (1982).

Habit: parasite. **Habitat:** forest-grassland transition. **Distribution:** Cameroon and S. Nigeria. **Chorology:** Lower Guinea (montane). **Star:** gold. **Alt. range:** 2000-2200m (1), 2201-2400m (1). **Flowering:** Nov (1). **Specimens:** - *Eastern slopes:* Cheek M. 5331. *No locality:* Mann G. 1210

Globimetula sp.

Specimens: - *Eastern slopes:* Tchouto P. 53

Helixanthera mannii (Oliv.) Danser
F.W.T.A. 1: 659 (1958); Fl. Cameroun 23: 23 (1982).

Habit: parasite. **Habitat:** forest. **Distribution:** SE Nigeria, Cameroon, S. Tomé, Gabon, Congo (Kinshasa), Angola and Uganda. **Chorology:** Guineo-Congolian. **Star:** green. **Alt. range:** 1-200m (1). **Flowering:** Oct (1). **Specimens:** - *Onge:* Tchouto P. 1020

Phragmanthera capitata (Spreng.) Balle
Fl. Cameroun 23: 29 (1982).
   Phragmanthera incana (Schum.) Balle, F.W.T.A. 1: 664 (1958).
   Phragmanthera lapathifolia (Engl. & K.Krause) Balle, F.W.T.A. 1: 664 (1958).

Habit: parasite. **Habitat:** forest. **Distribution:** Guinea to Bioko, Cameroon, Gabon, Congo (Brazzaville) and Congo (Kinshasa). **Chorology:** Guineo-Congolian (montane). **Star:** green. **Alt. range:** 1-200m (2), 601-800m (1), 801-1000m (2). **Flowering:** Mar (2), Apr (1), Oct (2). **Fruiting:** Mar (1), Apr (1). **Specimens:** - *Eastern slopes:* Tchouto P. 95. *Etinde:* Cable S. 1577; Dawson S. 38. *Limbe Botanic Garden:* Cheek M. 4956. *Mokoko River F.R.:* Acworth J. M. 225. *No locality:* Kalbreyer W. 95

Phragmanthera kamerunensis (Engl.) Balle
F.W.T.A. 1: 664 (1958); Fl. Cameroun 23: 34 (1982).

Habit: parasite. **Habitat:** forest. **Distribution:** Nigeria and Cameroon. **Chorology:** Lower Guinea. **Star:** blue. **Specimens:** - *No locality:* Winkler H. J. P. 1067

Tapinanthus apodanthus (Sprague) Danser
Verh. Kon. Wet. Akad. Amsterdam sect.2 Natk. 29: 107 (1933).
   Tapinanthus globiferus (A.Rich.) Tiegh. subsp. *apodanthus* (Sprague) Balle, Fl. Cameroun 23: 56 (1982).

Habit: parasite. **Habitat:** forest. **Distribution:** Nigeria to Zaire and CAR. **Chorology:** Guineo-Congolian. **Star:** green. **Alt. range:** 1-200m (1). **Flowering:** Jul (1), Oct (1). **Specimens:** - *Limbe Botanic Garden:* Cheek M. 4955. *Mabeta-Moliwe:* Cheek M. 3588. *Kokoko:* Acworth J.M. 225b

Tapinanthus globiferus (A.Rich.) Tiegh.

Specimens: - *Mabeta-Moliwe:* Cheek 3588. *Mokoko:* Acworth 2256

MALPIGHIACEAE

Acridocarpus longifolius (G.Don) Hook.f.
F.W.T.A. 1: 352 (1958); Fl. Cameroun 14: 6 (1972); Fl. Gabon 21: 6 (1973).

Habit: woody climber. **Habitat:** forest. **Distribution:** Liberia to Bioko, Cameroon, Gabon and Angola. **Chorology:** Guineo-Congolian. **Star:** green. **Alt. range:** 1-200m (3). **Flowering:** May (1), Jul (1). **Fruiting:** May (1), Jul (1). **Specimens:** - *Etinde:* Winkler H. J. P. 1228. *Mabeta-Moliwe:* Cheek M. 3555, 3580; Plot Voucher W 152, W 504, W 842; Wheatley J. I. 311

Heteropterys leona (Cav.) Exell
F.W.T.A. 1: 353 (1958); Fl. Cameroun 14: 20 (1972); Fl. Gabon 21: 18 (1973).

Habit: woody climber. **Habitat:** forest. **Distribution:** Guinea Bissau to Cameroon, Gabon, Congo (Kinshasa) and Angola. **Chorology:** Guineo-Congolian. **Star:** green. **Alt. range:** 1-200m (2). **Flowering:** Jul (2). **Specimens:** - *Etinde:* Preuss P. R. 1332. *Mabeta-Moliwe:* Cheek M. 3516, 3563; Maitland T. D. 1148

Triaspis stipulata Oliv.
F.W.T.A. 1: 354 (1958); Fl. Cameroun 14: 15 (1972); Fl. Gabon 21: 10 (1973).

Habit: woody climber. **Guild:** pioneer. **Habitat:** forest. **Distribution:** Togo, Nigeria and Cameroon. **Chorology:** Guineo-Congolian. **Star:** green. **Specimens:** - *Etinde:* Maitland T. D. 911

MALVACEAE

Abutilon mauritianum (Jacq.) Medic.
F.W.T.A. 1: 337 (1958); Fl. Senegal 6: 145 (1979).

Habit: herb to 1.5 m. **Guild:** pioneer. **Distribution:** tropical Africa. **Star:** excluded. **Specimens:** - *Mabeta-Moliwe:* Maitland T. D. 975a

Hibiscus manihot L. var. manihot
F.W.T.A. 1: 348 (1958).

Habit: shrub to 2 m. **Distribution:** introduced from Asia. **Star:** excluded. **Specimens:** - *No locality:* Maitland T. D. 142

Hibiscus rostellatus Guill. & Perr.
F.W.T.A. 1: 346 (1958).

Habit: shrub to 3 m. **Guild:** pioneer. **Distribution:** tropical Africa. **Star:** excluded. **Specimens:** - *Etinde:* Maitland T. D. 577

Hibiscus surattensis L.
F.W.T.A. 1: 346 (1958).

Habit: scrambling herb. **Habitat:** forest. **Distribution:** palaeotropical. **Star:** excluded. **Specimens:** - *Eastern slopes:* Hutchinson J. & Metcalfe C. R. 103

Hibiscus tiliaceus L.
F.W.T.A. 1: 345 (1958).

Habit: tree to 5 m. **Habitat:** sea shores. **Distribution:** pantropical. **Star:** excluded. **Alt. range:** 1-200m (1). **Flowering:** Mar (1). **Fruiting:** Mar (1). **Specimens:** - *Mabeta-Moliwe:* Sunderland T. C. H. 1103

Hibiscus vitifolius L.
Bol. Soc. Brot. (ser.2) 32: 69 (1958); F.W.T.A. 1: 346 (1958).
Hibiscus vitifolius L. var. *ricinifolius* (E.Mey. ex Harv.) Hochr., F.W.T.A. 1: 346 (1958).

Habit: shrub to 2 m. **Distribution:** Mount Cameroon, Uganda, Ethiopia and southern Africa. **Chorology:** Afromontane. **Star:** blue. **Alt. range:** 600-800m (1). **Specimens:** - *Etinde:* Maitland T. D. FHI 12332

Kosteletzkya adoensis (Hochst. ex A.Rich.) Mast.
F.W.T.A. 1: 349 (1958); Fl. Congo B., Rw. & Bur. 10: 139 (1963).

Habit: herb. **Guild:** pioneer. **Distribution:** tropical Africa. **Chorology:** Afromontane. **Star:** green. **Alt. range:** 1200-1400m (1). **Specimens:** - *No locality:* Dunlap 35

Pavonia urens Cav. var. glabrescens (Ulbr.) Brenan
F.W.T.A. 1: 341 (1958).

Habit: shrub to 3 m. **Habitat:** forest. **Distribution:** tropical Africa. **Chorology:** Afromontane. **Star:** green. **Specimens:** - *Eastern slopes:* Dunlap 126; Ekema N. 823

Pavonia urens Cav. var. urens
F.W.T.A. 1: 341 (1958).

Habit: shrub to 3 m. **Distribution:** tropical Africa. **Chorology:** Afromontane. **Star:** green. **Alt. range:** 1200-1400m (1). **Specimens:** - *No locality:* Maitland T. D. 840

Sida acuta Burm.f. subsp. carpinifolia (L.f.) Borss.Waalk.
Blumea 14: 188 (1966).
Sida stipulata Cav., F.W.T.A. 1: 339 (1958).

Habit: herb. **Guild:** pioneer. **Distribution:** tropical Africa. **Star:** excluded. **Specimens:** - *Etinde:* Maitland T. D. 49

Sida garckeana Pell.
Bol. Soc. Brot. (ser.2) 54: 104 (1980).
Sida corymbosa R.E.Fr., F.W.T.A. 1: 339 (1958).

Habit: shrub to 2 m. **Guild:** pioneer. **Habitat:** roadsides and villages. **Distribution:** tropical Africa and America. **Star:** excluded. **Alt. range:** 600-800m (1). **Flowering:** Mar (1). **Specimens:** - *Etinde:* Mbani J. M. 57

Sida pilosa Retz.
Bol. Soc. Brot. (ser.2) 54: 11 (1980).
Sida veronicifolia Lam., F.W.T.A. 1: 338 (1958).

Habit: herb. **Guild:** pioneer. **Distribution:** pantropical. **Star:** excluded. **Specimens:** - *Eastern slopes:* Maitland T. D. 905

Sida rhombifolia L. subsp. rhombifolia
F.W.T.A. 1: 339 (1958); Bol. Soc. Brot. (ser.2) 54: 70 (1980).
Sida rhombifolia L. var. *β.*, F.W.T.A. 1: 339 (1958).
Sida rhombifolia L. var. *μ.*, F.W.T.A. 1: 339 (1958).

Habit: herb. **Guild:** pioneer. **Habitat:** forest. **Distribution:** pantropical. **Chorology:** Montane. **Star:** excluded. **Alt. range:** 200-400m (1), 1601-1800m (1). **Specimens:** - *Eastern slopes:* Maitland T. D. 118. *Mokoko River F.R.:* Thomas D. W. 10018. *No locality:* Maitland T. D. 809

Urena lobata L.
F.W.T.A. 1: 341 (1958); Fl. Senegal 6: 266 (1979).

Habit: shrub to 3 m. **Guild:** pioneer. **Habitat:** lava flows. **Distribution:** pantropical. **Star:** excluded. **Specimens:** - *Eastern slopes:* Migeod F. 224. *Etinde:* Brodie 6

MEDUSANDRACEAE

Medusandra richardsiana Brenan
F.W.T.A. 1: 656 (1958).

Habit: tree to 18 m. **Habitat:** forest. **Distribution:** Cameroon and Gabon. **Chorology:** Lower Guinea. **Star:** gold. **Alt. range:** 1-200m (4), 201-400m (1). **Flowering:** Mar (1), May (1). **Fruiting:** Mar (1), May (4). **Specimens:** - *Mabeta-Moliwe:* W837, W852, W874, W942. *Mokoko River F.R.:* Batoum A. 35; Ekema N. 968; Fraser P. 342; Mbani J. M. 389, 390; Tchouto P. 575. *Southern Bakundu F.R.:* Brenan J. P. M. 9402

Soyauxia gabonensis Oliv.
F.W.T.A. 1: 653 (1958).

Habit: shrub to 3 m. **Habitat:** forest. **Distribution:** Nigeria, Cameroon and Gabon. **Chorology:** Lower Guinea. **Star:** blue. **Alt. range:** 1-200m (2), 201-400m (2). **Specimens:** - *Mokoko River F.R.:* Mbani J. M. 322; Ndam N. 1129, 1311. *Onge:* Harris D. J. 3699; Tchouto P. 895

MELASTOMATACEAE

Calvoa hirsuta Hook.f.
F.W.T.A. 1: 251 (1954); Fl. Cameroun 24: 81 (1983).

Habit: herb. **Habitat:** on rocks in forest. **Distribution:** Ghana, Bioko and Cameroon. **Chorology:** Guineo-Congolian (montane). **Star:** blue. **Alt. range:** 400-600m (2), 1001-1200m (3). **Flowering:** Oct (4). **Fruiting:** Oct (5). **Specimens:** - *Etinde:* Tchouto P. 437, 438; Thomas D. W. 9131; Watts J. 536; Wheatley J. I. 597

Dinophora spenneroides Benth.
F.W.T.A. 1: 252 (1954); Fl. Cameroun 24: 116 (1983).

Habit: shrub to 4 m. **Habitat:** forest. **Distribution:** Guinea to Bioko, Cameroon, Gabon, Congo (Brazzaville), Congo (Kinshasa) and Angola. **Chorology:** Guineo-Congolian. **Star:** green. **Alt. range:** 1-200m (1), 601-800m (1). **Flowering:** Nov (1). **Fruiting:** Jul (1). **Specimens:** - *Etinde:* Kwangue A. 133. *Mabeta-Moliwe:* Cheek M. 3529

Dissotis rotundifolia (Sm.) Triana var. rotundifolia

F.W.T.A. 1: 257 (1954); Fl. Cameroun 24: 37 (1983).

Habit: herb. **Habitat:** forest. **Distribution:** tropical Africa. **Star:** excluded. **Alt. range:** 1-200m (2). **Flowering:** May (1), Nov (1). **Fruiting:** Oct (1), Nov (1). **Specimens:** - *Eastern slopes:* Migeod F. 7. *Etinde:* Dawson S. 35. *Mabeta-Moliwe:* Nguembock F. 50; Sunderland T. C. H. 1355

Guyonia ciliata Hook.f.

F.W.T.A. 1: 246 (1954); Fl. Cameroun 24: 8 (1983).

Habit: terrestrial or epiphytic herb. **Habitat:** forest, by streams. **Distribution:** Guinea to Bioko, Congo (Kinshasa), Uganda and Tanzania. **Chorology:** Guineo-Congolian (montane). **Star:** green. **Alt. range:** 800-1000m (2), 1201-1400m (1), 1601-1800m (1). **Flowering:** Sep (1), Oct (2). **Fruiting:** Sep (1), Oct (1). **Specimens:** - *Eastern slopes:* Migeod F. 26. *Etinde:* Tchouto P. 447; Watts J. 573; Wheatley J. I. 572

Medinilla mannii Hook.f.

F.W.T.A. 1: 251 (1954); Fl. Cameroun 24: 112 (1983).

Habit: woody epiphyte. **Habitat:** forest. **Distribution:** Liberia to Bioko, Cameroon, Congo (Kinshasa) and Uganda. **Chorology:** Guineo-Congolian. **Star:** green. **Flowering:** Jun (1). **Specimens:** - *Etinde:* Winkler H. J. P. 91

Melastomastrum capitatum (Vahl) A. & R.Fern.

F.W.T.A. 1: 761 (1958); Fl. Cameroun 24: 48 (1983).
Dissotis erecta (Guill. & Perr.) Dandy, F.W.T.A. 1: 259 (1954).

Habit: shrub to 1.5 m. **Habitat:** forest and lava flows. **Distribution:** tropical Africa. **Star:** excluded. **Alt. range:** 1-200m (1). **Flowering:** Oct (1). **Specimens:** - *Etinde:* Brodie S. 11

Memecylon afzelii G.Don

Specimens: - *Mabeta-Moliwe:* Plot Voucher W418.

Memecylon dasyanthum Gilg & Ledermann ex Engl.

F.W.T.A. 1: 262 (1954); Fl. Cameroun 24: 134 (1983).

Habit: tree to 8 m. **Habitat:** forest. **Distribution:** Cameroon. **Chorology:** Lower Guinea. **Star:** gold. **Alt. range:** 1-200m (1), 601-800m (1). **Flowering:** Apr (1). **Specimens:** - *Mabeta-Moliwe:* Mildbraed G. W. J. 10534; Wheatley J. I. 141

Memecylon englerianum Cogn.

F.W.T.A. 1: 263 (1954); Fl. Cameroun 24: 147 (1983).
Memecylon obanense Baker f., F.W.T.A. 1: 262 (1954).

Habit: shrub to 4 m. **Habitat:** forest. **Distribution:** Guinea to Cameroon. **Chorology:** Guineo-Congolian. **Star:** green. **Alt. range:** 600-800m (2), 1001-1200m (1). **Flowering:** Jan (1).

Fruiting: Apr (1). **Specimens:** - *Eastern slopes:* Kwangue A. 56; Tekwe C. 13. *Mabeta-Moliwe:* Mildbraed G. W. J. 10757

Memecylon cf. englerianum Cogn.

Specimens: - *Mabeta-Moliwe:* Sunderland T. C. H. 1125

Memecylon cf. myrianthum Gilg

Specimens: - *Mabeta-Moliwe:* Plot Voucher W647

Memecylon zenkeri Gilg

F.W.T.A. 1: 263 (1954); Fl. Cameroun 24: 143 (1983).

Habit: shrub to 4 m. **Habitat:** forest. **Distribution:** SE Nigeria and Cameroon. **Chorology:** Lower Guinea. **Star:** blue. **Alt. range:** 1-200m (3), 201-400m (1), 401-600m (2), 601-800m (2). **Flowering:** Apr (3), May (1). **Fruiting:** Mar (1), Apr (1), Oct (1). **Specimens:** - *Eastern slopes:* Tchouto P. 140. *Etinde:* Kalbreyer W. 174; Tekwe C. 41; Watts J. 493. *Mabeta-Moliwe:* Jaff B. 52; Sunderland T. C. H. 1403; Watts J. 198. *Onge:* Wheatley J. I. 684

Memecylon sp.

Specimens: - *Onge:* Harris D. J. 3709, 3730; Watts J. 945

Preussiella kamerunensis Gilg

F.W.T.A. 1: 251 (1954); Fl. Cameroun 24: 92 (1983).
Preussiella chevalieri Jacq.-Fél., F.W.T.A. 1: 251 (1954).

Habit: woody epiphyte. **Habitat:** forest. **Distribution:** Guinea to Cameroon. **Chorology:** Guineo-Congolian (montane). **Star:** green. **Alt. range:** 1200-1400m (1), 1401-1600m (1). **Flowering:** Sep (1), Oct (1). **Fruiting:** Oct (1). **Specimens:** - *Etinde:* Tekwe C. 166; Wheatley J. I. 616

Preussiella cf. kamerunensis Gilg

Specimens: - *Etinde:* Tchouto P. 415; Tekwe C. 257; Wheatley J. I. 571

Tristemma albiflorum (G.Don) Benth.

Adansonia (ser.3) 28: 181 (1976).
Tristemma incompletum sensu Keay, F.W.T.A. 1: 250 (1954); F.W.T.A. 1: 761 (1958).

Habit: shrub to 1 m. **Guild:** pioneer. **Habitat:** forest. **Distribution:** Senegal to Bioko, Cameroon, Angola and Sudan. **Chorology:** Guineo-Congolian (montane). **Star:** green. **Alt. range:** 1200-1400m (1). **Flowering:** Nov (1). **Specimens:** - *Eastern slopes:* Migeod F. s.n.

Tristemma hirtum P.Beauv.

F.W.T.A. 1: 250 (1954); Fl. Cameroun 24: 68 (1983).

Habit: herb. **Guild:** pioneer. **Distribution:** Senegal to Bioko, Cameroon, Gabon and Congo (Kinshasa). **Chorology:** Guineo-Congolian. **Star:** green. **Specimens:** - *Etinde:* Mildbraed G. W. J. 10665

Tristemma leiocalyx Cogn.

Fl. Cameroun 24: 64 (1983).
Tetraphyllaster rosaceum Gilg, F.W.T.A. 1: 246 (1954).

Habit: shrub to 1 m. Habitat: forest. Distribution: SW Cameroon, Congo (Kinshasa) and Uganda. Chorology: Guineo-Congolian (montane). Star: green. Alt. range: 1200-1400m (1). Specimens: - *Eastern slopes:* Preuss P. R. FHI 70226

Tristemma littorale Benth. subsp. **biafranum** Jacq.-Fél.
Fl. Cameroun 24: 62 (1983).

Habit: shrub to 1 m. Guild: pioneer. Habitat: forest. Distribution: Ivory Coast to Cameroon, Gabon, Congo (Kinshasa) and CAR. Chorology: Guineo-Congolian. Star: green. Alt. range: 1-200m (5), 201-400m (1), 801-1000m (2). Flowering: Apr (1), May (2), Jun (2), Jul (1), Oct (2). Fruiting: Apr (1), May (1), Jun (1), Jul (1), Aug (1), Oct (1). Specimens: - *Eastern slopes:* Tekwe C. 109, 135. *Etinde:* Nkeng P. 78. *Mabeta-Moliwe:* Ndam N. 502, 506; Nguembock F. 6; Sidwell K. 10; Watts J. 397; Wheatley J. I. 163

Tristemma littorale Benth. subsp. **littorale**
F.W.T.A. 1: 250 (1954); Fl. Cameroun 24: 60 (1983).

Habit: shrub to 1 m. Habitat: forest. Distribution: Mount Cameroon and Bioko. Chorology: Western Cameroon. Star: black. Flowering: Jan (1), Oct (1). Specimens: - *Etinde:* Leeuwenberg A. J. M. 6929; Schlechter F. R. R. 12407

Tristemma oreophilum Gilg
Fl. Cameroun 24: 74 (1983).

Habit: shrub to 2 m. Habitat: Aframomum thicket. Distribution: Cameroon, Gabon and Congo (Kinshasa). Chorology: Guineo-Congolian (montane). Star: green. Alt. range: 600-800m (1), 1201-1400m (1). Flowering: Sep (2). Specimens: - *Eastern slopes:* Preuss P. R. 921. *Etinde:* Wheatley J. I. 590

Warneckea sp. nov. aff. **cinnamomoides** (G.Don) Jacq.-Fél.

Specimens: - *Mabeta-Moliwe:* Sunderland T. C. H. 1208; Wheatley J. I. 147

Warneckea membranifolia (Hook.f) Jacq.-Fél.
Fl. Cameroun 24: 172 (1983); Hawthorne W., F.G.F.T. Ghana: 47 (1990).
  Memecylon membranifolium Hook.f., F.W.T.A. 1: 263 (1954).

Habit: shrub to 3 m. Habitat: forest. Distribution: Ivory Coast to Bioko, Cameroon and Congo (Kinshasa). Chorology: Guineo-Congolian. Star: green. Alt. range: 1-200m (4), 201-400m (3). Flowering: Mar (1), Jun (1), Sep (1), Oct (1). Fruiting: Mar (1), Jun (1), Oct (3). Specimens: - *Bambuko F.R.:* Watts J. 629. *Mabeta-Moliwe:* Plot Voucher W 62, W 328, W 642, W 674; Wheatley J. I. 318. *Onge:* Tchouto P. 763; Watts J. 721, 851, 872; Wheatley J. I. 741

Warneckea memecyloides (Benth.) Jacq.-Fél.
Adansonia (ser.2) 18: 232 (1978); Fl. Cameroun 24: 163 (1983).
  Memecylon memecyloides (Benth.) Exell, F.W.T.A. 1: 263 (1954); Hawthorne W., F.G.F.T. Ghana: 47 (1990).

Habit: tree to 18 m. Guild: shade-bearer. Habitat: forest. Distribution: Ivory Coast to Cameroon. Chorology: Guineo-Congolian (montane). Star: green. Alt. range: 200-400m (1). Specimens: - *Mokoko River F.R.:* Ndam N. 1127

82

Warneckea pulcherrima (Gilg) Jacq.-Fél.
Fl. Cameroun 24: 163 (1983).

Habit: tree to 15 m. Habitat: forest. Distribution: Cameroon, Gabon and Congo (Kinshasa). Chorology: Guineo-Congolian. Star: green. Alt. range: 400-600m (1).Fruiting: Jan (1). Specimens: - *Etinde:* Tekwe C. 3

MELIACEAE

Carapa procera DC.
F.W.T.A. 1: 702 (1958); Keay R.W.J., Trees of Nigeria: 348 (1989); Hawthorne W., F.G.F.T. Ghana: 177 (1990); F.T.E.A. Meliaceae: 62 (1991).
  Carapa grandiflora Sprague, F.W.T.A. 1: 702 (1958).

Habit: tree to 30 m. Guild: shade-bearer. Habitat: forest. Distribution: Senegal to Bioko, Cameroon, Congo (Kinshasa) and Congo (Brazzaville). Chorology: Guineo-Congolian. Star: green. Alt. range: 1-200m (3), 201-400m (1). Flowering: Apr (2), May (1). Fruiting: Apr (1). Specimens: - *Etinde:* Maitland T. D. 430. *Mabeta-Moliwe:* Plot Voucher W 660, W 672; Wheatley J. I. 184. *Mokoko River F.R.:* Akogo M. 174; Ndam N. 1315; Watts J. 1123

Carapa sp.

Specimens: - *Bambuko F.R.:* Watts J. 632. *Onge:* Tchouto P. 733

Entandrophragma angolense (Welw.) C.DC.
F.W.T.A. 1: 700 (1958); Keay R.W.J., Trees of Nigeria: 342 (1989); Hawthorne W., F.G.F.T. Ghana: 174 (1990).

Habit: tree to 50 m. Guild: light demander. Habitat: forest and farmbush. Distribution: Guinea to Bioko, Cameroon, Congo (Kinshasa), Angola, Uganda and Sudan. Chorology: Guineo-Congolian. Star: scarlet. Specimens: - *Etinde:* Maitland T. D. 410

Entandrophragma candollei Harms
F.W.T.A. 1: 700 (1958); Keay R.W.J., Trees of Nigeria: 342 (1989); Hawthorne W., F.G.F.T. Ghana: 174 (1990).

Habit: tree to 50 m. Guild: light demander. Habitat: forest. Distribution: Guinea to Congo (Kinshasa) and Angola (Cabinda). Chorology: Guineo-Congolian. Star: red. Specimens: - *Southern Bakundu F.R.:* Rosevear D. R. Cam 85/37

Entandrophragma cylindricum (Sprague) Sprague
F.W.T.A. 1: 701 (1958); Keay R.W.J., Trees of Nigeria: 344 (1989); Hawthorne W., F.G.F.T. Ghana: 174 (1990).

Habit: tree to 50 m. Guild: light demander. Habitat: forest. Distribution: Sierra Leone to Congo (Kinshasa), Angola (Cabinda) and Uganda. Chorology: Guineo-Congolian. Star: red. Alt. range: 600-800m (1). Specimens: - *Mabeta-Moliwe:* Mildbraed G. W. J. 10770

Entandrophragma sp.

Specimens: - *Mokoko River F.R.:* Thomas D. W. 10084

Guarea glomerulata Harms
F.W.T.A. 1: 706 (1958).

Habit: tree or shrub to 6 m. Habitat: forest. Distribution: Nigeria, Bioko, Cameroon, Gabon and Congo (Kinshasa). Chorology: Guineo-Congolian. Star: green. Alt. range: 1-200m (5), 201-400m (3), 401-600m (3), 601-800m (1), 801-1000m (1), 1001-1200m (1), 1201-1400m (1). Flowering: Feb (1), Mar (3), May (1), Jun (1), Oct (3), Nov (1). Fruiting: Mar (7), Apr (1), May (1), Jun (2). Specimens: - *Bambuko F.R.:* Watts J. 633. *Etinde:* Cable S. 1629; Etuge M. 1233; Groves M. 298; Khayota B. 523, 576; Kwangue A. 42; Tchouto P. 32; Wheatley J. I. 630. *Mokoko River F.R.:* Ekema N. 988, 1203; Ndam N. 1260; Pouakouyou D. 114. *No locality:* Nkeng P. 9. *Onge:* Harris D. J. 3799; Tchouto P. 978; Watts J. 823

Heckeldora staudtii (Harms) Staner
F.W.T.A. 1: 707 (1958); Keay R.W.J., Trees of Nigeria: 354 (1989).

Habit: tree or shrub to 5 m. Habitat: forest. Distribution: Liberia to Cameroon, Gabon and Congo (Kinshasa). Chorology: Guineo-Congolian. Star: green. Alt. range: 1-200m (22), 201-400m (6), 401-600m (1), 601-800m (1). Flowering: Feb (1), Mar (2), May (2), Jun (1), Oct (6). Fruiting: Feb (1), Mar (7), Apr (5), May (3), Jun (1), Oct (2), Nov (2). Specimens: - *Bambuko F.R.:* Watts J. 666. *Etinde:* Cable S. 1518; Kalbreyer W. 4; Tchouto P. 72; Tekwe C. 69. *Mabeta-Moliwe:* Cheek M. 3224; Cable S. 1429; Groves M. 280; Khayota B. 457; Plot Voucher W 64, W 267, W 389, W 559, W 619, W 838; Sunderland T. C. H. 1084, 1165, 1231, 1335, 1337, 1463; Watts J. 108, 178, 286. *Mokoko River F.R.:* Acworth J. M. 170; Akogo M. 164; Ekema N. 1090; Pouakouyou D. 98; Watts J. 1090. *Onge:* Cheek M. 4996; Harris D. J. 3650, 3798; Ndam N. 662, 702, 736, 752; Tchouto P. 844; Watts J. 763

Khaya anthotheca C.DC.

Specimens: - *S. Bakundu F.R.:* Staudt 877

Khaya cf. anthotheca C.DC.

Specimens: - *Bambuko F.R.:* Keay in F.H.I. 37426

Trichilia gilgiana Harms
F.W.T.A. 1: 705 (1958); Meded. Land. Wag. 68(2): 76 (1968); Keay R.W.J., Trees of Nigeria: 351 (1989).

Habit: tree to 22 m. Habitat: forest. Distribution: Nigeria, Cameroon, Gabon, Congo (Kinshasa) and Angola. Chorology: Guineo-Congolian. Star: green. Alt. range: 1-200m (1). Flowering: Jul (1). Specimens: - *Mabeta-Moliwe:* Cheek M. 3522

Trichilia monadelpha (Thonn.) J.J.de Wilde
Meded. Land. Wag. 68(2): 108 (1968); Keay R.W.J., Trees of Nigeria: 351 (1989); Hawthorne W., F.G.F.T. Ghana: 178 (1990). *Trichilia heudelotii* Planch. ex Oliv., F.W.T.A. 1: 704 (1958).

Habit: tree to 16 m. Habitat: forest. Distribution: Guinea Bissau to Bioko, Cameroon, Gabon, Congo (Kinshasa) and Angola (Cabinda). Chorology: Guineo-Congolian. Star: green. Alt. range: 1-200m (3). Flowering: Apr (2). Fruiting: Jun (1). Specimens: - *Etinde:* Maitland T. D. 317; Tekwe C. 86. *Mabeta-Moliwe:* Plot Voucher W 529, W 1155; Sunderland T. C. H. 1227; Wheatley J. I. 180

Trichilia prieureana A.Juss. subsp. vermoesenii J.J.de Wilde
F.W.T.A. 1: 704 (1958); Meded. Land. Wag. 68(2): 139 (1968); Keay R.W.J., Trees of Nigeria: 350 (1989); Hawthorne W., F.G.F.T. Ghana: 177 (1990).

Habit: tree to 30 m. Habitat: forest. Distribution: Sierra Leone to Congo (Kinshasa), Angola, Uganda and Sudan. Chorology: Guineo-Congolian. Star: green. Specimens: - *Mabeta-Moliwe:* Dunlap 181

Trichilia cf. prieureana A.Juss.

Specimens: - *Mabeta-Moliwe:* Plot Voucher W 90

Trichilia rubescens Oliv.
F.W.T.A. 1: 704 (1958); Meded. Land. Wag. 68(2): 161 (1968); Keay R.W.J., Trees of Nigeria: 351 (1989).

Habit: tree to 18 m. Habitat: forest and Aframomum thicket. Distribution: Nigeria, Cameroon, Bioko, Gabon, Congo (Kinshasa), CAR, Uganda and Tanzania. Chorology: Guineo-Congolian (montane). Star: green. Alt. range: 1-200m (11), 201-400m (3), 401-600m (2), 601-800m (2), 801-1000m (3). Flowering: Feb (3), Mar (8), Apr (1), May (1). Fruiting: Feb (1), Mar (3), Apr (2), May (4), Jun (3). Specimens: - *Eastern slopes:* Groves M. 120; Nkeng P. 64; Tchouto P. 36. *Etinde:* Cable S. 1596; Groves M. 288; Khayota B. 511; Mbani J. M. 10, 45; Nkeng P. 76; Tchouto P. 12; Tekwe C. 42. *Mabeta-Moliwe:* Khayota B. 492; Plot Voucher W 391; Sunderland T. C. H. 1281, 1445; Watts J. 193; Wheatley J. I. 250, 369. *Mokoko River F.R.:* Batoum A. 32; Ekema N. 1047; Pouakouyou D. 4; Tchouto P. 623, 1245. *Onge:* Tchouto P. 674, 883

Trichilia sp.

Specimens: - *Mokoko River F.R.:* Tchouto P. 604, 1121

Turraea vogelii Hook.f. ex Benth.
F.W.T.A. 1: 708 (1958); F.T.E.A. Meliaceae: 15 (1991).

Habit: woody climber to 5 m. Guild: pioneer. Habitat: forest. Distribution: Ghana to Bioko, Cameroon, Gabon, Congo (Kinshasa), Angola, Uganda and Sudan. Chorology: Guineo-Congolian. Star: green. Alt. range: 1-200m (3), 201-400m (1), 601-800m (1). Flowering: Jun (1), Oct (1), Nov (2). Fruiting: Oct (1). Specimens: - *Eastern slopes:* Tekwe C. 95. *Etinde:* Maitland T. D. 269. *Limbe Botanic Garden:* Wheatley J. I. 662. *Onge:* Cheek M. 5100; Harris D. J. 3771; Tchouto P. 860

Turraeanthus africanus (Welw. ex C.DC.) Pellegr.
F.W.T.A. 1: 707 (1958); Keay R.W.J., Trees of Nigeria: 352 (1989); Hawthorne W., F.G.F.T. Ghana: 177 (1990).

Habit: tree to 35 m. Guild: shade-bearer. Habitat: forest. Distribution: Sierra Leone to Cameroon, Gabon, Congo (Kinshasa) and Angola. Chorology: Guineo-Congolian. Star: pink. Alt. range: 200-400m (1), 401-600m (1). Flowering: Apr (2). Specimens: - *Eastern slopes:* Dalziel J. M. 8209. *Etinde:* Tchouto P. 110; Tekwe C. 28. *Mokoko River F.R.:* Mbani J. M. 362

MELIANTHACEAE

Bersama abyssinica Fresen.
Kew Bull. 5: 233 (1950); Kew Bull. 10: 600 (1955); F.T.E.A. Melianthaceae: 2 (1958); F.W.T.A. 1: 726 (1958); Fl. Zamb. 2: 545 (1966); Keay R.W.J., Trees of Nigeria: 365 (1989); Hawthorne W., F.G.F.T. Ghana: 166 (1990).
    Bersama maxima Baker, F.W.T.A. 1: 726 (1958).
    Bersama acutidens Welw. ex Hiern, F.W.T.A. 1: 726 (1958).

Habit: tree to 15 m. **Guild:** pioneer. **Habitat:** forest. **Distribution:** tropical Africa. **Chorology:** Afromontane. **Star:** green. **Alt. range:** 1-200m (1), 201-400m (1), 401-600m (2), 601-800m (2), 1001-1200m (1). **Flowering:** Mar (1), Apr (1), May (1), Dec (1). **Fruiting:** May (1), Sep (1), Oct (1). **Specimens:** - *Eastern slopes:* Tchouto P. 45. *Etinde:* Cable S. 363; Tchouto P. 221; Tekwe C. 54; Wheatley J. I. 556. *Mabeta-Moliwe:* Cheek M. 5707. *Onge:* Akogo M. 29

MENISPERMACEAE

Cissampelos owariensis P.Beauv. ex DC.
F.W.T.A. 1: 75 (1954).

Habit: twining herb. **Guild:** pioneer. **Habitat:** forest. **Distribution:** tropical Africa. **Star:** excluded. **Alt. range:** 1-200m (3). **Flowering:** Jun (1). **Fruiting:** Jul (1). **Specimens:** - *Etinde:* Maitland T. D. 635. *Mabeta-Moliwe:* Cheek M. 3527. *Mokoko River F.R.:* Acworth J. M. 345; Ekema N. 1180; Mbani J. M. 513

Dioscoreophyllum cumminsii (Stapf) Diels
F.W.T.A. 1: 73 (1954).

Habit: woody climber. **Guild:** pioneer. **Habitat:** forest. **Distribution:** Guinea to Cameroon, Gabon, Congo (Kinshasa), CAR and Sudan. **Chorology:** Guineo-Congolian. **Star:** green. **Alt. range:** 200-400m (1). **Flowering:** Oct (1). **Specimens:** - *Onge:* Ndam N. 623

Dioscoreophyllum volksenii Engl. var. volksenii
F.W.T.A. 1: 758 (1958).
    Dioscoreophyllum tenerum Engl. var. *tenerum*, F.W.T.A. 1: 73 (1954).

Habit: woody climber. **Habitat:** forest. **Distribution:** tropical Africa. **Star:** excluded. **Alt. range:** 1-200m (4). **Flowering:** May (1), Jun (1). **Fruiting:** Jun (2). **Specimens:** - *Etinde:* Kwangue A. 110. *Mabeta-Moliwe:* Watts J. 256; Wheatley J. I. 350. *Mokoko River F.R.:* Ekema N. 1109

Jateorhiza macrantha (Hook.f.) Exell & Mendonça
F.W.T.A. 1: 74 (1954).

Habit: woody climber. **Habitat:** forest. **Distribution:** Nigeria, Cameroon, Bioko, Gabon, Congo (Kinshasa) and Angola (Cabinda). **Chorology:** Guineo-Congolian. **Star:** green. **Alt. range:** 1-200m (10), 201-400m (1). **Flowering:** Feb (1), Mar (1), Apr (3), Jun (1), Oct (1). **Fruiting:** Feb (1), Mar (1), Apr (3), May (1), Oct (1). **Specimens:** - *Etinde:* Maitland T. D. 412; Mbani J. M. 76; Nkeng P. 19; Tchouto P. 124; Tekwe C. 71. *Mabeta-Moliwe:* Jaff B. 183; Sunderland T. C. H. 1201; Watts J. 102;

Wheatley J. I. 217. *Mokoko River F.R.:* Acworth J. M. 47; Pouakouyou D. 94. *Onge:* Ndam N. 679

Kolobopetalum auriculatum Engl.
F.W.T.A. 1: 758 (1958).

Habit: woody climber. **Guild:** shade-bearer. **Habitat:** forest. **Distribution:** Ghana, Nigeria, Cameroon and Gabon. **Chorology:** Guineo-Congolian. **Star:** green. **Alt. range:** 1-200m (8), 201-400m (2). **Flowering:** Mar (1), Apr (2). **Fruiting:** Mar (4), Apr (2), May (2), Jun (2). **Specimens:** - *Bambuko F.R.:* Watts J. 621. *Etinde:* Kalbreyer W. 19; Tchouto P. 79. *Mabeta-Moliwe:* Jaff B. 23, 40; Plot Voucher W 215, W 537, W 731. *Mokoko River F.R.:* Akogo M. 183, 283; Ekema N. 919, 981; Fraser P. 448. *Onge:* Wheatley J. I. 801

Penianthus camerounensis A.Dekker
Bull. Jard. Bot. Nat. Belg. 53: 43 (1983).

Habit: tree or shrub to 8 m. **Habitat:** forest. **Distribution:** Cameroon. **Chorology:** Lower Guinea. **Star:** gold. **Specimens:** - *Mokoko River F.R.:* Fraser P. 405

Penianthus cf. camerounensis A.Dekker

Specimens: - *Etinde:* Mbani J. M. 154

Penianthus longifolius Miers
F.W.T.A. 1: 76 (1954); Bull. Jard. Bot. Nat. Belg. 53: 46 (1983).

Habit: shrub to 3 m. **Habitat:** forest. **Distribution:** Nigeria, Cameroon, Bioko, Gabon, Congo (Brazzaville), Congo (Kinshasa) and CAR. **Chorology:** Guineo-Congolian. **Star:** green. **Alt. range:** 1-200m (11). **Fruiting:** Apr (7), May (5), Jun (1). **Specimens:** - *Etinde:* Tchouto P. 122; Tekwe C. 68. *Mabeta-Moliwe:* Plot Voucher W 67, W 123, W 592; Sunderland T. C. H. 1246, 1271, 1397, 1455; Wheatley J. I. 205. *Mokoko River F.R.:* Acworth J. M. 69; Mbani J. M. 305, 324; Ndam N. 1139, 1292; Watts J. 1084. *No locality:* Mann G. 1205

Penianthus zenkeri (Engl.) Diels
F.W.T.A. 1: 76 (1954); Bull. Jard. Bot. Nat. Belg. 53: 56 (1983).

Habit: tree or shrub to 6 m. **Habitat:** stream banks in forest. **Distribution:** SE Nigeria, Cameroon and E. Congo (Kinshasa). **Chorology:** Guineo-Congolian. **Star:** green. **Alt. range:** 1-200m (2). **Flowering:** Mar (1), Nov (1). **Specimens:** - *Onge:* Harris D. J. 3875; Wheatley J. I. 773

Penianthus sp.

Specimens: - *Etinde:* Nkeng P. 39; Tchouto P. 31

Perichasma laetificata Miers
Adansonia (ser.2) 17: 223 (1977).
    Stephania laetificata (Miers) Benth., F.W.T.A. 1: 75 (1954).

Habit: woody climber. **Habitat:** forest. **Distribution:** Nigeria, Cameroon, Congo (Kinshasa) and Angola. **Chorology:** Guineo-Congolian. **Star:** green. **Alt. range:** 1-200m (1), 401-600m (1). **Flowering:** Mar (1), Apr (1). **Fruiting:** Mar (1). **Specimens:** - *Etinde:* Deistel H. 142; Tchouto P. 63, 116. *Mabeta-Moliwe:* Nguembock F. 83

Rhigiocarya racemifera Miers
F.W.T.A. 1: 72 (1954).

Habit: woody climber. **Habitat:** forest. **Distribution:** Sierra Leone to Cameroon, Gabon and Angola (Cabinda). **Chorology:** Guineo-Congolian. **Star:** green. **Alt. range:** 1-200m (1), 201-400m (1), 401-600m (1).**Fruiting:** Mar (1), Apr (1), Oct (1). **Specimens: -** *Mabeta-Moliwe:* Sunderland T. C. H. 1247. *Mokoko River F.R.:* Tchouto P. 591. *Onge:* Tchouto P. 782

Stephania abyssinica (Dill. & Rich.) Walp. var. abyssinica
F.W.T.A. 1: 75 (1954).

Habit: woody climber. **Habitat:** forest. **Distribution:** tropical Africa. **Chorology:** Afromontane. **Star:** green. **Alt. range:** 1-200m (2), 1201-1400m (1), 2001-2200m (1). **Flowering:** Apr (1). **Fruiting:** Jul (1), Oct (1). **Specimens: -** *Eastern slopes:* Watts J. 430. *Mabeta-Moliwe:* Cheek M. 3568; Watts J. 236. *No locality:* Mann G. 1242

Stephania dinklagei (Engl.) Diels
F.W.T.A. 1: 758 (1958).

Habit: woody climber. **Guild:** pioneer. **Habitat:** forest. **Distribution:** Guinea to Cameroon, Congo (Brazzaville) and Angola. **Chorology:** Guineo-Congolian. **Star:** green. **Alt. range:** 1-200m (5), 1001-1200m (1). **Flowering:** Apr (1), May (1), Oct (1). **Fruiting:** May (2), Oct (1). **Specimens: -** *Eastern slopes:* Lehmbach H. 199. *Etinde:* Cheek M. 3793; Tekwe C. 70. *Mabeta-Moliwe:* Plot Voucher W 639; Sunderland T. C. H. 1409. *Mokoko River F.R.:* Ekema N. 1087; Watts J. 1184. *Onge:* Watts J. 878

Stephania sp.

Specimens: - *Onge:* Watts J. 707

Synclisia scabrida Miers
F.W.T.A. 1: 68 (1954).

Habit: woody climber. **Habitat:** forest. **Distribution:** SE Nigeria, Cameroon, Gabon, Congo (Kinshasa) and Angola. **Chorology:** Guineo-Congolian. **Star:** green. **Specimens: -** *No locality:* Box H. E. 3558

Syntriandrium preussii Engl.
F.W.T.A. 1: 73 (1954).

Habit: woody climber. **Habitat:** forest. **Distribution:** Nigeria, Cameroon, Gabon and Congo (Kinshasa). **Chorology:** Guineo-Congolian. **Star:** green. **Alt. range:** 1-200m (2), 401-600m (2). **Flowering:** Apr (1), Jun (1). **Fruiting:** May (2). **Specimens: -** *Etinde:* Maitland T. D. 374; Mbani J. M. 151; Tchouto P. 114. *Mabeta-Moliwe:* Preuss P. R. 1273; Wheatley J. I. 353. *Mokoko River F.R.:* Ekema N. 918

Syntriandrium sp.

Specimens: - *Mabeta-Moliwe:* Jaff B. 247; Watts J. 390

Tiliacora lehmbachii Engl.
F.W.T.A. 1: 757 (1958).

Habit: woody climber. **Habitat:** forest. **Distribution:** Mount Cameroon and possibly a single collection from Congo (Kinshasa). **Chorology:** Endemic (montane). **Star:** black. **Alt. range:** 800-1000m (1). **Specimens: -** *Eastern slopes:* Lehmbach H. 90

Tinospora bakis (A.Rich.) Miers
F.W.T.A. 1: 74 (1954).

Habit: twining herb. **Habitat:** forest gaps. **Distribution:** tropical Africa in dry areas. **Star:** excluded. **Alt. range:** 1-200m (1). **Flowering:** Mar (1). **Specimens: -** *Mabeta-Moliwe:* Watts J. 101

Triclisia dictyophylla Diels
F.W.T.A. 1: 757 (1958).
   Triclisia gilletii (De Wild.) Staner, F.W.T.A. 1: 71 (1954).

Habit: woody climber. **Habitat:** forest. **Distribution:** Ivory Coast to Cameroon, Gabon, Congo (Kinshasa), Angola (Cabinda) and CAR. **Chorology:** Guineo-Congolian. **Star:** green. **Alt. range:** 1-200m (4), 201-400m (2). **Flowering:** Feb (1), Apr (1). **Fruiting:** Apr (2), Jun (2), Oct (1). **Specimens: -** *Etinde:* Wheatley J. I. 27. *Mabeta-Moliwe:* Cheek M. 3266, 3268; Plot Voucher W 68, W 206, W 232, W 361, W 407, W 767, W 846; Sunderland T. C. H. 1196; Watts J. 126; Wheatley J. I. 368. *Mokoko River F.R.:* Acworth J. M. 89. *Onge:* Ndam N. 639

Triclisia macrophylla Oliv.
F.W.T.A. 1: 71 (1954).

Habit: woody climber. **Habitat:** forest. **Distribution:** SE Nigeria, Bioko and SW Cameroon. **Chorology:** Western Cameroon. **Star:** gold. **Specimens: -** *No locality:* Jungner J. R. 144

MOLLUGINACEAE

Mollugo pentaphylla L.
F.W.T.A. 1: 134 (1954); Taxon 32: 124 (1983).

Habit: herb. **Distribution:** palaeotropical. **Star:** excluded. **Specimens: -** *Eastern slopes:* Maitland T. D. 1177

MONIMIACEAE

Glossocalyx longicuspis Benth.
F.W.T.A. 1: 55 (1954); Fl. Gabon 10: 107 (1965); Fl. Cameroun 18: 112 (1974).

Habit: tree to 12 m. **Habitat:** forest. **Distribution:** Cameroon, Bioko and Gabon. **Chorology:** Lower Guinea. **Star:** gold. **Alt. range:** 1-200m (9), 201-400m (4). **Flowering:** Mar (2), Apr (2), May (4), Nov (4). **Fruiting:** Apr (2), Nov (1). **Specimens: -** *Etinde:* Cheek M. 5386. *Mabeta-Moliwe:* Jaff B. 180; Wheatley J. I. 202. *Mokoko River F.R.:* Batoum A. 29; Ekema N. 835; Tchouto P. 561, 1167; Thomas D. W. 10071. *Onge:* Ndam N. 730; Tchouto P. 1030; Thomas D. W. 9923; Watts J. 963; Wheatley J. I. 740

Glossocalyx cf. longicuspis Benth.

Specimens: - *Onge:* Thomas D. W. 9924

Xymalos monospora (Harv.) Baill. ex Warb.
F.W.T.A. 1: 55 (1954); Fl. Gabon 10: 104 (1965); Fl. Cameroun 18: 109 (1974); Keay R.W.J., Trees of Nigeria: 34 (1989).

Habit: tree to 15 m. **Habitat:** forest and forest-grassland transition. **Distribution:** SE Nigeria to East and southern Africa. **Chorology:** Afromontane. **Star:** green. **Alt. range:** 1600-1800m (2), 1801-2000m (1).**Fruiting:** Sep (2), Oct (1). **Specimens:** - *Eastern slopes:* Cheek M. 3692; Tekwe C. 326. *Etinde:* Tekwe C. 231

MORACEAE

C.C. Berg (BG)

Antiaris toxicaria Lesch. subsp. **welwitschii** (Engl.) C.C.Berg var. **welwitschii**
Bull. Jard. Bot. Nat. Belg. 48: 466 (1978); Fl. Cameroun 28: 106 (1985); Keay R.W.J., Trees of Nigeria: 298 (1989).
Antiaris welwitschii Engl., F.W.T.A. 1: 613 (1958).
Antiaris toxicaria Lesch. subsp. *africana* (Engl.) C.C.Berg var. *welwitschii* (Engl.) Corner, Bull. Jard. Bot. Nat. Belg. 47: 316 (1977).

Habit: tree to 40 m. **Guild:** light demander. **Habitat:** forest. **Distribution:** Sierra Leone to Congo (Kinshasa), Angola, Uganda and Tanzania. **Chorology:** Guineo-Congolian. **Star:** pink. **Alt. range:** 800-1000m (1). **Specimens:** - *Mabeta-Moliwe:* Mildbraed G. W. J. 10556; Plot Voucher W 392, W 594

Dorstenia africana (Baill.) C.C.Berg
Fl. Gabon 26: 30 (1984); Fl. Cameroun 28: 29 (1985).
Craterogyne africana (Baill.) Lanjouw, F.W.T.A. 1: 599 (1958).

Habit: shrub to 2.5 m. **Habitat:** forest. **Distribution:** SE Nigeria, Cameroon and Gabon. **Chorology:** Lower Guinea. **Star:** blue. **Alt. range:** 1-200m (6), 201-400m (2). **Flowering:** Mar (1), Apr (2), May (4). **Fruiting:** Apr (2). **Specimens:** - *Bambuko F.R.:* Watts J. 643. *Mokoko River F.R.:* Acworth J. M. 122; Akogo M. 211; Ekema N. 965, 996; Tchouto P. 1080; Thomas D. W. 10033; Watts J. 1079. *Southern Bakundu F.R.:* Brenan J. P. M. 9258

Dorstenia angusticornis Engl.
F.W.T.A. 1: 599 (1958); Bull. Jard. Bot. Nat. Belg. 50: 341 (1980); Fl. Cameroun 28: 42 (1985).

Habit: shrub to 1 m. **Habitat:** forest. **Distribution:** Cameroon. **Chorology:** Lower Guinea. **Star:** gold. **Specimens:** - *Southern Bakundu F.R.:* Brenan J. P. M. 9489

Dorstenia barteri Bureau
F.W.T.A. 1: 599 (1958).**Alt. range:** 200-400m (1). **Flowering:** Jun (2), Oct (1). **Specimens:** - *Eastern slopes:* Brenan J. P. M. 9349. *Etinde:* Maitland T. D. 543. *Mabeta-Moliwe:* Plot Voucher W 548, W 801, W 961. *Onge:* Ndam N. 621

Dorstenia barteri Bureau var. **barteri**
F.W.T.A. 1: 599 (1958); Fl. Gabon 26: 72 (1984); Fl. Cameroun 28: 69 (1985).

Habit: herb. **Habitat:** forest. **Distribution:** SE Nigeria, Bioko and W. Cameroon. **Chorology:** Western Cameroon. **Star:** gold. **Alt. range:** 1-200m (11), 401-600m (2), 601-800m (1), 1001-1200m

(1). **Flowering:** Apr (2), May (6), Jun (2), Oct (3), Nov (2). **Fruiting:** Jun (1), Oct (1). **Specimens:** - *Etinde:* Tchouto P. 440; Watts J. 540; Williams S. 73. *Mabeta-Moliwe:* Brodie S. 17; Nguembock F. 47; Sunderland T. C. H. 1214, 1283, 1286, 1384, 1392, 1400; Wheatley J. I. 89, 284, 358. *Mokoko River F.R.:* Ekema N. 1154; Mbani J. M. 520; Ndam N. 1349. *Onge:* Cheek M. 4977

Dorstenia barteri Bureau var. **multiradiata** (Engl.) Hijman & C.C.Berg
Adansonia (ser.2) 16: 434 (1977); Fl. Gabon 26: 73 (1984); Fl. Cameroun 28: 69 (1985).
Dorstenia multiradiata Engl., F.W.T.A. 1: 597 (1958).

Habit: herb. **Habitat:** forest. **Distribution:** SE Nigeria and Cameroon. **Chorology:** Lower Guinea. **Star:** blue. **Alt. range:** 200-400m (6). **Flowering:** Oct (6). **Fruiting:** Oct (2). **Specimens:** - *Onge:* Akogo M. 139; Ndam N. 642, 691, 707, 719, 761. *Southern Bakundu F.R.:* Brenan J. P. M. 9436

Dorstenia barteri Bureau var. **subtriangularis** (Engl.) Hijman & C.C.Berg
Adansonia (ser.2) 16: 436 (1977); Fl. Cameroun 28: 70 (1985).
Dorstenia subtriangularis Engl., F.W.T.A. 1: 597 (1958).

Habit: herb. **Habitat:** forest. **Distribution:** SE Nigeria and Cameroon. **Chorology:** Lower Guinea. **Star:** blue. **Alt. range:** 1-200m (5), 201-400m (4), 401-600m (1). **Flowering:** Mar (2), Jun (1), Aug (1), Oct (4), Nov (2). **Fruiting:** Oct (2). **Specimens:** - *Bambuko F.R.:* Tchouto P. 547. *Etinde:* Cable S. 254; Mbani J. M. 60; Tchouto P. 460. *Mabeta-Moliwe:* Baker W. J. 213; Dunlap 184; Jaff B. 302; Mann G. 14. *Onge:* Akogo M. 132; Cheek M. 5142; Harris D. J. 3633; Watts J. 820

Dorstenia ciliata Engl.
F.W.T.A. 1: 597 (1958); Fl. Gabon 26: 57 (1984); Fl. Cameroun 28: 55 (1985).

Habit: herb. **Habitat:** forest. **Distribution:** SE Nigeria, Cameroon and Gabon. **Chorology:** Lower Guinea. **Star:** blue. **Specimens:** - *Southern Bakundu F.R.:* Brenan J. P. M. 9261

Dorstenia elliptica Bureau
F.W.T.A. 1: 597 (1958); Fl. Gabon 26: 37 (1984); Fl. Cameroun 28: 35 (1985).

Habit: shrub to 1 m. **Habitat:** forest. **Distribution:** SW Cameroon, Bioko, Gabon, Congo (Brazzaville) and Angola (Cabinda) **Chorology:** Guineo-Congolian. **Star:** green. **Alt. range:** 1-200m (3), 201-400m (3). **Flowering:** Mar (2), Apr (1), Jun (1), Oct (1). **Fruiting:** Mar (1), Apr (2), Oct (1). **Specimens:** - *Bambuko F.R.:* Tchouto P. 541, 546. *Mabeta-Moliwe:* Plot Voucher W 15, W 48, W 50, W 74; Watts J. 334; Wheatley J. I. 197. *Mokoko River F.R.:* Akogo M. 288. *Onge:* Tchouto P. 936. *Southern Bakundu F.R.:* Brenan J. P. M. 9266

Dorstenia kameruniana Engl.
Fl. Cameroun 28: 32 (1985).
Craterogyne kameruniana (Engl.) Lanjouw, F.W.T.A. 1: 599 (1958).

Habit: shrub to 3 m. **Guild:** shade-bearer. **Habitat:** forest. **Distribution:** tropical Africa. **Star:** excluded. **Alt. range:** 1-200m (1). **Flowering:** Mar (1). **Specimens:** - *Bambuko F.R.:* Watts J. 658

Dorstenia mannii Hook.f. var. mannii

F.W.T.A. 1: 597 (1958); Fl. Gabon 26: 50 (1984); Fl. Cameroun 28: 50 (1985).
> *Dorstenia ophiocoma* K.Schum. & Engl., F.W.T.A. 1: 597 (1958).

Habit: herb. **Habitat:** forest and Aframomum thicket. **Distribution:** SE Nigeria, Cameroon and Gabon. **Chorology:** Lower Guinea. **Star:** blue. **Alt. range:** 1-200m (17), 201-400m (8), 401-600m (3), 601-800m (1). **Flowering:** Feb (1), Mar (2), Apr (1), May (6), Jun (2), Aug (1), Oct (8), Nov (7), Dec (1). **Fruiting:** Aug (1), Dec (1). **Specimens:** - *Bambuko F.R.:* Watts J. 641. *Etinde:* Cable S. 168, 205, 220, 253; Cheek M. 5565; Mbani J. M. 20, 131; Mildbraed G. W. J. 10637; Watts J. 478, 515. *Mabeta-Moliwe:* Baker W. J. 321; Brodie S. 16; Cheek M. 5702; Plot Voucher W 114. *Mokoko River F.R.:* Akogo M. 176; Batoum A. 45; Ekema N. 846, 1031, 1046; Fraser P. 365, 443; Ndam N. 1231; Pouakouyou D. 77; Tchouto P. 631. *Onge:* Akogo M. 44; Cheek M. 4972, 5104; Harris D. J. 3706; Ndam N. 696, 791; Tchouto P. 859; Watts J. 929

Dorstenia mannii Hook.f. var. mungensis (Engl.) Hijman

Adansonia (ser.4) 3: 314 (1984); Fl. Gabon 26: 53 (1984); Fl. Cameroun 28: 54 (1985).
> *Dorstenia mungensis* Engl., F.W.T.A. 1: 597 (1958).

Habit: herb. **Habitat:** forest. **Distribution:** Nigeria, Cameroon and Gabon. **Chorology:** Lower Guinea. **Star:** blue. **Alt. range:** 1-200m (3), 1001-1200m (1). **Flowering:** May (2), Dec (1). **Specimens:** - *Eastern slopes:* Maitland T. D. 925. *Etinde:* Preuss P. R. 1107. *Mabeta-Moliwe:* Dunlap 255; Jaff B. 161; Sunderland T. C. H. 1020, 1402

Dorstenia cf. picta Bureau

Specimens: - *Etinde:* Thomas D. W. 9159

Dorstenia poinsettifolia Engl. var. angularis Hijman & C.C.Berg

Adansonia (ser.2) 16: 438 (1977); Fl. Cameroun 28: 87 (1985).
> *Dorstenia prorepens* Engl., F.W.T.A. 1: 597 (1958).

Habit: herb. **Habitat:** forest. **Distribution:** Nigeria, Bioko and SW Cameroon. **Chorology:** Lower Guinea. **Star:** blue. **Alt. range:** 200-400m (1), 1201-1400m (1). **Flowering:** Dec (1). **Specimens:** - *Etinde:* Cheek M. 5901. *Mokoko River F.R.:* Mbani J. M. 492. *No locality:* Mann G. 1956

Dorstenia poinsettifolia Engl. var. poinsettifolia

F.W.T.A. 1: 597 (1958); Fl. Gabon 26: 80 (1984); Fl. Cameroun 28: 80 (1985).

Habit: herb. **Habitat:** forest. **Distribution:** Cameroon and Gabon. **Chorology:** Lower Guinea. **Star:** blue. **Specimens:** - *Eastern slopes:* Deistel H. 435

Dorstenia poinsettifolia Engl. var. nov.

Specimens: - *Mabeta-Moliwe:* Sunderland 1060, Cheek M. 3413

Dorstenia psilurus Welw.

Fl. Gabon 26: 77 (1984); Fl. Cameroun 28: 74 (1985).
> *Dorstenia scabra* (Bureau) Engl., F.W.T.A. 1: 599 (1958).

> *Dorstenia tenuifolia* Engl., F.W.T.A. 1: 599 (1958).

Habit: herb. **Habitat:** forest. **Distribution:** Nigeria, Cameroon, Gabon, Congo (Kinshasa), Uganda and Zambia. **Chorology:** Guineo-Congolian. **Star:** green. **Alt. range:** 200-400m (1). **Flowering:** Dec (1). **Fruiting:** Dec (1). **Specimens:** - *Etinde:* Cheek M. 5878; Morton J. K. K 923

Dorstenia turbinata Engl.

F.W.T.A. 1: 599 (1958); Bull. Jard. Bot. Nat. Belg. 50: 341 (1980); Fl. Gabon 26: 38 (1984); Fl. Cameroun 28: 36 (1985).
> *Dorstenia buesgenii* Engl., F.W.T.A. 1: 599 (1958).
> *Dorstenia ledermannii* Engl., F.W.T.A. 1: 599 (1958).

Habit: shrub to 2 m. **Habitat:** forest. **Distribution:** Sierra Leone to Cameroon and Gabon. **Chorology:** Guineo-Congolian. **Star:** green. **Specimens:** - *Southern Bakundu F.R.:* Schlechter F. R. R. 12871

Dorstenia sp.

Specimens: - *Mabeta-Moliwe:* Tchouto P. 191

Ficus ardisioides Warb. subsp. camptoneura (Mildbr.) C.C.Berg

Fl. Cameroun 28: 238 (1985); Kew Bull. 43: 77 (1988); Keay R.W.J., Trees of Nigeria: 297 (1989); Hawthorne W., F.G.F.T. Ghana: 122 (1990); Kirkia 13(2): 271 (1990).
> *Ficus camptoneura* Mildbr., F.W.T.A. 1: 607 (1958).

Habit: tree to 10 m, or epiphytic shrub. **Guild:** strangler. **Habitat:** forest. **Distribution:** Ivory Coast, Nigeria, Bioko, Cameroon, Gabon, E. Congo (Kinshasa) and N. Zambia. **Chorology:** Guineo-Congolian. **Star:** green. **Alt. range:** 1-200m (2), 201-400m (2), 801-1000m (1), 1401-1600m (1). **Flowering:** Mar (1). **Fruiting:** Feb (1), Mar (1), Sep (1), Oct (2), Nov (1). **Specimens:** - *Eastern slopes:* Deistel H. s.n.; Tchouto P. 103. *Etinde:* Sunderland T. C. H. 1083; Tekwe C. 216. *Onge:* Tchouto P. 720, 865; Thomas D. W. 9930

Ficus asperifolia Miq.

F.W.T.A. 1: 606 (1958); Meded. Land. Wag. 82(3): 229 (1982); Fl. Gabon 26: 127 (1984); Fl. Cameroun 28: 124 (1985); Keay R.W.J., Trees of Nigeria: 293 (1989).
> *Ficus warburgii* Winkl., F.W.T.A. 1: 611 (1958).

Habit: scrambling shrub. **Guild:** pioneer. **Habitat:** forest. **Distribution:** Senegal to Bioko, Cameroon, Congo (Kinshasa), Angola, Zambia, Uganda and Tanzania. **Chorology:** Guineo-Congolian. **Star:** green. **Alt. range:** 1000-1200m (1). **Fruiting:** Apr (1). **Specimens:** - *Eastern slopes:* Kwangue A. 55. *Etinde:* Winkler H. J. P. 449

Ficus cf. camptoneura Mildbr.

Specimens: - *Etinde:* Tchouto P. 242

Ficus chlamydocarpa Mildbr. & Burret subsp. chlamydocarpa

F.W.T.A. 1: 608 (1958); Fl. Cameroun 28: 212 (1985); Kew Bull. 43: 80 (1988); Keay R.W.J., Trees of Nigeria: 298 (1989); Kirkia 13(2): 261 (1990).

Habit: tree to 35 m. **Habitat:** forest. **Distribution:** Bioko and Cameroon. **Chorology:** Lower Guinea (montane). **Star:** gold. **Alt.**

range: 1600-1800m (1). **Specimens: -** *No locality:* Maitland T. D. 843

Ficus conraui Warb.

F.W.T.A. 1: 607 (1958); Fl. Gabon 26: 222 (1984); Fl. Cameroun 28: 234 (1985); Keay R.W.J., Trees of Nigeria: 296 (1989); Hawthorne W., F.G.F.T. Ghana: 122 (1990); Kirkia 13(2): 270 (1990).
Ficus praticola Mildbr. & Hutch., F.W.T.A. 1: 607 (1958).

Habit: tree to 10 m, or epiphytic shrub. **Guild:** strangler. **Habitat:** forest. **Distribution:** Sierra Leone, Ivory Coast, Nigeria, Cameroon, Gabon, Congo (Kinshasa), CAR and Uganda. **Chorology:** Guineo-Congolian. **Star:** green. **Specimens: -** *Eastern slopes:* Reder H. 395 (partly)

Ficus cyathistipula Warb. subsp. **cyathistipula**

Fl. Gabon 26: 232 (1984); Fl. Cameroun 28: 246 (1985); Kew Bull. 43: 82 (1988); Hawthorne W., F.G.F.T. Ghana: 122 (1990); Kirkia 13(2): 271 (1990).
Ficus rederi Hutch., F.W.T.A. 1: 608 (1958).

Habit: tree to 10 m, or epiphytic shrub. **Guild:** strangler. **Habitat:** forest. **Distribution:** Ivory Coast to Angola, N. Zambia, Malawi and Tanzania. **Star:** green. **Alt. range:** 800-1000m (1).**Fruiting:** Jul (1). **Specimens: -** *Eastern slopes:* Reder H. 395; Tekwe C. 104. *Etinde:* Maitland T. D. 370

Ficus dryepondtiana Gentil ex De Wild.

F.W.T.A. 1: 611 (1958); Fl. Gabon 26: 206 (1984); Fl. Cameroun 28: 222 (1985); Fl. Cameroun 28: 222 (1985); Kirkia 13(2): 276 (1990).

Habit: tree to 30 m, or epiphytic shrub. **Guild:** strangler. **Habitat:** forest. **Distribution:** Cameroon, Gabon, Congo (Brazzaville), Congo (Kinshasa) and CAR. **Chorology:** Guineo-Congolian. **Star:** green. **Specimens: -** *Etinde:* Rosevear D. R. 64/37. *Mabeta-Moliwe:* Plot Voucher W 1191 a

Ficus exasperata Vahl

F.W.T.A. 1: 605 (1958); Fl. Gabon 26: 126 (1984); Fl. Cameroun 28: 121 (1985); Keay R.W.J., Trees of Nigeria: 291 (1989); Hawthorne W., F.G.F.T. Ghana: 155 (1990).

Habit: tree to 22 m. **Guild:** pioneer. **Habitat:** forest and farmbush. **Distribution:** tropical Africa to South India and Sri Lanka. **Star:** excluded. **Alt. range:** 1-200m (1).**Fruiting:** Jun (1), Aug (1). **Specimens: -** *Mabeta-Moliwe:* Plot Voucher W 1180, W 1207; von Rege I. 106

Ficus kamerunensis Mildbr. & Burret

F.W.T.A. 1: 608 (1958); Fl. Gabon 26: 165 (1984); Fl. Cameroun 28: 179 (1985); Keay R.W.J., Trees of Nigeria: 296 (1989); Hawthorne W., F.G.F.T. Ghana: 122 (1990); Kirkia 13(2): 269 (1990).

Habit: tree to 30 m, or epiphytic shrub. **Guild:** strangler. **Habitat:** forest. **Distribution:** Sierra Leone to Cameroon, Gabon and Congo (Kinshasa). **Chorology:** Guineo-Congolian. **Star:** green. **Specimens: -** *Etinde:* Winkler H. J. P. 1091

Ficus lutea Vahl

Kew Bull. 36: 597 (1981); Fl. Gabon 26: 193 (1984); Fl. Cameroun 28: 206 (1985); Keay R.W.J., Trees of Nigeria: 297 (1989); Kirkia 13(2): 261 (1990).
Ficus vogelii (Miq.) Miq., F.W.T.A. 1: 609 (1958).

88

Habit: tree to 30 m, or epiphytic shrub. **Guild:** strangler. **Habitat:** forest. **Distribution:** tropical and subtropical Africa. **Star:** excluded. **Alt. range:** 1-200m (2).**Fruiting:** Oct (1), Dec (1). **Specimens: -** *Etinde:* Banks H. 46; Brodie S. 13; Cheek M. 5944; Maitland T. D. 637; Rosevear D. R. 63/37. *Mabeta-Moliwe:* Plot Voucher W 1212

Ficus sp. aff. lutea Vahl

Specimens: - *Mabeta-Moliwe:* Cheek M. 3512, Plot Voucher W 1184

Ficus lyrata Warb.

F.W.T.A. 1: 607 (1958); Fl. Cameroun 28: 254 (1985); Keay R.W.J., Trees of Nigeria: 296 (1989); Hawthorne W., F.G.F.T. Ghana: 121 (1990); Kirkia 13(2): 273 (1990).

Habit: tree to 15 m, or epiphytic shrub. **Guild:** strangler. **Habitat:** forest and lava flows. **Distribution:** Sierra Leone to Cameroon. **Chorology:** Guineo-Congolian. **Star:** green. **Alt. range:** 1-200m (1).**Fruiting:** Jul (1). **Specimens: -** *Etinde:* Rosevear D. R. Cam 76/37. *Mabeta-Moliwe:* Cheek M. 3573

Ficus mucuso Welw. ex Ficalho

F.W.T.A. 1: 606 (1958); Fl. Gabon 26: 131 (1984); Fl. Cameroun 28: 132 (1985); Keay R.W.J., Trees of Nigeria: 291 (1989); Hawthorne W., F.G.F.T. Ghana: 118 (1990); Kirkia 13(2): 256 (1990).

Habit: tree to 40 m. **Guild:** pioneer. **Habitat:** forest and Aframomum thicket. **Distribution:** Guinea Bissau to Bioko, Cameroon, Gabon, Congo (Kinshasa), Angola, Uganda and Tanzania. **Chorology:** Guineo-Congolian. **Star:** green. **Alt. range:** 800-1000m (1). **Specimens: -** *Etinde:* Winkler H. J. P. 24. *Mabeta-Moliwe:* Mildbraed G. W. J. 10791; Plot Voucher W 1198

Ficus natalensis Hochst. subsp. **leprieurii** (Miq.) C.C.Berg

Meded. Land. Wag. 82(3): 235 (1982); Fl. Gabon 26: 173 (1984); Fl. Cameroun 28: 188 (1985); Kew Bull. 43: 87 (1988); Keay R.W.J., Trees of Nigeria: 296 (1989); Kirkia 13(2): 267 (1990).
Ficus leprieuri Miq., F.W.T.A. 1: 608 (1958); Hawthorne W., F.G.F.T. Ghana: 121 (1990).

Habit: tree to 30 m, or epiphytic shrub. **Guild:** strangler. **Habitat:** lava flows. **Distribution:** Senegal to Congo (Kinshasa), Angola, CAR and NW Zambia. **Chorology:** Guineo-Congolian. **Star:** green. **Alt. range:** 1-200m (1).**Fruiting:** Feb (1). **Specimens: -** *Etinde:* Rosevear D. R. 66/37; Tekwe C. 22

Ficus oreodryadum Mildbr.

Fl. Cameroun 28: 200 (1985); Kirkia 13(2): 269 (1990).

Habit: tree to 30 m, or epiphytic shrub. **Guild:** strangler. **Habitat:** forest. **Distribution:** SW Cameroon, Bioko, E. Congo (Kinshasa), Burundi and Rwanda. **Chorology:** Guineo-Congolian (montane). **Star:** green. **Alt. range:** 1400-1600m (1). **Specimens: -** *Eastern slopes:* Thomas D. W. 9481

Ficus ottoniifolia (Miq.) Miq.

F.W.T.A. 1: 611 (1958); Fl. Gabon 26: 201 (1984); Hawthorne W., F.G.F.T. Ghana: 122 (1990); Kirkia 13(2): 273 (1990).

Habit: tree to 15 m. **Habitat:** forest. **Distribution:** tropical Africa. **Star:** excluded. **Specimens: -** *Mabeta-Moliwe:* Plot Voucher W 1028

Ficus ovata Vahl
F.W.T.A. 1: 608 (1958); Meded. Land. Wag. 82(3): 236 (1982); Fl.
Gabon 26: 216 (1984); Fl. Cameroun 28: 230 (1985); Keay R.W.J.,
Trees of Nigeria: 294 (1989); Hawthorne W., F.G.F.T. Ghana: 121
(1990); Kirkia 13(2): 277 (1990).

Habit: tree to 30 m, or epiphytic shrub. **Guild:** strangler. **Habitat:**
forest. **Distribution:** tropical Africa. **Star:** excluded. **Specimens:** -
Etinde: Maitland T. D. 697

Ficus polita Vahl subsp. **polita**
F.W.T.A. 1: 611 (1958); Meded. Land. Wag. 82(3): 237 (1982); Fl.
Gabon 26: 214 (1984); Fl. Cameroun 28: 227 (1985); Kew Bull.
43: 93 (1988); Keay R.W.J., Trees of Nigeria: 294 (1989);
Hawthorne W., F.G.F.T. Ghana: 121 (1990); Kirkia 13(2): 275
(1990).

Habit: tree to 20 m. **Habitat:** forest. **Distribution:** Senegal to
Angola, Mozambique, Uganda, South Africa and Madagascar.
Star: excluded. **Alt. range:** 1-200m (1).**Fruiting:** Jul (1).
Specimens: - *Etinde:* Winkler H. J. P. 428. *Mabeta-Moliwe:*
Cheek M. 3572; Plot Voucher W 1189 b

Ficus saussureana DC.
Fl. Gabon 26: 113 (1984); Fl. Cameroun 28: 209 (1985); Keay
R.W.J., Trees of Nigeria: 296 (1989); Hawthorne W., F.G.F.T.
Ghana: 121 (1990); Kirkia 13(2): 261 (1990).
Ficus eriobotryoides Kunth & Bouché var. *eriobotryoides*,
F.W.T.A. 1: 608 (1958).

Habit: tree to 30 m, or epiphytic shrub. **Guild:** strangler. **Habitat:**
forest and forest-grassland transition. **Distribution:** Guinea to
Bioko and Cameroon. **Chorology:** Guineo-Congolian (montane).
Star: green. **Alt. range:** 1-200m (1), 601-800m (1), 1401-1600m
(1), 1601-1800m (1). **Flowering:** Nov (1). **Fruiting:** Nov (1).
Specimens: - *Eastern slopes:* Hutchinson J. & Metcalfe C. R. 110;
Thomas D. W. 9487. *Etinde:* Maitland T. D. 1081; Thomas D. W.
9217. *Onge:* Watts J. 906

Ficus sur Forssk.
F.W.T.A. 1: 606 (1958); Meded. Land. Wag. 82(3): 238 (1982); Fl.
Gabon 26: 134 (1984); Fl. Cameroun 28: 135 (1985); Keay R.W.J.,
Trees of Nigeria: 291 (1989); Hawthorne W., F.G.F.T. Ghana: 118
(1990); Kirkia 13(2): 256 (1990).
Ficus capensis Thunb., F.W.T.A. 1: 606 (1958).

Habit: tree to 25 m. **Guild:** pioneer. **Habitat:** forest and farmbush.
Distribution: tropical and subtropical Africa. **Star:** excluded. **Alt.
range:** 1-200m (2). **Flowering:** Jul (1). **Fruiting:** Jul (1), Oct (1).
Specimens: - *Eastern slopes:* Maitland T. D. 889. *Etinde:* Dawson
S. 34; Tchouto P. 246. *Mabeta-Moliwe:* Plot Voucher W 1201

Ficus cf. sur Forssk.

Specimens: - *Onge:* Harris D. J. 3848

Ficus tesselata Warb.
F.W.T.A. 1: 607 (1958); F.W.T.A. 1: 607 (1958); Kew Bull. 43: 83
(1988); Keay R.W.J., Trees of Nigeria: 297 (1989); Keay R.W.J.,
Trees of Nigeria: 297 (1989); Kirkia 13(2): 256 (1990).
Ficus winkleri Mildbr. & Burret, F.W.T.A. 1: 607 (1958).
Ficus camptoneuroides Hutch., F.W.T.A. 1: 607 (1958).
Habit: tree to 10 m, or epiphytic shrub. **Guild:** strangler. **Habitat:**
forest. **Distribution:** Sierra Leone to Bioko, Congo (Kinshasa) and
Rwanda. **Chorology:** Guineo-Congolian. **Star:** green. **Specimens:**
- *Etinde:* Winkler H. J. P. 1204

Ficus thonningii Blume
F.W.T.A. 1: 610 (1958); Meded. Land. Wag. 82(3): 239 (1982); Fl.
Gabon 26: 161 (1984); Fl. Cameroun 28: 175 (1985); Keay R.W.J.,
Trees of Nigeria: 297 (1989); Hawthorne W., F.G.F.T. Ghana: 115
(1990); Kirkia 13(2): 268 (1990).
Ficus dekdekena (Miq.) A.Rich., F.W.T.A. 1: 610 (1958).
Ficus iteophylla Miq., F.W.T.A. 1: 610 (1958).

Habit: tree to 20 m. **Habitat:** forest. **Distribution:** tropical and
subtropical Africa. **Chorology:** Afromontane. **Star:** green. **Alt.
range:** 1600-1800m (1). **Specimens:** - *Eastern slopes:* Thomas D.
W. 9493

Ficus vogeliana (Miq.) Miq.
F.W.T.A. 1: 606 (1958); Fl. Gabon 26: 138 (1984); Fl. Cameroun
28: 140 (1985); Keay R.W.J., Trees of Nigeria: 293 (1989);
Hawthorne W., F.G.F.T. Ghana: 118 (1990); Kirkia 13(2): 257
(1990).

Habit: tree to 30 m. **Habitat:** forest and stream banks.
Distribution: Guinea to Bioko, Cameroon, Gabon, Congo
(Brazzaville), Congo (Kinshasa) and Angola (Cabinda).
Chorology: Guineo-Congolian. **Star:** green. **Alt. range:** 1-200m
(15), 201-400m (1). **Flowering:** Jul (1). **Fruiting:** Mar (1), Apr
(4), May (2), Jun (1), Jul (1), Aug (1), Oct (1), Nov (3), Dec (2).
Specimens: - *Mabeta-Moliwe:* Cable S. 545, 610; Cheek M. 3434;
Mann G. 702; Ndam N. 478; Plot Voucher W 707; Sunderland T.
C. H. 1237, 1446; Watts J. 242; Wheatley J. I. 182; von Rege I. 24,
78. *Mokoko River F.R.:* Acworth J. M. 88; Ekema N. 836. *Onge:*
Harris D. J. 3768; Ndam N. 792; Thomas D. W. 9827; Watts J.
1007; Wheatley J. I. 767

Milicia excelsa (Welw.) C.C.Berg
Bull. Jard. Bot. Nat. Belg. 52: 227 (1982); Fl. Gabon 26: 6 (1984);
Fl. Cameroun 28: 9 (1985); Keay R.W.J., Trees of Nigeria: 298
(1989); Hawthorne W., F.G.F.T. Ghana: 118 (1990).
Chlorophora excelsa (Welw.) Benth., F.W.T.A. 1: 595 (1958).

Habit: tree to 50 m. **Guild:** pioneer. **Habitat:** forest. **Distribution:**
tropical Africa. **Star:** scarlet. **Alt. range:** 200-400m (1).
Specimens: - *Mabeta-Moliwe:* Mann G. 205. *Mokoko River F.R.:*
Thomas D. W. 10076

Treculia acuminata Baill.
Bull. Jard. Bot. Nat. Belg. 47: 396 (1977); Fl. Gabon 26: 17
(1984); Fl. Cameroun 28: 19 (1985).
Treculia zenkeri Engl., F.W.T.A. 1: 613 (1958).

Habit: shrub to 3 m. **Habitat:** forest. **Distribution:** SE Nigeria,
Cameroon and Gabon. **Chorology:** Lower Guinea. **Star:** blue. **Alt.
range:** 1-200m (1). **Flowering:** May (1). **Specimens:** - *Mabeta-
Moliwe:* Plot Voucher W 105; Sunderland T. C. H. 1399. *Southern
Bakundu F.R.:* Onochie C. F. A. FHI 32058

Treculia africana Decne. subsp. **africana** var.
africana
Bull. Jard. Bot. Nat. Belg. 47: 384 (1977); Fl. Cameroun 28: 16
(1985); Keay R.W.J., Trees of Nigeria: 300 (1989).
Treculia africana Decne. var. *africana*, F.W.T.A. 1: 613 (1958).

Habit: tree to 30 m. **Habitat:** forest. **Distribution:** tropical Africa.
Star: excluded. **Fruiting:** Feb (1). **Specimens:** - *Mabeta-Moliwe:*
Mann G. 773

Treculia obovoidea N.E.Br.

F.W.T.A. 1: 613 (1958); Bull. Jard. Bot. Nat. Belg. 47: 399 (1977);
Fl. Gabon 26: 18 (1984); Fl. Cameroun 28: 20 (1985); Keay
R.W.J., Trees of Nigeria: 302 (1989).

Habit: tree to 20 m. **Habitat:** forest and forest gaps. **Distribution:**
SE Nigeria, Cameroon, Congo (Brazzaville), Congo (Kinshasa)
and Angola (Cabinda). **Chorology:** Guineo-Congolian. **Star:**
green. **Alt. range:** 1-200m (19), 201-400m (3), 401-600m (1).
Flowering: Apr (7). **Fruiting:** Mar (1), Apr (10), May (6), Jun (1).
Specimens: - *Bambuko F.R.:* Watts J. 656. *Etinde:* Kwangue A.
71, 76; Mbani J. M. 161. *Mabeta-Moliwe:* Jaff B. 35, 46;
Sunderland T. C. H. 1251, 1260, 1368, 1412; Tchouto P. 157, 159;
Watts J. 187; Wheatley J. I. 223; Winkler H. J. P. 1283. *Mokoko
River F.R.:* Acworth J. M. 137; Akogo M. 259, 260; Ekema N.
1127a; Fraser P. 439; Mbani J. M. 304; Tchouto P. 1083, 1088;
Watts J. 1059. *Onge:* Tchouto P. 961. *Southern Bakundu F.R.:*
Brenan J. P. M. 9272

Trilepisium madagascariense DC.

Bull. Jard. Bot. Nat. Belg. 47: 299 (1977); Fl. Cameroun 28: 103
(1985); Keay R.W.J., Trees of Nigeria: 302 (1989).
   Bosqueia angolensis Ficalho, F.W.T.A. 1: 612 (1958);
Hawthorne W., F.G.F.T. Ghana: 117 (1990).

Habit: tree to 25 m. **Guild:** light demander. **Habitat:** forest.
Distribution: tropical Africa. **Star:** excluded. **Specimens:** -
Etinde: Maitland T. D. 186. *Mabeta-Moliwe:* Dunlap 247; Plot
Voucher W 773, W 1205

MYRICACEAE

Myrica arborea Hutch.

F.W.T.A. 1: 589 (1958).

Habit: tree to 10 m. **Habitat:** forest and forest-grassland
transition. **Distribution:** Bioko to Bamenda Highlands.
Chorology: Western Cameroon Uplands (montane). **Star:** black.
Alt. range: 1800-2000m (1). **Specimens:** - *No locality:* Mann G.
1203

MYRISTICACEAE

Coelocaryon preussii Warb.

F.W.T.A. 1: 61 (1954); Fl. Gabon 10: 94 (1965); Fl. Cameroun 18:
98 (1974); Keay R.W.J., Trees of Nigeria: 38 (1989).

Habit: tree to 30 m. **Habitat:** forest. **Distribution:** Nigeria,
Cameroon, Gabon and Congo (Brazzaville). **Chorology:** Guineo-
Congolian. **Star:** pink. **Alt. range:** 1-200m (3), 201-400m (2).
Flowering: May (1), Jun (1). **Fruiting:** Mar (1), Apr (1), May (2).
Specimens: - *Etinde:* Preuss P. R. 331. *Mabeta-Moliwe:* Jaff B.
198; Plot Voucher W 909, W 980, W 1178; Sunderland T. C. H.
1141; Wheatley J. I. 107, 356. *Mokoko River F.R.:* Thomas D. W.
10068

Pycnanthus angolensis (Welw.) Warb.

F.W.T.A. 1: 61 (1954); Fl. Gabon 10: 87 (1965); Fl. Cameroun 18:
91 (1974); Keay R.W.J., Trees of Nigeria: 36 (1989); Hawthorne
W., F.G.F.T. Ghana: 83 (1990).
   Pycnanthus microcephalus (Benth.) Warb., F.W.T.A. 1: 61
(1954).

Habit: tree to 35 m. **Guild:** light demander. **Habitat:** forest.
Distribution: Guinea to Angola and Uganda. **Chorology:** Guineo-
Congolian. **Star:** pink. **Alt. range:** 1-200m (1), 601-800m
(1).**Fruiting:** Apr (1). **Specimens:** - *Mabeta-Moliwe:* Mildbraed
G. W. J. 10699. *Mokoko River F.R.:* Akogo M. 256

Scyphocephalium mannii (Benth.) Warb.

F.W.T.A. 1: 61 (1954); Fl. Gabon 10: 89 (1965); Fl. Cameroun 18:
94 (1974); Keay R.W.J., Trees of Nigeria: 38 (1989).

Habit: tree to 30 m. **Habitat:** forest. **Distribution:** SE Nigeria,
Cameroon and Gabon. **Chorology:** Lower Guinea. **Star:** blue. **Alt.
range:** 1-200m (2), 201-400m (1). **Flowering:** Mar (1), Apr (1).
Fruiting: May (1). **Specimens:** - *Etinde:* Maitland T. D. 403.
Mabeta-Moliwe: Ndam N. 479; Plot Voucher W 686; Sunderland
T. C. H. 1209; Wheatley J. I. 61. *Southern Bakundu F.R.:* Brenan
J. P. M. 9491

Staudtia kamerunensis Warb. var. gabonensis (Warb.) Fouilloy

Fl. Cameroun 18: 104 (1974); Keay R.W.J., Trees of Nigeria: 38
(1989); Fl. Zamb. 9(2): 40 (1997).
   Staudtia stipitata Warb., F.W.T.A. 1: 62 (1954).
   Staudtia gabonensis Warb., Fl. Gabon 10: 99 (1965).

Habit: tree to 40 m. **Habitat:** forest. **Distribution:** Nigeria,
Cameroon, Gabon, Congo (Kinshasa) and Angola (Cabinda).
Chorology: Guineo-Congolian. **Star:** pink. **Specimens:** - *Mabeta-
Moliwe:* Plot Voucher W 384, W 554

Staudtia kamerunensis Warb. var. kamerunensis

F.W.T.A. 1: 62 (1954); Fl. Cameroun 18: 103 (1974).

Habit: tree to 40 m. **Habitat:** forest. **Distribution:** Cameroon,
Gabon and Congo (Kinshasa). **Chorology:** Guineo-Congolian.
Star: green. **Specimens:** - *Etinde:* Winkler H. J. P. 1090, 1335

MYRSINACEAE

Ardisia dolichocalyx Taton

Bull. Jard. Bot. Nat. Belg. 49: 98 (1979).

Habit: shrub to 3 m. **Habitat:** forest. **Distribution:** Cameroon.
Chorology: Lower Guinea (montane). **Star:** gold. **Alt. range:**
1000-1200m (1), 1201-1400m (1). **Flowering:** Jul (1), Oct (1).
Fruiting: Jul (1), Oct (1). **Specimens:** - *Etinde:* Tchouto P. 257,
398

Ardisia etindensis Taton

Bull. Jard. Bot. Nat. Belg. 49: 100 (1979).

Habit: shrub to 1 m. **Distribution:** Cameroon. **Chorology:** Lower
Guinea (montane). **Star:** black. **Alt. range:** 400-600m (1).
Flowering: May (1). **Specimens:** - *Etinde:* Letouzey R. 14962

Ardisia oligantha (Gilg & Schellenb.) Taton

Bull. Jard. Bot. Nat. Belg. 49: 109 (1979).
   Afrardisia oligantha Gilg & Schellenb., F.W.T.A. 2: 31 (1963).

Habit: shrub? **Distribution:** Mount Cameroon. **Chorology:** Endemic. **Star:** black. **Specimens:** - *No locality:* Weberbauer A. 48

Ardisia schlechteri Gilg
Bull. Jard. Bot. Nat. Belg. 49: 112 (1979).
Afrardisia schlechteri (Gilg) Mez, F.W.T.A. 2: 31 (1963).

Habit: shrub to 0.5 m. **Habitat:** forest. **Distribution:** Mount Cameroon. **Chorology:** Endemic. **Star:** black. **Specimens:** - *Etinde:* Schlechter F. R. R. 12417

Ardisia staudtii Gilg
Bull. Jard. Bot. Nat. Belg. 49: 112 (1979).
Afrardisia staudtii (Gilg) Mez, F.W.T.A. 2: 31 (1963).
Afrardisia cymosa (Bak.) Mez, F.W.T.A. 2: 31 (1963).

Habit: shrub to 2 m. **Habitat:** forest. **Distribution:** Nigeria, Bioko, Cameroon, Gabon, Congo (Brazzaville), Congo (Kinshasa) and CAR. **Chorology:** Guineo-Congolian (montane). **Star:** green. **Alt. range:** 1-200m (4), 201-400m (1), 1201-1400m (1). **Flowering:** Jun (2), Sep (1). **Fruiting:** Apr (1), May (3). **Specimens:** - *Etinde:* Nkeng P. 74; Tekwe C. 180. *Mabeta-Moliwe:* Jaff B. 152; Ndam N. 497; Plot Voucher W 372, W 792; Watts J. 123; Wheatley J. I. 279

Ardisia cf. staudtii Gilg

Specimens: - *Etinde:* Tchouto P. 139; Watts J. 487; Wheatley J. I. 527. *Mabeta-Moliwe:* Sunderland T. C. H. 1352; Watts J. 281; Wheatley J. I. 278

Ardisia cf. zenkeri Metz

Specimens: - *Mabeta-Moliwe:* Cheek M. 3253

Ardisia sp.

Specimens: - *Mokoko River F.R.:* Mbani J. M. 345a

Embelia sp. aff. guineensis Baker

Specimens: - *Etinde:* Thomas D. W. 9544

Embelia mildbraedii Gilg & Schellenb.
Bull. Jard. Bot. Nat. Belg. 50: 203 (1980).
Embelia sp. aff. welwitschii (Hiern) K.Schum., F.W.T.A. 2: 32 (1963).

Habit: woody climber. **Habitat:** forest. **Distribution:** SW Cameroon. **Chorology:** Western Cameroon. **Star:** black. **Specimens:** - *Eastern slopes:* Maitland T. D. 692

Embelia cf. mildbraedii Gilg & Schellenb.

Specimens: - *Eastern slopes:* Tchouto P. 377; Tekwe C. 320; Thomas D. W. 9458. *Etinde:* Tekwe C. 189

Maesa kamerunensis Mez
F.W.T.A. 2: 32 (1963); Bull. Jard. Bot. Nat. Belg. 50: 208 (1980).

Habit: shrub to 6 m. **Habitat:** forest. **Distribution:** Bioko, Cameroon, S. Tomé, Gabon and Congo (Brazzaville). **Chorology:**

Lower Guinea. **Star:** blue. **Alt. range:** 600-800m (1), 801-1000m (1).**Fruiting:** Mar (1). **Specimens:** - *Eastern slopes:* Tchouto P. 51. *No locality:* Mann G. 1198

Maesa lanceolata Forssk.
F.W.T.A. 2: 33 (1963); Bull. Jard. Bot. Nat. Belg. 50: 208 (1980); Meded. Land. Wag. 82(3): 243 (1982); Keay R.W.J., Trees of Nigeria: 399 (1989).

Habit: tree to 6 m. **Habitat:** forest-grassland transition. **Distribution:** tropical and subtropical Africa. **Chorology:** Afromontane. **Star:** green. **Alt. range:** 800-1000m (2), 1601-1800m (3), 1801-2000m (2), 2001-2200m (1). **Flowering:** May (1), Sep (1). **Fruiting:** Jul (1), Aug (1), Sep (4). **Specimens:** - *Eastern slopes:* Sunderland T. C. H. 1434; Tekwe C. 114, 140. *Etinde:* Tekwe C. 237, 260; Wheatley J. I. 573, 577b. *No locality:* Mann G. 1208

Rapanea melanophloeos (L.) Mez
Bull. Jard. Bot. Nat. Belg. 50: 222 (1980); Keay R.W.J., Trees of Nigeria: 399 (1989).
Rapanea neurophylla (Gilg) Mez, F.W.T.A. 2: 31 (1963).

Habit: tree or shrub to 15 m. **Habitat:** forest. **Distribution:** W. Cameroon to East and southern Africa. **Chorology:** Afromontane. **Star:** green. **Alt. range:** 1600-1800m (2), 2001-2200m (1). **Flowering:** Oct (1). **Fruiting:** Sep (1), Oct (1). **Specimens:** - *Eastern slopes:* Tchouto P. 301. *Etinde:* Tekwe C. 233. *No locality:* Mann G. 1200

MYRTACEAE

Eugenia kalbreyeri Engl. & Brehmer
F.W.T.A. 1: 238 (1954); Fl. Gabon 11: 22 (1966).

Habit: shrub to 3 m. **Guild:** shade-bearer. **Distribution:** Sierra Leone and Mount Cameroon. **Chorology:** Guineo-Congolian. **Star:** black. **Alt. range:** 600-800m (1). **Specimens:** - *Etinde:* Kalbreyer W. 157

Eugenia kameruniana Engl.
F.W.T.A. 1: 238 (1954).

Habit: shrub? **Distribution:** Mount Cameroon. **Chorology:** Endemic. **Star:** black. **Specimens:** - *Etinde:* Winkler H. J. P. 1110

Eugenia sp. nov? 1 aff. gilgii

Specimens: - *Mabeta-Moliwe:* Sunderland 1344; Wheatley 312, 330

Eugenia sp. nov? 2 aff. pobeguinii

Specimens: - *Mabeta-Moliwe:* Jaff 82; Watts 122; Wheatley 162

Eugenia sp. nov? 3 aff. zenkeri

Specimens: - *Mabeta-Moliwe:* Cheek M. 3515; Nguembock F. 29

Syzygium guineense (Willd.) DC

Specimens: - *Mabeta-Moliwe:* Cheek 3464

Syzygium staudtii (Engl.) Mildbr.

F.W.T.A. 1: 240 (1954); Acta. Bot. Neerl. 9: 404 (1960); Fl. Gabon 11: 8 (1966); Keay R.W.J., Trees of Nigeria: 76 (1989).

Habit: tree to 30 m. **Habitat:** forest-grassland transition. **Distribution:** Liberia to Bioko and Cameroon. **Chorology:** Guineo-Congolian (montane). **Star:** green. **Alt. range:** 2000-2200m (1). **Specimens:** - *No locality:* Maitland T. D. 987

OCHNACEAE

Campylospermum calanthum (Gilg) Farron

Bull. Jard. Bot. Brux. 35: 394 (1965).
Ouratea calantha Gilg, F.W.T.A. 1: 228 (1954); Keay R.W.J., Trees of Nigeria: 72 (1989); Hawthorne W., F.G.F.T. Ghana: 97 (1990).
Ouratea nigroviolacea Gilg ex De Wild., F.W.T.A. 1: 228 (1954).

Habit: tree or shrub to 5 m. **Guild:** shade-bearer. **Habitat:** forest. **Distribution:** Nigeria, Cameroon and Angola. **Chorology:** Lower Guinea. **Star:** blue. **Specimens:** - *Southern Bakundu F.R.:* Brenan J. P. M. 9303

Campylospermum elongatum (Oliv.) Tiegh.

Bull. Jard. Bot. Brux. 35: 396 (1965); Bot. Helv. 95: 67 (1985).
Ouratea elongata (Oliv.) Engl., F.W.T.A. 1: 226 (1954).

Habit: tree or shrub to 6 m. **Habitat:** forest. **Distribution:** Nigeria, Cameroon, Bioko, Gabon and Congo (Kinshasa). **Chorology:** Guineo-Congolian. **Star:** green. **Alt. range:** 1-200m (4), 201-400m (2). **Flowering:** Jun (3). **Fruiting:** Mar (2), Jun (1), Nov (1). **Specimens:** - *Bambuko F.R.:* Tchouto P. 553; Watts J. 647. *Mabeta-Moliwe:* Cheek M. 3246; Dunlap 92; Plot Voucher W 1018, W 1034; Sunderland T. C. H. 1461; Watts J. 411; Wheatley J. I. 327. *Onge:* Tchouto P. 1032. *Southern Bakundu F.R.:* Brenan J. P. M. 9271

Campylospermum flavum (Schum. & Thonn.) Farron

Bull. Jard. Bot. Brux. 35: 397 (1965).
Ouratea flava (Schum. & Thonn.) Hutch. & Dalziel, F.W.T.A. 1: 228 (1954); Hawthorne W., F.G.F.T. Ghana: 97 (1990).

Habit: tree or shrub to 4 m. **Habitat:** forest. **Distribution:** Guinea to Bioko, Cameroon and CAR. **Chorology:** Guineo-Congolian. **Star:** green. **Specimens:** - *Eastern slopes:* Maitland T. D. 298. *Mabeta-Moliwe:* Nguembock F. 63, 90; Sunderland 1111, 1121, 1171

Campylospermum letouzeyi Farron

Adansonia (ser.2) 9: 117 (1969).
Habit: tree or shrub to 4 m. **Habitat:** forest. **Distribution:** Cameroon. **Chorology:** Lower Guinea. **Star:** gold. **Alt. range:** 1-200m (1), 201-400m (2). **Flowering:** Mar (1). **Fruiting:** Mar (3). **Specimens:** - *Bambuko F.R.:* Tchouto P. 551; Watts J. 618. *Onge:* Wheatley J. I. 826

Campylospermum mannii (Oliv.) Tiegh.

Bot. Helv. 95: 68 (1985).
Ouratea mannii (Oliv.) Engl., F.W.T.A. 1: 226 (1954).
Habit: shrub to 3 m. **Habitat:** forest. **Distribution:** SE Nigeria, Bioko and SW Cameroon. **Chorology:** Western Cameroon. **Star:**

gold. **Alt. range:** 1-200m (6). **Flowering:** Feb (2). **Fruiting:** Feb (2), Mar (3), Aug (1). **Specimens:** - *Etinde:* Mbani J. M. 26; Sunderland T. C. H. 1077; Tekwe C. 125. *Mabeta-Moliwe:* Nguembock F. 59; Plot Voucher W 192, W 775, W 974; Sunderland T. C. H. 1100; Wheatley J. I. 69. *Onge:* Mbani J. M. 31

Campylospermum monticola (Gilg) Cheek comb. nov.

Basionym: *Ouratea monticola* Gilg in Engl. Jahrb. 33: 272 (1903).
Type: Cameroon, Buea, Deistel 283 (isosyntype, K!)

Specimens: - *Eastern Slopes:* Diestel 283; Lehmbach 111; Preuss 760, 829.

Campylospermum reticulatum (P.Beauv.) Farron var. reticulatum

Bull. Jard. Bot. Brux. 35: 400 (1965).
Ouratea reticulata (P.Beauv.) Engl., F.W.T.A. 1: 228 (1954).

Habit: tree or shrub to 4 m. **Guild:** shade-bearer. **Habitat:** forest. **Distribution:** Guinea to Cameroon and Gabon. **Chorology:** Guineo-Congolian. **Star:** green. **Specimens:** - *Etinde:* Kalbreyer W. 97. *Mabeta-Moliwe:* Plot Voucher W 19, W 913

Campylospermum sp. aff. reticulatum (P.Beauv.) Farron

Specimens: - *Etinde:* Kwangue A. 86; Mbani J. M. 85

Campylospermum subcordatum (Stapf) Farron

Bull. Jard. Bot. Brux. 35: 402 (1965).
Ouratea subcordata (Stapf) Engl., F.W.T.A. 1: 226 (1954).

Habit: shrub to 2.5 m. **Habitat:** forest. **Distribution:** Ivory Coast and Ghana. **Chorology:** Guineo-Congolian. **Star:** gold. **Alt. range:** 1-200m (1). **Flowering:** Dec (1). **Specimens:** - *Mabeta-Moliwe:* Plot Voucher W 149, W 783, W 922, W 995; Sunderland T. C. H. 1010

Campylospermum sulcatum (Tiegh.) Farron

Bull. Jard. Bot. Brux. 35: 403 (1965).
Ouratea sulcata (Tiegh.) Keay, F.W.T.A. 1: 229 (1954).

Habit: tree or shrub to 4 m. **Guild:** shade-bearer. **Habitat:** forest. **Distribution:** Ivory Coast to Cameroon, Gabon and Congo (Brazzaville). **Chorology:** Guineo-Congolian. **Star:** green. **Alt. range:** 1-200m (2). **Flowering:** Nov (2). **Specimens:** - *Onge:* Tchouto P. 1035; Watts J. 986. *Southern Bakundu F.R.:* Brenan J. P. M. 9268

Campylospermum zenkeri (Engl. ex Tiegh.) Farron

Bot. Helv. 95: 68 (1985).
Ouratea ambacensis Hutch. & Dalziel, F.W.T.A. 1: 229 (1954).

Habit: tree or shrub to 4 m. **Habitat:** forest. **Distribution:** Mount Cameroon. **Chorology:** Endemic. **Star:** black. **Alt. range:** 1-200m (1), 601-800m (1). **Flowering:** Apr (1). **Specimens:** - *Mabeta-Moliwe:* Dunlap 171; Mann G. 12; Mildbraed G. W. J. 10520; Watts J. 214

Campylospermum sp. C

Ouratea sp. C sensu Keay F.W.T.A. 1: 231 (1954).

Habit: tree or shrub to 4 m. **Distribution:** Mount Cameroon. **Chorology:** Endemic. **Star:** black. **Specimens:** - *Etinde:* Brenan J. P. M. 9589

Idertia axillaris (Oliv.) Farron

Ber. Schweiz. Bot. Ges. 73: 212 (1963); Bot. Helv. 95: 66 (1985). *Ouratea axillaris* (Oliv.) Engl., F.W.T.A. 1: 228 (1954).

Habit: shrub to 3 m. **Habitat:** forest. **Distribution:** Nigeria, Cameroon and Gabon. **Chorology:** Lower Guinea. **Star:** blue. **Specimens:** - *Southern Bakundu F.R.:* Olorunfemi J. FHI 30519

Lophira alata Banks ex Gaertn.f.

F.W.T.A. 1: 231 (1954); Keay R.W.J., Trees of Nigeria: 74 (1989); Hawthorne W., F.G.F.T. Ghana: 97 (1990).

Habit: tree to 40 m. **Guild:** pioneer. **Habitat:** forest. **Distribution:** Sierra Leone to Cameroon, Rio Muni, Gabon and Congo (Kinshasa). **Chorology:** Guineo-Congolian. **Star:** red. **Alt. range:** 600-800m (1). **Specimens:** - *Etinde:* Maitland T. D. 793. *Mabeta-Moliwe:* Mann G. 709; Mildbraed G. W. J. 10711; Plot Voucher W 720

Rhabdophyllum affine (Hook.f.) Tiegh. subsp. pauciflorum (Gilg) Farron

Bull. Jard. Bot. Brux. 35: 390 (1965). *Ouratea pauciflora* Gilg, F.W.T.A. 1: 229 (1954).

Habit: tree or shrub to 6 m. **Habitat:** forest. **Distribution:** Nigeria and SW Cameroon. **Chorology:** Lower Guinea. **Star:** gold. **Alt. range:** 600-800m (1). **Specimens:** - *Mabeta-Moliwe:* Mildbraed G. W. J. 10532

Rhabdophyllum cf. affine (Hook.f.) Tiegh. subsp. myrioneurum (Gilg) Farron

Specimens:- *Mabeta-Moliwe:* Jaff 51; Sunderland 1011.

Rhabdophyllum calophyllum (Hook.f.) Tiegh.

Bull. Jard. Bot. Brux. 35: 392 (1965). *Ouratea calophylla* (Hook.f.) Engl., F.W.T.A. 1: 229 (1954); Keay R.W.J., Trees of Nigeria: 72 (1989); Hawthorne W., F.G.F.T. Ghana: 97 (1990).

Habit: tree to 15 m. **Guild:** shade-bearer. **Habitat:** forest and forest gaps. **Distribution:** Sierra Leone to Cameroon. **Chorology:** Guineo-Congolian. **Star:** green. **Alt. range:** 1-200m (19), 201-400m (4). **Flowering:** Mar (5), Apr (7), May (5), Oct (3). **Fruiting:** Mar (2), Apr (2), May (4), Jun (4), Oct (5). **Specimens:** - *Etinde:* Kwangue A. 59, 80; Tchouto P. 130; Watts J. 467. *Mabeta-Moliwe:* Cheek M. 3483; Ndam N. 482; Plot Voucher W 79, W 352, W 494, W 1013; Sunderland T. C. H. 1130, 1142, 1169, 1225, 1257, 1310, 1366, 1393; Tchouto P. 158; Watts J. 109, 150, 270; Wheatley J. I. 166. *Onge:* Akogo M. 39; Ndam N. 666; Tchouto P. 923; Watts J. 837; Wheatley J. I. 852

OLACACEAE

Coula edulis Baill.

F.W.T.A. 1: 645 (1958); Fl. Gabon 20: 144 (1973); Fl. Cameroun 15: 144 (1973); Keay R.W.J., Trees of Nigeria: 313 (1989); Hawthorne W., F.G.F.T. Ghana: 146 (1990).

Habit: tree to 18 m. **Guild:** shade-bearer. **Habitat:** forest. **Distribution:** Sierra Leone to Cameroon, Gabon, Congo (Kinshasa) and Angola (Cabinda). **Chorology:** Guineo-Congolian. **Star:** green. **Alt. range:** 1-200m (1), 201-400m (2). **Flowering:** May (1). **Specimens:** - *Mokoko River F.R.:* Tchouto P. 1223; Thomas D. W. 10091, 10137. *Southern Bakundu F.R.:* Olorunfemi J. FHI 30539

Diogoa zenkeri (Engl.) Exell & Mendonça

F.W.T.A. 1: 648 (1958); Fl. Gabon 20: 150 (1973); Fl. Cameroun 15: 150 (1973); Keay R.W.J., Trees of Nigeria: 313 (1989).

Habit: tree to 20 m. **Habitat:** forest. **Distribution:** Nigeria, Cameroon, Gabon, Congo (Kinshasa) and Angola (Cabinda). **Chorology:** Guineo-Congolian. **Star:** green. **Alt. range:** 1-200m (2), 201-400m (1), 401-600m (2). **Flowering:** Mar (1), Jul (1). **Fruiting:** Mar (1), May (1), Jun (1), Jul (1). **Specimens:** - *Etinde:* Akogo M. 148; Khayota B. 562; Kwangue A. 94; Mbani J. M. 166; Tchouto P. 262

Heisteria parvifolia Sm.

F.W.T.A. 1: 645 (1958); Fl. Gabon 20: 126 (1973); Fl. Cameroun 15: 126 (1973); Keay R.W.J., Trees of Nigeria: 314 (1989); Hawthorne W., F.G.F.T. Ghana: 85 (1990).

Habit: tree or shrub to 10 m. **Guild:** shade-bearer. **Habitat:** forest. **Distribution:** Senegal to Bioko, Cameroon, Gabon, Congo (Kinshasa), Angola and Uganda. **Chorology:** Guineo-Congolian. **Star:** green. **Alt. range:** 1-200m (22), 201-400m (6), 601-800m (1). **Flowering:** Jan (1), Mar (4), Apr (3), May (6), Jul (1), Aug (1), Oct (3), Nov (2). **Fruiting:** Jan (1), Mar (4), Apr (2), May (9), Jun (2), Jul (1), Oct (2), Nov (3), Dec (2). **Specimens:** - *Bambuko F.R.:* Watts J. 586. *Etinde:* Khayota B. 519; Maitland T. D. 444; Mbani J. M. 130; Nkeng P. 4; Tekwe C. 35. *Mabeta-Moliwe:* Baker W. J. 320; Cable S. 1403, 1407; Cheek M. 5688; Hunt L. V. 9; Jaff B. 185, 197; Khayota B. 479; Ndam N. 507; Nguembock F. 62; Plot Voucher W 427, W 1162; Sunderland T. C. H. 1017, 1179, 1383, 1453; Watts J. 251, 393; Wheatley J. I. 292. *Mokoko River F.R.:* Akogo M. 159; Ekema N. 970; Fraser P. 413; Ndam N. 1089, 1142, 1307. *Onge:* Akogo M. 102; Harris D. J. 3678; Tchouto P. 718; Thomas D. W. 9758; Watts J. 777, 970

Octoknema affinis Pierre

F.W.T.A. 1: 656 (1958); Fl. Gabon 20: 187 (1973); Fl. Cameroun 15: 187 (1973); Keay R.W.J., Trees of Nigeria: 318 (1989). *Octoknema winkleri* Engl., F.W.T.A. 1: 656 (1958).

Habit: tree to 20 m. **Habitat:** forest. **Distribution:** Nigeria, Cameroon and Gabon. **Chorology:** Lower Guinea. **Star:** blue. **Alt. range:** 1-200m (11), 201-400m (1). **Flowering:** Mar (1), Apr (3), May (1). **Fruiting:** Apr (5). **Specimens:** - *Etinde:* Cheek M. 5881. *Mabeta-Moliwe:* Jaff B. 59, 80; Plot Voucher W 5, W 169, W 371, W 483, W 605, W 671, W 904, W 1052; Sunderland T. C. H. 1205; Watts J. 226; Wheatley J. I. 170; Winkler H. J. P. 1238. *Mokoko River F.R.:* Acworth J. M. 51, 100; Ekema N. 839; Tchouto P. 1098. *Onge:* Cheek M. 5119; Tchouto P. 507

Olax gambecola Baill.

F.W.T.A. 1: 647 (1958); Fl. Gabon 20: 111 (1973); Fl. Cameroun 15: 111 (1973).

Habit: shrub to 3 m. **Guild:** shade-bearer. **Habitat:** forest. **Distribution:** Guinea to Congo (Kinshasa), Angola and Uganda.

Chorology: Guineo-Congolian. **Star:** green. **Alt. range:** 1-200m (6). **Flowering:** May (3), Jun (1). **Fruiting:** Apr (1), May (2), Jun (1), Oct (1). **Specimens:** - *Mabeta-Moliwe:* Sunderland T. C. H. 1330, 1401; Tchouto P. 183. *Mokoko River F.R.:* Akogo M. 179; Ekema N. 1188; Ndam N. 1182. *Onge:* Watts J. 887

Olax latifolia Engl.
F.W.T.A. 1: 647 (1958); Fl. Gabon 20: 122 (1973); Fl. Cameroun 15: 122 (1973).

Habit: shrub to 3 m. **Habitat:** forest. **Distribution:** Cameroon, Gabon and Congo (Kinshasa). **Chorology:** Guineo-Congolian. **Star:** green. **Alt. range:** 1-200m (8), 201-400m (9), 601-800m (2), 801-1000m (1). **Flowering:** Mar (1), Apr (1), Jun (2), Aug (2), Sep (1), Oct (2). **Fruiting:** Jan (2), Mar (6), Apr (1), May (1), Jun (1), Aug (2), Nov (2), Dec (2). **Specimens:** - *Etinde:* Cheek M. 5894, 5914; Etuge M. 1245; Khayota B. 504, 506; Mbani J. M. 2; Nkeng P. 1, 43; Tchouto P. 24, 26, 475; Watts J. 466; Wheatley J. I. 515. *Mabeta-Moliwe:* Baker W. J. 220; Nguembock F. 30; Watts J. 235; Wheatley J. I. 325a, 325b; von Rege I. 33. *Mokoko River F.R.:* Ekema N. 967b. *Onge:* Harris D. J. 3681

Olax mannii Oliv.
F.W.T.A. 1: 647 (1958); Fl. Gabon 20: 113 (1973); Fl. Cameroun 15: 113 (1973).

Habit: shrub to 3 m. **Habitat:** forest. **Distribution:** Sierra Leone to Cameroon, Gabon and Congo (Kinshasa). **Chorology:** Guineo-Congolian. **Star:** green. **Alt. range:** 1-200m (3), 201-400m (5). **Flowering:** Mar (1), May (1), Nov (1). **Fruiting:** Mar (5), May (4). **Specimens:** - *Bambuko F.R.:* Tchouto P. 550; Watts J. 597, 664. *Mokoko River F.R.:* Mbani J. M. 361; Ndam N. 1104, 1203, 1278; Tchouto P. 529, 640. *Onge:* Harris D. J. 3689. *Southern Bakundu F.R.:* Brenan J. P. M. 9473

Olax staudtii Engl.
Fl. Gabon 20: 119 (1973); Fl. Cameroun 15: 119 (1973).

Habit: shrub to 1.5 m. **Habitat:** forest, sometimes on rocks. **Distribution:** Cameroon, Gabon and CAR. **Chorology:** Guineo-Congolian. **Star:** green. **Alt. range:** 1-200m (6), 201-400m (5). **Flowering:** Mar (1), Aug (1), Oct (6), Nov (2), Dec (1). **Fruiting:** Mar (1), Aug (1), Oct (2), Nov (2), Dec (1). **Specimens:** - *Bambuko F.R.:* Watts J. 588. *Mabeta-Moliwe:* Cheek M. 5760. *Onge:* Akogo M. 104; Cheek M. 5063, 5122; Harris D. J. 3638; Ndam N. 607; Watts J. 697, 737, 891, 971

Olax triplinerva Oliv.
Fl. Cameroun 15: 120 (1973).

Habit: shrub to 2.5 m. **Habitat:** forest and farmbush. **Distribution:** Cameroon, Rio Muni, Gabon and CAR. **Chorology:** Guineo-Congolian. **Star:** green. **Alt. range:** 1-200m (1). **Flowering:** Oct (1). **Fruiting:** Oct (1). **Specimens:** - *Mabeta-Moliwe:* Sidwell K. 113

Olax cf. triplinerva Oliv.

Specimens: - *Mokoko River F.R.:* Ekema N. 1025

Ongokea gore (Hua) Pierre
F.W.T.A. 1: 649 (1958); Fl. Gabon 20: 159 (1973); Fl. Cameroun 15: 159 (1973); Keay R.W.J., Trees of Nigeria: 314 (1989); Hawthorne W., F.G.F.T. Ghana: 83 (1990).

Habit: tree to 40 m. **Guild:** light demander. **Habitat:** forest. **Distribution:** Sierra Leone to Cameroon, Gabon, Congo (Kinshasa) and Angola. **Chorology:** Guineo-Congolian. **Star:** green. **Alt. range:** 1-200m (1), 801-1000m (1). **Flowering:** Jul (1). **Fruiting:** Jul (1). **Specimens:** - *Mabeta-Moliwe:* Cheek M. 3578; Mildbraed G. W. J. 10703

Ptychopetalum petiolatum Oliv.
F.W.T.A. 1: 647 (1958); Fl. Gabon 20: 147 (1973); Fl. Cameroun 15: 147 (1973).

Habit: tree or shrub to 10 m. **Habitat:** forest. **Distribution:** SE Nigeria, Cameroon, Rio Muni and Gabon. **Chorology:** Lower Guinea. **Star:** blue. **Alt. range:** 1-200m (9), 201-400m (2).**Fruiting:** Mar (1), Apr (1), May (4), Jun (2), Oct (1). **Specimens:** - *Mokoko River F.R.:* Acworth J. M. 292; Ekema N. 1204; Mbani J. M. 396, 410; Ndam N. 1166, 1361, 1363; Pouakouyou D. 38; Tchouto P. 615, 1089, 1141; Watts J. 1073. *Onge:* Tchouto P. 928. *Southern Bakundu F.R.:* Brenan J. P. M. 9443

Strombosia grandifolia Hook.f. ex Benth.
F.W.T.A. 1: 648 (1958); Fl. Gabon 20: 134 (1973); Fl. Cameroun 15: 134 (1973); Keay R.W.J., Trees of Nigeria: 314 (1989).

Habit: tree to 13 m. **Habitat:** forest. **Distribution:** Benin to Bioko, Cameroon, Gabon, Congo (Kinshasa) and Angola (Cabinda). **Chorology:** Guineo-Congolian. **Star:** green. **Alt. range:** 1-200m (4). **Flowering:** Apr (3). **Fruiting:** Apr (2), May (1). **Specimens:** - *Etinde:* Maitland T. D. 180. *Mabeta-Moliwe:* Tchouto P. 148, 165, 1055. *Mokoko River F.R.:* Akogo M. 166; Ekema N. 890

Strombosia pustulata Oliv.
F.W.T.A. 1: 648 (1958); Fl. Cameroun 15: 137 (1973); Keay R.W.J., Trees of Nigeria: 314 (1989).
  Strombosia glaucescens Engl., F.W.T.A. 1: 648 (1958).

Habit: tree to 20 m. **Habitat:** forest. **Distribution:** Senegal to Congo (Kinshasa). **Chorology:** Guineo-Congolian. **Star:** green. **Alt. range:** 1-200m (1). **Specimens:** - *Mokoko River F.R.:* Mbani J. M. 312

Strombosia scheffleri Engl.
F.W.T.A. 1: 648 (1958); Fl. Gabon 20: 139 (1973); Fl. Cameroun 15: 139 (1973); Keay R.W.J., Trees of Nigeria: 314 (1989).

Habit: tree to 30 m. **Habitat:** forest. **Distribution:** SE Nigeria, Bioko, Cameroon, Congo (Kinshasa), Uganda and East Africa. **Star:** excluded. **Alt. range:** 400-600m (1).**Fruiting:** May (1). **Specimens:** - *Etinde:* Mbani J. M. 155

Strombosiopsis tetrandra Engl.
F.W.T.A. 1: 649 (1958); Fl. Gabon 20: 152 (1973); Fl. Cameroun 15: 152 (1973); Keay R.W.J., Trees of Nigeria: 313 (1989).

Habit: tree 4-20 m tall. **Habitat:** forest. **Distribution:** SE Nigeria, Gabon, Congo (Kinshasa) and Uganda. **Chorology:** Guineo-Congolian. **Star:** green. **Alt. range:** 1-200m (3), 801-1000m (1). **Flowering:** Oct (1). **Fruiting:** May (1), Jun (1). **Specimens:** - *Mabeta-Moliwe:* Mildbraed G. W. J. 10541. *Mokoko River F.R.:* Ekema N. 1101; Watts J. 1116. *Onge:* Cheek M. 5136

Ximenia americana L.

F.W.T.A. 1: 646 (1958); Fl. Gabon 20: 106 (1973); Fl. Cameroun 15: 106 (1973); Meded. Land. Wag. 82(3): 251 (1982); Keay R.W.J., Trees of Nigeria: 311 (1989); Hawthorne W., F.G.F.T. Ghana: 95 (1990).

Habit: tree or shrub to 5 m. **Distribution:** tropical Africa. **Star:** excluded. **Specimens:** - *Mabeta-Moliwe:* Mann G. 764

OLEACEAE

P. Green (K)

Chionanthus africanus (Knobl.) Stearn

Bot. J. Linn. Soc. 80: 197 (1980); Keay R.W.J., Trees of Nigeria: 403 (1989); Hawthorne W., F.G.F.T. Ghana: 45 (1990).
Linociera africana (Knobl.) Knobl., F.W.T.A. 2: 48 (1963).

Habit: tree or shrub to 10 m. **Habitat:** forest. **Distribution:** Sierra Leone to Congo (Kinshasa), Angola, Uganda and Tanzania. **Chorology:** Guineo-Congolian. **Star:** green. **Alt. range:** 600-800m (1). **Flowering:** May (1). **Fruiting:** May (1). **Specimens:** - *Eastern slopes:* Deistel H. 85. *Etinde:* Tchouto P. 219

Chionanthus mannii (Soler.) Stearn subsp. congesta (Baker) Stearn

Bot. J. Linn. Soc. 80: 200 (1980).
Linociera congesta Baker, F.W.T.A. 2: 48 (1963).

Habit: tree or shrub to 4 m. **Habitat:** forest. **Distribution:** Cameroon, Rio Muni and Gabon. **Chorology:** Lower Guinea. **Star:** blue. **Specimens:** - *Southern Bakundu F.R.:* Binuyo A. & Daramola B. O. FHI 35187

Chionanthus mannii (Soler.) Stearn subsp. mannii

Bot. J. Linn. Soc. 80: 199 (1980); Hawthorne W., F.G.F.T. Ghana: 45 (1990).
Linociera mannii Soler., F.W.T.A. 2: 48 (1963).

Habit: tree to 10 m. **Habitat:** forest. **Distribution:** Sierra Leone to Cameroon and Gabon. **Chorology:** Guineo-Congolian. **Star:** green. **Specimens:** - *Mabeta-Moliwe:* Plot Voucher W 83

Jasminum pauciflorum Benth.

F.W.T.A. 2: 50 (1963); Meded. Land. Wag. 82(3): 253 (1982).

Habit: woody climber. **Guild:** light demander. **Habitat:** forest. **Distribution:** tropical Africa. **Chorology:** Afromontane. **Star:** green. **Alt. range:** 1-200m (2). **Flowering:** May (2). **Specimens:** - *Mabeta-Moliwe:* Sunderland T. C. H. 1354; Tchouto P. 180

Jasminum preussii Engl. & Knobl.

F.W.T.A. 2: 51 (1963); Meded. Land. Wag. 82(3): 253 (1982).

Habit: woody climber. **Guild:** pioneer. **Habitat:** forest and forest-grassland transition. **Distribution:** Ghana, Nigeria and Cameroon. **Chorology:** Guineo-Congolian (montane). **Star:** green. **Alt. range:** 200-400m (1), 1601-1800m (1), 2001-2200m (1). **Flowering:** Oct (1). **Fruiting:** Sep (1). **Specimens:** - *Eastern slopes:* Watts J. 464. *Etinde:* Wheatley J. I. 493. *No locality:* Brenan J. P. M. 9354

Olea capensis L. subsp. hochstetteri (Baker) Friis & P.S.Green

Meded. Land. Wag. 82(3): 253 (1982); Kew Bull. 41: 36 (1986); Keay R.W.J., Trees of Nigeria: 403 (1989).
Olea hochstetteri Baker, F.W.T.A. 2: 49 (1963).

Habit: tree to 26 m. **Habitat:** forest. **Distribution:** tropical and subtropical Africa. **Chorology:** Afromontane. **Star:** green. **Specimens:** - *No locality:* Maitland T. D. 498

ONAGRACEAE

Ludwigia octovalis (Jacq.) Raven subsp. brevisepala (Brenan) Raven

Fl. Cameroun 5: 104 (1966); Fl. Zamb. 4: 337 (1978).
Jussiaea suffruticosa L. var. *brevisepala* Brenan, F.W.T.A. 1: 169 (1954).
Jussiaea suffruticosa L. var. *linearis* (Willd.) Oliv. ex Kuntze, F.W.T.A. 1: 169 (1954).

Habit: herb. **Distribution:** tropical Africa. **Star:** excluded. **Specimens:** - *Etinde:* Maitland T. D. 55

OPILIACEAE

Rhopalopilia pallens Pierre

Fl. Cameroun 15: 176 (1973); Bot. Jahrb. Syst. 108: 271 (1987).

Habit: shrub to 2 m. **Habitat:** forest. **Distribution:** Cameroon, Gabon and Congo (Kinshasa). **Chorology:** Guineo-Congolian. **Star:** green. **Specimens:** - *Mabeta-Moliwe:* Plot Voucher W 112

Urobotrya congolana (Baill.) Hiepko subsp. congolana

Bot. Jahrb. Syst. 107: 142 (1985).
Urobotrya minutiflora Stapf, F.W.T.A. 1: 652 (1958).
Opilia congolana Baill., Fl. Cameroun 15: 172 (1973); Fl. Gabon 20: 172 (1973).

Habit: tree or shrub to 5 m. **Habitat:** forest. **Distribution:** SE Nigeria, W. Cameroon, Rio Muni, Gabon and Congo (Brazzaville). **Chorology:** Guineo-Congolian. **Star:** green. **Alt. range:** 1-200m (10), 201-400m (3), 601-800m (1), 1001-1200m (1). **Flowering:** Feb (1), Apr (3), May (1), Dec (1). **Fruiting:** Feb (1), Mar (2), Apr (7), May (3), Dec (1). **Specimens:** - *Bambuko F.R.:* Watts J. 652. *Eastern slopes:* Dalziel J. M. 8244; Kwangue A. 53. *Etinde:* Mbani J. M. 7; Wheatley J. I. 5. *Mabeta-Moliwe:* Cheek M. 5738; Maitland T. D. 707; Sunderland T. C. H. 1254; Watts J. 189. *Mokoko River F.R.:* Acworth J. M. 183, 298; Akogo M. 230, 257; Ekema N. 934, 1017; Tchouto P. 610, 1060

OXALIDACEAE

Biophytum umbraculum Welw.

Brittonia 33: 451 (1981).
Biophytum petersianum Klotzsch, F.W.T.A. 1: 158 (1954).

Habit: herb. Distribution: palaeotropical. Star: excluded. Specimens: - *Eastern slopes:* Maitland T. D. 1290

Biophytum zenkeri Guillaumin
F.W.T.A. 1: 159 (1954).

Habit: herb. Habitat: forest, by streams. Distribution: SE Nigeria, Cameroon, Gabon, Congo (Kinshasa) and Angola. Chorology: Guineo-Congolian. Star: green. Alt. range: 1-200m (1). Flowering: Jan (1). Specimens: - *Limbe Botanic Garden:* Tchouto P. 1056a

Oxalis barrelieri L.
Willdenowia 8: 17 (1977).

Habit: herb. Habitat: forest gaps. Distribution: introduced. Star: excluded. Alt. range: 1-200m (1). Flowering: Oct (1). Fruiting: Oct (1). Specimens: - *Onge:* Tchouto P. 672

Oxalis corniculata L.
F.W.T.A. 1: 159 (1954).

Habit: herb. Habitat: forest. Distribution: cosmopolitan. Chorology: Montane. Star: excluded. Alt. range: 1400-1600m (2), 2001-2200m (1). Flowering: Oct (2). Specimens: - *Eastern slopes:* Tchouto P. 319, 370. *No locality:* Mann G. 2020

Oxalis corymbosa DC.
F.W.T.A. 1: 159 (1954).

Habit: herb. Habitat: forest. Distribution: cosmopolitan. Chorology: Montane. Star: excluded. Specimens: - *Eastern slopes:* Deistel H. 150

PANDACEAE

S. Cable (K)

Microdesmis camerunensis J.Léonard
Fl. Gabon 22: 28 (1973); Fl. Cameroun 19: 56 (1975).

Habit: shrub to 3 m. Habitat: forest. Distribution: Cameroon and Gabon. Chorology: Lower Guinea. Star: blue. Alt. range: 200-400m (2). Flowering: Jul (1). Fruiting: Mar (1), Jul (1). Specimens: - *Bambuko F.R.:* Watts J. 644. *Etinde:* Tchouto P. 259

Microdesmis haumaniana J.Léonard
Fl. Gabon 22: 26 (1973); Fl. Cameroun 19: 54 (1975).

Habit: shrub 2-6 m tall. Habitat: forest. Distribution: Cameroon, Gabon, Congo (Kinshasa) and Angola. Chorology: Guineo-Congolian. Star: green. Alt. range: 1-200m (1). Flowering: Dec (1). Fruiting: Dec (1). Specimens: - *Etinde:* Sunderland T. C. H. 1004

Microdesmis puberula Hook.f. ex Planch.
F.W.T.A. 1: 392 (1958); Fl. Congo B., Rw. & Bur. 8: 104 (1962); Fl. Gabon 22: 23 (1973); Fl. Cameroun 19: 51 (1975); Keay R.W.J., Trees of Nigeria: 308 (1989); Hawthorne W., F.G.F.T. Ghana: 107 (1990).
Microdesmis zenkeri Pax, F.T.A. 6: 742 (1897).

Habit: tree or shrub to 15 m. Guild: shade-bearer. Habitat: forest. Distribution: Nigeria, Cameroon, Bioko, Gabon, Congo (Kinshasa), CAR and Uganda. Chorology: Guineo-Congolian. Star: green. Alt. range: 1-200m (21), 201-400m (6), 401-600m (2), 801-1000m (1). Flowering: Mar (1), Apr (1), May (1), Aug (1), Oct (1), Nov (2), Dec (1). Fruiting: Mar (9), Apr (6), May (5), Aug (1), Dec (3). Specimens: - *Bambuko F.R.:* Tchouto P. 534; Watts J. 623, 657. *Etinde:* Cable S. 250; Kalbreyer W. 28; Mbani J. M. 63, 70; Tchouto P. 71, 86, 109; Tekwe C. 62. *Mabeta-Moliwe:* Baker W. J. 323; Cable S. 516, 528, 548; Cheek M. 5697; Jaff B. 203; Mildbraed G. W. J. 10588; Nguembock F. 27a; Plot Voucher W 128, W 279, W 415, W 534, W 567, W 616, W 986, W 1030, W 1160; Sunderland T. C. H. 1006, 1284; Tchouto P. 205; Watts J. 246; Wheatley J. I. 116; von Rege I. 88. *Mokoko River F.R.:* Acworth J. M. 48, 224, 307; Akogo 149, 302; Fraser P. 364; Mbani J. M. 345b; Ndam N. 1159, 1328; Sonké 1111; Tchouto P. 586, 1051. *Onge:* Thomas D. W. 9854; Watts J. 857; Wheatley J. I. 743

Panda oleosa Pierre
F.W.T.A. 1: 636 (1958); Fl. Gabon 22: 15 (1973); Fl. Cameroun 19: 43 (1975); Keay R.W.J., Trees of Nigeria: 308 (1989); Hawthorne W., F.G.F.T. Ghana: 107 (1990).

Habit: tree to 15 m. Guild: shade-bearer. Habitat: forest. Distribution: Liberia to Cameroon, Gabon and Congo (Kinshasa). Chorology: Guineo-Congolian. Star: green. Alt. range: 1-200m (3), 201-400m (2). Flowering: Apr (2). Fruiting: Oct (2), Nov (1). Specimens: - *Etinde:* Maitland T. D. 474. *Mabeta-Moliwe:* Jaff B. 2; Plot Voucher W 958; Sunderland T. C. H. 1224; Wheatley J. I. 151. *Mokoko:* Thomas 10015. *Onge:* Akogo M. 145; Watts J. 798

PASSIFLORACEAE

Adenia cissampeloides (Planch. ex Benth.) Harms
F.W.T.A. 1: 203 (1954); Meded. Land. Wag. 71(19): 246 (1971).

Habit: woody climber. Guild: pioneer. Habitat: forest. Distribution: tropical Africa. Star: excluded. Alt. range: 1-200m (3). Flowering: Apr (2), May (1). Fruiting: Apr (1), May (1). Specimens: - *Eastern slopes:* Maitland T. D. 679. *Etinde:* Tekwe C. 59. *Mabeta-Moliwe:* Cheek M. 3422; Ndam N. 492; Wheatley J. I. 186

Adenia gracilis Harms
F.W.T.A. 1: 203 (1954); Meded. Land. Wag. 71(19): 255 (1971).

Habit: herbaceous climber. Guild: pioneer. Habitat: forest. Distribution: Senegal to Cameroon, CAR, Gabon, Congo (Brazzaville), Congo (Kinshasa), Angola (Cabinda), Uganda and Burundi. Chorology: Guineo-Congolian. Star: green. Alt. range: 1-200m (1). Flowering: Jun (1). Fruiting: Jun (1). Specimens: - *Eastern slopes:* Maitland T. D. 228. *Mabeta-Moliwe:* Wheatley J. I. 338

Adenia letouzeyi W.J.de Wilde
Meded. Land. Wag. 71(19): 148 (1971).

Habit: woody climber. Habitat: forest. Distribution: Cameroon, Gabon, Congo (Brazzaville), Congo (Kinshasa) and Angola. Chorology: Guineo-Congolian. Star: green. Alt. range: 1-200m (1), 201-400m (1). Flowering: Mar (1), Apr (1). Specimens: - *Etinde:* Mbani J. M. 65; Tekwe C. 30

Adenia cf. letouzeyi J.J.de Wilde

Specimens: - *Mabeta-Moliwe:* Wheatley J. I. 236

Adenia lobata (Jacq.) Engl.

F.W.T.A. 1: 203 (1954); Meded. Land. Wag. 71(19): 149 (1971).

Habit: woody climber. **Guild:** pioneer. **Habitat:** forest.
Distribution: Senegal to Cameroon. **Chorology:** Guineo-
Congolian. **Star:** green. **Alt. range:** 1-200m (2), 201-400m (1),
1401-1600m (1). **Flowering:** Apr (1). **Fruiting:** Jun (1), Jul (1),
Oct (1). **Specimens:** - *Eastern slopes:* Thomas D. W. 9509.
Etinde: Maitland T. D. 618; Tchouto P. 247. *Mabeta-Moliwe:* Plot
Voucher W 691; Sunderland T. C. H. 1212; Watts J. 399. *Southern
Bakundu F.R.:* Ejiofor M. C. FHI 29337

Adenia cf. lobata (Jacq.) Engl.

Specimens: - *Eastern slopes:* Nkeng P. 81. *Mabeta-Moliwe:*
Cheek M. 3581; Wheatley J. I. 51

Adenia mannii (Mast.) Engl.

F.W.T.A. 1: 203 (1954); Meded. Land. Wag. 71(19): 144 (1971).
Adenia oblongifolia Harms, F.W.T.A. 1: 202 (1954).

Habit: herbaceous climber. **Guild:** pioneer. **Habitat:** forest and
farmbush. **Distribution:** Sierra Leone to Cameroon, CAR, Gabon,
Congo (Brazzaville), Congo (Kinshasa) and Uganda. **Chorology:**
Guineo-Congolian. **Star:** green. **Alt. range:** 1-200m (1), 801-
1000m (1).**Fruiting:** Feb (1), Jun (1). **Specimens:** - *Etinde:*
Wheatley J. I. 11. *Mabeta-Moliwe:* Mann G. 17; Plot Voucher W
1059; Wheatley J. I. 361

Barteria fistulosa Mast.

F.W.T.A. 1: 201 (1954); Keay R.W.J., Trees of Nigeria: 69 (1989).
Barteria nigritana Mast. subsp. *fistulosa* (Mast.) Sleumer,
Blumea 22: 14 (1974).

Habit: tree to 15 m. **Habitat:** forest. **Distribution:** Nigeria,
Cameroon, Bioko, Gabon, Congo (Kinshasa), Uganda and
Tanzania. **Chorology:** Guineo-Congolian. **Star:** green. **Alt. range:**
1-200m (7). **Flowering:** Mar (1), Apr (1), May (4). **Fruiting:** May
(1), Jun (1). **Specimens:** - *Etinde:* Maitland T. D. 616; Mbani J.
M. 86; Tchouto P. 84; Tekwe C. 85. *Mabeta-Moliwe:* Ndam N.
490; Sunderland T. C. H. 1276; Tchouto P. 185; Wheatley J. I. 248

Barteria nigritana Hook.f.

F.W.T.A. 1: 201 (1954); F.T.E.A. Passifloraceae: 4 (1975); Keay
R.W.J., Trees of Nigeria: 69 (1989).

Habit: tree to 15 m. **Habitat:** forest. **Distribution:** Nigeria,
Cameroon, Bioko, Gabon and Congo (Kinshasa). **Chorology:**
Guineo-Congolian. **Star:** green. **Alt. range:** 800-1000m (1).
Flowering: Mar (1). **Specimens:** - *Eastern slopes:* Tchouto P.
100; *Mabeta-Moliwe:* Plot Voucher W356

Efulensia clematoides C.H.Wright

Blumea 22: 33 (1974).
Deidamia clematoides (C.H.Wright) Harms, F.W.T.A. 1: 761
(1958).

Habit: woody climber. **Habitat:** forest. **Distribution:** SE Nigeria,
Cameroon, Rio Muni, Gabon and Congo (Kinshasa). **Chorology:**
Guineo-Congolian. **Star:** green. **Alt. range:** 1-200m (1).**Fruiting:**
Jun (1). **Specimens:** - *Etinde:* Kwangue A. 115. *Mabeta-Moliwe:*
Plot Voucher W 549, W 679

Passiflora foetida L.

F.W.T.A. 1: 199 (1954); Fl. Zamb. 4: 410 (1978).

Habit: herbaceous climber. **Guild:** pioneer. **Habitat:** forest and
farmbush. **Distribution:** tropical Africa, introduced from tropical
America. **Star:** excluded. **Alt. range:** 1-200m (2). **Flowering:** Mar
(1). **Specimens:** - *Etinde:* Maitland T. D. 400. *Mokoko River F.R.:*
Pouakouyou D. 24. *Onge:* Mbani J. M. 69

PEDALIACEAE

Sesamum radiatum Schum. & Thonn.

F.W.T.A. 2: 391 (1963); Fl. Zamb. 8(3): 104 (1988).

Habit: herb. **Guild:** pioneer. **Distribution:** pantropical. **Star:**
excluded. **Specimens:** - *Etinde:* Ejiofor M. C. FHI 29334

PHYTOLACCACEAE

Hilleria latifolia (Lam.) H.Walt.

F.W.T.A. 1: 143 (1954); Fl. Gabon 7: 60 (1963).

Habit: herb. **Guild:** pioneer. **Distribution:** tropical Africa,
introduced from tropical America. **Star:** excluded. **Specimens:** -
Mabeta-Moliwe: Mann G. 775

Phytolacca dodecandra L'Hér.

F.W.T.A. 1: 143 (1954); Fl. Gabon 7: 56 (1963).

Habit: woody climber. **Guild:** pioneer. **Habitat:** forest.
Distribution: tropical and subtropical Africa. **Star:** excluded. **Alt.
range:** 600-800m (1). **Flowering:** Mar (1). **Fruiting:** Mar (1).
Specimens: - *Eastern slopes:* Tchouto P. 58

PIPERACEAE

Peperomia fernandopoiana C.DC.

F.W.T.A. 1: 81 (1954); F.T.E.A. Piperaceae: 19 (1996).

Habit: epiphyte. **Habitat:** grassland and lava flows. **Distribution:**
Guinea to Bioko and Cameroon. **Chorology:** Guineo-Congolian
(montane). **Star:** green. **Alt. range:** 1000-1200m (1). **Specimens:**
- *No locality:* Dalziel J. M. 8340

Peperomia kamerunana C.DC.

F.W.T.A. 1: 83 (1954).

Habit: epiphyte. **Habitat:** forest. **Distribution:** Mount Cameroon
and Bioko. **Chorology:** Western Cameroon Uplands (montane).
Star: black. **Alt. range:** 2000-2200m (1). **Specimens:** - *No
locality:* Brenan J. P. M. 9371

Peperomia molleri C.DC.

F.W.T.A. 1: 82 (1954); F.T.E.A. Piperaceae: 16 (1996); Fl. Zamb.
9(2): 35 (1997).

Habit: epiphyte. Habitat: forest. Distribution: tropical Africa. Star: excluded. Alt. range: 800-1000m (1). Specimens: - *Eastern slopes:* Preuss P. R. 1060

Peperomia pellucida (L.) H.B. & K.

F.W.T.A. 1: 82 (1954); F.T.E.A. Piperaceae: 11 (1996); Fl. Zamb. 9(2): 32 (1997).

Habit: herb. Habitat: stream banks in forest. Distribution: pantropical. Star: excluded. Alt. range: 1-200m (1). Fruiting: May (1). Specimens: - *Etinde:* Thorold C. A. 77. *Mabeta-Moliwe:* Wheatley J. I. 264

Peperomia retusa (L.f.) A.Dietr.

F.W.T.A. 1: 82 (1954); F.T.E.A. Piperaceae: 15 (1996); Fl. Zamb. 9(2): 33 (1997).
   Peperomia retusa (L.f.) A.Dietr. var. *mannii* (Hook.f.) Düll, Engl. Bot. Jahrb. 93: 90 (1973); F.T.E.A. Piperaceae: 13 (1996).
   Peperomia mannii Hook.f. ex C.DC., F.W.T.A. 1: 82 (1954).

Habit: epiphyte. Habitat: forest. Distribution: SE Nigeria and W. Cameroon to East and South Africa. Star: excluded. Alt. range: 1200-1400m (1). Specimens: - *No locality:* Mann G. 2016

Peperomia tetraphylla (G.Forst.) Hook. & Arn.

Engl. Bot. Jahrb. 93: 81 (1973); F.T.E.A. Piperaceae: 12 (1996); Fl. Zamb. 9(2): 29 (1997).
   Peperomia reflexa (L.f.) A.Dietr., F.W.T.A. 1: 83 (1954).

Habit: epiphyte. Habitat: forest. Distribution: pantropical. Chorology: Montane. Star: excluded. Alt. range: 1200-1400m (1). Specimens: - *No locality:* Dunlap 5

Peperomia thomeana C.DC.

Engl. Bot. Jahrb. 93: 104 (1974).
   Peperomia vaccinifolia C.DC., F.W.T.A. 1: 82 (1954).

Habit: epiphyte. Habitat: forest. Distribution: Bioko, W. Cameroon and S. Tomé. Chorology: Western Cameroon Uplands (montane). Star: gold. Alt. range: 1400-1600m (1). Specimens: - *Eastern slopes:* Brenan J. P. M. 9364. *No locality:* Preuss P. R. 875a

Peperomia vulcanica Baker & C.H.Wright

Engl. Bot. Jahrb. 93: 102 (1973).
   Peperomia hygrophila Engl., F.W.T.A. 1: 83 (1954).

Habit: epiphyte. Habitat: forest-grassland transition. Distribution: Sierra Leone to W. Cameroon. Chorology: Guineo-Congolian (montane). Star: green. Alt. range: 2400-2600m (2). Specimens: - *No locality:* Mann G. 1305; Mildbraed G. W. J. 3400

Peperomia cf. vulcanica Baker & C.H.Wright

Specimens: - *Etinde:* Cheek M. 3602

Piper capense L.f.

F.W.T.A. 1: 84 (1954); F.T.E.A. Piperaceae: 5 (1996); Fl. Zamb. 9(2): 27 (1997).

Habit: shrub to 3 m, sometimes scandent. Habitat: forest. Distribution: tropical and subtropical Africa. Chorology: Afromontane. Star: green. Alt. range: 1200-1400m (1), 1601-1800m (1). Flowering: Jul (1). Specimens: - *Etinde:* Cheek M.

3611. *Mabeta-Moliwe:* Plot Voucher W700. *No locality:* Mann G. 2015

Piper guineense Schum. & Thonn.

F.W.T.A. 1: 84 (1954); F.T.E.A. Piperaceae: 4 (1996); Fl. Zamb. 9(2): 28 (1997).

Habit: herbaceous climber, often reaching the canopy. Guild: shade-bearer. Habitat: forest. Distribution: tropical Africa. Star: excluded. Alt. range: 1-200m (1). Fruiting: Jun (1). Specimens: - *Mabeta-Moliwe:* Watts 361; Wheatley J. I. 355

Piper umbellatum L.

F.W.T.A. 1: 84 (1954); F.T.E.A. Piperaceae: 8 (1996); Fl. Zamb. 9(2): 25 (1997).

Habit: shrub to 4 m. Guild: pioneer. Habitat: forest, clearings and Aframomum thicket. Distribution: pantropical. Star: excluded. Alt. range: 1-200m (8), 801-1000m (1). Flowering: Apr (1), May (3), Jun (3). Specimens: - *Mabeta-Moliwe:* Sunderland T. C. H. 1230, 1450. *Mokoko River F.R.:* Akogo M. 177; Ekema N. 825, 895, 1004, 1106, 1202; Pouakouyou D. 84. *No locality:* Mann G. 1272

PITTOSPORACEAE

Pittosporum mannii Hook.f.

F.W.T.A. 1: 182 (1954); F.T.E.A. Pittosporaceae: 8 (1966); Kew Bull. 42: 328 (1987); Keay R.W.J., Trees of Nigeria: 53 (1989).
   Pittosporum viridiflorum Sims subsp. *dalzielii* (Hutch.) Cuf., Meded. Land. Wag. 82(3): 260 (1982).

Habit: tree to 10 m. Habitat: forest-grassland transition. Distribution: SE Nigeria, Bioko and Cameroon to East Africa. Chorology: Afromontane. Star: green. Alt. range: 1200-1400m (1), 1801-2000m (3), 2201-2400m (1). Flowering: May (1). Fruiting: Sep (1), Oct (2). Specimens: - *Eastern slopes:* Mbani J. M. 106; Thomas D. W. 9292, 9349. *Etinde:* Tekwe C. 259. *No locality:* Mann G. 1202

PLANTAGINACEAE

Plantago palmata Hook.f.

F.W.T.A. 2: 306 (1963).

Habit: herb. Habitat: forest. Distribution: Mount Cameroon, Bioko and East Africa. Chorology: Afromontane. Star: blue. Alt. range: 1800-2000m (1), 2201-2400m (1). Flowering: Oct (1). Specimens: - *Eastern slopes:* Tchouto P. 328. *No locality:* Mann G. 1962

PLUMBAGINACEAE

Plumbago zeylanica L.

F.W.T.A. 2: 306 (1963).

Habit: woody climber. **Guild:** pioneer. **Habitat:** forest.
Distribution: pantropical. **Star:** excluded. **Alt. range:** 600-800m
(1). **Specimens:** - *No locality:* Maitland T. D. 1113

PODOSTEMACEAE

Tristicha trifaria (Bory) Spreng.
F.W.T.A. 1: 124 (1954); Fl. Cameroun 30: 98 (1987).

Habit: rheophyte. **Habitat:** streams. **Distribution:** Africa, India
and tropical America. **Star:** excluded. **Specimens:** - *Etinde:*
Brenan J. P. M. 9495; Rosevear D. R. 131/36

POLYGALACEAE

Atroxima afzeliana (Oliv.) Stapf
F.W.T.A. 1: 109 (1954); Meded. Land. Wag. 77(18): 15 (1977).

Habit: woody climber. **Guild:** shade-bearer. **Habitat:** forest.
Distribution: Guinea Bissau to Cameroon, CAR, Congo
(Kinshasa) and Angola. **Chorology:** Guineo-Congolian. **Star:**
green. **Alt. range:** 1-200m (2). **Flowering:** Mar (2). **Fruiting:** Mar
(2). **Specimens:** - *Bambuko F.R.:* Watts J. 659. *Mokoko River
F.R.:* Tchouto P. 645

Atroxima liberica Stapf
F.W.T.A. 1: 109 (1954); Meded. Land. Wag. 77(18): 19 (1977).

Habit: woody climber. **Guild:** shade-bearer. **Habitat:** forest.
Distribution: Liberia to Cameroon, Gabon, Congo (Brazzaville),
Congo (Kinshasa) and Angola (Cabinda). **Chorology:** Guineo-
Congolian. **Star:** green. **Alt. range:** 1-200m (10), 201-400m (4),
601-800m (1). **Flowering:** Feb (1), Mar (1), Oct (2), Nov (1).
Fruiting: Mar (4), Apr (5), May (2), Oct (1), Nov (1). **Specimens:**
- *Bambuko F.R.:* Watts J. 582. *Etinde:* Wheatley J. I. 8. *Mabeta-
Moliwe:* Jaff B. 95, 278; Wheatley J. I. 165. *Mokoko River F.R.:*
Acworth J. M. 68, 113; Ndam N. 1299; Tchouto P. 636, 639, 1140.
Onge: Cheek M. 5050; Tchouto P. 713; Watts J. 864, 1018;
Wheatley J. I. 766

Carpolobia alba G.Don
F.W.T.A. 1: 108 (1954); Meded. Land. Wag. 77(18): 24 (1977).
Carpolobia glabrescens Hutch. & Dalziel, F.W.T.A. 1: 108
(1954).

Habit: tree or shrub to 5 m. **Habitat:** forest. **Distribution:** Guinea
Bissau to Bioko, Cameroon, Gabon, Congo (Brazzaville), Congo
(Kinshasa), Angola and CAR. **Chorology:** Guineo-Congolian.
Star: green. **Alt. range:** 1-200m (1), 201-400m (2). **Flowering:**
Mar (1), May (1). **Fruiting:** Mar (1). **Specimens:** - *Etinde:* Nkeng
P. 48; Tchouto P. 30. *Mabeta-Moliwe:* Nguembock F. 21b.
Mokoko River F.R.: Thomas D. W. 10158

Carpolobia cf. alba G.Don

Specimens: - *Mabeta-Moliwe:* Nguembock F. 67; Sunderland T.
C. H. 1126

Carpolobia lutea G.Don
F.W.T.A. 1: 109 (1954); Meded. Land. Wag. 77(18): 38 (1977).

Habit: shrub to 3 m. **Guild:** shade-bearer. **Habitat:** forest and
stream banks. **Distribution:** Guinea to SW Cameroon. **Chorology:**
Guineo-Congolian. **Star:** green. **Alt. range:** 1-200m (4), 201-
400m (1). **Flowering:** Apr (1). **Fruiting:** Apr (3), May (2).
Specimens: - *Mokoko River F.R.:* Acworth J. M. 49; Akogo M.
276; Ekema N. 887, 1023; Tchouto P. 1157. *Southern Bakundu
F.R.:* Brenan J. P. M. 9308

Polygala tenuicaulis Hook.f. subsp. tenuicaulis
Polygala tenuicaulis F.W.T.A. 1: 114 (1954).

Habit: herb. **Habitat:** grassland. **Distribution:** Mount Cameroon.
Chorology: Endemic (montane). **Star:** black. **Alt. range:** 2000-
2200m (2), 2201-2400m (1), 2401-2600m (2). **Flowering:** Sep (1),
Oct (3). **Specimens:** - *Eastern slopes:* Cheek M. 3704, 3705;
Tchouto P. 268. *Etinde:* Tekwe C. 304. *No locality:* Mann G. 1282

Securidaca welwitschii Oliv.
F.W.T.A. 1: 110 (1954).

Habit: woody climber. **Guild:** pioneer. **Habitat:** forest.
Distribution: Guinea to Cameroon, Congo (Kinshasa), Angola and
Uganda. **Chorology:** Guineo-Congolian. **Star:** green. **Alt. range:**
600-800m (1). **Specimens:** - *Eastern slopes:* Maitland T. D. 569

POLYGONACEAE

Polygonum salicifolium Brouss. ex Willd.
F.W.T.A. 1: 141 (1954); F.T.E.A. Polygonaceae: 17 (1958); Fl.
Gabon 7: 14 (1963).

Habit: herb. **Distribution:** pantropical, also southern Europe.
Star: excluded. **Specimens:** - *Mabeta-Moliwe:* Mann G. 762

Polygonum setosulum A.Rich.
F.T.E.A. Polygonaceae: 22 (1958); F.W.T.A. 1: 759 (1958).
Polygonum nyikense Baker, F.W.T.A. 1: 142 (1954).

Habit: herb. **Distribution:** Mount Cameroon and Bioko to East
Africa. **Chorology:** Afromontane. **Star:** green. **Alt. range:** 1200-
1400m (1). **Specimens:** - *No locality:* Maitland T. D. 1115

Rumex abyssinicus Jacq.
F.W.T.A. 1: 139 (1954); F.T.E.A. Polygonaceae: 7 (1958); Fl.
Gabon 7: 8 (1963).

Habit: herb. **Guild:** pioneer. **Distribution:** tropical Africa.
Chorology: Afromontane. **Star:** green. **Alt. range:** 400-600m (1).
Specimens: - *Eastern slopes:* Maitland T. D. 297. *No locality:*
Mann G. 1217

Rumex nepalensis Spreng.
Fragm. Florist. Geobot. Suppl. 2 (1): 93 (1993)
Rumex bequaertii De Wild. F.W.T.A. 1: 139 (1954); F.T.E.A.
Polygonaceae: 8 (1958).

Habit: herb. **Guild:** pioneer. **Distribution:** tropical and southern
Africa. **Chorology:** Afromontane. **Star:** green. **Alt. range:** 2800-
3000m (1). **Flowering:** Jan (1). **Specimens:** - *No locality:* Mann
G. 1235

PORTULACACEAE

Talinum triangulare (Jacq.) Willd.
F.W.T.A. 1: 136 (1954); Fl. Gabon 7: 69 (1963).

Habit: herb. **Guild:** pioneer. **Habitat:** stream banks in forest. **Distribution:** tropical America, naturalized in Africa. **Star:** excluded. **Alt. range:** 1-200m (1). **Flowering:** Jun (1). **Fruiting:** Jun (1). **Specimens: -** *Mabeta-Moliwe:* Hutchinson J. & Metcalfe C. R. 137; Watts J. 354

PRIMULACEAE

Anagallis minima (L.) E.H.L.Krause
F.W.T.A. 2: 303 (1963).

Habit: herb. **Habitat:** grassland. **Distribution:** cosmopolitan. **Chorology:** Montane. **Star:** excluded. **Alt. range:** 2800-3000m (1). **Specimens: -** *Eastern slopes:* Morton J. K. K 867

Ardisiandra sibthorpioides Hook.f.
F.W.T.A. 2: 304 (1963).

Habit: herb. **Habitat:** forest. **Distribution:** Mount Cameroon, Bioko and East Africa. **Chorology:** Afromontane. **Star:** blue. **Alt. range:** 1600-1800m (1), 1801-2000m (1). **Flowering:** Oct (1). **Specimens: -** *Eastern slopes:* Thomas D. W. 9450. *No locality:* Mann G. 2022

RANUNCULACEAE

Clematis grandiflora DC.
F.W.T.A. 1: 64 (1954).

Habit: herbaceous climber. **Guild:** pioneer. **Habitat:** forest. **Distribution:** Guinea to Cameroon, Congo (Kinshasa), Angola and Uganda. **Chorology:** Guineo-Congolian (montane). **Star:** green. **Specimens: -** *No locality:* Maitland T. D. 438, 1108

Clematis simensis Fresen.
F.W.T.A. 1: 64 (1954).

Habit: herbaceous climber. **Habitat:** grassland. **Distribution:** tropical and southern Africa. **Chorology:** Afromontane. **Star:** green. **Alt. range:** 1600-1800m (1), 2001-2200m (1), 2801-3000m (1). **Flowering:** Oct (2). **Specimens: -** *Eastern slopes:* Sidwell K. 42; Tchouto P. 276. *No locality:* Mann G. 1245

Ranunculus multifidus Forssk.
F.T.E.A. Ranunculaceae: 19 (1952).
　Ranunculus extensus (Hook.f.) Schube ex Engl., F.W.T.A. 1: 64 (1954).

Habit: herb. **Habitat:** forest. **Distribution:** W. Cameroon, Bioko, E. Congo (Kinshasa) and East Africa. **Chorology:** Afromontane. **Star:** green. **Alt. range:** 1200-1400m (1). **Specimens: -** *Etinde:* Maitland T. D. 1121

Thalictrum rhynchocarpum Dill. & Rich.
F.W.T.A. 1: 64 (1954).

Habit: herb. **Habitat:** forest and forest-grassland transition. **Distribution:** Bioko, Cameroon and East Africa. **Chorology:** Afromontane. **Star:** blue. **Alt. range:** 800-1000m (1), 1601-1800m (1), 2001-2200m (1), 2201-2400m (1). **Flowering:** Oct (1). **Fruiting:** Sep (1), Oct (2). **Specimens: -** *Eastern slopes:* Tchouto P. 299; Thomas D. W. 9339. *Etinde:* Tekwe C. 235. *No locality:* Mann G. 1999

RHAMNACEAE

Gouania longipetala Hemsl.
F.W.T.A. 1: 670 (1958); Fl. Gabon 4: 68 (1962); Fl. Cameroun 33: 13 (1991).

Habit: woody climber. **Guild:** pioneer. **Habitat:** forest. **Distribution:** Guinea to Bioko, Congo (Brazzaville), Congo (Kinshasa) and Angola. **Chorology:** Guineo-Congolian. **Star:** green. **Alt. range:** 400-600m (1). **Flowering:** Oct (1). **Specimens: -** *Etinde:* Maitland T. D. 114. *Onge:* Tchouto P. 791

Lasiodiscus fasciculiflorus Engl.
F.W.T.A. 1: 671 (1958); Fl. Gabon 4: 60 (1962); Keay R.W.J., Trees of Nigeria: 323 (1989); Hawthorne W., F.G.F.T. Ghana: 41 (1990); Fl. Cameroun 33: 10 (1991).

Habit: tree to 12 m. **Guild:** shade-bearer. **Habitat:** forest. **Distribution:** Sierra Leone to Congo (Kinshasa). **Chorology:** Guineo-Congolian. **Star:** green. **Alt. range:** 1-200m (1). **Flowering:** Nov (1). **Fruiting:** Nov (1). **Specimens: -** *Onge:* Watts J. 961

Lasiodiscus mannii Hook.f.
F.W.T.A. 1: 671 (1958); Adansonia (ser.2) 2: 132 (1962); Fl. Gabon 4: 63 (1962); Keay R.W.J., Trees of Nigeria: 321 (1989); Hawthorne W., F.G.F.T. Ghana: 41 (1990); Fl. Cameroun 33: 5 (1991).

Habit: tree to 10 m. **Guild:** shade-bearer. **Habitat:** forest. **Distribution:** Cameroon, Gabon and Congo (Kinshasa). **Chorology:** Guineo-Congolian. **Star:** green. **Specimens: -** *Southern Bakundu F.R.:* Brenan J. P. M. 9439

Lasiodiscus sp.

Specimens: - *Onge:* Tchouto P. 700

Maesopsis eminii Engl.
F.W.T.A. 1: 669 (1958); Fl. Gabon 4: 51 (1962); Keay R.W.J., Trees of Nigeria: 321 (1989); Hawthorne W., F.G.F.T. Ghana: 41 (1990).

Habit: tree to 30 m. **Guild:** pioneer. **Habitat:** forest. **Distribution:** tropical Africa. **Star:** excluded. **Specimens: -** *Mabeta-Moliwe:* Plot Voucher W 761

Ventilago africana Exell
F.W.T.A. 1: 670 (1958); Fl. Gabon 4: 56 (1962).

Habit: woody climber. Guild: light demander. Habitat: lava flows. Distribution: Guinea Bissau to Cameroon, Gabon, Congo (Kinshasa), Angola and Uganda. Chorology: Guineo-Congolian. Star: green. Alt. range: 1600-1800m (1). Specimens: - *Eastern slopes:* Banks H. 21

Ventilago diffusa (G.Don) Exell
F.W.T.A. 1: 670 (1958); Fl. Cameroun 33: 36 (1991).

Habit: woody climber. Habitat: forest. Distribution: Nigeria, Cameroon, S. Tomé and Congo (Kinshasa). Chorology: Guineo-Congolian. Star: green. Specimens: - *Etinde:* Winkler H. J. P. 606

RHIZOPHORACEAE

Anopyxis klaineana (Pierre) Engl.
F.W.T.A. 1: 286 (1954); Keay R.W.J., Trees of Nigeria: 99 (1989); Hawthorne W., F.G.F.T. Ghana: 41 (1990).

Habit: tree to 50 m. Guild: light demander. Habitat: forest. Distribution: Sierra Leone to Cameroon, Gabon, Congo (Kinshasa) and Angola (Cabinda). Chorology: Guineo-Congolian. Star: red. Alt. range: 200-400m (1). Specimens: - *Mokoko River F.R.:* Thomas D. W. 10087

Cassipourea cf. afzelii (Oliv.) Alston

Specimens: - *Mabeta-Moliwe:* Cheek M. 3565

Cassipourea gummiflua Tul. var. mannii (Hook.f. ex Oliv.) J.Lewis
Kew Bull. 10: 158 (1955); F.W.T.A. 1: 762 (1958); Fl. Afr. Cent. Rhizophoraceae: 19 (1987); Keay R.W.J., Trees of Nigeria: 96 (1989); Hawthorne W., F.G.F.T. Ghana: 41 (1990).
    Cassipourea glabra Alston, F.W.T.A. 1: 283 (1954).

Habit: tree to 10 m. Habitat: forest. Distribution: Sierra Leone to Angola. Chorology: Guineo-Congolian. Star: green. Alt. range: 1400-1600m (1). Flowering: Oct (1). Specimens: - *Etinde:* Wheatley J. I. 622

Rhizophora harrisonii Leechman
F.W.T.A. 1: 285 (1954).

Habit: tree or shrub to 6 m. Habitat: mangrove forest. Distribution: Senegal to Angola and eastern coasts of tropical America. Star: excluded. Flowering: Jun (1). Specimens: - *Mabeta-Moliwe:* Plot Voucher W 1183

Rhizophora racemosa G.F.W.Mey.
F.W.T.A. 1: 285 (1954); Keay R.W.J., Trees of Nigeria: 96 (1989); Hawthorne W., F.G.F.T. Ghana: 43 (1990).

Habit: tree to 30 m. Habitat: mangrove forest. Distribution: Senegal to Cameroon, Gabon, Congo (Brazzaville) and Congo (Kinshasa), also the east coast of tropical America. Star: excluded. Alt. range: 1-200m (1). Flowering: Mar (1). Specimens: - *Mabeta-Moliwe:* Maitland T. D. 972; Sunderland T. C. H. 1104

ROSACEAE

Alchemilla cryptantha Steud. ex A.Rich.
Fl. Congo B., Rw. & Bur. 3: 7 (1952); F.W.T.A. 1: 424 (1958); Fl. Cameroun 20: 197 (1978); Fl. Ethiopia 3: 39 (1989).

Habit: herb. Guild: pioneer. Habitat: grassland. Distribution: tropical and southern Africa. Chorology: Afromontane. Star: green. Alt. range: 2400-2600m (1). Specimens: - *No locality:* Mann G. 1984

Alchemilla kiwuensis Engl.
Fl. Congo B., Rw. & Bur. 3: 5 (1952); F.W.T.A. 1: 424 (1958); Fl. Cameroun 20: 204 (1978); Nord. J. Bot. 6: 576 (1986).

Habit: herb. Guild: pioneer. Habitat: forest. Distribution: tropical Africa. Chorology: Afromontane. Star: green. Alt. range: 2000-2200m (1). Specimens: - *Eastern slopes:* Brenan J. P. M. 9549

Prunus africana (Hook.f.) Kalkman
F.T.E.A. Rosaceae: 46 (1960); Blumea 13: 33 (1965); Fl. Cameroun 20: 209 (1978); Fl. Ethiopia 3: 32 (1989); Keay R.W.J., Trees of Nigeria: 181 (1989).
    Pygeum africanum Hook.f., F.W.T.A. 1: 426 (1958).

Habit: tree to 25 m. Habitat: forest-grassland transition. Distribution: tropical and subtropical Africa. Chorology: Afromontane. Star: scarlet. Alt. range: 2200-2400m (1). Specimens: - *No locality:* Mann G. 1207

Rubus pinnatus Willd. var. afrotropicus (Engl.) Gust.
F.W.T.A. 1: 426 (1958); Fl. Cameroun 20: 218 (1978).

Habit: scrambling shrub. Guild: pioneer. Habitat: forest-grassland transition. Distribution: tropical and southern Africa. Chorology: Afromontane. Star: green. Alt. range: 800-1000m (1), 1801-2000m (2), 2001-2200m (1). Flowering: Jul (1). Fruiting: Sep (1), Nov (1). Specimens: - *Eastern slopes:* Cheek M. 5364; Tekwe C. 108. *Etinde:* Tekwe C. 280. *No locality:* Mann G. 1221

Rubus sp. aff. rigidus Sm. var. camerunensis Letouzey

Specimens: - *Eastern slopes:* Thomas D. W. 9460

Rubus rosifolius Sm.
F.W.T.A. 1: 426 (1958); Fl. Cameroun 20: 229 (1978); Blumea 27: 81 (1981); Fl. Ethiopia 3: 33 (1989).

Habit: shrub to 1 m. Guild: pioneer. Habitat: forest-grassland transition. Distribution: introduced from Asia. Chorology: Montane. Star: excluded. Specimens: - *Eastern slopes:* Dundas J. FHI 15232

S. Dawson, S. Cable, K. Sidwell, M. Cheek &
D. Bridson (K)

Aidia genipiflora (DC.) Dandy
Bull. Jard. Bot. Brux. 28: 22 (1958); F.W.T.A. 2: 114 (1963); Fl.
Gabon 17: 164 (1970); Keay R.W.J., Trees of Nigeria: 427 (1989);
Hawthorne W., F.G.F.T. Ghana: 35 (1990).

Habit: tree to 12 m. **Guild:** shade-bearer. **Habitat:** forest.
Distribution: Guinea Bissau to Bioko, Cameroon and Sudan.
Chorology: Guineo-Congolian. **Star:** green. **Alt. range:** 1-200m
(2). **Flowering:** Apr (1), Jul (1). **Fruiting:** Jul (1). **Specimens:** -
Mabeta-Moliwe: Cheek M. 3561. *Mokoko River F.R.:* Tchouto P.
1074

Aidia micrantha (K.Schum.) White
Fl. Gabon 17: 165 (1970).

Habit: tree or shrub to 9 m. **Habitat:** forest. **Distribution:**
Cameroon, CAR, Rio Muni, Gabon, Congo (Kinshasa) and
Angola. **Chorology:** Guineo-Congolian. **Star:** green. **Specimens:** -
No locality: Cheek M. 3036

Aidia rhacodosepala (K.Schum.) Petit
Fl. Gabon 17: 164 (1970).

Habit: tree or shrub to 10 m. **Habitat:** forest. **Distribution:**
Cameroon. **Chorology:** Lower Guinea. **Star:** gold. **Alt. range:** 1-
200m (2). **Flowering:** Apr (2). **Fruiting:** Apr (1). **Specimens:** -
Mabeta-Moliwe: Watts J. 209. *Mokoko River F.R.:* Tchouto P.
1113

Aidia rubens (Hiern) G.Taylor
F.W.T.A. 2: 114 (1963); Fl. Gabon 17: 170 (1970).

Habit: tree to 6 m. **Habitat:** forest. **Distribution:** Nigeria and
Cameroon. **Chorology:** Lower Guinea. **Star:** blue. **Alt. range:** 1-
200m (1). **Flowering:** May (1). **Fruiting:** May (1). **Specimens:** -
Mokoko River F.R.: Batoum A. 46; Fraser P. 388

Aidia sp.

Specimens: - *Mokoko River F.R.:* Ndam N. 1116

Anthospermum asperuloides Hook.f.
F.W.T.A. 2: 223 (1963); Pl. Syst. Evol. Suppl. 3: 294 (1986).
Anthospermum cameroonense Hutch. & Dalziel, F.W.T.A. 2:
223 (1963).

Habit: shrub to 0.5 m. **Habitat:** forest-grassland transition.
Distribution: Mount Cameroon and Bioko. **Chorology:** Western
Cameroon Uplands (montane). **Star:** black. **Alt. range:** 2400-
2600m (1), 2601-2800m (1). **Specimens:** - *No locality:* Boughey
A. S. s.n.; Brenan J. P. M. 9531; Breteler F. J. MC 108; Keay R.
W. J. FHI 28592; Mann G. 1290; Morton J. K. K 796

Aoranthe cladantha (K.Schum.) Somers
Bull. Jard. Bot. Nat. Belg. 58: 47 (1988).
Porterandia cladantha (K.Schum.) Keay, F.W.T.A. 2: 114
(1963).

Habit: tree to 15 m. **Habitat:** forest. **Distribution:** Nigeria,
Cameroon, S. Tomé, CAR, Gabon, Congo (Brazzaville), Congo

(Kinshasa) and Angola. **Chorology:** Guineo-Congolian. **Star:**
green. **Specimens:** - *Mabeta-Moliwe:* Plot Voucher W 763

Argocoffeopsis afzelii (Hiern) Robbr.
Bull. Jard. Bot. Nat. Belg. 51: 365 (1981).
Coffea afzelii Hiern, F.W.T.A. 2: 156 (1963).

Habit: woody climber. **Guild:** shade-bearer. **Habitat:** forest.
Distribution: Sierra Leone to Cameroon and Congo (Kinshasa).
Chorology: Guineo-Congolian. **Star:** green. **Specimens:** - *Etinde:*
Maitland T. D. 527

Argocoffeopsis subcordata (Hiern) Lebrun
Bull. Jard. Bot. Nat. Belg. 51: 366 (1981).
Coffea subcordata Hiern, F.W.T.A. 2: 156 (1963).

Habit: woody climber. **Habitat:** forest. **Distribution:** SE Nigeria,
Cameroon, Gabon, Congo (Kinshasa) and Angola (Cabinda).
Chorology: Guineo-Congolian. **Star:** green. **Specimens:** -
Southern Bakundu F.R.: Onochie C. F. A. FHI 30854, FHI 35477

Argostemma africanum K.Schum.
F.W.T.A. 2: 208 (1963); Fl. Gabon 17: 6 (1970).

Habit: herb. **Habitat:** on wet rocks in forest. **Distribution:** SE
Nigeria, Cameroon and Rio Muni. **Chorology:** Lower Guinea.
Star: blue. **Alt. range:** 400-600m (2), 601-800m (1). **Flowering:**
Oct (3). **Fruiting:** Oct (1). **Specimens:** - *Etinde:* Cheek M. 3758;
Thomas D. W. 9140; Watts J. 528. *Mabeta-Moliwe:* Schlechter F.
R. R. 15793

Atractogyne bracteata (Wernham) Hutch. & Dalziel
F.W.T.A. 2: 158 (1963); Fl. Gabon 17: 184 (1970).

Habit: woody climber. **Guild:** pioneer. **Habitat:** forest.
Distribution: Ivory Coast, SE Nigeria, Cameroon and Gabon.
Chorology: Guineo-Congolian. **Star:** green. **Alt. range:** 1-200m
(2), 201-400m (1). **Flowering:** May (1), Jul (1), Sep (1). **Fruiting:**
May (1), Jul (1), Sep (1). **Specimens:** - *Etinde:* Wheatley J. I. 492.
Mabeta-Moliwe: Cheek M. 3524; Wheatley J. I. 268

Atractogyne gabonii Pierre
F.W.T.A. 2: 158 (1963); Fl. Gabon 17: 181 (1970).

Habit: woody climber. **Habitat:** forest. **Distribution:** Cameroon,
Gabon and Congo (Brazzaville). **Chorology:** Guineo-Congolian.
Star: green. **Specimens:** - *Etinde:* Maitland T. D. 393

Aulacocalyx caudata (Hiern) Keay
F.W.T.A. 2: 145 (1963); Fl. Gabon 17: 155 (1970); Kew Bull. 52:
641 (1997).

Habit: tree or shrub to 5 m. **Habitat:** forest. **Distribution:** SE
Nigeria, Cameroon, Rio Muni and Gabon. **Chorology:** Lower
Guinea. **Star:** blue. **Alt. range:** 1-200m (2), 201-400m (1).
Flowering: May (1), Jun (3). **Specimens:** - *Etinde:* Kwangue A.
116; Tchouto P. 228. *Mabeta-Moliwe:* Plot Voucher W 170, W
864, W 1027; Watts J. 400

Aulacocalyx jasminiflora Hook.f.
F.W.T.A. 2: 145 (1963); Keay R.W.J., Trees of Nigeria: 427
(1989); Kew Bull. 52: 647 (1997).

Habit: tree to 13 m. **Guild:** shade-bearer. **Habitat:** forest. **Distribution:** Liberia to Cameroon, Gabon, Congo (Brazzaville), Congo (Kinshasa), Angola, Zambia and CAR. **Chorology:** Guineo-Congolian (montane). **Star:** green. **Flowering:** Mar (1). **Specimens:** - *Eastern slopes:* Maitland T. D. 456

Aulacocalyx talbotii (Wernham) Keay

F.W.T.A. 2: 145 (1963); Fl. Gabon 17: 160 (1970); Kew Bull. 52: 652 (1997).

Habit: tree to 15 m. **Habitat:** forest. **Distribution:** SE Nigeria, Cameroon and Gabon. **Chorology:** Lower Guinea. **Star:** blue. **Alt. range:** 1-200m (1), 201-400m (1).**Fruiting:** Apr (2). **Specimens:** - *Mokoko River F.R.:* Acworth J. M. 108; Mbani J. M. 307; Ndam N. 1096

Belonophora coriacea Hoyle

F.W.T.A. 2: 158 (1963).

Habit: tree to 12 m. **Habitat:** forest. **Distribution:** SE Nigeria, Cameroon, CAR, Gabon and Congo (Kinshasa). **Chorology:** Guineo-Congolian. **Star:** green. **Alt. range:** 1-200m (4), 201-400m (2), 401-600m (2). **Flowering:** May (1), Nov (1). **Fruiting:** Mar (1), Apr (2), May (3). **Specimens:** - *Etinde:* Mbani J. M. 141; Tchouto P. 28; Williams S. 47. *Mabeta-Moliwe:* Jaff B. 260; Plot Voucher W 94, W 165; Tchouto P. 201; Watts J. 168; Wheatley J. I. 199. *No locality:* Thomas D. W. 4431

Belonophora wernhamii Hutch. & Dalziel

F.W.T.A. 2: 158 (1963).

Habit: shrub to 2.5 m. **Habitat:** forest. **Distribution:** SE Nigeria and Cameroon. **Chorology:** Lower Guinea. **Star:** blue. **Specimens:** - *Onge:* Cheek M. 5049; *Southern Bakundu F.R.:* Brenan J. P. M. 9292, 9437

Belonophora sp. nov.

Habit: tree or shrub 3-6 m tall. **Habitat:** forest. **Distribution:** SW Cameroon. **Chorology:** Western Cameroon. **Star:** black. **Alt. range:** 1-200m (2). **Flowering:** Oct (2). **Fruiting:** Oct (1). **Specimens:** - *Onge:* Sonké 667, 1169; Tchouto P. 497, 731, 916

Bertiera bracteolata Hiern

F.W.T.A. 2: 159 (1963); Fl. Gabon 17: 38 (1970).

Habit: woody climber. **Habitat:** forest. **Distribution:** Sierra Leone to Cameroon, Gabon and Congo (Kinshasa). **Chorology:** Guineo-Congolian. **Star:** green. **Specimens:** - *Southern Bakundu F.R.:* Binuyo A. & Daramola B. O. 3549

Bertiera breviflora Hiern

F.W.T.A. 2: 159 (1963); Fl. Gabon 17: 47 (1970).

Habit: tree or shrub to 4 m. **Habitat:** forest. **Distribution:** Sierra Leone to Bioko, Cameroon, Gabon and Congo (Kinshasa). **Chorology:** Guineo-Congolian. **Star:** green. **Alt. range:** 1-200m (2). **Flowering:** Dec (1). **Fruiting:** Mar (2), Dec (1). **Specimens:** - *Mabeta-Moliwe:* Etuge M. 1217; Groves M. 232; Sunderland T. C. H. 1019. *Southern Bakundu F.R.:* Ejiofor M. C. FHI 29344

Bertiera laxa Benth.

F.W.T.A. 2: 159 (1963); Fl. Gabon 17: 44 (1970).

Habit: tree or shrub to 5 m. **Habitat:** forest. **Distribution:** Nigeria, Bioko and Cameroon. **Chorology:** Lower Guinea. **Star:** blue. **Alt. range:** 1-200m (15), 201-400m (4), 401-600m (1), 601-800m (2), 1001-1200m (1), 1601-1800m (1). **Flowering:** Mar (2), Apr (1), May (3), Jun (2), Jul (1), Sep (1), Oct (1). **Fruiting:** Jan (1), Mar (3), Apr (3), May (7), Jun (3), Jul (1), Sep (1), Oct (1). **Specimens:** - *Bambuko F.R.:* Keay R. W. J. FHI 37540. *Etinde:* Cheek M. 3615; Khayota B. 521; Maitland T. D. 544; Mbani J. M. 128; Tchouto P. 397; Tekwe C. 2; Wheatley J. I. 486. *Mabeta-Moliwe:* Groves M. 236; Jaff B. 84, 175; Nguembock F. 18; Plot Voucher W 474; Sunderland T. C. H. 1161; Watts J. 240, 287; Wheatley J. I. 62, 363. *Mokoko River F.R.:* Acworth J. M. 124, 263, 285; Ekema N. 1001, 1111, 1186; Fraser P. 416, 433; Ndam N. 1314

Bertiera racemosa (G.Don) K.Schum. var. elephantina N.Hallé

Fl. Gabon 17: 56 (1970); F.T.E.A. Rubiaceae: 482 (1988).

Habit: tree or shrub to 8 m. **Guild:** pioneer. **Habitat:** forest. **Distribution:** Sierra Leone to Cameroon, Gabon and Congo (Kinshasa). **Chorology:** Guineo-Congolian. **Star:** green. **Alt. range:** 1-200m (5).**Fruiting:** Apr (1), May (2), Jun (2). **Specimens:** - *Mokoko River F.R.:* Akogo M. 227; Ekema N. 994, 1135; Pouakouyou D. 51; Watts J. 1093

Bertiera racemosa (G.Don) K.Schum. var. racemosa (G.Don) K.Schum.

F.W.T.A. 2: 160 (1963); Fl. Gabon 17: 54 (1970); F.T.E.A. Rubiaceae: 480 (1988); Hawthorne W., F.G.F.T. Ghana: 30 (1990).

Habit: tree or shrub to 8 m. **Habitat:** forest. **Distribution:** Sierra Leone to Bioko, Cameroon, Gabon, Congo (Kinshasa), Angola and Uganda. **Chorology:** Guineo-Congolian. **Star:** green. **Specimens:** - *Mokoko River F.R.:* Ndam N. 1193

Bertiera cf. racemosa (G.Don) K.Schum.

Specimens: - *Onge:* Watts J. 736

Bertiera retrofracta K.Schum.

F.W.T.A. 2: 160 (1963).

Habit: shrub to 3 m. **Habitat:** forest. **Distribution:** SE Nigeria, Bioko and SW Cameroon. **Chorology:** Western Cameroon. **Star:** gold. **Alt. range:** 1-200m (4), 201-400m (2), 401-600m (2). **Flowering:** Mar (2), May (2), Jun (3). **Fruiting:** Feb (1), Mar (2), Apr (1), May (1), Jun (1), Sep (1). **Specimens:** - *Etinde:* Khayota B. 585; Kwangue A. 101; Preuss P. R. 1279; Tchouto P. 111; Wheatley J. I. 26, 516. *Mabeta-Moliwe:* Cheek M. 3271, 3489; Plot Voucher W 453, W 468, W 527, W 625, W 824; Watts J. 264; Wheatley J. I. 64. *Mokoko River F.R.:* Watts J. 1154. *Southern Bakundu F.R.:* Binuyo A. & Daramola B. O. FHI 35452

Bertiera sp.

Specimens: - *Mokoko River F.R.:* Ekema N. 1120a

Brenania brieyi (De Wild.) Petit

F.W.T.A. 2: 116 (1963); Fl. Gabon 17: 255 (1970); Keay R.W.J., Trees of Nigeria: 424 (1989).

Habit: tree to 30 m. **Habitat:** forest. **Distribution:** Nigeria, Cameroon, Gabon, Congo (Kinshasa) and Angola (Cabinda). **Chorology:** Guineo-Congolian. **Star:** green. **Alt. range:** 1-200m

(1). **Flowering:** Apr (1). **Specimens:** - *Mokoko River F.R.:* Tchouto P. 1119

Calochone acuminata Keay
F.W.T.A. 2: 119 (1963); Fl. Gabon 17: 213 (1970).

Habit: woody climber to 10 m. **Habitat:** forest and Aframomum thicket. **Distribution:** Cameroon, Gabon and Angola (Cabinda). **Chorology:** Lower Guinea. **Star:** gold. **Alt. range:** 800-1000m (1).**Fruiting:** Oct (1). **Specimens:** - *Etinde:* Watts J. 567

Calycosiphonia macrochlamys (K.Schum.) Robbr.
Fl. Gabon 17: 146 (1970); Bull. Jard. Bot. Nat. Belg. 51: 377 (1981).
Coffea macrochlamys K.Schum., F.W.T.A. 2: 156 (1963).

Habit: tree to 5 m. **Habitat:** forest. **Distribution:** Ghana, Bioko, Cameroon, Gabon and Congo (Kinshasa). **Chorology:** Guineo-Congolian. **Star:** green. **Alt. range:** 600-800m (1).**Fruiting:** Mar (1). **Specimens:** - *Etinde:* Mbani J. M. 48

Calycosiphonia spathicalyx (K.Schum.) Robbr.
Bull. Jard. Bot. Nat. Belg. 51: 373 (1981); F.T.E.A. Rubiaceae: 727 (1988); Hawthorne W., F.G.F.T. Ghana: 35 (1990).
Coffea spathicalyx K.Schum., F.W.T.A. 2: 156 (1963).

Habit: tree or shrub to 8 m. **Guild:** shade-bearer. **Habitat:** forest. **Distribution:** tropical Africa. **Star:** excluded. **Specimens:** - *Bambuko F.R.:* Keay R. W. J. FHI 37463

Chassalia cristata (Hiern) Bremek.
F.W.T.A. 2: 192 (1963); F.T.E.A. Rubiaceae: 122 (1976).

Habit: shrub to 3 m. **Habitat:** forest. **Distribution:** SE Nigeria, Bioko, Cameroon, Cameroon, Congo (Kinshasa) and Angola. **Chorology:** Guineo-Congolian. **Star:** green. **Alt. range:** 800-1000m (1), 1201-1400m (1). **Specimens:** - *No locality:* Deistel H. 645; Maitland T. D. 489, 1114

Chassalia ischnophylla (K.Schum.) Hepper
F.W.T.A. 2: 192 (1963).

Habit: shrub to 0.5 m. **Habitat:** forest. **Distribution:** Nigeria and SW Cameroon. **Chorology:** Lower Guinea. **Star:** gold. **Alt. range:** 200-400m (1). **Flowering:** Oct (1). **Fruiting:** Oct (1). **Specimens:** - *Etinde:* Tchouto P. 476

Chassalia kolly (Schumach.) Hepper
F.W.T.A. 2: 192 (1963).

Habit: shrub to 3 m. **Guild:** pioneer. **Habitat:** forest. **Distribution:** Guinea to Bioko and SW Cameroon. **Chorology:** Guineo-Congolian. **Star:** green. **Alt. range:** 1-200m (1).**Fruiting:** May (1). **Specimens:** - *Mabeta-Moliwe:* Cheek M. 3241; Jaff B. 176

Chassalia sp.

Specimens: - *Mabeta-Moliwe:* Jaff B. 88; Sunderland T. C. H. 1199. *Mokoko River F.R.:* Batoum A. 37; Ekema N. 886, 1012; Fraser P. 381; Ndam N. 1295; Pouakouyou D. 47. *No locality:* Breteler F. J. 291; Hambler D. J. 144; Hutchinson J. & Metcalfe C. R. 70; Maitland T. D. 706

Chazaliella domatiicola (De Wild.) Petit & Verdc.
Kew Bull. 30: 269 (1975).
Psychotria abrupta sensu Hepper, F.W.T.A. 2: 201 (1963).
Psychotria domatiicola De Wild., Pl. Bequaert. 2: 362 (1924).

Habit: shrub to 2 m. **Habitat:** forest. **Distribution:** Ghana, Nigeria, Cameroon, Rio Muni and Congo (Kinshasa). **Chorology:** Guineo-Congolian. **Star:** green. **Alt. range:** 400-600m (1). **Specimens:** - *No locality:* Thomas D. W. 2771

Chazaliella insidens (Hiern) Petit & Verdc. subsp. insidens
Kew Bull. 30: 269 (1975).
Psychotria insidens Hiern, F.W.T.A. 2: 200 (1963).

Habit: shrub to 1 m. **Habitat:** forest. **Distribution:** Nigeria, Bioko and Cameroon. **Chorology:** Lower Guinea. **Star:** blue. **Alt. range:** 1-200m (1), 201-400m (4), 401-600m (1), 601-800m (2), 801-1000m (3), 1001-1200m (1). **Flowering:** Feb (1), Mar (2), Apr (1), Nov (3), Dec (2). **Fruiting:** Oct (1), Nov (1), Dec (1). **Specimens:** - *Bambuko F.R.:* Watts J. 584. *Eastern slopes:* Kwangue A. 54. *Etinde:* Cable S. 219; Cheek M. 5527; Lighava M. 11, 22; Tekwe C. 32; Wheatley J. I. 15; Williams S. 58. *No locality:* Letouzey R. 14988. *Onge:* Tchouto P. 786; Wheatley J. I. 683

Chazaliella obanensis (Wernham) Petit & Verdc.
Kew Bull. 31: 798 (1977).
Psychotria obanensis Wernham, F.W.T.A. 2: 200 (1963).

Habit: shrub to 3 m. **Habitat:** forest. **Distribution:** SE Nigeria and SW Cameroon. **Chorology:** Western Cameroon. **Star:** gold. **Alt. range:** 1-200m (3), 201-400m (3). **Flowering:** Aug (1), Sep (1), Oct (2). **Fruiting:** Oct (3), Nov (1). **Specimens:** - *Mabeta-Moliwe:* Baker W. J. 259. *Onge:* Cheek M. 5054; Harris D. J. 3727; Ndam N. 693; Watts J. 722, 757

Chazaliella oddonii (De Wild.) Petit & Verdc. var. cameroonensis Verdc.
Kew Bull. 31: 802 (1977).

Habit: shrub to 1 m. **Habitat:** forest. **Distribution:** Nigeria and Cameroon. **Chorology:** Lower Guinea. **Star:** black. **Alt. range:** 200-400m (1). **Specimens:** - *No locality:* Hutchinson J. & Metcalfe C. R. 158

Chazaliella oddonii (De Wild.) Petit & Verdc. var. oddonii
Kew Bull. 31: 801 (1977).

Habit: shrub to 1 m. **Habitat:** forest. **Distribution:** Cameroon, Congo (Brazzaville) and Congo (Kinshasa). **Chorology:** Guineo-Congolian. **Star:** green. **Flowering:** May (1). **Specimens:** - *Mokoko River F.R.:* Ndam N. 1138

Chazaliella sciadephora (Hiern) Petit & Verdc. var. condensata Verdc.
Kew Bull. 31: 793 (1977).

Habit: shrub to 3 m. **Guild:** shade-bearer. **Habitat:** forest. **Distribution:** Cameroon. **Chorology:** Lower Guinea. **Star:** gold. **Alt. range:** 200-400m (1).**Fruiting:** May (1). **Specimens:** - *Mokoko River F.R.:* Ndam N. 1180

Chazaliella sciadephora (Hiern) Petit & Verdc. var. sciadephora

Kew Bull. 31: 790 (1977).
Psychotria sciadephora Hiern, F.W.T.A. 2: 201 (1963).

Habit: shrub to 3 m. **Habitat:** forest. **Distribution:** Guinea to Cameroon. **Chorology:** Guineo-Congolian. **Star:** green. **Alt. range:** 1-200m (24), 201-400m (3), 401-600m (2), 601-800m (1), 801-1000m (4), 1001-1200m (1), 1201-1400m (1). **Flowering:** Feb (4), Mar (6), Apr (1), May (1), Jun (1), Jul (1), Oct (1), Nov (1), Dec (2). **Fruiting:** Apr (1), May (4), Jun (5), Jul (1), Sep (1), Oct (3), Nov (5), Dec (1). **Specimens:** - *Eastern slopes:* Cable S. 1350, 1361; Tekwe C. 101. *Etinde:* Cable S. 171, 197, 271, 424; Cheek M. 5385, 5470, 5832, 5837; Kwangue A. 82; Mbani J. M. 64; Nkeng P. 47; Tekwe C. 31; Wheatley J. I. 548. *Mabeta-Moliwe:* Cable S. 1364, 1405, 1436; Dawson S. 50; Hunt L. V. 6; Jaff B. 17, 226; Plot Voucher W 54, W 311, W 401, W 568, W 800, W 582 b; Sidwell K. 2; Sunderland T. C. H. 1115, 1290, 1448; Watts J. 250, 375. *No locality:* Cheek M. 3022, 3048; Leeuwenberg A. J. M. 6911; Maitland T. D. 741; Mann G. 729, 1192; Nkeng P. 7; Thompson S. 1558. *Onge:* Harris D. J. 3790; Tchouto P. 789

Chazaliella sp.

Specimens: - *No locality:* Hepper F. N. 8613; Thompson S. 1391, 1662

Coffea brevipes Hiern

F.W.T.A. 2: 156 (1963).

Habit: shrub to 2 m. **Habitat:** forest. **Distribution:** Cameroon and Congo (Kinshasa). **Chorology:** Guineo-Congolian. **Star:** green. **Alt. range:** 1-200m (1), 601-800m (1), 801-1000m (1). **Specimens:** - *Bambuko F.R.:* Keay R. W. J. FHI 37394, FHI 37420. *Etinde:* Preuss P. R. 1383; Winkler H. J. P. 539. *No locality:* Mann G. 2158; Thomas D. W. 5119

Coffea canephora Pierre ex Froehner

F.W.T.A. 2: 154 (1963); F.T.E.A. Rubiaceae: 710 (1988); Hawthorne W., F.G.F.T. Ghana: 32 (1990).

Habit: tree or shrub to 6 m. **Guild:** shade-bearer. **Habitat:** farmbush. **Distribution:** Guinea to Cameroon, Gabon, Congo (Brazzaville), Congo (Kinshasa), Angola, Uganda and Sudan. **Chorology:** Guineo-Congolian. **Star:** green. **Specimens:** - *Mabeta-Moliwe:* Mildbraed G. W. J. 10766

Coffea liberica Bull. ex Hiern

F.W.T.A. 2: 154 (1963); F.T.E.A. Rubiaceae: 706 (1988).

Habit: tree or shrub to 9 m. **Habitat:** forest. **Distribution:** Cameroon, S. Tomé, CAR, Gabon, Congo (Kinshasa), Sudan and East Africa. **Star:** pink. **Specimens:** - *Mabeta-Moliwe:* Plot Voucher W 584

Coffea sp.

Specimens: - *Mokoko River F.R.:* Ekema N. 957; Tchouto P. 1190, 1215

Corynanthe pachyceras K.Schum.

F.W.T.A. 2: 111 (1963); Fl. Gabon 12: 62 (1966); Keay R.W.J., Trees of Nigeria: 425 (1989); Hawthorne W., F.G.F.T. Ghana: 32 (1990); Bot. J. Linn. Soc. 120: 303 (1996).

Habit: tree to 22 m. **Guild:** light demander. **Habitat:** forest. **Distribution:** Sierra Leone to Cameroon, Gabon, Congo (Brazzaville) and Congo (Kinshasa). **Chorology:** Guineo-Congolian. **Star:** green. **Alt. range:** 200-400m (1). **Fruiting:** Mar (1). **Specimens:** - *Mokoko River F.R.:* Tchouto P. 566. *Southern Bakundu F.R.:* Binuyo A. & Daramola B. O. FHI 35478

Craterispermum aristatum Wernham

F.W.T.A. 2: 188 (1963).

Habit: tree or shrub to 4 m. **Habitat:** forest. **Distribution:** SE Nigeria and SW Cameroon. **Chorology:** Western Cameroon. **Star:** gold. **Alt. range:** 200-400m (1). **Fruiting:** May (1). **Specimens:** - *Mokoko River F.R.:* Thomas D. W. 10108

Craterispermum caudatum Hutch.

F.W.T.A. 2: 188 (1963); Hawthorne W., F.G.F.T. Ghana: 35 (1990).

Habit: tree or shrub to 5 m. **Guild:** shade-bearer. **Habitat:** forest. **Distribution:** Sierra Leone to Cameroon. **Chorology:** Guineo-Congolian. **Star:** green. **Alt. range:** 1-200m (7), 201-400m (1), 401-600m (1). **Flowering:** Apr (1), May (3), Jun (3). **Fruiting:** May (2), Jun (2). **Specimens:** - *Etinde:* Thomas D. W. 2795. *Mabeta-Moliwe:* Plot Voucher W 53; Watts J. 324. *Mokoko River F.R.:* Ekema N. 1174; Pouakouyou D. 64; Tchouto P. 1161; Watts J. 1041, 1139, 1181. *Southern Bakundu F.R.:* Mbani J. M. 413

Craterispermum sp. aff. cerinanthum Hiern

Specimens: - *Mokoko River F.R.:* Mbani J. M. 394, 404; Ndam N. 1124. *Southern Bakundu F.R.:* Binuyo A. & Daramola B. O. FHI 35620

Cremaspora triflora (Thonn.) K.Schum. subsp. triflora

F.W.T.A. 2: 148 (1963); F.T.E.A. Rubiaceae: 733 (1988).

Habit: woody climber. **Guild:** shade-bearer. **Habitat:** forest. **Distribution:** tropical Africa. **Star:** excluded. **Alt. range:** 1-200m (1), 201-400m (2), 401-600m (1). **Flowering:** Apr (1), Oct (1). **Fruiting:** Mar (1), Apr (1), Sep (1), Oct (1). **Specimens:** - *Etinde:* Tchouto P. 454; Wheatley J. I. 523. *Mabeta-Moliwe:* Nguembock F. 80; Wheatley J. I. 74, 123

Cremaspora sp.

Specimens: - *Mabeta-Moliwe:* Tchouto P. 1047. *Mokoko River F.R.:* Fraser P. 425; Pouakouyou D. 27; Tchouto P. 1239

Cuviera acutiflora DC.

F.W.T.A. 2: 177 (1963); Keay R.W.J., Trees of Nigeria: 429 (1989); Hawthorne W., F.G.F.T. Ghana: 39 (1990).

Habit: tree or shrub to 10 m. **Guild:** pioneer. **Habitat:** forest. **Distribution:** Guinea to Cameroon and Gabon. **Chorology:** Guineo-Congolian. **Star:** green. **Alt. range:** 1-200m (1). **Flowering:** Oct (1). **Specimens:** - *Etinde:* Kalbreyer W. 29; Maitland T. D. 784. *Mabeta-Moliwe:* Mann G. 770. *Onge:* Watts J. 893

Cuviera calycosa Wernham

F.W.T.A. 2: 177 (1963).

Habit: tree to 6 m. **Habitat:** forest. **Distribution:** SE Nigeria, Cameroon and Gabon. **Chorology:** Lower Guinea. **Star:** blue. **Specimens:** - *Southern Bakundu F.R.:* Binuyo A. & Daramola B. O. FHI 35613

Cuviera leniochlamys K.Schum.
Engl. Bot. Jahrb. 28: 79 (1901).

Habit: tree or shrub to 6 m. **Habitat:** forest. **Distribution:** SE Nigeria and SW Cameroon. **Chorology:** Western Cameroon. **Star:** gold. **Alt. range:** 1-200m (6), 201-400m (1). **Flowering:** May (2). **Fruiting:** Apr (1), May (3), Jun (2). **Specimens:** - *Etinde:* Gentry A. 5284a. *Mokoko River F.R.:* Ekema N. 852, 922, 1131b, 1212; Fraser P. 347; Ndam N. 1132, 1185, 1244; Pouakouyou D. 76; Tchouto P. 1162; Watts J. 1164

Cuviera longiflora Hiern
F.W.T.A. 2: 177 (1963).

Habit: tree to 15 m. **Habitat:** forest and Aframomum thicket. **Distribution:** Cameroon, Congo (Kinshasa) and Angola. **Chorology:** Guineo-Congolian. **Star:** green. **Alt. range:** 400-600m (1), 601-800m (2), 801-1000m (1). **Flowering:** Feb (1), Mar (1). **Specimens:** - *Eastern slopes:* Maitland T. D. 1212. *Etinde:* Mbani J. M. 55; Wheatley J. I. 14. *No locality:* Mann G. 1211; Thomas D. W. 2844

Cuviera wernhamii Cheek *nom. nov.*
Cuviera minor (Wernham) Verdc. Kew Bull. 42: 189 (1987) non *Cuviera minor* C.H. Wright, Bull, Misc. Inform. Kew 1906: 105.
Basionym: *Globulostylis minor* Wernham. Cat. Talbot's Nigerian Pl: 50 (1913).

Habit: tree to 10 m. **Habitat:** forest. **Distribution:** SE Nigeria and SW Cameroon. **Chorology:** Western Cameroon. **Star:** gold. **Alt. range:** 1-200m (2), 401-600m (1). **Flowering:** Oct (1), Nov (1), Dec (1). **Fruiting:** Oct (1). **Specimens:** - *Etinde:* Lighava M. 31. *Onge:* Tchouto P. 1007; Watts J. 969

Cuviera subuliflora Benth.
F.W.T.A. 2: 177 (1963); Hawthorne W., F.G.F.T. Ghana: 39 (1990).

Habit: tree to 8 m. **Guild:** pioneer. **Habitat:** forest. **Distribution:** Ghana, SE Nigeria, Bioko and SW Cameroon. **Chorology:** Guineo-Congolian. **Star:** green. **Alt. range:** 1-200m (3), 201-400m (3), 401-600m (3). **Flowering:** Feb (1), Jun (1), Oct (2), Nov (1). **Specimens:** - *Etinde:* Cable S. 247; Maitland T. D. 417, 740; Wheatley J. I. 19. *Mokoko River F.R.:* Fraser P. 469; Ndam N. 1111. *No locality:* Thomas D. W. 2542. *Onge:* Cheek M. 5172; Watts J. 880

Cuviera sp.

Specimens: - *Mokoko River F.R.:* Mbani J. M. 380, 465

Dictyandra arborescens Welw. ex Hook.f.
F.W.T.A. 2: 132 (1963); F.T.E.A. Rubiaceae: 686 (1988).

Habit: tree or shrub to 8 m. **Guild:** shade-bearer. **Habitat:** forest. **Distribution:** Guinea to Bioko, Cameroon, Congo (Kinshasa), Angola and Uganda. **Chorology:** Guineo-Congolian. **Star:** green. **Specimens:** - *No locality:* Maitland T. D. s.n.

Didymosalpinx abbeokutae Hiern
F.W.T.A. 2: 130 (1963); Bull. Jard. Bot. Nat. Belg. 56: 153 (1986).

Habit: scrambling shrub. **Guild:** shade-bearer. **Habitat:** forest. **Distribution:** Guinea to Cameroon and Congo (Kinshasa). **Chorology:** Guineo-Congolian. **Star:** green. **Specimens:** - *No locality:* Preuss P. R. 1259

Didymosalpinx lanciloba (S.Moore) Keay
F.W.T.A. 2: 130 (1963); Bull. Jard. Bot. Nat. Belg. 56: 152 (1986); F.T.E.A. Rubiaceae: 524 (1988).

Habit: shrub to 3 m. **Habitat:** forest. **Distribution:** Nigeria and Cameroon. **Chorology:** Lower Guinea. **Star:** blue. **Alt. range:** 1-200m (1).**Fruiting:** May (1). **Specimens:** - *Mokoko River F.R.:* Acworth J. M. 300

Didymosalpinx cf. lanciloba (S.Moore) Keay

Specimens: - *Onge:* Tchouto P. 971

Diodia sarmentosa Sw.
F.T.E.A. Rubiaceae: 336 (1976).
Diodia scandens sensu Hepper, F.W.T.A. 2: 216 (1963).

Habit: herb. **Guild:** pioneer. **Habitat:** forest. **Distribution:** pantropical. **Star:** excluded. **Alt. range:** 1-200m (1), 1401-1600m (1). **Flowering:** Oct (1). **Specimens:** - *Etinde:* Dawson S. 63. *No locality:* Maitland T. D. s.n.

Diodia serrulata (P.Beauv.) G.Taylor
F.W.T.A. 2: 216 (1963).

Habit: herb. **Guild:** pioneer. **Habitat:** roadsides and villages. **Distribution:** tropical Africa and Central America. **Star:** excluded. **Alt. range:** 1-200m (1). **Flowering:** Jul (1). **Specimens:** - *Mabeta-Moliwe:* Cheek M. 3538. *No locality:* Maitland T. D. 1347, 1347a

Ecpoma gigantostipulum (K.Schum.) N.Hallé
Fl. Gabon 12: 221 (1966).
Sabicea gigantistipula K.Schum., F.W.T.A. 2: 173 (1963).

Habit: shrub to 3 m. **Habitat:** forest. **Distribution:** SE Nigeria, Bioko and SW Cameroon. **Chorology:** Western Cameroon. **Star:** gold. **Alt. range:** 200-400m (1). **Flowering:** May (1), Dec (1). **Specimens:** - *Etinde:* Cheek M. 5846. *Mokoko River F.R.:* Fraser P. 401

Ecpoma hiernianum (Wernham) F. & N.Hallé
Fl. Gabon 12: 224 (1966).

Habit: shrub to 3 m. **Habitat:** forest. **Distribution:** Cameroon and Gabon. **Chorology:** Lower Guinea. **Star:** blue. **Alt. range:** 1-200m (7), 201-400m (2). **Flowering:** Jun (1), Oct (3), Nov (1). **Fruiting:** Jun (3), Oct (1). **Specimens:** - *Etinde:* Cable S. 159. *Mokoko River F.R.:* Ekema N. 1100, 1114b, 1193b; Pouakouyou D. 32. *Onge:* Akogo M. 68; Cheek M. 5060, 5168; Harris D. J. 3634

Ecpoma sp. aff. hiernianum (Wernham) F. & N.Hallé

Specimens: - *Onge:* Harris D. J. 3775; Watts J. 900

Euclinia longiflora Salisb.

F.W.T.A. 2: 121 (1963); Fl. Gabon 17: 205 (1970); F.T.E.A. Rubiaceae: 495 (1988); Hawthorne W., F.G.F.T. Ghana: 36 (1990).

Habit: tree to 6 m. **Guild:** shade-bearer. **Habitat:** forest. **Distribution:** Guinea to Cameroon, Congo (Kinshasa), Angola, Uganda and Sudan. **Chorology:** Guineo-Congolian. **Star:** green. **Alt. range:** 1-200m (4), 601-800m (1), 801-1000m (1), 1001-1200m (2). **Flowering:** Mar (1), May (1), Dec (1). **Fruiting:** May (2), Oct (1). **Specimens:** - *Eastern slopes:* Kwangue A. 52. *Etinde:* Cable S. 412; Kwangue A. 44; Tchouto P. 225. *Mokoko River F.R.:* Ekema N. 854. *No locality:* Mann G. 2164; Mildbraed G. W. J. 10567. *Onge:* Tchouto P. 963; Wheatley J. I. 842

Gaertnera fissistipula (K.Schum. & K.Krause) Petit

Bull. Jard. Bot. Brux. 29: 39 (1959).

Habit: shrub to 3 m. **Habitat:** forest. **Distribution:** Cameroon. **Chorology:** Lower Guinea. **Star:** gold. **Alt. range:** 1-200m (4), 201-400m (3). **Flowering:** May (6). **Fruiting:** May (7), Oct (3). **Specimens:** - *Mokoko River F.R.:* Acworth J. M. 309; Fraser P. 344; Ndam N. 1238, 1267, 1313, 1356a; Watts J. 1141, 1153. *Onge:* Cheek M. 5075; Ndam N. 793; Tchouto P. 912

Gaertnera paniculata Benth.

F.W.T.A. 2: 191 (1963); Hawthorne W., F.G.F.T. Ghana: 35 (1990).

Habit: tree or shrub to 6 m. **Guild:** shade-bearer. **Habitat:** forest. **Distribution:** Guinea to Cameroon, Gabon, Congo (Kinshasa) and Zambia. **Chorology:** Guineo-Congolian (montane). **Star:** green. **Alt. range:** 1-200m (2), 1601-1800m (4). **Flowering:** May (1), Jun (1), Aug (1), Oct (2). **Fruiting:** May (1), Jul (1). **Specimens:** - *Etinde:* Hunt L. V. 35; Tchouto P. 251, 417; Wheatley J. I. 612. *Mokoko River F.R.:* Akogo M. 301; Pouakouyou D. 88

Gaertnera sp.

Specimens: - *No locality:* Migeod F. 253

Galium simense Fresen.

F.W.T.A. 2: 223 (1963); F.T.E.A. Rubiaceae: 391 (1976).

Habit: herb. **Guild:** pioneer. **Habitat:** forest-grassland transition. **Distribution:** Bioko, W. Cameroon and Ethiopia. **Chorology:** Afromontane. **Star:** gold. **Alt. range:** 1400-1600m (3), 1601-1800m (3), 1801-2000m (3), 2001-2200m (1), 2401-2600m (2), 2601-2800m (4), 2801-3000m (1). **Flowering:** Oct (2), Nov (1). **Fruiting:** Mar (1), Oct (2), Nov (1). **Specimens:** - *Eastern slopes:* Dahl A. 614; Dawson S. 81; Tchouto P. 302; Watts J. 451. *No locality:* Brenan J. P. M. 9516; Breteler F. J. 99, 278; Dunlap 29; Keay R. W. J. FHI 28634; Linder 3449; Maitland T. D. 227, 1232; Migeod F. 46; Morton J. K. K 514, K 1048, GC 6701, GC 7063; Thomas D. W. 2623

Galium thunbergianum Eckl. & Zeyh.

F.W.T.A. 2: 223 (1963); F.T.E.A. Rubiaceae: 387 (1976).

Habit: herb. **Guild:** pioneer. **Habitat:** grassland. **Distribution:** tropical and southern Africa. **Chorology:** Afromontane. **Star:** green. **Alt. range:** 1-200m (2), 801-1000m (1), 1801-2000m (1), 2601-2800m (4), 3001-3200m (3). **Flowering:** Oct (1). **Fruiting:** Oct (2). **Specimens:** - *Eastern slopes:* Cheek M. 3665; Thomas D. W. 9305. *No locality:* Brenan J. P. M. 9560; Breteler F. J. 12;

Mann G. 1284, 2001; Morton J. K. K 789, GC 6944, GC 6979, GC 7075, GC 10663a

Geophila afzelii Hiern

Bull. Jard. Bot. Brux. 31: 317 (1961); F.W.T.A. 2: 206 (1963).

Habit: herb. **Guild:** shade-bearer. **Habitat:** forest. **Distribution:** Guinea to Cameroon, Gabon, Congo (Kinshasa) and Angola (Cabinda). **Chorology:** Guineo-Congolian. **Star:** green. **Alt. range:** 1-200m (6), 201-400m (1). **Flowering:** May (2), Oct (2). **Fruiting:** May (1), Oct (3), Dec (1). **Specimens:** - *Mabeta-Moliwe:* Cheek M. 5683; Dawson S. 51. *Mokoko River F.R.:* Ekema N. 848, 849; Watts J. 1173. *Onge:* Tchouto P. 999; Watts J. 769

Geophila obvallata (Schumach.) Didr.

F.W.T.A. 2: 206 (1963); F.T.E.A. Rubiaceae: 112 (1976).

Habit: herb. **Guild:** shade-bearer. **Habitat:** forest. **Distribution:** Guinea Bissau to Cameroon, Gabon and Congo (Kinshasa). **Chorology:** Guineo-Congolian. **Star:** green. **Alt. range:** 1-200m (2). **Flowering:** May (1). **Fruiting:** Mar (1), May (1). **Specimens:** - *Mabeta-Moliwe:* Tchouto P. 189; Wheatley J. I. 45

Hallea ledermannii (K.Krause) Verdc.

Kew Bull. 40: 508 (1985); Hawthorne W., F.G.F.T. Ghana: 29 (1990).
  Mitragyna ledermannii (K.Krause) Ridsdale, Blumea 24: 68 (1978); Keay R.W.J., Trees of Nigeria: 422 (1989).
  Mitragyna ciliata Aubrév. & Pellegr., F.W.T.A. 2: 161 (1963).

Habit: tree to 36 m. **Habitat:** forest. **Distribution:** Liberia to Bioko, Cameroon, Gabon and Congo (Brazzaville). **Chorology:** Guineo-Congolian. **Star:** red. **Alt. range:** 1-200m (1). **Flowering:** Dec (1). **Specimens:** - *Etinde:* Cheek M. 5916; Maitland T. D. 363

Hallea stipulosa (DC.) Leroy

Adansonia (ser.2) 15: 65 (1975); F.T.E.A. Rubiaceae: 447 (1988); Hawthorne W., F.G.F.T. Ghana: 29 (1990).
  Mitragyna stipulosa (DC.) Kuntze, F.W.T.A. 2: 161 (1963); Keay R.W.J., Trees of Nigeria: 422 (1989).

Habit: tree to 30 m. **Habitat:** forest. **Distribution:** Senegal to Cameroon, Gabon, Congo (Brazzaville), Angola, CAR, Uganda and Sudan. **Chorology:** Guineo-Congolian. **Star:** red. **Alt. range:** 1-200m (1). **Flowering:** May (1). **Specimens:** - *Mabeta-Moliwe:* Ndam N. 496; Plot Voucher W 724

Heinsia crinita (Afzel.) G.Taylor

F.W.T.A. 2: 161 (1963); Fl. Gabon 12: 132 (1966); F.T.E.A. Rubiaceae: 476 (1988).
  Heinsia scandens Mildbr. nomen., Wiss. Ergebn. Deutsch. Zentr.-Afr. Exped. 1910-11, 2: 64 (1922).

Habit: tree or shrub to 12 m. **Guild:** pioneer. **Habitat:** forest and stream banks. **Distribution:** Guinea to Bioko, Cameroon, CAR, Gabon, Congo (Kinshasa) and Angola. **Chorology:** Guineo-Congolian. **Star:** green. **Alt. range:** 1-200m (22), 201-400m (6), 401-600m (3). **Flowering:** Mar (4), Apr (5), May (5), Jun (1), Oct (2), Nov (1). **Fruiting:** Mar (3), Apr (3), May (8), Jun (4), Aug (2), Oct (4), Nov (1), Dec (1). **Specimens:** - *Bambuko F.R.:* Watts J. 619, 662. *Etinde:* Kalbreyer W. 12; Khayota B. 572; Tchouto P. 118, 474; Tekwe C. 88; Watts J. 522. *Mabeta-Moliwe:* Baker W. J. 222, 333; Jaff B. 55, 209; Plot Voucher W 190, W 706, W 1166; Sunderland T. C. H. 1164, 1215, 1405; Tchouto P. 176; Wheatley J. I. 50, 294. *Mokoko River F.R.:* Acworth J. M. 168, 321; Akogo M. 203; Ekema N. 893, 998, 1020, 1097; Ndam N. 1297, 1322;

Tchouto P. 1082. *Onge:* Akogo M. 76; Ndam N. 766; Tchouto P. 751; Watts J. 912. *Southern Bakundu F.R.:* Adebusuyi J. K. FHI 44033; Binuyo A. & Daramola B. O. FHI 35077; Latilo M. G. FHI 43820; Leeuwenberg A. J. M. 9820; Olorunfemi J. FHI 30535

Hekistocarpa minutiflora Hook.f.
F.W.T.A. 2: 213 (1963).

Habit: shrub to 1.5 m. **Habitat:** forest and forest gaps.
Distribution: Nigeria and Cameroon. **Chorology:** Lower Guinea.
Star: blue. **Alt. range:** 1-200m (11), 201-400m (2). **Flowering:**
Jan (1), Mar (2), Apr (1), May (1), Jun (3), Aug (2), Oct (2), Nov
(1). **Fruiting:** Mar (1), Oct (2). **Specimens:** - *Mabeta-Moliwe:*
Baker W. J. 265, 338; Cable S. 683; Cheek M. 3520; Groves M.
234; Sunderland 1019. *Mokoko River F.R.:* Akogo M. 252; Ekema
N. 1065, 1190; Fraser P. 446; Pouakouyou D. 96; Tchouto P. 564.
Onge: Cheek M. 5103; Harris D. J. 3861; Tchouto P. 1022

Hymenocoleus glaber Robbr.
Bull. Jard. Bot. Nat. Belg. 47: 14 (1977).

Habit: herb. **Habitat:** forest. **Distribution:** Cameroon.
Chorology: Lower Guinea. **Star:** gold. **Alt. range:** 400-600m
(1).**Fruiting:** Nov (1). **Specimens:** - *Etinde:* Cable S. 260. *No
locality:* Schlechter F. R. R. 12366

Hymenocoleus hirsutus (Benth.) Robbr.
Bull. Jard. Bot. Nat. Belg. 45: 288 (1975); F.T.E.A. Rubiaceae:
115 (1976).
Geophila hirsuta Benth., F.W.T.A. 2: 205 (1963).

Habit: herb. **Guild:** shade-bearer. **Habitat:** forest. **Distribution:**
Guinea to Cameroon, Gabon, Congo (Brazzaville), Congo
(Kinshasa), CAR, Uganda and Tanzania. **Chorology:** Guineo-
Congolian. **Star:** green. **Alt. range:** 1-200m (5), 601-800m (2).
Flowering: Mar (1), Jun (1), Aug (1). **Fruiting:** Mar (1), Nov (2).
Specimens: - *Etinde:* Cheek M. 5601; Khayota B. 496. *Mabeta-
Moliwe:* Baker W. J. 249; Plot Voucher W 653; Wheatley J. I. 352.
No locality: Mildbraed G. W. J. 10581, 10586. *Onge:* Thomas D.
W. 9821

Hymenocoleus libericus (A.Chev. ex Hutch. & Dalziel) Robbr.
F.T.E.A. Rubiaceae: 115 (1976); Bull. Jard. Bot. Nat. Belg. 47: 19
(1977).
Geophila liberica Hutch. & Dalziel, F.W.T.A. 2: 205 (1963).

Habit: herb. **Guild:** shade-bearer. **Habitat:** forest. **Distribution:**
Liberia, Ivory Coast, Ghana, Cameroon, Gabon, Congo
(Brazzaville), Congo (Kinshasa) and CAR. **Chorology:** Guineo-
Congolian. **Star:** green. **Alt. range:** 1-200m (1). **Flowering:** May
(1). **Specimens:** - *Mabeta-Moliwe:* Tchouto P. 188

Hymenocoleus neurodictyon (K.Schum.) Robbr. var. neurodictyon
Bull. Jard. Bot. Nat. Belg. 45: 291 (1975); F.T.E.A. Rubiaceae:
116 (1976).
Geophila neurodictyon (K.Schum.) Hepper, F.W.T.A. 2: 206
(1963).

Habit: herb. **Habitat:** forest. **Distribution:** Sierra Leone to
Cameroon, Gabon, Congo (Brazzaville) and CAR. **Chorology:**
Guineo-Congolian. **Star:** green. **Alt. range:** 1-200m (6), 201-
400m (1). **Flowering:** Mar (1), Apr (2), May (1). **Fruiting:** Apr
(2), Nov (1), Dec (1). **Specimens:** - *Etinde:* Cable S. 199. *Mabeta-
Moliwe:* Cable S. 597; Watts J. 133, 154; Wheatley J. I. 276.

Mokoko River F.R.: Tchouto P. 597. *No locality:* Mildbraed G. W.
J. 10589; Schlechter F. R. R. 12364

Hymenocoleus rotundifolius (A.Chev. ex Hepper) Robbr.
Bull. Jard. Bot. Nat. Belg. 45: 287 (1975); F.T.E.A. Rubiaceae:
114 (1976); Bull. Jard. Bot. Nat. Belg. 47: 19 (1977).
Geophila rotundifolia A.Chev. ex Hepper, F.W.T.A. 2: 206
(1963).

Habit: herb. **Guild:** shade-bearer. **Habitat:** forest. **Distribution:**
Sierra Leone to Cameroon, Gabon and Congo (Kinshasa).
Chorology: Guineo-Congolian. **Star:** green. **Alt. range:** 1-200m
(3), 401-600m (1). **Flowering:** Mar (1). **Fruiting:** Mar (1), Oct
(1), Nov (1). **Specimens:** - *Etinde:* Cheek M. 5552. *Mabeta-
Moliwe:* Cheek M. 3238; Sunderland 1007. *Onge:* Tchouto P.
920; Wheatley J. I. 732, 778

Hymenocoleus subipecacuanha (K.Schum.) Robbr.
Bull. Jard. Bot. Nat. Belg. 47: 20 (1977).
Hymenocoleus petitianus Robbr., Bull. Jard. Bot. Nat. Belg. 45:
286 (1975).

Habit: herb. **Habitat:** forest. **Distribution:** Nigeria, Cameroon
and Congo (Kinshasa). **Chorology:** Guineo-Congolian. **Star:**
green. **Alt. range:** 1-200m (13), 201-400m (6), 401-600m (1),
801-1000m (1), 1001-1200m (1). **Flowering:** Apr (2), May (2),
Jun (5), Aug (1), Sep (1), Oct (3), Nov (3). **Fruiting:** Apr (2), May
(1), Jun (1), Aug (1), Sep (1), Oct (6), Nov (4). **Specimens:** -
Etinde: Cheek M. 5445, 5526, 5558; Tchouto P. 428; Watts J. 483;
Wheatley J. I. 547. *Mabeta-Moliwe:* Baker W. J. 285; Dawson S.
54; Nguembock F. 5b; Plot Voucher W 376, W 408, W 726, W
967; Sunderland T. C. H. 1298, 1391, 1466; Watts J. 142, 327;
Wheatley J. I. 274; von Rege I. 44. *Mokoko River F.R.:* Akogo M.
263; Ndam N. 1091, 1317. *Onge:* Akogo M. 131; Cheek M. 5162;
Harris D. J. 3644; Ndam N. 718; Tchouto P. 947; Watts J. 989

Hymenodictyon biafranum Hiern
F.W.T.A. 2: 111 (1963).

Habit: tree or shrub to 10 m. **Habitat:** lava flows. **Distribution:**
SE Nigeria, Bioko, Cameroon and Gabon. **Chorology:** Lower
Guinea. **Star:** blue. **Alt. range:** 1-200m (5), 201-400m (2), 401-
600m (2), 601-800m (2), 801-1000m (1), 1201-1400m (2).
Flowering: May (3), Jul (1). **Fruiting:** Jan (1), Feb (1), Jul (1).
Specimens: - *Eastern slopes:* Dawson S. 57, 58; Maitland T. D.
201. *Etinde:* Keay R. W. J. FHI 28652; Rosevear D. R. 68/37;
Tchouto P. 238; Thomas D. W. 7057; Thompson S. 1603. *Mokoko
River F.R.:* Ndam N. 1168; Tchouto P. 1180; Thomas D. W.
10052. *No locality:* Cheek M. 3030; Maitland T. D. 1074; Mann
G. 1194; Upson T. 160. *Onge:* Harris D. J. 3701

Hymenodictyon floribundum (Steud. & Hochst.) B.L.Rob.
F.W.T.A. 2: 111 (1963); Meded. Land. Wag. 82(3): 281 (1982);
F.T.E.A. Rubiaceae: 452 (1988).

Habit: tree or shrub to 8 m. **Habitat:** lava flows and forest.
Distribution: tropical Africa. **Star:** excluded. **Alt. range:** 1-200m
(1), 401-600m (1). **Flowering:** Mar (1). **Fruiting:** Mar (1).
Specimens: - *Bambuko F.R.:* Watts J. 672. *Eastern slopes:*
Thomas D. W. 7042

Ixora delicatula Keay
F.W.T.A. 2: 142 (1963); Opera Botanica Belgica 9: 96 (1998).

Habit: shrub to 2 m. Habitat: forest. Distribution: Nigeria and Cameroon. Chorology: Lower Guinea. Star: blue. Alt. range: 800-1000m (1). Flowering: Dec (1). Specimens: - *Etinde:* Faucher P. 4

Ixora foliosa Hiern
F.W.T.A. 2: 144 (1963); Opera Botanica Belgica 9: 102 (1998).

Habit: tree to 15 m. Habitat: forest and forest-grassland transition. Distribution: W. Cameroon. Chorology: Western Cameroon Uplands (montane). Star: gold. Alt. range: 1200-1400m (2), 1401-1600m (2), 1601-1800m (1), 1801-2000m (5), 2001-2200m (1), 2201-2400m (1). Flowering: Sep (1). Fruiting: Sep (2). Specimens: - *Eastern slopes:* Brenan J. P. M. 9562; Maitland T. D. 198, 208, 514; Manning S. 1296; Mildbraed G. W. J. 10820; Thomas D. W. 2598; Thorold C. A. CM 42. *Etinde:* Tekwe C. 195; Wheatley J. I. 575. *No locality:* Breteler F. J. 156, 170; Keay R. W. J. FHI 28645

Ixora guineensis Benth.
F.W.T.A. 2: 142 (1963); Opera Botanica Belgica 9: 104 (1998).
  Ixora breviflora Hiern, F.W.T.A. 2: 142 (1963).
  Ixora talbotii Wernham, F.W.T.A. 2: 142 (1963).

Habit: tree or shrub to 9 m. Habitat: forest. Distribution: Nigeria, Bioko, Cameroon, Rio Muni, Gabon and Congo (Kinshasa). Chorology: Guineo-Congolian (montane). Star: green. Alt. range: 1-200m (35), 201-400m (5), 601-800m (2), 1201-1400m (4), 1401-1600m (2), 1601-1800m (1). Flowering: Mar (5), May (3), Jun (3), Jul (2), Aug (1), Sep (2), Oct (1), Nov (4), Dec (4). Fruiting: Feb (1), Mar (6), Apr (4), May (10), Jun (4), Jul (2), Aug (2), Sep (2), Oct (1), Nov (2), Dec (1). Specimens: - *Eastern slopes:* Cable S. 1393; Mbani J. M. 118. *Etinde:* Cable S. 172; Cheek M. 5396a; Faucher P. 28; Hunt L. V. 43; Khayota B. 541; Kwangue A. 98; Mbani J. M. 77; Sidwell K. 165; Tchouto P. 23, 127; Tekwe C. 153; Wheatley J. I. 563. *Mabeta-Moliwe:* Baker W. J. 235; Cable S. 529, 546, 1365, 1474; Cheek M. 3567, 5699; Groves M. 252; Hurst J. 8; Jaff B. 147, 174; Khayota B. 477; Ndam N. 480; Plot Voucher W 454, W 657, W 799, W 938, W 972; Sunderland T. C. H. 1021, 1172, 1336, 1338, 1346; Watts J. 165, 263. *Mokoko River F.R.:* Acworth J. M. 61; Akogo M. 167; Ekema N. 1038, 1050, 1163; Fraser P. 410, 450; Ndam N. 1148, 1355; Pouakouyou D. 58. *No locality:* Brenan J. P. M. 9592

Ixora hippoperifera Bremek.
F.W.T.A. 2: 142 (1963); Opera Botanica Belgica 9: 110 (1998).

Habit: tree or shrub to 4 m. Habitat: forest. Distribution: SE Nigeria, Cameroon, Rio Muni and Gabon. Chorology: Lower Guinea. Star: blue. Specimens: - *Southern Bakundu F.R.:* Binuyo A. & Daramola B. O. FHI 35163; Olorunfemi J. FHI 30747; Onochie C. F. A. FHI 30868

Ixora nematopoda K.Schum.
F.W.T.A. 2: 142 (1963); Opera Botanica Belgica 9: 138 (1998).

Habit: shrub to 3 m. Habitat: forest. Distribution: SE Nigeria, Bioko, Cameroon, Gabon and Congo (Kinshasa). Chorology: Guineo-Congolian. Star: green. Alt. range: 1-200m (9), 201-400m (4), 401-600m (3). Flowering: Jan (1), May (1), Jun (5), Jul (2), Sep (1), Nov (1). Fruiting: Jan (2), May (1), Jun (2), Oct (3), Nov (3), Dec (1). Specimens: - *Etinde:* Cable S. 173, 280, 386; Cheek M. 3058; Kwangue A. 114; Mbani J. M. 8; Nkeng P. 6, 100; Tchouto P. 811; Thompson S. 1804; Watts J. 489. *Mabeta-Moliwe:* Cheek, M. 3259; Hurst J. 15; Nguembock F. 8; Plot Voucher W 166, W 185, W 911, W 1035; von Rege I. 3. *Mokoko River F.R.:* Acworth J. M. 266a; Ekema N. 1161; Fraser P. 345. *No locality:* Hepper F. N. 8694. *Onge:* Watts J. 781

Ixora sp.

Specimens: - *Mokoko River F.R.:* Acworth J. M. 219, 259; Akogo M. 305, 310; Fraser P. 359; Ndam N. 1078, 1170

Keetia hispida (Benth.) Bridson
Kew Bull. 41: 986 (1986).
  Canthium hispidum Benth., F.W.T.A. 2: 182 (1963).
  Canthium setosum Hiern, F.T.A. 3: 141 (1877).
  Canthium rubrinerve (K.Krause) Hepper, F.W.T.A. 2: 182 (1963).

Habit: woody climber. Guild: pioneer. Habitat: forest. Distribution: Sierra Leone to Cameroon, Rio Muni, Congo (Kinshasa) and Angola. Chorology: Guineo-Congolian. Star: green. Alt. range: 1-200m (2), 201-400m (4), 401-600m (1). Flowering: Mar (1), Oct (2), Nov (1). Fruiting: Mar (1), Jun (2), Oct (2). Specimens: - *Eastern slopes:* Deistel H. 628. *Etinde:* Khayota B. 594; Mildbraed G. W. J. 10564, 10585. *Mabeta-Moliwe:* Maitland T. D. 392; Plot Voucher W 393; Wheatley J. I. 322. *No locality:* Mann G. 1190. *Onge:* Cheek M. 5036; Harris D. J. 3811; Ndam N. 711; Watts J. 784, 855

Keetia leucantha (K.Krause) Bridson
Kew Bull. 41: 987 (1986).
  Canthium setosum sensu Hepper, F.W.T.A. 2: 182 (1963).

Habit: woody climber. Habitat: forest. Distribution: Sierra Leone, Liberia, Ivory Coast, Ghana, Nigeria and Cameroon. Chorology: Guineo-Congolian. Star: green. Alt. range: 200-400m (1). Flowering: Mar (1). Fruiting: Mar (1). Specimens: - *Mokoko River F.R.:* Tchouto P. 605

Keetia mannii (Hiern) Bridson
Kew Bull. 41: 988 (1986).
  Canthium mannii Hiern, F.W.T.A. 2: 182 (1963).

Habit: woody climber. Guild: pioneer. Habitat: forest. Distribution: Guinea to Cameroon, CAR and Sudan. Chorology: Guineo-Congolian. Star: green. Specimens: - *Mabeta-Moliwe:* Maitland T. D. 974

Keetia molundensis (K.Krause) Bridson var. macrostipulata (De Wild.) Bridson
Kew Bull. 41: 976 (1986); F.T.E.A. Rubiaceae: 916 (1991).
  Canthium macrostipulatum (De Wild.) Evrard, Bull. Jard. Bot. Brux. 37: 459 (1967).

Habit: woody climber. Habitat: forest. Distribution: Nigeria, Cameroon, Congo (Kinshasa) and Uganda. Chorology: Guineo-Congolian. Star: green. Alt. range: 1-200m (1).Fruiting: Jan (1). Specimens: - *Mabeta-Moliwe:* Tchouto P. 1049

Keetia sp. aff. molundensis (K.Krause) Bridson
  Canthium sp. A sensu Hepper & Keay, F.W.T.A. 2: 182 (1963).
Specimens: - *Etinde:* Olorunfemi J. FHI 30513. *Mabeta-Moliwe:* Baker W. J. 348

Keetia venosa (Oliv.) Bridson
Kew Bull. 41: 974 (1986); F.T.E.A. Rubiaceae: 914 (1991); Fl. Zamb. 5(2): 365 (1998).
  Canthium venosum (Oliv.) Hiern, F.W.T.A. 2: 184 (1963).

Habit: woody climber, shrub or small tree. Guild: pioneer. Habitat: forest. Distribution: tropical Africa. Star: excluded. Alt.

range: 800-1000m (1). **Specimens:** - *Eastern slopes:* Maitland T. D. 194

Keetia sp. aff. venosa (Oliv.) Bridson

Specimens: - *Onge:* Thomas D. W. 9771; Watts J. 997

Keetia sp.

Specimens: - *Mokoko River F.R.:* Tchouto P. 583

Lasianthus batangensis K.Schum.
F.W.T.A. 2: 190 (1963); Bull. Jard. Bot. Nat. Belg. 51: 453 (1981).

Habit: shrub to 3 m. **Guild:** shade-bearer. **Habitat:** forest. **Distribution:** Sierra Leone to Cameroon, Gabon, Congo (Kinshasa) and Angola (Cabinda). **Chorology:** Guineo-Congolian. **Star:** green. **Alt. range:** 1-200m (6), 201-400m (5). **Flowering:** Mar (1), May (2), Jun (2), Oct (2), Nov (1). **Fruiting:** Mar (1), May (3), Jun (1), Oct (5). **Specimens:** - *Mokoko River F.R.:* Batoum A. 28; Ekema N. 1118b; Fraser P. 351; Ndam N. 1216; Pouakouyou D. 109. *Onge:* Harris D. J. 3749; Ndam N. 723, 762; Tchouto P. 986; Watts J. 765, 832; Wheatley J. I. 804

Leptactina involucrata Hook.f.
Pl. Syst. Evol. 145: 114 (1984).
    Dictyandra involucrata (Hook.f.) Hiern, F.W.T.A. 2: 132 (1963); Fl. Gabon 17: 90 (1970).

Habit: shrub to 6 m, often scandent. **Guild:** shade-bearer. **Habitat:** forest. **Distribution:** Sierra Leone to Cameroon, Rio Muni and CAR. **Chorology:** Guineo-Congolian. **Star:** green. **Alt. range:** 1-200m (1).**Fruiting:** Dec (1). **Specimens:** - *Mabeta-Moliwe:* Cheek M. 5732. *No locality:* Mann G. 2156

Massularia acuminata (G.Don) Bullock ex Hoyle
F.W.T.A. 2: 114 (1963); Fl. Gabon 17: 178 (1970); Hawthorne W., F.G.F.T. Ghana: 36 (1990).

Habit: tree to 10 m. **Guild:** shade-bearer. **Habitat:** forest. **Distribution:** Guinea to Cameroon, Gabon, Congo (Kinshasa) and Angola (Cabinda). **Chorology:** Guineo-Congolian. **Star:** green. **Alt. range:** 1-200m (31), 201-400m (4). **Flowering:** Jan (1), Feb (1), Mar (2), Apr (4), May (6), Jun (1), Jul (1), Oct (2), Nov (2), Dec (4). **Fruiting:** Feb (1), Mar (3), Apr (6), May (10), Jun (1), Jul (1), Aug (1), Oct (2), Nov (1), Dec (2). **Specimens:** - *Etinde:* Acworth J. M. 23, 24; Cheek M. 5497; Lighava M. 48; Nkeng P. 54; Tchouto P. 131; Tekwe C. 20, 72. *Mabeta-Moliwe:* Baker W. J. 234; Cable S. 515, 518, 577, 1440; Cheek M. 3258; Groves M. 262; Hurst J. 21; Jaff B. 192, 223; Ndam N. 494; Nguembock F. 1a, 89; Plot Voucher W 987; Sunderland T. C. H. 1018, 1217, 1294; Tchouto P. 182; Watts J. 177. *Mokoko River F.R.:* Acworth J. M. 177; Akogo M. 178, 303; Batoum A. 3; Ekema N. 873, 987, 1082; Pouakouyou D. 101. *No locality:* Cheek M. 3065. *Onge:* Harris D. J. 3760, 3788; Ndam N. 738; Watts J. 849

Mitriostigma barteri Hook.f. ex Hiern
F.W.T.A. 2: 130 (1963).

Habit: shrub to 1 m. **Habitat:** forest. **Distribution:** Bioko and SW Cameroon. **Chorology:** Western Cameroon. **Star:** black. **Alt. range:** 1-200m (3), 401-600m (1), 601-800m (1), 801-1000m (1). **Flowering:** May (1). **Fruiting:** Mar (1), Apr (1), May (1), Jun (1), Nov (2). **Specimens:** - *Etinde:* Cable S. 265; Etuge M. 1242; Kwangue A. 136. *Mabeta-Moliwe:* Sunderland T. C. H. 1469; Tchouto P. 181; Wheatley J. I. 210

Morinda geminata DC.
F.W.T.A. 2: 189 (1963).

Habit: tree to 10 m. **Distribution:** Senegal to Cameroon. **Chorology:** Guineo-Congolian. **Star:** green. **Alt. range:** 1-200m (1). **Flowering:** Feb (1), Jun (1). **Fruiting:** Feb (1), Jun (1), Dec (1). **Specimens:** - *Etinde:* Cheek M. 5934; Dalziel J. M. 8189; Thompson S. 1356

Morinda lucida Benth.
F.W.T.A. 2: 189 (1963); F.T.E.A. Rubiaceae: 146 (1976); Keay R.W.J., Trees of Nigeria: 418 (1989); Hawthorne W., F.G.F.T. Ghana: 29 (1990).

Habit: tree to 18 m. **Guild:** pioneer. **Habitat:** forest. **Distribution:** tropical Africa. **Star:** excluded. **Specimens:** - *Mabeta-Moliwe:* Dunlap 249

Morinda morindoides (Baker) Milne-Redh.
F.W.T.A. 2: 189 (1963).

Habit: woody climber. **Guild:** light demander. **Habitat:** forest. **Distribution:** Guinea Bissau to Bioko, Cameroon, Congo (Kinshasa), Angola and Sudan. **Chorology:** Guineo-Congolian. **Star:** green. **Alt. range:** 1-200m (1).**Fruiting:** Jun (1), Dec (1). **Specimens:** - *Mabeta-Moliwe:* Cheek M. 5754; Plot Voucher W 309

Mussaenda arcuata Lam. ex Poir.
F.W.T.A. 2: 165 (1963); Fl. Gabon 12: 152 (1966); Meded. Land. Wag. 82(3): 284 (1982); F.T.E.A. Rubiaceae: 461 (1988).

Habit: woody climber. **Habitat:** forest. **Distribution:** tropical Africa. **Star:** excluded. **Alt. range:** 1-200m (7), 201-400m (1), 401-600m (2), 801-1000m (2). **Flowering:** Oct (2), Nov (2), Dec (3). **Fruiting:** Mar (2), Apr (1), Oct (2), Dec (3). **Specimens:** - *Etinde:* Cable S. 148, 350, 399, 456; Ejiofor M. C. FHI 29367; Faucher P. 44; Leeuwenberg A. J. M. 6951; Mbani J. M. 75. *Mabeta-Moliwe:* Watts J. 233. *Onge:* Akogo M. 51; Harris D. J. 3829; Tchouto P. 520, 727

Mussaenda elegans Schum. & Thonn.
F.W.T.A. 2: 167 (1963); Fl. Gabon 12: 141 (1966); Meded. Land. Wag. 82(3): 284 (1982); F.T.E.A. Rubiaceae: 462 (1988).

Habit: woody climber. **Guild:** pioneer. **Habitat:** forest and farmbush. **Distribution:** Mali to Cameroon, Congo (Kinshasa), Uganda and Sudan. **Chorology:** Guineo-Congolian. **Star:** green. **Alt. range:** 1-200m (4). **Flowering:** May (2), Jun (1). **Fruiting:** Jun (1). **Specimens:** - *Etinde:* Mildbraed G. W. J. 10598; Wheatley J. I. 233. *Mabeta-Moliwe:* Sunderland T. C. H. 1316; Watts J. 412; Wheatley J. I. 364

Mussaenda erythrophylla Schum. & Thonn.
F.W.T.A. 2: 165 (1963); Fl. Gabon 12: 148 (1966); F.T.E.A. Rubiaceae: 463 (1988).

Habit: woody climber. **Guild:** pioneer. **Habitat:** forest. **Distribution:** tropical Africa. **Star:** excluded. **Alt. range:** 800-1000m (2). **Flowering:** Mar (1). **Fruiting:** Jul (1). **Specimens:** - *Eastern slopes:* Tekwe C. 111. *Etinde:* Cable S. 1592

Mussaenda polita Hiern
F.W.T.A. 2: 165 (1963); Fl. Gabon 12: 149 (1966).

Habit: woody climber. Habitat: forest. Distribution: SE Nigeria, Cameroon and Gabon. Chorology: Lower Guinea. Star: blue. Alt. range: 1-200m (6), 201-400m (1), 401-600m (1). Flowering: May (1), Oct (1), Nov (2), Dec (1). Fruiting: Mar (1), May (3), Oct (2), Nov (1). Specimens: - *Etinde:* Cable S. 388. *Mokoko River F.R.:* Ekema N. 915; Ndam N. 1131; Tchouto P. 580; Watts J. 1064. *Onge:* Harris D. J. 3787; Tchouto P. 915, 995; Watts J. 996

Mussaenda tenuiflora Benth.
F.W.T.A. 2: 167 (1963); Fl. Gabon 12: 145 (1966).

Habit: woody climber. Habitat: forest. Distribution: Guinea to Bioko, Cameroon, Congo (Kinshasa) and Angola. Chorology: Guineo-Congolian. Star: green. Alt. range: 1-200m (30), 201-400m (5), 401-600m (2). Flowering: Feb (1), Mar (5), Apr (3), May (3), Aug (1), Oct (1), Nov (3), Dec (1). Fruiting: Feb (1), Mar (2), Apr (2), May (6), Jun (2), Aug (3), Sep (1), Oct (4), Nov (5), Dec (2). Specimens: - *Etinde:* Cable S. 144, 182; Cheek M. 5507; Hutchinson J. & Metcalfe C. R. 141; Khayota B. 574; Marsden J. 20; Mbani J. M. 68, 138; Nkeng P. 17; Tchouto P. 62; Tekwe C. 129; Wheatley J. I. 511. *Mabeta-Moliwe:* Baker W. J. 307, 314, 343; Cable S. 555, 612, 1470, 1472; Cheek M. 5764; Khayota B. 486; Ndam N. 518; Nguembock F. 70; Sunderland T. C. H. 1193, 1203. *Mokoko River F.R.:* Akogo M. 175, 197; Ekema N. 984, 1060, 1200; Ndam N. 1196; Pouakouyou D. 100; Watts J. 1047. *Onge:* Akogo M. 66; Cheek M. 4965; Harris D. J. 3664, 3700, 3828; Ndam N. 694; Watts J. 768, 1001. *Southern Bakundu F.R.:* Binuyo A. & Daramola B. O. FHI 35174

Nauclea diderrichii (De Wild. & T.Durand) Merrill
F.W.T.A. 2: 164 (1963); Fl. Gabon 12: 44 (1966); F.T.E.A. Rubiaceae: 441 (1988); Keay R.W.J., Trees of Nigeria: 418 (1989); Hawthorne W., F.G.F.T. Ghana: 29 (1990).

Habit: tree to 40 m. Guild: pioneer. Habitat: forest. Distribution: Sierra Leone to Cameroon, Gabon, Congo (Kinshasa), CAR, Uganda, Mozambique. Star: scarlet. Alt. range: 1-200m (1).Fruiting: Nov (2). Specimens: - *Mabeta-Moliwe:* Mildbraed G. W. J. 10605. *Onge:* Harris D. J. 3832

Nauclea vanderguchtii (De Wild.) Petit
F.W.T.A. 2: 164 (1963); Fl. Gabon 12: 48 (1966); Keay R.W.J., Trees of Nigeria: 420 (1989).

Habit: tree to 15 m. Habitat: forest. Distribution: Nigeria, Cameroon, Gabon, Congo (Kinshasa) and Angola (Cabinda). Chorology: Guineo-Congolian. Star: green. Alt. range: 1-200m (1). Specimens: - *Onge:* Harris D. J. 3672

Nichallea soyauxii (Hiern) Bridson
Kew Bull. 33: 288 (1978).
Tarenna soyauxii (Hiern) Bremek., F.W.T.A. 2: 135 (1963).

Habit: tree or shrub to 5 m. Guild: light demander. Habitat: forest. Distribution: Sierra Leone to Cameroon, Gabon, Congo (Kinshasa) and Angola (Cabinda). Chorology: Guineo-Congolian. Star: green. Alt. range: 1-200m (6), 201-400m (2). Flowering: May (1), Sep (1), Oct (1), Nov (1). Fruiting: Mar (1), May (3), Jun (1), Oct (1). Specimens: - *Mokoko River F.R.:* Akogo M. 308; Ekema N. 920; Pouakouyou D. 56a; Tchouto P. 601; Watts J. 1081. *Onge:* Tchouto P. 825, 1033; Watts J. 852

Oldenlandia lancifolia (Schumach.) DC.
F.W.T.A. 2: 212 (1963); Fl. Gabon 12: 100 (1966); F.T.E.A. Rubiaceae: 292 (1976).

Habit: herb. Guild: pioneer. Habitat: wet areas in forest. Distribution: tropical Africa. Chorology: Afromontane. Star: green. Alt. range: 1-200m (3), 2401-2600m (1). Flowering: May (1). Fruiting: May (1), Jul (1), Oct (1), Nov (1). Specimens: - *Eastern slopes:* Maitland T. D. 580. *Etinde:* Dawson S. 62; Ogu 205; Thompson S. 1596. *Mokoko River F.R.:* Watts J. 1144

Otomeria cameronica (Bremek.) Hepper
F.W.T.A. 2: 215 (1963); Fl. Gabon 12: 114 (1966).

Habit: herb. Habitat: forest. Distribution: Sierra Leone to Bioko and Cameroon. Chorology: Guineo-Congolian (montane). Star: green. Alt. range: 800-1000m (3), 1001-1200m (1), 1201-1400m (1), 1401-1600m (1), 1601-1800m (1). Flowering: Oct (1), Nov (1). Specimens: - *Eastern slopes:* Breteler F. J. 305; Dawson S. 85; Dunlap 140; Etuge M. 432; Hambler D. J. 646; Hutchinson J. & Metcalfe C. R. 118; Maitland T. D. 277; Migeod F. 8, 10. *Etinde:* Adebusuyi J. K. FHI 44023; Sidwell K. 167

Oxyanthus formosus Hook.f. ex Planch.
F.W.T.A. 2: 129 (1963); Fl. Gabon 17: 188 (1970); F.T.E.A. Rubiaceae: 537 (1988); Hawthorne W., F.G.F.T. Ghana: 36 (1990).

Habit: tree or shrub to 5 m. Guild: shade-bearer. Habitat: forest. Distribution: Guinea to Bioko, Cameroon, Gabon, Congo (Kinshasa) and Sudan. Chorology: Guineo-Congolian. Star: green. Alt. range: 1-200m (5), 801-1000m (1), 1401-1600m (1). Flowering: May (3), Jun (2). Fruiting: Sep (1), Dec (1). Specimens: - *Etinde:* Cable S. 334; Tekwe C. 228. *Mokoko River F.R.:* Acworth J. M. 262; Ekema N. 1103; Pouakouyou D. 110; Tchouto P. 1204; Watts J. 1199. *No locality:* Hutchinson J. & Metcalfe C. R. 68

Oxyanthus gracilis Hiern
F.W.T.A. 2: 129 (1963); Fl. Gabon 17: 201 (1970).

Habit: shrub to 3 m. Habitat: forest. Distribution: Nigeria, Bioko and Cameroon. Chorology: Lower Guinea. Star: blue. Alt. range: 1-200m (3), 201-400m (5), 401-600m (2), 601-800m (2), 1001-1200m (1). Flowering: Mar (1), Oct (2). Fruiting: Oct (8), Nov (3), Dec (1). Specimens: - *Etinde:* Cheek M. 5499; Faucher P. 10; Tchouto P. 443; Watts J. 503, 550. *Onge:* Akogo M. 84; Cheek M. 4998, 5017; Ndam N. 648, 773; Watts J. 950, 987; Wheatley J. I. 840

Oxyanthus cf. gracilis Hiern

Specimens: - *Onge:* Cheek M. 5095; Ndam N. 770

Oxyanthus laxiflorus K.Schum. ex Hutch. & Dalziel
F.W.T.A. 2: 129 (1963); Fl. Gabon 17: 200 (1970).

Habit: shrub to 3 m. Habitat: forest. Distribution: SE Nigeria and Cameroon. Chorology: Lower Guinea. Star: blue. Alt. range: 1-200m (3), 201-400m (1). Flowering: Apr (1). Fruiting: Mar (1), Apr (1), May (2). Specimens: - *Bambuko F.R.:* Watts J. 645. *Mokoko River F.R.:* Akogo M. 219; Ekema N. 935; Tchouto P. 1212

Oxyanthus montanus Sonké
Bull. Jard. Bot. Nat. Belg. 63: 397 (1994).
Oxyanthus sp. A sensu Hepper & Keay, F.W.T.A. 2: 129 (1963).

Habit: tree or shrub to 4 m. Habitat: forest. Distribution: Cameroon. Chorology: Lower Guinea. Star: gold. Alt. range:

111

1400-1600m (2).**Fruiting:** Sep (2). **Specimens:** - *Eastern slopes:* Hutchinson J. & Metcalfe C. R. 71; Maitland T. D. 681. *Etinde:* Tekwe C. 192, 215. *No locality:* Dunlap 15; Etuge M. 117

Oxyanthus cf. pallidus Hiern

Specimens: - *Etinde:* Tekwe C. 174

Oxyanthus speciosus DC. subsp. globosus Bridson
Kew Bull. 34: 115 (1979); F.T.E.A. Rubiaceae: 529 (1988).

Habit: tree to 15 m. **Habitat:** forest. **Distribution:** Cameroon. **Chorology:** Lower Guinea. **Star:** gold. **Alt. range:** 200-400m (1), 401-600m (1).**Fruiting:** Mar (1), May (1). **Specimens:** - *Etinde:* Mbani J. M. 145. *Mokoko River F.R.:* Tchouto P. 582

Oxyanthus speciosus DC. subsp. speciosus
F.W.T.A. 2: 129 (1963); Fl. Gabon 17: 196 (1970); F.T.E.A. Rubiaceae: 529 (1988); Hawthorne W., F.G.F.T. Ghana: 36 (1990); Bull. Jard. Bot. Nat. Belg. 65: 117 (1996).

Habit: tree to 15 m. **Habitat:** forest. **Distribution:** Cameroon. **Chorology:** Lower Guinea. **Star:** gold. **Alt. range:** 400-600m (1). **Flowering:** Nov (1). **Fruiting:** Dec (1). **Specimens:** - *Etinde:* Cable S. 397. *Mabeta-Moliwe:* Nguembock F. 46

Oxyanthus unilocularis Hiern
F.W.T.A. 2: 129 (1963); Fl. Gabon 17: 193 (1970); F.T.E.A. Rubiaceae: 537 (1988); Hawthorne W., F.G.F.T. Ghana: 30 (1990).

Habit: tree or shrub to 8 m. **Guild:** shade-bearer. **Habitat:** forest. **Distribution:** Sierra Leone to Cameroon, CAR, Gabon, Congo (Kinshasa), Angola (Cabinda), Uganda and Sudan. **Chorology:** Guineo-Congolian. **Star:** green. **Alt. range:** 1-200m (5). **Flowering:** Mar (1). **Fruiting:** Mar (1), Apr (1), Dec (2). **Specimens:** - *Etinde:* Cheek M. 5919. *Mabeta-Moliwe:* Cable S. 1493; Cheek M. 5711; Groves M. 248; Jaff B. 121

Parapentas setigera (Hiern) Verdc.
F.W.T.A. 2: 214 (1963); Fl. Gabon 12: 108 (1966).

Habit: herb. **Guild:** pioneer. **Habitat:** forest. **Distribution:** Guinea to Bioko, Cameroon, Gabon, Congo (Kinshasa) and Angola (Cabinda). **Chorology:** Guineo-Congolian. **Star:** green. **Alt. range:** 1-200m (2), 601-800m (1). **Flowering:** Apr (1), May (1), Nov (1). **Specimens:** - *Etinde:* Cable S. 232. *Mokoko River F.R.:* Acworth J. M. 81; Watts J. 1145

Pauridiantha canthiiflora Hook.f.
F.W.T.A. 2: 168 (1963); Fl. Gabon 12: 239 (1966).
Habit: shrub to 5 m. **Habitat:** forest. **Distribution:** Nigeria, Bioko, Cameroon, Gabon and Congo (Kinshasa). **Chorology:** Guineo-Congolian. **Star:** green. **Alt. range:** 1-200m (12), 201-400m (1). **Flowering:** Apr (1), May (1), Aug (1), Nov (1). **Fruiting:** Apr (3), May (8), Nov (1), Dec (1). **Specimens:** - *Etinde:* Winkler H. J. P. 548. *Mabeta-Moliwe:* Cable S. 525; Jaff B. 72; Nguembock F. 19; Sunderland T. C. H. 1351; Tchouto P. 213; Wheatley J. I. 77; von Rege I. 95. *Mokoko River F.R.:* Akogo M. 244, 304; Ekema N. 956, 989, 1021; Fraser P. 378; Ndam N. 1178; Watts J. 1071. *Southern Bakundu F.R.:* Binuyo A. & Daramola B. O. FHI 35591

Pauridiantha divaricata (K.Schum.) Bremek.
Fl. Gabon 12: 236 (1966).

112

Habit: shrub to 2 m. **Habitat:** forest. **Distribution:** SE Nigeria and SW Cameroon. **Chorology:** Western Cameroon. **Star:** gold. **Flowering:** Mar (1). **Specimens:** - *Southern Bakundu F.R.:* Brenan J. P. M. 9452

Pauridiantha floribunda (K.Schum. & K.Krause) Bremek.
F.W.T.A. 2: 168 (1963); Fl. Gabon 12: 255 (1966).

Habit: tree or shrub to 8 m. **Habitat:** forest. **Distribution:** Nigeria, Cameroon, S. Tomé and Gabon. **Chorology:** Lower Guinea. **Star:** blue. **Alt. range:** 1-200m (5), 201-400m (1). **Flowering:** May (2), Jun (1). **Fruiting:** Oct (1). **Specimens:** - *Mabeta-Moliwe:* Jaff B. 181; Watts J. 389. *Mokoko River F.R.:* Ekema N. 979; Mbani J. M. 508; Tchouto P. 1184. *Onge:* Tchouto P. 913. *Southern Bakundu F.R.:* Akuo FHI 15170

Pauridiantha hirtella (Benth.) Bremek.
F.W.T.A. 2: 169 (1963); Hawthorne W., F.G.F.T. Ghana: 30 (1990).

Habit: shrub to 5 m. **Guild:** pioneer. **Habitat:** forest. **Distribution:** Guinea Bissau to Cameroon. **Chorology:** Guineo-Congolian. **Star:** green. **Alt. range:** 1-200m (1).**Fruiting:** Oct (1). **Specimens:** - *Mabeta-Moliwe:* Schlechter F. R. R. 12380. *Onge:* Akogo M. 61

Pauridiantha paucinervis (Hiern) Bremek.
F.W.T.A. 2: 168 (1963); F.T.E.A. Rubiaceae: 153 (1976).

Habit: tree or shrub to 6 m. **Habitat:** forest. **Distribution:** Bioko and SW Cameroon. **Chorology:** Western Cameroon Uplands (montane). **Star:** black. **Alt. range:** 1400-1600m (3), 1601-1800m (1), 1801-2000m (1). **Flowering:** Sep (2). **Fruiting:** Sep (1), Oct (2). **Specimens:** - *Eastern slopes:* Cheek M. 3694; Thomas D. W. 9470. *Etinde:* Tekwe C. 226, 296. *No locality:* Maitland T. D. 1308

Pauridiantha rubens (Benth.) Bremek.
F.W.T.A. 2: 168 (1963); Fl. Gabon 12: 250 (1966).

Habit: tree or shrub to 10 m. **Habitat:** forest and lava flows. **Distribution:** Cameroon, Bioko, S. Tomé and Congo (Kinshasa). **Chorology:** Guineo-Congolian. **Star:** green. **Alt. range:** 1-200m (1), 601-800m (1). **Flowering:** Feb (1). **Specimens:** - *Etinde:* Maitland T. D. 576; Tekwe C. 25

Pauridiantha sp. aff. rubens (Benth.) Bremek.

Specimens: - *Mabeta-Moliwe:* Watts J. 386

Pauridiantha venusta N.Hallé
Fl. Gabon 12: 237 (1966).

Habit: tree or shrub to 6 m. **Habitat:** forest. **Distribution:** Cameroon and Gabon. **Chorology:** Lower Guinea. **Star:** blue. **Alt. range:** 1-200m (7), 201-400m (3). **Flowering:** May (2), Jun (1), Nov (2). **Fruiting:** Mar (1), May (2), Oct (2), Nov (4). **Specimens:** - *Etinde:* Cheek M. 5419, 5479; Mbani J. M. 129. *Mokoko River F.R.:* Ndam N. 1187; Pouakouyou D. 115. *Onge:* Akogo M. 122; Harris D. J. 3721; Ndam N. 720; Thomas D. W. 9850; Watts J. 967; Wheatley J. I. 735

Pauridiantha viridiflora (Schweinf. ex Hiern) Hepper

F.W.T.A. 2: 168 (1963); Fl. Gabon 12: 237 (1966); F.T.E.A. Rubiaceae: 158 (1976).

Habit: tree or shrub to 6 m. **Habitat:** forest. **Distribution:** SE Nigeria, Cameroon, Congo (Kinshasa), CAR, Uganda and W. Tanzania. **Chorology:** Guineo-Congolian (montane). **Star:** green. **Alt. range:** 800-1000m (2). **Specimens:** - *Etinde:* Maitland T. D. 502, 1075

Pauridiantha sp.

Specimens: - *Onge:* Tchouto P. 769

Pausinystalia macroceras (K.Schum.) Pierre ex Beille

F.W.T.A. 2: 112 (1963); Keay R.W.J., Trees of Nigeria: 424 (1989); Bot. J. Linn. Soc. 120: 311 (1996).

Habit: tree 14-40 m tall. **Habitat:** forest. **Distribution:** Nigeria, Cameroon, Rio Muni, Gabon, Congo (Kinshasa) and Angola (Cabinda). **Chorology:** Guineo-Congolian. **Star:** green. **Alt. range:** 1-200m (1). **Specimens:** - *Southern Bakundu F.R.:* Binuyo A. & Daramola B. O. FHI 35654; Mambo 72

Pausinystalia talbotii Wernham

F.W.T.A. 2: 112 (1963); Bot. J. Linn. Soc. 120: 320 (1996).
Corynanthe dolichocarpa W.Brandt, F.W.T.A. 2: 111 (1963).

Habit: tree to 30 m. **Habitat:** forest. **Distribution:** Nigeria and Cameroon. **Chorology:** Lower Guinea. **Star:** blue. **Fruiting:** Dec (1). **Specimens:** - *Mabeta-Moliwe:* Mildbraed G. W. J. 10773

Pavetta bidentata Hiern var. bidentata

F.W.T.A. 2: 139 (1963); Ann. Missouri Bot. Gard. 83: 103 (1996).

Habit: shrub to 3 m. **Habitat:** forest. **Distribution:** SE Nigeria, Bioko, Cameroon and Congo (Kinshasa). **Chorology:** Guineo-Congolian. **Star:** green. **Alt. range:** 1-200m (2). **Flowering:** Mar (1), Apr (1). **Specimens:** - *Bambuko F.R.:* Watts J. 671. *Etinde:* Maitland T. D. 1180. *Mokoko River F.R.:* Acworth J. M. 154

Pavetta brachycalyx Hiern

F.W.T.A. 2: 139 (1963); Ann. Missouri Bot. Gard. 83: 105 (1996).

Habit: shrub to 3 m. **Habitat:** forest. **Distribution:** Cameroon. **Chorology:** Lower Guinea (montane). **Star:** gold. **Alt. range:** 400-600m (1), 801-1000m (1), 1201-1400m (3), 1401-1600m (1), 1601-1800m (1).**Fruiting:** Mar (1), Aug (1), Sep (1), Oct (1). **Specimens:** - *Eastern slopes:* Cable S. 1388; Maitland T. D. 213. *Etinde:* Cheek M. 3797; Hunt L. V. 38; Maitland T. D. 988; Tekwe C. 161; Thomas D. W. 2827. *No locality:* Dunlap 20; Mann G. 2159

Pavetta camerounensis S.Manning subsp. brevirama S.Manning

Ann. Missouri Bot. Gard. 83: 108 (1996).

Habit: shrub to 1 m. **Habitat:** forest. **Distribution:** SW Cameroon. **Chorology:** Western Cameroon. **Star:** black. **Specimens:** - *Southern Bakundu F.R.:* Brenan J. P. M. 9279a; Olorunfemi J. FHI 30719

Pavetta camerounensis S.Manning subsp. camerounensis

Ann. Missouri Bot. Gard. 83: 108 (1996).

Habit: shrub to 1 m. **Habitat:** forest. **Distribution:** Cameroon. **Chorology:** Lower Guinea. **Star:** gold. **Alt. range:** 1-200m (1). **Flowering:** Apr (1). **Fruiting:** Apr (1). **Specimens:** - *Mokoko River F.R.:* Akogo M. 242

Pavetta gabonica Bremek.

Ann. Missouri Bot. Gard. 83: 113 (1996).

Habit: shrub to 3 m. **Habitat:** stream banks in forest. **Distribution:** Cameroon and Gabon. **Chorology:** Lower Guinea. **Star:** blue. **Alt. range:** 1-200m (1). **Flowering:** Apr (1). **Fruiting:** May (1). **Specimens:** - *Mokoko River F.R.:* Fraser P. 383; Pouakouyou D. 1

Pavetta cf. gracilipes Hiern

Specimens: - *Mabeta-Moliwe:* Sidwell K. 127

Pavetta hispida Hiern

F.W.T.A. 2: 139 (1963).

Habit: shrub to 3 m. **Habitat:** forest. **Distribution:** SE Nigeria, Cameroon, Congo (Kinshasa) and Angola (Cabinda). **Chorology:** Guineo-Congolian. **Star:** green. **Alt. range:** 1-200m (2), 201-400m (3). **Flowering:** Mar (1), Apr (1), Nov (1). **Fruiting:** Oct (1), Nov (1). **Specimens:** - *Bambuko F.R.:* Watts J. 590. *Mokoko River F.R.:* Ndam N. 1133; Tchouto P. 1163. *Onge:* Harris D. J. 3792; Watts J. 788, 1034

Pavetta hookeriana Hiern var. hookeriana

F.W.T.A. 2: 139 (1963); Ann. Missouri Bot. Gard. 83: 116 (1996).

Habit: tree or shrub to 10 m. **Habitat:** forest-grassland transition. **Distribution:** Bioko and W. Cameroon. **Chorology:** Western Cameroon Uplands (montane). **Star:** gold. **Alt. range:** 2000-2200m (5), 2201-2400m (1). **Flowering:** Oct (1). **Fruiting:** Oct (1). **Specimens:** - *Bambuko F.R.:* Thomas D. W. 4608. *Eastern slopes:* Onochie C. F. A. FHI 9525; Tchouto P. 280; Upson T. 91. *No locality:* Maitland T. D. 992; Mann G. 2166

Pavetta hookeriana Hiern var. pubinervata S.Manning

Ann. Missouri Bot. Gard. 83: 117 (1996).

Habit: tree or shrub to 10 m. **Habitat:** forest. **Distribution:** Mount Cameroon. **Chorology:** Endemic (montane). **Star:** black. **Alt. range:** 400-600m (1). **Flowering:** Mar (1). **Specimens:** - *Eastern slopes:* Kalbreyer W. 94

Pavetta cf. ixorifolia Bremek.

Specimens: - *Mabeta-Moliwe:* Sunderland T. C. H. 1244

Pavetta longibrachiata Bremek.

Ann. Missouri Bot. Gard. 83: 123 (1996).

Habit: shrub to 4 m. **Habitat:** forest. **Distribution:** Cameroon. **Chorology:** Lower Guinea. **Star:** gold. **Alt. range:** 200-400m (1), 401-600m (1). **Flowering:** Feb (1). **Specimens:** - *Etinde:* Wheatley J. I. 23. *Onge:* Thomas D. W. 4494

Pavetta neurocarpa Benth.

F.W.T.A. 2: 139 (1963); Ann. Missouri Bot. Gard. 83: 133 (1996).

Habit: shrub to 1 m. **Habitat:** forest. **Distribution:** Bioko and SW Cameroon. **Chorology:** Western Cameroon. **Star:** black. **Alt. range:** 1-200m (6), 201-400m (7), 401-600m (2), 601-800m (4), 801-1000m (2). **Flowering:** Mar (4), May (2), Dec (3). **Fruiting:** May (1), Aug (1), Sep (1), Oct (3), Nov (4), Dec (2). **Specimens:** - *Etinde:* Cable S. 137, 393, 1513, 1580, 1605; Cheek M. 5483, 5605, 5850; Khayota B. 505; Lighava M. 28; Satabie B. 651; Tekwe C. 126; Thomas D. W. 5120, 9115; Watts J. 519. *Mabeta-Moliwe:* Mann G. s.n.. *Mokoko River F.R.:* Ndam N. 1072, 1119, 1255. *Onge:* Akogo M. 37; Harris D. J. 3708; Ndam N. 655; Thomas D. W. 9864

Pavetta owariensis P.Beauv. var. **glaucescens** (Hiern) S.Manning

Ann. Missouri Bot. Gard. 83: 136 (1996).
   Pavetta glaucescens Hiern, F.W.T.A. 2: 139 (1963).

Habit: tree or shrub to 5 m. **Habitat:** forest. **Distribution:** Cameroon. **Chorology:** Lower Guinea. **Star:** gold. **Alt. range:** 1-200m (1), 201-400m (2), 401-600m (2), 601-800m (1). **Flowering:** Jan (1), Mar (1), Apr (1), Jul (1). **Fruiting:** Mar (1), Jul (1), Aug (1), Oct (1), Nov (1). **Specimens:** - *Etinde:* Khayota B. 568; Kwangue A. 134; Tchouto P. 264; Tekwe C. 5; Watts J. 472. *Mokoko River F.R.:* Akogo M. 215. *Onge:* Baker W. J. 354. *Southern Bakundu F.R.:* Binuyo A. & Daramola B. O. FHI 35576; Ejiofor M. C. FHI 29306; Onochie C. F. A. FHI 31199

Pavetta plumosa Hutch.

F.W.T.A. 2: 139 (1963).

Habit: shrub to 1 m. **Distribution:** Mount Cameroon. **Chorology:** Endemic. **Star:** black. **Specimens:** - *Etinde:* Mildbraed G. W. J. 10743

Pavetta rigida Hiern

F.W.T.A. 2: 140 (1963); Ann. Missouri Bot. Gard. 78: 535 (1991).

Habit: shrub to 4 m. **Habitat:** forest. **Distribution:** Nigeria, Bioko and Cameroon. **Chorology:** Lower Guinea (montane). **Star:** blue. **Alt. range:** 1-200m (5), 201-400m (1), 401-600m (1), 801-1000m (4), 1001-1200m (2), 1201-1400m (3). **Flowering:** Feb (1), Mar (2), Apr (3), May (3). **Fruiting:** Jun (2), Oct (2), Dec (2). **Specimens:** - *Bambuko F.R.:* Keay R. W. J. FHI 37484. *Eastern slopes:* Cable S. 1352; Etuge M. 1152; Maitland T. D. 509. *Etinde:* Cable S. 321, 477, 1529, 1643; Cheek M. 3771; Nkeng P. 42; Thomas D. W. 9165. *Mabeta-Moliwe:* Jaff B. 316; Plot Voucher W 44, W 183, W 186; Sunderland T. C. H. 1326; Watts J. 167; Wheatley J. I. 222. *Mokoko River F.R.:* Acworth J. M. 299; Mbani J. M. 338; Ndam N. 1364. *Onge:* Thomas D. W. 4442

Pavetta cf. rigida Hiern

Specimens: - *Mokoko River F.R.:* Ekema N. 1040

Pavetta staudtii Hutch. & Dalziel

F.W.T.A. 2: 139 (1963); Ann. Missouri Bot. Gard. 83: 140 (1996).

Habit: shrub to 3 m. **Habitat:** forest. **Distribution:** Cameroon. **Chorology:** Lower Guinea. **Star:** gold. **Alt. range:** 1-200m (1), 401-600m (1). **Fruiting:** May (1), Dec (1). **Specimens:** - *Etinde:* Cable S. 298. *Mokoko River F.R.:* Ekema N. 940

Pavetta talbotii Wernham

F.W.T.A. 2: 140 (1963).

Habit: shrub to 1 m. **Habitat:** forest. **Distribution:** SE Nigeria and SW Cameroon. **Chorology:** Western Cameroon. **Star:** gold. **Alt. range:** 1-200m (1). **Specimens:** - *Southern Bakundu F.R.:* Thomas D. W. 2587a

Pavetta tenuissima S.Manning

Ann. Missouri Bot. Gard. 83: 142 (1996).

Habit: shrub to 2 m. **Habitat:** forest. **Distribution:** Cameroon. **Chorology:** Lower Guinea. **Star:** gold. **Specimens:** - *Southern Bakundu F.R.:* Brenan J. P. M. 9280, 9280a

Pavetta tetramera (Hiern) Bremek.

F.W.T.A. 2: 140 (1963).

Habit: shrub to 1 m. **Habitat:** forest. **Distribution:** Cameroon, Rio Muni, Gabon, Congo (Kinshasa) and Angola. **Chorology:** Guineo-Congolian. **Star:** green. **Specimens:** - *Southern Bakundu F.R.:* Binuyo A. & Daramola B. O. FHI 35156; Latilo M. G. FHI 43817; Olorunfemi J. FHI 30517

Pavetta sp.

Specimens: - *Mokoko River F.R.:* Acworth J. M. 134; Batoum A. 6; Ekema N. 986, 1105. *Southern Bakundu F.R.:* Binuyo A. & Daramola B. O. FHI 35164

Pentas schimperana (A.Rich.) Vatke subsp. **occidentalis** (Hook.f.) Verdc.

F.W.T.A. 2: 216 (1963); F.T.E.A. Rubiaceae: 188 (1976).

Habit: shrub to 2 m. **Habitat:** grassland. **Distribution:** Cameroon, Bioko, S. Tomé and Congo (Kinshasa). **Chorology:** Guineo-Congolian (montane). **Star:** green. **Alt. range:** 1-200m (1), 1201-1400m (1), 1601-1800m (1), 1801-2000m (1), 2001-2200m (3), 2201-2400m (1), 2401-2600m (5), 2601-2800m (3), 2801-3000m (1). **Specimens:** - *Eastern slopes:* Boughey A. S. GC 12582; Brenan J. P. M. 9385; Dundas J. FHI 20370; Hambler D. J. 186; Hutchinson J. & Metcalfe C. R. 49; Keay R. W. J. FHI 28596; Letouzey R. MS 89; Maitland T. D. 476, 683; Morton J. K. GC 6861. *Etinde:* Maitland T. D. 1079. *No locality:* Brenan J. P. M. 9385a; Dalziel J. M. 8306; Gregory H. 114; Maitland T. D. 988; Mann G. 1993; Meurillon A. 1159; Migeod F. 164

Pentodon pentandrus (Schum. & Thonn.) Vatke

F.W.T.A. 2: 213 (1963); Fl. Gabon 12: 105 (1966); F.T.E.A. Rubiaceae: 263 (1976).

Habit: herb. **Guild:** pioneer. **Habitat:** forest and stream banks. **Distribution:** tropical Africa and Arabia. **Star:** excluded. **Alt. range:** 1-200m (4). **Flowering:** Apr (2), May (2). **Fruiting:** May (1). **Specimens:** - *Mabeta-Moliwe:* Sunderland T. C. H. 1278; Tchouto P. 207. *Mokoko River F.R.:* Acworth J. M. 64, 162. *Southern Bakundu F.R.:* Brenan J. P. M. 9326

Petitiocodon parviflorum (Keay) Robbr.

Bull. Jard. Bot. Nat. Belg. 58: 117 (1988).
   Didymosalpinx parviflora Keay, F.W.T.A. 2: 130 (1963).

Habit: tree or shrub to 4 m. **Habitat:** forest. **Distribution:** Nigeria and Cameroon. **Chorology:** Lower Guinea. **Star:** gold. **Alt. range:**

1-200m (10), 201-400m (1), 401-600m (1). **Flowering:** Mar (2), Apr (3). **Fruiting:** May (4), Jun (2). **Specimens:** - *Etinde:* Akogo M. 150; Mbani J. M. 164; Nkeng P. 38. *Mabeta-Moliwe:* Jaff B. 83, 164, 216; Khayota B. 454; Ndam N. 484; Plot Voucher W 757, W 764; Sunderland T. C. H. 1185; Watts J. 382; Wheatley J. I. 146. *Mokoko River F.R.:* Tchouto P. 1097. *No locality:* Motuba I. M. 15067

Poecilocalyx schumannii Bremek.
Fl. Gabon 12: 230 (1966).

Habit: shrub to 2 m. **Habitat:** forest. **Distribution:** Cameroon and Rio Muni. **Chorology:** Lower Guinea. **Star:** blue. **Alt. range:** 1-200m (4), 201-400m (2). **Flowering:** May (3), Oct (1). **Fruiting:** Mar (1), May (3), Nov (1). **Specimens:** - *Mokoko River F.R.:* Batoum A. 21; Ekema N. 843; Fraser P. 361; Tchouto P. 1203. *Onge:* Watts J. 834, 983; Wheatley J. I. 752

Polysphaeria macrophylla K.Schum.
F.W.T.A. 2: 148 (1963).

Habit: tree or shrub to 5 m. **Guild:** shade-bearer. **Habitat:** forest and stream banks. **Distribution:** Ghana, Nigeria and Cameroon. **Chorology:** Guineo-Congolian (montane). **Star:** green. **Alt. range:** 1-200m (11), 201-400m (5), 2401-2600m (1). **Flowering:** Mar (1), Apr (1), May (2), Jun (1), Nov (3), Dec (1). **Fruiting:** Mar (1), Apr (2), May (3), Jun (1), Aug (1), Oct (4), Nov (2), Dec (1). **Specimens:** - *Etinde:* Cheek M. 5929. *Mabeta-Moliwe:* Tchouto P. 199. *Mokoko River F.R.:* Acworth J. M. 132, 282; Akogo M. 279; Ekema N. 1058; Fraser P. 453. *No locality:* Maitland T. D. 599; Thompson S. 1634. *Onge:* Akogo M. 32; Baker W. J. 355; Cheek M. 5182; Harris D. J. 3658, 3796; Tchouto P. 875; Watts J. 797, 920; Wheatley J. I. 685, 796

Polysphaeria sp. aff. macrophylla K.Schum.

Specimens: - *Mabeta-Moliwe:* Sunderland T. C. H. 1232; Wheatley J. I. 365

Preussiodora sulphurea (K.Schum.) Keay
F.W.T.A. 2: 119 (1963); Fl. Gabon 17: 208 (1970).

Habit: shrub to 2 m. **Habitat:** forest. **Distribution:** SE Nigeria, Bioko and SW Cameroon. **Chorology:** Western Cameroon. **Star:** gold. **Alt. range:** 1-200m (5), 201-400m (1). **Flowering:** Mar (1), Apr (2), Oct (1). **Fruiting:** May (1), Oct (2). **Specimens:** - *Etinde:* Tekwe C. 80. *Mokoko River F.R.:* Akogo M. 250; Tchouto P. 619, 1147. *Onge:* Cheek M. 5106; Watts J. 803

Pseudosabicea batesii (Wernham) N.Hallé
Fl. Gabon 12: 202 (1966).

Habit: woody climber. **Habitat:** forest. **Distribution:** Cameroon and Gabon. **Chorology:** Lower Guinea. **Star:** blue. **Alt. range:** 1-200m (6). **Flowering:** Apr (1), May (2), Jun (3). **Specimens:** - *Mokoko River F.R.:* Acworth J. M. 233; Batoum A. 15; Ekema N. 1005, 1172, 1177b; Pouakouyou D. 33

Pseudosabicea medusula (K.Schum.) N.Hallé
Fl. Gabon 12: 200 (1966).
Sabicea medusula K.Schum. ex Wernham, F.W.T.A. 2: 174 (1963).

Habit: creeping herb. **Habitat:** forest. **Distribution:** Cameroon. **Chorology:** Lower Guinea. **Star:** gold. **Alt. range:** 1-200m (3).

Flowering: May (3). **Fruiting:** May (2). **Specimens:** - *Mokoko River F.R.:* Batoum A. 14; Ekema N. 870; Tchouto P. 1220

Psilanthus mannii Hook.f.
F.W.T.A. 2: 157 (1963).

Habit: tree or shrub to 4 m. **Guild:** shade-bearer. **Habitat:** forest. **Distribution:** Guinea to Bioko, Cameroon, CAR, Gabon, Congo (Kinshasa) and Angola. **Chorology:** Guineo-Congolian. **Star:** green. **Alt. range:** 1-200m (6), 201-400m (6), 401-600m (3), 801-1000m (1). **Flowering:** Feb (1), Apr (1), May (1), Jul (1), Nov (1). **Fruiting:** Apr (1), May (2), Jul (1), Sep (1), Oct (3), Nov (3), Dec (3). **Specimens:** - *Bambuko F.R.:* Olorunfemi J. FHI 30769. *Etinde:* Cable S. 263, 433; Cheek M. 5467, 5897; Lighava M. 5; Mbani J. M. 122, 158; Tchouto P. 260. *Mabeta-Moliwe:* Maitland T. D. 612. *Mokoko River F.R.:* Akogo M. 261. *No locality:* Nkeng P. 13; Thompson S. 1805. *Onge:* Akogo M. 90; Ndam N. 593; Thomas D. W. 9804; Watts J. 709, 801, 982

Psilanthus sp. nov.
Psilanthus ebracteolatus sensu auct., non Hiern, F.T.A. 3: 186 (1877).
Coffea ebracteolata sensu Hepper & Keay p.p., F.W.T.A. 2: 157 (1963).

Habit: tree or shrub to 4 m. **Habitat:** forest. **Distribution:** Cameroon. **Chorology:** Lower Guinea. **Star:** gold. **Alt. range:** 1-200m (1), 201-400m (1). **Fruiting:** Sep (1), Nov (1). **Specimens:** - *Etinde:* Tchouto P. 810. *Mabeta-Moliwe:* Mann G. 740. *Onge:* Thomas D. W. 9851. *Southern Bakundu F.R.:* Latilo M. G. FHI 43830

Psychotria bifaria Hiern var. bifaria
F.W.T.A. 2: 199 (1963); Bull. Jard. Bot. Brux. 36: 152 (1966).

Habit: shrub to 1 m. **Habitat:** forest. **Distribution:** Bioko and SW Cameroon. **Chorology:** Western Cameroon. **Star:** black. **Alt. range:** 1-200m (5), 201-400m (4), 601-800m (3). **Flowering:** Feb (1), Mar (4), Apr (1), Oct (1). **Fruiting:** Oct (2), Nov (2). **Specimens:** - *Eastern slopes:* Maitland T. D. 541. *Etinde:* Cable S. 183, 1502; Etuge M. 1230; Mbani J. M. 61; Tchouto P. 88, 461; Thomas D. W. 5543; Wheatley J. I. 28. *Mabeta-Moliwe:* Plot Voucher W 154; Wheatley J. I. 149. *Onge:* Akogo M. 125; Thomas D. W. 9801

Psychotria calceata Petit
Bull. Jard. Bot. Brux. 34: 201 (1964).

Habit: shrub to 2 m. **Habitat:** forest. **Distribution:** Nigeria and Cameroon. **Chorology:** Lower Guinea. **Star:** blue. **Alt. range:** 200-400m (1), 401-600m (1). **Flowering:** Oct (1). **Fruiting:** Oct (2). **Specimens:** - *Etinde:* Tchouto P. 463; Watts J. 485

Psychotria calva Hiern
F.W.T.A. 2: 200 (1963); Bull. Jard. Bot. Brux. 36: 100 (1966); Meded. Land. Wag. 82(3): 291 (1982).

Habit: shrub to 2 m. **Guild:** pioneer. **Habitat:** forest. **Distribution:** Senegal to Bioko and Cameroon. **Chorology:** Guineo-Congolian. **Star:** green. **Alt. range:** 1-200m (1). **Flowering:** Apr (1). **Fruiting:** Apr (1). **Specimens:** - *Mokoko River F.R.:* Akogo M. 267

Psychotria camerunensis Petit
Bull. Jard. Bot. Brux. 36: 158 (1966).

Habit: shrub to 1.5 m. Habitat: forest. Distribution: Cameroon. Chorology: Lower Guinea. Star: gold. Specimens: - *Etinde:* Hepper F. N. 8613

Psychotria camptopus Verdc.
Kew Bull. 30: 259 (1975).
Cephaelis mannii (Hook.f.) Hiern, F.W.T.A. 2: 203 (1963).

Habit: tree to 5 m. Habitat: forest. Distribution: SE Nigeria, Bioko, SW Cameroon and Congo (Kinshasa). Chorology: Guineo-Congolian (montane). Star: gold. Alt. range: 1-200m (1), 201-400m (2), 1201-1400m (2), 1601-1800m (2), 1801-2000m (1). Flowering: May (1), Oct (3). Fruiting: Jun (1), Aug (1), Oct (3). Specimens: - *Eastern slopes:* Sunderland T. C. H. 1443. *Etinde:* Hunt L. V. 45; Nkeng P. 71; Thomas D. W. 9537; Wheatley J. I. 624. *No locality:* Maitland T. D. s.n.. *Onge:* Akogo M. 95; Ndam N. 643

Psychotria ceratalabastron K.Schum.
F.W.T.A. 2: 199 (1963); Bull. Jard. Bot. Brux. 34: 218 (1964).

Habit: shrub to 2 m. Habitat: forest. Distribution: SW Cameroon. Chorology: Western Cameroon. Star: black. Alt. range: 200-400m (1). Flowering: Apr (1), May (1), Jun (1). Specimens: - *Mabeta-Moliwe:* Preuss P. R. 1271. *Mokoko River F.R.:* Fraser P. 447; Ndam N. 1075

Psychotria chalconeura (K.Schum.) Petit
F.W.T.A. 2: 202 (1963); Bull. Jard. Bot. Brux. 34: 60 (1964); Bull. Jard. Bot. Nat. Belg. 42: 354 (1972).
Psychotria malchairei De Wild., F.W.T.A. 2: 201 (1963).

Habit: tree or shrub to 8 m. Habitat: forest. Distribution: Cameroon and Congo (Kinshasa). Chorology: Guineo-Congolian (montane). Star: green. Alt. range: 1-200m (2), 201-400m (2).Fruiting: May (6), Jun (1). Specimens: - *Etinde:* Cheek M. 3012. *Mabeta-Moliwe:* Cheek M. 3491. *Mokoko River F.R.:* Ekema N. 878, 1162; Fraser P. 349, 350; Ndam N. 1141, 1171, 1215

Psychotria dorotheae Wernham
F.W.T.A. 2: 200 (1963); Bull. Jard. Bot. Brux. 34: 67 (1964).

Habit: shrub to 2 m. Guild: shade-bearer. Habitat: forest. Distribution: Liberia, SE Nigeria and SW Cameroon. Chorology: Guineo-Congolian. Star: gold. Alt. range: 1-200m (11), 201-400m (3). Flowering: Apr (2). Fruiting: Mar (1), Apr (3), May (10), Jun (1), Nov (1). Specimens: - *Bambuko F.R.:* Watts J. 650. *Mokoko River F.R.:* Acworth J. M. 153, 296; Akogo M. 285; Batoum A. 43; Ekema N. 912, 914, 958, 1193a; Fraser P. 380; Ndam N. 1109, 1152, 1300; Tchouto P. 1075; Watts J. 1094. *Onge:* Harris D. J. 3781

Psychotria ebensis K.Schum.
Bull. Jard. Bot. Brux. 34: 157 (1964).
Psychotria coeruleo-violacea K.Schum., F.W.T.A. 2: 198 (1963).

Habit: shrub to 2 m. Guild: shade-bearer. Habitat: forest. Distribution: Cameroon and Gabon. Chorology: Lower Guinea. Star: blue. Alt. range: 1-200m (8), 201-400m (3). Flowering: Mar (2), Apr (2), May (4), Jun (2), Aug (1). Fruiting: Jul (1), Aug (1), Oct (1). Specimens: - *Etinde:* Maitland T. D. 631. *Mabeta-Moliwe:* Baker W. J. 266; Hurst J. 1; Ndam N. 476; Plot Voucher W 466, W 670; Sunderland T. C. H. 1273; Watts J. 164; Wheatley J. I. 71, 280. *Mokoko River F.R.:* Acworth J. M. 128; Fraser P. 422. *Onge:* Ndam N. 737; Wheatley J. I. 749

Psychotria cf. ebensis K.Schum.
Specimens: - *Mokoko River F.R.:* Ekema N. 1052; Tchouto P. 1227

Psychotria erythropus K.Schum.
F.W.T.A. 2: 202 (1963); Bull. Jard. Bot. Brux. 34: 219 (1964).

Habit: shrub to 4 m. Habitat: forest. Distribution: Mount Cameroon. Chorology: Endemic (montane). Star: black. Alt. range: 1800-2000m (1). Specimens: - *Eastern slopes:* Preuss P. R. 1044

Psychotria foliosa Hiern
Bull. Jard. Bot. Brux. 34: 204 (1964).

Habit: shrub to 1 m. Habitat: forest. Distribution: Cameroon and Gabon. Chorology: Lower Guinea. Star: blue. Alt. range: 1-200m (2), 201-400m (3), 401-600m (1). Flowering: May (1), Oct (1). Fruiting: Oct (4), Dec (1). Specimens: - *Etinde:* Faucher P. 26. *Mokoko River F.R.:* Batoum A. 50. *Onge:* Cheek M. 5023; Ndam N. 743; Tchouto P. 739; Watts J. 734. *Southern Bakundu F.R.:* Akuo FHI 15178

Psychotria gabonica Hiern
F.W.T.A. 2: 202 (1963); Bull. Jard. Bot. Brux. 34: 65 (1964); Bull. Jard. Bot. Nat. Belg. 42: 355 (1972).
Psychotria rowlandii Hutch. & Dalziel, F.W.T.A. 2: 201 (1963).

Habit: tree or shrub to 4 m. Habitat: forest. Distribution: Liberia to Cameroon and Gabon. Chorology: Guineo-Congolian (montane). Star: green. Alt. range: 1-200m (6), 201-400m (1), 401-600m (4), 601-800m (1), 801-1000m (4), 1001-1200m (5), 1201-1400m (3), 1401-1600m (3). Flowering: Feb (1), Mar (1), Apr (2), May (1), Aug (1), Oct (4), Nov (1), Dec (3). Fruiting: Feb (2), Apr (2), Jun (1), Aug (1), Oct (5), Nov (1), Dec (1). Specimens: - *Eastern slopes:* Breteler F. J. 237; Cable S. 1326; Groves M. 179; Keay R. W. J. FHI 28575; Maitland T. D. 455; Tchouto P. 39. *Etinde:* Cable S. 336, 429; Cheek M. 3796, 5599; Hunt L. V. 24; Lighava M. 10; Thomas D. W. 9524, 9543. *Mabeta-Moliwe:* Jaff B. 128; Watts J. 321. *Mokoko River F.R.:* Acworth J. M. 133; Ekema N. 959; Pouakouyou D. 16. *No locality:* Dunlap 16, 222; Hutchinson J. & Metcalfe C. R. 77; Maitland T. D. 1217; Mann G. 2160; Migeod F. 253. *Onge:* Cheek M. 4970, 4975, 5005; Ndam N. 667; Tchouto P. 816; Wheatley J. I. 742

Psychotria globiceps K.Schum.
F.W.T.A. 2: 200 (1963); Bull. Jard. Bot. Brux. 34: 172 (1964).

Habit: shrub to 1 m. Habitat: forest. Distribution: Cameroon and Congo (Kinshasa). Chorology: Guineo-Congolian. Star: green. Alt. range: 1-200m (11), 201-400m (5), 401-600m (3). Flowering: May (4), Jun (3), Aug (1), Sep (1), Oct (1), Nov (1). Fruiting: Jun (1), Aug (1), Sep (1), Oct (3), Nov (4), Dec (2). Specimens: - *Etinde:* Cable S. 269; Cheek M. 5412, 5904; Letouzey R. 14973; Lighava M. 33; Mbani J. M. 125; Schlechter F. R. R. 12374; Tekwe C. 124; Wheatley J. I. 528. *Mokoko River F.R.:* Ekema N. 1051, 1160; Fraser P. 444; Pouakouyou D. 107; Watts J. 1092. *Onge:* Cheek M. 5151; Harris D. J. 3653, 3867; Ndam N. 611; Tchouto P. 856; Watts J. 767

Psychotria cf. globiceps K.Schum.
Specimens: - *Mokoko River F.R.:* Batoum A. 42

Psychotria globosa Hiern

F.W.T.A. 2: 198 (1963).**Alt. range:** 1-200m (2). **Flowering:** Apr (1), May (1). **Fruiting:** Nov (1). **Specimens:** - *Mabeta-Moliwe:* Mann G. 1330; Ndam N. 503; Nguembock F. 32a, 78; Preuss P. R. 1159; Tchouto P. 145

Psychotria globosa Hiern var. ciliata (Hiern) Petit

Bull. Jard. Bot. Brux. 34: 160 (1964).
   Psychotria nigerica Hepper, F.W.T.A. 2: 198 (1963).

Habit: shrub to 1 m. **Habitat:** forest. **Distribution:** Nigeria and Cameroon. **Chorology:** Lower Guinea. **Star:** blue. **Alt. range:** 1-200m (5), 201-400m (1), 601-800m (2), 801-1000m (1), 1201-1400m (1). **Flowering:** Mar (2), Apr (2), May (2). **Fruiting:** May (1), Oct (2). **Specimens:** - *Etinde:* Cheek M. 3770; Kalbreyer W. 153; Maitland T. D. 1110; Mbani J. M. 96; Nkeng P. 51. *Mokoko River F.R.:* Acworth J. M. 46; Akogo M. 292; Ekema N. 1091. *Onge:* Tchouto P. 779; Wheatley J. I. 821

Psychotria globosa Hiern var. globosa

F.W.T.A. 2: 198 (1963); Bull. Jard. Bot. Brux. 34: 159 (1964).

Habit: shrub to 1 m. **Habitat:** forest. **Distribution:** Nigeria and Cameroon. **Chorology:** Lower Guinea. **Star:** blue. **Alt. range:** 1-200m (1). **Flowering:** Jun (1). **Specimens:** - *Mokoko River F.R.:* Ekema N. 1178

Psychotria humilis Hiern var. humilis

F.W.T.A. 2: 199 (1963); Bull. Jard. Bot. Brux. 36: 180 (1966).

Habit: herb. **Habitat:** forest. **Distribution:** Nigeria, Cameroon and Gabon. **Chorology:** Lower Guinea. **Star:** blue. **Alt. range:** 1-200m (1), 201-400m (1). **Flowering:** Oct (1). **Fruiting:** Mar (1). **Specimens:** - *Onge:* Tchouto P. 749; Wheatley J. I. 817. *Southern Bakundu F.R.:* Binuyo A. & Daramola B. O. FHI 35453

Psychotria humilis Hiern var. maior Petit

F.W.T.A. 2: 199 p.p. (1963); Bull. Jard. Bot. Brux. 36: 181 (1966).

Habit: herb. **Habitat:** forest. **Distribution:** Nigeria and Cameroon. **Chorology:** Lower Guinea. **Star:** blue. **Alt. range:** 1-200m (5), 201-400m (3). **Flowering:** Mar (1), Aug (1), Oct (1), Nov (4). **Fruiting:** May (1), Nov (1). **Specimens:** - *Etinde:* Cheek M. 5430, 5454. *Mabeta-Moliwe:* Baker W. J. 239; Cable S. 1446. *Mokoko River F.R.:* Ekema N. 997. *Onge:* Harris D. J. 3732, 3868; Ndam N. 645

Psychotria konguensis Hiern

F.W.T.A. 2: 199 (1963); Bull. Jard. Bot. Brux. 36: 160 (1966). **Habit:** shrub to 1 m. **Habitat:** forest. **Distribution:** Cameroon, Gabon, Congo (Brazzaville), Congo (Kinshasa) and Angola (Cabinda). **Chorology:** Guineo-Congolian. **Star:** green. **Alt. range:** 200-400m (1). **Flowering:** Oct (1). **Fruiting:** Oct (2). **Specimens:** - *Mabeta-Moliwe:* Nguembock F. 17. *Onge:* Tchouto P. 762

Psychotria latistipula Benth.

F.W.T.A. 2: 198 (1963); Bull. Jard. Bot. Brux. 34: 143 (1964).

Habit: tree or shrub to 4 m. **Habitat:** forest. **Distribution:** Nigeria, Bioko, Cameroon and Gabon. **Chorology:** Lower Guinea (montane). **Star:** blue. **Alt. range:** 1-200m (16), 201-400m (3), 401-600m (4), 601-800m (1). **Flowering:** Mar (1), May (2), Jun (3), Aug (3), Oct (3), Nov (1), Dec (2). **Fruiting:** May (1), Aug (1), Oct (4), Nov (2), Dec (4). **Specimens:** - *Etinde:* Cable S. 379; Cheek M. 5464, 5829; Khayota B. 573; Kwangue A. 140; Maitland T. D. 172, 721; Nkeng P. 72; Thompson S. 1365, 1493, 1548; Watts J. 529; Winkler H. J. P. 29. *Mabeta-Moliwe:* Baker W. J. 317; Cable S. 523; Leeuwenberg A. J. M. 6942; Ndam N. 510; Nguembock F. 55; Plot Voucher W 1189; Sidwell K. 109; Sunderland T. C. H. 1024, 1300, 1447; Watts J. 416; von Rege I. 36, 85, 104. *Onge:* Cheek M. 5133; Ndam N. 618; Tchouto P. 788

Psychotria sp. aff. lauracea (K.Schum.) Petit

Specimens: - *Mabeta-Moliwe:* Plot Voucher W6

Psychotria cf. ledermannii auct.

Specimens: - *Mabeta-Moliwe:* Sunderland 1395

Psychotria leptophylla Hiern

F.W.T.A. 2: 200 (1963); Bull. Jard. Bot. Brux. 36: 166 (1966).

Habit: shrub to 2 m. **Habitat:** forest. **Distribution:** SE Nigeria, Bioko, Cameroon and Congo (Kinshasa). **Chorology:** Guineo-Congolian. **Star:** green. **Alt. range:** 1-200m (8), 201-400m (2), 401-600m (1), 601-800m (1), 801-1000m (2), 1001-1200m (1), 1201-1400m (3). **Flowering:** Feb (1), Mar (2), Apr (1), May (2), Sep (1), Nov (2). **Fruiting:** Mar (2), Sep (2), Nov (4), Dec (2). **Specimens:** - *Eastern slopes:* Dalziel J. M. 8208; Maitland T. D. 704. *Etinde:* Cable S. 175, 1648; Cheek M. 5456, 5580, 5838; Khayota B. 498; Leeuwenberg A. J. M. 6904; Maitland T. D. 1166; Tekwe C. 157; Thompson S. 1418, 1557; Wheatley J. I. 517. *Mabeta-Moliwe:* Cable S. 531; Watts J. 196. *Mokoko River F.R.:* Ekema N. 1006. *No locality:* Dunlap 21; Mann G. 1191. *Onge:* Watts J. 952

Psychotria mannii (Hook.f.) Hiern

F.W.T.A. 2: 199 (1963); Bull. Jard. Bot. Brux. 36: 171 (1966).

Habit: shrub to 1.5 m. **Habitat:** forest. **Distribution:** Nigeria, Cameroon, Gabon and Congo (Kinshasa). **Chorology:** Guineo-Congolian. **Star:** green. **Specimens:** - *No locality:* Deistel H. 147; Lehmbach H. 21. *Southern Bakundu F.R.:* Daramola B. O. FHI 35504

Psychotria peduncularis (Salisb.) Steyerm. var. hypsophila (K.Schum. & K.Krause) Verdc.

Kew Bull. 30: 257 (1975).
   Cephaelis peduncularis Salisb. var. *hypsophila* (K.Schum. & K.Krause) Hepper, F.W.T.A. 2: 204 (1963).

Habit: shrub to 3 m. **Habitat:** forest. **Distribution:** Guinea to Bioko and Cameroon. **Chorology:** Guineo-Congolian (montane). **Star:** green. **Alt. range:** 1-200m (2), 801-1000m (2), 1201-1400m (1), 1401-1600m (1), 1601-1800m (1), 1801-2000m (3). **Flowering:** May (1), Sep (1). **Fruiting:** Jun (1), Aug (1), Sep (1), Dec (1). **Specimens:** - *Eastern slopes:* Maitland T. D. 205. *Etinde:* Cable S. 473; Hunt L. V. 42; Tekwe C. 164, 193. *Mabeta-Moliwe:* Nguembock F. 77; Plot Voucher W 1056; Wheatley J. I. 291. *No locality:* Brenan J. P. M. 9528; Dundas J. FHI 15336; Dunlap 1; Hutchinson J. & Metcalfe C. R. 3; Maitland T. D. 617; Migeod F. 136; Preuss P. R. 1138

Psychotria peduncularis (Salisb.) Steyerm. var. palmetorum (DC.) Verdc.

Kew Bull. 30: 257 (1975).

Cephaelis peduncularis Salisb. var. *palmetorum* (DC.) Hepper, F.W.T.A. 2: 204 (1963).

Habit: shrub to 3 m. **Habitat:** forest. **Distribution:** Senegal to Cameroon. **Chorology:** Guineo-Congolian (montane). **Star:** green. **Alt. range:** 1200-1400m (1), 1401-1600m (1). **Specimens:** - *No locality:* Dunlap 244; Maitland T. D. s.n.

Psychotria peduncularis (Salisb.) Steyerm. var. peduncularis

Kew Bull. 30: 257 (1975); F.T.E.A. Rubiaceae: 72 (1976).
   Cephaelis peduncularis Salisb. var. *A* sensu Keay, F.W.T.A. 2: 204 (1963).
   Cephaelis peduncularis Salisb. var. *B* sensu Keay, F.W.T.A. 2: 204 (1963).

Habit: shrub to 3 m. **Guild:** shade-bearer. **Habitat:** forest. **Distribution:** tropical Africa. **Chorology:** Afromontane. **Star:** green. **Alt. range:** 1-200m (4), 201-400m (1), 801-1000m (1), 1801-2000m (1). **Flowering:** Mar (2). **Fruiting:** Mar (1), Aug (1), Sep (1), Oct (1). **Specimens:** - *Eastern slopes:* Tekwe C. 141. *Etinde:* Mbani J. M. 78; Tchouto P. 78; Thompson S. 1553; Wheatley J. I. 487. *Mabeta-Moliwe:* Dawson S. 52. *No locality:* Breteler F. J. 177

Psychotria podocarpa Petit

F.W.T.A. 2: 198 (1963); Bull. Jard. Bot. Brux. 34: 145 (1964).

Habit: shrub to 3 m. **Habitat:** forest. **Distribution:** SE Nigeria and SW Cameroon. **Chorology:** Western Cameroon. **Star:** gold. **Alt. range:** 1-200m (3), 201-400m (1), 401-600m (1). **Fruiting:** Feb (1), Sep (1), Oct (1), Nov (2). **Specimens:** - *Etinde:* Cheek M. 5383, 5412a; Sunderland T. C. H. 1078; Wheatley J. I. 529. *Onge:* Akogo M. 31

Psychotria pteropetala K.Schum.

F.W.T.A. 2: 201 (1963).

Habit: shrub to 1.5 m. **Habitat:** forest. **Distribution:** Cameroon. **Chorology:** Lower Guinea. **Star:** gold. **Alt. range:** 200-400m (1). **Flowering:** Mar (1). **Specimens:** - *Mabeta-Moliwe:* Wheatley J. I. 57

Psychotria cf. sadebeckiana K.Schum. var. sadebeckiana

Specimens: - *Mabeta-Moliwe:* Cheek M. 3492

Psychotria subobliqua Hiern

F.W.T.A. 2: 200 (1963); Bull. Jard. Bot. Brux. 34: 168 (1964); Bull. Jard. Bot. Nat. Belg. 42: 359 (1972).

Habit: shrub to 3 m. **Guild:** shade-bearer. **Habitat:** forest. **Distribution:** Guinea to Bioko, Cameroon, Rio Muni, Gabon and Congo (Kinshasa). **Chorology:** Guineo-Congolian. **Star:** green. **Specimens:** - *Etinde:* Maitland T. D. 701, 1142

Psychotria cf. subobliqua Hiern

Specimens: - *Mabeta-Moliwe:* Wheatley J. I. 283

Psychotria venosa (Hiern) Petit

F.W.T.A. 2: 202 (1963); Bull. Jard. Bot. Brux. 34: 110 (1964).

Habit: tree or shrub to 5 m. **Habitat:** forest. **Distribution:** Nigeria, Bioko, Cameroon, Gabon, Congo (Brazzaville) and Congo (Kinshasa). **Chorology:** Guineo-Congolian (montane). **Star:** green. **Alt. range:** 1-200m (1), 201-400m (1). **Flowering:** May (1). **Fruiting:** May (1). **Specimens:** - *Mokoko River F.R.:* Mbani J. M. 392; Ndam N. 1113

Psychotria vogeliana Benth.

F.W.T.A. 2: 198 (1963); Bull. Jard. Bot. Brux. 34: 135 (1964); F.T.E.A. Rubiaceae: 63 (1976); Meded. Land. Wag. 82(3): 292 (1982).

Habit: tree or shrub to 5 m. **Habitat:** forest. **Distribution:** Mali to Cameroon, CAR, Gabon and Congo (Kinshasa). **Chorology:** Guineo-Congolian. **Star:** green. **Specimens:** - *Southern Bakundu F.R.:* Brenan J. P. M. 9325

Psychotria sp.

Specimens: - *Mabeta-Moliwe:* Sunderland T. C. H. 1395. *Mokoko River F.R.:* Fraser P. 389; Letouzey R. 15058. *No locality:* Manning S. 1276

Psychotria sp. nov. 1 aff. dorotheae Wernh.

Specimens: - *Mabeta-Moliwe:* Cable S. 593; Cheek M. 5709, 5761; Jaff B. 149; Watts J. 127. *Onge:* Akogo M. 94

Psychotria sp. nov. 2 aff. bidentata (Thunb. ex Roem. & K.Schult.) Hiern

Specimens: - *Mabeta-Moliwe:* Schlechter F.R.R. 12374; Watts J. 110; Plot Voucher W923

Psychotria sp. nov. 3

Specimens: - *Mabeta-Moliwe:* Jaff B. 222, 319; Watts J. 260, 374

Psychotria sp. nov. 4

Specimens: - *Mabeta-Moliwe:* Jaff B. 62; Sunderland T. C. H. 1361; Watts J. 294; Wheatley J. I. 275

Psydrax acutiflora (Hiern) Bridson

F.T.E.A. Rubiaceae: 906 (1991).
   Canthium acutiflorum Hiern, F.W.T.A. 2: 182 (1963).
   Canthium henriquesianum (K.Schum.) G.Taylor, Meded. Land. Wag. 82(3): 271 p.p. (1982).

Habit: woody climber. **Habitat:** forest. **Distribution:** SE Nigeria, Cameroon, S. Tomé, Gabon and Congo (Kinshasa). **Chorology:** Guineo-Congolian (montane). **Star:** green. **Alt. range:** 1-200m (1), 1401-1600m (4). **Fruiting:** Sep (1). **Specimens:** - *Eastern slopes:* Thomas D. W. 9508. *Etinde:* Kalbreyer W. 182; Tekwe C. 294. *No locality:* Maitland T. D. 1124; Mann G. 1179

Psydrax dunlapii (Hutch. & Dalziel) Bridson

Kew Bull. 40: 699 (1985).
   Canthium dunlapii Hutch. & Dalziel, F.W.T.A. 2: 184 (1963).

Habit: tree to 10 m. **Habitat:** forest. **Distribution:** Nigeria, Bioko and Cameroon. **Chorology:** Lower Guinea (montane). **Star:** blue. **Alt. range:** 1-200m (1), 801-1000m (1), 1201-1400m (2), 1401-

1600m (1), 1601-1800m (1), 1801-2000m (4). **Flowering:** Oct (1). **Fruiting:** Aug (1), Oct (1). **Specimens:** - *Eastern slopes:* Brenan J. P. M. 9553; Breteler F. J. 182; Hutchinson J. & Metcalfe C. R. 10; Maitland T. D. 239, 508, 795; Nkeng P. 85. *Etinde:* Thomas D. W. 9180; Watts J. 571. *No locality:* Breteler F. J. 196; Dunlap 93

Psydrax kraussioides (Hiern) Bridson
F.T.E.A. Rubiaceae: 907 (1991); Fl. Zamb. 5(2): 362 (1998).
   Canthium kraussioides Hiern, F.W.T.A. (ed.1) 2: 113 (1931).
   Canthium henriquezianum sensu Hepper, F.W.T.A. 2: 181 (1963).

Habit: woody climber. **Guild:** pioneer. **Habitat:** roadsides and villages. **Distribution:** Guinea to Angola. **Chorology:** Guineo-Congolian. **Star:** green. **Alt. range:** 1-200m (1).**Fruiting:** Apr (1). **Specimens:** - *Mabeta-Moliwe:* Jaff B. 116

Rothmannia hispida (K.Schum.) Fagerlind
F.W.T.A. 2: 125 (1963); Keay R.W.J., Trees of Nigeria: 416 (1989); Hawthorne W., F.G.F.T. Ghana: 36 (1990); Bull. Jard. Bot. Nat. Belg. 65: 232 (1996).

Habit: tree or shrub to 10 m. **Guild:** shade-bearer. **Habitat:** forest. **Distribution:** Guinea to Congo (Brazzaville), Congo (Kinshasa) and CAR. **Chorology:** Guineo-Congolian. **Star:** green. **Alt. range:** 1-200m (7), 201-400m (1). **Flowering:** May (1). **Fruiting:** Mar (2), Apr (2), May (1), Jun (2), Aug (1), Oct (1). **Specimens:** - *Etinde:* Tchouto P. 236. *Mabeta-Moliwe:* Cable S. 1492; Jaff B. 100; Khayota B. 451; Nguembock F. 5a; Plot Voucher W 111, W 281, W 562, W 566, W 618; Sunderland T. C. H. 1245, 1410; Watts J. 404; von Rege I. 94. *Mokoko River F.R.:* Mbani J. M. 348. *No locality:* Leeuwenberg A. J. M. 6963

Rothmannia cf. hispida (K.Schum.) Fagerlind

Specimens: - *Mabeta-Moliwe:* Watts J. 105

Rothmannia longiflora Salisb.
F.W.T.A. 2: 125 (1963); Fl. Gabon 17: 237 (1970); F.T.E.A. Rubiaceae: 515 (1988); Hawthorne W., F.G.F.T. Ghana: 36 (1990); Bull. Jard. Bot. Nat. Belg. 65: 235 (1996).

Habit: tree or shrub to 6 m. **Guild:** shade-bearer. **Habitat:** forest. **Distribution:** tropical Africa. **Star:** excluded. **Alt. range:** 1-200m (2). **Flowering:** Dec (1). **Fruiting:** May (1). **Specimens:** - *Mabeta-Moliwe:* Cable S. 606; Tchouto P. 198

Rothmannia cf. longiflora Salisb.

Specimens: - *Mabeta-Moliwe:* Cheek M. 3558; Sunderland T. C. H. 1194

Rothmannia lujae (De Wild.) Keay
F.W.T.A. 2: 125 (1963); Fl. Gabon 17: 244 (1970); Keay R.W.J., Trees of Nigeria: 416 (1989); Bull. Jard. Bot. Nat. Belg. 65: 236 (1996).

Habit: tree to 12 m. **Habitat:** forest. **Distribution:** Nigeria, Cameroon, Gabon and Congo (Brazzaville). **Chorology:** Lower Guinea. **Star:** blue. **Alt. range:** 200-400m (1).**Fruiting:** May (1). **Specimens:** - *Mokoko River F.R.:* Thomas D. W. 10103

Rothmannia octomera (Hook.f.) Fagerlind
F.W.T.A. 2: 125 (1963); Fl. Gabon 17: 236 (1970); Bull. Jard. Bot. Nat. Belg. 65: 230 (1996).

Habit: tree or shrub to 7 m. **Habitat:** forest. **Distribution:** Nigeria, Cameroon, Bioko, Gabon, Congo (Brazzaville), Congo (Kinshasa) and CAR. **Chorology:** Guineo-Congolian. **Star:** green. **Alt. range:** 1-200m (2). **Flowering:** Apr (1), May (1). **Fruiting:** Apr (1). **Specimens:** - *Mokoko River F.R.:* Ekema N. 875; Tchouto P. 1059

Rothmannia talbotii (Wernham) Keay
F.W.T.A. 2: 126 (1963); Fl. Gabon 17: 250 (1970); Bull. Jard. Bot. Nat. Belg. 65: 244 (1996).

Habit: tree to 5 m. **Habitat:** forest. **Distribution:** Nigeria, Cameroon, Gabon, Congo (Brazzaville) and Angola (Cabinda). **Chorology:** Guineo-Congolian. **Star:** green. **Alt. range:** 1-200m (9). **Flowering:** Apr (1), Jul (1), Oct (1). **Fruiting:** Apr (3), May (1), Jun (2). **Specimens:** - *Etinde:* Kwangue A. 64. *Mabeta-Moliwe:* Jaff B. 255; Sunderland T. C. H. 1256; von Rege I. 2. *Mokoko River F.R.:* Akogo M. 236, 265; Ekema N. 1118a; Pouakouyou D. 36. *Onge:* Akogo M. 124

Rothmannia urcelliformis (Hiern) Bullock ex Robyns
F.W.T.A. 2: 125 (1963); F.T.E.A. Rubiaceae: 514 (1988); Keay R.W.J., Trees of Nigeria: 416 (1989); Hawthorne W., F.G.F.T. Ghana: 36 (1990); Bull. Jard. Bot. Nat. Belg. 65: 226 (1996).

Habit: tree or shrub to 10 m. **Guild:** shade-bearer. **Habitat:** forest. **Distribution:** tropical Africa. **Chorology:** Afromontane. **Star:** green. **Alt. range:** 1000-1200m (1), 1201-1400m (1), 1401-1600m (1), 1601-1800m (1), 1801-2000m (2).**Fruiting:** Sep (1). **Specimens:** - *Eastern slopes:* Thomas D. W. 9482. *Etinde:* Tekwe C. 299. *No locality:* Breteler F. J. 195; Etuge M. 440; Hutchinson J. & Metcalfe C. R. 44; Thomas D. W. 2601

Rothmannia whitfieldii (Lindl.) Dandy
F.W.T.A. 2: 126 (1963); Fl. Gabon 17: 252 (1970); F.T.E.A. Rubiaceae: 518 (1988); Keay R.W.J., Trees of Nigeria: 416 (1989); Hawthorne W., F.G.F.T. Ghana: 36 (1990); Bull. Jard. Bot. Nat. Belg. 65: 240 (1996).

Habit: tree to 12 m. **Guild:** shade-bearer. **Habitat:** forest. **Distribution:** tropical Africa. **Star:** excluded. **Alt. range:** 1-200m (5), 201-400m (1). **Flowering:** Apr (1). **Fruiting:** Mar (2), Apr (2), May (1), Jun (1), Aug (1). **Specimens:** - *Mabeta-Moliwe:* Baker W. J. 276; Cable S. 1491; Ndam N. 500; Plot Voucher W 1037; Sunderland T. C. H. 1109; Watts J. 243. *Mokoko River F.R.:* Tchouto P. 1171. *No locality:* Maitland T. D. 463

Rothmannia cf. whitfieldii (Lindl.) Dandy

Specimens: - *Mabeta-Moliwe:* Wheatley J. I. 293

Rothmannia sp.

Specimens: - *Mokoko River F.R.:* Acworth J. M. 293; Mbani J. M. 388

Rutidea decorticata Hiern
F.W.T.A. 2: 146 (1963); Kew Bull. 33: 273 (1978).

Habit: woody climber. **Habitat:** forest. **Distribution:** Nigeria, Bioko, Cameroon, Gabon, Congo (Brazzaville) and Congo (Kinshasa). **Chorology:** Guineo-Congolian (montane). **Star:** green. **Alt. range:** 1-200m (10), 201-400m (1), 401-600m (2), 601-800m (1), 1001-1200m (1), 1201-1400m (1), 1401-1600m (2). **Flowering:** Feb (1), Oct (1). **Fruiting:** Jan (1), Feb (1), Mar

(1), Sep (1), Oct (5), Nov (4), Dec (2). **Specimens:** - *Eastern slopes:* Cable S. 1391. *Etinde:* Cable S. 196, 373, 492; Cheek M. 5478; Maitland T. D. 1178; Mbani J. M. 14; Nkeng P. 31; Sunderland T. C. H. 1072; Tekwe C. 190; Watts J. 561. *Mabeta-Moliwe:* Sidwell K. 11. *Onge:* Akogo M. 42; Cheek M. 5126; Harris D. J. 3676; Tchouto P. 690, 790; Thomas D. W. 9870

Rutidea glabra Hiern

F.W.T.A. 2: 146 (1963); Kew Bull. 33: 276 (1978).

Habit: woody climber. **Habitat:** forest and stream banks. **Distribution:** Nigeria, Cameroon, Gabon and Congo (Brazzaville). **Chorology:** Guineo-Congolian. **Star:** green. **Alt. range:** 1-200m (6), 401-600m (2), 601-800m (1). **Flowering:** Mar (2), Apr (1). **Fruiting:** Aug (3), Oct (1), Nov (1), Dec (1). **Specimens:** - *Eastern slopes:* Maitland T. D. 556. *Etinde:* Cable S. 272; Etuge M. 1239. *Mabeta-Moliwe:* Baker W. J. 280, 339; Cable S. 1451; Sunderland T. C. H. 1022, 1199. *Onge:* Akogo M. 114; Baker W. J. 360

Rutidea hispida Hiern

Kew Bull. 33: 260 (1978).

Habit: woody climber. **Habitat:** forest. **Distribution:** SE Nigeria, Cameroon and Gabon. **Chorology:** Lower Guinea. **Star:** blue. **Alt. range:** 200-400m (2). **Fruiting:** Oct (1), Nov (1). **Specimens:** - *Etinde:* Williams S. 55. *Onge:* Watts J. 838

Rutidea membranacea Hiern

F.W.T.A. 2: 146 (1963); Kew Bull. 33: 273 (1978).

Habit: woody climber. **Guild:** pioneer. **Habitat:** forest. **Distribution:** Sierra Leone to Cameroon, CAR, Congo (Brazzaville), Congo (Kinshasa), Angola and Tanzania. **Chorology:** Guineo-Congolian. **Star:** green. **Alt. range:** 800-1000m (1). **Specimens:** - *Eastern slopes:* Maitland T. D. 262

Rutidea olenotricha Hiern

F.W.T.A. 2: 146 (1963); Kew Bull. 33: 268 (1978).

Habit: woody climber. **Guild:** pioneer. **Habitat:** forest. **Distribution:** Sierra Leone to Cameroon, Gabon, Congo (Brazzaville), Congo (Kinshasa), Angola, Zambia and Sudan. **Chorology:** Guineo-Congolian. **Star:** green. **Alt. range:** 200-400m (1). **Specimens:** - *Mokoko River F.R.:* Ndam N. 1151

Rutidea smithii Hiern subsp. smithii

F.W.T.A. 2: 146 (1963); Kew Bull. 33: 270 (1978); F.T.E.A. Rubiaceae: 609 (1988).
Habit: woody climber. **Guild:** pioneer. **Habitat:** forest. **Distribution:** tropical Africa. **Star:** excluded. **Alt. range:** 200-400m (1), 801-1000m (2). **Fruiting:** Dec (1). **Specimens:** - *Eastern slopes:* Hutchinson J. & Metcalfe C. R. 94; Maitland T. D. 1016. *Etinde:* Cheek M. 5866

Rutidea sp. nov. aff. hispida Hiern

Specimens: - *Mabeta-Moliwe:* Sunderland T.C.H. 1012

Rytigynia neglecta (Hiern) Robyns

F.W.T.A. 2: 186 (1963); Kew Bull. 42: 158 (1987); F.T.E.A. Rubiaceae: 813 (1991).

Habit: woody climber. **Habitat:** forest. **Distribution:** Cameroon to East Africa. **Chorology:** Afromontane. **Star:** green. **Alt. range:** 1-200m (1), 1401-1600m (1), 1601-1800m (1). **Specimens:** -

Eastern slopes: Hutchinson J. & Metcalfe C. R. 30; Maitland T. D. 648; Morton J. K. GC 6717

Rytigynia umbellulata (Hiern) Robyns

F.W.T.A. 2: 186 (1963); Meded. Land. Wag. 82(3): 293 (1982); Kew Bull. 42: 152 (1987); Hawthorne W., F.G.F.T. Ghana: 36 (1990); F.T.E.A. Rubiaceae: 808 (1991); Fl. Zamb. 5(2): 286 (1998).
Vangueria ituriensis De Wild., Bull. Jard. Bot. Brux. 8: 56 (1922).

Habit: tree to 10 m. **Guild:** pioneer. **Habitat:** forest. **Distribution:** tropical Africa. **Chorology:** Afromontane. **Star:** green. **Alt. range:** 1400-1600m (2).**Fruiting:** Oct (1). **Specimens:** - *Eastern slopes:* Thomas D. W. 4600, 9466

Sabicea calycina Benth.

F.W.T.A. 2: 172 (1963); F.T.E.A. Rubiaceae: 472 (1988).

Habit: woody climber. **Guild:** pioneer. **Habitat:** forest and stream banks. **Distribution:** Sierra Leone to Congo (Kinshasa). **Chorology:** Guineo-Congolian. **Star:** green. **Alt. range:** 1-200m (9), 201-400m (1), 401-600m (1), 601-800m (2). **Flowering:** Feb (1), Mar (2), Apr (2), May (5), Aug (1), Dec (2). **Fruiting:** Feb (1), Apr (2), May (4), Aug (1), Nov (1). **Specimens:** - *Etinde:* Cable S. 389; Hutchinson J. & Metcalfe C. R. 147; Khayota B. 493; Maitland T. D. 54; Tekwe C. 34. *Mabeta-Moliwe:* Cable S. 564, 1447; Ndam N. 530; Sunderland T. C. H. 1312; von Rege I. 84; Wheatley J.I. 135, 334. *Mokoko River F.R.:* Ekema N. 991, 1085; Ndam N. 1197; Pouakouyou D. 18. *No locality:* Nkeng P. 16. *Onge:* Harris D. J. 3762

Sabicea capitellata Benth.

F.W.T.A. 2: 174 (1963).

Habit: woody climber. **Habitat:** forest. **Distribution:** Nigeria and Cameroon. **Chorology:** Lower Guinea. **Star:** blue. **Alt. range:** 1-200m (3), 201-400m (1). **Flowering:** Mar (1), May (1), Oct (1), Nov (2). **Fruiting:** Mar (1), May (1), Oct (1), Nov (1). **Specimens:** - *Etinde:* Cheek M. 5455. *Mokoko River F.R.:* Ndam N. 1135. *Onge:* Harris D. J. 3838; Tchouto P. 994; Wheatley J. I. 776. *Southern Bakundu F.R.:* Brenan J. P. M. 9299

Sabicea gabonica (Hiern) Hepper

F.W.T.A. 2: 172 (1963); Bull. Jard. Bot. Nat. Belg. 50: 255 (1980). *Sabicea efulenensis* (Hutch.) Hepper, Kew Bull. 13: 292 (1958).

Habit: woody climber. **Habitat:** forest. **Distribution:** Togo to Cameroon and Gabon. **Chorology:** Guineo-Congolian. **Star:** green. **Alt. range:** 1-200m (2), 201-400m (1), 401-600m (2).**Fruiting:** Apr (1), Oct (1), Dec (3). **Specimens:** - *Etinde:* Cable S. 347, 381; Lighava M. 57. *Mokoko River F.R.:* Akogo M. 173. *Onge:* Tchouto P. 1006

Sabicea pilosa Hiern

F.W.T.A. 2: 172 (1963).

Habit: woody climber. **Habitat:** forest. **Distribution:** Nigeria, Cameroon and Gabon. **Chorology:** Lower Guinea. **Star:** blue. **Alt. range:** 1-200m (12), 201-400m (1), 601-800m (1), 801-1000m (2). **Flowering:** Feb (1), Mar (3), Apr (4), Aug (1), Nov (1), Dec (2). **Fruiting:** Apr (2), May (2), Jun (1), Aug (1), Nov (1), Dec (2). **Specimens:** - *Etinde:* Cable S. 145, 1554; Faucher P. 46; Khayota B. 497; Mbani J. M. 16. *Mabeta-Moliwe:* Baker W. J. 217; Cable S. 616, 1463; Sunderland T. C. H. 1202; Watts J. 125. *Mokoko River F.R.:* Acworth J. M. 212, 235; Akogo M. 275; Ekema N. 1084; Fraser P. 430; Ndam N. 1194; Pouakouyou D. 41

Sabicea speciosa K.Schum.
F.W.T.A. 2: 172 (1963).

Habit: woody climber. Habitat: forest. Distribution: Togo to Cameroon. Chorology: Guineo-Congolian. Star: green. Alt. range: 1-200m (6). Flowering: Apr (1), May (2), Jun (1). Fruiting: Aug (2). Specimens: - *Etinde:* Winkler H. J. P. 27. *Mabeta-Moliwe:* Baker W. J. 227; Watts J. 132; Wheatley J. I. 257; von Rege I. 42. *Mokoko River F.R.:* Ekema N. 980, 1183

Sabicea sp. aff. tchapensis Krause

Specimens: - *Mokoko River F.R.:* Ekema N. 869; Ndam N. 1188; Tchouto P. 1221. *Onge:* Watts J. 1044

Sabicea venosa Benth.
F.W.T.A. 2: 172 (1963).

Habit: woody climber. Guild: pioneer. Habitat: forest and farmbush. Distribution: Guinea Bissau to Congo (Kinshasa). Chorology: Guineo-Congolian. Star: green. Alt. range: 800-1000m (1). Flowering: Aug (1). Fruiting: Aug (1). Specimens: - *Eastern slopes:* Maitland T. D. 32; Tekwe C. 139

Sabicea xanthotricha Wernham
F.W.T.A. 2: 172 (1963).

Habit: tree or shrub to 4 m. Habitat: forest. Distribution: Nigeria and Cameroon. Chorology: Lower Guinea. Star: blue. Alt. range: 1-200m (2), 201-400m (1), 401-600m (1), 601-800m (1), 801-1000m (1). Flowering: Feb (1), Mar (2), Apr (1), May (1), Jun (1). Fruiting: Feb (1), May (1), Jun (1). Specimens: - *Etinde:* Cable S. 1500, 1606; Tekwe C. 40; Wheatley J. I. 25. *Mokoko River F.R.:* Batoum A. 27; Pouakouyou D. 35

Sabicea sp. nov.

Specimens: - *Mabeta-Moliwe:* Sunderland T.C.H. 1274a

Sabicea sp.

Specimens: - *Mokoko River F.R.:* Acworth J. M. 230

Sacosperma paniculatum (Benth.) G.Taylor
F.W.T.A. 2: 213 (1963).
Habit: woody climber. Guild: pioneer. Habitat: forest. Distribution: tropical Africa. Star: excluded. Alt. range: 1-200m (2), 801-1000m (2). Flowering: Mar (1). Fruiting: Mar (1), May (1). Specimens: - *Eastern slopes:* Deistel H. 624; Maitland T. D. 240. *Mabeta-Moliwe:* Wheatley J. I. 67. *Mokoko River F.R.:* Tchouto P. 1219

Sacosperma sp. nov.

Specimens: - *Mabeta-Moliwe:* Cheek M. 3577

Schumanniophyton magnificum (K.Schum.) Harms
F.W.T.A. 2: 213 (1963); Fl. Gabon 17: 24 (1970).

Habit: tree to 6 m. Habitat: forest. Distribution: Nigeria to Angola (Cabinda). Chorology: Guineo-Congolian. Star: green. Alt. range: 1-200m (2), 401-600m (2), 801-1000m (1).

Flowering: Oct (1). Fruiting: Feb (1), Nov (2). Specimens: - *Eastern slopes:* Etuge M. 1181. *Etinde:* Cheek M. 5537. *Mabeta-Moliwe:* Plot Voucher W 131, W 470, W 770. *No locality:* Thomas D. W. 2857. *Onge:* Tchouto P. 1044; Watts J. 886

Sericanthe auriculata (Keay) Robbr.
Bull. Jard. Bot. Nat. Belg. 51: 172 (1981).
Tricalysia auriculata Keay, F.W.T.A. 2: 152 (1963).

Habit: tree or shrub to 4 m. Habitat: forest. Distribution: Nigeria and Cameroon. Chorology: Lower Guinea. Star: blue. Alt. range: 1-200m (6), 201-400m (3). Flowering: Jun (1), Oct (2). Fruiting: Mar (1), May (5), Jun (1), Oct (2). Specimens: - *Bambuko F.R.:* Tchouto P. 536. *Mokoko River F.R.:* Batoum A. 4; Ekema N. 883, 1113; Ndam N. 1150; Tchouto P. 1189; Watts J. 1105. *Onge:* Tchouto P. 979; Watts J. 842

Sherbournia bignoniiflora (Welw.) Hua
F.W.T.A. 2: 127 (1963); F.T.E.A. Rubiaceae: 521 (1988).

Habit: woody climber. Guild: pioneer. Habitat: forest. Distribution: tropical Africa. Chorology: Afromontane. Star: green. Alt. range: 800-1000m (1), 1201-1400m (2). Flowering: Feb (2). Fruiting: Feb (1), Oct (1). Specimens: - *Eastern slopes:* Cable S. 1351; Groves M. 148. *Etinde:* Cheek M. 3774

Sherbournia zenkeri Hua
F.W.T.A. 2: 127 (1963).

Habit: woody climber. Habitat: forest. Distribution: Nigeria to Angola (Cabinda). Chorology: Guineo-Congolian. Star: green. Alt. range: 1-200m (6), 201-400m (1), 401-600m (1). Flowering: Mar (2), Apr (2), May (2), Aug (1), Nov (1). Fruiting: Mar (1), Apr (1), May (2), Aug (1). Specimens: - *Etinde:* Cable S. 190; Mbani J. M. 35; Tchouto P. 120. *Mabeta-Moliwe:* Baker W. J. 212; Wheatley J. I. 39, 96, 302. *Mokoko River F.R.:* Batoum A. 34

Sherbournia sp.

Specimens: - *Mokoko River F.R.:* Akogo M. 226; Ekema N. 866, 1015; Mbani J. M. 341; Ndam N. 1291, 1327

Spermacoce mauritiana Gideon
F.T.E.A. Rubiaceae: 927 (1991).

Habit: herb. Guild: pioneer. Habitat: roadsides and villages. Distribution: tropical Africa. Star: excluded. Alt. range: 1-200m (1). Flowering: Jul (1). Specimens: - *Mabeta-Moliwe:* Cheek M. 3543. *No locality:* Thompson S. 1469

Spermacoce ocymoides Burm.f.
F.W.T.A. 2: 220 (1963); F.T.E.A. Rubiaceae: 361 (1976).
Borreria ocymoides (Burm.f.) DC., Opera Bot. Belg. 7: 317 (1996).

Habit: herb. Guild: pioneer. Distribution: tropical Africa. Star: excluded. Alt. range: 1-200m (1), 201-400m (1). Specimens: - *No locality:* Chuml H. 225; Leeuwenberg A. J. M. 6898

Spermacoce princeae (K.Schum.) Verdc. var. princeae
F.T.E.A. Rubiaceae: 362 (1976).
Borreria princeae K.Schum. var. *princeae*, F.W.T.A. 2: 222 (1963).

Habit: herb. Guild: pioneer. Habitat: forest-grassland transition. Distribution: Cameroon to Tanzania. Chorology: Guineo-Congolian (montane). Star: green. Alt. range: 1-200m (1), 1801-2000m (1), 2201-2400m (1). Specimens: - *No locality:* Brenan J. P. M. 9552; Maitland T. D. 1327; Morton J. K. GC 7095

Spermacoce princeae (K.Schum.) Verdc. var. pubescens (Hepper) Verdc.

F.T.E.A. Rubiaceae: 364 (1976).
Borreria princeae K.Schum. var. *pubescens* Hepper, F.W.T.A. 2: 222 (1963).

Habit: herb. Guild: pioneer. Habitat: forest. Distribution: Cameroon to Tanzania. Chorology: Guineo-Congolian (montane). Star: green. Alt. range: 1200-1400m (1). Specimens: - *No locality:* Dundas J. FHI 15316

Spermacoce tenuior L.

F.T.E.A. Rubiaceae: 349 (1976); Kew Bull. 37: 545 (1983).
Borreria laevis (Lam.) Griseb., Goett. Abh. 7: 231 (1857).

Habit: herb. Habitat: forest. Distribution: tropical Africa. Star: excluded. Specimens: - *No locality:* Chuml H. 186; Maitland T. D. 51

Stelechantha sp. nov.

Habit: shrub to 2.5 m. Habitat: forest. Distribution: Cameroon. Chorology: Lower Guinea. Star: black. Alt. range: 1-200m (1), 201-400m (1). Flowering: May (1), Jun (1). Specimens: - *Mabeta-Moliwe:* Wheatley J. I. 298. *Mokoko River F.R.:* Tchouto P. 1246

Stipularia africana P.Beauv.

Fl. Gabon 12: 158 (1966).
Sabicea africana (P.Beauv.) Hepper, F.W.T.A. 2: 173 (1963).

Habit: shrub to 3 m. Guild: pioneer. Habitat: forest. Distribution: Sierra Leone to Cameroon, Gabon and Congo (Kinshasa). Chorology: Guineo-Congolian. Star: green. Alt. range: 1-200m (1), 201-400m (1). Flowering: Feb (1). Fruiting: Sep (1). Specimens: - *Etinde:* Nkeng P. 27; Wheatley J. I. 514

Tarenna baconoides Wernham var. baconoides

F.W.T.A. 2: 135 (1963); Fl. Gabon 17: 99 (1970).

Habit: woody climber. Habitat: forest. Distribution: SE Nigeria and SW Cameroon. Chorology: Western Cameroon. Star: gold. Alt. range: 200-400m (1). Fruiting: Oct (1). Specimens: - *Onge:* Ndam N. 601

Tarenna bipindensis (K.Schum.) Bremek.

F.W.T.A. 2: 135 (1963); Fl. Gabon 17: 103 (1970).

Habit: woody climber. Habitat: forest. Distribution: Guinea to Cameroon and Gabon. Chorology: Guineo-Congolian. Star: green. Alt. range: 200-400m (1). Specimens: - *Etinde:* Thomas D. W. 3456

Tarenna conferta (Benth.) Hiern

F.W.T.A. 2: 135 (1963); Fl. Gabon 17: 108 (1970).

Habit: woody climber. Habitat: forest. Distribution: Nigeria, Cameroon and Gabon. Chorology: Lower Guinea. Star: blue. Alt.

range: 200-400m (1), 601-800m (1). Flowering: Mar (1), Nov (1). Fruiting: Mar (1). Specimens: - *Eastern slopes:* Tchouto P. 33. *Etinde:* Banks 110; Brodie 28; Cheek M. 5574; Etuge M. 1231; Letouzey R. 14957. *Mokoko River F.R.:* Letouzey R. 15083

Tarenna eketensis Wernham

F.W.T.A. 2: 135 (1963); Fl. Gabon 17: 114 (1970); Kew Bull. 34: 379 (1979).

Habit: woody climber. Guild: shade-bearer. Habitat: forest. Distribution: Liberia to Cameroon and Gabon. Chorology: Guineo-Congolian. Star: green. Alt. range: 1-200m (3), 201-400m (2), 401-600m (1). Flowering: May (1), Jun (2), Oct (1). Fruiting: Sep (1), Oct (2). Specimens: - *Etinde:* Kwangue A. 112; Wheatley J. I. 488. *Mabeta-Moliwe:* Wheatley J. I. 246, 333. *Mokoko River F.R.:* Letouzey R. 15092. *Onge:* Tchouto P. 781, 885

Tarenna fusco-flava (K.Schum.) N.Hallé

Adansonia (ser.2) 7: 506 (1967); Kew Bull. 34: 379 (1979); F.T.E.A. Rubiaceae: 588 (1988).
Tarenna flavo-fusca (K.Schum.) S.Moore, F.W.T.A. 2: 134 (1963).

Habit: woody climber. Habitat: forest and Aframomum thicket. Distribution: Liberia to Cameroon, Gabon, Congo (Kinshasa), Angola and Uganda. Chorology: Guineo-Congolian. Star: green. Alt. range: 1-200m (2), 801-1000m (1). Flowering: May (1). Fruiting: Jan (1), Oct (1), Dec (1). Specimens: - *Etinde:* Cable S. 457. *Mokoko River F.R.:* Ekema N. 891. *Onge:* Tchouto P. 996. *Southern Bakundu F.R.:* Binuyo A. & Daramola B. O. FHI 35479

Tarenna grandiflora Hiern

F.W.T.A. 2: 134 (1963).

Habit: tree or shrub to 8 m. Habitat: forest. Distribution: Nigeria and Cameroon. Chorology: Lower Guinea. Star: blue. Alt. range: 1-200m (7), 201-400m (3), 401-600m (2). Flowering: Aug (1), Nov (1). Fruiting: Mar (4), Apr (2), May (1), Nov (1), Dec (4). Specimens: - *Etinde:* Cable S. 409; Cheek M. 5893; Mbani J. M. 147; Olorunfemi J. FHI 30745; Tchouto P. 68. *Mabeta-Moliwe:* Cheek M. 5797; Jaff B. 90; Sunderland T. C. H. 1131; Wheatley J. I. 171; von Rege I. 93. *Mokoko River F.R.:* Etuge M. 385; Tchouto P. 567. *Onge:* Thomas D. W. 9853. *Southern Bakundu F.R.:* Binuyo A. & Daramola B. O. FHI 35643; Latilo M. G. FHI 43809

Tarenna lasiorachis (K.Schum. & K.Krause) Bremek.

F.W.T.A. 2: 135 (1963); Fl. Gabon 17: 110 (1970).

Habit: tree or shrub to 6 m. Habitat: forest. Distribution: Cameroon. Chorology: Lower Guinea. Star: gold. Alt. range: 1-200m (2), 201-400m (1), 601-800m (1). Flowering: Oct (2), Nov (1). Fruiting: Apr (1). Specimens: - *Etinde:* Maitland T. D. 87, 620. *Onge:* Harris D. J. 3860; Tchouto P. 1001; Watts J. 841

Tarenna thomasii Hutch. & Dalziel

F.W.T.A. 2: 135 (1963).

Habit: shrub to 3 m. Habitat: lava flows. Distribution: Guinea to Cameroon. Chorology: Guineo-Congolian. Star: green. Alt. range: 1-200m (1). Flowering: Jul (1). Specimens: - *Etinde:* Tchouto P. 237

Tarenna sp.

Specimens: - *Etinde:* Dawson 64. *Mokoko River F.R.:* Mbani J. M. 315

Tricalysia atherura N.Hallé
Fl. Gabon 17: 292 (1970); Bull. Jard. Bot. Nat. Belg. 57: 76 (1987).

Habit: tree or shrub 1.5-8 m tall. Habitat: forest. Distribution: Cameroon and Gabon. Chorology: Lower Guinea. Star: blue. Alt. range: 400-600m (1).Fruiting: Oct (1). Specimens: - *Etinde:* Watts J. 498

Tricalysia biafrana Hiern
F.W.T.A. 2: 151 (1963); Fl. Gabon 17: 315 (1970); Bull. Jard. Bot. Nat. Belg. 57: 129 (1987); Hawthorne W., F.G.F.T. Ghana: 35 (1990).

Habit: tree or shrub to 10 m. Guild: shade-bearer. Habitat: forest. Distribution: Liberia to Bioko, Cameroon, CAR, Congo (Brazzaville), Congo (Kinshasa) and Angola (Cabinda). Chorology: Guineo-Congolian. Star: green. Alt. range: 1-200m (1), 401-600m (1), 1201-1400m (2), 1401-1600m (1). Flowering: Oct (1). Fruiting: May (1). Specimens: - *Etinde:* Cheek M. 3798; Maitland T. D. 981; Mbani J. M. 156. *Mabeta-Moliwe:* Mann G. 7; Preuss P. R. 1310, 1311. *No locality:* Maitland T. D. 1117, 1319; Mann G. 1212

Tricalysia discolor Brenan
F.W.T.A. 2: 151 (1963); Bull. Jard. Bot. Nat. Belg. 49: 294 (1979); Keay R.W.J., Trees of Nigeria: 429 (1989); Hawthorne W., F.G.F.T. Ghana: 32 (1990).
Tricalysia mildbraedii Keay, F.W.T.A. 2: 151 (1963).

Habit: tree to 15 m. Guild: shade-bearer. Habitat: forest. Distribution: Liberia to Cameroon. Chorology: Guineo-Congolian. Star: green. Alt. range: 200-400m (1). Specimens: - *Mokoko River F.R.:* Akuo FHI 15158

Tricalysia gossweileri S.Moore
Bull. Jard. Bot. Nat. Belg. 49: 300 (1979).

Habit: tree or shrub 1-10 m tall. Habitat: forest. Distribution: Cameroon, Gabon, Congo (Brazzaville), Congo (Kinshasa) and Angola. Chorology: Guineo-Congolian (montane). Star: green. Alt. range: 1-200m (7), 201-400m (1), 401-600m (1), 601-800m (3), 801-1000m (1), 1401-1600m (1), 2401-2600m (1). Flowering: Feb (1), Apr (3). Fruiting: Jan (1), Mar (1), Apr (3), May (3), Jun (1), Oct (2), Nov (1). Specimens: - *Eastern slopes:* Tekwe C. 11. *Etinde:* Cheek M. 3769; Mbani J. M. 52; Tchouto P. 112; Tekwe C. 36; Wheatley J. I. 10. *Mabeta-Moliwe:* Jaff B. 178, 187; Nguembock F. 22; Sidwell K. 105; Watts J. 137, 325; Wheatley J. I. 135, 295. *No locality:* Maitland T. D. 555, 893

Tricalysia oligoneura K.Schum.
F.W.T.A. 2: 151 (1963); Fl. Gabon 17: 312 (1970); Bull. Jard. Bot. Nat. Belg. 57: 135 (1987).

Habit: tree or shrub to 7 m. Guild: shade-bearer. Habitat: forest. Distribution: Nigeria, Cameroon, CAR and Congo (Kinshasa). Chorology: Guineo-Congolian. Star: green. Alt. range: 400-600m (1). Specimens: - *No locality:* Thomas D. W. 2838

Tricalysia pallens Hiern
F.W.T.A. 2: 152 (1963); Bull. Jard. Bot. Nat. Belg. 57: 114 (1987); F.T.E.A. Rubiaceae: 548 (1988); Hawthorne W., F.G.F.T. Ghana: 35 (1990).
Tricalysia pallens Hiern var. *gabonica* (Hiern) N.Hallé, Fl. Gabon 17: 310 (1970).
Tricalysia sp. B sensu N.Hallé, Fl. Gabon 17: 320 (1970).

Habit: tree or shrub to 10 m. Guild: shade-bearer. Habitat: forest. Distribution: tropical Africa. Star: excluded. Alt. range: 400-600m (1), 801-1000m (1), 1401-1600m (1). Flowering: Jan (1). Fruiting: Feb (1). Specimens: - *Eastern slopes:* Maitland T. D. 492. *Etinde:* Sunderland T. C. H. 1071; Tchouto P. 16. *No locality:* Thomas D. W. 2861

Tricalysia reflexa Hutch. var. reflexa
F.W.T.A. 2: 151 (1963); Bull. Jard. Bot. Nat. Belg. 49: 302 (1979).

Habit: shrub 1-6 m tall. Guild: shade-bearer. Habitat: forest. Distribution: Guinea to Cameroon. Chorology: Guineo-Congolian. Star: green. Specimens: - *Bambuko F.R.:* Olorunfemi J. FHI 30778

Tricalysia soyauxii K.Schum.
F.W.T.A. 2: 152 (1963); Fl. Gabon 17: 302 (1970); Bull. Jard. Bot. Nat. Belg. 57: 106 (1987).

Habit: tree to 14 m. Habitat: forest. Distribution: Cameroon and Gabon. Chorology: Lower Guinea. Star: blue. Alt. range: 1-200m (1). Specimens: - *Etinde:* Kwangue A. 61

Tricalysia sp.

Specimens: - *Mokoko River F.R.:* Batoum A. 33; Fraser P. 462; Mbani J. M. 374, 504; Pouakouyou D. 45. *No locality:* Thomas D. W. 5127

Trichostachys aurea Hiern
F.W.T.A. 2: 207 (1963); Mem. Mus. Natl. Hist. Nat. B. Bot. 24: 1334 (1975).
Trichostachys zenkeri De Wild., Pl. Bequeart. 6: 83 (1932).

Habit: shrub to 0.5 m. Guild: shade-bearer. Habitat: forest. Distribution: Sierra Leone to Cameroon and Gabon. Chorology: Guineo-Congolian. Star: green. Alt. range: 1-200m (6). Flowering: May (4), Jun (1), Nov (1). Fruiting: Oct (1), Nov (2). Specimens: - *Mokoko River F.R.:* Ekema N. 850, 1117a; Fraser P. 390, 394; Watts J. 1165. *Onge:* Harris D. J. 3733; Tchouto P. 985; Watts J. 1032

Trichostachys lehmbachii K.Schum.
F.W.T.A. 2: 207 (1963).

Habit: herb. Distribution: Cameroon. Chorology: Lower Guinea. Star: gold. Specimens: - *Etinde:* Lehmbach H. s.n.

Trichostachys sp. A sensu Hepper & Keay
F.W.T.A. 2: 207 (1963).

Habit: shrub to 1 m. Habitat: forest. Distribution: Cameroon. Chorology: Lower Guinea. Star: gold. Alt. range: 1-200m (4). Flowering: Apr (2), May (2). Fruiting: May (2). Specimens: - *Mokoko River F.R.:* Batoum A. 16; Ekema N. 974; Fraser P. 382; Tchouto P. 1058, 1136

Uncaria donisii Petit
Bull. Jard. Bot. Brux. 27: 443 (1957).

Habit: woody climber. **Habitat:** forest. **Distribution:** Congo (Kinshasa). **Chorology:** Guineo-Congolian. **Star:** green. **Alt. range:** 1-200m (1). **Flowering:** Mar (1). **Fruiting:** Mar (1). **Specimens:** - *Mabeta-Moliwe:* Watts J. 103

Vangueriella campylacantha (Mildbr.) Verdc.
Kew Bull. 42: 196 (1987).
Vangueriopsis subulata Robyns, F.W.T.A. 2: 180 (1963).

Habit: woody climber. **Habitat:** forest. **Distribution:** Guinea to Cameroon. **Chorology:** Guineo-Congolian. **Star:** green. **Specimens:** - *Southern Bakundu F.R.:* Binuyo A. & Daramola B. O. FHI 35649

Vangueriella laxiflora (K.Schum.) Verdc.
Kew Bull. 42: 192 (1987).
Vangueriopsis calycophila (K.Schum.) Robyns, F.W.T.A. 2: 179 (1963).

Habit: woody climber. **Habitat:** forest. **Distribution:** Cameroon. **Chorology:** Lower Guinea. **Star:** gold. **Alt. range:** 1-200m (1), 201-400m (1).**Fruiting:** Apr (1). **Specimens:** - *Etinde:* Thomas D. W. 5544. *Mabeta-Moliwe:* Plot Voucher W 208; Watts J. 173

Virectaria angustifolia (Hiern) Bremek.
F.W.T.A. 2: 209 (1963).

Habit: herb. **Habitat:** swamp forest. **Distribution:** Nigeria, Cameroon and Gabon. **Chorology:** Lower Guinea. **Star:** blue. **Alt. range:** 1-200m (4). **Flowering:** Oct (2), Nov (2). **Fruiting:** Oct (1), Nov (2). **Specimens:** - *Onge:* Akogo M. 105; Harris D. J. 3663; Tchouto P. 724; Watts J. 917

Virectaria procumbens (Sm.) Bremek.
F.W.T.A. 2: 208 (1963).

Habit: herb. **Guild:** pioneer. **Habitat:** forest. **Distribution:** Guinea Bissau to Bioko, Cameroon, Congo (Kinshasa) and Angola (Cabinda). **Chorology:** Guineo-Congolian. **Star:** green. **Alt. range:** 1-200m (2), 201-400m (1). **Flowering:** Apr (1), May (1), Nov (1). **Fruiting:** Apr (1), May (1), Nov (1). **Specimens:** - *Etinde:* Cheek M. 5399. *Mabeta-Moliwe:* Cheek M. 3440. *Mokoko River F.R.:* Akogo M. 214; Ekema N. 902

RUTACEAE

Araliopsis soyauxii Engl.
Fl. Cameroun 1: 99 (1963).

Habit: tree to 15 m. **Habitat:** forest. **Distribution:** Nigeria, Cameroon and Gabon. **Chorology:** Lower Guinea. **Star:** blue. **Alt. range:** 200-400m (1). **Specimens:** - *Mabeta-Moliwe:* Plot Voucher W 348. *Mokoko River F.R.:* Ndam N. 1153

Citropsis articulata (Spreng.) Swingle & M.Kellerm.
F.W.T.A. 1: 688 (1958); Fl. Gabon 6: 92 (1963); F.T.E.A. Rutaceae: 33 (1982); Hawthorne W., F.G.F.T. Ghana: 166 (1990).

Habit: tree or shrub to 6 m. **Guild:** shade-bearer. **Habitat:** forest. **Distribution:** Sierra Leone to Cameroon and Gabon. **Chorology:** Guineo-Congolian. **Star:** green. **Specimens:** - *Mabeta-Moliwe:* Plot Voucher W 465

Citropsis sp.

Specimens: - *Mabeta-Moliwe:* Wheatley J. I. 329

Clausena anisata (Willd.) Hook.f. ex Benth.
F.W.T.A. 1: 686 (1958); Fl. Gabon 6: 87 (1963); Fl. Cameroun 1: 127 (1963); Meded. Land. Wag. 82(3): 299 (1982); F.T.E.A. Rutaceae: 49 (1982); Fl. Ethiopia 3: 429 (1989); Hawthorne W., F.G.F.T. Ghana: 165 (1990).

Habit: tree or shrub to 6 m. **Guild:** pioneer. **Habitat:** forest-grassland transition. **Distribution:** tropical Africa. **Chorology:** Afromontane. **Star:** green. **Alt. range:** 1800-2000m (1). **Specimens:** - *No locality:* Mann G. 1187

Oricia (?) lecomteana Pierre
Fl. Cameroun 1: 94 (1963).
Oricia sp. B sensu Keay, F.W.T.A. 1: 688 (1958).

Habit: tree to 8 m. **Habitat:** forest. **Distribution:** Nigeria, Cameroon and Gabon. **Chorology:** Lower Guinea. **Star:** blue. **Specimens:** - *Mabeta-Moliwe:* Plot Voucher W 31, W 82, W 199. *Southern Bakundu F.R.:* Olorunfemi J. FHI 30709

Oricia sp. aff. lecomteana Pierre

Specimens: - *Mabeta-Moliwe:* Sunderland T. C. H. 1291; Watts J. 238; Wheatley J. I. 253. *Mokoko River F.R.:* Ekema N. 1014

Oricia trifoliata (Engl.) Verdoorn
F.W.T.A. 1: 688 (1958); Fl. Gabon 6: 56 (1963); Fl. Cameroun 1: 90 (1963).
Araliopsis trifoliata Engl.,

Habit: large tree? **Habitat:** forest. **Distribution:** Mount Cameroon. **Chorology:** Endemic (montane). **Star:** black. **Flowering:** Mar (2). **Specimens:** - *Etinde:* Zahn 499. *No locality:* Maitland T. D. 586

Teclea afzelii Engl.
F.W.T.A. 1: 689 (1958); Fl. Gabon 6: 82 (1963); Fl. Cameroun 1: 122 (1963).
Habit: tree or shrub to 5 m. **Habitat:** forest. **Distribution:** Guinea to Cameroon. **Chorology:** Guineo-Congolian. **Star:** green. **Specimens:** - *Etinde:* Schlechter F. R. R. 12410

Zanthoxylum dinklagei (Engl.) P.G.Waterman
Taxon 24: 363 (1975).
Fagara dinklagei Engl., F.W.T.A. 1: 686 (1958).

Habit: woody climber. **Habitat:** forest. **Distribution:** Nigeria, Cameroon and Gabon. **Chorology:** Lower Guinea. **Star:** blue. **Alt. range:** 1-200m (1), 201-400m (1). **Flowering:** May (1), Jun (1), Oct (1). **Fruiting:** Oct (1). **Specimens:** - *Mokoko River F.R.:* Ndam N. 1237; Pouakouyou D. 73. *Onge:* Tchouto P. 836

Zanthoxylum gilletii (De Wild.) P.G.Waterman

Taxon 24: 363 (1975); F.T.E.A. Rutaceae: 38 (1982); Keay R.W.J., Trees of Nigeria: 324 (1989); Hawthorne W., F.G.F.T. Ghana: 169 (1990).
Fagara macrophylla Engl., F.W.T.A. 1: 685 (1958).
Fagara melanorhachis Hoyle, F.W.T.A. 1: 685 (1958).
Fagara tessmannii Engl., Fl. Cameroun 1: 55 (1963).
Fagara inaequalis Engl., F.W.T.A. 1: 685 (1958).

Habit: tree to 40 m. **Habitat:** forest and farmbush. **Distribution:** Sierra Leone to Cameroon, Gabon, Congo (Kinshasa), Angola, Uganda and Sudan. **Chorology:** Guineo-Congolian. **Star:** green. **Alt. range:** 1-200m (2). **Flowering:** Apr (1). **Fruiting:** Jul (1). **Specimens:** - *Etinde:* Maitland T. D. 202. *Mabeta-Moliwe:* Cheek M. 3557; Watts J. 199

Zanthoxylum cf. heitzii (Aubrév. & Pellegr.) P.G.Waterman

Specimens: - *Southern Bakundu F.R.:* Mbani J. M. 421

Zanthoxylum cf. leprieurii Guill. & Perr.

Specimens: - *Eastern slopes:* Thomas D. W. 9436

Zanthoxylum poggei (Engl.) P.G.Waterman

Taxon 24: 363 (1975).
Fagara poggei Engl., Fl. Cameroun 1: 70 (1963).

Habit: woody climber. **Habitat:** forest. **Distribution:** Cameroon, CAR and Congo (Kinshasa). **Chorology:** Guineo-Congolian. **Star:** green. **Alt. range:** 200-400m (1). **Fruiting:** Oct (1). **Specimens:** - *Onge:* Akogo M. 128

SANTALACEAE

Thesium tenuissimum Hook.f.

F.W.T.A. 1: 666 (1958); Fl. Cameroun 14: 67 (1972); Boissiera 32: 275 (1980).

Habit: shrub to 0.4 m. **Habitat:** grassland and forest-grassland transition. **Distribution:** Bioko and Cameroon. **Chorology:** Lower Guinea (montane). **Star:** gold. **Alt. range:** 2200-2400m (1), 2801-3000m (1). **Flowering:** Oct (1). **Specimens:** - *Eastern slopes:* Thomas D. W. 9346. *No locality:* Mann G. 1223

SAPINDACEAE

S. Cable (K)

Allophylus africanus P.Beauv.

F.W.T.A. 1: 713 (1958); Fl. Gabon 23: 47 (1973); Fl. Cameroun 16: 47 (1973); Keay R.W.J., Trees of Nigeria: 357 (1989); Hawthorne W., F.G.F.T. Ghana: 165 (1990).

Habit: tree to 8 m. **Guild:** pioneer. **Habitat:** forest. **Distribution:** tropical Africa. **Star:** excluded. **Alt. range:** 1-200m (2). **Flowering:** May (1), Jun (1). **Specimens:** - *Etinde:* Maitland T. D. 748. *Mabeta-Moliwe:* Plot Voucher W 721, W 1213; Wheatley J. I. 372. *Mokoko River F.R.:* Watts J. 1126a

Allophylus bullatus Radlk.

F.W.T.A. 1: 713 (1958); Fl. Gabon 23: 40 (1973); Fl. Cameroun 16: 40 (1973); Keay R.W.J., Trees of Nigeria: 357 (1989).

Habit: tree to 11 m. **Habitat:** forest. **Distribution:** SE Nigeria and W. Cameroon. **Chorology:** Western Cameroon Uplands (montane). **Star:** gold. **Alt. range:** 800-1000m (1), 1401-1600m (2), 1601-1800m (1), 2001-2200m (1). **Flowering:** Oct (1). **Fruiting:** Aug (1), Sep (2). **Specimens:** - *Eastern slopes:* Tekwe C. 136; Thomas D. W. 9311. *Etinde:* Tekwe C. 229; Wheatley J. I. 586. *No locality:* Mann G. 1184

Allophylus grandifolius (Baker) Radlk.

F.W.T.A. 1: 713 (1958); Fl. Gabon 23: 38 (1973); Fl. Cameroun 16: 38 (1973).

Habit: tree or shrub to 4 m. **Habitat:** forest and farmbush. **Distribution:** SE Nigeria, SW Cameroon and S. Tomé. **Chorology:** Western Cameroon. **Star:** gold. **Specimens:** - *Etinde:* Preuss P. R. 1326

Allophylus hirtellus (Hook.f.) Radlk.

F.W.T.A. 1: 714 (1958); Fl. Gabon 23: 29 (1973); Fl. Cameroun 16: 29 (1973).

Habit: tree or shrub to 6 m. **Habitat:** forest. **Distribution:** SE Nigeria, Bioko and SW Cameroon. **Chorology:** Western Cameroon. **Star:** gold. **Alt. range:** 1-200m (3), 201-400m (7), 401-600m (1). **Flowering:** May (1), Jun (1), Sep (1), Oct (3). **Fruiting:** Oct (4), Nov (2). **Specimens:** - *Etinde:* Cheek M. 5421; Tchouto P. 809; Thomas D. W. 9142. *Mabeta-Moliwe:* Mann G. 727. *Mokoko River F.R.:* Pouakouyou D. 93; Watts J. 1060. *Onge:* Akogo M. 21, 99; Cheek M. 5059; Ndam N. 602; Watts J. 751, 1030

Allophylus longicuneatus Vermoesen ex Hauman

Fl. Cameroun 16: 44 (1973).

Habit: tree or shrub to 15 m. **Habitat:** forest. **Distribution:** Cameroon, Gabon and Congo (Kinshasa). **Chorology:** Guineo-Congolian. **Star:** green. **Alt. range:** 1-200m (1). **Flowering:** Apr (1). **Specimens:** - *Mokoko River F.R.:* Akogo M. 245

Allophylus megaphyllus Hutch. & Dalziel

F.W.T.A. 1: 714 (1958); Fl. Gabon 23: 28 (1973); Fl. Cameroun 16: 28 (1973).

Habit: shrub to 3 m. **Habitat:** forest. **Distribution:** SE Nigeria and SW Cameroon. **Chorology:** Western Cameroon. **Star:** gold. **Alt. range:** 1-200m (4), 201-400m (1). **Flowering:** Oct (1), Nov (1). **Fruiting:** Oct (3), Nov (2). **Specimens:** - *Onge:* Harris D. J. 3836; Tchouto P. 735, 990; Thomas D. W. 9865; Watts J. 867. *Southern Bakundu F.R.:* Brenan J. P. M. 9275

Blighia unijugata Baker

F.W.T.A. 1: 723 (1958); Fl. Gabon 23: 184 (1973); Fl. Cameroun 16: 184 (1973); Keay R.W.J., Trees of Nigeria: 358 (1989); Hawthorne W., F.G.F.T. Ghana: 187 (1990).

Habit: tree to 20 m. **Guild:** shade-bearer. **Habitat:** forest. **Distribution:** tropical Africa. **Star:** excluded. **Alt. range:** 200-400m (1), 601-800m (1). **Flowering:** Mar (1). **Specimens:** - *Mabeta-Moliwe:* Mann G. 760; Mildbraed G. W. J. 10553; Plot Voucher W 340, W 1041, W 1190 b. *Mokoko River F.R.:* Tchouto P. 563

Blighia welwitschii (Hiern) Radlk.

F.W.T.A. 1: 722 (1958); Fl. Gabon 23: 183 (1973); Fl. Cameroun 16: 183 (1973); Keay R.W.J., Trees of Nigeria: 358 (1989); Hawthorne W., F.G.F.T. Ghana: 187 (1990).

Habit: tree to 50 m. **Guild:** light demander. **Habitat:** forest. **Distribution:** Sierra Leone to Cameroon, Gabon, Congo (Kinshasa), Angola, CAR and Angola. **Chorology:** Guineo-Congolian. **Star:** green. **Alt. range:** 1-200m (1).**Fruiting:** Feb (1). **Specimens:** - *Etinde:* Mbani J. M. 11

Cardiospermum grandiflorum Sw.

F.W.T.A. 1: 711 (1958); Fl. Gabon 23: 17 (1973); Fl. Cameroun 16: 17 (1973).

Habit: herbaceous climber. **Guild:** pioneer. **Habitat:** forest. **Distribution:** Guinea to Cameroon, Gabon, Congo (Kinshasa), Angola, Sudan and Uganda. **Chorology:** Guineo-Congolian. **Star:** green. **Alt. range:** 1-200m (1), 201-400m (1), 601-800m (1). **Flowering:** Oct (1), Nov (1). **Fruiting:** Nov (1). **Specimens:** - *Etinde:* Dundas J. FHI 20502. *Limbe Botanic Garden:* Wheatley J. I. 661. *No locality:* Mann G. 1299. *Onge:* Tchouto P. 933

Chytranthus angustifolius Exell

Fl. Gabon 23: 100 (1973); Fl. Cameroun 16: 100 (1973).
   Chytranthus bracteosus Radlk., F.W.T.A. 1: 717 (1958).

Habit: tree to 8 m. **Habitat:** forest. **Distribution:** Cameroon and Gabon. **Chorology:** Lower Guinea. **Star:** blue. **Specimens:** - *Southern Bakundu F.R.:* Brenan J. P. M. 9487

Chytranthus atroviolaceus Baker f. ex Hutch. & Dalziel

F.W.T.A. 1: 717 (1958); Fl. Gabon 23: 102 (1973); Fl. Cameroun 16: 102 (1973); Keay R.W.J., Trees of Nigeria: 363 (1989); Hawthorne W., F.G.F.T. Ghana: 188 (1990).
   Chytranthus brunneo-tomentosus Gilg ex Radlk., F.W.T.A. 1: 718 (1958).

Habit: tree to 10 m. **Guild:** shade-bearer. **Habitat:** forest. **Distribution:** Sierra Leone, Ghana, Nigeria, Cameroon and Congo (Kinshasa). **Chorology:** Guineo-Congolian. **Star:** green. **Alt. range:** 1-200m (4), 601-800m (1), 801-1000m (2), 1001-1200m (2). **Flowering:** Feb (1), Mar (4), Apr (3). **Fruiting:** Feb (1), Apr (1), Aug (1). **Specimens:** - *Eastern slopes:* Etuge M. 1171; Tchouto P. 141. *Etinde:* Cable S. 1598, 1645; Hunt L. V. 31. *Mabeta-Moliwe:* Jaff B. 26, 77; Watts J. 140. *Onge:* Wheatley J. I. 677

Chytranthus carneus Radlk.

Fl. Gabon 23: 94 (1973); Fl. Cameroun 16: 94 (1973); Hawthorne W., F.G.F.T. Ghana: 188 (1990).
   Chytranthus longiracemosus Gilg, Engl. Monogr. Afr. Pfanzenfam. Sapindaceae: 788 (1932).

Habit: tree to 10 m. **Guild:** shade-bearer. **Habitat:** forest. **Distribution:** Sierra Leone to Cameroon, Gabon, Congo (Kinshasa), Angola and CAR. **Chorology:** Guineo-Congolian. **Star:** green. **Alt. range:** 1-200m (1). **Flowering:** Mar (1), Jun (1). **Fruiting:** Mar (1). **Specimens:** - *Mabeta-Moliwe:* Khayota B. 480; Plot Voucher W 155, W 367, W 469, W 1050; Sunderland T. C. H. 1464

Chytranthus macrobotrys (Gilg) Exell & Mendonça

F.W.T.A. 1: 717 (1958); Fl. Gabon 23: 111 (1973); Fl. Cameroun 16: 111 (1973); Keay R.W.J., Trees of Nigeria: 363 (1989); Hawthorne W., F.G.F.T. Ghana: 188 (1990).

Habit: tree to 11 m. **Habitat:** forest. **Distribution:** Ghana to Cameroon, Gabon, Congo (Kinshasa) and Angola (Cabinda). **Chorology:** Guineo-Congolian. **Star:** green. **Alt. range:** 1-200m (1). **Flowering:** Jul (1). **Specimens:** - *Mabeta-Moliwe:* Plot Voucher W 626, W 1031, W 1040; Winkler H. J. P. 1457; von Rege I. 1. *Southern Bakundu F.R.:* Dundas J. FHI 8368

Chytranthus setosus Radlk.

F.W.T.A. 1: 717 (1958); Fl. Gabon 23: 106 (1973); Fl. Cameroun 16: 106 (1973).

Habit: tree or shrub to 5 m. **Guild:** shade-bearer. **Habitat:** forest. **Distribution:** Togo to Cameroon, Congo (Kinshasa) and Angola (Cabinda). **Chorology:** Guineo-Congolian. **Star:** green. **Alt. range:** 200-400m (3). **Flowering:** Oct (2), Nov (1). **Specimens:** - *Etinde:* Cheek M. 5488. *Onge:* Tchouto P. 838; Watts J. 815

Chytranthus talbotii (Baker f.) Keay

F.W.T.A. 1: 717 (1958); Fl. Gabon 23: 108 (1973); Fl. Cameroun 16: 108 (1973); Keay R.W.J., Trees of Nigeria: 365 (1989).

Habit: tree to 10 m. **Habitat:** forest. **Distribution:** Ivory Coast to Cameroon and Gabon. **Chorology:** Guineo-Congolian. **Star:** green. **Alt. range:** 1-200m (8), 201-400m (1). **Flowering:** Apr (3), May (4), Jun (1). **Specimens:** - *Mabeta-Moliwe:* Cheek M. 3408; Jaff B. 123; Ndam N. 477; Plot Voucher W1031; Sunderland T. C. H. 1267, 1287, 1465; Watts J. 230; Wheatley J. I. 249, 321. *Mokoko River F.R.:* Ekema N. 1062; Ndam N. 1106. *Southern Bakundu F.R.:* Brenan J. P. M. 9417

Deinbollia sp. aff. macrantha Radlk.

Specimens: - *Bambuko F.R.:* Keay R. W. J. FHI 37428

Deinbollia maxima Gilg

F.W.T.A. 1: 716 (1958); Fl. Gabon 23: 66 (1973); Fl. Cameroun 16: 66 (1973).

Habit: tree or shrub to 6 m. **Habitat:** forest. **Distribution:** Sierra Leone, Cameroon and Gabon. **Chorology:** Guineo-Congolian. **Star:** green. **Alt. range:** 1-200m (7), 1401-1600m (1). **Flowering:** Feb (2), Mar (1), May (1), Jun (1), Nov (1). **Fruiting:** Mar (1), May (3). **Specimens:** - *Etinde:* Kwangue A. 88; Mbani J. M. 12, 28, 133; Tchouto P. 227. *Mokoko River F.R.:* Ekema N. 1107. *Onge:* Wheatley J. I. 791. *Southern Bakundu F.R.:* Keay R. W. J. FHI 37489

Deinbollia pinnata (Poir.) Schumach. & Thonn.

F.W.T.A. 1: 715 (1958); Fl. Cameroun 16: 63 (1973).

Habit: shrub to 3 m. **Habitat:** forest. **Distribution:** Senegal to SW Cameroon. **Chorology:** Guineo-Congolian. **Star:** green. **Alt. range:** 400-600m (1). **Flowering:** Feb (1). **Specimens:** - *Eastern slopes:* Etuge M. 1199

Deinbollia sp. A sensu Keay

F.W.T.A. 1: 716 (1958).

Habit: tree to 5 m. **Distribution:** Nigeria and Cameroon. **Chorology:** Lower Guinea. **Star:** blue. **Fruiting:** Feb (1).

Specimens: - *Southern Bakundu F.R.:* Binuyo A. & Daramola B. O. FHI 35524

Eriocoelum macrocarpum Gilg

F.W.T.A. 1: 724 (1958); Fl. Gabon 23: 178 (1973); Fl. Cameroun 16: 178 (1973); Keay R.W.J., Trees of Nigeria: 358 (1989).

Habit: tree to 26 m. **Habitat:** forest. **Distribution:** Ivory Coast to Cameroon, Gabon and Congo (Kinshasa). **Chorology:** Guineo-Congolian. **Star:** green. **Alt. range:** 1-200m (1).**Fruiting:** Jul (1). **Specimens:** - *Etinde:* Maitland T. D. 101. *Mabeta-Moliwe:* Cheek M. 3592; Plot Voucher W 705

Laccodiscus ferrugineus (Baker) Radlk.

F.W.T.A. 1: 721 (1958); Fl. Gabon 23: 159 (1973); Fl. Cameroun 16: 159 (1973).

Habit: tree or shrub to 6 m. **Habitat:** forest and forest gaps. **Distribution:** Nigeria, Bioko, Cameroon and Gabon. **Chorology:** Lower Guinea. **Star:** blue. **Alt. range:** 1-200m (17), 201-400m (2), 401-600m (1), 601-800m (1). **Flowering:** Jan (1), Mar (4), Apr (2), Jun (1). **Fruiting:** Jan (1), Mar (9), Apr (4), May (6), Jun (1). **Specimens:** - *Etinde:* Cable S. 1519; Maitland T. D. 587; Mbani J. M. 59, 88; Nkeng P. 5, 41; Tchouto P. 107. *Mabeta-Moliwe:* Plot Voucher W 588; Sunderland T. C. H. 1168, 1180, 1299; Watts J. 115, 162; Wheatley J. I. 68. *Mokoko River F.R.:* Acworth J. M. 50; Akogo M. 224; Ekema N. 948, 966, 967a, 1056; Fraser P. 452; Mbani J. M. 343b; Watts J. 1053. *Onge:* Mbani J. M. 33

Lepisanthes senegalensis (Juss. ex Poir.) Leenh.

Blumea 17: 85 (1969); Fl. Cameroun 16: 75 (1973); Keay R.W.J., Trees of Nigeria: 361 (1989).
Aphania senegalensis (Juss. ex Poir.) Radlk., F.W.T.A. 1: 716 (1958).

Habit: tree to 15 m. **Habitat:** forest. **Distribution:** palaeotropical. **Star:** excluded. **Specimens:** - *Etinde:* Maitland T. D. 186a

Lychnodiscus cerospermus Radlk.

Fl. Gabon 23: 168 (1973); Fl. Cameroun 16: 168 (1973).

Habit: tree to 20 m. **Habitat:** forest. **Distribution:** Cameroon, Gabon and Congo (Kinshasa). **Chorology:** Guineo-Congolian. **Star:** green. **Alt. range:** 1-200m (1), 201-400m (1). **Flowering:** Mar (1). **Fruiting:** May (1). **Specimens:** - *Mokoko River F.R.:* Mbani J. M. 376; Tchouto P. 626; Thomas D. W. 10072

Lychnodiscus grandifolius Radlk.

Fl. Gabon 23: 165 (1973); Fl. Cameroun 16: 165 (1973).

Habit: tree to 10 m. **Habitat:** forest. **Distribution:** Cameroon. **Chorology:** Lower Guinea. **Star:** gold. **Alt. range:** 1-200m (1). **Flowering:** Apr (1). **Specimens:** - *Mabeta-Moliwe:* Sunderland T. C. H. 1221

Majidea fosteri (Sprague) Radlk.

F.W.T.A. 1: 725 (1958); Fl. Gabon 23: 195 (1973); Fl. Cameroun 16: 195 (1973); Keay R.W.J., Trees of Nigeria: 360 (1989); Hawthorne W., F.G.F.T. Ghana: 188 (1990).

Habit: tree to 30 m. **Guild:** light demander. **Habitat:** forest. **Distribution:** Ivory Coast to Congo (Kinshasa), Uganda and Sudan. **Chorology:** Guineo-Congolian. **Star:** green. **Specimens:** - *Etinde:* Maitland T. D. 386

Namataea simplicifolia D.W.Thomas & D.J.Harris, ined.

Kew Bull. (in prep.).

Habit: tree or shrub to 6 m. **Habitat:** forest. **Distribution:** SE Nigeria and SW Cameroon. **Chorology:** Western Cameroon. **Star:** gold. **Alt. range:** 1-200m (1). **Specimens:** - *Onge:* Thomas D. W. 9768

Pancovia sp. nov.

Specimens: - *Bambuko F.R.:* Tchouto P. 538; Watts J. 594, 612, 613

Paullinia pinnata L.

F.W.T.A. 1: 710 (1958); Fl. Gabon 23: 13 (1973); Fl. Cameroun 16: 13 (1973); Meded. Land. Wag. 82(3): 306 (1982).

Habit: woody climber. **Guild:** pioneer. **Habitat:** forest and farmbush. **Distribution:** tropical Africa. **Star:** excluded. **Alt. range:** 1-200m (8), 201-400m (2), 401-600m (1), 601-800m (1), 801-1000m (1), 1001-1200m (1). **Flowering:** Feb (1), Mar (4), Apr (2), Nov (2), Dec (2). **Fruiting:** Feb (1), Mar (1), Apr (2), May (1), Dec (1). **Specimens:** - *Eastern slopes:* Etuge M. 1156; Groves M. 113; Nkeng P. 63; Tchouto P. 40, 46. *Etinde:* Cable S. 506; Cheek M. 5917; Tekwe C. 16. *Limbe Botanic Garden:* Wheatley J. I. 659. *Mabeta-Moliwe:* Cheek M. 5728; Mann G. 13; Nguembock F. 71; Sunderland T. C. H. 1134, 1347; Wheatley J. I. 124. *Mokoko River F.R.:* Akogo M. 193; Thomas D. W. 10035. *Onge:* Harris D. J. 3871

Placodiscus glandulosus Radlk.

F.W.T.A. 1: 720 (1958); Fl. Gabon 23: 134 (1973); Fl. Cameroun 16: 134 (1973); Keay R.W.J., Trees of Nigeria: 363 (1989).

Habit: tree to 12 m. **Habitat:** forest. **Distribution:** SE Nigeria, Cameroon and Gabon. **Chorology:** Lower Guinea. **Star:** blue. **Alt. range:** 1-200m (2). **Flowering:** Aug (1). **Fruiting:** Aug (1). **Specimens:** - *Etinde:* Maitland T. D. 766. *Mabeta-Moliwe:* Baker W. J. 270, 271

Placodiscus leptostachys Radlk.

F.W.T.A. 1: 720 (1958); Fl. Gabon 23: 130 (1973); Fl. Cameroun 16: 130 (1973).

Habit: tree to 7 m. **Habitat:** forest. **Distribution:** Liberia, Ghana, Nigeria and Cameroon. **Chorology:** Guineo-Congolian. **Star:** green. **Alt. range:** 600-800m (1). **Specimens:** - *No locality:* Mann G. 2150

Placodiscus turbinatus Radlk.

F.W.T.A. 1: 720 (1958); Fl. Gabon 23: 134 (1973); Fl. Cameroun 16: 134 (1973); Keay R.W.J., Trees of Nigeria: 363 (1989).

Habit: tree to 5 m. **Habitat:** forest. **Distribution:** SE Nigeria, Cameroon and Gabon. **Chorology:** Lower Guinea. **Star:** blue. **Specimens:** - *Southern Bakundu F.R.:* Ejiofor M. C. FHI 29348

Placodiscus sp. nov.

Specimens: - *Onge:* Akogo M. 89; Ndam N. 755; Tchouto P. 878; Watts J. 800, 873

Placodiscus sp.

Specimens: - *Mokoko River F.R.:* Mbani J. M. 331; Tchouto P. 1130. *Onge:* Tchouto P. 702

SAPOTACEAE

Autranella congolensis (De Wild.) A.Chev.
F.W.T.A. 2: 21 (1963); Fl. Cameroon 2: 39 (1964); Keay R.W.J., Trees of Nigeria: 392 (1989).

Habit: tree to 50 m. Habitat: forest. Distribution: SE Nigeria, Cameroon, Gabon and Congo (Kinshasa). Chorology: Guineo-Congolian. Star: green. Specimens: - *Etinde:* Ogu FHI 50311

Baillonella toxisperma Pierre

Specimens: - *S. Bakundu:* Ejiofor FHI 15115

Chrysophyllum delevoyi De Wild.
F.W.T.A. 2: 28 (1963); Kew Bull. 20: 461 (1966); Keay R.W.J., Trees of Nigeria: 397 (1989); Hawthorne W., F.G.F.T. Ghana: 62 (1990).
  Chrysophyllum africanum sensu Baker, F.T.A. 3: 500 (1877).
  Gambeya africana (Baker) Pierre, Fl. Cameroon 2: 114 (1964).

Habit: tree to 21 m. Habitat: forest. Distribution: tropical Africa. Star: excluded. Alt. range: 1-200m (3). Flowering: Apr (1). Fruiting: Nov (1). Specimens: - *Etinde:* Maitland T. D. 416. *Mabeta-Moliwe:* Jaff B. 3; Wheatley J. I. 196. *Mokoko River F.R.:* Mbani J. M. 308

Chrysophyllum gorungosanum Engl.
F.W.T.A. 2: 28 (1963).

Habit: tree to 30 m. Habitat: forest. Distribution: SW Cameroon, Angola and East Africa. Chorology: Afromontane. Star: green. Alt. range: 800-1000m (1). Specimens: - *Eastern slopes:* Maitland T. D. 462

Chrysophyllum subnudum Baker
F.W.T.A. 2: 27 (1963); Keay R.W.J., Trees of Nigeria: 395 (1989); Hawthorne W., F.G.F.T. Ghana: 62 (1990).
  Gambeya subnuda (Baker) Pierre, Fl. Cameroon 2: 113 (1964).
Habit: tree to 10 m. Guild: shade-bearer. Habitat: forest. Distribution: Sierra Leone to Cameroon, Gabon and Congo (Kinshasa). Chorology: Guineo-Congolian. Star: green. Specimens: - *Southern Bakundu F.R.:* Ejiofor M. C. FHI 29324

Chrysophyllum sp.

Specimens: - *Etinde:* Tekwe C. 214

Englerophytum cf. kennedyi Aubrév. & Pellegr.

Specimens: - *Mabeta-Moliwe:* Plot Voucher WA3 3

Omphalocarpum elatum Miers

Specimens: - *No locality:* Mann G. 712

128

Omphalocarpum procerum P.Beauv.

Specimens: - *S. Bakundu:* Preuss P.R. 282b

Pouteria aningeri Baehni
Candollea 9: 289 (1942)
  Aningeria robusta (A.Chev.) Aubrev. & Pellegr. F.W.T.A. 2: 24 (1963)

Specimens: - *S. Bakundu:* Nemba & Thomas 40

Synsepalum brenanii (Heine) T.D.Penn.
Pennington T., Gen. of Sapot.: 249 (1991).
  Vincentella brenanii Heine, F.W.T.A. 2: 23 (1963); Fl. Cameroon 2: 101 (1964).

Habit: tree or shrub to 5 m. Habitat: forest. Distribution: Mount Cameroon. Chorology: Endemic. Star: black. Flowering: Mar (1). Specimens: - *Southern Bakundu F.R.:* Brenan J. P. M. 9273

Synsepalum brevipes (Baker) T.D.Penn.
Pennington T., Gen. of Sapot.: 249 (1991).
  Pachystela brevipes (Baker) Baill. ex Engl., F.W.T.A. 2: 28 (1963); Fl. Cameroun 2: 83 (1964); Meded. Land. Wag. 82(3): 310 (1982); Keay R.W.J., Trees of Nigeria: 397 (1989).

Habit: tree to 15 m. Habitat: forest. Distribution: tropical Africa. Star: excluded. Alt. range: 200-400m (1). Specimens: - *Mokoko River F.R.:* Ndam N. 1290

Synsepalum dulcificum (Schum. & Thonn.) Daniell
F.W.T.A. 2: 22 (1963); Pennington T., Gen. of Sapot.: 248 (1991).

Habit: tree to 10 m. Habitat: forest. Distribution: Ghana, Cameroon, Gabon and Congo (Kinshasa), sometimes cultivated. Chorology: Guineo-Congolian. Star: green. Specimens: - *Etinde:* Dalziel J. M. 8174

Synsepalum msolo (Engl.) T.D.Penn.
Pennington T., Gen. of Sapot.: 249 (1991).
  Pachystela msolo (Engl.) Engl., F.W.T.A. 2: 28 (1963); Fl. Cameroun 2: 86 (1964); Keay R.W.J., Trees of Nigeria: 397 (1989).

Habit: tree to 45 m. Habitat: forest. Distribution: tropical Africa. Star: excluded. Specimens: - *Bambuko F.R.:* Keay R. W. J. FHI 37452

Synsepalum revolutum (Baker) T.D.Penn.
Pennington T., Gen. of Sapot.: 249 (1991).
  Vincentella revoluta (Baker) Pierre, F.W.T.A. 2: 23 (1963); Fl. Cameroun 2: 98 (1964); Keay R.W.J., Trees of Nigeria: 398 (1989); Hawthorne W., F.G.F.T. Ghana: 65 (1990).

Habit: tree to 20 m. Habitat: forest. Distribution: Ivory Coast to Bioko, Cameroon, Gabon, Congo (Brazzaville) and Congo (Kinshasa). Chorology: Guineo-Congolian. Star: green. Specimens: - *Eastern slopes:* Maitland T. D. 44

Synsepalum sp.

Specimens: - *Eastern slopes:* Nkeng P. 93

Tridesmostemon omphalocarpoides Engl.

F.W.T.A. 2: 19 (1963); Fl. Cameroun 2: 68 (1964); Pennington T., Gen. of Sapot.: 261 (1991).

Habit: tree to 35 m. **Habitat:** forest. **Distribution:** Cameroon, Gabon, Congo (Kinshasa) and Angola (Cabinda). **Chorology:** Guineo-Congolian. **Star:** green. **Specimens:** - *Etinde:* Ejiofor M. C. FHI 15254

SCROPHULARIACEAE

S. Bidgood (K) & E. Fischer (KOBLENZ)

Alectra sessiliflora (Vahl) Kuntze var. monticola (Engl.) Melch.

F.W.T.A. 2: 368 (1963); Fl. Zamb. 8(2): 90 (1990).

Habit: herb. **Habitat:** grassland. **Distribution:** tropical Africa and Asia. **Chorology:** Montane. **Star:** excluded. **Alt. range:** 800-1000m (1). **Flowering:** Nov (1). **Specimens:** - *Eastern slopes:* Banks H. 83; Migeod F. 56

Bacopa procumbens (Mill.) Greenman

F.W.T.A. 2: 359 (1963).

Habit: herb. **Distribution:** Mount Cameroon, tropical America, India and Australia. **Chorology:** Montane. **Star:** gold. **Alt. range:** 1200-1400m (1). **Specimens:** - *Eastern slopes:* Morton J. K. K 947

Buchnera leptostachya Benth.

F.W.T.A. 2: 369 (1963); Fl. Zamb. 8(2): 115 (1990).

Habit: herb. **Distribution:** tropical Africa. **Chorology:** Afromontane. **Star:** green. **Alt. range:** 1600-1800m (1). **Specimens:** - *No locality:* Keay R. W. J. FHI 37505

Diclis ovata Benth.

F.W.T.A. 2: 345 (1963); Fl. Zamb. 8(2): 12 (1990).

Habit: herb. **Distribution:** tropical Africa. **Star:** excluded. **Specimens:** - *Eastern slopes:* Morton J. K. K 943

Hedbergia abyssinica (Benth.) Molau

Fl. Zamb. 8(2): 156 (1990).
Bartsia mannii Hemsl., F.W.T.A. 2: 367 (1963).
Bartsia petitiana (A.Rich.) Hemsl., F.W.T.A. 2: 367 (1963).
Habit: herb. **Habitat:** grassland. **Distribution:** Nigeria, Cameroon, Congo (Kinshasa), Sudan, Ethiopia, Zambia, Malawi and East Africa. **Chorology:** Afromontane. **Star:** green. **Alt. range:** 2200-2400m (1), 3001-3200m (1). **Specimens:** - *No locality:* Johnston H. H. 80; Mann G. 1264

Limosella africana Glück

F.W.T.A. 2: 366 (1963).

Habit: herb. **Habitat:** grassland. **Distribution:** tropical and subtropical Africa. **Chorology:** Afromontane. **Star:** green. **Alt. range:** 3000-3200m (1). **Specimens:** - *No locality:* Mann G. s.n.

Lindernia nummulariifolia (D.Don) Wettst.

F.W.T.A. 2: 364 (1963); Fl. Zamb. 8(2): 65 (1990).

Habit: herb. **Distribution:** tropical Africa and Asia. **Star:** excluded. **Specimens:** - *Eastern slopes:* Dundas J. FHI 15338

Scoparia dulcis L.

F.W.T.A. 2: 356 (1963); Fl. Zamb. 8(2): 77 (1990).

Habit: herb. **Guild:** pioneer. **Habitat:** farmbush and roadsides. **Distribution:** pantropical. **Star:** excluded. **Alt. range:** 1-200m (1). **Flowering:** Apr (1). **Specimens:** - *Mokoko River F.R.:* Acworth J. M. 240

Sibthorpia europaea L.

F.W.T.A. 2: 356 (1963); Fl. Zamb. 8(2): 77 (1990).

Habit: herb. **Distribution:** Bioko, W. Cameroon, Zambia, Zimbabwe, Ethiopia, Tanzania and Europe. **Chorology:** Montane. **Star:** excluded. **Alt. range:** 2200-2400m (1). **Specimens:** - *No locality:* Mann G. 1963

Sopubia mannii Skan var. mannii

F.W.T.A. 2: 369 (1963); Fl. Zamb. 8(2): 144 (1990).

Habit: herb. **Habitat:** grassland. **Distribution:** tropical Africa. **Chorology:** Afromontane. **Star:** green. **Alt. range:** 2000-2200m (1). **Specimens:** - *No locality:* Mann G. 1287

Torenia thouarsii (Cham. & Schltdl.) Kuntze

F.W.T.A. 2: 363 (1963); Fl. Zamb. 8(2): 57 (1990).

Habit: herb. **Habitat:** forest. **Distribution:** tropical Africa and America. **Star:** excluded. **Alt. range:** 1-200m (1). **Flowering:** May (1). **Fruiting:** May (1). **Specimens:** - *Etinde:* Nning 22. *Mokoko River F.R.:* Watts J. 1150. *Mabeta-Moliwe:* Sunderland T.C.H. 12776

Rhabdotosperma densifolia (Hook.f.) Hartl.

Verbascum densifolium (Hook.f.) Hub.-Mor. Bauhinia 5: 12 (1973).
Celsia densifolia Hook.f., F.W.T.A. 2: 355 (1963).

Habit: shrub to 1 m. **Habitat:** grassland. **Distribution:** Bioko and W. Cameroon. **Chorology:** Western Cameroon Uplands (montane). **Star:** gold. **Alt. range:** 2800-3000m (1). **Specimens:** - *No locality:* Johnston H. H. 9

Veronica abyssinica Fresen.

F.W.T.A. 2: 355 (1963); Fl. Zamb. 8(2): 82 (1990).
Habit: herb. **Habitat:** forest-grassland transition. **Distribution:** Nigeria, W. Cameroon, Sudan, Ethiopia, Zambia, Zimbabwe, Malawi and East Africa. **Chorology:** Afromontane. **Star:** green. **Alt. range:** 2000-2200m (1). **Specimens:** - *No locality:* Mann G. 1263

Veronica mannii Hook.f.

F.W.T.A. 2: 355 (1963).

Habit: herb. **Habitat:** grassland. **Distribution:** Mount Cameroon and Bioko. **Chorology:** Western Cameroon Uplands (montane). **Star:** black. **Alt. range:** 3200-3400m (1), 3401-3600m (1). **Flowering:** Nov (1). **Specimens:** - *Eastern slopes:* Banks H. 68. *No locality:* Mann G. 1312

SCYTOPETALACEAE

Brazzeia soyauxii (Oliv.) Tiegh. var. soyauxii
F.W.T.A. 1: 300 (1958); Fl. Gabon 24: 146 (1978); Fl. Cameroun 20: 146 (1978).

Habit: tree or shrub to 10 m. **Habitat:** forest. **Distribution:** Cameroon and Gabon. **Chorology:** Lower Guinea. **Star:** blue. **Alt. range:** 1-200m (4), 201-400m (3). **Flowering:** Jun (1). **Fruiting:** Mar (1), Apr (2), May (1), Jun (1). **Specimens:** - *Mokoko River F.R.:* Acworth J. M. 173; Akogo M. 221; Ekema N. 1115a; Ndam N. 1140, 1181, 1247; Pouakouyou D. 48; Tchouto P. 594; Thomas D. W. 10145. *Southern Bakundu F.R.:* Brenan J. P. M. 9479

Oubanguia africana Baill.
Fl. Congo B., Rw. & Bur. 10: 329 (1963); Fl. Gabon 24: 152 (1978); Fl. Cameroun 20: 152 (1978).

Habit: tree 10-20 m tall. **Habitat:** forest. **Distribution:** Cameroon, Rio Muni, Gabon, Congo (Brazzaville) and Congo (Kinshasa). **Chorology:** Guineo-Congolian. **Star:** green. **Alt. range:** 1-200m (8), 201-400m (1). **Flowering:** Apr (4), Jun (1). **Specimens:** - *Mabeta-Moliwe:* Sunderland T. C. H. 1240; Tchouto P. 171; Wheatley J. I. 200. *Mokoko River F.R.:* Akogo M. 201; Ekema N. 1175; Mbani J. M. 320, 357. *Onge:* Tchouto P. 917; Watts J. 944

Oubanguia alata Baker f.
F.W.T.A. 1: 300 (1958); Fl. Gabon 24: 156 (1978); Fl. Cameroun 20: 156 (1978); Keay R.W.J., Trees of Nigeria: 112 (1989); Keay R.W.J., Trees of Nigeria: 112 (1989).

Habit: tree 10-20 m tall. **Habitat:** forest. **Distribution:** SE Nigeria, Cameroon and Gabon. **Chorology:** Lower Guinea. **Star:** blue. **Alt. range:** 1-200m (1), 201-400m (1), 401-600m (1).**Fruiting:** Sep (1). **Specimens:** - *Etinde:* Mildbraed G. W. J. 10649. *Onge:* Cheek M. 4959, 5089; Watts J. 723

Rhaptopetalum coriaceum Oliv.
F.W.T.A. 1: 300 (1958); Fl. Gabon 24: 170 (1978); Fl. Cameroun 20: 170 (1978); Keay R.W.J., Trees of Nigeria: 112 (1989).

Habit: tree to 10 m. **Habitat:** forest. **Distribution:** SE Nigeria, Bioko, Cameroon and Gabon. **Chorology:** Lower Guinea. **Star:** blue. **Alt. range:** 1-200m (6), 201-400m (5), 401-600m (4), 801-1000m (1). **Flowering:** Feb (2), Apr (1), Oct (1), Nov (1), Dec (2). **Fruiting:** Feb (1), Apr (1), Jun (1), Sep (1), Oct (6), Nov (2), Dec (1). **Specimens:** - *Etinde:* Cheek M. 5473; Faucher P. 35; Lighava M. 40; Nkeng P. 26; Tekwe C. 24, 65; Watts J. 526; Wheatley J. I. 539. *Mokoko River F.R.:* Ekema N. 1099. *Onge:* Akogo M. 20; Cheek M. 5109; Ndam N. 627, 663, 768; Tchouto P. 792; Watts J. 918

Rhaptopetalum sp. nov.
Rhaptopetalum pachyphyllum (Gürke) Engler sensu Fl. Cameroun 20: 178 (1978) pro parte

Specimens: - *S. Bakundu:* Binuyo A.& Daramola B.O. FHI 35094, 35577

Rhaptopetalum sp.

Specimens: - *Mokoko River F.R.:* Mbani J. M. 317

Scytopetalum klaineanum Pierre ex Engl.
Adansonia (ser.2) 1: 118 (1961); Fl. Cameroun 20: 187 (1978).

Habit: tree to 40 m. **Habitat:** forest. **Distribution:** Cameroon, Congo (Brazzaville) and Angola (Cabinda). **Chorology:** Lower Guinea. **Star:** blue. **Alt. range:** 1-200m (1).**Fruiting:** Nov (1). **Specimens:** - *Onge:* Thomas D. W. 9861

SIMAROUBACEAE

Brucea antidysenterica J.F.Mill.
F.W.T.A. 1: 692 (1958).

Habit: tree to 10 m. **Habitat:** forest. **Distribution:** tropical Africa. **Chorology:** Afromontane. **Star:** green. **Alt. range:** 2200-2400m (1). **Specimens:** - *No locality:* Mann G. 1188

Brucea cf. antidysenterica J.F.Mill.

Specimens: - *Eastern slopes:* Thomas D. W. 9426

Brucea guineensis G.Don
F.W.T.A. 1: 692 (1958); Hawthorne W., F.G.F.T. Ghana: 171 (1990).

Habit: tree or shrub to 5 m. **Guild:** pioneer. **Habitat:** forest. **Distribution:** Sierra Leone to Cameroon and Rio Muni. **Chorology:** Guineo-Congolian. **Star:** green. **Alt. range:** 1-200m (4), 801-1000m (1). **Flowering:** Feb (1), Mar (1), Jun (1), Nov (1), Dec (1). **Fruiting:** Mar (1), Apr (1), Jun (1), Nov (1). **Specimens:** - *Eastern slopes:* Tchouto P. 97. *Etinde:* Tekwe C. 17, 91. *Mabeta-Moliwe:* Cheek M. 3264; Mann G. 774; Nguembock F. 34; Sunderland T. C. H. 1008. *Mokoko River F.R.:* Pouakouyou D. 29

Hannoa klaineana Pierre & Engl.
F.W.T.A. 1: 691 (1958); Fl. Gabon 3: 40 (1962); Keay R.W.J., Trees of Nigeria: 330 (1989); Hawthorne W., F.G.F.T. Ghana: 171 (1990).

Habit: tree to 35 m. **Guild:** pioneer. **Habitat:** forest. **Distribution:** Guinea to Congo (Kinshasa) and Angola. **Chorology:** Guineo-Congolian. **Star:** green. **Specimens:** - *Mabeta-Moliwe:* Plot Voucher W 582, W 617, W 741, W 1204

Hannoa sp.

Specimens: - *Mokoko River F.R.:* Ndam N. 1098

Nothospondias staudtii Engl.
F.W.T.A. 1: 729 (1958); Fl. Gabon 3: 50 (1962); Keay R.W.J., Trees of Nigeria: 330 (1989); Hawthorne W., F.G.F.T. Ghana: 171 (1990).

Habit: tree to 26 m. **Habitat:** forest. **Distribution:** Nigeria, Cameroon, Gabon and Congo (Kinshasa). **Chorology:** Guineo-Congolian. **Star:** green. **Alt. range:** 200-400m (1).**Fruiting:** Oct (1). **Specimens:** - *Etinde:* Maitland T. D. 673. *Mabeta-Moliwe:* Mann G. 2148. *Onge:* Watts J. 859

Brugmansia suaveolens (Humb. & Bonpl. ex Willd.) Bercht. & Presl.

Fl. Rwanda 3: 359 (1985).
Datura suaveolens Humb. & Bonpl. ex Willd., F.W.T.A. 2: 326 (1963).

Habit: tree or shrub 3-5 m tall. **Guild:** pioneer. **Distribution:** native of Brazil, introduced throughout the tropics. **Star:** excluded. **Specimens:** - *Eastern slopes:* Migeod F. 113

Brugmansia x candida Pers.

Fl. Rwanda 3: 359 (1985).
Datura candida (Pers.) Safford, F.W.T.A. 2: 326 (1963).

Habit: tree or shrub 3-5 m tall. **Distribution:** native of the Andes, cultivated throughout the tropics. **Star:** excluded. **Specimens:** - *Eastern slopes:* Migeod F. 148

Discopodium penninervium Hochst.

F.W.T.A. 2: 328 (1963); Fl. Rwanda 3: 368 (1985).

Habit: tree or shrub to 6 m. **Habitat:** forest-grassland transition. **Distribution:** Bioko, Cameroon, Congo (Kinshasa), Ethiopia and East Africa. **Chorology:** Afromontane. **Star:** green. **Alt. range:** 2200-2400m (1), 2401-2600m (1). **Flowering:** Oct (1). **Fruiting:** Oct (1). **Specimens:** - *Eastern slopes:* Thomas D. W. 9427. *No locality:* Mann G. 1236

Physalis peruviana L.

F.W.T.A. 2: 329 (1963); Fl. Rwanda 3: 370 (1985).

Habit: herb. **Habitat:** forest. **Distribution:** tropical America, naturalized in West Africa. **Chorology:** Montane. **Star:** excluded. **Alt. range:** 2200-2400m (1). **Flowering:** Oct (1). **Fruiting:** Oct (1). **Specimens:** - *Eastern slopes:* Tchouto P. 331

Solanum aculeatissimum Jacq.

F.W.T.A. 2: 334 (1963); Bothalia 25: 51 (1995).

Habit: shrub to 1 m. **Guild:** pioneer. **Habitat:** forest. **Distribution:** tropical and southern Africa. **Chorology:** Afromontane. **Star:** green. **Alt. range:** 1200-1400m (1). **Specimens:** - *Eastern slopes:* Migeod F. s.n.

Solanum anomalum Thonn.

F.W.T.A. 2: 334 (1963).

Habit: shrub to 2 m. **Guild:** pioneer. **Habitat:** forest. **Distribution:** Liberia to Cameroon. **Chorology:** Guineo-Congolian (montane). **Star:** green. **Alt. range:** 2000-2200m (1). **Flowering:** Oct (1). **Fruiting:** Oct (1). **Specimens:** - *Eastern slopes:* Tchouto P. 309

Solanum distichum Thonn.

F.T.A. 4: 223 (1906).
Solanum indicum L. subsp. *distichum* (Thonn.) Bitter, F.W.T.A. 2: 333 (1963).

Habit: shrub to 2 m. **Guild:** pioneer. **Habitat:** forest. **Distribution:** tropical Africa. **Chorology:** Afromontane. **Star:** green. **Alt. range:** 1800-2000m (2). **Specimens:** - *No locality:* Hutchinson J. & Metcalfe C. R. 1322; Mann G. 1322

Solanum giganteum Jacq.

F.W.T.A. 2: 332 (1963); Fl. Rwanda 3: 378 (1985); Bothalia 25: 56 (1995).

Habit: tree or shrub to 8 m. **Guild:** pioneer. **Distribution:** tropical and subtropical Africa and India. **Star:** excluded. **Alt. range:** 1200-1400m (1). **Specimens:** - *Eastern slopes:* Mildbraed G. W. J. 9507

Solanum mauritianum Scop.

F.W.T.A. 2: 332 (1963); Fl. Rwanda 3: 380 (1985); Bothalia 25: 50 (1995).

Habit: tree or shrub to 6 m. **Guild:** pioneer. **Habitat:** forest. **Distribution:** tropical America, naturalized in the Palaeotropics. **Star:** excluded. **Alt. range:** 800-1000m (1), 1001-1200m (1). **Flowering:** Feb (1), Mar (1). **Fruiting:** Feb (1), Mar (1). **Specimens:** - *Eastern slopes:* Dundas J. FHI 20389; Nkeng P. 57. *Etinde:* Tchouto P. 13

Solanum nigrum L.

F.W.T.A. 2: 335 (1963); Fl. Rwanda 3: 380 (1985); Bothalia 25: 46 (1995).

Habit: herb. **Guild:** pioneer. **Distribution:** cosmopolitan. **Chorology:** Montane. **Star:** excluded. **Alt. range:** 800-1000m (1), 2201-2400m (1). **Flowering:** Oct (1). **Specimens:** - *Eastern slopes:* Migeod F. 88; Thomas D. W. 9348

Solanum pseudospinosum C.H.Wright

F.W.T.A. 2: 335 (1963).

Habit: herb. **Guild:** pioneer. **Habitat:** grassland. **Distribution:** Mount Cameroon. **Chorology:** Endemic (montane). **Star:** black. **Alt. range:** 2400-2600m (1), 2601-2800m (1). **Flowering:** Oct (1). **Fruiting:** Oct (1). **Specimens:** - *Eastern slopes:* Cheek M. 3664. *No locality:* Mann G. 1321

Solanum terminale Forssk.

Kew Bull. 14: 247 (1960); F.W.T.A. 2: 331 (1963); Fl. Rwanda 3: 382 (1985); Ann. Missouri Bot. Gard. 79: 35 (1992).
Solanum terminale Forssk. subsp. *inconstans* (C.H.Wright) Heine, Bothalia 25: 49 (1995).
Solanum terminale Forssk. subsp. *sanaganum* (Bitter) Heine, Fl. Rwanda 3: 382 (1985).

Habit: herbaceous climber. **Guild:** pioneer. **Habitat:** forest. **Distribution:** tropical Africa. **Star:** excluded. **Alt. range:** 1-200m (2), 201-400m (2), 401-600m (1), 1201-1400m (1). **Flowering:** Apr (1), May (3), Aug (1), Oct (1). **Fruiting:** Apr (1), May (2), Aug (1). **Specimens:** - *Eastern slopes:* Mbani J. M. 119. *Etinde:* Mbani J. M. 142; Tchouto P. 478; Tekwe C. 128. *Mabeta-Moliwe:* Jaff B. 39; Plot Voucher W 369; Sunderland T. C. H. 1348. *Mokoko:* Ndam 1347. *No locality:* Kalbreyer W. 172

Solanum torvum Sw.

F.W.T.A. 2: 333 (1963); Fl. Rwanda 3: 382 (1985).

Habit: shrub to 3 m. **Guild:** pioneer. **Habitat:** forest. **Distribution:** pantropical. **Star:** excluded. **Alt. range:** 1-200m (2), 201-400m (1), 601-800m (1), 801-1000m (2). **Flowering:** Feb (1), Mar (2), Apr (1), May (1). **Fruiting:** Feb (1), Mar (2), Apr (1), May (1). **Specimens:** - *Etinde:* Mbani J. M. 41, 51; Tchouto P. 11.

131

Mabeta-Moliwe: Sunderland T. C. H. 1308; Tchouto P. 164. *No locality:* Maitland T. D. 281

Solanum welwitschii C.H.Wright

Bothalia 25: 49 (1995).
   Solanum terminale Forssk. subsp. *welwitschii* (C.H.Wright) Heine, F.W.T.A. 2: 331 (1963).

Habit: woody climber. **Guild:** pioneer. **Habitat:** forest. **Distribution:** Guinea to Bioko, Cameroon, Gabon, Congo (Brazzaville), Congo (Kinshasa) and Angola. **Chorology:** Guineo-Congolian. **Star:** green. **Alt. range:** 200-400m (1). **Flowering:** Mar (1). **Fruiting:** Mar (1). **Specimens:** - *Etinde:* Tchouto P. 21. *Mabeta-Moliwe:* Mann G. s.n.

STERCULIACEAE

Cola argentea Mast.

F.W.T.A. 1: 326 (1958); Fl. Gabon 2: 86 (1961); Keay R.W.J., Trees of Nigeria: 132 (1989).
   Cola sp. D sensu Keay, F.W.T.A. 1: 332 (1958).

Habit: tree 5-12 m tall. **Habitat:** forest. **Distribution:** SE Nigeria, Cameroon and Gabon. **Chorology:** Lower Guinea. **Star:** blue. **Alt. range:** 600-800m (3). **Flowering:** Mar (1), Dec (1). **Specimens:** - *Etinde:* Cable S. 413, 1511; Maitland T. D. 1069. *Southern Bakundu F.R.:* Clayton W. D. 226

Cola brevipes K.Schum.

F.W.T.A. 1: 328 (1958); Fl. Gabon 2: 57 (1961).

Habit: tree or shrub to 4 m. **Habitat:** forest. **Distribution:** SE Nigeria and Cameroon. **Chorology:** Lower Guinea. **Star:** blue. **Specimens:** - *Etinde:* Braun J. 25

Cola cauliflora Mast.

F.W.T.A. 1: 329 (1958); Fl. Gabon 2: 68 (1961).

Habit: shrub 3-5 m tall. **Habitat:** forest. **Distribution:** SE Nigeria, Cameroon and Gabon. **Chorology:** Lower Guinea. **Star:** blue. **Alt. range:** 1-200m (7), 201-400m (5), 401-600m (4). **Flowering:** Feb (1), Mar (3), Apr (2), Oct (1). **Fruiting:** Jun (2), Jul (1), Sep (1), Oct (5), Nov (1). **Specimens:** - *Etinde:* Cheek M. 5608; Groves M. 282; Mbani J. M. 19; Nkeng P. 34, 68; Tchouto P. 25; Tekwe C. 45; Thomas D. W. 9146; Watts J. 730. *Mabeta-Moliwe:* Hunt L. V. 2; Jaff B. 103; Mann G. 772; Plot Voucher W 96, W 455, W 478, W 479; Watts J. 376. *Onge:* Akogo M. 93; Ndam N. 689, 709; Tchouto P. 676. *Southern Bakundu F.R.:* Brenan J. P. M. 9304

Cola chlamydantha K.Schum.

F.W.T.A. 1: 328 (1958); Fl. Gabon 2: 32 (1961); Keay R.W.J., Trees of Nigeria: 132 (1989); Hawthorne W., F.G.F.T. Ghana: 157 (1990).
   Chlamydocola chlamydantha (K.Schum.) Bodard., Fl. Gabon 2: 32 (1961).
   Sterculia mirabilis (A.Chev.) Roberty, Bull. Inst. Franc. Afr. Noire 15: 1402 (1953).

Habit: tree to 20 m. **Guild:** shade-bearer. **Habitat:** forest and farmbush. **Distribution:** Guinea to Cameroon, Gabon, Congo (Kinshasa) and CAR. **Chorology:** Guineo-Congolian. **Star:** blue. **Alt. range:** 1-200m (2), 201-400m (1). **Flowering:** May (1).

Specimens: - *Mabeta-Moliwe:* Baker W. J. 316; Cheek M. 3407; Sunderland T. C. H. 1306. *Mokoko River F.R.:* Ndam N. 1108

Cola digitata Mast.

F.W.T.A. 1: 326 (1958); Fl. Gabon 2: 101 (1961); Keay R.W.J., Trees of Nigeria: 133 (1989); Hawthorne W., F.G.F.T. Ghana: 157 (1990).

Habit: tree to 10 m. **Guild:** pioneer. **Habitat:** forest. **Distribution:** Liberia to Congo (Kinshasa). **Chorology:** Guineo-Congolian. **Star:** green. **Alt. range:** 1-200m (8). **Flowering:** Apr (1), May (2). **Fruiting:** Apr (3), May (3), Jun (1), Nov (1). **Specimens:** - *Mabeta-Moliwe:* Sunderland T. C. H. 1331; Watts J. 145; Wheatley J. I. 164, 307. *Mokoko River F.R.:* Akogo M. 220; Batoum A. 49; Ekema N. 1179; Ndam N. 1207. *No locality:* Box H. E. 3567. *Onge:* Watts J. 990

Cola ficifolia Mast.

F.W.T.A. 1: 328 (1958); Fl. Gabon 2: 79 (1961); Keay R.W.J., Trees of Nigeria: 133 (1989).

Habit: tree to 6 m. **Habitat:** forest. **Distribution:** SE Nigeria, Bioko, Cameroon, Gabon and Congo (Kinshasa). **Chorology:** Guineo-Congolian. **Star:** gold. **Alt. range:** 1-200m (11), 201-400m (2). **Flowering:** Mar (3), Apr (2), May (1), Jun (2), Aug (1). **Fruiting:** May (2), Jun (3). **Specimens:** - *Etinde:* Maitland T. D. 601; Tchouto P. 233. *Mabeta-Moliwe:* Baker W. J. 214; Cheek M. 3415, 3419; Plot Voucher W 95, W 486, W 533, W 542, W 788, W 835; Watts J. 119, 136, 296, 378; Wheatley J. I. 70, 209, 289, 360. *Mokoko River F.R.:* Thomas D. W. 10055. *Onge:* Tchouto P. 901; Wheatley J. I. 738. *Southern Bakundu F.R.:* Brenan J. P. M. 9295

Cola flaviflora Engl. & K.Krause

F.W.T.A. 1: 328 (1958).

Habit: shrub 2-8 m tall. **Habitat:** forest. **Distribution:** SE Nigeria and W. Cameroon. **Chorology:** Western Cameroon. **Star:** gold. **Alt. range:** 1-200m (1). **Flowering:** Mar (1). **Specimens:** - *Bambuko F.R.:* Watts J. 678

Cola flavo-velutina K.Schum.

F.W.T.A. 1: 329 (1958); Fl. Gabon 2: 72 (1961); Keay R.W.J., Trees of Nigeria: 133 (1989); Hawthorne W., F.G.F.T. Ghana: 145 (1990).

Habit: tree 4-10 m tall. **Guild:** shade-bearer. **Habitat:** forest. **Distribution:** Ghana, Nigeria, Cameroon and Gabon. **Chorology:** Guineo-Congolian. **Star:** blue. **Alt. range:** 1-200m (19). **Flowering:** Mar (2), Apr (8), May (1), Jun (1). **Fruiting:** Apr (1), May (1), Jun (6), Jul (1). **Specimens:** - *Etinde:* Kwangue A. 91; Nkeng P. 53; Tchouto P. 64, 232. *Mabeta-Moliwe:* Cheek M. 3414; Hunt L. V. 7; Jaff B. 129, 263; Plot Voucher W 65, W 285, W 298, W 416, W 438, W 439, W 991, W 1049; Sunderland T. C. H. 1249, 1311; Watts J. 170, 175, 241, 323, 343, 363; Wheatley J. I. 129, 193, 323. *Mokoko River F.R.:* Akogo M. 161; Tchouto P. 1064. *Southern Bakundu F.R.:* Brenan J. P. M. 9416; Ejiofor M. C. FHI 29328; Olorunfemi J. FHI 30503

Cola cf. gigantea A.Chev.

Specimens: - *Mabeta-Moliwe:* Plot Voucher W 164, W 1194 b

Cola lateritia K.Schum. var. lateritia

F.W.T.A. 1: 330 (1958); Fl. Gabon 2: 47 (1961); Keay R.W.J., Trees of Nigeria: 133 (1989); Hawthorne W., F.G.F.T. Ghana: 151 (1990).

Habit: tree 15-30 m tall. Habitat: forest. Distribution: SE Nigeria, Cameroon, Gabon and Congo (Kinshasa). Chorology: Guineo-Congolian. Star: green. Alt. range: 1-200m (2).Fruiting: Apr (2). Specimens: - *Etinde:* Maitland T. D. 530. *Mabeta-Moliwe:* Jaff B. 92. *Mokoko River F.R.:* Acworth J. M. 127

Cola lepidota K.Schum.

F.W.T.A. 1: 326 (1958); Fl. Gabon 2: 88 (1961); Keay R.W.J., Trees of Nigeria: 132 (1989).

Habit: tree 5-15 m tall. Habitat: forest and farmbush. Distribution: SE Nigeria, Cameroon and Gabon. Chorology: Lower Guinea. Star: gold. Alt. range: 1-200m (2), 201-400m (1). Flowering: Apr (3). Specimens: - *Mabeta-Moliwe:* Wheatley J. I. 231. *Mokoko River F.R.:* Akogo M. 254, 280. *Southern Bakundu F.R.:* Brenan J. P. M. 9444

Cola letouzeyana Nkongm.

Adansonia (ser.2) 7: 337 (1986).

Habit: shrub 2-4 m tall. Habitat: forest. Distribution: Cameroon. Chorology: Lower Guinea. Star: black. Alt. range: 200-400m (1).Fruiting: Mar (1). Specimens: - *Mokoko River F.R.:* Tchouto P. 525

Cola marsupium K.Schum.

F.W.T.A. 1: 328 (1958); Fl. Gabon 2: 51 (1961); Keay R.W.J., Trees of Nigeria: 135 (1989).

Habit: tree or shrub to 10 m. Habitat: forest. Distribution: SE Nigeria, Cameroon, Rio Muni, Gabon and Congo (Kinshasa). Chorology: Guineo-Congolian. Star: green. Alt. range: 1-200m (10), 201-400m (1). Flowering: Mar (1), Apr (2), May (3), Jun (3), Dec (1). Fruiting: Apr (2), May (1), Jun (1), Jul (1). Specimens: - *Etinde:* Maitland T. D. 603; Mbani J. M. 39. *Mabeta-Moliwe:* Hunt L. V. 17; Jaff B. 53; Plot Voucher W 260, W 444; Sunderland T. C. H. 1027, 1263, 1295; Watts J. 332; Wheatley J. I. 78. *Mokoko River F.R.:* Batoum A. 8; Ekema N. 1122; Fraser P. 353; Ndam N. 1303; Watts J. 1157. *No locality:* Box H. E. 3566. *Southern Bakundu F.R.:* Brenan J. P. M. 9451

Cola nigerica Brenan & Keay

F.W.T.A. 1: 329 (1958); Keay R.W.J., Trees of Nigeria: 133 (1989).

Habit: tree to 8 m. Habitat: forest. Distribution: Nigeria and SW Cameroon. Chorology: Lower Guinea. Star: gold. Alt. range: 1-200m (2).Fruiting: Apr (1), Oct (1). Specimens: - *Mabeta-Moliwe:* Watts J. 244. *Onge:* Tchouto P. 694

Cola nitida (Vent.) Schott & Endl.

F.W.T.A. 1: 329 (1958); Fl. Gabon 2: 44 (1961); Keay R.W.J., Trees of Nigeria: 136 (1989); Hawthorne W., F.G.F.T. Ghana: 145 (1990).

Habit: tree to 25 m. Guild: shade-bearer. Habitat: farmbush and around villages. Distribution: native from Guinea to Ghana, introduced in Cameroon. Star: pink. Specimens: - *Mabeta-Moliwe:* Cheek M. 3463. Mann G. 8

Cola pachycarpa K.Schum.

F.W.T.A. 1: 326 (1958); Fl. Gabon 2: 94 (1961); Keay R.W.J., Trees of Nigeria: 132 (1989).

Habit: tree 4-10 m tall. Habitat: forest. Distribution: SE Nigeria, Cameroon, Gabon and Congo (Kinshasa). Chorology: Guineo-Congolian. Star: gold. Alt. range: 1-200m (10), 201-400m (1). Flowering: Mar (1), Apr (5), May (3), Aug (1), Nov (1). Fruiting: Aug (1), Oct (1). Specimens: - *Mabeta-Moliwe:* Baker W. J. 351; Groves M. 275; Jaff B. 91; Plot Voucher W 265, W 283, W 784; Sunderland T. C. H. 1255; Watts J. 176; Wheatley J. I. 126, 211. *Mokoko River F.R.:* Ekema N. 859, 1019; Fraser P. 402. *Onge:* Tchouto P. 1027, 1045. *Southern Bakundu F.R.:* Dundas J. FHI 8356, FHI 8486

Cola praeacuta Brenan & Keay

F.W.T.A. 1: 329 (1958).

Habit: tree or shrub 3-7 m tall. Habitat: forest. Distribution: Mount Cameroon. Chorology: Endemic. Star: black. Alt. range: 1-200m (4), 201-400m (4).Fruiting: Jun (2), Oct (5). Specimens: - *Etinde:* Kwangue A. 92, 109. *Onge:* Ndam N. 701, 739, 798; Tchouto P. 678, 884; Thomas D. W. 9863

Cola ricinifolia Engl. & K.Krause

F.W.T.A. 1: 328 (1958).

Habit: tree or shrub to 4 m. Habitat: swamp forest. Distribution: SE Nigeria, SW Cameroon and Rio Muni. Chorology: Lower Guinea. Star: gold. Alt. range: 1-200m (1). Flowering: Apr (1). Specimens: - *Mokoko River F.R.:* Acworth J. M. 83

Cola rostrata K.Schum.

F.W.T.A. 1: 326 (1958); Fl. Gabon 2: 89 (1961); Keay R.W.J., Trees of Nigeria: 133 (1989).

Habit: tree 10-20 m tall. Habitat: forest. Distribution: SE Nigeria, Cameroon and Gabon. Chorology: Lower Guinea. Star: gold. Alt. range: 1-200m (6), 201-400m (2). Flowering: Apr (2), May (2). Fruiting: Apr (1), Oct (2). Specimens: - *Mabeta-Moliwe:* Jaff B. 94; Plot Voucher W 688, W 748; Wheatley J. I. 300, 328. *Mokoko River F.R.:* Akogo M. 253; Tchouto P. 1101; Watts J. 1107. *Onge:* Cheek M. 5070; Tchouto P. 886. *Southern Bakundu F.R.:* Brenan J. P. M. 9433

Cola semecarpophylla K.Schum.

F.W.T.A. 1: 329 (1958); Keay R.W.J., Trees of Nigeria: 133 (1989).

Habit: tree to 8 m. Habitat: forest and stream banks. Distribution: SE Nigeria and Cameroon. Chorology: Lower Guinea. Star: gold. Alt. range: 1-200m (4), 201-400m (1).Fruiting: Nov (1). Specimens: - *Onge:* Tchouto P. 900; Thomas D. W. 9781, 9782, 9783, 9890

Cola verticillata (Thonn.) Stapf ex A.Chev.

F.W.T.A. 1: 330 (1958); Fl. Gabon 2: 45 (1961); Keay R.W.J., Trees of Nigeria: 135 (1989); Hawthorne W., F.G.F.T. Ghana: 145 (1990).

Habit: tree 6-25 m tall. Guild: shade-bearer. Habitat: forest. Distribution: Ghana to Congo (Kinshasa) and Angola (Cabinda). Chorology: Guineo-Congolian. Star: green. Alt. range: 1-200m (1), 401-600m (2), 601-800m (1), 801-1000m (1). Flowering: Mar (2), Apr (2). Fruiting: Aug (1). Specimens: - *Etinde:* Cable S. 1578; Etuge M. 1240; Maitland T. D. 592; Tchouto P. 119; Tekwe C. 48. *Mabeta-Moliwe:* Baker W. J. 251

133

Cola sp. nov. aff. flavo-velutina K.Schum.

Specimens: - *Bambuko F.R.:* Watts J. 670. *Mokoko:* Ndam 1160

Cola sp. nov. 1 aff. philipi-jonesii Brenan & Keay

Habit: shrub to 1 m. Habitat: forest. Distribution: Mount Cameroon. Chorology: Endemic. Star: black. Alt. range: 1-200m (1), 601-800m (1).Fruiting: Jun (1), Nov (1). Specimens: - *Etinde:* Kwangue A. 97, 132. *Mokoko:* Watts J. 1169

Cola sp. nov. 2 aff. philipi-jonesii Brenan & Keay

Specimens: - *Mokoko:* Ndam 1240. *Onge:* Ndam N. 797; Thomas D. W. 9869; Wheatley J. I. 792

Cola sp.

Specimens: - *Onge:* Harris D. J. 3753

Leptonychia batangensis (C.H.Wright) Burret
F.W.T.A. 1: 316 (1958).

Habit: tree 5-7 m tall. Habitat: forest. Distribution: Bioko and Cameroon. Chorology: Lower Guinea. Star: black. Flowering: Jan (1). Specimens: - *Mabeta-Moliwe:* Mann G. 13

Leptonychia echinocarpa K.Schum.
F.W.T.A. 1: 316 (1958); Fl. Gabon 2: 130 (1961).

Habit: tree or shrub to 4 m. Habitat: forest. Distribution: Cameroon and Gabon. Chorology: Lower Guinea. Star: blue. Alt. range: 1-200m (2), 201-400m (2). Flowering: Mar (1). Fruiting: Mar (1), Oct (2), Nov (1). Specimens: - *Onge:* Cheek M. 5038; Harris D. J. 3791; Tchouto P. 771; Wheatley J. I. 748

Leptonychia sp. aff. echinocarpa K.Schum.
F.W.T.A. 1: 316 (1958).

Habit: tree or shrub to 6 m. Habitat: forest. Distribution: SE Nigeria, Cameroon and Gabon. Chorology: Lower Guinea. Star: gold. Alt. range: 1-200m (1), 201-400m (2). Flowering: Aug (1). Fruiting: Aug (1), Nov (2). Specimens: - *Etinde:* Cheek M. 5493, 5572. *Mabeta-Moliwe:* Baker W. J. 230

Leptonychia multiflora K.Schum.
F.W.T.A. 1: 316 (1958); Fl. Congo B., Rw. & Bur. 10: 237 (1963).

Habit: shrub to 2 m. Habitat: forest. Distribution: Cameroon and Congo (Kinshasa). Chorology: Guineo-Congolian. Star: green. Alt. range: 1-200m (2), 401-600m (1). Flowering: Mar (1), Dec (1). Fruiting: Dec (1). Specimens: - *Etinde:* Cheek M. 5817; Khayota B. 579. *Mabeta-Moliwe:* Cable S. 586

Leptonychia cf. multiflora K.Schum.

Specimens: - *Eastern slopes:* Kwangue A. 50; Tchouto P. 37. *Mabeta-Moliwe:* Jaff B. 20; Sunderland T. C. H. 1340; Wheatley J. I. 228

Leptonychia sp. aff. multiflora K.Schum.

Habit: shrub to 3 m. Habitat: forest. Distribution: Cameroon, Gabon and Congo (Kinshasa). Chorology: Guineo-Congolian. Star: green. Alt. range: 1-200m (2), 201-400m (3). Flowering: Mar (1), Oct (2). Fruiting: Mar (2), Oct (1). Specimens: - *Mokoko River F.R.:* Tchouto P. 521a, 528. *Onge:* Cheek M. 5140; Tchouto P. 988; Wheatley J. I. 823

Leptonychia pallida K.Schum.
F.W.T.A. 1: 316 (1958); Keay R.W.J., Trees of Nigeria: 126 (1989).

Habit: tree to 10 m. Habitat: forest. Distribution: SE Nigeria, Bioko and Cameroon. Chorology: Lower Guinea. Star: blue. Alt. range: 1-200m (13), 201-400m (1), 401-600m (2). Flowering: Jul (1), Oct (2), Nov (6). Fruiting: Feb (1), Mar (4), Apr (1), Jul (1), Nov (5), Dec (1). Specimens: - *Bambuko F.R.:* Olorunfemi J. FHI 30759. *Etinde:* Lighava M. 42; Nkeng P. 50; Tekwe C. 47. *Mabeta-Moliwe:* Hunt L. V. 5, 13; Jaff B. 21; Khayota B. 487; Plot Voucher W 32. *No locality:* Nkeng P. 10. *Onge:* Cheek M. 5113; Harris D. J. 3659, 3874; Thomas D. W. 9784, 9868, 9886; Watts J. 858, 914, 965; Wheatley J. I. 678

Leptonychia pubescens Keay
F.W.T.A. 1: 316 (1958).

Habit: tree to 10 m. Habitat: forest. Distribution: Ivory Coast to Cameroon. Chorology: Guineo-Congolian. Star: green. Specimens: - *Mabeta-Moliwe:* Plot Voucher W 253

Mansonia altissima (A.Chev.) A.Chev. var. kamerunica Jacq.-Fél.
F.W.T.A. 1: 313 (1958); Hawthorne W., F.G.F.T. Ghana: 149 (1990).

Habit: tree to 30 m. Habitat: forest. Distribution: Cameroon. Chorology: Lower Guinea. Star: gold. Specimens: - *Southern Bakundu F.R.:* Smith J. 129/36

Melochia melissifolia Benth. var. bracteosa (F.Hoffm.) K.Schum.
F.W.T.A. 1: 318 (1958).

Habit: herb. Guild: pioneer. Habitat: forest. Distribution: tropical Africa. Star: excluded. Alt. range: 1-200m (1). Flowering: Dec (1). Specimens: - *Mabeta-Moliwe:* Cheek M. 5784

Octolobus spectabilis Welw.
Fl. Gabon 2: 107 (1961).
Octolobus angustatus Hutch., F.W.T.A. 1: 319 (1958).

Habit: shrub 2-4 m tall. Guild: shade-bearer. Habitat: forest. Distribution: Sierra Leone to Angola. Chorology: Guineo-Congolian. Star: green. Alt. range: 800-1000m (1). Flowering: Mar (1). Specimens: - *Etinde:* Etuge M. 1244

Pterygota macrocarpa K.Schum.
F.W.T.A. 1: 320 (1958); Keay R.W.J., Trees of Nigeria: 123 (1989); Hawthorne W., F.G.F.T. Ghana: 149 (1990).

Habit: tree to 40 m. Guild: light demander. Habitat: forest. Distribution: Sierra Leone to Congo (Kinshasa). Chorology: Guineo-Congolian. Star: red. Alt. range: 800-1000m (2).

Specimens: - *Etinde:* Maitland T. D. 90. *Mabeta-Moliwe:*
Mildbraed G. W. J. 10701, 10778; Plot Voucher W 262, W 495, W
532, W 661, W 844, W 982

Scaphopetalum cf. zenkeri K.Schum.
F.W.T.A. 1: 315 (1958); Keay R.W.J., Trees of Nigeria: 126
(1989).

Habit: tree to 14 m. Habitat: forest. Distribution: SE Nigeria and
Cameroon. Chorology: Lower Guinea. Star: gold. Alt. range: 1-
200m (2). Flowering: Apr (1). Specimens: - *Mabeta-Moliwe:*
Cheek M. 3406. *Onge:* Harris D. J. 3713; Wheatley J. I. 863.
Southern Bakundu F.R.: Brenan J. P. M. 9291, 9427

Sterculia oblonga Mast.
F.W.T.A. 1: 321 (1958); Hawthorne W., F.G.F.T. Ghana: 145
(1990).
   Eribroma oblonga (Mast.) Pierre ex A.Chev., Fl. Gabon 2: 19
(1961); Keay R.W.J., Trees of Nigeria: 129 (1989).

Habit: tree to 40 m. Guild: light demander. Habitat: forest.
Distribution: Ivory Coast to Bioko and Gabon. Chorology:
Guineo-Congolian. Star: pink. Alt. range: 800-1000m (1).
Specimens: - *Mabeta-Moliwe:* Mildbraed G. W. J. 10616

Sterculia rhinopetala K.Schum.
F.W.T.A. 1: 321 (1958); Hawthorne W., F.G.F.T. Ghana: 145
(1990).

Habit: tree to 40 m. Guild: light demander. Habitat: forest.
Distribution: Ivory Coast, Ghana, Nigeria and Cameroon.
Chorology: Guineo-Congolian. Star: pink. Specimens: - *Mabeta-
Moliwe:* Plot Voucher W 1007

Sterculia tragacantha Lindl.
Kew Bull. 10: 607 (1955); F.W.T.A. 1: 321 (1958); Fl. Gabon 2:
13 (1961); Keay R.W.J., Trees of Nigeria: 129 (1989); Hawthorne
W., F.G.F.T. Ghana: 149 (1990).

Habit: tree to 25 m. Guild: pioneer. Habitat: forest. Distribution:
tropical Africa. Star: green. Alt. range: 1-200m (2), 201-400m
(1). Flowering: Oct (1). Fruiting: Oct (1), Nov (1). Specimens: -
Etinde: Cheek M. 5512; Maitland T. D. 366. *Mabeta-Moliwe:*
Cheek M. 3260; Mann G. 759; Plot Voucher W 496, W 752, W
948, W 1174, W 1199. *Mokoko River F.R.:* Ndam N. 1257. *Onge:*
Ndam N. 783

Triplochiton scleroxylon K.Schum.

Specimens: - Common in *S. Bakundu,* but no specimens seen

THYMELAEACEAE

Craterosiphon scandens Engl. & Gilg
F.W.T.A. 1: 174 (1954); Fl. Cameroun 5: 38 (1966).

Habit: woody climber. Habitat: forest. Distribution: Guinea to
Cameroon and Uganda. Chorology: Guineo-Congolian (montane).
Star: green. Alt. range: 1200-1400m (1). Specimens: - *Eastern
slopes:* Maitland T. D. 657

Dicranolepis buchholzii Engl. & Gilg
F.W.T.A. 1: 173 (1954); Fl. Gabon 11: 78 (1966); Fl. Cameroun 5:
30 (1966).

Habit: shrub to 2 m. Habitat: forest. Distribution: Cameroon,
Gabon and Congo (Kinshasa). Chorology: Guineo-Congolian
(montane). Star: green. Alt. range: 1-200m (1), 601-800m (1).
Flowering: Dec (1). Specimens: - *Mabeta-Moliwe:* Cheek M.
5695; Mildbraed G. W. J. 10691

Dicranolepis disticha Planch.
F.W.T.A. 1: 173 (1954); Fl. Gabon 11: 76 (1966); Fl. Cameroun 5:
27 (1966).

Habit: tree or shrub to 5 m. Habitat: forest. Distribution: Guinea
to Bioko, Cameroon, Gabon and Congo (Kinshasa). Chorology:
Guineo-Congolian. Star: green. Alt. range: 1-200m (8), 401-
600m (1), 801-1000m (1). Flowering: May (1). Fruiting: Mar (1),
Apr (3), May (3), Jun (3), Aug (1), Oct (1). Specimens: - *Eastern
slopes:* Tekwe C. 142. *Etinde:* Watts J. 521. *Mabeta-Moliwe:* Jaff
B. 140, 167; Nguembock F. 61; Plot Voucher W 40, W 325, W
819, W 896, W 945; Sunderland T. C. H. 1242; Tchouto P. 193;
Watts J. 156, 206, 369; Wheatley J. I. 31. *Mokoko:* Acworth J.M.
274; Akogo 170, 228, 258; Ekema 1028, 1123, 1164; Fraser 375;
Pouakouyou 61; Watts 1103

Dicranolepis glandulosa H.H.W.Pearson
F.W.T.A. 1: 173 (1954); Fl. Gabon 11: 67 (1966); Fl. Cameroun 5:
12 (1966).

Habit: tree or shrub to 4 m. Habitat: forest and stream banks.
Distribution: Cameroon. Chorology: Lower Guinea. Star: gold.
Alt. range: 1-200m (8), 201-400m (4), 401-600m (3), 601-800m
(1). Flowering: Feb (1), Mar (1), Oct (1), Nov (3), Dec (3).
Fruiting: Feb (1), Mar (3), Apr (3), May (2). Specimens: -
Bambuko F.R.: Watts J. 630. *Etinde:* Cable S. 166, 291; Cheek M.
5435; Faucher P. 29; Lighava M. 54; Mbani J. M. 159; Mildbraed
G. W. J. 10652; Sunderland T. C. H. 1081; Tchouto P. 29.
Mabeta-Moliwe: Plot Voucher W 834; Watts J. 134. *Mokoko River
F.R.:* Acworth J. M. 175, 217; Ekema 841, 1125; Pouakouyou 86;
Watts J. 1110. *Onge:* Cheek M. 4993; Harris D. J. 3757; Wheatley
J. I. 708

Dicranolepis grandiflora Engl.
F.W.T.A. 1: 173 (1954); Fl. Cameroun 5: 19 (1966).

Habit: tree or shrub to 6 m. Guild: shade-bearer. Habitat: forest.
Distribution: Ghana, Nigeria, Bioko and Cameroon. Chorology:
Guineo-Congolian. Star: green. Alt. range: 1200-1400m
(1).Fruiting: Aug (1). Specimens: - *Eastern slopes:* Kwangue A.
117

Dicranolepis vestita Engl.
F.W.T.A. 1: 173 (1954); Fl. Cameroun 5: 22 (1966).

Habit: tree or shrub to 4 m. Habitat: forest. Distribution: Nigeria,
Bioko and Cameroon. Chorology: Lower Guinea (montane). Star:
blue. Alt. range: 1200-1400m (4), 1401-1600m (2). Flowering:
Jul (1). Fruiting: May (2), Jul (1), Sep (1). Specimens: - *Eastern
slopes:* Mbani J. M. 103, 116. *Etinde:* Tchouto P. 256; Tekwe C.
207; Wheatley J. I. 562. *No locality:* Mann G. 1214

Dicranolepis sp.

Specimens: - *Etinde:* Cheek M. 5549; Hunt L. V. 39. *Mabeta-
Moliwe:* Hurst J. 27; von Rege I. 40. *Mokoko River F.R.:* Tchouto
P. 612. *Onge:* Wheatley J. I. 798

Gnidia glauca (Fres.) Gilg

Fl. Gabon 11: 95 (1966); Fl. Cameroun 5: 69 (1966); Fl. Afr. Cent. Thymelaeaceae: 61 (1975).

Lasiosiphon glaucus Fresen., F.W.T.A. 1: 176 (1954).

Habit: tree to 5 m. **Habitat:** forest-grassland transition. **Distribution:** Mount Cameroon, Ethiopia, Zambia and East Africa. **Chorology:** Afromontane. **Star:** gold. **Alt. range:** 1800-2000m (2), 2201-2400m (1). **Flowering:** Oct (1). **Specimens:** - *Eastern slopes:* Tchouto P. 345. *No locality:* Mann G. 1201; Mildbraed G. W. J. 10822

Octolepis casearia Oliv.

F.W.T.A. 1: 174 (1954); Fl. Gabon 11: 43 (1966); Fl. Cameroun 5: 79 (1966); Keay R.W.J., Trees of Nigeria: 50 (1989).

Habit: tree or shrub to 4 m. **Habitat:** forest. **Distribution:** SE Nigeria, Cameroon and Rio Muni. **Chorology:** Lower Guinea. **Star:** blue. **Alt. range:** 1-200m (2), 201-400m (4). **Flowering:** Oct (1), Nov (2). **Fruiting:** Mar (3), Nov (1). **Specimens:** - *Bambuko F.R.:* Tchouto P. 543; Watts J. 616, 649. *Mokoko:* Akogo 217; Ekema 1211, 1148. *Onge:* Tchouto P. 1040; Watts J. 827, 894. *Southern Bakundu F.R.:* Brenan J. P. M. 9270.

Octolepis sp.

Specimens: - *Mabeta-Moliwe:* Plot Voucher W 1007

Peddiea africana Harv.

F.W.T.A. 1: 174 (1954); Bull. Jard. Bot. Nat. Belg. 63: 206 (1994).

Peddiea fischeri Engl., F.W.T.A. 1: 174 (1954).

Peddiea parviflora Hook.f., F.W.T.A. 1: 174 (1954).

Habit: tree to 6 m. **Habitat:** forest and forest-grassland transition. **Distribution:** tropical and subtropical Africa. **Chorology:** Afromontane. **Star:** green. **Alt. range:** 1600-1800m (2). **Flowering:** Sep (1), Oct (1). **Fruiting:** Oct (1). **Specimens:** - *Etinde:* Tchouto P. 416; Tekwe C. 232

TILIACEAE

Ancistrocarpus densispinosus Oliv.

F.W.T.A. 1: 310 (1958).

Habit: tree or shrub to 4 m. **Habitat:** forest. **Distribution:** Nigeria, Cameroon, Gabon, Congo (Kinshasa) and Angola (Cabinda). **Chorology:** Guineo-Congolian. **Star:** green. **Alt. range:** 1-200m (6), 201-400m (1), 401-600m (1). **Flowering:** Mar (2), Apr (3), May (1). **Fruiting:** Mar (1), Apr (1), May (1), Oct (1). **Specimens:** - *Etinde:* Maitland T. D. 672; Tchouto P. 129. *Mabeta-Moliwe:* Jaff B. 115; Sunderland T. C. H. 1235, 1313. *Mokoko River F.R.:* Tchouto P. 589. *Onge:* Tchouto P. 519, 830; Wheatley J. I. 782

Christiana africana DC.

F.W.T.A. 1: 301 (1958); Keay R.W.J., Trees of Nigeria: 113 (1989); Hawthorne W., F.G.F.T. Ghana: 149 (1990).

Habit: tree to 12 m. **Guild:** shade-bearer. **Habitat:** forest. **Distribution:** Senegal to Cameroon, Congo (Kinshasa), Angola and Sudan, also tropical S. America. **Star:** excluded. **Alt. range:** 800-1000m (1). **Specimens:** - *Mabeta-Moliwe:* Mildbraed G. W. J. 10748

Desplatsia chrysochlamys (Mildbr. & Burret) Mildbr. & Burret

F.W.T.A. 1: 307 (1958); Hawthorne W., F.G.F.T. Ghana: 125 (1990).

Habit: tree or shrub to 6 m. **Guild:** shade-bearer. **Habitat:** forest. **Distribution:** Sierra Leone to Cameroon, Congo (Kinshasa) and Angola. **Chorology:** Guineo-Congolian. **Star:** green. **Alt. range:** 600-800m (1), 801-1000m (1). **Flowering:** Feb (1). **Specimens:** - *Eastern slopes:* Etuge M. 1180. *Etinde:* Kalbreyer W. 107; Maitland T. D. 562

Desplatsia dewevrei (De Wild. & T.Durand) Burret

F.W.T.A. 1: 305 (1958); Keay R.W.J., Trees of Nigeria: 114 (1989); Hawthorne W., F.G.F.T. Ghana: 125 (1990).

Habit: tree to 14 m. **Guild:** shade-bearer. **Habitat:** forest. **Distribution:** Ivory Coast to Congo (Kinshasa) and Uganda. **Chorology:** Guineo-Congolian. **Star:** green. **Alt. range:** 200-400m (1). **Specimens:** - *Etinde:* Maitland T. D. 797. *Mokoko River F.R.:* Mbani J. M. 435

Desplatsia subericarpa Bocq.

F.W.T.A. 1: 307 (1958); Keay R.W.J., Trees of Nigeria: 114 (1989); Hawthorne W., F.G.F.T. Ghana: 125 (1990).

Habit: tree to 10 m. **Guild:** shade-bearer. **Habitat:** forest. **Distribution:** Sierra Leone to Cameroon, Gabon, Congo (Kinshasa) and Angola (Cabinda). **Chorology:** Guineo-Congolian. **Star:** green. **Alt. range:** 1-200m (3), 201-400m (1), 401-600m (1). **Flowering:** Mar (1), Apr (1), Oct (1), Nov (1). **Fruiting:** Mar (2), Apr (1), Oct (1), Nov (1). **Specimens:** - *Bambuko F.R.:* Watts J. 585. *Etinde:* Watts J. 524. *Mabeta-Moliwe:* Cable S. 1442; Nguembock F. 48; Wheatley J. I. 158. *Onge:* Cheek M. 5147

Duboscia macrocarpa Bocq.

F.W.T.A. 1: 305 (1958); Keay R.W.J., Trees of Nigeria: 114 (1989).

Duboscia viridiflora (K.Schum.) Mildbr., F.W.T.A. 1: 305 (1958); Hawthorne W., F.G.F.T. Ghana: 125 (1990).

Habit: tree to 30 m. **Guild:** light demander. **Habitat:** forest. **Distribution:** Ivory Coast to Congo (Kinshasa). **Chorology:** Guineo-Congolian. **Star:** green. **Specimens:** - *Eastern slopes:* Maitland T. D. 584. *Etinde:* Winkler H. J. P. 1089

Glyphaea brevis (Spreng.) Monach.

F.W.T.A. 1: 308 (1958); Keay R.W.J., Trees of Nigeria: 117 (1989); Hawthorne W., F.G.F.T. Ghana: 125 (1990).

Habit: tree or shrub to 6 m. **Guild:** shade-bearer. **Habitat:** forest. **Distribution:** tropical Africa. **Star:** excluded. **Alt. range:** 1-200m (6), 201-400m (1), 601-800m (1). **Flowering:** Mar (2), Apr (2), Jun (1), Nov (1), Dec (1). **Fruiting:** Mar (1), Jun (2). **Specimens:** - *Bambuko F.R.:* Watts J. 661. *Eastern slopes:* Maitland T. D. 578; Tchouto P. 57. *Etinde:* Cheek M. 5573, 5927. *Mabeta-Moliwe:* Dunlap 183; Nguembock F. 81; Watts J. 224, 367; Wheatley J. I. 144, 362

Microcos coriacea (Mast.) Burret

Notizbl. Bot. Gart. Berlin 9: 759 (1926).

Grewia coriacea Mast., F.W.T.A. 1: 303 (1958).

Habit: tree to 15 m. **Habitat:** forest. **Distribution:** Nigeria, Cameroon, Gabon, Congo (Kinshasa) and Angola. **Chorology:**

Guineo-Congolian. **Star:** green. **Alt. range:** 1-200m (1), 801-1000m (1). **Flowering:** Nov (1). **Specimens:** - *Mabeta-Moliwe:* Mildbraed G. W. J. 10608; Plot Voucher W 175, W 395, W 628, W 809, W 940, W 993. *Onge:* Thomas D. W. 9835. *Southern Bakundu F.R.:* Ejiofor M. C. FHI 14065

Microcos malacocarpa (Mast.) Burret
Notizbl. Bot. Gart. Berlin 9: 760 (1926).
Grewia malacocarpa Mast., F.W.T.A. 1: 303 (1958).

Habit: woody climber. **Guild:** light demander. **Habitat:** forest. **Distribution:** Sierra Leone to Bioko and Cameroon. **Chorology:** Guineo-Congolian. **Star:** green. **Alt. range:** 800-1000m (1). **Flowering:** Mar (1). **Specimens:** - *Etinde:* Etuge M. 1243; Winkler H. J. P. 22a

Microcos oligoneura (Sprague) Burret
Notizbl. Bot. Gart. Berlin 9: 763 (1926); Keay R.W.J., Trees of Nigeria: 113 (1989).
Grewia oligoneura Sprague, F.W.T.A. 1: 303 (1958).

Habit: tree to 10 m. **Habitat:** forest. **Distribution:** SE Nigeria, Cameroon, Gabon, Congo (Kinshasa) and Sudan. **Chorology:** Guineo-Congolian. **Star:** green. **Alt. range:** 1-200m (2). **Flowering:** Oct (1), Nov (1). **Specimens:** - *Onge:* Tchouto P. 957; Watts J. 896

Microcos sp.

Specimens: - *Onge:* Watts J. 796

Triumfetta cordifolia A.Rich.
F.W.T.A. 1: 310 (1958); Fl. Zamb. 2: 76 (1963).

Habit: shrub to 2 m. **Habitat:** forest and farmbush. **Distribution:** tropical Africa. **Star:** excluded. **Alt. range:** 1-200m (1), 201-400m (1), 401-600m (1), 1001-1200m (1). **Flowering:** Feb (1), Oct (1), Nov (1), Dec (1). **Fruiting:** Nov (1). **Specimens:** - *Eastern slopes:* Cable S. 1314; Maitland T. D. 322. *Etinde:* Cheek M. 5585. *Mabeta-Moliwe:* Cheek M. 5782. *Onge:* Ndam N. 606

Triumfetta eriophlebia Hook.f.
F.W.T.A. 1: 309 (1958).

Habit: herb. **Distribution:** Nigeria, Bioko and Cameroon. **Chorology:** Lower Guinea. **Star:** blue. **Specimens:** - *Eastern slopes:* Hutchinson J. & Metcalfe C. R. 114; Maitland T. D. 279

Triumfetta pentandra A.Rich.

Specimens: - *Mabeta-Moliwe:* Cheek M. 3549

Triumfetta rhomboidea Jacq.
F.W.T.A. 1: 309 (1958); Fl. Zamb. 2: 73 (1963).

Habit: herb. **Habitat:** forest and forest-grassland transition. **Distribution:** pantropical. **Chorology:** Montane. **Star:** excluded. **Alt. range:** 1-200m (1). **Flowering:** May (1). **Specimens:** - *Eastern slopes:* Maitland T. D. 134. *Mokoko River F.R.:* Ekema N. 983

Triumfetta tomentosa Boj.
F.W.T.A. 1: 309 (1958); Fl. Zamb. 2: 72 (1963).

Habit: shrub to 3 m. **Distribution:** tropical Africa and America. **Star:** excluded. **Specimens:** - *Eastern slopes:* Maitland T. D. 135

ULMACEAE

Celtis philippensis Blanco
Fl. Cameroun 8: 26 (1968); Keay R.W.J., Trees of Nigeria: 284 (1989).
Celtis brownii Rendle, F.W.T.A. 1: 592 (1958).

Habit: tree to 15 m. **Habitat:** forest. **Distribution:** tropical Africa, Asia and Australia. **Star:** excluded. **Specimens:** - *Etinde:* Mildbraed G. W. J. 10749

Celtis zenkeri Engl.
F.W.T.A. 1: 592 (1958); Fl. Cameroun 8: 19 (1968); Keay R.W.J., Trees of Nigeria: 284 (1989); Hawthorne W., F.G.F.T. Ghana: 113 (1990).

Habit: tree to 30 m. **Guild:** light demander. **Habitat:** forest. **Distribution:** Guinea to Congo (Kinshasa), Angola, Uganda, Sudan and Tanzania. **Chorology:** Guineo-Congolian. **Star:** green. **Alt. range:** 800-1000m (1). **Specimens:** - *Mabeta-Moliwe:* Mildbraed G. W. J. 10764

Trema orientalis (L.) Blume
F.W.T.A. 1: 593 (1958); Kew Bull. 19: 143 (1964); Fl. Cameroun 8: 48 (1968); Meded. Land. Wag. 82(3): 324 (1982); Keay R.W.J., Trees of Nigeria: 285 (1989); Hawthorne W., F.G.F.T. Ghana: 113 (1990).
Trema guineensis (Schum. & Thonn.) Ficalho, F.W.T.A. 1: 592 (1958).

Habit: tree to 12 m. **Guild:** pioneer. **Habitat:** forest and farmbush. **Distribution:** palaeotropical. **Star:** excluded. **Alt. range:** 1-200m (6). **Flowering:** Apr (2), May (1), Jul (1), Aug (1), Dec (1). **Fruiting:** Apr (2), Aug (1). **Specimens:** - *Eastern slopes:* Maitland T. D. 275. *Mabeta-Moliwe:* Baker W. J. 295; Cheek M. 3556, 5780; Ndam N. 533; Wheatley J. I. 152, 178

UMBELLIFERAE

Agrocharis melanantha Hochst.
Fl. Cameroun 10: 52 (1970); F.T.E.A. Umbelliferae: 33 (1989).
Caucalis melanantha (Hochst.) Hiern, F.W.T.A. 1: 753 (1958).

Habit: herb. **Habitat:** grassland. **Distribution:** Bioko, Cameroon, Congo (Kinshasa), Rwanda, Sudan, Ethiopia and East Africa. **Chorology:** Afromontane. **Star:** green. **Alt. range:** 1800-2000m (1), 2001-2200m (1), 2601-2800m (1). **Flowering:** Oct (2). **Specimens:** - *Eastern slopes:* Cheek M. 3675; Watts J. 426. *No locality:* Mann G. 1243

Centella asiatica (L.) Urb.
F.W.T.A. 1: 753 (1958); Fl. Cameroun 10: 38 (1970); Fl. Zamb. 4: 562 (1978); F.T.E.A. Umbelliferae: 15 (1989).

Habit: herb. **Habitat:** forest-grassland transition. **Distribution:** pantropical. **Chorology:** Montane. **Star:** excluded. **Alt. range:** 800-1000m (1). **Specimens:** - *Eastern slopes:* Maitland T. D. 218

Cryptotaenia africana (Hook.f.) Drude

F.W.T.A. 1: 755 (1958); Fl. Cameroun 10: 83 (1970); F.T.E.A. Umbelliferae: 54 (1989).

Habit: herb. **Habitat:** forest-grassland transition. **Distribution:** tropical Africa. **Chorology:** Afromontane. **Star:** green. **Alt. range:** 1600-1800m (2), 2201-2400m (1). **Flowering:** Sep (1), Oct (1). **Specimens:** - *Eastern slopes:* Thomas D. W. 9395. *Etinde:* Thomas D. W. 9185. *No locality:* Brenan J. P. M. 9351

Eryngium foetidum L.

F.W.T.A. 1: 753 (1958); Fl. Cameroun 10: 45 (1970); F.T.E.A. Umbelliferae: 22 (1989).

Habit: herb. **Guild:** pioneer. **Habitat:** stream banks in forest. **Distribution:** tropical America, introduced in West Africa. **Star:** excluded. **Specimens:** - *Etinde:* Brenan J. P. M. 9264

Hydrocotyle mannii Hook.f.

F.W.T.A. 1: 753 (1958); Fl. Cameroun 10: 34 (1970); Fl. Zamb. 4: 561 (1978); F.T.E.A. Umbelliferae: 12 (1989).

Habit: herb. **Habitat:** forest-grassland transition. **Distribution:** tropical Africa. **Chorology:** Afromontane. **Star:** green. **Alt. range:** 1200-1400m (1), 1801-2000m (1). **Flowering:** Oct (1). **Specimens:** - *Eastern slopes:* Thomas D. W. 9439. *No locality:* Maitland T. D. 215

Peucedanum angustisectum (Engl.) Norman

F.W.T.A. 1: 755 (1958); Fl. Cameroun 10: 92 (1970); F.T.E.A. Umbelliferae: 101 (1989).

Habit: herb. **Habitat:** forest-grassland transition. **Distribution:** Cameroon. **Chorology:** Lower Guinea (montane). **Star:** gold. **Alt. range:** 2000-2200m (1), 2401-2600m (1). **Flowering:** Oct (1). **Specimens:** - *Etinde:* Thomas D. W. 9362. *No locality:* Maitland T. D. 819

Peucedanum camerunensis Jacq.-Fél.

Fl. Cameroun 10: 93 (1970).

Habit: herb. **Habitat:** grassland. **Distribution:** Cameroon. **Chorology:** Lower Guinea (montane). **Star:** gold. **Alt. range:** 2800-3000m (1). **Flowering:** Oct (1). **Fruiting:** Oct (1). **Specimens:** - *Eastern slopes:* Sidwell K. 36

Peucedanum winkleri H.Wolff

F.W.T.A. 1: 755 (1958); Fl. Cameroun 10: 87 (1970).

Habit: herb. **Habitat:** forest-grassland transition. **Distribution:** Bioko and Cameroon. **Chorology:** Lower Guinea (montane). **Star:** gold. **Alt. range:** 2200-2400m (1), 2401-2600m (1). **Flowering:** Feb (1). **Fruiting:** Feb (1), Oct (1). **Specimens:** - *Eastern slopes:* Thomas D. W. 9338. *No locality:* Maitland T. D. 1315

Pimpinella oreophila Hook.f.

F.W.T.A. 1: 754 (1958); Fl. Cameroun 10: 67 (1970); F.T.E.A. Umbelliferae: 66 (1989).

Habit: herb. **Habitat:** grassland. **Distribution:** Bioko, Cameroon, Sudan, Ethiopia, Uganda and Kenya. **Chorology:** Afromontane. **Star:** green. **Alt. range:** 2600-2800m (1), 3201-3400m (1). **Flowering:** Oct (1). **Specimens:** - *Eastern slopes:* Sidwell K. 56. *No locality:* Mann G. 1291

Sanicula elata D.Don

F.W.T.A. 1: 753 (1958); Fl. Cameroun 10: 42 (1970); Fl. Zamb. 4: 565 (1978); F.T.E.A. Umbelliferae: 17 (1989).

Habit: herb. **Habitat:** forest-grassland transition. **Distribution:** palaeotropical. **Chorology:** Montane. **Star:** excluded. **Alt. range:** 1200-1400m (1), 1601-1800m (1), 1801-2000m (1), 2001-2200m (1). **Flowering:** Oct (1). **Fruiting:** Sep (1), Oct (1). **Specimens:** - *Eastern slopes:* Cheek M. 3695; Thomas D. W. 9309. *Etinde:* Thomas D. W. 9215. *No locality:* Mann G. 1233

URTICACEAE

I. Friis (C) & M. Thomas (K)

Boehmeria macrophylla Hornem.

F.T.E.A. Urticaceae: 44 (1989); Fl. Zamb. 9(6): 108 (1991).
Boehmeria platyphylla D.Don, F.W.T.A. 1: 622 (1958).

Habit: herb to 3 m. **Guild:** pioneer. **Habitat:** forest. **Distribution:** palaeotropical. **Chorology:** Montane. **Star:** excluded. **Alt. range:** 1-200m (1), 201-400m (4), 801-1000m (3), 1001-1200m (1). **Flowering:** Mar (1), Aug (1), Sep (1), Oct (3), Nov (2), Dec (1). **Fruiting:** Dec (1). **Specimens:** - *Bambuko F.R.:* Olorunfemi J. FHI 30761. *Eastern slopes:* Tekwe C. 137. *Etinde:* Cable S. 1542; Cheek M. 3781, 5402, 5848; Watts J. 570; Wheatley J. I. 521. *Onge:* Akogo M. 19; Watts J. 1011

Droguetia iners (Forssk.) Schweinf.

F.W.T.A. 1: 622 (1958); Fl. Cameroun 8: 213 (1968); F.T.E.A. Urticaceae: 56 (1989); Fl. Zamb. 9(6): 112 (1991).

Habit: herb. **Distribution:** Bioko and SW Cameroon to East and South Africa. **Chorology:** Afromontane. **Star:** green. **Alt. range:** 1200-1400m (1). **Specimens:** - *No locality:* Adams C. D. GC 11728

Elatostema mannii Wedd.

F.W.T.A. 1: 620 (1958); Fl. Cameroun 8: 138 (1968).

Habit: herb. **Habitat:** on rocks and by streams in forest. **Distribution:** SE Nigeria, Bioko and SW Cameroon. **Chorology:** Western Cameroon. **Star:** gold. **Alt. range:** 1-200m (5), 201-400m (4), 401-600m (1). **Flowering:** Mar (2), Apr (1), May (2), Sep (1), Oct (1), Nov (3). **Specimens:** - *Bambuko F.R.:* Tchouto P. 537. *Etinde:* Cable S. 164; Cheek M. 5427, 5880; Tchouto P. 126; Wheatley J. I. 526. *Mokoko River F.R.:* Fraser P. 367; Ndam N. 1230. *Onge:* Cheek M. 5187; Harris D. J. 3807; Ndam N. 712; Wheatley J. I. 694

Elatostema monticolum Hook.f.

F.W.T.A. 1: 620 (1958); Fl. Cameroun 8: 141 (1968); F.T.E.A. Urticaceae: 41 (1989); Fl. Zamb. 9(6): 104 (1991).

Habit: herb. **Habitat:** forest and forest-grassland transition. **Distribution:** Bioko, W. Cameroon, Congo (Kinshasa), Rwanda, Burundi, Sudan, Ethiopia, Zimbabwe and East Africa. **Chorology:** Afromontane. **Star:** green. **Alt. range:** 1200-1400m (1), 1801-2000m (1). **Flowering:** Oct (1). **Specimens:** - *Eastern slopes:* Thomas D. W. 9517. *No locality:* Adams C. D. GC 11737

Elatostema paivaeanum Wedd.

F.W.T.A. 1: 620 (1958); Fl. Cameroun 8: 147 (1968); F.T.E.A. Urticaceae: 39 (1989); Fl. Zamb. 9(6): 104 (1991).

Habit: herb. Habitat: forest, sometimes on rocks. Distribution: tropical Africa. Star: excluded. Alt. range: 1200-1400m (1), 1401-1600m (1), 1601-1800m (1). Flowering: Jul (1), Sep (1). Specimens: - *Eastern slopes:* Dalziel J. M. 8206. *Etinde:* Cheek M. 3610; Wheatley J. I. 588. *No locality:* Maitland T. D. 1083

Elatostema welwitschii Engl.

F.W.T.A. 1: 620 (1958); Fl. Cameroun 8: 151 (1968); F.T.E.A. Urticaceae: 39 (1989); Fl. Zamb. 9(6): 104 (1991).

Habit: herb. Habitat: forest, often by streams and on rocks. Distribution: Guinea to Bioko, Cameroon, Gabon, Congo (Kinshasa), Angola, Malawi, Uganda and Tanzania. Chorology: Afromontane. Star: green. Alt. range: 1000-1200m (3), 1401-1600m (1), 1601-1800m (1). Flowering: Sep (1), Oct (2). Fruiting: Sep (1). Specimens: - *Eastern slopes:* Brenan J. P. M. 9344c; Tekwe C. 329; Thomas D. W. 9498. *Etinde:* Cheek M. 3782; Tchouto P. 425

Laportea aestuans (L.) Chew

Fl. Cameroun 8: 121 (1968); F.T.E.A. Urticaceae: 23 (1989); Fl. Zamb. 9(6): 92 (1991).
Fleurya aestuans (L.) Gaudich., F.W.T.A. 1: 619 (1958).

Habit: herb. Guild: pioneer. Habitat: forest. Distribution: tropical Africa. Star: excluded. Alt. range: 1-200m (1). Flowering: Apr (1). Specimens: - *Mokoko River F.R.:* Pouakouyou D. 11

Laportea alatipes Hook.f.

F.W.T.A. 1: 620 (1958); Fl. Cameroun 8: 117 (1968); F.T.E.A. Urticaceae: 16 (1989); Fl. Zamb. 9(6): 93 (1991).

Habit: herb. Guild: pioneer. Habitat: forest. Distribution: Cameroon to East and South Africa. Chorology: Afromontane. Star: green. Alt. range: 800-1000m (1), 1401-1600m (1), 1601-1800m (1), 2201-2400m (1). Flowering: Sep (1), Oct (2). Fruiting: Sep (1). Specimens: - *Eastern slopes:* Tchouto P. 386; Thomas D. W. 9394. *Etinde:* Wheatley J. I. 585. *No locality:* Mann G. 1973

Laportea ovalifolia (Schumach.) Chew

Fl. Cameroun 8: 131 (1968); F.T.E.A. Urticaceae: 18 (1989); Fl. Zamb. 9(6): 89 (1991).
Fleurya ovalifolia (Schumach.) Dandy, F.W.T.A. 1: 619 (1958).

Habit: herb. Guild: pioneer. Habitat: forest, often by streams. Distribution: tropical and subtropical Africa. Chorology: Afromontane. Star: green. Alt. range: 1-200m (2), 1201-1400m (1), 1601-1800m (2). Flowering: Mar (1), Apr (1), Sep (1). Fruiting: Apr (1). Specimens: - *Eastern slopes:* Thomas D. W. 9490. *Etinde:* Dalziel J. M. 8201; Wheatley J. I. 574. *Mabeta-Moliwe:* Cheek M. 3222; Sunderland T. C. H. 1262. *No locality:* Mann G. 1950. *Onge:* Wheatley J. I. 703

Lecanthus peduncularis (Wall. ex Royle) Wedd.

F.W.T.A. 1: 621 (1958); Fl. Cameroun 8: 180 (1968); F.T.E.A. Urticaceae: 2 (1989); Fl. Zamb. 9(6): 101 (1991).

Habit: herb. Habitat: forest. Distribution: Cameroon and Ethiopia, also India and SE Asia. Chorology: Montane. Star: excluded. Alt. range: 1800-2000m (1). Specimens: - *Eastern slopes:* Banks H. 79. *No locality:* Adams C. D. GC 11766

Parietaria debilis G.Forster

F.W.T.A. 1: 622 (1958); Fl. Cameroun 8: 208 (1968); Fl. Rwanda 1: 158 (1978); F.T.E.A. Urticaceae: 52 (1989).
Parietaria laxiflora Engl., F.W.T.A. 1: 622 (1958).

Habit: herb. Habitat: forest-grassland transition. Distribution: pantropical. Chorology: Montane. Star: excluded. Alt. range: 1800-2000m (1), 2001-2200m (1), 2201-2400m (1), 2401-2600m (1), 3001-3200m (1), 3601-3800m (1). Flowering: Oct (4). Specimens: - *Eastern slopes:* Cheek M. 3640, 3670; Thomas D. W. 9308, 9335. *No locality:* Brenan J. P. M. 9509; Morton J. K. GC 6902

Pilea microphylla (L.) Liebm.

F.W.T.A. 1: 621 (1958); Fl. Cameroun 8: 178 (1968); F.T.E.A. Urticaceae: 37 (1989).

Habit: herb. Habitat: forest, by streams. Distribution: pantropical. Star: excluded. Alt. range: 1400-1600m (1). Flowering: Oct (1). Specimens: - *Eastern slopes:* Tchouto P. 373

Pilea rivularis Wedd.

Fl. Cameroun 8: 163 (1968); F.T.E.A. Urticaceae: 29 (1989); Fl. Zamb. 9(6): 98 (1991).
Pilea ceratomera Wedd., F.W.T.A. 1: 621 (1958).

Habit: herb. Habitat: forest-grassland transition. Distribution: tropical and subtropical Africa. Chorology: Afromontane. Star: green. Alt. range: 800-1000m (1), 1401-1600m (2), 2201-2400m (2). Flowering: Sep (2), Oct (2). Specimens: - *Eastern slopes:* Thomas D. W. 9352, 9461. *Etinde:* Tekwe C. 223; Wheatley J. I. 558. *No locality:* Johnston H. H. 66

Pilea sublucens Wedd.

F.W.T.A. 1: 621 (1958); Fl. Cameroun 8: 171 (1968); F.T.E.A. Urticaceae: 31 (1989).

Habit: herb. Habitat: forest and Aframomum thicket. Distribution: Liberia to Bioko, Cameroon, Gabon, Congo (Kinshasa), Rwanda and Uganda. Chorology: Guineo-Congolian (montane). Star: green. Alt. range: 1-200m (1), 201-400m (1), 401-600m (3), 801-1000m (1), 1001-1200m (1), 1201-1400m (1), 1401-1600m (3), 1601-1800m (1), 1801-2000m (1), 2201-2400m (1). Flowering: Sep (3), Oct (7), Nov (2). Specimens: - *Eastern slopes:* Adams C. D. GC 11296; Cheek M. 5338; Sidwell K. 85. *Etinde:* Cheek M. 3603, 5524; Tchouto P. 413; Tekwe C. 191; Thomas D. W. 9134; Watts J. 532; Wheatley J. I. 545, 581, 594. *Onge:* Ndam N. 690; Tchouto P. 797

Pilea tetraphylla (Steud.) Blume

F.W.T.A. 1: 621 (1958); Fl. Cameroun 8: 173 (1968); F.T.E.A. Urticaceae: 27 (1989); Fl. Zamb. 9(6): 96 (1991).

Habit: herb. Habitat: forest-grassland transition. Distribution: tropical Africa. Chorology: Afromontane. Star: green. Alt. range: 1200-1400m (1), 2001-2200m (2), 2201-2400m (1). Flowering: Sep (1), Oct (2). Fruiting: Sep (1). Specimens: - *Eastern slopes:* Tchouto P. 329. *Etinde:* Thomas D. W. 9366; Wheatley J. I. 561. *No locality:* Mann G. 2012

Pouzolzia guineensis Benth.

F.W.T.A. 1: 622 (1958); Fl. Cameroun 8: 184 (1968); F.T.E.A. Urticaceae: 47 (1989).

Habit: herb. Guild: pioneer. Habitat: forest. Distribution: Senegal to Bioko, S. Tomé, Cameroon, Congo (Kinshasa), Angola and CAR. Chorology: Guineo-Congolian. Star: green. Alt. range:

1-200m (1). **Flowering:** Jul (1). **Specimens:** - *Etinde:* Maitland T. D. 550. *Mabeta-Moliwe:* Cheek M. 3537

Pouzolzia parasitica (Forssk.) Schweinf.
F.W.T.A. 1: 763 (1958); Fl. Cameroun 8: 194 (1968); F.T.E.A. Urticaceae: 51 (1989); Fl. Zamb. 9(6): 111 (1991).

Habit: herb to 2 m. **Habitat:** forest. **Distribution:** tropical Africa and South America. **Chorology:** Montane. **Star:** excluded. **Alt. range:** 2000-2200m (1). **Flowering:** Oct (1). **Specimens:** - *Etinde:* Thomas D. W. 9368

Procris crenata C.B.Robinson
F.W.T.A. 1: 621 (1958); Fl. Cameroun 8: 155 (1968); F.T.E.A. Urticaceae: 42 (1989); Fl. Zamb. 9(6): 101 (1991).

Habit: epiphyte. **Habitat:** forest. **Distribution:** Bioko, SW Cameroon, Congo (Kinshasa), Rwanda, Burundi, Sudan, Ethiopia, East Africa and Zimbabwe. **Chorology:** Afromontane. **Star:** green. **Alt. range:** 1000-1200m (3), 1201-1400m (1). **Flowering:** Sep (1), Oct (2). **Fruiting:** Aug (1). **Specimens:** - *Eastern slopes:* Preuss P. R. 963. *Etinde:* Hunt L. V. 32; Tekwe C. 162; Thomas D. W. 9523; Wheatley J. I. 598

Urera batesii Rendle
F.W.T.A. 1: 618 (1958); Fl. Cameroun 8: 102 (1968).

Habit: woody climber. **Habitat:** forest. **Distribution:** Bioko, Cameroon and Congo (Kinshasa). **Chorology:** Guineo-Congolian. **Star:** green. **Alt. range:** 1-200m (1). **Flowering:** Mar (1). **Specimens:** - *Etinde:* Tchouto P. 85

Urera cordifolia Engl.
F.W.T.A. 1: 618 (1958); Fl. Cameroun 8: 90 (1968).

Habit: woody climber. **Habitat:** forest. **Distribution:** Nigeria, Cameroon, Gabon and CAR. **Chorology:** Guineo-Congolian. **Star:** green. **Alt. range:** 1-200m (6). **Flowering:** Feb (1), Mar (3), Apr (1), Jul (1). **Fruiting:** Apr (1). **Specimens:** - *Etinde:* Maitland T. D. 411; Nkeng P. 23; Tchouto P. 69. *Mabeta-Moliwe:* Cheek M. 3539; Sunderland T. C. H. 1190. *Onge:* Tchouto P. 510; Wheatley J. I. 832

Urera gabonensis Pierre ex Letouzey
Fl. Cameroun 8: 95 (1968).

Habit: woody climber. **Habitat:** forest. **Distribution:** Cameroon and Gabon. **Chorology:** Lower Guinea. **Star:** blue. **Alt. range:** 1-200m (1). **Flowering:** Apr (1). **Fruiting:** Apr (1). **Specimens:** - *Mokoko River F.R.:* Pouakouyou D. 17

Urera gravenreuthi Engl.
F.W.T.A. 1: 618 (1958); Fl. Cameroun 8: 85 (1968).

Habit: woody climber. **Habitat:** forest. **Distribution:** Cameroon. **Chorology:** Lower Guinea (montane). **Star:** gold. **Specimens:** - *Eastern slopes:* Dundas J. FHI 15340

Urera mannii (Wedd.) Benth. & Hook.f. ex Rendle
F.W.T.A. 1: 618 (1958); Fl. Cameroun 8: 86 (1968).

Habit: woody climber. **Guild:** light demander. **Habitat:** forest. **Distribution:** Liberia to Bioko and SW Cameroon. **Chorology:**

140

Guineo-Congolian (montane). **Star:** green. **Alt. range:** 800-1000m (1). **Flowering:** Jul (1). **Specimens:** - *Eastern slopes:* Tekwe C. 110

Urera cf. mannii (Wedd.) Benth. & Hook.f. ex Rendle

Specimens: - *Onge:* Ndam N. 692

Urera repens (Wedd.) Rendle
F.W.T.A. 1: 618 (1958); Fl. Cameroun 8: 104 (1968).

Habit: woody climber. **Habitat:** forest. **Distribution:** Liberia to Bioko, Cameroon, Gabon, Congo (Brazzaville) and Congo (Kinshasa). **Chorology:** Guineo-Congolian. **Star:** green. **Alt. range:** 1-200m (4), 201-400m (3), 401-600m (1). **Flowering:** Mar (2), Apr (1), May (1), Jun (2), Dec (1). **Fruiting:** May (1). **Specimens:** - *Bambuko F.R.:* Watts J. 639. *Etinde:* Cable S. 285. *Mokoko River F.R.:* Acworth J. M. 339; Akogo M. 274; Fraser P. 459; Ndam N. 1348; Tchouto P. 574; Watts J. 1049, 1082

Urera rigida (Benth.) Keay
F.W.T.A. 1: 618 (1958); Fl. Cameroun 8: 89 (1968).

Habit: woody climber. **Habitat:** forest. **Distribution:** Sierra Leone to SW Cameroon. **Chorology:** Guineo-Congolian. **Star:** green. **Specimens:** - *Mabeta-Moliwe:* Dalziel J. M. 8184

Urera sp. aff. rigida (Benth.) Keay

Specimens: - *Mabeta-Moliwe:* Jaff B. 108; Plot Voucher W 364

Urera trinervis (Hochst.) Friis & Immelmann
F.T.E.A. Urticaceae: 6 (1989); Fl. Zamb. 9(6): 83 (1991); Fl. Zamb. 9(6): 81 (1991).
Urera cameroonensis Wedd., F.W.T.A. 1: 618 (1958).

Habit: woody climber. **Guild:** shade-bearer. **Habitat:** forest. **Distribution:** tropical and subtropical Africa. **Star:** excluded. **Alt. range:** 1-200m (3), 1001-1200m (1), 1401-1600m (1). **Flowering:** Apr (2), May (1), Oct (1). **Fruiting:** Apr (1), May (1). **Specimens:** - *Eastern slopes:* Maitland T. D. 312. *Etinde:* Wheatley J. I. 646. *Mabeta-Moliwe:* Plot Voucher W 796, W 1192; Sunderland T. C. H. 1325; Wheatley J. I. 183. *Mokoko River F.R.:* Acworth J. M. 139. *No locality:* Mann G. 2173

Urera sp.

Specimens: - *Etinde:* Cable S. 1623. *Mokoko River F.R.:* Acworth J. M. 85, 142, 348

VERBENACEAE

Clerodendrum bipindense Gürke
F.W.T.A. 2: 444 (1963).

Habit: woody climber. **Guild:** pioneer. **Habitat:** forest. **Distribution:** Nigeria, Bioko and Cameroon. **Chorology:** Lower Guinea. **Star:** blue. **Flowering:** Oct (1). **Specimens:** - *Etinde:* Maitland T. D. 783. *Mabeta-Moliwe:* Banks H. 14

Clerodendrum buettneri Gürke
F.W.T.A. 2: 442 (1963).

Habit: woody climber. **Habitat:** forest. **Distribution:** Nigeria, Cameroon, Rio Muni and Gabon. **Chorology:** Lower Guinea. **Star:** blue. **Alt. range:** 1-200m (3). **Flowering:** Jun (2). **Fruiting:** Aug (1). **Specimens:** - *Mabeta-Moliwe:* Watts J. 395; Wheatley J. I. 346; von Rege I. 70

Clerodendrum capitatum (Willd.) Schum. & Thonn. var. captitatum
F.W.T.A. 2: 443 (1963); Meded. Land. Wag. 82(3): 327 (1982).

Habit: woody climber. **Guild:** pioneer. **Habitat:** forest. **Distribution:** tropical and subtropical Africa. **Chorology:** Afromontane. **Star:** green. **Alt. range:** 800-1000m (2). **Flowering:** Dec (1). **Specimens:** - *Etinde:* Lighava M. 3. *No locality:* Mann G. 1975

Clerodendrum chinense (Osbeck) Mabb.
F.T.E.A. Verbenaceae: 94 (1992).
   Clerodendrum japonicum sensu Huber, Hepper & Meikle, F.W.T.A. 2: 443 (1963).

Habit: shrub to 2 m. **Guild:** pioneer. **Habitat:** villages. **Distribution:** introduced to tropical Africa, native to China and Japan. **Star:** excluded. **Specimens:** - *Etinde:* Ndi FHI 50351

Clerodendrum dusenii Gürke
F.W.T.A. 2: 444 (1963).
   Clerodendrum barteri Baker, F.T.A. 5: 298 (1900).

Habit: scrambling shrub. **Habitat:** forest. **Distribution:** Nigeria, Cameroon, S. Tomé and Angola. **Chorology:** Guineo-Congolian. **Star:** green. **Alt. range:** 1-200m (5). **Flowering:** Jan (1), Apr (1), May (2), Aug (1). **Fruiting:** Jan (1), Apr (1). **Specimens:** - *Mabeta-Moliwe:* Baker W. J. 296; Maitland T. D. 980; Ndam N. 488; Sunderland T. C. H. 1373; Tchouto P. 1048; Wheatley J. I. 179

Clerodendrum globuliflorum B.Thomas
F.W.T.A. 2: 443 (1963).

Habit: shrub to 2 m. **Guild:** pioneer. **Habitat:** forest. **Distribution:** Nigeria, Bioko and Cameroon. **Chorology:** Lower Guinea. **Star:** blue. **Alt. range:** 1-200m (7), 201-400m (3). **Flowering:** Mar (1), Apr (2), May (3), Jun (1), Oct (2). **Fruiting:** Mar (1), Apr (2), May (2), Jun (1). **Specimens:** *Bambuko F.R.:* Watts J. 603. *Mokoko River F.R.:* Acworth J. M. 176; Akogo M. 169; Ekema N. 830, 992, 1089; Fraser P. 417; Pouakouyou D. 74. *No locality:* Weberbauer A. 58. *Onge:* Tchouto P. 987; Watts J. 740

Clerodendrum cf. globuliflorum B.Thomas

Specimens: - *Mokoko River F.R.:* Ndam N. 1214

Clerodendrum melanocrater Gürke
F.W.T.A. 2: 444 (1963); F.T.E.A. Verbenaceae: 113 (1992).

Habit: woody climber. **Guild:** pioneer. **Habitat:** forest. **Distribution:** Bioko, Cameroon, Congo (Kinshasa), Uganda and Tanzania. **Chorology:** Guineo-Congolian. **Star:** green. **Alt. range:** 1-200m (2), 601-800m (1). **Flowering:** Sep (1). **Fruiting:** Sep (1), Nov (1). **Specimens:** - *Mabeta-Moliwe:* Mildbraed G. W. J. 10510. *Onge:* Harris D. J. 3779; Tchouto P. 822

Clerodendrum silvanum Henriq. var. buchholzii (Gürke) Verdc.
Mem. Mus. Natl. Hist. Nat. B. Bot. 25: 1555 (1975); F.T.E.A. Verbenaceae: 110 (1992).
   Clerodendrum buchholzii Gürke, F.W.T.A. 2: 443 (1963).

Habit: woody climber. **Guild:** pioneer. **Habitat:** forest. **Distribution:** tropical Africa. **Star:** excluded. **Alt. range:** 1-200m (7), 201-400m (3), 1201-1400m (1). **Flowering:** Mar (2), May (1), Aug (2), Oct (1), Nov (1). **Fruiting:** Mar (1), May (2), Nov (1). **Specimens:** - *Bambuko F.R.:* Watts J. 640. *Etinde:* Cable S. 130; Cheek M. 5448. *Mabeta-Moliwe:* Baker W. J. 224, 228, 315. *Mokoko River F.R.:* Ekema N. 842, 982; Fraser P. 362. *No locality:* Maitland T. D. 311. *Onge:* Ndam N. 620; Wheatley J. I. 713

Clerodendrum splendens G.Don
F.W.T.A. 2: 444 (1963).

Habit: woody climber. **Guild:** pioneer. **Habitat:** forest. **Distribution:** Senegal to Cameroon, Rio Muni, Gabon, Congo (Kinshasa), Angola and CAR. **Chorology:** Guineo-Congolian. **Star:** green. **Alt. range:** 1-200m (2), 401-600m (1), 601-800m (3), 801-1000m (2). **Flowering:** Mar (2), Oct (1), Dec (1). **Fruiting:** Mar (4). **Specimens:** - *Eastern slopes:* Groves M. 192. *Etinde:* Cable S. 462, 1501, 1579; Groves M. 295; Hutchinson J. & Metcalfe C. R. 143; Khayota B. 507. *Onge:* Harris D. J. 3837; Tchouto P. 1021

Clerodendrum umbellatum Poir.
F.W.T.A. 2: 442 (1963); F.T.E.A. Verbenaceae: 97 (1992).

Habit: woody climber. **Guild:** pioneer. **Habitat:** forest and farmbush. **Distribution:** Senegal to Cameroon, Bioko, Gabon, Congo (Kinshasa), Uganda and Tanzania. **Chorology:** Guineo-Congolian. **Star:** green. **Alt. range:** 1-200m (2). **Flowering:** Mar (1). **Fruiting:** Dec (1). **Specimens:** - *Etinde:* Dundas J. FHI 20501.

Mabeta-Moliwe: Cheek M. 5682; Nguembock F. 72, 73; Sunderland T. C. H. 1146

Clerodendrum violaceum Gürke
F.W.T.A. 2: 441 (1963); F.T.E.A. Verbenaceae: 125 (1992).

Habit: woody climber. **Guild:** pioneer. **Habitat:** forest. **Distribution:** Guinea to Cameroon, Congo (Kinshasa) and Zimbabwe. **Star:** green. **Specimens:** - *Etinde:* Maitland T. D. 554

Clerodendrum sp.

Specimens: *Mokoko River F.R.:* Acworth J. M. 147; Fraser P. 392

Premna angolensis Gürke
F.W.T.A. 2: 438 (1963); Keay R.W.J., Trees of Nigeria: 437 (1989); Hawthorne W., F.G.F.T. Ghana: 45 (1990); F.T.E.A. Verbenaceae: 70 (1992).

Habit: tree to 12 m. **Guild:** pioneer. **Habitat:** forest. **Distribution:** tropical Africa. **Chorology:** Afromontane. **Star:** green. **Alt. range:** 800-1000m (1). **Specimens:** - *Eastern slopes:* Maitland T. D. 661

Vitex grandifolia Gürke

F.W.T.A. 2: 446 (1963); Keay R.W.J., Trees of Nigeria: 439 (1989); Hawthorne W., F.G.F.T. Ghana: 161 (1990).

Habit: tree to 10 m. **Guild:** shade-bearer. **Habitat:** forest. **Distribution:** Sierra Leone to Cameroon, Rio Muni and Gabon. **Chorology:** Guineo-Congolian. **Star:** green. **Specimens:** - *Etinde:* Maitland T. D. 361. *Mabeta-Moliwe:* Cheek M. 3420; Plot Voucher W 544

Vitex sp. aff. grandifolia Gürke

Specimens: - *Mabeta-Moliwe:* Cheek M. 3586

Vitex cf. phaeotricha Mildbr. & Pieper

Specimens: - *Mabeta-Moliwe:* Wheatley J. I. 174

Vitex rivularis Gürke

F.W.T.A. 2: 446 (1963); Keay R.W.J., Trees of Nigeria: 439 (1989); Hawthorne W., F.G.F.T. Ghana: 161 (1990).

Habit: tree to 15 m. **Guild:** light demander. **Habitat:** forest. **Distribution:** Liberia to Congo (Kinshasa). **Chorology:** Guineo-Congolian. **Star:** green. **Alt. range:** 600-800m (1). **Specimens:** - *Mabeta-Moliwe:* Mildbraed G. W. J. 10535

Vitex thyrsiflora Baker

F.W.T.A. 2: 446 (1963).

Habit: woody climber. **Habitat:** forest. **Distribution:** Guinea to Cameroon and Congo (Kinshasa). **Chorology:** Guineo-Congolian. **Star:** green. **Alt. range:** 1-200m (1). **Flowering:** Jun (1). **Specimens:** - *Etinde:* Maitland T. D. 565. *Mabeta-Moliwe:* Wheatley J. I. 337

Vitex zenkeri Gürke

Engl. Bot. Jahrb. 33: 293 (1903).
Habit: tree or shrub to 10 m. **Habitat:** forest. **Distribution:** Cameroon, Gabon and Congo (Kinshasa). **Chorology:** Guineo-Congolian. **Star:** green. **Alt. range:** 1-200m (1). **Fruiting:** May (1). **Specimens:** - *Mokoko River F.R.:* Watts J. 1159

Vitex sp. A sensu Huber, Hepper & Meikle

F.W.T.A. 2: 448 (1963).

Habit: tree to 20 m. **Distribution:** Nigeria and Cameroon. **Chorology:** Lower Guinea. **Star:** blue. **Specimens:** - *Bambuko F.R.:* Olorunfemi J. FHI 30751

Vitex sp. B sensu Huber, Hepper & Meikle

F.W.T.A. 2: 448 (1963).

Habit: tree to 15 m. **Habitat:** forest. **Distribution:** Mount Cameroon. **Chorology:** Endemic. **Star:** black. **Specimens:** - *Southern Bakundu F.R.:* Olorunfemi J. FHI 30712

VIOLACEAE

G. Achoundong (YA)

Rinorea adnata Chipp

F.W.T.A. 1: 104 (1954).

Habit: shrub to 2 m. **Guild:** shade-bearer. **Habitat:** forest. **Distribution:** Cameroon. **Chorology:** Lower Guinea. **Star:** gold. **Specimens:** - *Southern Bakundu F.R.:* Brenan J. P. M. 9260

Rinorea afzelii Engl.

F.W.T.A. 1: 103 (1954); Fl. Afr. Cent. Violaceae: 20 (1969).

Habit: tree or shrub to 10 m. **Guild:** shade-bearer. **Habitat:** forest. **Distribution:** Sierra Leone, Liberia, Cameroon and Congo (Kinshasa). **Chorology:** Guineo-Congolian. **Star:** green. **Alt. range:** 1-200m (1). **Flowering:** Apr (1). **Specimens:** - *Mokoko River F.R.:* Tchouto P. 1057

Rinorea angustifolia (Thou.) Baill. subsp. engleriana (De Wild. & T.Durand) Grey-Wilson

Kew Bull. 16: 412 (1963); Kew Bull. 36: 111 (1981).
   Rinorea gracilipes Engl., Fl. Afr. Cent. Violaceae: 17 (1969).
   Rinorea ardisiiflora sensu Keay, F.W.T.A. 1: 101 (1954).

Habit: tree or shrub to 8 m. **Guild:** shade-bearer. **Habitat:** forest. **Distribution:** Cameroon, Gabon, Congo (Kinshasa), Angola and Uganda. **Chorology:** Guineo-Congolian. **Star:** green. **Specimens:** - *No locality:* Box H. E. 3563

Rinorea breviracemosa Chipp

F.W.T.A. 1: 103 (1954).

Habit: shrub to 3 m. **Guild:** shade-bearer. **Habitat:** swamp forest. **Distribution:** Sierra Leone to Cameroon. **Chorology:** Guineo-Congolian. **Star:** green. **Alt. range:** 1-200m (3), 201-400m (3). **Flowering:** Oct (1), Nov (2). **Fruiting:** Mar (1), Oct (3). **Specimens:** - *Onge:* Akogo M. 101; Harris D. J. 3680; Ndam N. 787; Tchouto P. 868; Watts J. 1026; Wheatley J. I. 712

Rinorea cf. crassifolia (Baker f.) De Wild.

Specimens: - *Etinde:* Tchouto P. 92

Rinorea dentata P.Beauv.

F.W.T.A. 1: 104 (1954); Fl. Afr. Cent. Violaceae: 35 (1969); Kew Bull. 36: 116 (1981); Keay R.W.J., Trees of Nigeria: 46 (1989); Hawthorne W., F.G.F.T. Ghana: 102 (1990).

Habit: tree to 10 m. **Guild:** shade-bearer. **Habitat:** forest. **Distribution:** Liberia to Bioko, Angola, Uganda and Tanzania. **Chorology:** Guineo-Congolian. **Star:** green. **Alt. range:** 1-200m (11), 201-400m (1), 401-600m (1), 601-800m (1). **Flowering:** Feb (2), Mar (1), Oct (3), Nov (1), Dec (1). **Fruiting:** Mar (1), Sep (1), Oct (2), Nov (1), Dec (1). **Specimens:** - *Etinde:* Groves M. 291; Maitland T. D. 519; Mbani J. M. 13, 27; Nkeng P. 97; Sunderland T. C. H. 1005; Wheatley J. I. 499. *Mabeta-Moliwe:* Cable S. 1443; Mildbraed G. W. J. 10733. *Mokoko River F.R.:* Mbani J. M. 384; Tchouto P. 1105, 1176. *Onge:* Cheek M. 5096; Tchouto P. 768, 1025

Rinorea cf. dentata P.Beauv.

Specimens: - *Etinde:* Mbani J. M. 80

Rinorea exappendiculata Engl. ex Brandt
Engl. Bot. Jahrb. 49: 127 (1913).

Habit: tree or shrub to 4 m. **Guild:** shade-bearer. **Habitat:** forest.
Distribution: Cameroon. **Chorology:** Lower Guinea. **Star:** gold.
Alt. range: 1-200m (1). **Flowering:** Oct (1). **Fruiting:** Oct (1).
Specimens: - *Onge:* Tchouto P. 697

Rinorea cf. ilicifolia (Welw. ex Oliv.) Kuntze

Specimens: - *Mabeta-Moliwe:* Plot Voucher W 517

Rinorea kamerunensis Engl.
F.W.T.A. 1: 103 (1954).

Habit: shrub to 1.5 m. **Guild:** shade-bearer. **Habitat:** forest.
Distribution: Cameroon. **Chorology:** Lower Guinea. **Star:** gold.
Alt. range: 1-200m (6), 201-400m (2). **Flowering:** Mar (4), May
(1), Jun (1). **Fruiting:** Mar (1), May (1), Jun (1). **Specimens:** -
Etinde: Nkeng P. 49. *Mabeta-Moliwe:* Cable S. 1456; Plot
Voucher W 11, W 34, W 249, W 745, W 1167; Preuss P. R. 1221.
Mokoko River F.R.: Ekema N. 1138; Pouakouyou D. 52; Tchouto
P. 595; Watts J. 1115, 1138. *Onge:* Wheatley J. I. 820. *Southern
Bakundu F.R.:* Brenan J. P. M. 9289

Rinorea cf. kamerunensis Engl.

Specimens: - *Mabeta-Moliwe:* Plot Voucher W 76, W 255

Rinorea ledermannii Engl.
F.W.T.A. 1: 103 (1954).

Habit: shrub to 1.5 m. **Guild:** shade-bearer. **Habitat:** forest.
Distribution: Cameroon. **Chorology:** Lower Guinea. **Star:** gold.
Specimens: - *Southern Bakundu F.R.:* Brenan J. P. M. 9290

Rinorea oblongifolia (C.H.Wright) Marqua
F.W.T.A. 1: 104 (1954); Fl. Afr. Cent. Violaceae: 27 (1969); Keay
R.W.J., Trees of Nigeria: 48 (1989); Hawthorne W., F.G.F.T.
Ghana: 132 (1990).

Habit: tree to 16 m. **Guild:** shade-bearer. **Habitat:** forest and
farmbush. **Distribution:** Sierra Leone to Cameroon, Gabon,
Uganda and Sudan. **Chorology:** Guineo-Congolian. **Star:** green.
Alt. range: 1-200m (1), 601-800m (1). **Flowering:** Mar (1).
Fruiting: Mar (1). **Specimens:** - *Etinde:* Tchouto P. 73. *Mabeta-
Moliwe:* Mildbraed G. W. J. 10765

Rinorea preussii Engl.
F.W.T.A. 1: 104 (1954); Fl. Afr. Cent. Violaceae: 39 (1969).

Habit: tree or shrub to 5 m. **Guild:** shade-bearer. **Habitat:** forest
and farmbush. **Distribution:** SE Nigeria, SW Cameroon and
Congo (Kinshasa). **Chorology:** Guineo-Congolian. **Star:** green.
Specimens: - *Etinde:* Preuss P. R. 1154

Rinorea subintegrifolia (P.Beauv.) Kuntze
F.W.T.A. 1: 104 (1954); Fl. Afr. Cent. Violaceae: 30 (1969);
Hawthorne W., F.G.F.T. Ghana: 132 (1990).

Habit: shrub to 2 m. **Guild:** shade-bearer. **Habitat:** forest.
Distribution: Guinea to Cameroon, Gabon, Congo (Kinshasa) and
Tanzania. **Chorology:** Guineo-Congolian. **Star:** green. **Alt. range:**
1-200m (9). **Flowering:** Jan (1), Mar (1), Jul (1), Nov (5).

Fruiting: Mar (3), Nov (1). **Specimens:** - *Etinde:* Maitland T. D.
FHI 7850. *Mabeta-Moliwe:* Cable S. 684, 1417; Cheek M. 3589;
Groves M. 220; Khayota B. 467. *Onge:* Harris D. J. 3671, 3679;
Watts J. 925, 942, 1015. *Southern Bakundu F.R.:* Keay R. W. J.
FHI 28565

Rinorea cf. subintegrifolia (P.Beauv.) Kuntze

Specimens: - *Mabeta-Moliwe:* Plot Voucher W 33, W 104, W 423,
W 442, W 895

Rinorea thomasii Achoundong in ed.

Habit: shrub to 1.5 m. **Guild:** shade-bearer. **Habitat:** forest.
Distribution: Cameroon. **Chorology:** Lower Guinea. **Star:** gold.
Alt. range: 1-200m (1), 201-400m (2). **Flowering:** Oct (1), Nov
(1). **Fruiting:** Mar (1). **Specimens:** - *Bambuko F.R.:* Tchouto P.
554. *Onge:* Harris D. J. 3690; Watts J. 799

Rinorea welwitschii (Oliv.) Kuntze
Fl. Zamb. 1: 251 (1960); Kew Bull. 16: 418 (1963); Fl. Afr. Cent.
Violaceae: 59 (1969); Kew Bull. 36: 121 (1981); Keay R.W.J.,
Trees of Nigeria: 46 (1989); Hawthorne W., F.G.F.T. Ghana: 132
(1990).
    Rinorea elliotii Engl., F.W.T.A. 1: 103 (1954).
    Rinorea longicuspis Engl., F.W.T.A. 1: 103 (1954).

Habit: tree to 10 m. **Guild:** shade-bearer. **Habitat:** forest, by
streams. **Distribution:** tropical Africa. **Star:** excluded. **Alt. range:**
1-200m (2), 201-400m (1), 401-600m (2). **Flowering:** Mar (2),
Nov (1). **Fruiting:** Mar (2), Apr (2). **Specimens:** - *Bambuko F.R.:*
Watts J. 587. *Etinde:* Tchouto P. 115; Tekwe C. 39. *Onge:* Harris
D. J. 3859; Wheatley J. I. 799

Rinorea sp.

Specimens: - *Etinde:* Tchouto P. 87; Tekwe C. 38. *Onge:*
Wheatley J. I. 816

Viola abyssinica Staud. ex Oliv.
F.W.T.A. 1: 107 (1954); Fl. Afr. Cent. Violaceae: 69 (1969).

Habit: herb. **Habitat:** forest-grassland transition. **Distribution:**
tropical and subtropical Africa. **Chorology:** Afromontane. **Star:**
green. **Alt. range:** 1000-1200m (1), 1601-1800m (1), 1801-2000m
(1), 2001-2200m (2). **Flowering:** Mar (1), Nov (1). **Specimens:** -
Eastern slopes: Cheek M. 5365; Dahl A. 618. *Etinde:* Thomas D.
W. 9173. *No locality:* Brenan J. P. M. 9358; Maitland T. D. 1013

VISCACEAE

R.M. Polhill (K)

Viscum congolense De Wild.
F.W.T.A. 1: 665 (1958); Fl. Cameroun 23: 68 (1982).

Habit: parasite. **Habitat:** forest. **Distribution:** Ivory Coast,
Ghana, Nigeria, Cameroon, Gabon, Congo (Kinshasa) and Angola
(Cabinda). **Chorology:** Guineo-Congolian (montane). **Star:** green.
Alt. range: 1-200m (1), 1601-1800m (1). **Flowering:** Sep (1).
Fruiting: May (1). **Specimens:** - *Etinde:* Tekwe C. 242; Winkler
H. J. P. 23a. *Mokoko River F.R.:* Watts J. 1109

Viscum decurrens (Engl.) Baker & Sprague
F.W.T.A. 1: 665 (1958); Fl. Cameroun 23: 70 (1982).

Habit: parasite. Habitat: forest. Distribution: SE Nigeria, Cameroon, Gabon, Congo (Kinshasa) and Angola. Chorology: Guineo-Congolian. Star: green. Specimens: - *Etinde:* Winkler H. J. P. 550

VITACEAE

Ampelocissus abyssinica (A.Rich.) Planch.
Fl. Cameroun 13: 14 (1972); F.T.E.A. Vitaceae: 5 (1993).
Ampelocissus cavicaulis (Baker) Planch., F.W.T.A. 1: 682 (1958).

Habit: woody climber. Habitat: forest. Distribution: Cameroon, Gabon, Sudan, Ethiopia, Uganda and Tanzania. Chorology: Guineo-Congolian (montane). Star: green. Alt. range: 1-200m (4). Flowering: Mar (1). Fruiting: Apr (2), Jun (1). Specimens: - *Etinde:* Maitland T. D. 583; Tchouto P. 82; Tekwe C. 55, 89. *Mabeta-Moliwe:* Tchouto P. 149

Ampelocissus sp. aff. africana (Lour.) Merr.

Specimens: - *Mabeta-Moliwe:* Nguembock F. 74; Sunderland T. C. H. 1458; Watts J. 255; Wheatley J. I. 60

Ampelocissus bombycina (Baker) Planch.
F.W.T.A. 1: 682 (1958); Fl. Cameroun 13: 8 (1972).

Habit: woody climber. Habitat: forest. Distribution: Guinea to Cameroon, Congo (Kinshasa), CAR, Sudan and Ethiopia. Chorology: Afromontane. Star: green. Specimens: - *Eastern slopes:* Dalziel J. M. 8231

Cayratia debilis (Baker) Suesseng.
Fl. Cameroun 13: 22 (1972); F.T.E.A. Vitaceae: 139 (1993).
Cissus debilis (Baker) Planch., F.W.T.A. 1: 679 (1958).

Habit: herbaceous climber. Habitat: forest. Distribution: Liberia to Bioko, Cameroon, Gabon, Congo (Kinshasa), Sudan and Uganda. Chorology: Guineo-Congolian. Star: green. Specimens: - *Etinde:* Preuss P. R. 1128

Cayratia gracilis (Guill. & Perr.) Suesseng.
Fl. Cameroun 13: 20 (1972); F.T.E.A. Vitaceae: 138 (1993).
Cissus gracilis Guill. & Perr., F.W.T.A. 1: 679 (1958).

Habit: herbaceous climber. Habitat: forest. Distribution: tropical and subtropical Africa, also Yemen. Star: excluded. Specimens: - *Eastern slopes:* Deistel H. 178

Cissus amoena Gilg & Brandt
F.W.T.A. 1: 677 (1958); Fl. Gabon 14: 87 (1968); Fl. Cameroun 13: 108 (1972).

Habit: woody climber. Habitat: forest. Distribution: Cameroon and Gabon. Chorology: Lower Guinea. Star: blue. Alt. range: 1-200m (1).Fruiting: Jun (1). Specimens: - *Mabeta-Moliwe:* Wheatley J. I. 345. *Mokoko River F.R.:* Dusen P. K. H. 263

Cissus barbeyana De Wild. & T.Durand
F.W.T.A. 1: 678 (1958); Fl. Gabon 14: 96 (1968); Fl. Cameroun 13: 118 (1972).

Habit: woody climber. Habitat: forest. Distribution: Nigeria, Cameroon, Gabon and Congo (Kinshasa). Chorology: Guineo-Congolian. Star: green. Alt. range: 1-200m (6). Flowering: Apr (1), May (2), Jun (2). Fruiting: May (3), Jun (1). Specimens: - *Eastern slopes:* Maitland T. D. 33. *Mokoko River F.R.:* Acworth J. M. 86; Ekema N. 924, 943, 1177a; Pouakouyou D. 39; Watts J. 1066. *Southern Bakundu F.R.:* Onochie C. F. A. FHI 30859

Cissus barteri (Baker) Planch.
F.W.T.A. 1: 678 (1958); Fl. Gabon 14: 102 (1968); Fl. Cameroun 13: 126 (1972).

Habit: herbaceous climber. Habitat: forest. Distribution: Nigeria, Cameroon, Bioko, Gabon, Congo (Kinshasa) and Angola (Cabinda). Chorology: Guineo-Congolian. Star: green. Alt. range: 1-200m (4), 201-400m (3), 401-600m (1). Flowering: Jan (1), Mar (1), Oct (2), Nov (2). Fruiting: Jan (1), Apr (1), Oct (1), Nov (2). Specimens: - *Etinde:* Kalbreyer W. 5; Mbani J. M. 43. *Mabeta-Moliwe:* Nguembock F. 38; Sunderland T. C. H. 1057. *Mokoko River F.R.:* Acworth J. M. 214; Ndam N. 1120. *Onge:* Cheek M. 4963; Harris D. J. 3774; Ndam N. 636; Watts J. 1027

Cissus cf. barteri (Baker) Planch.

Specimens: - *Etinde:* Wheatley J. I. 512

Cissus diffusiflora (Baker) Planch.
F.W.T.A. 1: 678 (1958); Fl. Gabon 14: 101 (1968); Fl. Cameroun 13: 122 (1972).

Habit: herbaceous climber. Guild: pioneer. Habitat: forest. Distribution: Guinea to Bioko, Cameroon, Gabon, Congo (Kinshasa), Angola, Sudan and Uganda. Chorology: Guineo-Congolian. Star: green. Alt. range: 1-200m (1), 201-400m (3). Flowering: Oct (1). Fruiting: Apr (1), Sep (1), Oct (3). Specimens: - *Etinde:* Tchouto P. 806. *Mokoko River F.R.:* Ndam N. 1094. *Onge:* Akogo M. 91; Ndam N. 715; Tchouto P. 851

Cissus dinklagei Gilg & Brandt
Fl. Cameroun 13: 114 (1972).

Habit: woody climber. Habitat: forest. Distribution: Cameroon, Gabon, Congo (Kinshasa) and Angola. Chorology: Guineo-Congolian. Star: green. Alt. range: 1-200m (2).Fruiting: Apr (1), Jun (1). Specimens: - *Mokoko River F.R.:* Acworth J. M. 98; Ekema N. 1156

Cissus glaucophylla Hook.f.
F.W.T.A. 1: 678 (1958); Fl. Gabon 14: 89 (1968).

Habit: woody climber. Guild: pioneer. Habitat: forest. Distribution: Sierra Leone to Bioko, Cameroon, Gabon, Congo (Kinshasa) and Angola. Chorology: Guineo-Congolian. Star: green. Alt. range: 1-200m (1). Flowering: Apr (1). Fruiting: Apr (1). Specimens: - *Etinde:* Maitland T. D. 402. *Mabeta-Moliwe:* Wheatley J. I. 129

Cissus leonardii Dewit
Fl. Gabon 14: 95 (1968); Fl. Cameroun 13: 116 (1972).

Habit: herbaceous climber. **Habitat:** forest. **Distribution:**
Cameroon, Gabon, Congo (Brazzaville), Congo (Kinshasa) and
CAR. **Chorology:** Guineo-Congolian. **Star:** green. **Alt. range:** 1-
200m (7), 201-400m (1). **Flowering:** Mar (2), May (2), Jul (1),
Oct (1). **Fruiting:** Mar (1), May (1), Jul (1), Oct (1), Nov (1).
Specimens: - *Mabeta-Moliwe:* Cheek M. 3535; Plot Voucher W
501; Watts J. 284; Wheatley J. I. 317. *Onge:* Harris D. J. 3773;
Ndam N. 713; Tchouto P. 508; Watts J. 879; Wheatley J. I. 733

Cissus oreophila Gilg & Brandt
F.W.T.A. 1: 677 (1958); Fl. Gabon 14: 89 (1968); Fl. Cameroun
13: 110 (1972).

Habit: herbaceous climber. **Guild:** pioneer. **Habitat:** forest.
Distribution: Liberia to Cameroon and Gabon. **Chorology:**
Guineo-Congolian. **Star:** green. **Alt. range:** 600-800m
(1).**Fruiting:** Mar (1). **Specimens:** - *Eastern slopes:* Tchouto P.
50. *No locality:* Mann G. 1279

Cissus petiolata Hook.f.
F.W.T.A. 1: 678 (1958); Fl. Gabon 14: 92 (1968); Fl. Cameroun
13: 96 (1972); F.T.E.A. Vitaceae: 16 (1993).

Habit: woody climber. **Guild:** light demander. **Habitat:** stream
banks in forest. **Distribution:** tropical Africa. **Star:** excluded. **Alt.
range:** 1-200m (1). **Flowering:** Mar (1). **Specimens:** - *Etinde:*
Ejiofor M. C. FHI 29332; Maitland T. D. 552. *Onge:* Wheatley J. I.
838

Cissus polyantha Gilg & Brandt
F.W.T.A. 1: 678 (1958); F.T.E.A. Vitaceae: 30 (1993).

Habit: herbaceous climber. **Guild:** pioneer. **Habitat:** forest.
Distribution: Sierra Leone to Cameroon and CAR. **Chorology:**
Guineo-Congolian. **Star:** green. **Alt. range:** 1-200m (1).**Fruiting:**
Apr (1). **Specimens:** - *Mokoko River F.R.:* Acworth J. M. 90

Cissus producta Afzel.
F.W.T.A. 1: 678 (1958); Fl. Gabon 14: 103 (1968); Fl. Cameroun
13: 128 (1972); F.T.E.A. Vitaceae: 21 (1993).

Habit: herbaceous climber. **Guild:** light demander. **Habitat:**
forest. **Distribution:** Senegal to Cameroon, Gabon, Congo
(Kinshasa), Angola, Zambia, Zimbabwe, Uganda and Tanzania.
Star: excluded. **Alt. range:** 1-200m (2), 201-400m (1), 801-
1000m (1). **Flowering:** Mar (1), Oct (1). **Fruiting:** Mar (1), Sep
(1), Oct (1), Nov (1). **Specimens:** - *Eastern slopes:* Tchouto P. 96.
Etinde: Wheatley J. I. 504. *Mokoko River F.R.:* Ndam N. 1130.
Onge: Akogo M. 146; Harris D. J. 3852

Cissus cf. producta Afzel.

Specimens: - *Mokoko River F.R.:* Ekema N. 885; Ndam N. 1184

Cissus ruginosicarpa Desc.
Fl. Gabon 14: 86 (1968); Fl. Cameroun 13: 102 (1972).

Habit: herbaceous climber. **Habitat:** forest. **Distribution:**
Cameroon, Gabon and Congo (Kinshasa). **Chorology:** Guineo-
Congolian. **Star:** green. **Alt. range:** 1-200m (1). **Flowering:** Oct
(1). **Specimens:** - *Onge:* Tchouto P. 992

Cissus sp.

Specimens: - *Onge:* Thomas D. W. 9833

Cyphostemma adenopodum (Sprague) Desc.
Fl. Cameroun 13: 46 (1972); F.T.E.A. Vitaceae: 131 (1993).
Cissus adenopoda Sprague, F.W.T.A. 1: 679 (1958).

Habit: herbaceous climber. **Habitat:** forest. **Distribution:** Ivory
Coast to Bioko, Cameroon, Gabon, Congo (Kinshasa), Angola
(Cabinda) and Uganda. **Chorology:** Guineo-Congolian. **Star:**
green. **Specimens:** - *Etinde:* Winkler H. J. P. 127

Cyphostemma mannii (Baker) Desc.
Fl. Cameroun 13: 68 (1972).
Cissus mannii (Baker) Planch., F.W.T.A. 1: 680 (1958).

Habit: herbaceous climber. **Habitat:** forest-grassland transition.
Distribution: Bioko and Cameroon. **Chorology:** Lower Guinea
(montane). **Star:** gold. **Alt. range:** 1400-1600m (1), 1801-2000m
(1), 2001-2200m (2). **Flowering:** Sep (2), Oct (1). **Specimens:** -
Eastern slopes: Thomas D. W. 9283. *Etinde:* Tekwe C. 278;
Thomas D. W. 9213. *No locality:* Dunlap 94

VOCHYSIACEAE

Erismadelphus exsul Mildbr. var. platyphyllus
Keay & Stafleu
F.W.T.A. 1: 114 (1954); Keay R.W.J., Trees of Nigeria: 48 (1989).

Habit: tree to 35 m. **Habitat:** forest. **Distribution:** SE Nigeria,
Cameroon and Gabon. **Chorology:** Lower Guinea. **Star:** blue. **Alt.
range:** 200-400m (1). **Specimens:** - *S. Bakundu F.R.:* Obiorah FHI
29508. *Mokoko River F.R.:* Thomas D. W. 10098

MONOCOTYLEDONS

AMARYLLIDACEAE
I. Nordal (O)

Crinum natans Baker
F.W.T.A. 3: 134 (1968); Fl. Gabon 28: 34 (1986); Fl. Cameroun
30: 15 (1987).

Habit: herb. **Habitat:** aquatic. **Distribution:** Guinea to Bioko,
Cameroon, Gabon and Congo (Kinshasa). **Chorology:** Guineo-
Congolian. **Star:** green. **Specimens:** - *Eastern slopes:* Schlechter
F. R. R. 12851

Crinum purpurascens Herbert
Fl. Cameroun 30: 14 (1987)

Specimens: - *Bambuko F.R.:* Letouzey 15045

Scadoxus cinnabarinus (Decne.) Friis &
Nordal
Fl. Gabon 28: 25 (1986); Fl. Cameroun 30: 5 (1987).
Haemanthus cinnabarinus Decne., F.W.T.A. 3: 132 (1968).

Habit: herb. **Guild**: shade-bearer. **Habitat**: forest. **Distribution**: Sierra Leone to Bioko, Cameroon, Angola and Uganda. **Chorology**: Guineo-Congolian. **Star**: green. **Alt. range**: 1-200m (7), 201-400m (4), 601-800m (1), 801-1000m (2), 1201-1400m (2). **Flowering**: Jan (1), Feb (1), Mar (1), Oct (2), Nov (2), Dec (2). **Fruiting**: Mar (1), Apr (1), Aug (1), Sep (1), Oct (3), Nov (2), Dec (3). **Specimens**: - *Eastern slopes*: Kwangue A. 121. *Etinde*: Cable S. 135, 469; Cheek M. 5911; Faucher P. 18; Lighava M. 20; Mbani J. M. 21, 92; Sunderland T. C. H. 1052; Tekwe C. 151. *Mabeta-Moliwe*: Mann G. 779; Plot Voucher W 1214. *Mokoko River F.R.*: Akogo 145; Tchouto P. 521b. *Onge*: Akogo M. 126, 127; Ndam N. 714; Watts J. 915, 1002

Scadoxus pseudocaulus (Bjornst. & Friis) Friis & Nordal

Fl. Cameroun 30: 6 (1987).
Haemanthus sp. A sensu Hepper, F.W.T.A. 3: 132 (1968).

Habit: herb. **Distribution**: Nigeria, Cameroon and Gabon. **Chorology**: Lower Guinea. **Star**: blue. **Specimens**: - *Eastern slopes*: Dundas FHI 20387; Gregory 91; Hambler D. J. 178; Swarbrick 2278

ANTHERICACEAE

I. Nordal (O)

Chlorophytum cauliferum Poelln.

Portug. Acta. Biol. (ser.B) 1: 222 (1945).

Habit: herb. **Habitat**: forest. **Distribution**: Cameroon. **Chorology**: Lower Guinea. **Star**: gold. **Alt. range**: !-200m (2), 801-1000m (1). **Flowering**: Mar (4). **Fruiting**: Mar (1). **Specimens**: - *Etinde*: Cable S. 1634. *Mabeta-Moliwe*: Cable S. 1453, 1489; Khayota B. 481

Chlorophytum comosum (Thunb.) Jacq.

Journ. Soc. Imp. Cent. Hort. 8: 345 (1862); Bothalia 7: 698 (1962); Fl. Rwanda 4: 43 (1988); F.T.E.A. Anthericaceae: 55 (1997).
Chlorophytum sparsiflorum Baker F.W.T.A. 3: 100 (1968).
Chlorophytum deistelianum Engl. & K.Krause F.W.T.A. 3: 102 (1968).

Habit: herb. **Habitat**: forest and farmbush. **Distribution**: tropical and southern Africa. **Chorology**: Afromontane. **Star**: green. **Alt. range**: 800-1000m (1), 1001-1200m (1), 1601-1800m (2). **Flowering**: Feb (2), Mar (2). **Fruiting**: Feb (1). **Specimens**: - *Eastern slopes*: Dahl A. 617, 623; Deistel H. 649; Etuge M. 1197; Groves M. 112; Tekwe C. 297; Wheatley J. I. 587. *Mabeta-Moliwe*: Plot Voucher W 743, W 760; Watts J. 279, 331. *No locality*: Dalziel J. M. 8345
Note: This material may in future be recognised as a subspecies of *C. comosum*

Chlorophytum cf. comusum

Specimens: - *Etinde*: Cable S. 167, 488; Cheek M. 5413, 5431; Tchouto P. 445, 473; Watts J. 538; Wheatley J. I. 495, 536. *Mabeta-Moliwe*: Baker W. J. 347; Cable S. 590. *Onge*: Harris D. J. 3767; Ndam N. 756; Tchouto P. 741; Thomas D. W. 9806; Watts J. 902, 964; Wheatley J. I. 695

Chlorophytum orchidastrum Lindl.

F.W.T.A. 3: 100 (1968).
Habit: herb. **Habitat**: forest. **Distribution**: Guinea to Cameroon, Congo (Kinshasa) and Zambia. **Chorology**: Guineo-Congolian. **Star**: green. **Alt. range**: 1-200m (1), 201-400m (3), 601-800m (2). **Flowering**: Mar (1), Nov (3), Dec (1). **Fruiting**: Mar (1), Nov (1). **Specimens**: - *Etinde*: Cable S. 165; Cheek M. 5432, 5437, 5857; Khayota B. 500. *No locality*: Mann G. 2132

Chlorophytum petrophilum K.Krause

Engl. Bot. Jahrb. 51: 441 (1914).

Habit: herb. **Habitat**: forest. **Distribution**: Cameroon. **Chorology**: Lower Guinea. **Star**: black. **Alt. range**: 200-400m (1). **Flowering**: Nov (1). **Fruiting**: Nov (1). **Specimens**: - *Etinde*: Cheek M. 5451

ARACEAE

P. Boyce (K)

Amorphophallus preussii (Engl.) N.E.Br.

F.W.T.A. 3: 118 (1968); Fl. Cameroun 31: 37 (1988).

Habit: herb. **Habitat**: forest, sometimes on rocks. **Distribution**: Mount Cameroon. **Chorology**: Endemic (montane). **Star**: black. **Alt. range**: 800-1000m (1). **Specimens**: - *No locality*: Preuss P. R. 588

Amorphophallus zenkeri (Engl.) N.E.Br.

F.W.T.A. 3: 119 (1968); Fl. Cameroun 31: 29 (1988).

Habit: herb. **Habitat**: forest. **Distribution**: SE Nigeria and Cameroon. **Chorology**: Lower Guinea. **Star**: black. **Alt. range**: 800-1000m (1), 1001-1200m (1). **Flowering**: Dec (2). **Specimens**: - *Etinde*: Cable S. 470, 500. *Mabeta-Moliwe*: Cheek M. s.n.

Amorphophallus sp.

Specimens: - *Etinde*: Mbani J. M. 93. *Mabeta-Moliwe*: Baker W. J. 340

Anchomanes difformis (Blume) Engl.

F.W.T.A. 3: 121 (1968); Fl. Cameroun 31: 23 (1988).
Anchomanes difformis (Blume) Engl. var. *pallidus* (Hook.) Hepper, F.W.T.A. 3: 122 (1968).
Anchomanes welwitschii Rendle, F.W.T.A. 3: 121 (1968).

Habit: herb. **Habitat**: forest and forest-grassland transition. **Distribution**: Sierra Leone to Congo (Kinshasa), Angola and Sudan. **Chorology**: Guineo-Congolian (montane). **Star**: green. **Alt. range**: 1-200m (1). **Flowering**: Apr (1). **Specimens**: - *Mabeta-Moliwe*: Wheatley J. I. 168

Anchomanes sp.

Specimens: - *Eastern slopes*: Cable S. 1331. *Etinde*: Lighava M. 46; Thomas D. W. 9166. *Mabeta-Moliwe*: Khayota B. 471. *Onge*: Ndam N. 728

Anubias barteri Schott

F.W.T.A. 3: 120 (1968); Meded. Land. Wag. 79(14): 8 (1979); Fl. Cameroun 31: 16 (1988).

Anubias lanceolata N.E.Br., F.W.T.A. 3: 120 (1968).
Anubias minima A.Chev., F.W.T.A. 3: 120 (1968).

Habit: herb. **Habitat:** forest, on rocks and by streams. **Distribution:** Guinea to Bioko, Cameroon, Gabon and Congo (Brazzaville). **Chorology:** Guineo-Congolian. **Star:** green. **Alt. range:** 1-200m (12), 201-400m (4), 401-600m (1), 801-1000m (3). **Flowering:** Mar (2), May (1), Aug (1), Sep (1), Oct (6), Nov (3), Dec (2). **Fruiting:** Mar (1), Sep (1), Oct (1). **Specimens:** - *Etinde:* Cable S. 140, 337, 1638; Faucher P. 49; Maitland T. D. 1306; Wheatley J. I. 518. *Mabeta-Moliwe:* Baker W. J. 292; Cable S. 591; Ndam N. 512; Sunderland T. C. H. 1135; Wheatley J. I. 73. *Mokoko River F.R.:* Acworth J. M. 202; Ndam N. 1143. *Onge:* Akogo M. 77; Cheek M. 5014; Tchouto P. 742, 799, 955; Watts J. 746, 910, 922; Wheatley J. I. 697

Cercestis camerunensis (Ntepe-Nyame) Bogner

Aroideana 8(3): 73 (1986); Fl. Cameroun 31: 60 (1988).
Rhektophyllum camerunense Ntepe-Nyame, Fl. Cameroun 31: 60 (1988).

Habit: hemi-epiphyte. **Habitat:** forest. **Distribution:** Nigeria, Cameroon and Gabon. **Chorology:** Lower Guinea. **Star:** blue. **Alt. range:** 1-200m (2), 201-400m (4). **Flowering:** Oct (2). **Fruiting:** Dec (1). **Specimens:** - *Etinde:* Cheek M. 5876, 5909. *Mabeta-Moliwe:* Plot Voucher W 696. *Onge:* Ndam N. 686, 687; Tchouto P. 682; Thomas D. W. 9889

Cercestis congensis Engl.

Fl. Cameroun 31: 65 (1988).

Habit: herbaceous climber. **Habitat:** forest. **Distribution:** Guinea, Cameroon, Gabon, Congo (Brazzaville), Congo (Kinshasa), Angola and CAR. **Chorology:** Guineo-Congolian. **Star:** green. **Alt. range:** 1-200m (1).**Fruiting:** Feb (1). **Specimens:** - *Etinde:* Mbani J. M. 17

Cercestis dinklagei Engl.

Fl. Cameroun 31: 66 (1988).
Cercestis stigmaticus N.E.Br., F.W.T.A. 3: 126 (1968).

Habit: hemi-epiphyte. **Habitat:** forest. **Distribution:** Guinea to Congo (Brazzaville). **Chorology:** Guineo-Congolian. **Star:** green. **Alt. range:** 1-200m (6), 201-400m (2). **Flowering:** Aug (1), Oct (1). **Fruiting:** Aug (1), Oct (3), Nov (3). **Specimens:** - *Mabeta-Moliwe:* Baker W. J. 260; Khayota B. 456; Plot Voucher W 14, W 1044. *Onge:* Cheek M. 5186; Tchouto P. 848; Watts J. 761, 897, 981, 1038

Cercestis ivorensis A.Chev.

F.W.T.A. 3: 127 (1968); Fl. Cameroun 31: 68 (1988).
Cercestis lanceolatus Engl., F.W.T.A. 3: 127 (1968).

Habit: hemi-epiphyte. **Habitat:** forest. **Distribution:** Sierra Leone to Cameroon and Gabon. **Chorology:** Guineo-Congolian. **Star:** green. **Alt. range:** 200-400m (1). **Specimens:** - *Eastern slopes:* Reder H. 1742. *Onge:* Ndam N. 680

Cercestis kamerunianus (Engl.) N.E.Br.

F.W.T.A. 3: 127 (1968); Fl. Cameroun 31: 69 (1988).

Habit: hemi-epiphyte. **Habitat:** forest. **Distribution:** Nigeria, Cameroon and Gabon. **Chorology:** Lower Guinea. **Star:** blue. **Alt.**

range: 1-200m (2), 201-400m (4). **Flowering:** Jul (1), Oct (1). **Fruiting:** Jul (1), Oct (2), Nov (1), Dec (1). **Specimens:** - *Etinde:* Cheek M. 5845. *Mabeta-Moliwe:* von Rege I. 22. *Onge:* Akogo M. 134; Ndam N. 644; Tchouto P. 888, 1028. *Southern Bakundu F.R.:* Brenan J. P. M. 9493

Cercestis mirabilis (N.E.Br.) Bogner

Aroideana 8(3): 73 (1986); Fl. Cameroun 31: 59 (1988).
Rhektophyllum mirabile N.E.Br., F.W.T.A. 3: 122 (1968).

Habit: herbaceous climber. **Habitat:** forest. **Distribution:** Benin to Bioko, Cameroon, Congo (Kinshasa), Angola and Uganda. **Chorology:** Guineo-Congolian. **Star:** green. **Alt. range:** 1-200m (1), 201-400m (1). **Specimens:** - *Mabeta-Moliwe:* Plot Voucher W 103, W 153, W 234, W 331, W 362, W 402, W 434, W 557, W 687, W 715, W 766, W 802, W 1010. *Onge:* Cheek M. 5197; Ndam N. 685

Culcasia annetii Ntepe-Nyame

Fl. Cameroun 31: 91 (1988).

Habit: herbaceous climber. **Habitat:** forest. **Distribution:** Guinea to Cameroon. **Chorology:** Guineo-Congolian. **Star:** green. **Alt. range:** 400-600m (2).**Fruiting:** Dec (2). **Specimens:** - *Etinde:* Cable S. 398, 513. *Mabeta-Moliwe:* Plot Voucher W 970

Culcasia barombensis N.E.Br.

F.W.T.A. 3: 124 (1968); Fl. Cameroun 31: 85 (1988).
Culcasia angolensis Welw. ex Schott, F.W.T.A. 3: 124 (1968).

Habit: herbaceous climber. **Habitat:** forest. **Distribution:** Guinea to Bioko, Cameroon, Gabon, Congo (Brazzaville), Congo (Kinshasa) and Angola. **Chorology:** Guineo-Congolian. **Star:** green. **Alt. range:** 1-200m (1), 401-600m (1), 601-800m (1), 1201-1400m (1).**Fruiting:** Apr (3), Oct (1). **Specimens:** - *Eastern slopes:* Tchouto P. 143. *Etinde:* Cheek M. 3799; Tchouto P. 113. *Mabeta-Moliwe:* Jaff B. 86; Plot Voucher W 200, W 224

Culcasia dinklagei Engl.

Fl. Cameroun 31: 80 (1988).

Habit: herbaceous climber. **Habitat:** forest. **Distribution:** Guinea to Cameroon, Gabon and Congo (Brazzaville). **Chorology:** Guineo-Congolian. **Star:** green. **Alt. range:** 1-200m (12), 201-400m (5). **Flowering:** Feb (1), Mar (2), Oct (1), Nov (1). **Fruiting:** Mar (2), Aug (1), Oct (4), Nov (4). **Specimens:** - *Etinde:* Cheek M. 5397; Mbani J. M. 62; Nkeng P. 33. *Mabeta-Moliwe:* Cable S. 578, 1431; Khayota B. 461, 483; von Rege I. 19, 34. *Mokoko River F.R.:* Mbani J. M. 459. *Onge:* Akogo M. 15, 54; Harris D. J. 3677; Tchouto P. 899, 977, 1037; Watts J. 755, 937; Wheatley J. I. 699

Culcasia ekongoloi Ntepe-Nyame

Fl. Cameroun 31: 87 (1988).

Habit: herbaceous climber. **Habitat:** stream banks in forest. **Distribution:** Nigeria, Cameroon, Gabon, Congo (Kinshasa) and CAR. **Chorology:** Guineo-Congolian. **Star:** green. **Alt. range:** 1-200m (1). **Flowering:** Apr (1). **Fruiting:** Apr (1). **Specimens:** - *Mabeta-Moliwe:* Watts J. 210

Culcasia insulana N.E.Br.

F.W.T.A. 3: 124 (1968); Fl. Cameroun 31: 96 (1988).

Habit: herbaceous climber. **Habitat:** forest. **Distribution:** Bioko, Cameroon, Congo (Kinshasa) and CAR. **Chorology:** Guineo-Congolian (montane). **Star:** green. **Alt. range:** 1200-1400m (1).**Fruiting:** Sep (1). **Specimens:** - *Etinde:* Tekwe C. 177

Culcasia mannii (Hook.f.) Engl.

F.W.T.A. 3: 124 (1968); Fl. Cameroun 31: 82 (1988).

Habit: herb. **Habitat:** forest. **Distribution:** SE Nigeria, Cameroon and Rio Muni. **Chorology:** Lower Guinea. **Star:** blue. **Alt. range:** 1-200m (1), 201-400m (4), 401-600m (3), 601-800m (1). **Flowering:** Mar (1), Oct (1). **Fruiting:** Aug (1), Sep (1), Oct (4), Dec (2). **Specimens:** - *Etinde:* Cable S. 352, 1510; Cheek M. 5883; Tekwe C. 123; Wheatley J. I. 531. *Mabeta-Moliwe:* Mann G. s.n.. *No locality:* Mann G. s.n.. *Onge:* Cheek M. 4985; Ndam N. 678, 772; Watts J. 738

Culcasia obliquifolia Engl.

Fl. Cameroun 31: 95 (1988).

Habit: herbaceous climber. **Habitat:** forest. **Distribution:** Cameroon and Gabon. **Chorology:** Lower Guinea (montane). **Star:** blue. **Alt. range:** 1600-1800m (1).**Fruiting:** Aug (1). **Specimens:** - *Etinde:* Hunt L. V. 44

Culcasia panduriformis Engl. & Krause

Fl. Cameroun 31: 80 (1988).

Habit: herb. **Habitat:** forest and forest-grassland transition. **Distribution:** Cameroon. **Chorology:** Lower Guinea. **Star:** gold. **Alt. range:** 1-200m (1).**Fruiting:** Jul (1). **Specimens:** - *Mabeta-Moliwe:* Hunt L. V. 18

Culcasia parviflora N.E.Br.

F.W.T.A. 3: 124 (1968); Fl. Cameroun 31: 84 (1988).

Habit: herbaceous climber. **Guild:** shade-bearer. **Habitat:** forest. **Distribution:** Guinea to Cameroon, Gabon, Congo (Brazzaville) and Congo (Kinshasa). **Chorology:** Guineo-Congolian. **Star:** green. **Alt. range:** 1-200m (14), 201-400m (11), 401-600m (2), 601-800m (1), 1001-1200m (1). **Flowering:** Mar (1), Apr (1), May (2), Aug (1), Sep (1), Oct (2), Nov (1). **Fruiting:** Mar (2), Apr (2), May (2), Aug (1), Sep (1), Oct (10), Nov (5), Dec (4). **Specimens:** - *Eastern slopes:* Kwangue A. 49. *Etinde:* Banks H. 40; Cable S. 202, 408; Cheek M. 5480, 5882, 5903; Kwangue A. 143; Tekwe C. 82; Watts J. 479, 497, 726; Wheatley J. I. 494. *Mabeta-Moliwe:* Baker W. J. 210; Cheek M. 5703; Khayota B. 470; Plot Voucher W 375, W 406, W 560, W 633, W 1025; Sidwell K. 108; Sunderland T. C. H. 1265; Wheatley J. I. 273; von Rege I. 15, 100. *Onge:* Akogo M. 25; Cheek M. 5057, 5090, 5161; Thomas D. W. 9774; Watts J. 747, 808, 1022, 1023; Wheatley J. I. 819. *Southern Bakundu F.R.:* Brenan J. P. M. 9287

Culcasia cf. parviflora N.E.Br.

Specimens: - *Etinde:* Mbani J. M. 163. *Onge:* Thomas D. W. 9762

Culcasia sanagensis Ntepe-Nyame

Fl. Cameroun 31: 87 (1988).

Habit: herbaceous climber. **Habitat:** forest. **Distribution:** Cameroon. **Chorology:** Lower Guinea. **Star:** gold. **Alt. range:** 1-200m (1). **Flowering:** May (1). **Specimens:** - *Mabeta-Moliwe:* Jaff B. 231

Culcasia sapinii De Wild.

F.W.T.A. 3: 126 (1968); Fl. Cameroun 31: 94 (1988).
Culcasia seretii De Wild., F.W.T.A. 3: 126 (1968).

Habit: herbaceous climber. **Habitat:** forest. **Distribution:** Guinea to Congo (Kinshasa). **Chorology:** Guineo-Congolian. **Star:** green. **Alt. range:** 1-200m (2), 801-1000m (1). **Flowering:** Mar (1). **Fruiting:** Mar (2), Dec (1). **Specimens:** - *Etinde:* Faucher P. 15. *Mabeta-Moliwe:* Cable S. 1419, 1421

Culcasia scandens P.Beauv.

F.W.T.A. 3: 124 (1968); Fl. Cameroun 31: 88 (1988).
Culcasia lancifolia N.E.Br., F.W.T.A. 3: 124 (1968).
Culcasia saxatilis A.Chev., F.W.T.A. 3: 124 (1968).

Habit: herbaceous climber. **Guild:** shade-bearer. **Habitat:** forest and farmbush. **Distribution:** Senegal to Congo (Kinshasa). **Chorology:** Guineo-Congolian. **Star:** green. **Alt. range:** 1-200m (2), 201-400m (1), 401-600m (1). **Flowering:** Oct (1). **Fruiting:** Mar (1), Aug (1), Oct (1), Nov (1). **Specimens:** - *Eastern slopes:* Deistel H. 148. *Etinde:* Cable S. 273. *Mabeta-Moliwe:* Wheatley J. I. 41. *Onge:* Watts J. 688, 807

Culcasia cf. scandens P.Beauv.

Specimens: - *Eastern slopes:* Cable S. 1375

Culcasia simiarum Ntepe-Nyame

Fl. Cameroun 31: 92 (1988).

Habit: herbaceous climber. **Habitat:** forest. **Distribution:** Ivory Coast to Cameroon. **Chorology:** Guineo-Congolian. **Star:** green. **Alt. range:** 1-200m (1).**Fruiting:** Mar (2), Jul (1). **Specimens:** - *Mabeta-Moliwe:* Etuge M. 1220; Hurst J. 14; Khayota B. 472

Culcasia striolata Engl.

F.W.T.A. 3: 124 (1968); Fl. Cameroun 31: 79 (1988).

Habit: herb. **Guild:** shade-bearer. **Habitat:** forest. **Distribution:** Guinea to Cameroon and Gabon. **Chorology:** Guineo-Congolian. **Star:** green. **Alt. range:** 1-200m (11), 201-400m (2). **Flowering:** Apr (1), May (2), Jun (1), Aug (2). **Fruiting:** Mar (4), May (1), Jun (1), Aug (2), Oct (1), Dec (1). **Specimens:** - *Etinde:* Preuss P. R. 1161. *Mabeta-Moliwe:* Baker W. J. 209; Cable S. 535, 1418, 1432; Jaff B. 49; Khayota B. 476; Plot Voucher W 8, W 147, W 163, W 223, W 233, W 404, W 413, W 569, W 813, W 877, W 971; Sunderland T. C. H. 1349, 1381; Watts J. 254; Wheatley J. I. 58; von Rege I. 37, 72. *Onge:* Tchouto P. 902

Culcasia cf. striolata Engl.

Specimens: - *Onge:* Wheatley J. I. 698, 746

Culcasia tenuifolia Engl.

F.W.T.A. 3: 126 (1968); Fl. Cameroun 31: 90 (1988).

Habit: hemi-epiphyte. **Habitat:** forest. **Distribution:** Sierra Leone to Cameroon, Gabon and CAR. **Chorology:** Guineo-Congolian. **Star:** green. **Alt. range:** 1-200m (7), 201-400m (1). **Flowering:** Jul (1), Aug (2). **Fruiting:** Jul (2), Aug (3), Oct (1). **Specimens:** - *Etinde:* Kwangue A. 63. *Mabeta-Moliwe:* Baker W. J. 211; Hurst J. 24; Plot Voucher W 280; von Rege I. 21, 41. *Onge:* Ndam N. 677; Tchouto P. 911; Watts J. 698. *Southern Bakundu F.R.:* Brenan J. P. M. 9485

Culcasia sp.

Specimens: - *Onge:* Tchouto P. 980

Culcasia sp. nov. 1 of Ntepe-Nyame
Fl. Cameroun 31: 97 (1988).

Specimens: - *Mabeta-Moliwe:* Plot Voucher W 269

Culcasia sp. nov. 2 of Ntepe-Nyame
Fl. Cameroun 31: 97 (1988).

Specimens: - *Mabeta-Moliwe:* Plot Voucher W 351

Lasimorpha senegalensis Schott
Mayo S. et al., Gen. of Araceae: 138 (1997).
Cyrtosperma senegalense (Schott) Engl., Fl. Cameroun 31: 8 (1988).

Habit: herb. Habitat: swamp forest. Distribution: Senegal to Bioko, Cameroon and Congo (Kinshasa). Chorology: Guineo-Congolian. Star: green. Alt. range: 1-200m (1). Flowering: Mar (1). Specimens: - *Mabeta-Moliwe:* Jaff B. 31

Nephthytis poissonii (Engl.) N.E.Br.
F.W.T.A. 3: 121 (1968); Fl. Cameroun 31: 52 (1988).
Nephthytis gravenreuthii (Engl.) Engl., Fl. Cameroun 31: 50 (1988).
Nephthytis constricta N.E.Br., F.W.T.A. 3: 121 (1968).
Nephthytis poissonii (Engl.) N.E.Br. var. *constricta* (N.E.Br.) Ntepe-Nyame, Fl. Cameroun 31: 56 (1988).

Habit: herb. Habitat: forest. Distribution: Sierra Leone to Bioko, Cameroon and Gabon. Chorology: Guineo-Congolian. Star: green. Alt. range: 1-200m (16), 201-400m (7), 401-600m (1), 601-800m (5), 801-1000m (4). Flowering: Jan (1), Feb (1), Mar (7), May (2), Oct (2), Nov (3), Dec (2). Fruiting: Jan (1), Feb (1), Mar (9), May (1), Aug (1), Sep (1), Oct (4), Nov (3), Dec (6). Specimens: - *Etinde:* Cable S. 184, 328, 400, 463, 1556, 1584; Cheek M. 5410, 5855, 5895; Faucher P. 14; Khayota B. 528, 529, 539; Lighava M. 16; Mbani J. M. 5, 126; Nkeng P. 32; Wheatley J. I. 489. *Mabeta-Moliwe:* Baker W. J. 252; Cable S. 549, 1433; Jaff B. 32, 200; Khayota B. 460, 466; Sunderland T. C. H. 1127. *Onge:* Akogo M. 67; Cheek M. 5105; Harris D. J. 3649, 3840; Ndam N. 619; Watts J. 739, 941; Wheatley J. I. 711. *Southern Bakundu F.R.:* Brenan J. P. M. 9302

Rhaphidophora africana N.E.Br.
F.W.T.A. 3: 114 (1968); Fl. Cameroun 31: 75 (1988).

Habit: herbaceous climber. Guild: shade-bearer. Habitat: forest. Distribution: Sierra Leone to Bioko, Cameroon, Gabon, Congo (Brazzaville), Congo (Kinshasa) and CAR. Chorology: Guineo-Congolian. Star: green. Alt. range: 1-200m (2). Flowering: Mar (1). Specimens: - *Etinde:* Mbani J. M. 25. *Mabeta-Moliwe:* Nguembock F. 79. *Onge:* Wheatley J. I. 830

Stylochaeton zenkeri Engl.
F.W.T.A. 3: 114 (1968); Fl. Cameroun 31: 43 (1988).

Habit: herb. Habitat: forest. Distribution: Sierra Leone, Bioko, Cameroon, Rio Muni, Gabon and Congo (Brazzaville). Chorology: Guineo-Congolian (montane). Star: blue. Alt. range: 1-200m (4), 601-800m (1). Flowering: Apr (3), Jun (1). Specimens: - *Mabeta-Moliwe:* Mildbraed G. W. J. 10593; Plot Voucher W 313, W 461; Watts J. 139, 219, 351; Wheatley J. I. 192

BURMANNIACEAE

Afrothismia pachyantha Schltr.
F.W.T.A. 3: 180 (1968).

Habit: saprophyte. Habitat: forest. Distribution: SW Cameroon. Chorology: Western Cameroon. Star: black. Fruiting: Sep (1). Specimens: - *Mabeta-Moliwe:* Schlechter F. R. R. 15789

Afrothismia winkleri (Engl.) Schltr.
F.W.T.A. 3: 180 (1968).

Habit: saprophyte. Habitat: forest. Distribution: SE Nigeria, Cameroon and Uganda. Chorology: Guineo-Congolian. Star: black. Flowering: Jul (1), Sep (1). Specimens: - *Eastern slopes:* Winkler H. J. P. 225. *Mabeta-Moliwe:* Schlechter F. R. R. 15788

Burmannia congesta (Wright) Jonker
F.W.T.A. 3: 179 (1968).

Habit: saprophyte. Habitat: forest. Distribution: Liberia, Ghana, Cameroon, CAR, Gabon, Congo (Kinshasa) and Angola. Chorology: Guineo-Congolian. Star: gold. Alt. range: 1-200m (1). Specimens: - *Mabeta-Moliwe:* Cheek M. 3980; Schlechter F. R. R. 15787

Burmannia densiflora Schltr.
F.W.T.A. 3: 179 (1968).

Habit: saprophyte. Habitat: forest. Distribution: Cameroon. Chorology: Lower Guinea. Star: black. Specimens: - *Mabeta-Moliwe:* Stammler s.n.

Burmannia hexaptera Schltr.
F.W.T.A. 3: 179 (1968).

Habit: saprophyte. Habitat: forest. Distribution: Nigeria and Cameroon. Chorology: Lower Guinea. Star: black. Alt. range: 1-200m (1). Flowering: Oct (2). Fruiting: Nov (1). Specimens: - *Mabeta-Moliwe:* Cheek M. 3979; Schlechter F. R. R. 15785, 15786

Oxygyne triandra Schltr.
F.W.T.A. 3: 179 (1968).

Habit: saprophyte. Habitat: forest. Distribution: Mount Cameroon. Chorology: Endemic. Star: black. Flowering: Sep (1). Specimens: - *Mabeta-Moliwe:* Schlechter F. R. R. 15790

Oxygyne sp. nov.

Habit: saprophyte. Habitat: forest. Distribution: Mount Cameroon. Chorology: Endemic (montane). Star: black. Alt. range: 1200-1400m (1). Specimens: - *Etinde:* Cheek M. 3816

COLCHICACEAE

Gloriosa superba L.
F.W.T.A. 3: 106 (1968); Kew Bull. 25: 243 (1971).
Gloriosa simplex L., F.W.T.A. 3: 106 (1968).

Habit: herb. **Guild:** pioneer. **Habitat:** forest and farmbush. **Distribution:** palaeotropical. **Star:** excluded. **Alt. range:** 1-200m (3). **Flowering:** Dec (1). **Specimens:** - *Mabeta-Moliwe:* Cheek M. 5706, 5721; Ndam N. 184

Wurmbea tenuis (Hook.f.) Baker subsp. tenuis
F.W.T.A. 3: 107 (1968).

Habit: herb. **Habitat:** grassland. **Distribution:** Bioko and W. Cameroon. **Chorology:** Western Cameroon Uplands (montane). **Star:** gold. **Alt. range:** 2400-2600m (1). **Flowering:** May (1). **Specimens:** - *Eastern slopes:* Sunderland T. C. H. 1423

COMMELINACEAE
R. Faden (US)

Amischotolype tenuis (C.B.Clarke) R.S.Rao
Lebrun J.-P. & Stork A., E.P.F.A.T. 3: 21 (1995).
  Forrestia tenuis (C.B.Clarke) Benth., F.W.T.A. 3: 24 (1968).

Habit: herb. **Habitat:** forest. **Distribution:** Nigeria, Cameroon, Gabon and Congo (Kinshasa). **Chorology:** Guineo-Congolian. **Star:** green. **Alt. range:** 1-200m (1), 601-800m (1), 801-1000m (2), 1001-1200m (1). **Flowering:** Mar (2), Sep (1), Oct (2). **Specimens:** - *Etinde:* Cable S. 1617; Tchouto P. 431; Watts J. 543; Wheatley J. I. 546. *Mabeta-Moliwe:* Cable S. 1415; Khayota B. 484; Maitland T. D. 669; Plot Voucher W 365, W 414. *Mokoko:* Thomas 10114

Aneilema beniniense (P.Beauv.) Kunth
F.W.T.A. 3: 31 (1968).
Habit: herb. **Guild:** pioneer. **Habitat:** forest, stream banks and Aframomum thicket. **Distribution:** tropical Africa. **Star:** excluded. **Alt. range:** 1-200m (4), 401-600m (1), 801-1000m (1), 1001-1200m (3). **Flowering:** Apr (2), Jun (1), Oct (5). **Fruiting:** Oct (2). **Specimens:** - *Etinde:* Cable S. 1555; Morton J. K. K 902; Sidwell K. 162; Thomas D. W. 9148, 9522; Wheatley J. I. 601. *Mabeta-Moliwe:* Sidwell K. 7; Sunderland T. C. H. 1449; Watts J. 184; Wheatley J. I. 203. *Mokoko:* Acworth J.M. 62; Akogo 189; Pouakouyou 103

Aneilema dispermum Brenan
F.W.T.A. 3: 31 (1968).

Habit: herb. **Habitat:** forest-grassland transition. **Distribution:** Bioko, SW Cameroon, Malawi and Tanzania. **Chorology:** Afromontane. **Star:** blue. **Alt. range:** 1200-1400m (2), 1401-1600m (1), 1601-1800m (2). **Flowering:** Oct (2). **Specimens:** - *Eastern slopes:* Tchouto P. 366; Thomas D. W. 9504, 9519. *Etinde:* Sidwell K. 166. *No locality:* Morton J. K. K 694

Aneilema silvaticum Brenan
F.W.T.A. 3: 31 (1968).

Habit: herb. **Habitat:** forest. **Distribution:** Nigeria, Cameroon and Congo (Kinshasa). **Chorology:** Guineo-Congolian. **Star:** green. **Alt. range:** 200-400m (1). **Flowering:** Oct (1). **Specimens:** - *Etinde:* Tchouto P. 470

Aneilema umbrosum (Vahl) Kunth subsp. ovato-oblongum (P.Beauv.) J.K.Morton
F.W.T.A. 3: 30 (1968).

Habit: herb. **Guild:** pioneer. **Habitat:** forest gaps. **Distribution:** tropical Africa and South America. **Chorology:** Montane. **Star:** excluded. **Alt. range:** 1200-1400m (1). **Flowering:** Oct (1). **Specimens:** - *Eastern slopes:* Thomas D. W. 9514. *Mokoko:* Watts J. 1143

Aneilema umbrosum (Vahl) Kunth subsp. umbrosum
F.W.T.A. 3: 30 (1968).

Habit: herb. **Habitat:** forest. **Distribution:** Sierra Leone to Bioko, Cameroon, Gabon, Congo (Kinshasa) and Angola. **Chorology:** Guineo-Congolian (montane). **Star:** green. **Alt. range:** 1-200m (1), 201-400m (1), 601-800m (1). **Flowering:** Oct (2), Nov (1). **Specimens:** - *Etinde:* Nkeng P. 98; Tchouto P. 469; Watts J. 541

Aneilema sp.

Specimens: - *Eastern slopes:* Dahl A. 610

Buforrestia mannii C.B.Clarke
F.W.T.A. 3: 40 (1968).

Habit: herb. **Habitat:** forest and stream banks. **Distribution:** Nigeria, Bioko and Cameroon. **Chorology:** Lower Guinea. **Star:** blue. **Alt. range:** 1-200m (3), 201-400m (1), 401-600m (1), 601-800m (1). **Flowering:** Feb (1), Mar (2), Apr (2), Jun (1), Sep (1). **Specimens:** - *Etinde:* Khayota B. 530, 595; Wheatley J. I. 498. *Mabeta-Moliwe:* Jaff B. 66; Plot Voucher W 712; Sunderland T. C. H. 1353; Watts J. 163. *Mokoko:* Acworth J.M. 120. *No locality:* Nkeng P. 12

Coleotrype laurentii K.Schum

Specimens: - *Mokoko:* Ekema 1155

Commelina africana L. var. mannii (C.B.Clarke) Brenan
F.W.T.A. 3: 45 (1968).

Habit: herb. **Habitat:** grassland. **Distribution:** SW Cameroon and Ethiopia. **Chorology:** Afromontane. **Star:** gold. **Alt. range:** 2000-2200m (1), 2201-2400m (1), 2401-2600m (1). **Flowering:** Oct (2). **Specimens:** - *Eastern slopes:* Morton J. K. K 844; Tchouto P. 286; Thomas D. W. 9411

Commelina benghalensis L. var. hirsuta C.B.Clarke
F.W.T.A. 3: 48 (1968).

Habit: herb. **Distribution:** Guinea to Cameroon, Congo (Kinshasa), Rwanda, Uganda, Tanzania, Zambia and Malawi. **Chorology:** Afromontane. **Star:** green. **Alt. range:** 800-1000m (1). **Specimens:** - *Eastern slopes:* Migeod F. 2

Commelina cameroonensis J.K.Morton
F.W.T.A. 3: 49 (1968).

Habit: herb. **Habitat:** forest-grassland transition. **Distribution:** SE Nigeria, Bioko and W. Cameroon. **Chorology:** Western Cameroon Uplands (montane). **Star:** gold. **Alt. range:** 800-1000m (1), 1001-1200m (1), 1401-1600m (2), 1601-1800m (1), 1801-2000m (1). **Flowering:** Sep (2), Oct (2). **Specimens:** - *Eastern slopes:* Breteler

F. J. MC 229; Sidwell K. 74; Tchouto P. 392; Tekwe C. 318; Thomas D. W. 9496. *Etinde:* Tekwe C. 188; Thomas D. W. 9169

Commelina capitata Benth.
F.W.T.A. 3: 47 (1968).

Habit: herb. **Guild:** pioneer. **Habitat:** forest. **Distribution:** Senegal to Bioko, Cameroon, Uganda and Angola. **Chorology:** Guineo-Congolian (montane). **Star:** green. **Alt. range:** 1-200m (1), 601-800m (2), 801-1000m (1), 1001-1200m (3). **Flowering:** Mar (1), Oct (4), Nov (2). **Fruiting:** Mar (2). **Specimens:** - *Eastern slopes:* Maitland T. D. 1193. *Etinde:* Khayota B. 533; Kwangue A. 126; Tchouto P. 436; Watts J. 579; Wheatley J. I. 595. *Mabeta-Moliwe:* Khayota B. 469; Nguembock F. 45; Sidwell K. 118

Commelina congesta C.B.Clarke
F.W.T.A. 3: 49 (1968).

Habit: herb. **Guild:** pioneer. **Habitat:** roadsides and villages. **Distribution:** Guinea to Bioko, Cameroon, Gabon, Congo (Kinshasa) and CAR. **Chorology:** Guineo-Congolian. **Star:** green. **Alt. range:** 1-200m (1). **Flowering:** Oct (1). **Specimens:** - *Mabeta-Moliwe:* Sidwell K. 9

Commelina diffusa Burm.f.
F.W.T.A. 3: 47 (1968). **Alt. range:** 1-200m (4), 1001-1200m (1), 1601-1800m (1). **Flowering:** Apr (1), May (2), Oct (2). **Specimens:** - *Eastern slopes:* Dahl A. 611. *Etinde:* Brodie S. 14; Thomas D. W. 9174. *Mabeta-Moliwe:* Ndam N. 540; Sunderland T. C. H. 1357; Watts J. 190

Commelina diffusa Burm.f. subsp. diffusa
F.W.T.A. 3: 47 (1968).

Habit: herb. **Guild:** pioneer. **Habitat:** forest and farmbush. **Distribution:** pantropical. **Star:** excluded. **Specimens:** - *Eastern slopes:* Morton J. K. K 945

Commelina diffusa Burm.f. subsp. montana J.K.Morton
F.W.T.A. 3: 47 (1968).

Habit: herb. **Distribution:** Nigeria, Bioko and W. Cameroon. **Chorology:** Lower Guinea. **Star:** blue. **Specimens:** - *Eastern slopes:* Morton J. K. K 811

Commelina longicapsa C.B.Clarke
F.W.T.A. 3: 47 (1968).

Habit: herb. **Habitat:** forest. **Distribution:** Liberia to Cameroon, Gabon, Congo (Kinshasa) and Angola (Cabinda). **Chorology:** Guineo-Congolian. **Star:** green. **Alt. range:** 200-400m (1). **Flowering:** Oct (1). **Specimens:** - *Etinde:* Tchouto P. 472. *Mokoko:* Fraser 366; Sonké 1126

Cyanotis barbata D.Don
F.W.T.A. 3: 38 (1968).

Habit: herb. **Habitat:** grassland. **Distribution:** tropical Africa and the Himalayas. **Chorology:** Afromontane. **Star:** excluded. **Alt. range:** 2000-2200m (2), 2201-2400m (1), 2401-2600m (1), 2801-3000m (1). **Flowering:** Sep (1), Oct (3). **Specimens:** - *Eastern*

slopes: Sidwell K. 13; Tchouto P. 267; Thomas D. W. 9382. *Etinde:* Tekwe C. 273. *No locality:* Mann G. 1310

Floscopa africana (P.Beauv.) C.B.Clarke subsp. africana
F.W.T.A. 3: 28 (1968).

Habit: herb. **Distribution:** Gambia to Congo (Kinshasa) and Uganda. **Chorology:** Guineo-Congolian. **Star:** green. **Specimens:** - *Etinde:* Maitland T. D. 1304

Floscopa africana (P.Beauv.) C.B.Clarke subsp. majuscula (C.B.Clarke) Brenan
F.W.T.A. 3: 28 (1968).

Habit: herb. **Habitat:** forest. **Distribution:** Guinea Bissau to Cameroon, Gabon, Rio Muni and Congo (Kinshasa). **Chorology:** Guineo-Congolian. **Star:** green. **Alt. range:** 600-800m (1). **Flowering:** Nov (1). **Specimens:** - *Etinde:* Kwangue A. 138

Floscopa africana (P.Beauv.) C.B.Clarke subsp. petrophila J.K.Morton
F.W.T.A. 3: 28 (1968).

Habit: herb. **Distribution:** Liberia to Cameroon, Gabon, Congo (Kinshasa) and Uganda. **Chorology:** Guineo-Congolian. **Star:** green. **Specimens:** - *Mabeta-Moliwe:* Mildbraed G. W. J. 10775

Palisota ambigua (P.Beauv.) C.B.Clarke
F.W.T.A. 3: 35 (1968).

Habit: herb. **Habitat:** forest. **Distribution:** Nigeria, Cameroon, Gabon and Congo (Kinshasa). **Chorology:** Guineo-Congolian. **Star:** green. **Alt. range:** 1-200m (6). **Flowering:** Mar (1), Apr (1), May (1), Jun (3), Oct (1). **Fruiting:** Mar (1), Apr (1), Jun (2), Oct (1). **Specimens:** - *Mabeta-Moliwe:* Plot Voucher W 17, W 39, W 294, W 452, W 729; Sidwell K. 126; Sunderland T. C. H. 1252, 1457; Watts J. 338, 408; Wheatley J. I. 30. *Mokoko River F.R.:* Ekema 824, 1039; Fraser P. 371; Ndam 1271; Pouyoukou 79; Watts J. 1065

Palisota barteri Hook.
F.W.T.A. 3: 35 (1968).

Habit: herb. **Guild:** pioneer. **Habitat:** forest. **Distribution:** Sierra Leone to Bioko, Cameroon and Congo (Kinshasa). **Chorology:** Guineo-Congolian. **Star:** green. **Alt. range:** 1-200m (9), 601-800m (1), 801-1000m (3). **Flowering:** Mar (6), Apr (1), May (5), Jun (3). **Specimens:** - *Etinde:* Cable S. 1549, 1595, 1633; Khayota B. 515; Ndi FHI 50345. *Mabeta-Moliwe:* Cable S. 1488; Jaff B. 76, 154, 227; Khayota B. 463; Ndam N. 531; Plot Voucher W 355, W 477, W 776; Sunderland T. C. H. 1389; Watts J. 322; Wheatley J. I. 282. *Mokoko:* Acworth J.M. 283; Ekema 1054; Fraser 424; Sonké 1071

Palisota bracteosa C.B.Clarke
F.W.T.A. 3: 35 (1968).

Habit: herb. **Guild:** pioneer. **Habitat:** forest. **Distribution:** Guinea to Cameroon and S. Tomé. **Chorology:** Guineo-Congolian. **Star:** green. **Alt. range:** 200-400m (1). **Flowering:** May (1). **Specimens:** - *Mokoko River F.R.:* Ekema 824, 857, 858; Fraser P. 373; Ndam N. 1210, 1319; Sonké 1124; Watts J. 1083. *Southern Bakundu F.R.:* Richards P. W. s.n.

Palisota hirsuta (Thunb.) K.Schum.
F.W.T.A. 3: 35 (1968).

Habit: herb. **Guild:** pioneer. **Habitat:** forest. **Distribution:** Senegal to Bioko, Cameroon, Rio Muni, Congo (Kinshasa) and Angola (Cabinda). **Chorology:** Guineo-Congolian. **Star:** green. **Alt. range:** 1-200m (9), 601-800m (1). **Flowering:** Mar (4), Apr (4), May (2). **Fruiting:** Mar (1), Apr (2), May (2), Jun (1). **Specimens:** - *Eastern slopes:* Tekwe C. 99. *Etinde:* Mbani J. M. 90; Tchouto P. 77; Tekwe C. 56. *Mabeta-Moliwe:* Cable S. 1448; Etuge M. 1225; Groves M. 222; Jaff B. 132; Ndam N. 508; Plot Voucher W 60, W 558, W 840; Sunderland T. C. H. 1228; Wheatley J. I. 134. *Mokoko:* Batoum 48; Ekema 1069; Fraser 434; Poukouyouko 12; Sonké 1045; 1157, 1228

Palisota lagopus Mildbr.
F.W.T.A. 3: 35 (1968).

Habit: herb. **Guild:** pioneer. **Habitat:** forest. **Distribution:** Cameroon. **Chorology:** Lower Guinea. **Star:** gold. **Flowering:** May (1). **Specimens:** - *Mokoko River F.R.:* Ndam N. 1272. *Mokoko:* Acworth J.M. 288; Batoum 19; Ekema 929; Ndam 1272, 1282 *Southern Bakundu F.R.:* Morton J. K. K 922

Palisota mannii C.B.Clarke
F.W.T.A. 3: 35 (1968).

Habit: herb. **Guild:** pioneer. **Habitat:** forest and forest-grassland transition. **Distribution:** Senegal to Bioko, Cameroon, Rio Muni, Congo (Kinshasa), Angola (Cabinda), Sudan and Uganda. **Chorology:** Guineo-Congolian (montane). **Star:** green. **Alt. range:** 1200-1400m (3), 1401-1600m (1), 1801-2000m (1). **Flowering:** Sep (1). **Fruiting:** Aug (1), Sep (2), Oct (1). **Specimens:** - *Eastern slopes:* Dalziel J. M. 8349; Kwangue A. 120; Sidwell K. 77. *Etinde:* Tekwe C. 146; Thomas D. W. 9200

Palisota preussiana K.Schum. ex C.B.Clarke
F.W.T.A. 3: 35 (1968).

Habit: herb. **Guild:** pioneer. **Habitat:** forest and Aframomum thicket. **Distribution:** Mount Cameroon and Bioko. **Chorology:** Western Cameroon Uplands (montane). **Star:** black. **Alt. range:** 800-1000m (1), 1001-1200m (3), 1401-1600m (1). **Flowering:** Feb (1), Oct (4). **Fruiting:** Oct (4). **Specimens:** - *Eastern slopes:* Cable S. 1327; Migeod F. 87; Sidwell K. 100; Tchouto P. 394. *Etinde:* Thomas D. W. 9167; Wheatley J. I. 636

Palisota schweinfurthii C.B.Clarke
F.W.T.A. 3: 35 (1968).

Habit: herb. **Guild:** pioneer. **Habitat:** forest. **Distribution:** Cameroon, Rio Muni, C.A.R. Congo (Kinshasa), Angola, Sudan, Uganda, Tanzania and Zambia. **Chorology:** Congolian. **Star:** green. **Alt. range:** 600-800m (1), 1201-1400m (1). **Flowering:** Mar (1). **Specimens:** - *Eastern slopes:* Etuge M. 1205. *No locality:* Maitland T. D. 1071

Pollia condensata C.B.Clarke
F.W.T.A. 3: 33 (1968).

Habit: herb. **Guild:** shade-bearer. **Habitat:** forest and stream banks. **Distribution:** tropical Africa. **Chorology:** Tropical Africa. **Star:** green. **Alt. range:** 1-200m (3), 201-400m (1), 401-600m (1). **Flowering:** May (1), Aug (1), Oct (1). **Fruiting:** Mar (2), May (2), Aug (1), Oct (1), Dec (1). **Specimens:** - *Etinde:* Groves M. 302; Mildbraed G. W. J. 10528; Tchouto P. 67; Tekwe C. 132; Watts J.

480. *Mabeta-Moliwe:* Ndam N. 504; Nguembock F. 51. *Mokoko River F.R.:* Acworth J.M. 258; Akogo 184; Ekema 826, 834, 1098, 1133. Fraser P. 404; Ndam 1293; Watts J. 1043, 1050

Pollia mannii C.B.Clarke
F.W.T.A. 3: 32 (1968).

Habit: herb. **Guild:** shade-bearer. **Distribution:** Ivory Coast to Cameroon, S. Tomé, Uganda and W. Tanzania. **Chorology:** Guineo-Congolian. **Star:** green. **Specimens:** - *Bambuko F.R.:* Olorunfemi J. FHI 30779

Polyspatha paniculata Benth.
F.W.T.A. 3: 42 (1968).

Habit: herb. **Guild:** shade-bearer. **Habitat:** forest. **Distribution:** Guinea to Bioko, Cameroon, Rio Muni, Congo (Kinshasa), Angola (Cabinda) and Uganda. **Chorology:** Guineo-Congolian. **Star:** green. **Alt. range:** 800-1000m (1). **Specimens:** - *No locality:* Mann G. 2138

Stanfieldiella brachycarpa (Gilg & Lederm. ex Mildbr.) var. brachycarpa

Specimens: - *Mokoko:* Mbani 494; Sonké 1128

Stanfieldiella imperforata (C.B.Clarke) Brenan
F.W.T.A. 3: 23 (1968).

Habit: herb. **Guild:** pioneer. **Habitat:** forest. **Distribution:** tropical Africa. **Chorology:** Tropical Africa. **Star:** green. **Alt. range:** 600-800m (1), 801-1000m (2), 1201-1400m (1). **Flowering:** Feb (1), Mar (1), Oct (2). **Specimens:** - *Eastern slopes:* Etuge M. 1185. *Etinde:* Cable S. 1541; Mildbraed G. W. J. 10562; Watts J. 552; Wheatley J. I. 607

Stanfieldiella oligantha (Mildbr.) Brenan
F.W.T.A. 3: 24 (1968).

Habit: herb. **Guild:** pioneer. **Habitat:** forest. **Distribution:** Liberia to Cameroon. **Chorology:** Guineo-Congolian. **Star:** green. **Alt. range:** 1-200m (1), 201-400m (1), 601-800m (1). **Flowering:** Mar (1), Oct (2). **Specimens:** - *Etinde:* Khayota B. 502; Tchouto P. 458. *Mabeta-Moliwe:* Morton J. K. K 1199; Sidwell K. 12

COSTACEAE

Costus afer Ker-Gawl.

Specimens: - *S. Bakundu:* Daramola B.O. FHI 35537

Costus dinklagei K.Schum.
Fl. Gabon 9: 76 (1964); Fl. Cameroun 4: 76 (1965); Adansonia (ser.2) 7: 78 (1967).

Habit: herb to 2 m. **Habitat:** forest. **Distribution:** Nigeria and Cameroon. **Chorology:** Lower Guinea. **Star:** blue. **Alt. range:** 1-200m (4), 201-400m (1). **Flowering:** May (2), Jul (1), Nov (2). **Fruiting:** Jul (1). **Specimens:** - *Etinde:* Cable S. 133; Cheek M. 5441. *Mabeta-Moliwe:* Ndam N. 514; Sunderland T. C. H. 1268; von Rege I. 28

Costus cf. dinklagei K.Schum.

Specimens: - *Mabeta-Moliwe:* Nguembock F. 4a

Costus dubius (Afzel.) K.Schum.
Fl. Cameroun 4: 92 (1965); F.W.T.A. 3: 78 (1968); F.T.E.A. Zingiberaceae: 9 (1985).
   Costus albus A.Chev. ex J.Koechlin, Fl. Cameroun 4: 78 (1965).

Habit: herb. Guild: pioneer. Habitat: forest. Distribution: tropical Africa. Star: excluded. Alt. range: 1-200m (2). Flowering: Apr (1), Jun (1). Specimens: - *Mabeta-Moliwe:* Watts J. 228, 357

Costus englerianus K.Schum.
Fl. Gabon 9: 66 (1964); Fl. Cameroun 4: 75 (1965); F.W.T.A. 3: 78 (1968).

Habit: herb. Habitat: forest. Distribution: Sierra Leone to Bioko, Cameroon and Gabon. Chorology: Guineo-Congolian. Star: green. Alt. range: 1-200m (1), 201-400m (1). Flowering: Jun (1), Oct (1). Specimens: - *Etinde:* Tchouto P. 456. *Mabeta-Moliwe:* Wheatley J. I. 370

Costus letestui Pellegr.
Fl. Cameroun 4: 86 (1965).

Habit: epiphyte. Habitat: forest. Distribution: Cameroon and Gabon. Chorology: Lower Guinea. Star: blue. Specimens: - *Mokoko River F.R.:* Ndam N. 1201, 1265

Costus littoralis K.Schum.
Fl. Cameroun 4: 94 (1965).

Habit: herb. Habitat: forest. Distribution: Liberia and Cameroon. Chorology: Guineo-Congolian. Star: green. Fruiting: Jan (1). Specimens: - *Mabeta-Moliwe:* Acworth J. M. 19

Costus lucanusianus J.Braun & K.Schum.
Fl. Gabon 9: 83 (1964); Fl. Cameroun 4: 90 (1965); F.W.T.A. 3: 78 (1968); F.T.E.A. Zingiberaceae: 8 (1985).

Habit: herb to 3 m. Guild: pioneer. Habitat: forest and farmbush. Distribution: Sierra Leone to Bioko, Cameroon, Gabon and Uganda. Chorology: Guineo-Congolian (montane). Star: green. Alt. range: 800-1000m (1). Flowering: Mar (1). Specimens: - *Eastern slopes:* Migeod F. 105; Nkeng P. 62

Costus schlechteri Winkl.
Fl. Cameroun 4: 95 (1965); F.W.T.A. 3: 78 (1968).

Habit: herb. Guild: pioneer. Distribution: Liberia, Ivory Coast, SE Nigeria and SW Cameroon. Chorology: Guineo-Congolian. Star: green. Specimens: - *Etinde:* Winkler H. J. P. 25a

CYANASTRACEAE

Cyanastrum cordifolium Oliv.
F.W.T.A. 3: 108 (1968).

Habit: herb. Habitat: forest. Distribution: Nigeria, Cameroon and Gabon. Chorology: Lower Guinea. Star: blue. Alt. range: 1-200m (14), 201-400m (2), 401-600m (2). Flowering: Mar (3), Apr (5), May (1), Nov (1), Dec (3). Fruiting: Apr (2), May (3), Dec (1). Specimens: - *Etinde:* Cable S. 257, 507; Lighava M. 45; Mbani J. M. 72; Tchouto P. 70. *Mabeta-Moliwe:* Cable S. 584; Jaff B. 61; Nguembock F. 64; Sunderland T. C. H. 1207; Watts J. 157, 257; Wheatley J. I. 52, 290. *Mokoko River F.R.:* Acworth J. M. 169; Akogo M. 182; Batoum A. 38; Ekema N. 1026; Sonké 1160; Watts J. 1135. *Onge:* Wheatley J. I. 692

CYPERACEAE

K. Lye (NLH)

Bulbostylis densa (Wall.) Hand.-Mazz. var. cameroonensis Hooper
F.W.T.A. 3: 318 (1972).

Habit: herb. Habitat: grassland. Distribution: Mount Cameroon. Chorology: Endemic (montane). Star: black. Alt. range: 2400-2600m (1), 2601-2800m (1). Specimens: - *Eastern slopes:* Thomas D. W. 9407. *No locality:* Mann G. 1360b

Bulbostylis densa (Wall.) Hand.-Mazz. var. densa
F.W.T.A. 3: 318 (1972).

Habit: herb. Habitat: grassland. Distribution: palaeotropical. Chorology: Montane. Star: excluded. Alt. range: 1800-2000m (1), 2001-2200m (3), 2401-2600m (1). Flowering: Oct (3). Specimens: - *Eastern slopes:* Tchouto P. 271, 293, 294; Thomas D. W. 9307. *No locality:* Mann G. 2093

Bulbostylis hispidula (Vahl) R.W.Haines subsp. hispidula
Haines R. & Lye K., Sedges of E Africa: 104 (1983).
   Fimbristylis hispidula (Vahl) Kunth subsp. *hispidula* F.W.T.A. 3: 324 (1972).

Habit: herb. Habitat: grassland. Distribution: tropical Africa. Chorology: Afromontane. Star: green. Alt. range: 2000-2200m (1). Flowering: Oct (1). Specimens: - *Eastern slopes:* Tchouto P. 291

Bulbostylis pilosa (Willd.) Cherm.
F.W.T.A. 3: 316 (1972); Haines R. & Lye K., Sedges of E Africa: 96 (1983).

Habit: herb. Habitat: forest-grassland transition. Distribution: tropical Africa. Chorology: Afromontane. Star: green. Alt. range: 2400-2600m (1). Flowering: Oct (1). Specimens: - *Eastern slopes:* Watts J. 450

Bulbostylis pusilla (A.Rich.) C.B.Clarke subsp. yalingensis (Cherm.) R.W.Haines
Haines R. & Lye K., Sedges of E Africa: 115 (1983).

Habit: herb. Habitat: grassland. Distribution: Cameroon, CAR and East Africa. Chorology: Afromontane. Star: green. Alt. range: 1800-2000m (2). Specimens: - *Eastern slopes:* Thomas D. W. 9318, 9441

Bulbostylis schoenoides (Kunth) C.B.CL.
subsp. **erratica** (Hook.f.) Lye
Haines R. & Lye K., Sedges of E Africa: 101 (1983).

Habit: herb. **Habitat:** grassland. **Distribution:** Sierra Leone to Bioko and Cameroon. **Chorology:** Guineo-Congolian (montane). **Star:** green. **Alt. range:** 1800-2000m (1), 2801-3000m (2). **Flowering:** Oct (1). **Specimens:** - *Eastern slopes:* Thomas D. W. 9270, 9397. *No locality:* Mann G. 1344

Carex chlorosaccus C.B.Clarke
F.W.T.A. 3: 349 (1972); Haines R. & Lye K., Sedges of E Africa: 375 (1983).

Habit: herb. **Habitat:** forest. **Distribution:** Bioko, Mount Cameroon, E. Congo (Kinshasa), Sudan, Ethiopia and East Africa. **Chorology:** Afromontane. **Star:** green. **Alt. range:** 1400-1600m (1), 1601-1800m (1), 1801-2000m (1). **Flowering:** Oct (1). **Specimens:** - *Eastern slopes:* Cheek M. 3693. *Etinde:* Thomas D. W. 9258; Wheatley J. I. 621

Carex echinochloë Kunze
F.W.T.A. 3: 349 (1972); Haines R. & Lye K., Sedges of E Africa: 374 (1983).

Habit: herb. **Habitat:** grassland and forest-grassland transition. **Distribution:** W. Cameroon and East and NE Africa. **Chorology:** Afromontane. **Star:** blue. **Alt. range:** 1800-2000m (2), 2001-2200m (2). **Flowering:** Oct (1). **Fruiting:** Oct (1). **Specimens:** - *Eastern slopes:* Cheek M. 3698; Thomas D. W. 9265; Watts J. 442. *No locality:* Mann G. 1359

Carex mannii E.A.Bruce
F.W.T.A. 3: 349 (1972); Haines R. & Lye K., Sedges of E Africa: 380 (1983).

Habit: herb. **Distribution:** Bioko, Mount Cameroon and East Africa. **Chorology:** Afromontane. **Star:** blue. **Alt. range:** 1800-2000m (1). **Specimens:** - *No locality:* Maitland T. D. 1341

Carex preussii K.Schum.
F.W.T.A. 3: 347 (1972).

Habit: herb. **Habitat:** grassland. **Distribution:** Nigeria and W. Cameroon. **Chorology:** Lower Guinea (montane). **Star:** blue. **Alt. range:** 2200-2400m (1). **Specimens:** - *Eastern slopes:* Thomas D. W. 9372. *No locality:* Mann G. 2099

Cyperus atroviridis C.B.Clarke
F.W.T.A. 3: 288 (1972); Haines R. & Lye K., Sedges of E Africa: 199 (1983).

Habit: herb. **Habitat:** forest-grassland transition. **Distribution:** tropical Africa. **Chorology:** Afromontane. **Star:** green. **Alt. range:** 2000-2200m (1). **Specimens:** - *Eastern slopes:* Boughey A. S. GC 10634

Cyperus brevifolius (Rottb.) Hasskn.
Haines R. & Lye K., Sedges of E Africa: 236 (1983).
   Kyllinga brevifolia Rottb. F.W.T.A. 3: 307 (1972).

Habit: herb. **Distribution:** pantropical. **Star:** excluded. **Specimens:** - *Eastern slopes:* Chuml H. CDG 374

Cyperus difformis L.
F.W.T.A. 3: 290 (1972); Nord. J. Bot. 1: 58 (1981).

Habit: herb. **Habitat:** farmbush, roadsides and wet areas. **Distribution:** pantropical. **Star:** excluded. **Alt. range:** 1-200m (1). **Flowering:** Mar (1). **Specimens:** - *Onge:* Wheatley J. I. 762

Cyperus distans L.f. subsp. **distans**
F.W.T.A. 3: 287 (1972).

Habit: herb. **Habitat:** stream banks in forest. **Distribution:** pantropical. **Chorology:** Montane. **Star:** excluded. **Alt. range:** 1-200m (1). **Flowering:** May (1). **Specimens:** - *Eastern slopes:* Migeod F. 59. *Mabeta-Moliwe:* Wheatley J. I. 262

Cyperus distans L.f. subsp. **longibracteatus** (Cherm.) Lye var. **longibracteatus**
Nord. J. Bot. 3: 231 (1983).
   Cyperus longibracteatus (Cherm.) Kük., Feddes Rep. 26: 250 (1929).
   Mariscus longibracteatus Cherm., F.W.T.A. 3: 295 (1972).

Habit: herb. **Guild:** pioneer. **Distribution:** tropical Africa. **Star:** excluded. **Specimens:** - *Etinde:* Maitland T. D. 320

Cyperus dubius Rottb.
Haines R. & Lye K., Sedges of E Africa: 221 (1983).
   Mariscus dubius (Rottb.) C.E.C.Fischer, F.W.T.A. 3: 295 (1972).

Habit: herb. **Guild:** pioneer. **Habitat:** grassland. **Distribution:** palaeotropical. **Chorology:** Montane. **Star:** excluded. **Alt. range:** 2000-2200m (1), 2401-2600m (1). **Specimens:** - *Eastern slopes:* Brenan J. P. M. 9380; Sunderland T. C. H. 1431

Cyperus fertilis Boeck.
F.W.T.A. 3: 289 (1972); Haines R. & Lye K., Sedges of E Africa: 163 (1983).

Habit: herb. **Guild:** pioneer. **Habitat:** roadsides and villages. **Distribution:** Liberia to Angola and Uganda. **Chorology:** Guineo-Congolian. **Star:** green. **Alt. range:** 1-200m (1), 601-800m (1). **Flowering:** Apr (1). **Specimens:** - *Mabeta-Moliwe:* Mildbraed G. W. J. 10772. *Mokoko River F.R.:* Acworth J. M. 241

Cyperus laxus Lam. subsp. **buchholzii** (Boeck.) Lye
Nord. J. Bot. 3: 231 (1983); Haines R. & Lye K., Sedges of E Africa: 163 (1983).
   Cyperus diffusus Vahl subsp. *buchholzii* (Boeck.) Kük., F.W.T.A. 3: 289 (1972).

Habit: herb. **Habitat:** open areas in forest. **Distribution:** tropical Africa. **Star:** excluded. **Alt. range:** 1-200m (2), 401-600m (1). **Flowering:** Mar (1). **Fruiting:** Mar (1), Dec (1). **Specimens:** - *Etinde:* Khayota B. 553. *Mabeta-Moliwe:* Cheek M. 5777; Groves M. 217

Cyperus ligularis L.
Syst. Nat. 10(2): 867 (1759).
   Mariscus ligularis (L.) Urb., F.W.T.A. 3: 295 (1972).

Habit: herb. **Distribution:** Senegal to Bioko and Congo (Kinshasa), also tropical America. **Star:** excluded. **Specimens:** - *Etinde:* Maitland T. D. 31

Cyperus luteus Boeck.

Haines R. & Lye K., Sedges of E Africa: 203 (1983).
Mariscus foliosus C.B.Clarke, F.W.T.A. 3: 296 (1972).

Habit: herb. **Distribution:** tropical Africa. **Chorology:** Afromontane. **Star:** green. **Alt. range:** 800-1000m (1). **Specimens:** - *Eastern slopes:* Migeod F. 103

Cyperus mannii C.B.Clarke

F.W.T.A. 3: 289 (1972).

Habit: herb. **Habitat:** forest-grassland transition. **Distribution:** Sierra Leone to Bioko and W. Cameroon. **Chorology:** Guineo-Congolian (montane). **Star:** green. **Alt. range:** 1800-2000m (1), 2001-2200m (3). **Flowering:** Oct (1). **Specimens:** - *Eastern slopes:* Thomas D. W. 9296; Watts J. 425, 438. *No locality:* Mann G. 1358

Cyperus peruvianus (Lam.) F.N. Williams

Kyllinga peruviana Lam. F.W.T.A. 3: 305 (1972).

Habit: herb. **Distribution:** coasts of West Africa and eastern tropical America. **Star:** excluded. **Specimens:** - *Etinde:* Domke W. D. 711

Cyperus pinguis (C.B.CL.) Kük.

Haines R. & Lye K., Sedges of E Africa: 239 (1983).
Kyllinga elatior Kunth F.W.T.A. 3: 305 (1972);

Habit: herb. **Distribution:** tropical and subtropical Africa. **Chorology:** Afromontane. **Star:** green. **Specimens:** - *No locality:* Dunlap 60

Cyperus pseudoleptocladus Kük.

F.W.T.A. 3: 289 (1972); Haines R. & Lye K., Sedges of E Africa: 156 (1983).

Habit: herb. **Distribution:** W. Cameroon and East Africa. **Chorology:** Afromontane. **Star:** blue. **Alt. range:** 1800-2000m (1). **Specimens:** - *Eastern slopes:* Migeod F. 127

Cyperus renschii Boeck.

F.W.T.A. 3: 289 (1972); Haines R. & Lye K., Sedges of E Africa: 161 (1983).

Habit: herb. **Guild:** pioneer. **Habitat:** forest, by streams. **Distribution:** tropical Africa. **Chorology:** Afromontane. **Star:** green. **Alt. range:** 600-800m (1). **Specimens:** - *No locality:* Mann G. 2103

Cyperus sesquiflorus (Torr.) Mattf. & Kük. subsp. appendiculatus (K.Schum.) Lye, Haines R. & Lye K., Sedges of E Africa: 242 (1983).

Kyllinga odorata Vahl subsp. *appendiculatus* (K.Schum.) Lye Nord. J. Bot. 1: 746 (1981).
Kyllinga appendiculata K.Schum., F.W.T.A. 3: 307 (1972).

Habit: herb. **Habitat:** grassland and forest-grassland transition. **Distribution:** Bioko, Cameroon and Ethiopia. **Chorology:** Afromontane. **Star:** blue. **Alt. range:** 2000-2200m (1), 2201-2400m (2). **Flowering:** Oct (1), Nov (1). **Specimens:** - *Eastern slopes:* Cheek M. 5300; Sidwell K. 62. *No locality:* Mann G. 2104

Cyperus sesquiflorus (Torr.) Mattf. & Kük. subsp. cylindricus (Nees) Koyama, Haines R. & Lye K., Sedges of E Africa: 241 (1983).

Kyllinga odorata Vahl subsp. *cyllindrica* (Nees) Koyama Gard. Bull. Singapore 30: 161 (1977).

Habit: herb. **Habitat:** grassland. **Distribution:** pantropical. **Chorology:** Montane. **Star:** excluded. **Alt. range:** 1800-2000m (1). **Specimens:** - *Eastern slopes:* Thomas D. W. 9271

Cyperus sesquiflorus (Torr.) Mattf. & Kük. subsp. sesquiflorus

Haines R. & Lye K., Sedges of E Africa: 241 (1983).
Kyllinga odorata Vahl subsp. *odorata* F.W.T.A. 3: 304 (1972);

Habit: herb. **Habitat:** grassland. **Distribution:** tropical Africa and America. **Chorology:** Montane. **Star:** excluded. **Alt. range:** 2600-2800m (1). **Specimens:** - *No locality:* Johnston H. H. s.n.

Cyperus smithianus Ridl.

Haines R. & Lye K., Sedges of E Africa: 278 (1983).
Pycreus smithianus (Ridley) C.B.Clarke F.W.T.A. 3: 300 (1972).

Habit: herb. **Habitat:** forest clearings and river banks. **Distribution:** tropical Africa. **Star:** excluded. **Alt. range:** 1-200m (1). **Flowering:** Oct (1). **Specimens:** - *Onge:* Cheek M. 5171

Cyperus subumbellatus Kük.

Engl. Monogr. Afr. Pfanzenfam. Cyperaceae: 523 (1936).
Mariscus alternifolius sensu Hooper, F.W.T.A. 3: 296 (1972).

Habit: herb. **Guild:** pioneer. **Habitat:** clearings in forest and grassland. **Distribution:** Senegal to Cameroon, CAR, Gabon, Congo (Kinshasa), Ethiopia, Somalia and Sudan. **Chorology:** Afromontane. **Star:** green. **Alt. range:** 1-200m (1), 601-800m (1). **Flowering:** Jul (1). **Specimens:** - *Mabeta-Moliwe:* Cheek M. 3569; Mildbraed G. W. J. 10787

Cyperus tenuis Sw.

Prod. Ind. Occ. 20 (1788).
Mariscus flabelliformis Kunth var. *flabelliformis*, F.W.T.A. 3: 296 (1972).

Habit: herb. **Habitat:** farmbush and open areas. **Distribution:** Senegal to Bioko and Cameroon, also tropical America. **Star:** excluded. **Alt. range:** 1-200m (1). **Flowering:** Mar (1). **Specimens:** - *Onge:* Wheatley J. I. 764

Cyperus tomaiophyllus K.Schum.

Haines R. & Lye K., Sedges of E Africa: 207 (1983).
Mariscus tomaiophyllus (K.Schum.) C.B.Clarke, F.W.T.A. 3: 295 (1972).

Habit: herb. **Habitat:** forest and forest-grassland transition. **Distribution:** tropical Africa. **Chorology:** Afromontane. **Star:** green. **Alt. range:** 1400-1600m (1), 1601-1800m (1). **Specimens:** - *Etinde:* Thomas D. W. 9526. *No locality:* Maitland T. D. s.n.

Cyperus sp.

Specimens: - *Eastern slopes:* Tchouto P. 379. *Mabeta-Moliwe:* Cheek M. 3541

Cyperus sp. nov.

Habit: rheophyte. **Habitat:** streams. **Distribution:** Mount Cameroon. **Chorology:** Endemic. **Star:** black. **Alt. range:** 1-200m (1), 201-400m (2), 401-600m (1). **Flowering:** Nov (2). **Specimens:** - *Etinde:* Cable S. 203; Cheek M. 5542; Thomas D. W. 9713; Williams S. 53

Fimbristylis dichotoma (L.) Vahl
F.W.T.A. 3: 320 (1972); Haines R. & Lye K., Sedges of E Africa: 85 (1983).

Habit: herb. **Habitat:** grassland and stream banks. **Distribution:** pantropical. **Star:** excluded. **Alt. range:** 1-200m (1). **Specimens:** - *Onge:* Wheatley J. I. 761

Hypolytrum heteromorphum Nelmes
F.W.T.A. 3: 336 (1972); Haines R. & Lye K., Sedges of E Africa: 326 (1983).

Habit: herb. **Habitat:** forest. **Distribution:** tropical Africa. **Star:** excluded. **Alt. range:** 1-200m (2). **Flowering:** Nov (1), Dec (1). **Fruiting:** Nov (1). **Specimens:** - *Etinde:* Cheek M. 5915. *Onge:* Watts J. 968

Isolepis setacea (L.) R.Br.
Haines R. & Lye K., Sedges of E Africa: 134 (1983); Fl. Rwanda 4: 455 (1988).
Scirpus setaceus L., F.W.T.A. 3: 309 (1972).

Habit: herb. **Habitat:** grassland. **Distribution:** Mount Cameroon, East and southern Africa, Europe and Australia. **Chorology:** Montane. **Star:** blue. **Alt. range:** 2800-3000m (1). **Specimens:** - *Eastern slopes:* Morton J. K. K 865

Lipocarpha barteri C.B.Clarke
Haines R. & Lye K., Sedges of E Africa: 298 (1983).
Lipocarpha atra Ridl. var. *barteri* (C.B.Clarke) J. Raynal, F.W.T.A. 3: 328 (1972).

Habit: herb. **Distribution:** Ivory Coast to Cameroon. **Chorology:** Guineo-Congolian. **Star:** green. **Specimens:** - *Bambuko F.R.:* Eldin 68

Mapania amplivaginata K.Schum.
F.W.T.A. 3: 335 (1972); Simpson D., Rev. of Mapania: 139 (1992).

Habit: herb. **Habitat:** forest. **Distribution:** SE Nigeria, Cameroon, Gabon and Angola (Cabinda). **Chorology:** Guineo-Congolian. **Star:** green. **Alt. range:** 400-600m (3). **Flowering:** Sep (1), Oct (2). **Fruiting:** Oct (1). **Specimens:** - *Etinde:* Watts J. 486; Wheatley J. I. 532. *Onge:* Cheek M. 4976. *Southern Bakundu F.R.:* Brenan J. P. M. 9421

Mapania macrantha (Boeck) H.Pfeiffer
F.W.T.A. 3: 335 (1972); Simpson D., Rev. of Mapania: 131 (1992).

Habit: herb. **Habitat:** forest. **Distribution:** Cameroon and Gabon. **Chorology:** Lower Guinea (montane). **Star:** blue. **Alt. range:** 1200-1400m (2). **Flowering:** May (1), Sep (1). **Specimens:** - *Eastern slopes:* Mbani J. M. 105. *Etinde:* Thomas D. W. 9232

Mapania soyauxii (Boeck) K.Schum.
Simpson D., Rev. of Mapania: 138 (1992).

Habit: herb. **Habitat:** forest. **Distribution:** Cameroon, Gabon, Congo (Kinshasa) and Angola. **Chorology:** Guineo-Congolian. **Star:** green. **Alt. range:** 1-200m (2), 201-400m (1). **Flowering:** Apr (1), Aug (1), Oct (1). **Specimens:** - *Mabeta-Moliwe:* Baker W. J. 336; Watts J. 144. *Onge:* Watts J. 812

Mapania cf. soyauxii (Boeck) K.Schum.

Specimens: - *Mabeta-Moliwe:* Wheatley J. I. 304

Scleria naumanniana Boeck.
F.W.T.A. 3: 342 (1972).

Habit: herb. **Guild:** pioneer. **Habitat:** forest. **Distribution:** Senegal to Cameroon, Congo (Kinshasa) and Angola. **Chorology:** Guineo-Congolian. **Star:** green. **Alt. range:** 1-200m (1).**Fruiting:** Dec (1). **Specimens:** - *Mabeta-Moliwe:* Cheek M. 5723

Scleria verrucosa Willd.
F.W.T.A. 3: 340 (1972); Haines R. & Lye K., Sedges of E Africa: 359 (1983).

Habit: herb. **Guild:** pioneer. **Distribution:** Senegal to Congo (Kinshasa), Uganda and Tanzania. **Chorology:** Guineo-Congolian. **Star:** green. **Specimens:** - *Etinde:* Maitland T. D. 901

Scleria vogelii C.B.Clarke
F.W.T.A. 3: 340 (1972).

Habit: herb. **Guild:** pioneer. **Distribution:** Liberia to Cameroon and Gabon. **Chorology:** Guineo-Congolian. **Star:** green. **Specimens:** - *Eastern slopes:* Maitland T. D. 553

DIOSCOREACEAE

P. Wilkin (K)

Dioscorea bulbifera L.
F.W.T.A. 3: 152 (1968).

Habit: herbaceous climber. **Guild:** light demander. **Habitat:** farmbush. **Distribution:** cultivated throughout the tropics. **Star:** excluded. **Alt. range:** 1-200m (1).**Fruiting:** Jul (1). **Specimens:** - *Mabeta-Moliwe:* Cheek M. 3528. *Mokoko:* Watts J. 1196

Dioscorea hirtiflora Benth.
F.W.T.A. 3: 152 (1968).

Habit: herbaceous climber. **Guild:** pioneer. **Habitat:** forest. **Distribution:** tropical Africa. **Star:** excluded. **Specimens:** - *Eastern slopes:* Migeod F. 95

Dioscorea mangenotiana J.Miege
F.W.T.A. 3: 153 (1968).

Habit: herbaceous climber, often reaching the canopy. **Guild:** pioneer. **Habitat:** forest. **Distribution:** Liberia, Ivory Coast, Nigeria and Cameroon. **Chorology:** Guineo-Congolian. **Star:**

green. **Alt. range:** 1-200m (1). **Flowering:** Jun (1). **Specimens:** - *Mabeta-Moliwe:* Wheatley J. I. 335

Dioscorea minutiflora Engl.
F.W.T.A. 3: 153 (1968).

Habit: herbaceous climber. **Guild:** light demander. **Habitat:** forest. **Distribution:** Senegal to Cameroon, Congo (Kinshasa), Angola and Uganda. **Chorology:** Guineo-Congolian. **Star:** green. **Alt. range:** 1-200m (1), 801-1000m (1).**Fruiting:** Dec (2). **Specimens:** - *Eastern slopes:* Maitland T. D. 289. *Etinde:* Cable S. 466. *Mabeta-Moliwe:* Cheek M. 5693

Dioscorea preussii Pax
F.W.T.A. 3: 152 (1968).

Habit: herbaceous climber. **Guild:** pioneer. **Habitat:** forest and farmbush. **Distribution:** Senegal to Cameroon, Gabon, Congo (Kinshasa) and CAR. **Chorology:** Guineo-Congolian (montane). **Star:** green. **Alt. range:** 1400-1600m (1). **Flowering:** Jul (1). **Specimens:** - *Etinde:* Maitland T. D. 68; Tchouto P. 254

Dioscorea smilacifolia De Wild.
F.W.T.A. 3: 153 (1968).

Habit: herbaceous climber. **Guild:** light demander. **Habitat:** forest gaps. **Distribution:** Senegal to Bioko, Cameroon, Congo (Kinshasa), Angola and Uganda. **Chorology:** Guineo-Congolian. **Star:** green. **Alt. range:** 200-400m (1).**Fruiting:** Mar (1). **Specimens:** - *Mabeta-Moliwe:* Dunlap 180. *Onge:* Wheatley J. I. 682

DRACAENACEAE
J.J. Bos (WAG)

Dracaena sp. aff. aubryana Brongn. ex C.J.Morren

Specimens: - *Bambuko F.R.:* Watts J. 611, 614. *Mokoko River F.R.:* Tchouto P. 599. *Onge:* Wheatley J. I. 853

Dracaena sp. cf. bicolor Hook.

Specimens: - *Mabeta-Moliwe:* Plot Voucher W 184

Dracaena bueana Engl.
Engl. Jahrb. lix. Beibl. 131, 20 (1924)

Specimens: - *Eastern Slopes:* Deistel H. 461.

Dracaena camerooniana Baker
F.W.T.A. 3: 157 (1968); Agric. Univ. Wag. Papers 84(1): 45 (1984).

Habit: shrub to 8 m. **Guild:** shade-bearer. **Habitat:** forest. **Distribution:** tropical Africa. **Star:** excluded. **Alt. range:** 1-200m (12), 201-400m (10), 401-600m (3), 601-800m (1), 801-1000m (1). **Flowering:** Mar (1), May (2), Jun (1), Oct (2), Dec (1). **Fruiting:** Mar (3), May (3), Jun (1), Aug (2), Sep (2), Oct (6), Nov (3), Dec (3). **Specimens:** - *Bambuko F.R.:* Tchouto P. 539; Watts J. 609. *Etinde:* Cable S. 294, 425; Cheek M. 5443; Kwangue A. 135; Mbani J. M. 148; Nkeng P. 35, 75; Tchouto P. 477; Tekwe C. 131; Wheatley J. I. 530. *Mabeta-Moliwe:* Cable S. 550; Jaff B. 250; Plot Voucher W 817; Sidwell K. 107; Sunderland T. C. H. 1028, 1333; Watts J. 271; Wheatley J. I. 316; von Rege I. 108. *Onge:* Ndam N. 653; Tchouto P. 864; Thomas D. W. 9893; Watts J. 718, 794, 817; Wheatley J. I. 734

Dracaena sp. aff. cristula W.Bull
Agric. Univ. Wag. Papers 84(1): 64 (1984).
  Dracaena elliotii Baker, F.W.T.A. 3: 156 (1968).

Habit: shrub to 1 m. **Habitat:** forest. **Distribution:** Sierra Leone to Cameroon. **Chorology:** Guineo-Congolian. **Star:** green. **Specimens:** - *Mabeta-Moliwe:* Plot Voucher W 221

Dracaena fragrans (L.) Ker-Gawl.
F.W.T.A. 3: 157 (1968); Agric. Univ. Wag. Papers 84(1): 69 (1984).
  Dracaena deisteliana Engl., F.W.T.A. 3: 157 (1968).

Habit: tree or shrub to 15 m. **Guild:** shade-bearer. **Habitat:** forest. **Distribution:** tropical Africa. **Chorology:** Afromontane. **Star:** green. **Alt. range:** 1600-1800m (1). **Flowering:** Oct (1). **Specimens:** - *Eastern Slopes:* Lehmbach 16; Maitland 357; Breteler et al in MC 178; Dalziel 8348. *Etinde:* Thomas D. W. 9253

Dracaena laxissima Engl.
F.W.T.A. 3: 157 (1968); Agric. Univ. Wag. Papers 84(1): 79 (1984).

Habit: shrub to 2 m. **Habitat:** forest. **Distribution:** Nigeria to East Africa, Sudan, Zambia and Malawi. **Star:** excluded. **Alt. range:** 1-200m (7), 201-400m (6). **Flowering:** May (1), Jun (3). **Fruiting:** Mar (2), Apr (1), May (1), Aug (1), Oct (5). **Specimens:** - *Bambuko F.R.:* Tchouto P. 540; Watts J. 615. *Etinde:* Banks H. 38. *Mabeta-Moliwe:* Baker W. J. 262; Plot Voucher W 1054; Sunderland T. C. H. 1341; Watts J. 213, 350, 379; Wheatley J. I. 288, 320. *Onge:* Ndam N. 608, 748; Tchouto P. 863; Watts J. 816

Dracaena mannii Baker
Agric. Univ. Wag. Papers 84 (1): 82; F.W.T.A. 3: 156 (1968).

Specimens: - *Onge:* D. Thomas 4507

Dracaena mildbraedii K.Krause
Agric. Univ. Wag. Papers 84 (1): 89
  Dracaena viridiflora Engl. & K.Krause, F.W.T.A. 3: 157 (1968).
  Dracaena vaginata Hutch.; F.W.T.A. 2: 383 (1936).

Specimens:- *Mabeta-Moliwe:* Mildbraed G. 10571

Dracaena phrynioides Hook.
F.W.T.A. 3: 156 (1968); Agric. Univ. Wag. Papers 84(1): 97 (1984).

Habit: herb. **Habitat:** forest. **Distribution:** Liberia to Bioko, Rio Muni and Gabon. **Chorology:** Guineo-Congolian (montane). **Star:** green. **Alt. range:** 1-200m (3), 201-400m (5), 401-600m (1), 801-1000m (1), 1001-1200m (1). **Flowering:** May (1), Jun (3), Oct (1). **Fruiting:** Mar (2), Jul (1), Oct (4), Dec (3). **Specimens:** - *Bambuko F.R.:* Tchouto P. 542. *Etinde:* Cable S. 362, 499; Faucher P. 8. *Mabeta-Moliwe:* Banks H. 15; Hurst J. 20; Plot Voucher W 266, W 353, W 578, W 841, W 934, W 1032;

Wheatley J. I. 235. *Mokoko River F.R.:* Mbani J. M. 515. *Onge:* Akogo M. 71; Ndam N. 646, 654; Watts J. 735; Wheatley J. I. 772

Dracaena sp. aff. praetermissa Bos

Specimens: - *Eastern slopes:* Kwangue A. 51; Tekwe C. 103. *Etinde:* Cable S. 323; Faucher P. 22

Dracaena cf. preussii Engl.

Specimens: - *Onge:* Cheek M. 5155; Watts J. 1019

Dracaena cf. surculosa Lindl.

Specimens: - *Mabeta-Moliwe:* Watts J. 180

Sansevieria metallica Hort. ex Gerome & Labroy var. longituba N.E.Br.
Kew Bull. 1915: 245 (1915).

Habit: herb. Habitat: forest. Distribution: not known. Chorology: Guineo-Congolian. Star: green. Alt. range: 1-200m (1). Flowering: Jul (1). Specimens: - *Mabeta-Moliwe:* Cheek M. 3583

GRAMINEAE

T. Cope (K)

Acritochaete volkensii Pilg.
F.W.T.A. 3: 448 (1972); Agric. Univ. Wag. Papers 92(1): 309 (1992).

Habit: herb. Habitat: forest-grassland transition. Distribution: Nigeria, W. Cameroon and East Africa. Chorology: Afromontane. Star: green. Alt. range: 2000-2200m (1). Specimens: - *Eastern slopes:* Boughey A. S. GC 10716

Acroceras zizanioides (Kunth) Dandy
Fl. Gabon 5: 22 (1962); F.W.T.A. 3: 435 (1972); Agric. Univ. Wag. Papers 92(1): 241 (1992).

Habit: herb. Guild: pioneer. Habitat: forest. Distribution: Mostly Guineo-Congolian, but also in East Africa, tropical America and India. Star: excluded. Alt. range: 1-200m (2), 601-800m (1). Flowering: Oct (2). Specimens: - *Etinde:* Maitland T. D. 1291. *Mabeta-Moliwe:* Marsden J. 24. *Onge:* Marsden J. 30b; Thomas D. W. 9789

Agrostis mannii (Hook.f.) Stapf
F.W.T.A. 3: 372 (1972); Agric. Univ. Wag. Papers 92(1): 74 (1992).

Habit: herb. Guild: pioneer. Habitat: grassland. Distribution: Bioko and Cameroon. Chorology: Lower Guinea (montane). Star: gold. Alt. range: 2200-2400m (2), 2601-2800m (1). Flowering: Oct (2). Specimens: - *Eastern slopes:* Thomas D. W. 9322, 9387. *No locality:* Mann G. 2096

Agrostis quinqueseta (Steud.) Hochst.
F.W.T.A. 3: 372 (1972); Agric. Univ. Wag. Papers 92(1): 74 (1992).

Habit: herb. Guild: pioneer. Habitat: grassland. Distribution: Cameroon, Congo (Kinshasa) and East Africa. Chorology: Afromontane. Star: green. Alt. range: 2600-2800m (1). Specimens: - *No locality:* Mann G. 2086

Aira caryophyllea L.
F.W.T.A. 3: 372 (1972); Agric. Univ. Wag. Papers 92(1): 69 (1992).

Habit: herb. Habitat: grassland. Distribution: tropical and subtropical Africa, Europe and Asia. Chorology: Montane. Star: excluded. Alt. range: 2000-2200m (3), 2401-2600m (1), 2601-2800m (1), 3001-3200m (1). Flowering: Oct (3). Specimens: - *Eastern slopes:* Tchouto P. 274, 292; Thomas D. W. 9323, 9409; Watts J. 449. *No locality:* Mann G. 1356

Andropogon amethystinus Steud.
F.W.T.A. 3: 485 (1972); Agric. Univ. Wag. Papers 92(1): 427 (1992).

Habit: herb. Habitat: forest-grassland transition. Distribution: Bioko, Cameroon and East Africa. Chorology: Afromontane. Star: blue. Alt. range: 2000-2200m (1), 2401-2600m (1), 2601-2800m (1). Flowering: Oct (2). Specimens: - *Eastern slopes:* Tchouto P. 314; Thomas D. W. 9398. *No locality:* Migeod F. 170

Andropogon distachyos L.
F.W.T.A. 3: 485 (1972); Agric. Univ. Wag. Papers 92(1): 426 (1992).

Habit: herb. Distribution: tropical and subtropical Africa, Europe and Asia. Chorology: Montane. Star: excluded. Alt. range: 2200-2400m (1). Specimens: - *Eastern slopes:* Boughey A. S. GC 10621. *No locality:* Mann G. 2095

Andropogon lacunosus J.G.Anderson
F.W.T.A. 3: 485 (1972); Agric. Univ. Wag. Papers 92(1): 425 (1992).

Habit: herb. Distribution: Cameroon, Zimbabwe, Tanzania and South Africa. Chorology: Afromontane. Star: green. Specimens: - *Eastern slopes:* Maitland T. D. 351

Andropogon lima (Hack.) Stapf
F.W.T.A. 3: 485 (1972); Agric. Univ. Wag. Papers 92(1): 429 (1992).

Habit: herb. Habitat: grassland. Distribution: W. Cameroon and East Africa. Chorology: Afromontane. Star: blue. Alt. range: 2200-2400m (1), 2401-2600m (1). Flowering: Oct (1). Specimens: - *Eastern slopes:* Thomas D. W. 9373. *No locality:* Boughey A. S. GC 12543; Mildbraed G. W. J. 10856

Andropogon mannii Hook.f.
F.W.T.A. 3: 485 (1972); Agric. Univ. Wag. Papers 92(1): 427 (1992).

Habit: herb. Habitat: grassland. Distribution: tropical and subtropical Africa. Chorology: Afromontane. Star: green. Alt. range: 2600-2800m (1), 3201-3400m (1). Flowering: Oct (1).

Specimens: - *Eastern slopes:* Thomas D. W. 9402. *No locality:* Mildbraed G. W. J. 10892

Andropogon pusillus Hook.f.

F.W.T.A. 3: 485 (1972); Agric. Univ. Wag. Papers 92(1): 426 (1992).
   Hyparrhenia pusilla (Hook.f.) Stapf, F.W.T.A. (ed.1) 2: 591 (1936).

Habit: herb. **Habitat:** forest-grassland transition. **Distribution:** Nigeria and W. Cameroon. **Chorology:** Lower Guinea (montane). **Star:** blue. **Alt. range:** 1800-2000m (1), 2201-2400m (1). **Specimens:** - *Eastern slopes:* Thomas D. W. 9376, 9438

Andropogon schirensis A.Rich.

Fl. Gabon 5: 168 (1962); F.W.T.A. 3: 486 (1972); F.T.E.A. Gramineae: 781 (1982); Agric. Univ. Wag. Papers 92(1): 437 (1992).
   Andropogon dummeri Stapf, F.W.T.A. 3: 486 (1972).

Habit: herb. **Habitat:** grassland. **Distribution:** tropical and subtropical Africa. **Chorology:** Afromontane. **Star:** green. **Alt. range:** 1800-2000m (1), 2001-2200m (1), 2801-3000m (1). **Flowering:** Oct (3). **Specimens:** - *Eastern slopes:* Tchouto P. 273; Thomas D. W. 9267, 9400. *Etinde:* Brenan J. P. M. 9588. *No locality:* Mildbraed G. W. J. 10857

Andropogon tectorum Schum. & Thonn.

F.W.T.A. 3: 488 (1972); Agric. Univ. Wag. Papers 92(1): 436 (1992).

Habit: herb. **Guild:** pioneer. **Distribution:** Senegal to Cameroon and CAR. **Chorology:** Guineo-Congolian (montane). **Star:** green. **Specimens:** - *Bambuko F.R.:* Argent G. C. G. 977

Arthraxon hispidus (Thunb.) Makino var. hispidus

Blumea 27: 273 (1981); Agric. Univ. Wag. Papers 92(1): 461 (1992).
   Arthraxon quartinianus (A.Rich.) Nash, F.W.T.A. 3: 470 (1972).
   Arthraxon micans (Nees) Hochst., F.T.E.A. Gramineae: 742 (1982).

Habit: herb. **Guild:** pioneer. **Habitat:** forest. **Distribution:** pantropical. **Chorology:** Montane. **Star:** excluded. **Alt. range:** 1600-1800m (2). **Flowering:** Nov (2). **Specimens:** - *Eastern slopes:* Maitland T. D. 103; Marsden J. 37, 39

Axonopus compressus (Sw.) P.Beauv.

Fl. Gabon 5: 36 (1962); F.W.T.A. 3: 448 (1972); Agric. Univ. Wag. Papers 92(1): 286 (1992).

Habit: herb. **Guild:** pioneer. **Habitat:** roadsides and villages. **Distribution:** native to South America and introduced throughout the tropics. **Star:** excluded. **Alt. range:** 1-200m (1). **Flowering:** Oct (1). **Specimens:** - *Etinde:* Brunt M. A. 1038; Marsden J. 28. *No locality:* Maitland T. D. s.n.

Bambusa vulgaris Schrad. ex Wendell

Specimens:- *Mabeta-Moliwe:* No specimens collected.

Brachiaria mutica (Forssk.) Stapf

Fl. Gabon 5: 31 (1962); F.W.T.A. 3: 443 (1972); Agric. Univ. Wag. Papers 92(1): 264 (1992).

Habit: herb. **Distribution:** pantropical. **Star:** excluded. **Specimens:** - *Etinde:* Maitland T. D. 10

Brachiaria plantaginea (Link) Hitchc.

F.W.T.A. 3: 443 (1972); Agric. Univ. Wag. Papers 92(1): 263 (1992).

Habit: herb. **Distribution:** Burkina, Ghana, W. Cameroon, Congo (Kinshasa) and tropical America. **Star:** excluded. **Specimens:** - *Eastern slopes:* Maitland T. D. 26

Bromus leptoclados Nees

F.T.E.A. Gramineae: 68 (1970); F.W.T.A. 3: 369 (1972); Agric. Univ. Wag. Papers 92(1): 79 (1992).

Habit: herb to 2 m. **Habitat:** grassland and forest-grassland transition. **Distribution:** tropical and subtropical Africa. **Chorology:** Afromontane. **Star:** green. **Alt. range:** 2200-2400m (2), 2601-2800m (1). **Flowering:** Oct (2). **Specimens:** - *Eastern slopes:* Thomas D. W. 9374, 9434. *No locality:* Mann G. 2085

Centotheca lappacea (L.) Desv.

Fl. Gabon 5: 214 (1962); F.W.T.A. 3: 381 (1972); Agric. Univ. Wag. Papers 92(1): 83 (1992).

Habit: herb. **Guild:** pioneer. **Habitat:** forest. **Distribution:** palaeotropical. **Star:** excluded. **Alt. range:** 1-200m (3), 601-800m (1). **Flowering:** Oct (1), Dec (3). **Specimens:** - *Etinde:* Cheek M. 5830, 5836; Maitland T. D. 12. *Mabeta-Moliwe:* Cheek M. 5781. *Onge:* Marsden J. 30a

Chloris pilosa Schumach.

F.W.T.A. 3: 400 (1972); Agric. Univ. Wag. Papers 92(1): 172 (1992).

Habit: herb. **Guild:** pioneer. **Distribution:** tropical Africa. **Star:** excluded. **Specimens:** - *Etinde:* Maitland T. D. 9

Chrysopogon aciculatus (Retz.) Trin.

Fl. Gabon 5: 150 (1962); F.W.T.A. 3: 468 (1972); Agric. Univ. Wag. Papers 92(1): 412 (1992).

Habit: herb. **Guild:** pioneer. **Distribution:** introduced in West Africa from Asia. **Star:** excluded. **Specimens:** - *Etinde:* Maitland T. D. 16

Coix lacryma-jobi L.

F.W.T.A. 3: 511 (1972); Agric. Univ. Wag. Papers 92(1): 542 (1992).

Habit: herb. **Guild:** pioneer. **Distribution:** pantropical. **Star:** excluded. **Specimens:** - *Mabeta-Moliwe:* Maitland T. D. 978

Cyrtococcum chaetophoron (Roem. & Schult.) Dandy

Fl. Gabon 5: 76 (1962); F.W.T.A. 3: 426 (1972); Agric. Univ. Wag. Papers 92(1): 239 (1992).

Habit: herb. **Guild:** pioneer. **Habitat:** forest. **Distribution:** Senegal to Congo (Kinshasa), Angola and Sudan. **Chorology:**

Guineo-Congolian. **Star:** green. **Alt. range:** 1-200m (2), 601-800m (1). **Flowering:** Apr (2), Oct (1). **Fruiting:** Apr (1). **Specimens:** - *Mokoko River F.R.:* Acworth J. M. 198; Pouakouyou D. 20. *Onge:* Marsden J. 32

Deschampsia mildbraedii Pilg.

F.W.T.A. 3: 372 (1972); Agric. Univ. Wag. Papers 92(1): 73 (1992).

Habit: herb. **Habitat:** grassland. **Distribution:** SE Nigeria and W. Cameroon. **Chorology:** Western Cameroon Uplands (montane). **Star:** gold. **Alt. range:** 2600-2800m (1), 3201-3400m (1). **Specimens:** - *Eastern slopes:* Thomas D. W. 9399. *No locality:* Maitland T. D. 1246

Digitaria horizontalis Willd.

Fl. Gabon 5: 47 (1962); F.W.T.A. 3: 453 (1972); Agric. Univ. Wag. Papers 92(1): 325 (1992).

Habit: herb. **Guild:** pioneer. **Distribution:** tropical Africa and America. **Star:** excluded. **Specimens:** - *Eastern slopes:* Migeod F. 90

Digitaria radicosa (Presl) Miq.

F.T.E.A. Gramineae: 651 (1982); Agric. Univ. Wag. Papers 92(1): 324 (1992).
    Digitaria timorensis (Kunth) Bal., F.W.T.A. 3: 452 (1972).

Habit: herb. **Distribution:** SW Cameroon, Tanzania, Mauritius, Seychelles and Asia. **Star:** excluded. **Specimens:** - *Eastern slopes:* Maitland T. D. 2

Echinochloa pyramidalis (Lam.) Hitchc. & Chase

Fl. Gabon 5: 53 (1962); F.W.T.A. 3: 439 (1972); Agric. Univ. Wag. Papers 92(1): 251 (1992).

Habit: herb. **Distribution:** tropical Africa. **Star:** excluded. **Specimens:** - *Etinde:* Maitland T. D. 11

Eleusine indica (L.) Gaertn.

Fl. Gabon 5: 231 (1962); F.W.T.A. 3: 395 (1972); Agric. Univ. Wag. Papers 92(1): 142 (1992).

Habit: herb. **Guild:** pioneer. **Distribution:** pantropical. **Star:** excluded. **Specimens:** - *Eastern slopes:* Migeod F. 94

Eragrostis amabilis (L.) Hook. & Arn.

Hook. & Arn., Bot. Beech. Voy.: 251 (1838).
    Eragrostis tenella (L.) P.Beauv. ex Roem. & Schult., Fl. Gabon 5: 223 (1962); F.W.T.A. 3: 386 (1972); Agric. Univ. Wag. Papers 92(1): 114 (1992).

Habit: herb. **Distribution:** pantropical. **Star:** excluded. **Specimens:** - *Limbe Botanic Garden:* Brunt M. A. 1043

Eragrostis ciliaris (L.) R.Br.

Fl. Gabon 5: 224 (1962); F.W.T.A. 3: 386 (1972); Agric. Univ. Wag. Papers 92(1): 113 (1992).

Habit: herb. **Distribution:** pantropical. **Star:** excluded. **Flowering:** Nov (1). **Specimens:** - *Etinde:* Maitland T. D. 83

Eragrostis gangetica (Roxb.) Steud.

Fl. Gabon 5: 225 (1962); F.W.T.A. 3: 389 (1972); Agric. Univ. Wag. Papers 92(1): 132 (1992).

Habit: herb. **Distribution:** tropical Africa and India. **Star:** excluded. **Specimens:** - *Etinde:* Maitland T. D. 13

Festuca abyssinica A.Rich.

F.T.E.A. Gramineae: 60 (1970); F.W.T.A. 3: 369 (1972); Agric. Univ. Wag. Papers 92(1): 59 (1992).
    Festuca schimperiana A.Rich., F.W.T.A. (ed.1) 2: 508 (1936).

Habit: herb. **Habitat:** grassland. **Distribution:** Bioko, W. Cameroon and East Africa. **Chorology:** Afromontane. **Star:** blue. **Alt. range:** 2600-2800m (1), 3601-3800m (1). **Flowering:** Oct (1). **Specimens:** - *Eastern slopes:* Thomas D. W. 9403. *No locality:* Mann G. 1349

Festuca camusiana St.-Yves subsp. chodatiana St.-Yves

F.T.E.A. Gramineae: 57 (1970); F.W.T.A. 3: 369 (1972).
    Festuca chodatiana (St.-Yves) Alexeev, Agric. Univ. Wag. Papers 92(1): 60 (1992).

Habit: herb. **Habitat:** grassland. **Distribution:** Bioko, W. Cameroon and East Africa. **Chorology:** Afromontane. **Star:** blue. **Alt. range:** 2000-2200m (1). **Specimens:** - *No locality:* Mann G. 2069

Festuca mekiste W.D.Clayton

F.W.T.A. 3: 369 (1972); Agric. Univ. Wag. Papers 92(1): 60 (1992).
Habit: herb. **Habitat:** grassland. **Distribution:** Bioko, W. Cameroon and Kenya. **Chorology:** Afromontane. **Star:** blue. **Alt. range:** 2200-2400m (1). **Specimens:** - *No locality:* Mann G. 2078

Guaduella densiflora Pilg.

Fl. Gabon 5: 200 (1962); F.W.T.A. 3: 360 (1972); Agric. Univ. Wag. Papers 92(1): 25 (1992).
    Guaduella ledermannii Pilg., F.W.T.A. (ed.1) 2: 503 (1936).

Habit: herb. **Habitat:** forest. **Distribution:** SE Nigeria, Cameroon and Gabon. **Chorology:** Lower Guinea. **Star:** blue. **Alt. range:** 1-200m (1), 201-400m (1), 401-600m (1). **Flowering:** Sep (1), Oct (2). **Specimens:** - *Etinde:* Banks H. 28; Watts J. 488; Wheatley J. I. 506. *Southern Bakundu F.R.:* Brenan J. P. M. 9411

Guaduella humilis W.D.Clayton

F.W.T.A. 3: 360 (1972); Agric. Univ. Wag. Papers 92(1): 24 (1992).

Habit: herb. **Habitat:** forest. **Distribution:** SE Nigeria and SW Cameroon. **Chorology:** Western Cameroon. **Star:** gold. **Alt. range:** 1-200m (1). **Specimens:** - *Onge:* Thomas D. W. 9759

Guaduella macrostachys (K.Schum.) Pilger

F.W.T.A. 3: 360 (1972); Agric. Univ. Wag. Papers 92(1): 26 (1992).

Habit: herb. **Guild:** shade-bearer. **Habitat:** forest. **Distribution:** Ghana, Nigeria and Cameroon. **Chorology:** Guineo-Congolian. **Star:** green. **Alt. range:** 1-200m (6), 201-400m (2). **Flowering:** Apr (1), May (1), Oct (2), Nov (3). **Specimens:** - *Etinde:* Preuss P. R. s.n.. *Mokoko River F.R.:* Acworth J. M. 161, 294. *Onge:* Harris

D. J. 3716, 3756, 3858; Tchouto P. 847; Thomas D. W. 9761; Watts J. 795

Guaduella cf. macrostachys (K.Schum.) Pilger

Specimens: - *Etinde:* Cheek M. 5395

Guaduella oblonga Hutch. ex W.D.Clayton

Fl. Gabon 5: 206 (1962); F.W.T.A. 3: 360 (1972); Agric. Univ. Wag. Papers 92(1): 25 (1992).

Habit: herb. Habitat: forest. Distribution: Guinea to Cameroon and Gabon. Chorology: Guineo-Congolian. Star: green. Alt. range: 200-400m (1). Flowering: Oct (1). Specimens: - *Etinde:* Tchouto P. 466

Guaduella sp.

Specimens: - *Etinde:* Tchouto P. 467

Helictotrichon elongatum (Hochst ex A.Rich) C.E.Hubb.

F.W.T.A. 3: 372 (1972); Agric. Univ. Wag. Papers 92(1): 72 (1992).

Habit: herb. Habitat: grassland and forest-grassland transition. Distribution: Nigeria, Cameroon, Congo (Brazzaville), Sudan, Ethiopia, Zimbabwe and East Africa. Chorology: Afromontane. Star: green. Alt. range: 1800-2000m (1), 2001-2200m (2), 2201-2400m (1). Flowering: Oct (3). Specimens: - *Eastern slopes:* Tchouto P. 383; Thomas D. W. 9310. *Etinde:* Thomas D. W. 9367. *No locality:* Mann G. 2068

Helictotrichon mannii (Pilger) C.E.Hubb.

F.W.T.A. 3: 372 (1972); Agric. Univ. Wag. Papers 92(1): 71 (1992).

Habit: herb. Habitat: forest and forest-grassland transition. Distribution: Mount Cameroon and Bioko. Chorology: Western Cameroon Uplands (montane). Star: black. Alt. range: 800-1000m (1), 1601-1800m (1), 2001-2200m (1). Flowering: Oct (1), Nov (1). Specimens: - *Eastern slopes:* Marsden J. 36; Tchouto P. 382. *No locality:* Mann G. 2089

Hyparrhenia rufa (Nees) Stapf

Fl. Gabon 5: 181 (1962); F.W.T.A. 3: 492 (1972); Agric. Univ. Wag. Papers 92(1): 477 (1992).

Habit: herb. Guild: pioneer. Distribution: pantropical. Star: excluded. Specimens: - *Etinde:* Maitland T. D. 150a

Hyparrhenia smithiana (Hook.f.) Stapf var. major W.D.Clayton

F.W.T.A. 3: 491 (1972); Agric. Univ. Wag. Papers 92(1): 473 (1992).
Hyparrhenia chrysangyrea (Stapf) Stapf, F.W.T.A. (ed.1) 2: 591 (1936).

Habit: herb. Habitat: grassland. Distribution: Guinea to Congo (Brazzaville). Chorology: Guineo-Congolian (montane). Star: green. Alt. range: 1800-2000m (1), 2401-2600m (1). Flowering: Oct (2). Specimens: - *Eastern slopes:* Thomas D. W. 9264; Watts J. 448

Hypseochloa cameroonensis C.E.Hubb.

F.W.T.A. 3: 374 (1972); Agric. Univ. Wag. Papers 92(1): 87 (1992).

Habit: herb. Habitat: grassland. Distribution: Mount Cameroon. Chorology: Endemic (montane). Star: black. Alt. range: 2600-2800m (1). Specimens: - *No locality:* Mildbraed G. W. J. 10881

Ichnanthus vicinus (Baill.) Merr.

F.W.T.A. 3: 436 (1972); Agric. Univ. Wag. Papers 92(1): 194 (1992).

Habit: herb. Habitat: swamp forest. Distribution: pantropical. Star: excluded. Alt. range: 200-400m (1). Flowering: Oct (1). Specimens: - *Onge:* Tchouto P. 903

Imperata cylindrica (L.) Raeuschel

F.W.T.A. 3: 464 (1972); F.T.E.A. Gramineae: 700 (1982); Agric. Univ. Wag. Papers 92(1): 389 (1992).
Imperata cylindrica (L.) Raeuschel var. *africana* (Anderss.) C.E.Hubb., F.W.T.A. 3: 464 (1972).

Habit: herb. Guild: pioneer. Distribution: tropical and subtropical Africa. Star: excluded. Specimens: - *Eastern slopes:* Maitland T. D. 325

Isachne buettneri Hack.

Fl. Gabon 5: 109 (1962); F.W.T.A. 3: 420 (1972); Agric. Univ. Wag. Papers 92(1): 353 (1992).

Habit: herb. Guild: pioneer. Habitat: forest. Distribution: Guinea Bissau to Bioko, Cameroon, Congo (Brazzaville), Congo (Kinshasa), Uganda and Zambia. Chorology: Guineo-Congolian (montane). Star: green. Alt. range: 1-200m (1), 201-400m (1), 401-600m (1), 601-800m (1), 1601-1800m (1). Flowering: May (1), Jul (1), Oct (1), Nov (2). Specimens: - *Etinde:* Cable S. 215; Cheek M. 3598. *Mokoko River F.R.:* Fraser P. 384, 460. *Onge:* Cheek M. 4967; Thomas D. W. 9760

Isachne mauritiana Kunth

F.W.T.A. 3: 420 (1972); Agric. Univ. Wag. Papers 92(1): 354 (1992).

Habit: herb. Guild: pioneer. Habitat: forest and forest-grassland transition. Distribution: Ghana, Nigeria, Cameroon and Congo (Kinshasa) to East Africa. Chorology: Afromontane. Star: green. Alt. range: 1400-1600m (1), 1601-1800m (2). Flowering: Oct (2). Specimens: - *Eastern slopes:* Boughey A. S. GC 6762; Thomas D. W. 9507. *Etinde:* Thomas D. W. 9534; Wheatley J. I. 627

Koeleria capensis (Steud.) Nees

F.W.T.A. 3: 371 (1972); Agric. Univ. Wag. Papers 92(1): 66 (1992).

Habit: herb. Habitat: grassland. Distribution: Mount Cameroon, then Ethiopia to East and southern Africa. Chorology: Afromontane. Star: blue. Alt. range: 2400-2600m (1), 2601-2800m (1). Flowering: Oct (1). Specimens: - *Eastern slopes:* Thomas D. W. 9405. *No locality:* Mann G. 1357

Leptaspis zeylanica Nees

Agric. Univ. Wag. Papers 92(1): 39 (1992).
Leptaspis cochleata Thw., F.W.T.A. 3: 362 (1972).

Habit: herb. **Guild**: shade-bearer. **Habitat**: forest. **Distribution**: tropical Africa. **Chorology**: Afromontane. **Star**: green. **Alt. range**: 800-1000m (1), 1001-1200m (4), 1201-1400m (2). **Flowering**: Feb (1), May (1), Oct (1), Dec (2). **Fruiting**: Sep (1). **Specimens**: - *Eastern slopes*: Mbani J. M. 115. *Etinde*: Cable S. 496; Cheek M. 3778; Lighava M. 4; Sunderland T. C. H. 1086; Tekwe C. 176. *No locality*: Mann G. 1361

Loudetia simplex (Nees) C.E.Hubb.

Fl. Gabon 5: 262 (1962); F.W.T.A. 3: 419 (1972); F.T.E.A. Gramineae: 418 (1974); Agric. Univ. Wag. Papers 92(1): 377 (1992).
Loudetia camerunensis (Stapf) C.E.Hubb., F.W.T.A. (ed.1) 2: 544 (1936).

Habit: herb. **Habitat**: grassland and forest-grassland transition. **Distribution**: tropical Africa. **Chorology**: Afromontane. **Star**: green. **Alt. range**: 1600-1800m (1), 1801-2000m (1), 2001-2200m (1), 2201-2400m (2). **Flowering**: Oct (2), Nov (1). **Fruiting**: Oct (1). **Specimens**: - *Eastern slopes*: Marsden J. 42; Tchouto P. 266, 353; Thomas D. W. 9266. *No locality*: Mann G. 2080

Megastachya mucronata (Poir.) P.Beauv.

Fl. Gabon 5: 213 (1962); F.W.T.A. 3: 381 (1972); Agric. Univ. Wag. Papers 92(1): 84 (1992).

Habit: herb. **Guild**: pioneer. **Habitat**: forest. **Distribution**: tropical and subtropical Africa. **Star**: excluded. **Alt. range**: 1-200m (1). **Specimens**: - *Mokoko River F.R.*: Acworth J. M. 231

Melinis minutiflora P.Beauv.

F.W.T.A. 3: 455 (1972); F.T.E.A. Gramineae: 506 (1982); Agric. Univ. Wag. Papers 92(1): 304 (1992).
Melinis minutiflora P.Beauv. var. *setigera* W.D.Clayton, F.W.T.A. 3: 455 (1972).
Habit: herb. **Distribution**: native in tropical Africa and introduced throughout the tropics. **Star**: excluded. **Specimens**: - *Eastern slopes*: Maitland T. D. 102

Melinis repens (Willd.) Zizka

Fl. Zamb. 10: 116 (1989); Agric. Univ. Wag. Papers 92(1): 302 (1992).
Rhynchelytrum repens (Willd.) C.E.Hubb., F.W.T.A. 3: 454 (1972).

Habit: herb. **Distribution**: tropical and subtropical Africa. **Star**: excluded. **Specimens**: - *Eastern slopes*: Maitland T. D. 324

Microcalamus barbinodis Franch.

Fl. Gabon 5: 18 (1962); F.W.T.A. 3: 436 (1972); Agric. Univ. Wag. Papers 92(1): 244 (1992).

Habit: herb. **Habitat**: forest. **Distribution**: SW Cameroon, Congo (Brazzaville) and Congo (Kinshasa). **Chorology**: Guineo-Congolian. **Star**: green. **Alt. range**: 1-200m (2), 201-400m (2). **Flowering**: Apr (1), Oct (2). **Fruiting**: Oct (2). **Specimens**: - *Mokoko River F.R.*: Tchouto P. 1158. *Onge*: Tchouto P. 745, 981; Thomas D. W. 9852. *Southern Bakundu F.R.*: Brenan J. P. M. 9442

Microchloa indica (L.f.) P.Beauv.

F.W.T.A. 3: 403 (1972); Agric. Univ. Wag. Papers 92(1): 183 (1992).

Habit: herb. **Habitat**: forest-grassland transition. **Distribution**: pantropical. **Chorology**: Montane. **Star**: excluded. **Alt. range**:

1800-2000m (1). **Specimens**: - *Eastern slopes*: Thomas D. W. 9445

Microstegium vimineum (Trin.) A.Camus

F.W.T.A. 3: 477 (1972); Agric. Univ. Wag. Papers 92(1): 397 (1992).

Habit: herb. **Guild**: pioneer. **Distribution**: Mount Cameroon, Congo (Kinshasa) and Asia. **Chorology**: Montane. **Star**: blue. **Alt. range**: 1200-1400m (1). **Specimens**: - *Etinde*: Maitland T. D. 1086

Olyra latifolia L.

Fl. Gabon 5: 278 (1962); F.W.T.A. 3: 362 (1972); Agric. Univ. Wag. Papers 92(1): 38 (1992).

Habit: herb. **Guild**: shade-bearer. **Habitat**: forest. **Distribution**: tropical Africa and America. **Chorology**: Montane. **Star**: excluded. **Alt. range**: 400-600m (1). **Flowering**: Nov (1). **Fruiting**: Nov (2). **Specimens**: - *Eastern slopes*: Maitland T. D. 567. *Etinde*: Cable S. 262. *Mabeta-Moliwe*: Cheek M. 3551; Nguembock F. 41

Oplismenus burmannii (Retz.) P.Beauv.

Fl. Gabon 5: 58 (1962); F.W.T.A. 3: 437 (1972); Agric. Univ. Wag. Papers 92(1): 192 (1992).

Habit: herb. **Guild**: pioneer. **Distribution**: pantropical. **Star**: excluded. **Specimens**: - *Etinde*: Maitland T. D. 69

Oplismenus hirtellus (L.) P.Beauv.

Fl. Gabon 5: 56 (1962); F.W.T.A. 3: 437 (1972); Agric. Univ. Wag. Papers 92(1): 193 (1992).

Habit: herb. **Guild**: pioneer. **Habitat**: forest. **Distribution**: pantropical. **Chorology**: Montane. **Star**: excluded. **Alt. range**: 1-200m (4), 601-800m (1), 1401-1600m (1), 2001-2200m (1). **Flowering**: Oct (2), Dec (2). **Fruiting**: Dec (1). **Specimens**: - *Eastern slopes*: Thomas D. W. 9488. *Etinde*: Cable S. 238; Cheek M. 5834; Maitland T. D. 83; Marsden J. 19; Thomas D. W. 9357. *Mabeta-Moliwe*: Cable S. 532; Cheek M. 5778

Panicum acrotrichum Hook.f.

F.W.T.A. 3: 433 (1972); Agric. Univ. Wag. Papers 92(1): 219 (1992).

Habit: herb. **Guild**: pioneer. **Habitat**: forest. **Distribution**: W. Cameroon and Bioko. **Chorology**: Western Cameroon Uplands (montane). **Star**: gold. **Alt. range**: 600-800m (2). **Flowering**: Oct (2). **Specimens**: - *No locality*: Mann G. 2100. *Onge*: Marsden J. 34, 35

Panicum brevifolium L.

Fl. Gabon 5: 62 (1962); F.W.T.A. 3: 429 (1972); Agric. Univ. Wag. Papers 92(1): 207 (1992).

Habit: herb. **Guild**: pioneer. **Habitat**: forest and farmbush. **Distribution**: palaeotropical. **Star**: excluded. **Alt. range**: 1-200m (2), 601-800m (2). **Flowering**: Oct (3). **Fruiting**: Dec (1). **Specimens**: - *Eastern slopes*: Maitland T. D. 20; Marsden J. 16. *Mabeta-Moliwe*: Cheek M. 5779. *Onge*: Cheek M. 5158; Marsden J. 31

Panicum calvum Stapf
F.W.T.A. 3: 433 (1972); Agric. Univ. Wag. Papers 92(1): 222 (1992).

Habit: herb. Guild: pioneer. Habitat: forest and farmbush. Distribution: tropical Africa. Chorology: Afromontane. Star: green. Alt. range: 1400-1600m (1), 1601-1800m (1). Flowering: Nov (2). Specimens: - *Eastern slopes:* Marsden J. 41, 45. *No locality:* Hinds J. H. C 63

Panicum comorense Mez
F.W.T.A. 3: 434 (1972); Agric. Univ. Wag. Papers 92(1): 228 (1992).

Habit: herb. Guild: pioneer. Habitat: forest. Distribution: tropical Africa. Chorology: Afromontane. Star: green. Alt. range: 1400-1600m (1). Specimens: - *Eastern slopes:* Thomas D. W. 9506

Panicum hochstetteri Steud.
F.W.T.A. 3: 433 (1972); Agric. Univ. Wag. Papers 92(1): 221 (1992).

Habit: herb. Guild: pioneer. Habitat: forest-grassland transition. Distribution: tropical Africa. Chorology: Afromontane. Star: green. Alt. range: 1400-1600m (1), 1601-1800m (1), 1801-2000m (1), 2201-2400m (1). Flowering: Oct (2), Nov (2). Specimens: - *Eastern slopes:* Marsden J. 43, 44; Thomas D. W. 9297, 9385. *No locality:* Mann G. 2082

Panicum hymeniochilum Nees
F.W.T.A. 3: 429 (1972); Agric. Univ. Wag. Papers 92(1): 205 (1992).

Habit: herb. Guild: pioneer. Distribution: tropical Africa. Star: excluded. Specimens: - *No locality:* Boughey A. S. GC 12694

Panicum laxum Sw.
F.W.T.A. 3: 429 (1972); Agric. Univ. Wag. Papers 92(1): 206 (1992).

Habit: herb. Guild: pioneer. Habitat: forest. Distribution: tropical America, naturalized in Africa. Star: excluded. Alt. range: 1-200m (4). Flowering: Oct (1), Dec (1). Specimens: - *Mabeta-Moliwe:* Cheek M. 5786; Marsden J. 21. *Mokoko River F.R.:* Acworth J. M. 75. *Onge:* Thomas D. W. 9830

Panicum maximum Jacq.
Fl. Gabon 5: 64 (1962); F.W.T.A. 3: 429 (1972); Agric. Univ. Wag. Papers 92(1): 208 (1992).

Habit: herb. Guild: pioneer. Distribution: pantropical. Star: excluded. Specimens: - *Etinde:* Maitland T. D. 15

Panicum monticola Hook.f.
F.W.T.A. 3: 433 (1972); Agric. Univ. Wag. Papers 92(1): 221 (1992).

Habit: herb. Guild: pioneer. Habitat: forest. Distribution: tropical Africa. Chorology: Afromontane. Star: green. Alt. range: 800-1000m (1). Flowering: Dec (1). Specimens: - *Etinde:* Cable S. 501. *No locality:* Mann G. 1353

Panicum pusillum Hook.f.
F.W.T.A. 3: 433 (1972); Agric. Univ. Wag. Papers 92(1): 223 (1992).

Habit: herb. Guild: pioneer. Habitat: grassland. Distribution: Sierra Leone to Congo (Kinshasa), Ethiopia, Malawi and East Africa. Chorology: Afromontane. Star: green. Alt. range: 2600-2800m (1). Flowering: Oct (1). Specimens: - *Eastern slopes:* Thomas D. W. 9401. *No locality:* Boughey A. S. A 307

Panicum repens L.
Fl. Gabon 5: 72 (1962); F.W.T.A. 3: 434 (1972); Agric. Univ. Wag. Papers 92(1): 225 (1992).

Habit: herb. Guild: pioneer. Distribution: pantropical. Star: excluded. Specimens: - *Etinde:* Maitland T. D. 84

Paspalum conjugatum Berg
Fl. Gabon 5: 39 (1962); F.W.T.A. 3: 445 (1972); Agric. Univ. Wag. Papers 92(1): 279 (1992).

Habit: herb. Guild: pioneer. Habitat: forest. Distribution: pantropical. Star: excluded. Alt. range: 1-200m (3), 601-800m (2). Flowering: Apr (2), Oct (2), Dec (1). Specimens: - *Eastern slopes:* Maitland T. D. 151; Marsden J. 16a. *Mabeta-Moliwe:* Cheek M. 3441; 5789. *Mokoko River F.R.:* Acworth J. M. 73; Akogo M. 247. *Onge:* Marsden J. 33

Paspalum lamprocaryon K.Schum.
Fl. Zamb. 10(3): 90 (1989); Agric. Univ. Wag. Papers 92(1): 283 (1992).
> *Paspalum auriculatum* sensu Clayton, F.W.T.A. 3: 445 (1972).
> *Paspalum scrobiculatum* L. var. *lanceolatum* Konig & Sosef, Blumea 30: 312 (1985).

Habit: herb. Guild: pioneer. Habitat: forest. Distribution: tropical Africa. Chorology: Afromontane. Star: green. Alt. range: 1-200m (2), 1401-1600m (1). Flowering: Oct (1). Fruiting: Oct (1). Specimens: - *Eastern slopes:* Tchouto P. 371. *Onge:* Cheek M. 5209; Thomas D. W. 9788

Paspalum paniculatum L.
Fl. Gabon 5: 42 (1962); F.W.T.A. 3: 446 (1972); Agric. Univ. Wag. Papers 92(1): 280 (1992).

Habit: herb. Habitat: forest. Distribution: Liberia to Angola and Uganda, also in Polynesia, South America, New Guinea and Australia. Chorology: Montane. Star: excluded. Alt. range: 1-200m (5), 201-400m (1), 1401-1600m (1). Flowering: Oct (4), Dec (1). Specimens: - *Eastern slopes:* Tchouto P. 372. *Etinde:* Marsden J. 17, 29. *Mabeta-Moliwe:* Cheek M. 5785; Maitland T. D. 976; Marsden J. 23. *Mokoko River F.R.:* Acworth J. M. 232; Thomas D. W. 10019

Paspalum scrobiculatum L.
F.T.E.A. Gramineae: 610 (1982); Agric. Univ. Wag. Papers 92(1): 284 (1992).
> *Paspalum orbiculare* G.Forst., F.W.T.A. 3: 446 (1972).
> *Paspalum polystachyum* R.Br., F.W.T.A. 3: 446 (1972).

Habit: herb. Distribution: palaeotropical. Star: excluded. Specimens: - *Etinde:* Maitland T. D. 5. *Mabeta-Moliwe:* Cheek M. 3547

Paspalum vaginatum Sw.

Fl. Gabon 5: 40 (1962); F.W.T.A. 3: 445 (1972); Agric. Univ. Wag. Papers 92(1): 282 (1992).

Habit: herb. **Distribution:** pantropical. **Star:** excluded. **Specimens: -** *Etinde:* Brunt M. A. 1042

Pennisetum hordeoides (Lam.) Steud.

Fl. Gabon 5: 99 (1962); F.W.T.A. 3: 461 (1972); Agric. Univ. Wag. Papers 92(1): 336 (1992).

Habit: herb. **Guild:** pioneer. **Habitat:** roadsides and villages. **Distribution:** Senegal to Bioko and Congo (Kinshasa), also India. **Star:** excluded. **Alt. range:** 1-200m (2). **Flowering:** Oct (2). **Specimens: -** *Eastern slopes:* Marsden J. 26. *Etinde:* Marsden J. 16b

Pennisetum laxior (W.D.Clayton) W.D.Clayton

Kew Bull. 32: 580 (1978); Agric. Univ. Wag. Papers 92(1): 338 (1992).
Beckeropsis laxior W.D.Clayton, F.W.T.A. 3: 459 (1972).

Habit: herb. **Guild:** pioneer. **Distribution:** Ghana, Nigeria, Cameron, S. Tomé, Annobon and Sudan. **Chorology:** Guineo-Congolian. **Star:** green. **Specimens: -** *Eastern slopes:* Migeod F. 107

Pennisetum monostigma Pilg.

F.W.T.A. 3: 461 (1972); Agric. Univ. Wag. Papers 92(1): 343 (1992).

Habit: herb. **Guild:** pioneer. **Habitat:** grassland. **Distribution:** Sierra Leone to Cameroon and Bioko. **Chorology:** Guineo-Congolian (montane). **Star:** green. **Alt. range:** 1600-1800m (1), 1801-2000m (1), 2001-2200m (1), 2201-2400m (1), 2401-2600m (1). **Flowering:** Sep (1), Oct (2), Nov (1). **Specimens: -** *Eastern slopes:* Brenan J. P. M. 9535; Marsden J. 40; Thomas D. W. 9278, 9418. *Etinde:* Tekwe C. 276

Pennisetum polystachion (L.) Schult. subsp. polystachion

Fl. Gabon 5: 97 (1962); F.W.T.A. 3: 460 (1972); F.T.E.A. Gramineae: 679 (1982); Agric. Univ. Wag. Papers 92(1): 335 (1992).
Pennisetum subangustum (Schumach.) Stapf & C.E.Hubbard, F.W.T.A. 3: 461 (1972).

Habit: herb. **Guild:** pioneer. **Distribution:** pantropical. **Star:** excluded. **Specimens: -** *Etinde:* Maitland T. D. 153

Pennisetum purpureum Schumach.

Fl. Gabon 5: 94 (1962); F.W.T.A. 3: 461 (1972); Agric. Univ. Wag. Papers 92(1): 338 (1992).

Habit: herb. **Guild:** pioneer. **Habitat:** stream banks in forest. **Distribution:** tropical Africa, but introduced throughout the tropics. **Star:** excluded. **Alt. range:** 1-200m (3). **Flowering:** Oct (1), Nov (1). **Fruiting:** Oct (1). **Specimens: -** *Eastern slopes:* Marsden J. 27; Migeod F. 3. *Onge:* Cheek M. 5208; Thomas D. W. 9899

Pennisetum trachyphyllum Pilg.

F.W.T.A. 3: 461 (1972); Agric. Univ. Wag. Papers 92(1): 328 (1992).

Habit: herb. **Guild:** pioneer. **Distribution:** Cameroon, Congo (Kinshasa), Sudan, Ethiopia and East Africa. **Chorology:** Afromontane. **Star:** green. **Specimens: -** *Eastern slopes:* Maitland T. D. 341

Pennisetum unisetum (Nees) Benth.

F.T.E.A. Gramineae: 681 (1982); Agric. Univ. Wag. Papers 92(1): 338 (1992).
Beckeropsis uniseta (Nees) K.Schum., F.W.T.A. 3: 457 (1972).

Habit: herb. **Distribution:** tropical Africa. **Star:** excluded. **Specimens: -** *Bambuko F.R.:* Argent G. C. G. 974

Pentaschistis mannii C.E.Hubb.

F.T.E.A. Gramineae: 127 (1970); Agric. Univ. Wag. Papers 92(1): 87 (1992).
Pentaschistis pictigluma (Steud.) Pilg., F.W.T.A. 3: 374 (1972).

Habit: herb. **Habitat:** grassland. **Distribution:** Mount Cameroon. **Chorology:** Endemic (montane). **Star:** black. **Alt. range:** 3000-3200m (1). **Specimens: -** *No locality:* Mildbraed G. W. J. 10894

Perotis indica (L.) Kuntze

F.W.T.A. 3: 411 (1972).

Habit: herb. **Habitat:** open areas in forest. **Distribution:** tropical Asia, introduced to Africa. **Star:** excluded. **Alt. range:** 1-200m (1). **Flowering:** Apr (1). **Specimens: -** *Mokoko River F.R.:* Acworth J. M. 238

Poa annua L.

F.W.T.A. 3: 369 (1972); Agric. Univ. Wag. Papers 92(1): 63 (1992).

Habit: herb. **Habitat:** grassland. **Distribution:** worldwide in temperate regions and on tropical mountains. **Chorology:** Montane. **Star:** excluded. **Alt. range:** 1800-2000m (1). **Specimens: -** *No locality:* Maitland T. D. 1262

Poa leptoclada Hochst. ex A.Rich.

F.W.T.A. 3: 369 (1972); Agric. Univ. Wag. Papers 92(1): 61 (1992).

Habit: herb. **Habitat:** grassland. **Distribution:** tropical Africa. **Chorology:** Afromontane. **Star:** green. **Alt. range:** 2400-2600m (1). **Specimens: -** *No locality:* Mann G. 2071

Poa schimperiana Hochst. ex A.Rich.

F.W.T.A. 3: 369 (1972); Agric. Univ. Wag. Papers 92(1): 62 (1992).

Habit: herb. **Habitat:** grassland. **Distribution:** tropical Africa. **Chorology:** Afromontane. **Star:** green. **Alt. range:** 2600-2800m (1). **Specimens: -** *No locality:* Migeod F. 173

Poecilostachys oplismenoides (Hack.) W.D.Clayton

Agric. Univ. Wag. Papers 92(1): 189 (1992).
Chloachne oplismenoides (Hack.) Stapf ex Robyns, F.W.T.A. 3: 436 (1972).

Habit: herb. **Habitat:** forest. **Distribution:** Nigeria, Bioko, Cameroon, Sudan, Ethiopia, East Africa, Malawi and Zimbabwe. **Chorology:** Afromontane. **Star:** green. **Alt. range:** 1200-1400m

(1), 1601-1800m (1). **Specimens:** - *Eastern slopes:* Thomas D. W. 9491. *No locality:* Mann G. 1354

Polytrias indica (Houtt.) Veldkamp
Blumea 36: 180 (1991); Agric. Univ. Wag. Papers 92(1): 396 (1992).
Polytrias amaura (Büse) Kuntze, F.W.T.A. 3: 468 (1972).

Habit: herb. **Guild:** pioneer. **Distribution:** introduced. **Star:** excluded. **Specimens:** - *Limbe Botanic Garden:* Keay R. W. J. FHI 28661; Maitland T. D. 98, 155

Pseudechinolaena polystachya (Kunth) Stapf
Fl. Gabon 5: 25 (1962); F.W.T.A. 3: 436 (1972); Agric. Univ. Wag. Papers 92(1): 191 (1992).

Habit: herb. **Guild:** pioneer. **Habitat:** forest. **Distribution:** pantropical. **Chorology:** Montane. **Star:** excluded. **Alt. range:** 1-200m (1), 401-600m (1), 1401-1600m (1), 1601-1800m (1). **Flowering:** Oct (2), Nov (2). **Specimens:** - *Eastern slopes:* Keay R. W. J. FHI 28577; Marsden J. 38; Thomas D. W. 9459. *Onge:* Cheek M. 4979; Thomas D. W. 9871

Rottboellia cochinchinensis (Lour.) W.D.Clayton
Kew Bull. 35: 815 (1981); F.T.E.A. Gramineae: 853 (1982); Agric. Univ. Wag. Papers 92(1): 534 (1992).
Rottboellia exaltata L., F.W.T.A. 3: 506 (1958).

Habit: herb. **Guild:** pioneer. **Distribution:** palaeotropical. **Star:** excluded. **Specimens:** - *Etinde:* Maitland T. D. 900

Schizachyrium maclaudii (Jac.-Fél.) S.T.Blake
F.W.T.A. 3: 479 (1972); Agric. Univ. Wag. Papers 92(1): 453 (1992).

Habit: herb. **Distribution:** Guinea to Cameroon, also South America. **Star:** excluded. **Specimens:** - *Etinde:* Maitland T. D. 1292 (partly)

Schizachyrium platyphyllum (Franch.) Stapf
Fl. Gabon 5: 157 (1962); F.W.T.A. 3: 478 (1972); Agric. Univ. Wag. Papers 92(1): 452 (1992).

Habit: herb. **Guild:** pioneer. **Distribution:** Senegal to Cameroon, Congo (Brazzaville), Congo (Kinshasa), Sudan, Uganda and Malawi. **Star:** green. **Specimens:** - *Bambuko F.R.:* Argent G. C. G. 973

Schizachyrium pulchellum (Don ex Benth.) Stapf
F.W.T.A. 3: 481 (1972); Agric. Univ. Wag. Papers 92(1): 459 (1992).

Habit: herb. **Habitat:** coastal forest. **Distribution:** Senegal to Congo (Brazzaville). **Chorology:** Guineo-Congolian. **Star:** green. **Specimens:** - *Mabeta-Moliwe:* Plot Voucher W 1210

Schizachyrium sp.

Specimens: - *Etinde:* Williams S. 92

Setaria barbata (Lam.) Kunth
F.W.T.A. 3: 424 (1972); Agric. Univ. Wag. Papers 92(1): 289 (1992).

Habit: herb. **Guild:** pioneer. **Habitat:** forest. **Distribution:** Senegal to Bioko, Cameroon, Congo (Kinshasa), Angola and Sudan, introduced throughout the tropics. **Star:** excluded. **Alt. range:** 1-200m (2). **Flowering:** Oct (2). **Specimens:** - *Eastern slopes:* Marsden J. 25. *Etinde:* Maitland T. D. 78. *Mabeta-Moliwe:* Marsden J. 22

Setaria megaphylla (Steud.) T.Durand & Schinz
Fl. Gabon 5: 81 (1962); F.W.T.A. 3: 424 (1972); F.T.E.A. Gramineae: 539 (1982); Agric. Univ. Wag. Papers 92(1): 292 (1992).
Setaria chevalieri Stapf, F.W.T.A. 3: 424 (1972).

Habit: herb. **Guild:** pioneer. **Habitat:** forest and farmbush. **Distribution:** tropical Africa. **Chorology:** Afromontane. **Star:** green. **Alt. range:** 1-200m (1), 1401-1600m (1). **Flowering:** Nov (1). **Fruiting:** Nov (1). **Specimens:** - *Eastern slopes:* Marsden J. 46; Migeod F. 1. *Etinde:* Maitland T. D. 82. *Onge:* Harris D. J. 3872

Setaria poiretiana (Schult.) Kunth
Fl. Rwanda 4: 371 (1988); Agric. Univ. Wag. Papers 92(1): 293 (1992).
Setaria caudula Stapf, F.W.T.A. 3: 424 (1972).

Habit: herb. **Guild:** pioneer. **Distribution:** Nigeria, Cameroon, Bioko, Congo (Kinshasa), Sudan, Ethiopia and East Africa, introduced in tropical Asia and America. **Chorology:** Montane. **Star:** excluded. **Alt. range:** 600-800m (1). **Specimens:** - *No locality:* Mann G. 2102

Sorghastrum bipennatum (Hack.) Pilger
F.W.T.A. 3: 468 (1972); Agric. Univ. Wag. Papers 92(1): 406 (1992).

Habit: herb. **Guild:** pioneer. **Distribution:** tropical Africa. **Star:** excluded. **Specimens:** - *Bambuko F.R.:* Argent G. C. G. 976

Sorghum arundinaceum (Desv.) Stapf
Fl. Gabon 5: 144 (1962); F.W.T.A. 3: 467 (1972); F.T.E.A. Gramineae: 728 (1982); Agric. Univ. Wag. Papers 92(1): 402 (1992).
Sorghum vogelianum (Piper) Stapf, F.W.T.A. 3: 467 (1972).

Habit: herb to 3 m. **Guild:** pioneer. **Distribution:** tropical Africa, Asia and Australia, introduced in South America. **Star:** excluded. **Specimens:** - *Etinde:* Maitland T. D. 977

Sporobolus africanus (Poir.) Robyns & Tournay
F.W.T.A. 3: 410 (1972); F.T.E.A. Gramineae: 375 (1974).
Sporobolus indicus (L.) R.Br. var. *capensis* Engl., Agric. Univ. Wag. Papers 92(1): 157 (1992).

Habit: herb. **Habitat:** forest-grassland transition. **Distribution:** tropical and subtropical Africa, Asia and Australia. **Chorology:** Montane. **Star:** excluded. **Alt. range:** 2000-2200m (1). **Specimens:** - *Eastern slopes:* Morton J. K. K 856

Sporobolus molleri Hack.

Fl. Gabon 5: 253 (1962); F.W.T.A. 3: 408 (1972); Agric. Univ. Wag. Papers 92(1): 151 (1992).

Habit: herb. **Distribution:** tropical Africa. **Star:** excluded. **Specimens:** - *No locality:* Brenan J. P. M. 9262

Sporobolus montanus Engl.

F.W.T.A. 3: 407 (1972); Agric. Univ. Wag. Papers 92(1): 149 (1992).

Habit: herb. **Distribution:** W. Cameroon and NE Nigeria. **Chorology:** Western Cameroon Uplands (montane). **Star:** gold. **Flowering:** Jan (1), Dec (1). **Specimens:** - *No locality:* Boughey A. S. GC 12543; Maitland T. D. 1269

Sporobolus pyramidalis P.Beauv.

Fl. Gabon 5: 254 (1962); F.W.T.A. 3: 408 (1972); F.T.E.A. Gramineae: 373 (1974).
  Sporobolus indicus (L.) R.Br. var. *pyramidalis* (P.Beauv.) Veldkamp, Agric. Univ. Wag. Papers 92(1): 155 (1992).

Habit: herb. **Guild:** pioneer. **Habitat:** roadsides and villages. **Distribution:** tropical and subtropical Africa. **Star:** excluded. **Alt. range:** 1-200m (1). **Flowering:** Oct (1). **Specimens:** - *Etinde:* Maitland T. D. 173; Marsden J. 18

Stenotaphrum secundatum (Walt.) Kuntze

Fl. Gabon 5: 28 (1962); F.W.T.A. 3: 435 (1972); Agric. Univ. Wag. Papers 92(1): 300 (1992).

Habit: herb. **Distribution:** Atlantic coasts of tropical America and Africa and the Pacific. **Star:** excluded. **Specimens:** - *Etinde:* Maitland T. D. 25

Streblochaete longiarista (A.Rich.) Pilg.

F.W.T.A. 3: 371 (1972); Agric. Univ. Wag. Papers 92(1): 65 (1992).

Habit: herb. **Habitat:** forest-grassland transition. **Distribution:** tropical and subtropical Africa, also SE Asia. **Chorology:** Montane. **Star:** excluded. **Alt. range:** 1800-2000m (1), 2401-2600m (1). **Flowering:** Oct (1). **Specimens:** - *Eastern slopes:* Thomas D. W. 9424. *No locality:* Mann G. 2077

Trichopteryx elegantula (Hook.f.) Stapf

F.W.T.A. 3: 420 (1972); Agric. Univ. Wag. Papers 92(1): 365 (1992).

Habit: herb. **Habitat:** grassland and forest-grassland transition. **Distribution:** tropical and subtropical Africa. **Chorology:** Afromontane. **Star:** green. **Alt. range:** 1800-2000m (2), 2001-2200m (2), 2201-2400m (2). **Flowering:** Oct (3). **Specimens:** - *Eastern slopes:* Tchouto P. 288, 295, 342; Thomas D. W. 9437, 9444. *No locality:* Migeod F. 204

Tripogon major Hook.f.

F.W.T.A. 3: 393 (1972); Agric. Univ. Wag. Papers 92(1): 139 (1992).

Habit: herb. **Habitat:** grassland and forest-grassland transition. **Distribution:** tropical Africa. **Chorology:** Afromontane. **Star:** green. **Alt. range:** 2000-2200m (2), 2201-2400m (1), 2601-2800m (1). **Flowering:** Oct (2), Nov (1). **Specimens:** - *Eastern slopes:*

Boughey A. S. GC 12512; Cheek M. 5332; Tchouto P. 289; Thomas D. W. 9417

Vulpia bromoides (L.) S.F.Gray

F.W.T.A. 3: 369 (1972); Agric. Univ. Wag. Papers 92(1): 58 (1992).

Habit: herb. **Habitat:** grassland. **Distribution:** Europe and tropical African mountains. **Chorology:** Montane. **Star:** excluded. **Alt. range:** 2400-2600m (1), 2601-2800m (2). **Flowering:** Oct (2). **Specimens:** - *Eastern slopes:* Thomas D. W. 9406, 9410. *No locality:* Mann G. 2091

HYPOXIDACEAE

I. Nordal (O)

Hypoxis camerooniana Baker

Fl. Cameroun 30: 37 (1987).
  Hypoxis recurva Nel, F.W.T.A. 3: 172 (1968).

Habit: herb. **Habitat:** grassland and forest-grassland transition. **Distribution:** SE Nigeria and Cameroon. **Chorology:** Lower Guinea montane. **Star:** gold. **Alt. range:** 2200-2400m (1). **Specimens:** - *No locality:* Mann G. 1224. *Eastern Slopes:* Brenan J. P. M. 9581; Cheek M. 3623; Sidwell K. 58; Sunderland T.C.H. 1427; Tekwe C. 301; Thomas 9286

IRIDACEAE

Aristea ecklonii Baker

F.W.T.A. 3: 139 (1968); Fl. Zamb. 12(4): 5 (1993); F.T.E.A. Iridaceae: 4 (1996).

Habit: herb. **Distribution:** Cameroon, Congo (Kinshasa), Rwanda, Burundi, Uganda, Zimbabwe and South Africa. **Chorology:** Afromontane. **Star:** green. **Alt. range:** 1200-1400m (1). **Specimens:** - *Eastern slopes:* Maitland T. D. 27

Hesperantha petitiana (A.Rich.) Baker

Fl. Zamb. 12(4): 59 (1993); F.T.E.A. Iridaceae: 33 (1996).
  Hesperantha alpina (Hook.f.) Pax ex Engl., F.W.T.A. 3: 141 (1968).

Habit: herb. **Habitat:** grassland. **Distribution:** Cameroon, Ethiopia, Uganda, Malawi, Zimbabwe, Kenya and Tanzania. **Chorology:** Afromontane. **Star:** green. **Alt. range:** 2200-2400m (2), 2601-2800m (1). **Flowering:** Oct (1). **Fruiting:** Nov (1). **Specimens:** - *Eastern slopes:* Cheek M. 5357; Thomas D. W. 9315. *No locality:* Mann G. 2134

Romulea camerooniana Baker

F.W.T.A. 3: 139 (1968); Fl. Zamb. 12(4): 63 (1993); F.T.E.A. Iridaceae: 36 (1996).

Habit: herb. **Habitat:** grassland. **Distribution:** Cameroon to East and South Africa. **Chorology:** Afromontane. **Star:** green. **Alt. range:** 2000-2200m (1), 2201-2400m (1), 2801-3000m (1). **Flowering:** Sep (1), Oct (1). **Fruiting:** Sep (1). **Specimens:** - *Eastern slopes:* Thomas D. W. 9344. *Etinde:* Tekwe C. 302. *No locality:* Mann G. 2135

JUNCACEAE

Juncus capitatus Weig.
F.W.T.A. 3: 278 (1972); Haines R. & Lye K., Sedges of E Africa: 33 (1983).

Habit: herb. **Habitat:** grassland. **Distribution:** Mount Cameroon, North and East Africa, Atlantic Islands and southern Europe. **Chorology:** Montane. **Star:** blue. **Alt. range:** 2000-2200m (1). **Specimens:** - *No locality:* Mann G. 2094

Luzula campestris (L.) DC. var. mannii Buchenau
F.W.T.A. 3: 278 (1972).

Habit: herb. **Habitat:** grassland. **Distribution:** Bioko, Mount Cameroon, E. Congo (Kinshasa) and Uganda. **Chorology:** Guineo-Congolian (montane). **Star:** green. **Alt. range:** 3400-3600m (1). **Specimens:** - *No locality:* Mann G. 2108

MARANTACEAE

Afrocalathea rhizantha (K.Schum.) K.Schum.
Fl. Gabon 9: 134 (1964); Fl. Cameroun 4: 137 (1965); F.W.T.A. 3: 82 (1968).
Marantochloa sp. A sensu Hepper, F.W.T.A. 1: 83 (1958).

Habit: herb. **Habitat:** forest. **Distribution:** Nigeria, Cameroon and Gabon. **Chorology:** Lower Guinea. **Star:** blue. **Alt. range:** 1-200m (2). **Flowering:** Oct (1), Nov (1). **Specimens:** - *Onge:* Cheek M. 5128; Thomas D. W. 9755

Ataenidia conferta (Benth.) Milne-Redh.
Fl. Gabon 9: 131 (1964); Fl. Cameroun 4: 136 (1965); F.W.T.A. 3: 89 (1968).

Habit: herb. **Guild:** shade-bearer. **Habitat:** forest. **Distribution:** Ivory Coast to Cameroon, Gabon, Congo (Brazzaville), Congo (Kinshasa), Angola (Cabinda), CAR, Sudan and Uganda. **Chorology:** Guineo-Congolian. **Star:** green. **Alt. range:** 1-200m (4), 401-600m (1), 601-800m (1), 801-1000m (1). **Flowering:** Feb (1), May (2), Nov (2), Dec (1). **Specimens:** - *Etinde:* Cable S. 143, 383; Cheek M. 5602; Sunderland T. C. H. 1079. *Mabeta-Moliwe:* Sunderland T. C. H. 1398. *Mokoko River F.R.:* Watts J. 1130. *No locality:* Mann G. 2144

Halopegia azurea (K.Schum.) K.Schum.
Fl. Gabon 9: 136 (1964); Fl. Cameroun 4: 140 (1965); F.W.T.A. 3: 85 (1968).

Habit: herb. **Habitat:** forest, by streams. **Distribution:** Sierra Leone to Cameroon, Gabon, Congo (Kinshasa) and Angola (Cabinda). **Chorology:** Guineo-Congolian. **Star:** green. **Alt. range:** 1-200m (4), 201-400m (2). **Flowering:** Apr (1), Oct (2), Nov (1), Dec (1). **Specimens:** - *Etinde:* Cheek M. 5381. *Mabeta-Moliwe:* Cable S. 604; Sunderland T. C. H. 1230. *Onge:* Harris D. J. 3668; Ndam N. 724; Tchouto P. 890

Hypselodelphys scandens Louis & Mullend.
Fl. Gabon 9: 106 (1964); Fl. Cameroun 4: 113 (1965); F.W.T.A. 3: 89 (1968).

Habit: herb. **Guild:** pioneer. **Habitat:** forest. **Distribution:** Ivory Coast to Cameroon, Gabon and Congo (Kinshasa). **Chorology:** Guineo-Congolian. **Star:** green. **Alt. range:** 1-200m (1), 401-600m (1). **Flowering:** Mar (1), May (1). **Fruiting:** Mar (1), May (1), Oct (1). **Specimens:** - *Eastern slopes:* Tchouto P. 34. *Etinde:* Kwangue A. 87; Thomas D. W. 9150

Hypselodelphys violacea (Ridl.) Milne-Redh.
Fl. Gabon 9: 104 (1964); Fl. Cameroun 4: 112 (1965); F.W.T.A. 3: 88 (1968).

Habit: herb. **Guild:** pioneer. **Habitat:** forest. **Distribution:** Sierra Leone to Cameroon, Gabon, Congo (Kinshasa) and Angola. **Chorology:** Guineo-Congolian. **Star:** green. **Alt. range:** 1-200m (1). **Flowering:** Nov (1). **Specimens:** - *Onge:* Harris D. J. 3843

Hypselodelphys zenkeriana (K.Schum.) Milne-Redh.
Fl. Gabon 9: 103 (1964); Fl. Cameroun 4: 109 (1965); F.W.T.A. 3: 88 (1968).

Habit: herb. **Distribution:** Cameroon. **Chorology:** Lower Guinea. **Star:** gold. **Specimens:** - *Eastern slopes:* Deistel H. 206

Marantochloa congensis (K.Schum.) J.Léonard & Mullend. var. congensis
Fl. Gabon 9: 120 (1964); Fl. Cameroun 4: 123 (1965); F.W.T.A. 3: 83 (1968); Bull. Jard. Bot. Nat. Belg. 65: 376 (1996).

Habit: herb. **Guild:** pioneer. **Habitat:** forest. **Distribution:** Sierra Leone to Cameroon, Gabon, Congo (Brazzaville), Congo (Kinshasa) and Burundi. **Chorology:** Guineo-Congolian. **Star:** green. **Alt. range:** 1-200m (1). **Flowering:** Jun (1). **Specimens:** - *Mabeta-Moliwe:* Plot Voucher W 1157. *Onge:* Harris D. J. 3704

Marantochloa filipes (Benth.) Hutch.
Fl. Gabon 9: 122 (1964); Fl. Cameroun 4: 126 (1965); F.W.T.A. 3: 83 (1968); Bull. Jard. Bot. Nat. Belg. 65: 381 (1996).

Habit: herb to 2.5 m. **Guild:** pioneer. **Habitat:** forest and stream banks. **Distribution:** Guinea to Bioko, Cameroon, Gabon, Congo (Brazzaville), Congo (Kinshasa) and Angola. **Chorology:** Guineo-Congolian. **Star:** green. **Alt. range:** 1-200m (8), 201-400m (4), 401-600m (3), 601-800m (1). **Flowering:** Apr (1), Sep (1), Oct (1). **Fruiting:** Apr (1), Jun (1), Sep (1), Oct (6), Nov (3), Dec (4). **Specimens:** - *Etinde:* Cable S. 387; Cheek M. 5449, 5826; Kwangue A. 142; Lighava M. 39; Watts J. 473. *Mabeta-Moliwe:* Cable S. 603; Watts J. 188. *Mokoko River F.R.:* Ekema N. 1095. *Onge:* Akogo M. 119; Cheek M. 4990; Harris D. J. 3776; Ndam N. 640; Tchouto P. 951; Watts J. 725, 871

Marantochloa leucantha (K.Schum.) Milne-Redh.
Fl. Gabon 9: 124 (1964); Fl. Cameroun 4: 127 (1965); F.W.T.A. 3: 83 (1968); Bull. Jard. Bot. Nat. Belg. 65: 387 (1996).

Habit: scrambling herb to 3 m. **Guild:** pioneer. **Habitat:** forest. **Distribution:** tropical Africa. **Chorology:** Afromontane. **Star:** green. **Alt. range:** 1-200m (16), 201-400m (1), 401-600m (1), 601-800m (1), 1001-1200m (1). **Flowering:** Mar (1), Apr (2), May (1), Jul (1). **Fruiting:** Mar (1), Apr (3), May (6), Jun (2), Jul (1), Aug (1), Oct (3), Nov (2), Dec (1). **Specimens:** - *Eastern slopes:* Deistel H. 170; Tekwe C. 98. *Etinde:* Cable S. 299; Thomas D. W. 9522. *Mabeta-Moliwe:* Baker W. J. 225; Jaff B. 81, 269; Sunderland T. C. H. 1114; von Rege I. 27. *Mokoko River F.R.:* Acworth J. M. 261; Akogo M. 172, 289; Ekema N. 910, 913,

1088; Pouakouyou D. 71; Watts J. 1048. *Onge:* Harris D. J. 3761, 3823; Ndam N. 647; Tchouto P. 930

Marantochloa mannii (Benth.) Milne-Redh.

Fl. Gabon 9: 128 (1964); Fl. Cameroun 4: 130 (1965); F.W.T.A. 3: 83 (1968); Bull. Jard. Bot. Nat. Belg. 65: 391 (1996).

Habit: herb to 2.5 m. **Guild:** pioneer. **Habitat:** forest. **Distribution:** tropical Africa. **Star:** excluded. **Alt. range:** 1-200m (1). **Flowering:** Nov (1). **Specimens:** - *Mabeta-Moliwe:* Cheek M. 3256; Plot Voucher W 49, W 361. *Onge:* Harris D. J. 3849

Marantochloa monophylla (K.Schum.) D'Orey

Bull. Jard. Bot. Nat. Belg. 65: 371 (1996).
Marantochloa holostachya (Baker) Hutch., Fl. Gabon 9: 116 (1964); Fl. Cameroun 4: 120 (1965); F.W.T.A. 3: 83 (1968).

Habit: herb. **Habitat:** forest. **Distribution:** SE Nigeria, Cameroon, Gabon, Congo (Brazzaville) and Congo (Kinshasa). **Chorology:** Guineo-Congolian. **Star:** green. **Alt. range:** 1-200m (6), 201-400m (4), 801-1000m (1). **Flowering:** May (1), Aug (2), Oct (4), Nov (2), Dec (1). **Fruiting:** Aug (1), Oct (4), Nov (1). **Specimens:** - *Etinde:* Faucher P. 43; Williams S. 57. *Mabeta-Moliwe:* Baker W. J. 283; Plot Voucher W 25, W 49, W 631; Sidwell K. 119; von Rege I. 74. *Mokoko River F.R.:* Ekema N. 931. *Onge:* Akogo M. 79; Harris D. J. 3825; Ndam N. 671; Tchouto P. 696; Watts J. 778

Marantochloa sp. aff. monophylla (K.Schum.) D'Orey

Habit: herb. **Distribution:** Mount Cameroon. **Chorology:** Endemic. **Star:** black. **Alt. range:** 1-200m (2), 201-400m (1). **Flowering:** Mar (1). **Fruiting:** Oct (2). **Specimens:** - *Onge:* Ndam N. 767; Tchouto P. 921; Wheatley J. I. 809

Marantochloa purpurea (Ridl.) Milne-Redh.

Fl. Gabon 9: 125 (1964); Fl. Cameroun 4: 129 (1965); F.W.T.A. 3: 83 (1968); Bull. Jard. Bot. Nat. Belg. 65: 384 (1996).

Habit: herb to 2.5 m. **Habitat:** wet areas in forest. **Distribution:** tropical Africa. **Star:** excluded. **Flowering:** Jan (1). **Fruiting:** Jan (1). **Specimens:** - *Southern Bakundu F.R.:* Keay R. W. J. FHI 37433

Marantochloa ramosissima (Benth.) Hutch.

Fl. Gabon 9: 130 (1964); Fl. Cameroun 4: 133 (1965); F.W.T.A. 3: 85 (1968); Bull. Jard. Bot. Nat. Belg. 65: 393 (1996).

Habit: herb to 2.5 m. **Habitat:** forest, stream banks and Aframomum thicket. **Distribution:** Ivory Coast, SE Nigeria, Bioko and Cameroon. **Chorology:** Guineo-Congolian. **Star:** green. **Alt. range:** 1-200m (8), 201-400m (2), 401-600m (2), 601-800m (1). **Flowering:** May (1), Jun (1), Aug (1), Sep (1), Oct (3), Nov (4), Dec (2). **Fruiting:** Aug (1), Sep (1), Nov (1), Dec (2). **Specimens:** - *Etinde:* Cable S. 210, 380; Cheek M. 5922; Watts J. 530; Wheatley J. I. 507. *Mabeta-Moliwe:* Plot Voucher W 678; Sunderland T. C. H. 1378; von Rege I. 105. *Onge:* Cheek M. 5199; Harris D. J. 3707, 3802, 3804, 3856; Watts J. 741

Megaphrynium macrostachyum (Benth.) Milne-Redh.

Fl. Gabon 9: 156 (1964); Fl. Cameroun 4: 154 (1965); F.W.T.A. 3: 87 (1968).

Habit: herb. **Guild:** pioneer. **Habitat:** forest. **Distribution:** Sierra Leone to Cameroon, Gabon, Congo (Kinshasa), Angola (Cabinda), Sudan and Uganda. **Chorology:** Guineo-Congolian. **Star:** green. **Alt. range:** 1-200m (6), 201-400m (2), 401-600m (3). **Flowering:** May (2), Jun (1), Aug (2). **Fruiting:** May (2), Aug (1), Oct (4), Dec (2). **Specimens:** - *Etinde:* Cable S. 346. *Mabeta-Moliwe:* Baker W. J. 318, 328; Cable S. 547; Mann G. 1335; Ndam N. 515. *Mokoko River F.R.:* Batoum A. 25; Ekema N. 1102. *Onge:* Cheek M. 4964; Ndam N. 625, 801; Tchouto P. 795

Megaphrynium trichogynum Koechlin

Specimens: - *Mabeta-Moliwe:* Cheek M. 3248

Sarcophrynium brachystachyum (Benth.) K.Schum. var. brachystachyum

Fl. Gabon 9: 145 (1964); Fl. Cameroun 4: 146 (1965); F.W.T.A. 3: 87 (1968).

Habit: herb. **Habitat:** forest. **Distribution:** Senegal to Cameroon, Congo (Brazzaville) and CAR. **Chorology:** Guineo-Congolian. **Star:** green. **Alt. range:** 1-200m (16), 201-400m (4), 401-600m (1). **Flowering:** Mar (1), May (6), Jul (1), Oct (2). **Fruiting:** Mar (1), Apr (2), May (10), Jul (1), Oct (4), Nov (2), Dec (2). **Specimens:** - *Etinde:* Cable S. 293; Cheek M. 5398; Tchouto P. 66. *Mabeta-Moliwe:* Cable S. 533; Hurst J. 13; Ndam N. 486; Nguembock F. 32b; Sunderland T. C. H. 1272, 1379; Watts J. 280; Wheatley J. I. 133. *Mokoko River F.R.:* Acworth J. M. 273; Akogo M. 231; Ekema N. 828, 932, 993, 1075; Ndam N. 1294, 1308. *Onge:* Harris D. J. 3702; Tchouto P. 873, 962; Watts J. 742, 743

Sarcophrynium prionogonium (K.Schum.) K.Schum. var. puberulifolium Schnell

Fl. Gabon 9: 148 (1964); Fl. Cameroun 4: 149 (1965); F.W.T.A. 3: 88 (1968).

Habit: herb. **Guild:** shade-bearer. **Habitat:** forest. **Distribution:** Ivory Coast to Cameroon and CAR. **Chorology:** Guineo-Congolian. **Star:** green. **Alt. range:** 1-200m (2), 201-400m (2), 801-1000m (1). **Flowering:** Jun (2). **Fruiting:** Mar (1), Jun (2), Oct (1), Dec (1). **Specimens:** - *Etinde:* Cheek M. 5872; Mbani J. M. 36; Tekwe C. 92; Watts J. 581. *Mokoko River F.R.:* Ekema N. 1151

Sarcophrynium schweinfurthianum (Kuntze) Milne-Redh.

Fl. Gabon 9: 150 (1964); Fl. Cameroun 4: 150 (1965).

Habit: herb. **Habitat:** forest. **Distribution:** Cameroon, Gabon, Congo (Brazzaville), CAR, Sudan and Uganda. **Chorology:** Guineo-Congolian. **Star:** green. **Alt. range:** 1-200m (6), 201-400m (4), 401-600m (2). **Flowering:** Apr (1), May (1), Oct (1). **Fruiting:** Apr (1), Jun (1), Oct (5), Nov (3). **Specimens:** - *Etinde:* Watts J. 469. *Mokoko River F.R.:* Acworth J. M. 126; Akogo M. 210; Ekema N. 1124; Thomas D. W. 10132. *Onge:* Cheek M. 4966, 5051; Harris D. J. 3632, 3696, 3703; Tchouto P. 819; Watts J. 744

Thaumatococcus daniellii (Benn.) Benth.

Fl. Gabon 9: 141 (1964); Fl. Cameroun 4: 144 (1965); F.W.T.A. 3: 81 (1968).

Habit: herb. **Guild:** pioneer. **Habitat:** forest. **Distribution:** Sierra Leone to Cameroon, Gabon, Congo (Brazzaville), Congo (Kinshasa), Angola (Cabinda) and CAR. **Chorology:** Guineo-Congolian. **Star:** red. **Alt. range:** 1-200m (6). **Flowering:** Feb (1),

May (1), Jun (1), Aug (1). **Fruiting:** Feb (1), Apr (1), May (1), Jun (1), Aug (1), Oct (1). **Specimens:** - *Etinde:* Mbani J. M. 18. *Mabeta-Moliwe:* Baker W. J. 232; Mann G. 2145. *Mokoko River F.R.:* Akogo M. 235; Ekema N. 1074; Pouakouyou D. 112. *Onge:* Tchouto P. 1009

Trachyphrynium braunianum (K.Schum.) Baker
Fl. Gabon 9: 98 (1964); Fl. Cameroun 4: 106 (1965); F.W.T.A. 3: 89 (1968).

Habit: herb. **Guild:** pioneer. **Distribution:** Guinea to Cameroon, Gabon, Congo (Brazzaville), Congo (Kinshasa), Sudan and Uganda. **Chorology:** Guineo-Congolian. **Star:** green. **Specimens:** - *Bambuko F.R.:* Keay R. W. J. FHI 37469

ORCHIDACEAE
P.J. Cribb (K)

Aerangis biloba (Lindl.) Schltr.
F.W.T.A. 3: 265 (1968); Kew Bull. 34:281 (1979).

Habit: epiphyte. **Habitat:** forest. **Distribution:** Senegal to Cameroon. **Chorology:** Guineo-Congolian. **Star:** green. **Specimens:** - *Mokoko River F.R.:* Wright J. O. 58/12

Aerangis gravenreuthii (Kraenzl.) Schltr.
F.W.T.A. 3: 265 (1968); Kew Bull. 34:277 (1979).

Habit: epiphyte. **Habitat:** forest-grassland transition. **Distribution:** Bioko, Cameroon and Tanzania. **Chorology:** Guineo-Congolian (montane). **Star:** gold. **Alt. range:** 1800-2000m (1). **Specimens:** - *Eastern slopes:* Preuss P. R. 891

Ancistrochilus rothschildianus O'Brien
F.W.T.A. 3: 226 (1968); F.T.E.A. Orchidaceae: 285 (1984).
Habit: epiphyte. **Habitat:** forest. **Distribution:** Guinea to Cameroon and Uganda. **Chorology:** Guineo-Congolian. **Star:** green. **Specimens:** - *Mokoko River F.R.:* Wright J. O. 58/4

Ancistrochilus thomsonianus (Rchb.f.) Rolfe
F.W.T.A. 3: 226 (1968).

Habit: epiphyte. **Habitat:** forest. **Distribution:** SE Nigeria and Cameroon. **Chorology:** Lower Guinea. **Star:** blue. **Specimens:** - *Etinde:* Schlechter F. R. R. 15762

Ancistrorhynchus clandestinus (Lindl.) Schltr.
F.W.T.A. 3: 272 (1968); F.T.E.A. Orchidaceae: 588 (1989).

Habit: epiphyte. **Habitat:** forest. **Distribution:** Sierra Leone to Cameroon, Gabon and Congo (Kinshasa). **Chorology:** Guineo-Congolian. **Star:** green. **Specimens:** - *Eastern slopes:* Schlechter F. R. R. 12843

Ancistrorhynchus recurvus Finet
F.W.T.A. 3: 272 (1968); F.T.E.A. Orchidaceae: 589 (1989).

Habit: epiphyte. **Habitat:** forest. **Distribution:** Guinea to Cameroon, Gabon and Uganda. **Chorology:** Guineo-Congolian. **Star:** green. **Specimens:** - *Mokoko River F.R.:* Wright J. O. 58/50

Ancistrorhynchus serratus Summerh.
Kew Bull. 20: 195 (1966); F.W.T.A. 3: 272 (1968).

Habit: epiphyte. **Habitat:** forest. **Distribution:** SE Nigeria, Mount Cameroon and Bioko. **Chorology:** Western Cameroon Uplands (montane). **Star:** gold. **Alt. range:** 1400-1600m (1). **Specimens:** - *No locality:* Mann G. 2123

Ancistrorhynchus straussii (Schltr.) Schltr.
F.W.T.A. 3: 272 (1968).

Habit: epiphyte. **Habitat:** forest. **Distribution:** Ivory Coast to Congo (Kinshasa) and Uganda. **Chorology:** Guineo-Congolian. **Star:** green. **Specimens:** - *Mabeta-Moliwe:* Schlechter F. R. R. 15771

Angraecopsis cryptantha P.J.Cribb
Kew Bull. 51: 361 (1996).

Habit: epiphyte. **Habitat:** forest-grassland transition. **Distribution:** Mount Cameroon. **Chorology:** Endemic (montane). **Star:** black. **Alt. range:** 1800-2000m (1). **Flowering:** Oct (1). **Specimens:** - *Eastern slopes:* Thomas D. W. 9443

Angraecopsis elliptica Summerh.
F.W.T.A. 3: 273 (1968).

Habit: epiphyte. **Habitat:** forest. **Distribution:** Nigeria and W. Cameroon. **Chorology:** Lower Guinea. **Star:** blue. **Specimens:** - *No locality:* Gregory H. 194

Angraecopsis ischnopus (Schltr.) Schltr.
F.W.T.A. 3: 273 (1968).

Habit: epiphyte. **Habitat:** forest. **Distribution:** Sierra Leone to W. Cameroon. **Chorology:** Guineo-Congolian. **Star:** green. **Specimens:** - *Eastern slopes:* Deistel H. s.n., s.n.

Angraecopsis parviflora (Thouars) Schltr.
F.W.T.A. 3: 273 (1968); F.T.E.A. Orchidaceae: 598 (1989).

Habit: epiphyte. **Habitat:** forest. **Distribution:** Mount Cameroon and East Africa. **Chorology:** Afromontane. **Star:** gold. **Specimens:** - *Eastern slopes:* Gregory H. 870a

Angraecopsis tridens (Lindl.) Schltr.
F.W.T.A. 3: 273 (1968).

Habit: epiphyte. **Habitat:** forest. **Distribution:** Bioko and Cameroon. **Chorology:** Lower Guinea (montane). **Star:** black. **Alt. range:** 1200-1400m (1). **Specimens:** - *Eastern slopes:* Preuss P. R. 965

Angraecum aporoides Summerh.
F.W.T.A. 3: 254 (1968); Fl. Afr. Cent. Orchidaceae: 481 (1992).

Habit: epiphyte. **Habitat:** forest. **Distribution:** Nigeria, Bioko, Cameroon and Congo (Kinshasa). **Chorology:** Guineo-Congolian. **Star:** green. **Specimens:** - *Mokoko River F.R.:* Wright J. O. 58/49

169

Angraecum birrimense Rolfe
F.W.T.A. 3: 257 (1968).

Habit: epiphyte. Habitat: lava flows and forest. Distribution: Sierra Leone to Cameroon. Chorology: Guineo-Congolian. Star: green. Alt. range: 800-1000m (1). Flowering: Oct (2). Specimens: - *Eastern slopes:* Dundas J. FHI 15301. *Etinde:* Banks H. 50, 55

Angraecum distichum Lindl.
F.W.T.A. 3: 254 (1968); F.T.E.A. Orchidaceae: 494 (1989).

Habit: epiphyte. Habitat: mangrove forest and lava flows. Distribution: Guinea to Cameroon, Gabon, Congo (Brazzaville), Congo (Kinshasa), Angola and Uganda. Chorology: Guineo-Congolian. Star: green. Alt. range: 400-600m (1). Flowering: Oct (1). Fruiting: Oct (1). Specimens: - *Eastern slopes:* Brodie S. 18. *Etinde:* Ngongi FHI 15090. *Mabeta-Moliwe:* Cheek M. 3480; Sunderland T.C.H. 1372

Angraecum eichlerianum Kraenzl.
F.W.T.A. 3: 257 (1968); Fl. Afr. Cent. Orchidaceae: 470 (1992).

Habit: epiphyte. Habitat: forest. Distribution: SE Nigeria, Cameroon, Gabon, Congo (Kinshasa) and Angola. Chorology: Guineo-Congolian. Star: green. Specimens: - *Eastern slopes:* Keay R. W. J. FHI 25375

Angraecum infundibulare Lindl.
F.W.T.A. 3: 257 (1968); F.T.E.A. Orchidaceae: 503 (1989).

Habit: epiphyte. Habitat: forest. Distribution: Nigeria, Cameroon, Ethiopia, Kenya and Tanzania. Chorology: Afromontane. Star: green. Specimens: - *Etinde:* Rosevear D. R. 55/37

Angraecum podochiloides Schltr.
F.W.T.A. 3: 254 (1968); Fl. Afr. Cent. Orchidaceae: 478 (1992).

Habit: epiphyte. Habitat: forest. Distribution: Liberia to Cameroon and Congo (Kinshasa). Chorology: Guineo-Congolian. Star: green. Specimens: - *Etinde:* Schlechter F. R. R. 15769

Angraecum pungens Schltr.
F.W.T.A. 3: 254 (1968); Fl. Afr. Cent. Orchidaceae: 477 (1992).

Habit: epiphyte. Habitat: forest. Distribution: SE Nigeria, Cameroon and Congo (Kinshasa). Chorology: Guineo-Congolian. Star: green. Alt. range: 800-1000m (1). Specimens: - *Etinde:* Cole J. A. 7. *Mabeta-Moliwe:* Schlechter F. R. R. 15774

Angraecum sacciferum Lindl.
F.W.T.A. 3: 254 (1968); F.T.E.A. Orchidaceae: 497 (1989).

Habit: epiphyte. Habitat: forest. Distribution: Cameroon to East and South Africa. Chorology: Afromontane. Star: green. Alt. range: 800-1000m (1). Specimens: - *Eastern slopes:* Gregory H. 582

Angraecum subulatum Lindl.
F.W.T.A. 3: 254 (1968); Fl. Afr. Cent. Orchidaceae: 477 (1992).

Habit: epiphyte. Habitat: lava flows. Distribution: Sierra Leone to Bioko, Cameroon, Congo (Brazzaville) and Congo (Kinshasa). Chorology: Guineo-Congolian. Star: green. Alt. range: 400-

600m (1). Flowering: Oct (1). Specimens: - *Eastern slopes:* Brodie S. 19. *Southern Bakundu F.R.:* Daramola B. O. FHI 29819

Auxopus kamerunensis Schltr.
F.W.T.A. 3: 207 (1968).

Habit: saprophyte. Habitat: forest. Distribution: Ivory Coast, Ghana, Nigeria, Cameroon, Congo (Kinshasa) and Uganda. Chorology: Guineo-Congolian. Star: gold. Specimens: - *Southern Bakundu F.R.:* Keay R. W. J. FHI 28570

Auxopus macranthus Summerh.
F.W.T.A. 3: 207 (1968); F.T.E.A. Orchidaceae: 265 (1984).

Habit: saprophyte. Habitat: forest. Distribution: Liberia, Ivory Coast, Ghana, Nigeria and Cameroon. Chorology: Guineo-Congolian. Star: green. Alt. range: 200-400m (1). Flowering: Dec (1). Specimens: - *Etinde:* Cheek M. 5860

Brachycorythis kalbreyeri Rchb.f.
F.T.E.A. Orchidaceae: 24 (1968); F.W.T.A. 3: 187 (1968).

Habit: epiphyte. Habitat: forest. Distribution: Guinea to Cameroon, Congo (Brazzaville), Congo (Kinshasa), Uganda and Kenya. Chorology: Guineo-Congolian (montane). Star: green. Alt. range: 1600-1800m (1). Specimens: - *No locality:* Kalbreyer W. 145

Brownleea parviflora Harv. ex Lindl.
Kew Bull. 20: 169 (1966); F.T.E.A. Orchidaceae: 179 (1968); F.W.T.A. 3: 201 (1968); J. S. African Bot. 47: 34 (1981).

Habit: herb. Habitat: grassland. Distribution: Mount Cameroon and East to South Africa. Chorology: Afromontane. Star: gold. Alt. range: 2000-2200m (1), 2401-2600m (1). Flowering: Oct (1). Specimens: - *Eastern slopes:* Cheek M. 3624. *No locality:* Mann G. 2120

Bulbophyllum barbigerum Lindl.
F.W.T.A. 3: 236 (1968); Orchid Monographs 2: 36 (1987); Sanford ms. for Fl. Cameroun.

Habit: epiphyte. Habitat: forest. Distribution: Sierra Leone to Cameroon, Gabon, Congo (Brazzaville), Congo (Kinshasa) and CAR. Chorology: Guineo-Congolian. Star: green. Flowering: May (2). Specimens: - *Etinde:* Keay R. W. J. FHI 43914. *Mokoko River F.R.:* Ndam N. 1189

Bulbophyllum bibundiense Schltr.
F.W.T.A. 3: 241 (1968); Sanford ms. for Fl. Cameroun.

Habit: epiphyte. Habitat: forest. Distribution: Mount Cameroon. Chorology: Endemic. Star: black. Flowering: Oct (1). Specimens: - *Mabeta-Moliwe:* Schlechter F. R. R. 15784

Bulbophyllum bifarium Hook.f.
F.W.T.A. 3: 239 (1968); Orchid Monographs 2: 87 (1987); Sanford ms. for Fl. Cameroun.

Habit: epiphyte. Habitat: farmbush and gaps in forest. Distribution: Ivory Coast, SW Cameroon and Kenya. Chorology: Afromontane. Star: black. Alt. range: 1000-1200m (1), 1401-1600m (1). Flowering: Feb (1). Fruiting: Feb (1). Specimens: - *Eastern slopes:* Cable S. 1322. *No locality:* Mann G. 2121

Bulbophyllum bufo (Lindl.) Rchb.f.

F.W.T.A. 3: 241 (1968); Sanford ms. for Fl. Cameroun.
Bulbophyllum falcatum (Lindl.) Rchb.f. var. *bufo* (Lindl.)
Vermeulen, Orchid Monographs 2: 125 (1987).

Habit: epiphyte. **Habitat:** forest. **Distribution:** Guinea to
Cameroon, Bioko and Congo (Kinshasa). **Chorology:** Guineo-
Congolian. **Star:** green. **Specimens:** - *Eastern slopes:* Deistel H.
s.n.

Bulbophyllum calvum Summerh.

F.W.T.A. 3: 234 (1968); Orchid Monographs 2: 71 (1987).

Habit: epiphyte. **Habitat:** forest. **Distribution:** Mount Cameroon
and E. Nigeria. **Chorology:** Western Cameroon Uplands
(montane). **Star:** black. **Alt. range:** 1400-1600m (1). **Flowering:**
Sep (1). **Fruiting:** Sep (1). **Specimens:** - *Etinde:* Tekwe C. 200

Bulbophyllum calyptratum Kraenzl.

F.W.T.A. 3: 241 (1968); Orchid Monographs 2: 128 (1987);
Sanford ms. for Fl. Cameroun.

Habit: epiphyte. **Habitat:** forest. **Distribution:** Guinea to
Cameroon, Gabon, Congo (Brazzaville) and Congo (Kinshasa).
Chorology: Guineo-Congolian. **Star:** green. **Alt. range:** 1-200m
(2), 1001-1200m (1). **Flowering:** Feb (1), Mar (3), Apr (1).
Specimens: - *Eastern slopes:* Cable S. 1347. *Etinde:* Rosevear D.
R. 114/35; Schlechter F. R. R. 12369. *Mabeta-Moliwe:* Cable S.
1408; Groves M. 250

Bulbophyllum cochleatum Lindl.

F.W.T.A. 3: 236 (1968); F.T.E.A. Orchidaceae: 315 (1984);
Orchid Monographs 2: 41 (1987); Fl. Rwanda 4: 530 (1988);
Sanford ms. for Fl. Cameroun.

Habit: epiphyte. **Habitat:** forest, forest-grassland transition and
Aframomum thicket. **Distribution:** Guinea to Bioko, Cameroon,
Gabon, Sudan and Uganda to East and South Africa. **Chorology:**
Afromontane. **Star:** green. **Alt. range:** 1000-1200m (2), 1801-
2000m (1). **Flowering:** Oct (3). **Specimens:** - *Eastern slopes:*
Thomas D. W. 9435. *Etinde:* Wheatley J. I. 639, 640

Bulbophyllum cocoinum Batem. ex Lindl.

F.W.T.A. 3: 234 (1968); Sanford ms. for Fl. Cameroun.

Habit: epiphyte. **Habitat:** forest. **Distribution:** Sierra Leone to
Cameroon, Gabon, Congo (Kinshasa) and Uganda. **Chorology:**
Guineo-Congolian. **Star:** green. **Flowering:** Apr (1). **Specimens:**
- *Etinde:* Schlechter F. R. R. 12361

Bulbophyllum distans Lindl.

F.W.T.A. 3: 236 (1968); Sanford ms. for Fl. Cameroun.

Habit: epiphyte. **Habitat:** forest. **Distribution:** Liberia to Bioko,
Cameroon, Gabon, CAR, Congo (Kinshasa), Angola and Uganda.
Chorology: Guineo-Congolian. **Star:** green. **Specimens:** -
Mabeta-Moliwe: Cole J. A. 35

Bulbophyllum falcatum (Lindl.) Rchb.f. var. falcatum

F.W.T.A. 3: 241 (1968); F.T.E.A. Orchidaceae: 317 (1984);
Orchid Monographs 2: 124 (1987); Sanford ms. for Fl. Cameroun.

Habit: epiphyte. **Habitat:** forest. **Distribution:** Guinea to Bioko,
SW Cameroon, Congo (Kinshasa) and Uganda. **Chorology:**
Guineo-Congolian. **Star:** green. **Alt. range:** 1-200m (1), 201-
400m (1). **Flowering:** Jan (1), Nov (1). **Specimens:** - *Mabeta-
Moliwe:* Schlechter F. R. R. 12992; Wheatley J. I. 654. *No
locality:* Brenan J. P. M. FHI 28295

Bulbophyllum falcipetalum Lindl.

F.W.T.A. 3: 239 (1968); Orchid Monographs 2: 113 (1987);
Sanford ms. for Fl. Cameroun.

Habit: epiphyte. **Habitat:** forest. **Distribution:** Ivory Coast,
Ghana, Nigeria, Cameroon and Gabon. **Chorology:** Guineo-
Congolian. **Star:** green. **Specimens:** - *Mabeta-Moliwe:* Cole J. A.
33

Bulbophyllum filiforme Kraenzl.

F.W.T.A. 3: 239 (1968); Sanford ms. for Fl. Cameroun.
Bulbophyllum resupinatum Ridl. var. *filiforme* (Kraenzl.)
Vermeulen, Orchid Monographs 2: 120 (1987).

Habit: epiphyte. **Habitat:** forest. **Distribution:** Mount Cameroon.
Chorology: Endemic. **Star:** black. **Alt. range:** 1-200m (3).
Flowering: Mar (3). **Specimens:** - *Etinde:* Preuss P. R. 1242.
Mabeta-Moliwe: Cable S. 1406; Groves M. 251; Khayota B. 464

Bulbophyllum flavidum Lindl.

F.W.T.A. 3: 234 (1968); Sanford ms. for Fl. Cameroun.

Habit: epiphyte. **Habitat:** forest. **Distribution:** Guinea to Bioko,
Cameroon, S. Tomé and Gabon. **Chorology:** Guineo-Congolian.
Star: green. **Specimens:** - *Mabeta-Moliwe:* Schlechter F. R. R.
15757

Bulbophyllum fractiflexum Kraenzl.

Engl. Bot. Jahrb. 48: 392 (1912); Sanford ms. for Fl. Cameroun.
Bulbophyllum simonii Summerh., F.W.T.A. 3: 239 (1968).

Habit: epiphyte. **Habitat:** forest. **Distribution:** Nigeria, Bioko,
Cameroon and Gabon. **Chorology:** Lower Guinea. **Star:** blue.
Flowering: Mar (1), Apr (1). **Specimens:** - *Etinde:* Rosevear D. R.
56/37; Schlechter F. R. R. 12373; Thorold C. A. CM 15

Bulbophyllum fuscoides J.B.Petersen

F.W.T.A. 3: 239 (1968).

Habit: epiphyte. **Habitat:** forest. **Distribution:** Bioko and
Cameroon. **Chorology:** Lower Guinea. **Star:** gold. **Specimens:** -
Mokoko River F.R.: Wright J. O. 58/1

Bulbophyllum fuscum Lindl. subsp. fuscum

F.W.T.A. 3: 236 (1968); Orchid Monographs 2: 147 (1987);
Sanford ms. for Fl. Cameroun.

Habit: epiphyte. **Habitat:** forest. **Distribution:** Guinea to
Cameroon, Gabon, Congo (Brazzaville), Congo (Kinshasa), CAR
and Tanzania. **Chorology:** Guineo-Congolian. **Star:** green. **Alt.
range:** 800-1000m (1). **Specimens:** - *Etinde:* Cole J. A. 36

Bulbophyllum gravidum Lindl.

F.W.T.A. 3: 236 (1968); Sanford ms. for Fl. Cameroun.
Bulbophyllum cochleatum Lindl. var. *gravidum* (Lindl.)
Vermeulen, Orchid Monographs 2: 42 (1987).

Habit: epiphyte. **Habitat:** forest. **Distribution:** Mount Cameroon
and Bioko. **Chorology:** Western Cameroon Uplands (montane).

Star: black. **Alt. range:** 1400-1600m (1). **Flowering:** Nov (1).
Specimens: - *No locality:* Mann G. 2126

Bulbophyllum imbricatum Lindl.

F.W.T.A. 3: 242 (1968); Orchid Monographs 2: 103 (1987);
Sanford ms. for Fl. Cameroun.
Bulbophyllum leucorhachis (Rolfe) Schltr., F.W.T.A. 3: 242
(1968).

Habit: epiphyte. **Habitat:** forest. **Distribution:** Sierra Leone to
Cameroon, CAR, Gabon, Congo (Brazzaville) and Congo
(Kinshasa). **Chorology:** Guineo-Congolian. **Star:** green. **Alt.
range:** 1400-1600m (1). **Flowering:** Apr (1). **Fruiting:** Sep (1).
Specimens: - *Etinde:* Preuss P. R. 1241; Tekwe C. 211. *Mokoko
River F.R.:* Wright J. O. 58/44

Bulbophyllum intertextum Lindl.

F.W.T.A. 3: 234 (1968); F.T.E.A. Orchidaceae: 309 (1984);
Orchid Monographs 2: 54 (1987); Sanford ms. for Fl. Cameroun.

Habit: epiphyte. **Habitat:** lava flows and forest. **Distribution:**
tropical Africa. **Star:** excluded. **Alt. range:** 1000-1200m (1).
Flowering: Aug (1), Sep (1), Oct (2). **Specimens:** - *Etinde:* Banks
H. 52, 53; Keay R. W. J. FHI 42336; Wheatley J. I. 559. *Mabeta-
Moliwe:* Schlechter F. R. R. 15756

Bulbophyllum josephii (Kuntze) Summerh. var. josephii

F.W.T.A. 3: 234 (1968); Orchid Monographs 2: 68 (1987); Fl.
Rwanda 4: 532 (1988); Sanford ms. for Fl. Cameroun.

Habit: epiphyte. **Habitat:** forest. **Distribution:** Guinea to
Cameroon. **Chorology:** Guineo-Congolian (montane). **Star:** green.
Alt. range: 1400-1600m (2), 1601-1800m (1). **Flowering:** Oct
(2). **Specimens:** - *Etinde:* Thomas D. W. 9245; Wheatley J. I. 614.
No locality: Mann G. 2124

Bulbophyllum josephii (Kuntze) Summerh. var. mahonii (Rolfe) Verm.

Orchid Monographs 2: 69 (1987).

Habit: epiphyte. **Habitat:** forest. **Distribution:** Guinea to Bioko,
W. Cameroon, Congo (Kinshasa), Zambia and Malawi.
Chorology: Afromontane. **Star:** green. **Alt. range:** 1000-1200m
(1), 1601-1800m (3), 1801-2000m (1). **Flowering:** Sep (1), Oct
(3). **Fruiting:** Oct (1). **Specimens:** - *Eastern slopes:* Cheek M.
3647, 3702; Thomas D. W. 9483a, 9485. *Etinde:* Wheatley J. I.
579, 593

Bulbophyllum kamerunense Schltr.

F.W.T.A. 3: 242 (1968); Sanford ms. for Fl. Cameroun.

Habit: epiphyte. **Habitat:** forest. **Distribution:** Nigeria, Cameroon
and Gabon. **Chorology:** Lower Guinea. **Star:** blue. **Flowering:**
Apr (1). **Specimens:** - *Etinde:* Schlechter F. R. R. 12430

Bulbophyllum lupulinum Lindl.

F.W.T.A. 3: 239 (1968); Orchid Monographs 2: 138 (1987);
Sanford ms. for Fl. Cameroun.
Bulbophyllum wrightii Summerh., F.W.T.A. 3: 239 (1968).

Habit: epiphyte. **Habitat:** forest and farmbush. **Distribution:**
Guinea to Cameroon, Congo (Kinshasa), Ethiopia and Zambia.
Chorology: Guineo-Congolian. **Star:** green. **Alt. range:** 1000-

1200m (1). **Flowering:** Jan (1). **Specimens:** - *Eastern slopes:*
Schlechter F. R. R. 12844. *Mokoko River F.R.:* Wright J. O. 58/18

Bulbophyllum mannii Hook.f.

F.W.T.A. 3: 236 (1969); Sanford ms. for Fl. Cameroun.

Habit: epiphyte. **Habitat:** forest. **Distribution:** Sierra Leone to
Bioko and Cameroon. **Chorology:** Guineo-Congolian (montane).
Star: green. **Alt. range:** 1400-1600m (1). **Flowering:** Dec (1).
Specimens: - *No locality:* Mann G. 1337

Bulbophyllum maximum (Lindl.) Rchb.f.

F.W.T.A. 3: 241 (1968); F.T.E.A. Orchidaceae: 317 (1984);
Orchid Monographs 2: 97 (1987); Sanford ms. for Fl. Cameroun.
Bulbophyllum oxypterum (Lindl.) Rchb.f., F.W.T.A. 3: 241
(1968).

Habit: epiphyte. **Habitat:** forest. **Distribution:** tropical Africa.
Star: excluded. **Specimens:** - *Mabeta-Moliwe:* Schlechter F. R. R.
15758

Bulbophyllum modicum Summerh.

F.W.T.A. 3: 234 (1968); Sanford ms. for Fl. Cameroun.

Habit: epiphyte. **Habitat:** forest. **Distribution:** Mount Cameroon.
Chorology: Endemic (montane). **Star:** black. **Alt. range:** 800-
1000m (1). **Flowering:** Oct (1). **Specimens:** - *Eastern slopes:*
Gregory H. 193

Bulbophyllum oreonastes Rchb.f.

F.W.T.A. 3: 236 (1968); F.T.E.A. Orchidaceae: 312 (1984);
Orchid Monographs 2: 93 (1987); Fl. Rwanda 4: 533 (1988);
Sanford ms. for Fl. Cameroun.
Bulbophyllum zenkerianum Kraenzl., F.W.T.A. 3: 239 (1968).
Bulbophyllum melinostachyum Schltr., Engl. Bot. Jahrb. 26: 342
(1899).

Habit: epiphyte. **Habitat:** forest. **Distribution:** tropical Africa.
Chorology: Afromontane. **Star:** green. **Alt. range:** 800-1000m
(1), 1401-1600m (2). **Flowering:** Oct (1). **Specimens:** - *Eastern
slopes:* Schlechter F. R. R. 12377. *Etinde:* Wheatley J. I. 613. *No
locality:* Mann G. 2122

Bulbophyllum oxychilum Schltr.

Orchid Monographs 2: 76 (1987); Sanford ms. for Fl. Cameroun.
Bulbophyllum buntingii Rendle, F.W.T.A. 3: 234 (1968).

Habit: epiphyte. **Habitat:** forest. **Distribution:** Guinea to
Cameroon, CAR, Congo (Brazzaville), Congo (Kinshasa) and
Uganda. **Chorology:** Guineo-Congolian. **Star:** green. **Specimens:**
- *Mokoko River F.R.:* Wright J. O. 58/56

Bulbophyllum phaeopogon Schltr.

F.W.T.A. 3: 236 (1968); Sanford ms. for Fl. Cameroun.
Bulbophyllum schinzianum Kraenzl. var. *phaeopogon* (Schltr.)
Vermeulen, Orchid Monographs 2: 29 (1987).

Habit: epiphyte. **Habitat:** forest. **Distribution:** Ivory Coast to
Cameroon, Gabon and Congo (Kinshasa). **Chorology:** Guineo-
Congolian. **Star:** green. **Alt. range:** 1-200m (2). **Flowering:** Apr
(3). **Specimens:** - *Mabeta-Moliwe:* Preuss P. R. 1225; Wheatley J.
I. 84, 212. *Onge:* Wheatley J. I. 191

Bulbophyllum porphyroglossum Kraenzl.

F.W.T.A. 3: 234 (1968); Sanford ms. for Fl. Cameroun.

Habit: epiphyte. **Habitat:** forest. **Distribution:** SE Nigeria and SW Cameroon. **Chorology:** Western Cameroon. **Star:** gold. **Specimens:** - *Etinde:* Schlechter F. R. R. 12361

Bulbophyllum porphyrostachys Summerh.
F.W.T.A. 3: 239 (1968); Orchid Monographs 2: 137 (1987); Sanford ms. for Fl. Cameroun.

Habit: epiphyte. **Habitat:** forest. **Distribution:** Nigeria and SW Cameroon. **Chorology:** Lower Guinea. **Star:** black. **Specimens:** - *No locality:* Keay R. W. J. & Brenan J. P. M. FHI 28133

Bulbophyllum recurvum Lindl.
F.W.T.A. 3: 234 (1968); Sanford ms. for Fl. Cameroun.

Habit: epiphyte. **Habitat:** forest. **Distribution:** Sierra Leone, Liberia, Bioko and SW Cameroon. **Chorology:** Guineo-Congolian. **Star:** blue. **Specimens:** - *Mokoko River F.R.:* Wright J. O. 58/57

Bulbophyllum rhizophorae Lindl.
F.W.T.A. 3: 239 (1968); Sanford ms. for Fl. Cameroun.

Habit: epiphyte. **Habitat:** forest. **Distribution:** Sierra Leone to Cameroon, Gabon and Congo (Kinshasa). **Chorology:** Guineo-Congolian. **Star:** green. **Specimens:** - *Mokoko River F.R.:* Wright J. O. 58/5

Bulbophyllum sandersonii (Hook.f.) Rchb.f. subsp. sandersonii
F.T.E.A. Orchidaceae: 320 (1984); Orchid Monographs 2: 108 (1987); Fl. Rwanda 4: 533 (1988); Sanford ms. for Fl. Cameroun.
Bulbophyllum tentaculigerum Rchb.f., F.W.T.A. 3: 239 (1968).

Habit: epiphyte. **Habitat:** forest. **Distribution:** tropical Africa. **Chorology:** Afromontane. **Star:** green. **Alt. range:** 1000-1200m (1), 1601-1800m (1). **Flowering:** Feb (1), Mar (1), Apr (1). **Specimens:** - *Eastern slopes:* Etuge M. 1200. *Etinde:* Kalbreyer W. 25/3; Preuss P. R. 1217; Schlechter F. R. R. 12358. *Mabeta-Moliwe:* Jaff B. 213; Sunderland T.C.H. 1206; Wheatley J.I. 48, 83

Bulbophyllum scaberulum (Rolfe) Bolus var. scaberulum
F.T.E.A. Orchidaceae: 320 (1984); Orchid Monographs 2: 116 (1987); Fl. Rwanda 4: 630 (1988); Sanford ms. for Fl. Cameroun.

Habit: epiphyte. **Habitat:** forest. **Distribution:** Liberia to Cameroon, CAR, Gabon, Congo (Kinshasa) and Uganda. **Chorology:** Guineo-Congolian (montane). **Star:** green. **Alt. range:** 1600-1800m (1). **Specimens:** - *Etinde:* Keay R. W. J. FHI 37664

Bulbophyllum schimperianum Kraenzl.
F.W.T.A. 3: 234 (1968); F.T.E.A. Orchidaceae: 311 (1984); Orchid Monographs 2: 75 (1987); Sanford ms. for Fl. Cameroun.

Habit: epiphyte. **Habitat:** forest. **Distribution:** Liberia, Cameroon, Gabon, Congo (Brazzaville), Congo (Kinshasa) and CAR. **Chorology:** Guineo-Congolian. **Star:** green. **Specimens:** - *Etinde:* Schimper A. F. W. 341. *Mabeta-Moliwe:* Schlechter F.R.R. 15755

Bulbophyllum schinzianum Kraenzl.
F.W.T.A. 3: 236 (1968); Sanford ms. for Fl. Cameroun.

Bulbophyllum schinzianum Kraenzl. var. *schinzianum*, Orchid Monographs 2: 26 (1987).

Habit: epiphyte. **Habitat:** forest and farmbush. **Distribution:** Liberia to Cameroon, Gabon and Congo (Kinshasa). **Chorology:** Guineo-Congolian. **Star:** green. **Alt. range:** 1-200m (1). **Flowering:** Apr (1). **Fruiting:** Apr (1). **Specimens:** - *Mabeta-Moliwe:* Wheatley J. I. 84

Bulbophyllum tenuicaule Lindl.
F.W.T.A. 3: 236 (1968); Sanford ms. for Fl. Cameroun.
Bulbophyllum cochleatum Lindl. var. *tenuicaule* (Lindl.) Vermeulen, Fl. Cameroun 4: 36 (1965); Orchid Monographs 2: 45 (1987).

Habit: epiphyte. **Habitat:** forest. **Distribution:** Nigeria, SW Cameroon, Bioko and S. Tomé. **Chorology:** Lower Guinea (montane). **Star:** blue. **Alt. range:** 800-1000m (1), 1401-1600m (2), 1601-1800m (1), 1801-2000m (1). **Flowering:** Sep (3), Oct (2). **Specimens:** - *Eastern slopes:* Gregory H. 612; Thomas D. W. 9442. *Etinde:* Tekwe C. 199; Thomas D. W. 9246; Wheatley J. I. 580

Bulbophyllum teretifolium Schltr.
F.W.T.A. 3: 234 (1968); Orchid Monographs 2: 160 (1987); Sanford ms. for Fl. Cameroun.

Habit: epiphyte. **Habitat:** forest. **Distribution:** Cameroon. **Chorology:** Lower Guinea. **Star:** black. **Specimens:** - *Etinde:* Schlechter F. R. R. 12362

Bulbophyllum velutinum Lindl.
F.W.T.A. 3: 239 (1968); Sanford ms. for Fl. Cameroun.
Bulbophyllum falcatum (Lindl.) Rchb.f. var. *velutinum* (Lindl.) Vermeulen, Orchid Monographs 2: 126 (1987).

Habit: epiphyte. **Habitat:** forest. **Distribution:** Sierra Leone to Bioko, Cameroon, Gabon and Congo (Kinshasa). **Chorology:** Guineo-Congolian. **Star:** green. **Specimens:** - *Mokoko River F.R.:* Wright J. O. 58/39. *Mabeta-Moliwe:* Wheatley J.I. 47, 82, 177

Bulbophyllum winkleri Schltr.
F.W.T.A. 3: 234 (1968); Sanford ms. for Fl. Cameroun.

Habit: epiphyte. **Habitat:** forest. **Distribution:** Liberia to Cameroon and Congo (Kinshasa). **Chorology:** Guineo-Congolian. **Star:** green. **Fruiting:** Jun (1). **Specimens:** - *Eastern slopes:* Winkler H. J. P. 157. *Mokoko River F.R.:* Letouzey R. 15055

Bulbophyllum sp.

Specimens: - *Etinde:* Tekwe C. 209

Calanthe sylvatica (Thouars) Lindl.
F.T.E.A. Orchidaceae: 282 (1984).
Calanthe corymbosa Lindl., F.W.T.A. 3: 226 (1968).

Habit: herb. **Habitat:** forest. **Distribution:** tropical and subtropical Africa. **Chorology:** Afromontane. **Star:** green. **Alt. range:** 1400-1600m (2). **Flowering:** Mar (1). **Fruiting:** Mar (1). **Specimens:** - *Eastern slopes:* Etuge M. 1215. *No locality:* Johnston H. H. 107

Calyptrochilum emarginatum (Sw.) Schltr.
F.W.T.A. 3: 251 (1968); Fl. Afr. Cent. Orchidaceae: 594 (1992).

Habit: epiphyte. **Habitat:** forest. **Distribution:** Guinea to Cameroon, Gabon, Angola and CAR. **Chorology:** Guineo-Congolian. **Star:** green. **Specimens:** - *Etinde:* Preuss P. R. 1240

Chamaeangis odoratissima (Rchb.f.) Schltr.
F.W.T.A. 3: 264 (1968); F.T.E.A. Orchidaceae: 546 (1989).

Habit: epiphyte. **Habitat:** forest. **Distribution:** Sierra Leone to Cameroon, Congo (Kinshasa), Angola, CAR, Uganda, Rwanda, Malawi and East Africa. **Star:** excluded. **Specimens:** - *Eastern slopes:* Gregory H. 153a

Cheirostylis lepida (Rchb.f.) Rolfe
F.W.T.A. 3: 208 (1968); Fl. Afr. Cent. Orchidaceae: 22 (1984); F.T.E.A. Orchidaceae: 247 (1984).

Habit: herb. **Distribution:** SE Nigeria, SW Cameroon, S. Tomé, Congo (Kinshasa), Uganda, Rwanda and East Africa. **Chorology:** Afromontane. **Star:** green. **Alt. range:** 1200-1400m (1). **Specimens:** - *No locality:* Dunlap 95

Corymborkis corymbis Thouars
Bot. Tidsskr. 71: 161 (1977); F.T.E.A. Orchidaceae: 243 (1984); Fl. Afr. Cent. Orchidaceae: 12 (1984).
Corymborkis corymbosa Thouars, F.W.T.A. 3: 211 (1968).

Habit: herb. **Habitat:** stream banks in forest. **Distribution:** tropical and subtropical Africa. **Star:** excluded. **Specimens:** - *Mabeta-Moliwe:* Plot Voucher W 366, W 430, W 963, W 1033; Wheatley J.I. 340. *Southern Bakundu F.R.:* Akuo FHI 15162

Cynorkis debilis (Hook.f.) Summerh.
F.W.T.A. 3: 189 (1968); Fl. Afr. Cent. Orchidaceae: 166 (1984).

Habit: herb. **Habitat:** forest and forest-grassland transition. **Distribution:** Mount Cameroon, Bioko, S. Tomé, Congo (Kinshasa), Angola, Rwanda, Burundi, Malawi, Zimbabwe and East Africa. **Chorology:** Afromontane. **Star:** green. **Alt. range:** 1400-1600m (2), 1801-2000m (1). **Flowering:** Sep (1), Oct (1). **Specimens:** - *Etinde:* Thomas D. W. 9212, 9247. *No locality:* Mann G. 2127

Cynorkis gabonensis Summerh.
Kew Bull. s.n.: 143 (1938).

Habit: herb. **Habitat:** forest. **Distribution:** Cameroon and Gabon. **Chorology:** Lower Guinea. **Star:** blue. **Alt. range:** 400-600m (1). **Flowering:** Sep (1). **Specimens:** - *Etinde:* Wheatley J. I. 533

Cyrtorchis chailluana (Hook.f.) Schltr.
F.W.T.A. 3: 267 (1968); F.T.E.A. Orchidaceae: 576 (1989).

Habit: epiphyte. **Habitat:** forest. **Distribution:** Nigeria, Cameroon, Gabon, Congo (Brazzaville), Congo (Kinshasa), Burundi and Uganda. **Chorology:** Guineo-Congolian. **Star:** green. **Flowering:** Oct (3). **Specimens:** - *Etinde:* Banks H. 5, 48, 49

Cyrtorchis monteiroae (Rchb.f.) Schltr.
F.W.T.A. 3: 267 (1968); Fl. Afr. Cent. Orchidaceae: 493 (1992).

Habit: epiphyte. **Habitat:** lava flows. **Distribution:** Sierra Leone to Congo (Kinshasa), Angola and Uganda. **Chorology:** Guineo-Congolian. **Star:** green. **Flowering:** Oct (1). **Specimens:** - *Eastern slopes:* Dundas J. FHI 15302. *Etinde:* Banks H. 58

Cyrtorchis ringens (Rchb.f.) Summerh.
F.W.T.A. 3: 267 (1968); F.T.E.A. Orchidaceae: 580 (1989).

Habit: epiphyte. **Habitat:** forest. **Distribution:** tropical Africa. **Chorology:** Afromontane. **Star:** green. **Alt. range:** 1200-1400m (1). **Specimens:** - *No locality:* Mann G. 2114

Diaphananthe bidens (Sw.) Schltr.
F.W.T.A. 3: 261 (1968); F.T.E.A. Orchidaceae: 526 (1989).

Habit: epiphyte. **Habitat:** stream banks in forest. **Distribution:** Sierra Leone to Cameroon, Congo (Kinshasa), Angola and Uganda. **Chorology:** Guineo-Congolian. **Star:** green. **Specimens:** - *Etinde:* Maitland T. D. 725. *Mabeta-Moliwe:* Wheatley J.T. 266

Diaphananthe bueae (Schltr.) Schltr.
F.W.T.A. 3: 261 (1968); F.T.E.A. Orchidaceae: 524 (1989).

Habit: epiphyte. **Habitat:** forest. **Distribution:** Cameroon. **Chorology:** Lower Guinea (montane). **Star:** black. **Alt. range:** 1000-1200m (1). **Specimens:** - *Eastern slopes:* Deistel H. s.n.

Diaphananthe kamerunensis (Schltr.) Schltr.
F.W.T.A. 3: 261 (1968); F.T.E.A. Orchidaceae: 528 (1989).

Habit: epiphyte. **Habitat:** forest. **Distribution:** Cameroon, Congo (Kinshasa), Zambia and Uganda. **Chorology:** Guineo-Congolian (montane). **Star:** gold. **Alt. range:** 1000-1200m (1). **Specimens:** - *Eastern slopes:* Deistel H. s.n.

Diaphananthe obanensis (Rendle) Summerh.
F.W.T.A. 3: 263 (1968).

Habit: epiphyte. **Habitat:** forest. **Distribution:** SE Nigeria and SW Cameroon. **Chorology:** Western Cameroon Uplands (montane). **Star:** gold. **Specimens:** - *Mokoko River F.R.:* Wright J. O. 58/66

Diaphananthe plehniana (Schltr.) Schltr.
F.W.T.A. 3: 261 (1968).

Habit: epiphyte. **Habitat:** forest. **Distribution:** Nigeria and Cameroon. **Chorology:** Lower Guinea. **Star:** blue. **Specimens:** - *Mokoko River F.R.:* Wright J. O. 58/19

Dinklageella liberica Mansf.
F.W.T.A. 3: 270 (1968); Fl. Afr. Cent. Orchidaceae: 500 (1992).

Habit: epiphyte. **Habitat:** forest. **Distribution:** Liberia to Cameroon and Congo (Kinshasa). **Chorology:** Guineo-Congolian. **Star:** green. **Specimens:** - *Eastern slopes:* Chew C. W. 25

Disperis kamerunensis Schltr.
F.W.T.A. 3: 205 (1968).

Habit: herb. **Habitat:** forest. **Distribution:** Mount Cameroon. **Chorology:** Endemic (montane). **Star:** black. **Alt. range:** 1000-1200m (1). **Specimens:** - *Eastern slopes:* Preuss P. R. 609

Epipogium roseum (D.Don) Lindl.
F.W.T.A. 3: 207 (1968); F.T.E.A. Orchidaceae: 238 (1984).

Habit: saprophyte. **Guild:** pioneer. **Habitat:** forest. **Distribution:** Ghana, Nigeria, Cameroon, Rio Muni, Congo (Kinshasa), Uganda,

and Kenya, also in Asia. **Star:** black. **Alt. range:** 1000-1200m (1).
Flowering: Feb (1). **Specimens:** - *Eastern slopes:* Richards P. W.
9346. *Mabeta-Moliwe:* Dalziel J. M. 8205; Mann G. 784

Eulophia buettneri (Kraenzl.) Summerh.
F.W.T.A. 3: 250 (1968).

Habit: herb. **Distribution:** Guinea to Cameroon. **Chorology:**
Guineo-Congolian. **Star:** green. **Specimens:** - *Bambuko F.R.:*
Keay R. W. J. FHI 37460

Eulophia horsfallii (Batem.) Summerh.
F.W.T.A. 3: 250 (1968); F.T.E.A. Orchidaceae: 428 (1989).

Habit: herb. **Guild:** pioneer. **Distribution:** tropical and
subtropical Africa. **Star:** excluded. **Specimens:** - *Eastern slopes:*
Brown R. H. 105a

Gastrodia africana Kraenzl.
Sanford ms. for Fl. Cameroun.

Habit: saprophyte. **Habitat:** forest. **Distribution:** Cameroon.
Chorology: Lower Guinea (montane). **Star:** black. **Flowering:**
Apr (1). **Specimens:** - *No locality:* Dusen P. K. H. 397

Genyorchis macrantha Summerh.
Kew Bull. 11: 124 (1957); F.W.T.A. 3: 242 (1968).

Habit: epiphyte. **Habitat:** forest-grassland transition.
Distribution: Mount Cameroon. **Chorology:** Endemic (montane).
Star: black. **Alt. range:** 2000-2200m (1). **Specimens:** - *Eastern
slopes:* Brenan J. P. M. 9570

Genyorchis platybulbon Schltr.
F.W.T.A. 3: 242 (1968).

Habit: herb. **Distribution:** Mount Cameroon. **Chorology:**
Endemic. **Star:** black. **Specimens:** - *Mabeta-Moliwe:* Stammler
s.n.

Graphorkis lurida (Sw.) Kuntze
F.W.T.A. 3: 251 (1968); F.T.E.A. Orchidaceae: 415 (1989).

Habit: epiphyte. **Habitat:** forest. **Distribution:** Senegal to Bioko,
Cameroon, Gabon, Congo (Kinshasa), Burundi, Uganda and
Tanzania. **Chorology:** Guineo-Congolian. **Star:** green.
Specimens: - *Etinde:* Rosevear Cam. 52/37. *Mabeta-Moliwe:*
Mann G. 782

Habenaria amoena Summerh.
F.T.E.A. Orchidaceae: 75 (1968); Fl. Afr. Cent. Orchidaceae: 135
(1984).

Habit: herb. **Habitat:** forest-grassland transition. **Distribution:**
Cameroon, Congo (Kinshasa), Zambia, Zimbabwe and Tanzania.
Chorology: Afromontane. **Star:** green. **Alt. range:** 2000-2200m
(1). **Flowering:** Oct (1). **Specimens:** - *Etinde:* Thomas D. W. 9364

Habenaria attenuata Hook.f.
F.T.E.A. Orchidaceae: 52 (1968); F.W.T.A. 3: 193 (1968); Kew
Bull. 33: 660 (1979); Fl. Afr. Cent. Orchidaceae: 84 (1984).

Habit: herb. **Habitat:** grassland and forest-grassland transition.
Distribution: Mount Cameroon, Bioko, Congo (Kinshasa),
Uganda and Ethiopia. **Chorology:** Afromontane. **Star:** blue. **Alt.
range:** 2000-2200m (1), 2601-2800m (1). **Flowering:** Oct (2),
Nov (1). **Specimens:** - *Eastern slopes:* Banks H. 103; Cheek M.
3622, 3667

Habenaria bracteosa A.Rich.
F.T.E.A. Orchidaceae: 53 (1968); F.W.T.A. 3: 193 (1968).

Habit: herb. **Habitat:** grassland. **Distribution:** W. Cameroon,
Bioko, Uganda, Ethiopia, Kenya and Tanzania. **Chorology:**
Afromontane. **Star:** green. **Alt. range:** 2600-2800m (1).
Flowering: Oct (1). **Fruiting:** Oct (1). **Specimens:** - *Eastern
slopes:* Cheek M. 3666

Habenaria macrandra Lindl.
F.T.E.A. Orchidaceae: 63 (1968); F.W.T.A. 3: 193 (1968).

Habit: herb. **Habitat:** forest. **Distribution:** Guinea to Congo
(Kinshasa), Angola, Uganda and Tanzania. **Chorology:** Guineo-
Congolian. **Star:** green. **Alt. range:** 800-1000m (2), 1401-1600m
(1). **Flowering:** Oct (2). **Specimens:** - *Eastern slopes:* Sidwell K.
96. *Etinde:* Cheek M. 3768. *No locality:* Mann G. 2117

Habenaria malacophylla Rchb.f.
F.T.E.A. Orchidaceae: 73 (1968); F.W.T.A. 3: 196 (1968); Fl. Afr.
Cent. Orchidaceae: 135 (1984).

Habit: herb. **Habitat:** forest. **Distribution:** tropical and
subtropical Africa. **Chorology:** Afromontane. **Star:** green. **Alt.
range:** 1200-1400m (1), 1801-2000m (1). **Flowering:** Oct (1).
Specimens: - *Eastern slopes:* Dundas J. FHI 15309; Sidwell K. 92

Habenaria mannii Hook.f.
F.W.T.A. 3: 194 (1968).

Habit: herb. **Habitat:** grassland and forest-grassland transition.
Distribution: Nigeria, Bioko and Cameroon. **Chorology:** Lower
Guinea (montane). **Star:** blue. **Alt. range:** 2000-2200m (1), 2601-
2800m (1). **Flowering:** Oct (1). **Specimens:** - *Eastern slopes:*
Sidwell K. 55. *No locality:* Mann G. 2119

Habenaria microceras Hook.f.
F.W.T.A. 3: 193 (1968).

Habit: terrestrial or epiphytic herb. **Habitat:** grassland and forest-
grassland transition. **Distribution:** W. Cameroon and Bioko.
Chorology: Western Cameroon Uplands (montane). **Star:** black.
Alt. range: 2400-2600m (2). **Flowering:** Oct (1). **Fruiting:** Oct
(1). **Specimens:** - *Eastern slopes:* Cheek M. 3641. *No locality:*
Mann G. 2116

Habenaria obovata Summerh.
F.W.T.A. 3: 194 (1968).

Habit: herb. **Habitat:** grassland. **Distribution:** Mount Cameroon.
Chorology: Endemic (montane). **Star:** black. **Alt. range:** 2400-
2600m (1), 2801-3000m (1). **Flowering:** Oct (1), Nov (1).
Specimens: - *Eastern slopes:* Banks H. 104; Sidwell K. 39. *No
locality:* Maitland T. D. 804

Habenaria procera (Sw.) Lindl. var. gabonensis (Rchb.f.) Geerinck

Fl. Afr. Cent. Orchidaceae: 89 (1984).
Habenaria gabonensis Rchb.f., F.W.T.A. 3: 194 (1968).

Habit: herb. **Habitat:** lava flows. **Distribution:** Sierra Leone to Bioko, Cameroon, Gabon and Congo (Kinshasa). **Chorology:** Guineo-Congolian. **Star:** green. **Flowering:** Oct (1). **Specimens:** - *Etinde:* Banks H. 59

Habenaria procera (Sw.) Lindl. var. procera

F.T.E.A. Orchidaceae: 60 (1968); F.W.T.A. 3: 194 (1968); Fl. Afr. Cent. Orchidaceae: 89 (1984).

Habit: epiphyte. **Habitat:** forest. **Distribution:** Sierra Leone to Cameroon, Gabon and Uganda. **Chorology:** Guineo-Congolian. **Star:** green. **Specimens:** - *Etinde:* Ngongi FHI 15084

Habenaria weileriana Schltr.

F.W.T.A. 3: 194 (1968).

Habit: herb. **Habitat:** forest and stream banks. **Distribution:** SE Nigeria, Mount Cameroon and Gabon. **Chorology:** Lower Guinea. **Star:** blue. **Alt. range:** 800-1000m (1). **Specimens:** - *Etinde:* Weiler s.n.. *Mabeta-Moliwe:* Mildbraed G. W. J. 10600

Hetaeria heterosepala (Rchb.f.) Summerh.

F.W.T.A. 3: 210 (1968).
Zeuxine heterosepala (Rchb.f.) Geerinck Bull. Jard. Bot. Nat. Belg. 50: 120 (1980); Fl. Afr. Cent. Orchidaceae: 21 (1984).

Habit: herb. **Distribution:** Liberia, Ivory Coast, Cameroon, S. Tomé, Congo (Kinshasa) and Tanzania. **Chorology:** Guineo-Congolian (montane). **Star:** green. **Alt. range:** 800-1000m (1). **Specimens:** - *No locality:* Mann G. 2130a

Hetaeria stammleri (Schltr.) Summerh.

F.W.T.A. 3: 210 (1968).
Zeuxine stammleri Schltr. Bull. Jard. Bot. Nat. Belg. 50: 122 (1980); Fl. Afr. Cent. Orchidaceae: 19 (1984).

Habit: herb. **Distribution:** Ivory Coast, Nigeria, Cameroon, Bioko, Congo (Kinshasa) and CAR. **Chorology:** Guineo-Congolian. **Star:** green. **Specimens:** - *Mabeta-Moliwe:* Dunlap 245; Hepper N. 8621; Sunderland T.C.H. 1061; Stammler s.n.

Hetaeria tetraptera (Rchb.f.) Summerh.

F.W.T.A. 3: 210 (1968).
Zeuxine tetraptera (Rchb.f.) T.Durand & Schinz Bull. Jard. Bot. Nat. Belg. 50: 122 (1980); Fl. Afr. Cent. Orchidaceae: 19 (1984).

Habit: herb. **Habitat:** forest. **Distribution:** Nigeria, Cameroon, Gabon and Congo (Kinshasa). **Chorology:** Guineo-Congolian. **Star:** green. **Specimens:** - *Mabeta-Moliwe:* Wheatley J.I. 56. *Southern Bakundu F.R.:* Brenan J. P. M. 9447

Holothrix tridentata (Hook.f.) Rchb.f.

F.W.T.A. 3: 186 (1968).

Habit: herb. **Habitat:** forest-grassland transition. **Distribution:** Mount Cameroon and Ethiopia. **Chorology:** Afromontane. **Star:** gold. **Alt. range:** 1800-2000m (1), 2401-2600m (1), 2601-2800m (1). **Flowering:** Oct (2). **Specimens:** - *Eastern slopes:* Tchouto P. 336; Thomas D. W. 9269. *No locality:* Mann G. 2128

Liparis deistelii Schltr.

F.W.T.A. 3: 214 (1968); Fl. Afr. Cent. Orchidaceae: 279 (1984); F.T.E.A. Orchidaceae: 299 (1984).
Liparis kamerunensis Schltr., F.W.T.A. 3: 214 (1968).

Habit: epiphytic or terrestrial herb. **Habitat:** forest and forest-grassland transition. **Distribution:** Cameroon, Gabon, Congo (Kinshasa), Ethiopia, Malawi, Uganda, Kenya and Tanzania. **Chorology:** Afromontane. **Star:** green. **Alt. range:** 1800-2000m (2). **Specimens:** - *No locality:* Deistel H. s.n.; Mann G. 2129

Liparis goodyeroides Schltr.

F.W.T.A. 3: 214 (1968).

Habit: herb. **Distribution:** Cameroon. **Chorology:** Lower Guinea. **Star:** black. **Specimens:** - *Mabeta-Moliwe:* Stammler s.n.

Liparis nervosa (Thunb.) Lindley

Fl. Afr. Cent. Orchidaceae: 275 (1984); F.T.E.A. Orchidaceae: 298 (1984).
Liparis guineensis Lindl., F.W.T.A. 3: 214 (1968).

Habit: herb. **Habitat:** forest, amongst rocks and on lava flows. **Distribution:** tropical Africa. **Star:** excluded. **Alt. range:** 1-200m (1). **Flowering:** Oct (1). **Fruiting:** Oct (1). **Specimens:** - *Eastern slopes:* Banks H. 18. *Etinde:* Brodie S. 15

Liparis platyglossa Schltr.

F.W.T.A. 3: 214 (1968); F.T.E.A. Orchidaceae: 296 (1984).

Habit: epiphyte. **Habitat:** forest. **Distribution:** Ivory Coast, Nigeria, Bioko, Cameroon and Uganda. **Chorology:** Guineo-Congolian. **Star:** green. **Specimens:** - *Etinde:* Stossel 5b

Malaxis weberbaueriana (Kraenzl.) Summerh.

F.W.T.A. 3: 211 (1968); F.T.E.A. Orchidaceae: 289 (1984).

Habit: herb. **Distribution:** Bioko, Cameroon, Congo (Kinshasa), Zambia, Malawi, Zimbabwe and East Africa. **Star:** excluded. **Specimens:** - *No locality:* Weberbauer A. 42

Manniella gustavi Rchb.f.

F.W.T.A. 3: 207 (1968); Fl. Afr. Cent. Orchidaceae: 24 (1984).

Habit: herb. **Habitat:** forest. **Distribution:** Sierra Leone to Bioko, Cameroon, Congo (Brazzaville), Congo (Kinshasa), Uganda and Tanzania. **Chorology:** Guineo-Congolian (montane). **Star:** green. **Alt. range:** 1200-1400m (1). **Specimens:** - *No locality:* Mann G. 1336

Plectrelminthus caudatus (Lindl.) Summerh.

F.W.T.A. 3: 264 (1968); Fl. Afr. Cent. Orchidaceae: 497 (1992).

Habit: epiphyte. **Habitat:** forest. **Distribution:** Guinea to Cameroon, Gabon, Congo (Kinshasa) and CAR. **Chorology:** Guineo-Congolian. **Star:** green. **Specimens:** - *Etinde:* Nditapah J. K. FHI 52406

Podangis dactyloceras (Rchb.f.) Schltr.

F.W.T.A. 3: 251 (1968); Fl. Afr. Cent. Orchidaceae: 604 (1992).

Habit: epiphyte. **Habitat:** forest. **Distribution:** tropical Africa. **Chorology:** Afromontane. **Star:** green. **Alt. range:** 1200-1400m (1). **Specimens:** - *Eastern slopes:* Maitland T. D. 699

Polystachya adansoniae Rchb.f.
F.W.T.A. 3: 223 (1968); F.T.E.A. Orchidaceae: 367 (1984).

Habit: epiphyte. **Habitat:** on tree or rocks in forest. **Distribution:** tropical Africa. **Star:** excluded. **Specimens:** - *Eastern slopes:* Maitland T. D. 727

Polystachya affinis Lindl.
F.W.T.A. 3: 224 (1968); F.T.E.A. Orchidaceae: 391 (1984).

Habit: epiphyte. **Habitat:** lava flows and forest. **Distribution:** Guinea to Cameroon, Gabon, Congo (Kinshasa), Angola, CAR and Uganda. **Chorology:** Guineo-Congolian. **Star:** green. **Alt. range:** 1-200m (1). **Flowering:** Feb (1), Oct (1). **Specimens:** - *Etinde:* Brodie S. 8; Tchouto P. 9

Polystachya albescens Ridl. subsp. **angustifolia** (Summerh.) Summerh.
F.W.T.A. 3: 221 (1968).

Habit: epiphyte. **Habitat:** forest. **Distribution:** Mount Cameroon. **Chorology:** Endemic (montane). **Star:** black. **Alt. range:** 800-1000m (2). **Flowering:** Jul (1). **Fruiting:** Jul (1). **Specimens:** - *Eastern slopes:* Gregory H. 165; Tekwe C. 107

Polystachya alpina Lindl.
F.W.T.A. 3: 223 (1968).

Habit: epiphyte. **Habitat:** forest-grassland transition. **Distribution:** SE Nigeria, W. Cameroon and Bioko. **Chorology:** Western Cameroon Uplands (montane). **Star:** gold. **Alt. range:** 1400-1600m (1), 1601-1800m (1), 1801-2000m (1), 2001-2200m (1). **Flowering:** Sep (1), Oct (3). **Specimens:** - *Eastern slopes:* Cheek M. 3645; Thomas D. W. 9282; Watts J. 439. *Etinde:* Wheatley J. I. 582. *No locality:* Preuss P. R. 934

Polystachya cf. alpina Lindl.

Specimens: - *Eastern slopes:* Thomas D. W. 9432

Polystachya bennettiana Rchb.f.
F.T.E.A. Orchidaceae: 357 (1984); Fl. Rwanda 4: 604 (1988).
Polystachya stricta Rolfe, F.W.T.A. 3: 221 (1968).

Habit: epiphyte. **Habitat:** grassland. **Distribution:** Nigeria, Cameroon, Congo (Kinshasa), Rwanda, Uganda, Zambia, Sudan, Ethiopia, Kenya and Tanzania. **Chorology:** Afromontane. **Star:** green. **Alt. range:** 1200-1400m (1). **Flowering:** May (1). **Specimens:** - *Eastern slopes:* Mbani J. M. 113

Polystachya bicalcarata Kraenzl.
F.W.T.A. 3: 225 (1968).

Habit: epiphyte. **Habitat:** forest-grassland transition. **Distribution:** Mount Cameroon, Bioko and Bamboutos Mountains. **Chorology:** Western Cameroon Uplands (montane). **Star:** black. **Alt. range:** 1200-1400m (1), 1601-1800m (1). **Flowering:** Oct (1). **Specimens:** - *Eastern slopes:* Deistel H. 62c, 79. *Etinde:* Thomas D. W. 9178

Polystachya bifida Lindl.
F.W.T.A. 3: 221 (1968); Fl. Rwanda 4: 604 (1988); Fl. Afr. Cent. Orchidaceae: 423 (1992).

Habit: epiphyte. **Habitat:** forest and forest-grassland transition. **Distribution:** SE Nigeria, Cameroon, S. Tomé, Gabon, Congo (Kinshasa) and Rwanda. **Chorology:** Guineo-Congolian (montane). **Star:** green. **Alt. range:** 1400-1600m (2), 1601-1800m (1), 1801-2000m (1). **Flowering:** Oct (3). **Fruiting:** Oct (1). **Specimens:** - *Eastern slopes:* Cheek M. 3699. *Etinde:* Tchouto P. 404; Thomas D. W. 9242; Wheatley J. I. 623

Polystachya calluniflora Kraenzl.
F.W.T.A. 3: 223 (1968); F.T.E.A. Orchidaceae: 364 (1984).

Habit: epiphyte. **Habitat:** forest. **Distribution:** Nigeria, Cameroon, Bioko, Rwanda and Uganda. **Chorology:** Guineo-Congolian (montane). **Star:** green. **Alt. range:** 800-1000m (1). **Specimens:** - *Eastern slopes:* Deistel H. 75

Polystachya caloglossa Rchb.f.
F.W.T.A. 3: 221 (1968); F.T.E.A. Orchidaceae: 356 (1984).

Habit: epiphyte. **Habitat:** forest. **Distribution:** Mount Cameroon, Bioko, Congo (Kinshasa) and Uganda. **Chorology:** Guineo-Congolian (montane). **Star:** green. **Alt. range:** 400-600m (1), 1201-1400m (1), 1401-1600m (1), 1601-1800m (2). **Flowering:** Mar (1), Sep (1), Oct (2). **Fruiting:** Oct (1). **Specimens:** - *Eastern slopes:* Cable S. 1377. *Etinde:* Tchouto P. 410; Tekwe C. 222; Watts J. 533. *No locality:* Mann G. 2110

Polystachya cf. caloglossa Rchb.f.

Specimens: - *Etinde:* Mbani J. M. 24

Polystachya coriscensis Rchb.f.
F.W.T.A. 3: 223 (1968); Fl. Afr. Cent. Orchidaceae: 437 (1992).

Habit: epiphyte. **Habitat:** forest. **Distribution:** Nigeria, Cameroon, Gabon and Congo (Kinshasa). **Chorology:** Guineo-Congolian. **Star:** green. **Specimens:** - *Mabeta-Moliwe:* Schlechter F. R. R. 15780

Polystachya cultriformis (Thou.) Spreng.
F.W.T.A. 3: 225 (1968); F.T.E.A. Orchidaceae: 341 (1984).

Habit: epiphyte. **Habitat:** forest. **Distribution:** Bioko and Cameroon to East and South Africa. **Chorology:** Afromontane. **Star:** green. **Alt. range:** 1000-1200m (3), 1201-1400m (1). **Flowering:** Sep (1), Oct (1). **Fruiting:** Feb (1). **Specimens:** - *Eastern slopes:* Etuge M. 1167; Preuss P. R. 1009. *Etinde:* Thomas D. W. 9233; Wheatley J. I. 592

Polystachya dolichophylla Schltr.
F.W.T.A. 3: 224 (1968).

Habit: epiphyte. **Habitat:** forest. **Distribution:** Guinea to Cameroon and Gabon. **Chorology:** Guineo-Congolian. **Star:** green. **Alt. range:** 800-1000m (1). **Specimens:** - *Eastern slopes:* Schlechter F. R. R. 12837. *Mabeta-Moliwe:* Baum s.n.

Polystachya elegans Rchb.f.
F.W.T.A. 3: 223 (1968).

Habit: epiphyte. **Habitat:** forest. **Distribution:** Nigeria, W. Cameroon and Bioko. **Chorology:** Lower Guinea (montane). **Star:** blue. **Alt. range:** 1200-1400m (1), 1401-1600m (4), 1601-1800m

(1). **Flowering:** Sep (2), Oct (3). **Fruiting:** Oct (1). **Specimens:** - *Etinde:* Tekwe C. 198; Thomas D. W. 9250, 9545; Wheatley J. I. 555, 620. *No locality:* Mann G. 1338

Polystachya fusiformis (Thou.) Lindl.
F.W.T.A. 3: 225 (1968); F.T.E.A. Orchidaceae: 371 (1984).

Habit: epiphyte. **Habitat:** forest. **Distribution:** Ghana to East and South Africa. **Chorology:** Afromontane. **Star:** green. **Alt. range:** 800-1000m (1). **Specimens:** - *Eastern slopes:* Preuss P. R. 1072

Polystachya laxiflora Lindl.
F.W.T.A. 3: 221 (1968); Fl. Afr. Cent. Orchidaceae: 415 (1992).

Habit: epiphyte. **Habitat:** forest. **Distribution:** Guinea to Bioko, Cameroon, Gabon, Congo (Kinshasa) and Zambia. **Chorology:** Guineo-Congolian. **Star:** green. **Specimens:** - *Etinde:* Rosevear Cam. 51/37. *Mabeta-Moliwe:* Schlechter F. R. R. 15778

Polystachya leonensis Rchb.f.
F.W.T.A. 3: 224 (1968).

Habit: epiphyte. **Habitat:** forest. **Distribution:** Guinea to Cameroon. **Chorology:** Guineo-Congolian. **Star:** green. **Specimens:** - *Mabeta-Moliwe:* Lord Scarborough ?

Polystachya mystacioides De Wild.
F.W.T.A. 3: 225 (1968); Kew Bull. 32: 746 (1978).
Polystachya crassifolia Schltr., F.W.T.A. 3: 225 (1968).

Habit: epiphyte. **Habitat:** forest. **Distribution:** Ivory Coast, Cameroon and Congo (Kinshasa). **Chorology:** Guineo-Congolian. **Star:** green. **Specimens:** - *Mabeta-Moliwe:* Schlechter F. R. R. 12841

Polystachya odorata Lindl. var. odorata
F.W.T.A. 3: 224 (1968); F.T.E.A. Orchidaceae: 360 (1984).

Habit: epiphyte. **Habitat:** forest. **Distribution:** Ivory Coast to Congo (Kinshasa), Angola, Uganda and Tanzania. **Chorology:** Guineo-Congolian. **Star:** green. **Specimens:** - *Etinde:* Maitland T. D. 729

Polystachya paniculata (Sw.) Rolfe
F.W.T.A. 3: 221 (1968); F.T.E.A. Orchidaceae: 354 (1984).

Habit: epiphyte. **Habitat:** forest. **Distribution:** Sierra Leone to Cameroon, Gabon, Congo (Brazzaville), Congo (Kinshasa) and Uganda. **Chorology:** Guineo-Congolian. **Star:** green. **Specimens:** - *Etinde:* Ngongi FHI 15341

Polystachya polychaete Kraenzl.
F.W.T.A. 3: 223 (1968); F.T.E.A. Orchidaceae: 366 (1984).

Habit: epiphyte. **Habitat:** forest. **Distribution:** Sierra Leone to Bioko, Cameroon, Gabon, Congo (Kinshasa), Uganda, Kenya and Tanzania. **Chorology:** Guineo-Congolian. **Star:** green. **Alt. range:** 800-1000m (1). **Specimens:** - *Eastern slopes:* Maitland T. D. 731

Polystachya rhodoptera Rchb.f.
F.W.T.A. 3: 221 (1968); Fl. Afr. Cent. Orchidaceae: 420 (1992).

Habit: epiphyte. **Habitat:** Aframomum thicket. **Distribution:** Sierra Leone to Cameroon, Gabon and Congo (Kinshasa). **Chorology:** Guineo-Congolian. **Star:** green. **Alt. range:** 600-800m (1). **Flowering:** Oct (1). **Fruiting:** Oct (1). **Specimens:** - *Etinde:* Watts J. 558

Polystachya seticaulis Rendle
F.W.T.A. 3: 223 (1968); Fl. Afr. Cent. Orchidaceae: 440 (1992).

Habit: epiphyte. **Habitat:** forest. **Distribution:** Nigeria, Cameroon, Gabon and Congo (Kinshasa). **Chorology:** Guineo-Congolian. **Star:** green. **Specimens:** - *Mokoko River F.R.:* Wright J. O. 164

Polystachya superposita Rchb.f.
F.W.T.A. 3: 225 (1968).

Habit: epiphyte. **Habitat:** forest. **Distribution:** Mount Cameroon and Bioko. **Chorology:** Western Cameroon Uplands (montane). **Star:** black. **Alt. range:** 1400-1600m (1). **Specimens:** - *No locality:* Mann G. 2125

Polystachya supfiana Schltr.
F.W.T.A. 3: 225 (1968).

Habit: epiphyte. **Habitat:** forest. **Distribution:** SE Nigeria, Cameroon and Gabon. **Chorology:** Lower Guinea. **Star:** blue. **Specimens:** - *Etinde:* Schlechter F. R. R. 12415

Polystachya tessellata Lindl.
F.W.T.A. 3: 224 (1968).
Polystachya concreta (Jacq.) Garay & H.R.Sweet Orquideologia 9: 206 (1974); Kew Bull. 42: 724 (1987).

Habit: epiphyte. **Habitat:** forest. **Distribution:** tropical Africa. **Star:** excluded. **Alt. range:** 1-200m (1), 801-1000m (1). **Flowering:** Nov (1). **Fruiting:** Oct (1). **Specimens:** - *Eastern slopes:* Maitland T. D. 724. *Etinde:* Banks H. 51; Cable S. 147

Polystachya victoriae Kraenzl.
F.W.T.A. 3: 223 (1968).

Habit: epiphyte. **Habitat:** forest. **Distribution:** Cameroon. **Chorology:** Lower Guinea. **Star:** black. **Specimens:** - *Etinde:* Simon G. 14

Polystachya sp.

Specimens: - *Mabeta-Moliwe:* Wheatley J. I. 314

Solenangis scandens (Schltr.) Schltr.
F.W.T.A. 3: 269 (1968).

Habit: epiphyte. **Habitat:** lava flows. **Distribution:** Sierra Leone to Cameroon, Gabon, Congo (Brazzaville), Congo (Kinshasa) and CAR. **Chorology:** Guineo-Congolian. **Star:** green. **Alt. range:** 1-200m (1). **Specimens:** - *Etinde:* Dawson S. 39

Tridactyle anthomaniaca (Rchb.f.) Summerh.
F.W.T.A. 3: 274 (1968); Fl. Afr. Cent. Orchidaceae: 529 (1992).

Habit: epiphyte. **Habitat:** forest. **Distribution:** tropical Africa. **Chorology:** Afromontane. **Star:** green. **Alt. range:** 1800-2000m

(1). **Flowering:** Oct (1). **Specimens:** - *Eastern slopes:* Cheek M. 3703. *No locality:* King D. E. S. 123

Tridactyle tridactylites (Rolfe) Schltr.
F.W.T.A. 3: 274 (1968); Fl. Afr. Cent. Orchidaceae: 543 (1992).

Habit: epiphyte. **Habitat:** farmbush. **Distribution:** tropical Africa. **Chorology:** Afromontane. **Star:** green. **Specimens:** - *Eastern slopes:* Schlechter F. R. R. 12840

Vanilla africana LIndl.

Specimens:- *Mabeta-Moliwe:* Plot Voucher W 1215

Vanilla ramosa Rolfe
F.W.T.A. 3: 205 (1968); F.T.E.A. Orchidaceae: 261 (1984).

Habit: climbing or epiphyte herb. **Habitat:** forest. **Distribution:** Ghana to Cameroon, Gabon, Congo (Brazzaville), Congo (Kinshasa) and Tanzania. **Chorology:** Guineo-Congolian. **Star:** green. **Specimens:** - *Mabeta-Moliwe:* Cheek M. 3486. *Southern Bakundu F.R.:* Binuyo A. & Daramola B. O. FHI 35648

Zeuxine elongata Rolfe
F.W.T.A. 3: 208 (1968); F.T.E.A. Orchidaceae: 251 (1984); Fl. Afr. Cent. Orchidaceae: 20 (1984).

Habit: herb. **Habitat:** forest. **Distribution:** Sierra Leone to Cameroon, Congo (Kinshasa), CAR, Uganda, Kenya and Tanzania. **Chorology:** Guineo-Congolian. **Star:** green. **Alt. range:** 1200-1400m (1). **Flowering:** Feb (1). **Specimens:** - *Bambuko F.R.:* Keay R. W. J. FHI 37512. *Eastern slopes:* Etuge M. 1161

PALMAE
T.C.H. Sunderland & J. Dransfield (K)

Borassus aethiopum Mart.

Specimens: - *Bambuko and Mokoko:* Common in savanna but no specimens made.

Cocos nucifera L.

Specimens: - *Etinde, Mabeta-Moliwe:* Commonly naturalized along the sea-shore, but no collections available.

Elaeis guineensis Jacq.
F.W.T.A. 3: 161 (1968).

Habit: tree to 15 m. **Guild:** pioneer. **Habitat:** plantations and farmbush. **Distribution:** cultivated throughout the tropics. **Star:** excluded. **Specimens:** - *Mabeta-Moliwe:* Plot Voucher W 751, W 861

Eremospatha hookeri (Mann & H.Wendl.) H.Wendl.

Specimens: - *Mokoko:* Thomas, D.W. 10059. *Onge:* Harris 3738; Thomas 9733

Eremospatha laurentii de Wilde

Specimens: - *Etinde:* Cheek M. 5554

Eremospatha macrocarpa (Mann & H.Wendl.) H.Wendl.

Specimens: - *Mokoko:* Thomas D.W. 10058

Eremospatha wendlandiana Dammer ex Becc.
F.W.T.A. 3: 168 (1968).

Habit: rattan. **Habitat:** forest. **Distribution:** SE Nigeria and Cameroon. **Chorology:** Lower Guinea. **Star:** red. **Specimens:** - *Southern Bakundu F.R.:* Dransfield J. 7004; Dundas J. FHI 8381. *Mokoko:* Sunderland T.C.H. 1640

Eremospatha sp.

Specimens: - *S. Bakundu:* Dransfield J. 7005, 7008

Laccosperma acutiflorum (Becc.) J.Dransf.

Specimens: *S. Bakundu:* Dransfield J. 7006

Laccosperma opacum (G.Mann & H.Wendl.) Drude
Kew Bull. 37: 456 (1982); Hawthorne W., F.G.F.T. Ghana: 225 (1990); Hawthorne W., F.G.F.T. Ghana: 225 (1990).
Ancistrophyllum opacum (G.Mann & H.Wendl.) Drude, F.W.T.A. 3: 167 (1968).

Habit: rattan. **Guild:** light demander. **Habitat:** forest. **Distribution:** Ghana, Nigeria, Bioko, Cameroon and Gabon. **Chorology:** Guineo-Congolian. **Star:** pink. **Alt. range:** 1-200m (1). **Flowering:** Apr (1). **Specimens:** - *Etinde:* Dransfield J. 6998; Cheek M. 591; Maitland T. D. 761. *Mabeta-Moliwe:* Dransfield J. 6996, 6997; Plot Voucher W 145, W 305, W 320, W 473, W 749, W 923, W 1005; Wheatley J. I. 154. *Mokoko:* Mbani 497. *No locality:* Watts J. 821

Laccospermum secundiflorum (P. de Beauv.) O.Kuntze

Specimens: *Mokoko:* Sunderland T.C.H. 1645. *No locality:* Thomas 9738

Laccosperma sp.

Specimen: - *Onge:* Cheek M. 5062; Harris D. 3660, 3742; Thomas 9726

Oncocalamus mannii (H.Wendl.) H.Wendl. & Drude

Specimens: - *S. Bakundu:* Dransfield J. 7007. *Onge:* Thomas 9732

Oncocalamus sp.

Specimens: - *Onge*: Harris 3739

Nypa fruticans Wurmb.

Specimens: - *Mabeta-Moliwe & Etinde:* Naturalized along the shore, but no specimens available.

Phoenix reclinata Jacq.

Specimens:- *Etinde:* Cheek M. 5937

PANDANACEAE

Pandanus satabiei Huynh
Bull. Mus. Nat. Hist. Nat. Paris 4, Sér. 6, Sect. B, Adansonia 3: 335-358 (1984)

Habit: tree to 9 m. **Habitat:** swamp forest. **Distribution:** Senegal to W. Cameroon. **Chorology:** Guineo-Congolian. **Star:** green. **Specimens:** - *Mabeta-Moliwe:* Cheek K. 3433

SMILACACEAE

Smilax anceps Willd.
Meded. Land. Wag. 82(3): 219 (1982).
  Smilax kraussiana Meisn., F.W.T.A. 3: 112 (1968).

Habit: woody climber. **Guild:** pioneer. **Habitat:** forest. **Distribution:** tropical and subtropical Africa. **Star:** excluded. **Alt. range:** 1-200m (4), 401-600m (1), 1001-1200m (1). **Flowering:** Jun (1). **Fruiting:** Jan (1), Jun (1), Aug (1), Oct (1), Dec (1). **Specimens:** - *Mabeta-Moliwe:* Baker W. J. 346; Cable S. 563; Plot Voucher W 723; Sunderland T. C. H. 1063; Wheatley J. I. 373. *No locality:* Mann G. 1271. *Onge:* Tchouto P. 815

TRIURIDACEAE

Sciaphila ledermannii Engl.
F.W.T.A. 3: 15 (1968); Fl. Cameroun 26: 72 (1984).

Habit: saprophyte. **Habitat:** forest. **Distribution:** SE Nigeria and Cameroon. **Chorology:** Lower Guinea. **Star:** gold. **Alt. range:** 1-200m (2). **Specimens:** - *Onge:* Harris D. J. 3737; Tchouto P. 1014

ZINGIBERACEAE
David Harris (E)

Aframomum flavum Lock
Bull. Jard. Bot. Nat. Belg. 48: 393 (1978).
  Aframomum hanburyi sensu Koechlin, Fl. Cameroun 4: 65 (1965).

Habit: herb to 4 m. **Habitat:** forest, clearings and Aframomum thicket. **Distribution:** Cameroon and Rio Muni. **Chorology:** Lower Guinea. **Star:** blue. **Alt. range:** 1-200m (3), 401-600m (3), 601-800m (1). **Flowering:** Mar (1), Apr (1), Oct (2), Nov (1), Dec (1). **Fruiting:** Mar (2), Oct (2), Nov (1). **Specimens:** - *Etinde:* Cable S. 345, 1657; Cheek M. 5545. *Mabeta-Moliwe:* Sunderland T. C. H. 1166. *Mokoko River F.R.:* Acworth J. M. 234. *Onge:* Cheek M. 4971; Tchouto P. 970

Aframomum leptolepis (K.Schum.) K.Schum.
F.W.T.A. 3: 76 (1968); Kew Bull. 28: 444 (1973); Kew Bull. 35: 309 (1980).
  Aframomum sp. A sensu Hepper, F.W.T.A. 3: 76 (1968).
  Aframomum dalzielii Hutch., ined., F.W.T.A. (ed.1) 2: 331 (1936).

Habit: herb. **Distribution:** Bioko and Cameroon. **Chorology:** Lower Guinea. **Star:** gold. **Alt. range:** 1000-1200m (1). **Flowering:** Feb (1). **Specimens:** - *Eastern slopes:* Cable S. 1355, 1395; Dalziel J. M. 8235; Deistel H. 455; Groves M. 111; Mbani J. M. 121; Nkeng P. 61. *Etinde:* Cheek M. 5544; Sunderland T. C. H. 1065; Wheatley J. I. 505. *Mokoko River F.R.:* Acworth J. M. 332. *Onge:* Cheek M. 5000; Harris D. J. 3770, 3822; Ndam N. 624; Tchouto P. 906, 991; Thomas D. W. 9919

Aframomum limbatum (Oliv. & Hanb.) K.Schum.
Fl. Gabon 9: 37 (1964); F.W.T.A. 3: 76 (1968); Kew Bull. 35: 307 (1980).

Habit: herb. **Habitat:** forest. **Distribution:** Nigeria, Bioko, Cameroon and Gabon. **Chorology:** Lower Guinea. **Star:** blue. **Alt. range:** 1-200m (2). **Flowering:** Mar (2). **Specimens:** - *Onge:* Tchouto P. 505; Wheatley J. I. 813

Aframomum pilosum (Oliv. & Hanb.) K.Schum.
F.W.T.A. 3: 75 (1968); Kew Bull. 35: 302 (1980).

Habit: herb. **Habitat:** forest. **Distribution:** Nigeria, Bioko and Cameroon. **Chorology:** Lower Guinea. **Star:** blue. **Alt. range:** 200-400m (1).**Fruiting:** Oct (1). **Specimens:** - *Onge:* Ndam N. 675

Aframomum subsericeum (Oliv. & Hanb.) K.Schum. subsp. glaucophyllum (K.Schum.) Lock
Fl. Gabon 9: 36 (1964); Fl. Cameroun 4: 48 (1965); F.W.T.A. 3: 75 (1968); Kew Bull. 35: 306 (1980).

Habit: herb to 4 m. **Habitat:** forest. **Distribution:** Ivory Coast, Nigeria, Cameroon, Gabon, Congo (Kinshasa) and Angola (Cabinda). **Chorology:** Guineo-Congolian. **Star:** green. **Alt. range:** 1-200m (2). **Flowering:** Apr (1). **Specimens:** - *Mabeta-Moliwe:* Jaff B. 57. *Onge:* Harris D. J. 3834. *Southern Bakundu F.R.:* Binuyo A. & Daramola B. O. FHI 35633

Aframomum zambesiacum (Baker) K.Schum.
F.W.T.A. 3: 75 (1968); Kew Bull. 35: 301 (1980); F.T.E.A. Zingiberaceae: 25 (1985).
  Aframomum chlamydanthum Loes. & Mildbr., F.W.T.A. 3: 75 (1968).

Habit: herb to 2 m. **Habitat:** forest. **Distribution:** Malawi, Nigeria, Cameroon, Congo (Kinshasa) and East Africa. **Chorology:** Afromontane. **Star:** green. **Specimens:** - *Etinde:* Mildbraed G. W. J. 10712; Thomas D. W. 9187

Aframomum sp. 2

Specimens: - *Mokoko River F.R.:* Batoum A. 23. *Onge:* Harris D. J. 3826; Tchouto P. 512, 922

Aframomum sp. 5

Specimens: - *Etinde:* Cable S. 329, 1654; Faucher P. 42. *Mokoko River F.R.:* Akogo M. 232; Ekema N. 863; Ndam N. 1213; Watts J. 1120. *Onge:* Harris D. J. 3736, 3821; Ndam N. 669; Tchouto P. 509

Aframomum sp. 6

Specimens: - *Mabeta-Moliwe:* Nguembock F. 31; Watts J. 111. *Onge:* Harris D. J. 3735, 3819; Ndam N. 668, 746; Watts J. 717

Aframomum sp. 7

Specimens: - *Mokoko River F.R.:* Watts J. 1121

Aframomum sp. 8

Specimens: - *Mokoko River F.R.:* Acworth J. M. 203; Akogo M. 223; Ekema N. 860; Watts J. 1170. *Onge:* Harris D. J. 3740; Tchouto P. 506, 993

Aframomum sp. 9

Specimens: - *Mokoko River F.R.:* Acworth J. M. 322. *Onge:* Ndam N. 799; Tchouto P. 504

Aframomum sp. 10

Specimens: - *Etinde:* Cable S. 1644

Aframomum sp. 11

Specimens: - *Etinde:* Thomas D. W. 9188

Aframomum sp. 12

Specimens: - *Onge:* Harris D. J. 3824

Aframomum sp. 13

Specimens: - *Onge:* Akogo M. 140; Tchouto P. 511

Aframomum sp. 14

Specimens: - *Eastern slopes:* Thomas D. W. 9511

Renealmia africana (K.Schum.) Benth.
Fl. Gabon 9: 30 (1964); Fl. Cameroun 4: 38 (1965); F.W.T.A. 3: 70 (1968).

Habit: herb to 1.5 m. **Guild:** shade-bearer. **Habitat:** forest. **Distribution:** Nigeria, Bioko, Cameroon, Gabon and Congo (Kinshasa). **Chorology:** Guineo-Congolian (montane). **Star:** green. **Alt. range:** 1400-1600m (1). **Specimens:** - *No locality:* Johnston H. H. 106

Renealmia cf. africana (K.Schum.) Benth.

Specimens: - *Eastern slopes:* Mbani J. M. 107; Tekwe C. 312. *Etinde:* Tekwe C. 196; Thomas D. W. 9117, 9227; Watts J. 494

Renealmia albo-rosea K.Schum.
Fl. Cameroun 4: 41 (1965); F.W.T.A. 3: 71 (1968).

Habit: herb. **Habitat:** forest. **Distribution:** SW Cameroon. **Chorology:** Western Cameroon. **Star:** black. **Specimens:** - *Etinde:* Winkler H. J. P. 357

Renealmia cincinnata (K.Schum.) Baker
Fl. Gabon 9: 23 (1964); Fl. Cameroun 4: 30 (1965); F.W.T.A. 3: 70 (1968).

Habit: herb. **Habitat:** forest. **Distribution:** Nigeria, Cameroon and Gabon. **Chorology:** Lower Guinea. **Star:** blue. **Alt. range:** 1-200m (1). **Flowering:** May (1). **Fruiting:** May (1). **Specimens:** - *Etinde:* Preuss P. R. 1348. *Mabeta-Moliwe:* Watts J. 278

Renealmia sp.

Specimens: - *Mabeta-Moliwe:* Jaff B. 153

GYMNOSPERMS

GNETACEAE

Gnetum buchholzianum Engl.
F.W.T.A. 1: 33 (1954).

Habit: woody climber. **Habitat:** forest. **Distribution:** Cameroon. **Chorology:** Lower Guinea. **Star:** gold. **Alt. range:** 400-600m (1).**Fruiting:** Jan (1). **Specimens:** - *Etinde:* Maitland T. D. 749; Tekwe C. 6

PODOCARPACEAE

Podocarpus mannii Hook.f.
F.W.T.A. 1: 33 (1954).

Habit: tree to 15 m. **Habitat:** forest. **Distribution:** endemic to S. Tomé, planted in Cameroon. **Chorology:** Lower Guinea. **Star:** blue. **Specimens:** - *Eastern slopes:* Fairchild D. SPD 74656; Maitland T. D. 343

FERN ALLIES

P.J. Edwards (K)

LYCOPODIACEAE

Huperzia brachystachys (Baker) Pic.Serm.

Webbia 23: 162 (1968); Acta Botanica Barcinonensia 31: 9 (1978).
  Lycopodium brachystachys (Baker) Alston, F.W.T.A. Suppl.: 12 (1959).

Habit: epiphyte. **Habitat:** forest-grassland transition. **Distribution:** Guinea to SW Cameroon and Bioko. **Chorology:** Guineo-Congolian (montane). **Star:** green. **Alt. range:** 1400-1600m (1), 1601-1800m (3). **Specimens:** - *Eastern slopes:* Nkeng P. 84. *Etinde:* Tekwe C. 236; Thomas D. W. 9255; Wheatley J. I. 619. *No locality:* Mann G. 2041

Huperzia mildbraedii (Herter) Pic.Serm.

Webbia 23: 163 (1968); Acta Botanica Barcinonensia 31: 10 (1978); Bull. Jard. Bot. Nat. Belg. 53: 184 (1983).
  Lycopodium mildbraedii Herter, F.W.T.A. Suppl.: 11 (1959).

Habit: epiphyte. **Habitat:** forest-grassland transition. **Distribution:** Guinea to Bioko, W. Cameroon, S. Tomé and Congo (Kinshasa). **Chorology:** Guineo-Congolian (montane). **Star:** green. **Alt. range:** 1800-2000m (1), 2001-2200m (1). **Specimens:** - *Eastern slopes:* Thomas D. W. 9276, 9451. *No locality:* Johnston H. H. s.n.

Huperzia ophioglossoides (Lam.) Rothm.

Acta Botanica Barcinonensia 31: 8 (1978).
  Lycopodium ophioglossoides Lam., F.W.T.A. Suppl.: 12 (1959).

Habit: epiphyte. **Habitat:** forest. **Distribution:** tropical Africa. **Chorology:** Afromontane. **Star:** green. **Alt. range:** 2200-2400m (1). **Fertile:** Oct (1). **Specimens:** - *Eastern slopes:* Thomas D. W. 9391. *No locality:* Boughey A. S. GC 12581

Huperzia phlegmaria (L.) Rothm. var. phlegmaria

Acta Botanica Barcinonensia 31: 9 (1978).
  Lycopodium phlegmaria L., F.W.T.A. Suppl.: 12 (1959).

Habit: epiphyte. **Habitat:** forest. **Distribution:** palaeotropical. **Star:** excluded. **Specimens:** - *No locality:* Mann G. 2042 (partly)

Huperzia saururus (Lam.) Trevis.

Biologiske Skrifter 34: 20 (1989).
  Lycopodium saururus Lam., Fl. Zamb. Pteridophyta: 17 (1970).

Habit: herb. **Habitat:** grassland. **Distribution:** tropical Africa and America. **Chorology:** Montane. **Star:** excluded. **Specimens:** - *No locality:* Mann G. 2039

Lycopodiella cernua (L.) Pic.Serm.

Acta Botanica Barcinonensia 31: 11 (1978); Bull. Jard. Bot. Nat. Belg. 53: 187 (1983).
  Lycopodium cernuum L., F.W.T.A. Suppl.: 12 (1959).

Habit: herb. **Guild:** pioneer. **Habitat:** roadsides, farmbush and forest gaps. **Distribution:** pantropical. **Star:** excluded. **Alt. range:**

1-200m (3). **Fertile:** Oct (1). **Specimens:** - *Eastern slopes:* Fraser J. 37. *Mokoko River F.R.:* Acworth J. M. 228; Ekema N. 833. *Onge:* Tchouto P. 1000

SELAGINELLACEAE

Selaginella abyssinica Spring

F.W.T.A. Suppl.: 16 (1959); Fl. Cameroun 3: 31 (1964); Fl. Zamb. Pteridophyta: 27 (1970).

Habit: herb. **Habitat:** on rocks in forest. **Distribution:** Bioko and Cameroon to East and southern Africa. **Chorology:** Afromontane. **Star:** green. **Specimens:** - *Eastern slopes:* Preuss P. R. 978

Selaginella kalbreyeri Baker

F.W.T.A. Suppl.: 16 (1959); Fl. Cameroun 3: 28 (1964); Acta Botanica Barcinonensia 31: 17 (1978).

Habit: herb. **Habitat:** Aframomum thicket. **Distribution:** Guinea to Bioko, Cameroon, Gabon, Congo (Brazzaville), Congo (Kinshasa), Uganda and Ethiopia. **Star:** green. **Alt. range:** 600-800m (1). **Specimens:** - *Eastern slopes:* Kalbreyer W. 164

Selaginella kraussiana (Kunze) A.Braun

F.W.T.A. Suppl.: 16 (1959); Fl. Cameroun 3: 31 (1964); Fl. Zamb. Pteridophyta: 26 (1970); Acta Botanica Barcinonensia 31: 14 (1978); Bull. Jard. Bot. Nat. Belg. 53: 188 (1983).

Habit: herb. **Habitat:** forest. **Distribution:** tropical and subtropical Africa, naturalized in southern Europe. **Star:** excluded. **Specimens:** - *Eastern slopes:* Migeod F. 23

Selaginella molliceps Spring

F.W.T.A. Suppl.: 17 (1959); Fl. Cameroun 3: 41 (1964); Acta Botanica Barcinonensia 31: 20 (1978); Bull. Jard. Bot. Nat. Belg. 53: 189 (1983).

Habit: herb. **Habitat:** forest. **Distribution:** Guinea to Bioko, Cameroon, CAR, Gabon, Congo (Brazzaville), Congo (Kinshasa), Angola and Zimbabwe. **Star:** green. **Specimens:** - *Southern Bakundu F.R.:* Richards P. W. 4028

Selaginella myosurus (Sw.) Alston

F.W.T.A. Suppl.: 15 (1959); Fl. Cameroun 3: 23 (1964); Acta Botanica Barcinonensia 31: 13 (1978); Bull. Jard. Bot. Nat. Belg. 53: 189 (1983).

Habit: herbaceous climber. **Habitat:** forest regrowth, roadsides and farmbush. **Distribution:** Senegal to Bioko, Cameroon, Gabon, Congo (Brazzaville), Congo (Kinshasa), Angola and Kenya. **Chorology:** Guineo-Congolian. **Star:** green. **Alt. range:** 1-200m (2). **Fertile:** Oct (1). **Specimens:** - *Etinde:* Schlechter F. R. R. 12414. *Onge:* Cheek M. 5166; Tchouto P. 997

Selaginella soyauxii Hieron.

F.W.T.A. Suppl.: 16 (1959); Fl. Cameroun 3: 33 (1964); Acta Botanica Barcinonensia 31: 19 (1978); Bull. Jard. Bot. Nat. Belg. 53: 189 (1983).

Habit: herb. **Habitat:** forest. **Distribution:** Guinea to Bioko, Cameroon, Gabon, Congo (Brazzaville), Congo (Kinshasa) and Uganda. **Chorology:** Guineo-Congolian. **Star:** green. **Specimens:** - *Eastern slopes:* Migeod F. 6

Selaginella squarrosa Baker

F.W.T.A. Suppl.: 16 (1959); Fl. Cameroun 3: 34 (1964).

Habit: herb. **Habitat:** forest. **Distribution:** Cameroon and Rio Muni. **Chorology:** Lower Guinea. **Star:** blue. **Specimens:** - *No locality:* Mann G. 1407

Selaginella versicolor Spring

F.W.T.A. Suppl.: 16 (1959); Fl. Cameroun 3: 27 (1964); Acta Botanica Barcinonensia 31: 15 (1978); Bull. Jard. Bot. Nat. Belg. 53: 190 (1983).

Habit: terrestrial herb, sometimes epiphytic. **Guild:** shade-bearer. **Habitat:** forest. **Distribution:** Gambia to Bioko, Cameroon, Gabon, Congo (Brazzaville), Congo (Kinshasa), Angola, Malawi, Uganda and Sudan. **Star:** green. **Alt. range:** 200-400m (1), 401-600m (2). **Specimens:** - *Eastern slopes:* Rosevear D. R. 40/37. *Etinde:* Thomas D. W. 9121. *Onge:* Cheek M. 4980, 5021

Selaginella vogelii Spring

F.W.T.A. Suppl.: 15 (1959); Fl. Cameroun 3: 24 (1964); Acta Botanica Barcinonensia 31: 18 (1978).

Habit: herb. **Guild:** shade-bearer. **Habitat:** forest. **Distribution:** tropical Africa. **Star:** excluded. **Alt. range:** 1-200m (4), 201-400m (2), 601-800m (2), 801-1000m (1). **Fertile:** Mar (1), Oct (1), Nov (1), Dec (1). **Specimens:** - *Etinde:* Cable S. 476; Cheek M. 3753, 3755, 5874; Wheatley J. I. 496. *Mabeta-Moliwe:* Cheek M. 5710. *No locality:* Johnston H. H. 144. *Onge:* Harris D. J. 3656; Thomas D. W. 9894; Wheatley J. I. 785

FERNS

P.J. Edwards (K)

ADIANTACEAE

Adiantum philippense L.

F.W.T.A. Suppl.: 39 (1959); Fl. Cameroun 3: 143 (1964); Fl. Zamb. Pteridophyta: 110 (1970); Acta Botanica Barcinonensia 32: 26 (1980); Bull. Jard. Bot. Nat. Belg. 53: 235 (1983).

Habit: lithophyte. **Habitat:** on rocks and lava flows. **Distribution:** Palaeotropics and Central America. **Alt. range:** 1-200m (1). **Fertile:** Nov (1). **Specimens:** - *Etinde:* Richards P. W. 4066. *Limbe Botanic Garden:* Wheatley J. I. 649

Adiantum poiretii Wikstr. var. poiretii

F.W.T.A. Suppl.: 39 (1959); Fl. Cameroun 3: 148 (1964); Fl. Zamb. Pteridophyta: 112 (1970); Acta Botanica Barcinonensia 32: 27 (1980).

Habit: herb. **Habitat:** forest and forest-grassland transition. **Distribution:** pantropical. **Chorology:** Montane. **Star:** excluded. **Alt. range:** 2000-2200m (1), 2201-2400m (2). **Fertile:** Oct (2), Nov (1). **Specimens:** - *Eastern slopes:* Cheek M. 5337; Sidwell K. 61; Thomas D. W. 9274. *No locality:* Kalbreyer W. 113

Adiantum vogelii Mett. ex Keyserl.

F.W.T.A. Suppl.: 39 (1959); Fl. Cameroun 3: 147 (1964); Acta Botanica Barcinonensia 32: 26 (1980).

Habit: herb. **Habitat:** forest, plantations and stream banks. **Distribution:** Senegal to Bioko, Cameroon, CAR, Gabon, Congo (Brazzaville), Congo (Kinshasa), Angola and Zanzibar. **Chorology:** Guineo-Congolian. **Star:** green. **Alt. range:** 1-200m (4), 201-400m (1). **Fertile:** Apr (2), May (2), Oct (1). **Specimens:** - *Etinde:* Kalbreyer W. 2. *Mabeta-Moliwe:* Sunderland T. C. H. 1279; Wheatley J. I. 238. *Mokoko River F.R.:* Acworth J. M. 80; Tchouto P. 1073. *Onge:* Tchouto P. 891

Anogramma leptophylla (L.) Link

Fl. Zamb. Pteridophyta: 99 (1970).

Habit: herb. **Habitat:** grassland. **Distribution:** cosmopolitan. **Chorology:** Montane. **Star:** excluded. **Alt. range:** 2000-2200m (1). **Fertile:** Oct (1). **Specimens:** - *Eastern slopes:* Watts J. 422

Cheilanthes farinosa (Forssk.) Kaulf.

F.W.T.A. Suppl.: 43 (1959); Fl. Cameroun 3: 136 (1964); Fl. Zamb. Pteridophyta: 122 (1970).

Habit: herb. **Habitat:** grassland and lava flows. **Distribution:** palaeotropical. **Chorology:** Montane. **Star:** excluded. **Alt. range:** 1800-2000m (1), 3001-3200m (1). **Fertile:** Oct (2). **Specimens:** - *Eastern slopes:* Cheek M. 3672; Thomas D. W. 9295. *No locality:* Johnston H. H. 114

Coniogramme africana Hieron.

F.W.T.A. Suppl.: 38 (1959); Fl. Cameroun 3: 132 (1964); Fl. Zamb. Pteridophyta: 102 (1970); Acta Botanica Barcinonensia 32: 29 (1980); Bull. Jard. Bot. Nat. Belg. 53: 236 (1983).
Habit: herb. **Habitat:** rocky streams in forest. **Distribution:** tropical Africa. **Chorology:** Afromontane. **Star:** green. **Specimens:** - *No locality:* Johnston H. H. 127

Pellaea quadripinnata (Forssk.) Prantl

F.W.T.A. Suppl.: 44 (1959); Fl. Cameroun 3: 142 (1964); Fl. Zamb. Pteridophyta: 133 (1970); Bull. Jard. Bot. Nat. Belg. 53: 216 (1983).

Habit: herb. **Habitat:** forest-grassland transition and lava flows. **Distribution:** tropical and subtropical Africa, Middle East and Australia. **Chorology:** Montane. **Star:** excluded. **Alt. range:** 1800-2000m (1), 2401-2600m (1). **Fertile:** Oct (2). **Specimens:** - *Eastern slopes:* Cheek M. 3681; Thomas D. W. 9452. *No locality:* Maitland T. D. 964; Mann G. 1382

Pityrogramma calomelanos (L.) Link var. calomelanos

F.W.T.A. Suppl.: 38 (1959); Fl. Cameroun 3: 134 (1964); Fl. Zamb. Pteridophyta: 107 (1970); Acta Botanica Barcinonensia 32: 30 (1980); Bull. Jard. Bot. Nat. Belg. 53: 237 (1983).

Habit: herb. **Guild:** pioneer. **Habitat:** stream banks, forest clearings and roadsides. **Distribution:** pantropical, introduced from South America. **Star:** excluded. **Alt. range:** 1-200m (8), 201-400m (1), 401-600m (1), 1201-1400m (1). **Fertile:** Mar (2), May (1), Aug (1), Nov (2), Dec (3). **Specimens:** - *Etinde:* Cable S. 396; Cheek M. 5868; Maitland T. D. 1090. *Limbe Botanic Garden:* Wheatley J. I. 656. *Mabeta-Moliwe:* Cheek M. 5747; Sunderland T. C. H. 1140. *Mokoko River F.R.:* Acworth J. M. 78; Watts J. 1134. *Onge:* Watts J. 690, 1005; Wheatley J. I. 775

Asplenium abyssinicum Fée

F.W.T.A. Suppl.: 57 (1959); Fl. Cameroun 3: 216 (1964); Bull. Jard. Bot. Nat. Belg. 55: 125 (1985); Acta Botanica Barcinonensia 40: 24 (1991).

Habit: herb. **Habitat:** forest-grassland transition. **Distribution:** Cameroon, Bioko and Congo (Kinshasa) to East and southern Africa. **Chorology:** Afromontane. **Star:** green. **Alt. range:** 400-600m (1), 2001-2200m (1), 2201-2400m (2), 2401-2600m (1). **Fertile:** Oct (4). **Specimens: -** *Eastern slopes:* Sidwell K. 67; Thomas D. W. 9392; Watts J. 457. *Etinde:* Thomas D. W. 9358. *No locality:* Kalbreyer W. 111; Maitland T. D. 1047

Asplenium adamsii Alston

F.W.T.A. Suppl.: 59 (1959); Fl. Cameroun 3: 210 (1964).

Habit: lithophyte. **Habitat:** on rocks and lava flows. **Distribution:** Mount Cameroon and East Africa. **Chorology:** Afromontane. **Star:** gold. **Specimens: -** *No locality:* Adams C. D. 1271

Asplenium adiantum-nigrum L.

F.W.T.A. Suppl.: 59 (1959); Fl. Cameroun 3: 215 (1964).

Habit: lithophyte. **Habitat:** on rocks and lava flows. **Distribution:** Africa, Europe and North America. **Alt. range:** 3000-3200m (1), 3601-3800m (1). **Specimens: -** *Eastern slopes:* Brenan J. P. M. 4244. *No locality:* Adams C. D. 1294; Mann G. 1373

Asplenium aethiopicum (Burm.f.) Bech.

F.W.T.A. Suppl.: 59 (1959); Fl. Cameroun 3: 212 (1964); Fl. Zamb. Pteridophyta: 181 (1970); Bull. Jard. Bot. Nat. Belg. 55: 126 (1985); Acta Botanica Barcinonensia 40: 24 (1991).

Habit: epiphyte. **Habitat:** forest-grassland transition. **Distribution:** pantropical. **Chorology:** Montane. **Star:** excluded. **Alt. range:** 1400-1600m (1), 1601-1800m (1), 2001-2200m (1), 2201-2400m (1), 2601-2800m (1), 2801-3000m (1). **Fertile:** May (1), Sep (1), Oct (4). **Specimens: -** *Eastern slopes:* Sidwell K. 44; Tchouto P. 283; Thomas D. W. 9389; Upson T. 88. *Etinde:* Thomas D. W. 9203, 9260. *No locality:* Johnston H. H. 109

Asplenium aethiopicum (Burm.f.) Bech. var. 1

Specimens: - *Eastern slopes:* Upson T. 71

Asplenium aethiopicum (Burm.f.) Bech. var. 2

Specimens: - *Eastern slopes:* Sidwell K. 27

Asplenium aethiopicum (Burm.f.) Bech. var. 4

Specimens: - *Eastern slopes:* Thomas D. W. 9320

Asplenium anisophyllum Kunze

Fl. Zamb. Pteridophyta: 170 (1970).
   Asplenium geppii Carruth., F.W.T.A. Suppl.: 56 (1959).

Habit: epiphyte. **Habitat:** forest. **Distribution:** Guinea to Bioko, Cameroon, Gabon and Angola. **Chorology:** Guineo-Congolian. **Star:** green. **Specimens: -** *No locality:* Adams C. D. 1236

Asplenium sp. aff. anisophyllum Kunze

Specimens: - *Eastern slopes:* Watts J. 460. *Etinde:* Thomas D. W. 9186

Asplenium barteri Hook.

F.W.T.A. Suppl.: 56 (1959); Fl. Cameroun 3: 192 (1964); Bull. Jard. Bot. Nat. Belg. 55: 127 (1985); Acta Botanica Barcinonensia 40: 13 (1991).

Habit: epiphyte. **Habitat:** forest. **Distribution:** Guinea to Bioko, Cameroon, Gabon, Congo (Brazzaville), Congo (Kinshasa), Angola, CAR, Uganda, Sudan and Kenya. **Chorology:** Guineo-Congolian. **Star:** green. **Alt. range:** 1-200m (5), 201-400m (3), 401-600m (1), 801-1000m (1). **Fertile:** Mar (1), Apr (1), May (1), Jun (2), Oct (1), Nov (5), Dec (2). **Specimens: -** *Etinde:* Cable S. 201, 276, 420; Cheek M. 5491, 5871; Kalbreyer W. 199; Watts J. 474. *Mabeta-Moliwe:* Watts J. 130. *Mokoko River F.R.:* Mbani J. M. 471, 512; Ndam N. 1266. *Onge:* Thomas D. W. 9764; Watts J. 973; Wheatley J. I. 723

Asplenium cf. barteri Hook.

Specimens: - *Etinde:* Cable S. 178

Asplenium biafranum Alston & Ballard

F.W.T.A. Suppl.: 57 (1959); Fl. Cameroun 3: 200 (1964); Acta Botanica Barcinonensia 40: 23 (1991).

Habit: epiphyte. **Habitat:** grassland. **Distribution:** Benin, Nigeria, Bioko, Cameroon and S. Tomé. **Chorology:** Guineo-Congolian (montane). **Star:** blue. **Alt. range:** 1800-2000m (1). **Fertile:** Oct (1). **Specimens: -** *Eastern slopes:* Cheek M. 3706; Migeod F. 73

Asplenium cancellatum Alston

F.W.T.A. Suppl.: 59 (1959); Fl. Cameroun 3: 206 (1964); Acta Botanica Barcinonensia 40: 21 (1991).

Habit: epiphyte. **Habitat:** forest. **Distribution:** Ghana, Bioko, Cameroon and Gabon. **Chorology:** Guineo-Congolian (montane). **Star:** green. **Alt. range:** 1200-1400m (1). **Specimens: -** *Etinde:* Maitland T. D. 1094

Asplenium dregeanum Kunze

F.W.T.A. Suppl.: 59 (1959); Fl. Cameroun 3: 220 (1964); Fl. Zamb. Pteridophyta: 184 (1970); Bull. Jard. Bot. Nat. Belg. 55: 130 (1985); Acta Botanica Barcinonensia 40: 28 (1991).

Habit: epiphyte. **Habitat:** forest and Aframomum thicket. **Distribution:** tropical and subtropical Africa. **Chorology:** Afromontane. **Star:** green. **Alt. range:** 600-800m (1), 801-1000m (4), 1001-1200m (2), 1201-1400m (2), 1601-1800m (1), 1801-2000m (1), 2001-2200m (1). **Fertile:** Jul (1), Sep (2), Oct (5), Dec (2). **Specimens: -** *Eastern slopes:* Richards P. W. 4380; Sidwell K. 93; Tchouto P. 387. *Etinde:* Cable S. 421; Cheek M. 3616; Faucher P. 40; Tekwe C. 171; Thomas D. W. 9160, 9229; Watts J. 557, 576

Asplenium elliottii C.H.Wright

Kew Bull. 1908: 262 (1908).

Habit: epiphyte. **Habitat:** forest. **Distribution:** Ghana, Bioko and W. Cameroon. **Chorology:** Guineo-Congolian (montane). **Star:** blue. **Alt. range:** 1800-2000m (2), 2201-2400m (2). **Specimens: -** *No locality:* Dunlap 121, 231; Mann G. 1414, 2055

Asplenium emarginatum P.Beauv.

F.W.T.A. Suppl.: 56 (1959); Fl. Cameroun 3: 192 (1964); Bull. Jard. Bot. Nat. Belg. 55: 133 (1985); Acta Botanica Barcinonensia 40: 10 (1991).

Habit: herb. **Guild:** shade-bearer. **Habitat:** forest. **Distribution:** Guinea to Bioko, Cameroon, Gabon, Congo (Brazzaville), Congo (Kinshasa), Angola, CAR, Sudan and Uganda. **Chorology:** Guineo-Congolian. **Star:** green. **Specimens:** - *Eastern slopes:* Migeod F. 140

Asplenium erectum Bory ex Willd. var. usambarense (Hieron.) Schelpe

Fl. Zamb. Pteridophyta: 176 (1970); Bull. Jard. Bot. Nat. Belg. 55: 133 (1985); Acta Botanica Barcinonensia 40: 16 (1991).
Asplenium quintasii Gandoger, F.W.T.A. Suppl.: 57 (1959).

Habit: epiphytic or terrestrial herb. **Habitat:** forest and forest-grassland transition. **Distribution:** tropical Africa. **Chorology:** Afromontane. **Star:** green. **Alt. range:** 800-1000m (1), 1401-1600m (1), 1801-2000m (1). **Fertile:** Sep (1), Oct (2). **Specimens:** - *Eastern slopes:* Tchouto P. 388; Thomas D. W. 9453. *Etinde:* Thomas D. W. 9192b. *No locality:* Johnston H. H. 133

Asplenium friesiorum C.Chr.

F.W.T.A. Suppl.: 57 (1959); Fl. Cameroun 3: 202 (1964); Fl. Zamb. Pteridophyta: 178 (1970); Bull. Jard. Bot. Nat. Belg. 55: 134 (1985); Acta Botanica Barcinonensia 40: 22 (1991).
Habit: epiphyte. **Habitat:** forest-grassland transition. **Distribution:** tropical and subtropical Africa. **Chorology:** Afromontane. **Star:** green. **Alt. range:** 400-600m (1), 1201-1400m (1), 2201-2400m (1). **Fertile:** Sep (1), Oct (1). **Specimens:** - *Eastern slopes:* Thomas D. W. 9334. *Etinde:* Thomas D. W. 9223. *No locality:* Maitland T. D. s.n.; Mann G. 1402

Asplenium sp. aff. friesiorum C.Chr.

Specimens: - *Etinde:* Cable S. 158

Asplenium gemmascens Alston

F.W.T.A. Suppl.: 59 (1959); Fl. Cameroun 3: 207 (1964); Bull. Jard. Bot. Nat. Belg. 55: 135 (1985).

Habit: herb. **Habitat:** forest. **Distribution:** Nigeria, Cameroon and Congo (Kinshasa). **Chorology:** Guineo-Congolian. **Star:** green. **Alt. range:** 600-800m (2). **Fertile:** Jan (1), May (1). **Specimens:** - *Eastern slopes:* Tekwe C. 10. *Etinde:* Mbani J. M. 100

Asplenium gemmiferum Schrad.

F.W.T.A. Suppl.: 56 (1959); Fl. Cameroun 3: 188 (1964); Fl. Zamb. Pteridophyta: 173 (1970); Acta Botanica Barcinonensia 40: 11 (1991).

Habit: epiphyte. **Habitat:** forest. **Distribution:** SW Cameroon and Bioko to East and southern Africa. **Chorology:** Afromontane. **Star:** green. **Specimens:** - *No locality:* Mann G. 1394

Asplenium hemitomum Hieron.

F.W.T.A. Suppl.: 59 (1959); Bull. Jard. Bot. Nat. Belg. 55: 136 (1985); Acta Botanica Barcinonensia 40: 20 (1991).

Habit: herb. **Habitat:** forest. **Distribution:** Guinea to Bioko, Cameroon and Congo (Kinshasa). **Chorology:** Guineo-Congolian.

Star: green. **Alt. range:** 200-400m (1), 401-600m (1). **Fertile:** Oct (1), Dec (1). **Specimens:** - *Etinde:* Lighava M. 52; Watts J. 516

Asplenium hypomelas Kuhn

F.W.T.A. Suppl.: 60 (1959); Fl. Cameroun 3: 226 (1964); Fl. Zamb. Pteridophyta: 187 (1970); Acta Botanica Barcinonensia 40: 31 (1991).

Habit: epiphyte. **Habitat:** forest, sometimes on rocks. **Distribution:** tropical Africa. **Chorology:** Afromontane. **Star:** green. **Alt. range:** 1600-1800m (3). **Fertile:** Sep (1), Oct (1). **Specimens:** - *Eastern slopes:* Richards P. W. 4124; Tekwe C. 322; Thomas D. W. 9472. *No locality:* Maitland T. D. 1048

Asplenium inequilaterale Bory ex Willd.

F.W.T.A. Suppl.: 57 (1959); Fl. Zamb. Pteridophyta: 176 (1970); Bull. Jard. Bot. Nat. Belg. 55: 136 (1985); Acta Botanica Barcinonensia 40: 17 (1991).

Habit: herb. **Habitat:** forest. **Distribution:** tropical Africa. **Chorology:** Afromontane. **Star:** green. **Alt. range:** 1400-1600m (1). **Specimens:** - *Eastern slopes:* Thomas D. W. 9500

Asplenium lividum Mett. ex Kuhn

F.W.T.A. Suppl.: 59 (1959); Fl. Cameroun 3: 210 (1964); Fl. Zamb. Pteridophyta: 181 (1970).

Habit: epiphyte. **Habitat:** forest. **Distribution:** tropical Africa and America. **Star:** excluded. **Specimens:** - *No locality:* Adams C. D. 1220a

Asplenium macrophlebium Baker

F.W.T.A. Suppl.: 56 (1959); Fl. Cameroun 3: 193 (1964); Bull. Jard. Bot. Nat. Belg. 55: 143 (1985); Acta Botanica Barcinonensia 40: 14 (1991).

Habit: epiphytic or terrestrial herb. **Habitat:** forest. **Distribution:** tropical Africa. **Star:** excluded. **Alt. range:** 600-800m (1). **Specimens:** - *Etinde:* Kalbreyer W. 151

Asplenium mannii Hook.

F.W.T.A. Suppl.: 60 (1959); Fl. Cameroun 3: 224 (1964); Fl. Zamb. Pteridophyta: 187 (1970); Bull. Jard. Bot. Nat. Belg. 55: 144 (1985); Acta Botanica Barcinonensia 40: 30 (1991).

Habit: epiphyte. **Habitat:** forest-grassland transition. **Distribution:** tropical Africa. **Chorology:** Afromontane. **Star:** green. **Alt. range:** 1400-1600m (1), 2001-2200m (1). **Fertile:** Sep (1). **Specimens:** - *Eastern slopes:* Richards P. W. 4284. *Etinde:* Thomas D. W. 9209

Asplenium paucijugum Ballard

Bull. Jard. Bot. Nat. Belg. 55: 147 (1985); Acta Botanica Barcinonensia 40: 8 (1991).
Asplenium variabile Hook. var. *paucijugum* (Ballard) Alston, F.W.T.A. Suppl.: 56 (1959).

Habit: epiphytic or terrestrial herb. **Habitat:** forest, sometimes in stream beds. **Distribution:** tropical Africa. **Star:** excluded. **Specimens:** - *Eastern slopes:* Adams C. D. 1247

Asplenium preussii Hieron.

F.W.T.A. Suppl.: 60 (1959); Fl. Cameroun 3: 221 (1964); Acta Botanica Barcinonensia 40: 30 (1991).

Habit: terrestrial herb, sometimes epiphytic. Habitat: forest. Distribution: tropical and subtropical Africa. Chorology: Afromontane. Star: green. Alt. range: 800-1000m (1), 1401-1600m (1), 1601-1800m (1), 2001-2200m (1). Fertile: Oct (1), Dec (1). Specimens: - *Eastern slopes:* Richards P. W. 4311; Thomas D. W. 9476. *Etinde:* Cable S. 419; Thomas D. W. 9256

Asplenium protensum Schrad.

F.W.T.A. Suppl.: 57 (1959); Fl. Cameroun 3: 194 (1964); Fl. Zamb. Pteridophyta: 179 (1970); Bull. Jard. Bot. Nat. Belg. 55: 147 (1985); Acta Botanica Barcinonensia 40: 19 (1991).

Habit: epiphyte. Habitat: forest and forest-grassland transition. Distribution: tropical Africa. Chorology: Afromontane. Star: green. Alt. range: 1800-2000m (1), 2001-2200m (1), 2201-2400m (1), 2401-2600m (2). Fertile: Feb (1), May (2), Oct (2). Specimens: - *Eastern slopes:* Sidwell K. 94; Upson T. 72, 98; Watts J. 459. *No locality:* Maitland T. D. 1052; Mann G. 2043

Asplenium sandersonii Hook.

Fl. Zamb. Pteridophyta: 183 (1970); Bull. Jard. Bot. Nat. Belg. 55: 151 (1985); Acta Botanica Barcinonensia 40: 27 (1991).
   Asplenium sandersonii Hook. var. *vagans* (Baker) C.Chr., Ind. Fil. Suppl. 3: 39 (1934).
   Asplenium vagans Baker, F.W.T.A. Suppl.: 59 (1959).

Habit: epiphyte. Habitat: forest. Distribution: SE Nigeria, Bioko, Cameroon, Gabon and East Africa. Star: excluded. Alt. range: 200-400m (1). Fertile: Dec (1). Specimens: - *Etinde:* Cheek M. 5885

Asplenium subintegrum C.Chr.

F.W.T.A. Suppl.: 56 (1959); Fl. Cameroun 3: 181 (1964).

Habit: epiphyte. Habitat: forest. Distribution: SW Cameroon and Congo (Brazzaville). Chorology: Guineo-Congolian (montane). Star: blue. Specimens: - *Etinde:* Kalbreyer W. 167

Asplenium theciferum (Kunth) Mett. var. cornutum (Alston) Benl

Acta Botanica Barcinonensia 40: 32 (1991).
   Asplenium cornutum Alston, F.W.T.A. Suppl.: 60 (1959).

Habit: epiphyte. Habitat: forest, grassland and Aframomum thicket. Distribution: W. Cameroon and Bioko. Chorology: Western Cameroon. Star: gold. Alt. range: 600-800m (1), 1801-2000m (1), 2001-2200m (2). Fertile: Oct (2). Specimens: - *Eastern slopes:* Richards P. W. 4284; Thomas D. W. 9281; Watts J. 436. *Etinde:* Kalbreyer W. 185

Asplenium uhligii Hieron.

F.W.T.A. Suppl.: 59 (1959); Fl. Cameroun 3: 214 (1964).

Habit: epiphyte. Habitat: forest. Distribution: Togo, W. Cameroon, Bioko, Congo (Kinshasa) and East Africa. Star: excluded. Specimens: - *No locality:* Adams C. D. 1289

Asplenium unilaterale Lam.

F.W.T.A. Suppl.: 56 (1959); Fl. Cameroun 3: 195 (1964); Fl. Zamb. Pteridophyta: 174 (1970); Acta Botanica Barcinonensia 40: 15 (1991).

Habit: herb. Habitat: forest-grassland transition and stream banks. Distribution: palaeotropical. Star: excluded. Alt. range: 1-200m (1), 801-1000m (2), 1001-1200m (1), 1201-1400m (1).

Fertile: Sep (1), Oct (2), Dec (2). Specimens: - *Etinde:* Cable S. 471, 479; Thomas D. W. 9221, 9703; Wheatley J. I. 599. *No locality:* Savory H. J. UCI 561

Asplenium variabile Hook.

F.W.T.A. Suppl.: 56 (1959); Fl. Cameroun 3: 182 (1964); Acta Botanica Barcinonensia 40: 7 (1991).

Habit: herb. Habitat: forest and Aframomum thicket. Distribution: Guinea to Bioko, Cameroon, Gabon, Congo (Brazzaville) and Congo (Kinshasa). Chorology: Guineo-Congolian. Star: green. Alt. range: 1-200m (2), 201-400m (2). Fertile: Apr (1), May (1), Jun (1), Nov (1), Dec (1). Specimens: - *Etinde:* Cheek M. 5577. *Mabeta-Moliwe:* Cable S. 544; Wheatley J. I. 121, 239. *Mokoko River F.R.:* Mbani J. M. 443

Asplenium ? sp. nov.

Specimens: - *Etinde:* Wheatley J. I. 513.

Asplenium sp.

Specimens: - *Etinde:* Banks H. 36. *Mokoko River F.R.:* Mbani J. M. 511

Asplenium sp. 1

Habit: epiphyte. Habitat: forest and farmbush. Distribution: Cameroon and Uganda. Chorology: Guineo-Congolian (montane). Star: blue. Alt. range: 800-1000m (1). Fertile: Feb (1). Specimens: - *Eastern slopes:* Groves M. 126

Asplenium sp. 6

Habit: epiphyte. Habitat: forest. Distribution: Ivory Coast, Ghana, Cameroon and Angola. Chorology: Guineo-Congolian. Star: green. Alt. range: 1-200m (1), 201-400m (1). Fertile: Apr (1), Oct (1). Specimens: - *Mabeta-Moliwe:* Watts J. 131. *Onge:* Tchouto P. 893

Asplenium sp. 7

Habit: epiphyte. Habitat: forest. Distribution: Liberia to Cameroon. Chorology: Guineo-Congolian. Star: green. Alt. range: 1-200m (2). Fertile: Apr (1), May (1). Specimens: - *Mabeta-Moliwe:* Sunderland T. C. H. 1239; Watts J. 290

Asplenium sp. 8

Habit: epiphyte. Habitat: forest. Distribution: Mount Cameroon. Chorology: Endemic. Star: black. Alt. range: 1-200m (1). Fertile: Mar (1). Specimens: - *Mabeta-Moliwe:* Groves M. 230

Asplenium sp. 9

Habit: epiphyte. Habitat: forest. Distribution: Mount Cameroon. Chorology: Endemic. Star: black. Alt. range: 200-400m (1). Fertile: Nov (1). Specimens: - *Etinde:* Cheek M. 5576

BLECHNACEAE

Blechnum attenuatum (Sw.) Mett. var. attenuatum
F.W.T.A. Suppl.: 74 (1959); Fl. Zamb. Pteridophyta: 236 (1970); Acta Botanica Barcinonensia 38: 63 (1988).

Habit: bole epiphyte. **Habitat:** forest. **Distribution:** Mount Cameroon, Bioko, Uganda, East and southern Africa. **Chorology:** Afromontane. **Star:** green. **Alt. range:** 1-200m (1), 1401-1600m (3), 1601-1800m (1). **Fertile:** Sep (1), Oct (2). **Specimens:** - *Eastern slopes:* Box H. E. 3607; Thomas D. W. 9471. *Etinde:* Maitland T. D. 1045; Tekwe C. 186; Thomas D. W. 9240. *No locality:* Maitland T. D. 1931

CYATHEACEAE

Cyathea camerooniana Hook. var. camerooniana
F.W.T.A. Suppl.: 27 (1959); Fl. Cameroun 3: 66 (1964).
Alsophila camerooniana (Hook.) R.M.Tryon var. *camerooniana*, Acta Botanica Barcinonensia 31: 26 (1978).

Habit: tree fern to 3 m. **Habitat:** forest and stream banks. **Distribution:** Guinea to Bioko, Cameroon and Gabon. **Chorology:** Guineo-Congolian (montane). **Star:** blue. **Alt. range:** 1-200m (4), 201-400m (1). **Fertile:** Apr (3), May (3), Jun (4). **Specimens:** - *Mokoko River F.R.:* Acworth J. M. 93, 327, 328; Ekema N. 829; Fraser P. 429; Mbani J. M. 487, 488; Ndam N. 1083, 1092, 1235. *No locality:* Mann G. 2059

Cyathea camerooniana Hook. var. cf. zenkeri (Diels) Tardieu

Specimens: - *Mokoko River F.R.:* Akogo M. 273

Cyathea manniana Hook.
F.W.T.A. Suppl.: 29 (1959); Fl. Cameroun 3: 70 (1964); Fl. Zamb. Pteridophyta: 72 (1970); Hawthorne W., F.G.F.T. Ghana: 223 (1990).
Alsophila manniana (Hook.) R.M.Tryon, Acta Botanica Barcinonensia 31: 27 (1978).

Habit: tree fern to 10 m. **Habitat:** forest and farmbush. **Distribution:** tropical Africa. **Chorology:** Afromontane. **Star:** blue. **Specimens:** - *Eastern slopes:* Migeod F. 13

DENNSTAEDTIACEAE

Blotiella glabra (Bory) R.M.Tryon
Acta Botanica Barcinonensia 38: 31 (1988).
Lonchitis gracilis Alston, F.W.T.A. Suppl.: 34 (1959).

Habit: herb. **Habitat:** forest. **Distribution:** Cameroon, Bioko, S. Tomé and Gabon. **Chorology:** Lower Guinea. **Star:** blue. **Alt. range:** 2000-2200m (1). **Specimens:** - *Eastern slopes:* Richards P. W. 4245. *No locality:* Mann G. 2052

Blotiella mannii (Baker) Pic.Serm.
Acta Botanica Barcinonensia 38: 30 (1988).

Habit: herb. **Habitat:** forest. **Distribution:** Guinea, Liberia, Nigeria, Bioko, Cameroon and S. Tomé. **Chorology:** Guineo-Congolian (montane). **Star:** green. **Alt. range:** 200-400m (1), 1601-1800m (1). **Fertile:** Apr (2), Oct (1). **Specimens:** - *Etinde:* Tchouto P. 411. *Mokoko River F.R.:* Ndam N. 1093; Tchouto P. 1159

Hypolepis sparsisora (Schrad.) Kuhn
F.W.T.A. Suppl.: 33 (1959); Fl. Cameroun 3: 95 (1964); Fl. Zamb. Pteridophyta: 92 (1970); Bull. Jard. Bot. Nat. Belg. 53: 260 (1983); Acta Botanica Barcinonensia 38: 27 (1988).

Habit: herb. **Habitat:** forest-grassland transition. **Distribution:** tropical and subtropical Africa. **Chorology:** Afromontane. **Star:** green. **Alt. range:** 2000-2200m (1). **Specimens:** - *Eastern slopes:* Richards P. W. 4283

Lonchitis occidentalis Baker
Fl. Zamb. Pteridophyta: 86 (1970); Taxon 26: 578 (1977); Acta Botanica Barcinonensia 38: 34 (1988).
Anisosorus occidentalis (Baker) C.Chr., F.W.T.A. Suppl.: 33 (1959).

Habit: herb. **Guild:** shade-bearer. **Habitat:** forest. **Distribution:** tropical Africa. **Chorology:** Afromontane. **Star:** green. **Alt. range:** 1200-1400m (1). **Specimens:** - *No locality:* Mann G. 1386

Microlepia speluncae (L.) T.Moore var. speluncae
F.W.T.A. Suppl.: 33 (1959); Fl. Zamb. Pteridophyta: 89 (1970); Bull. Jard. Bot. Nat. Belg. 53: 258 (1983); Acta Botanica Barcinonensia 38: 23 (1988).

Habit: herb. **Habitat:** forest and farmbush. **Distribution:** pantropical. **Star:** excluded. **Alt. range:** 1-200m (1), 401-600m (1). **Fertile:** Oct (2). **Specimens:** - *Etinde:* Thomas D. W. 9144. *Mabeta-Moliwe:* Sidwell K. 125

Pteridium aquilinum (L.) Kuhn subsp. aquilinum
F.W.T.A. Suppl.: 33 (1959); Fl. Cameroun 3: 96 (1964); Bull. Jard. Bot. Nat. Belg. 53: 260 (1983); Acta Botanica Barcinonensia 38: 35 (1988).

Habit: herb. **Guild:** pioneer. **Habitat:** grassland. **Distribution:** cosmopolitan. **Chorology:** Montane. **Star:** excluded. **Alt. range:** 2200-2400m (1). **Fertile:** Oct (1). **Specimens:** - *Eastern slopes:* Box H. E. 3611; Thomas D. W. 9379

DRYOPTERIDACEAE

Ctenitis cirrhosa (Schum.) Ching
F.W.T.A. Suppl.: 71 (1959); Fl. Cameroun 3: 266 (1964); Fl. Zamb. Pteridophyta: 232 (1970); Bull. Jard. Bot. Nat. Belg. 55: 176 (1985); Acta Botanica Barcinonensia 40: 41 (1991).

Habit: herb. **Habitat:** forest. **Distribution:** tropical Africa. **Chorology:** Afromontane. **Star:** green. **Alt. range:** 1200-1400m (1). **Specimens:** - *No locality:* Mann G. 1390

Didymochlaena truncatula (Sw.) J.Sm.

F.W.T.A. Suppl.: 69 (1959); Fl. Cameroun 3: 254 (1964); Fl. Zamb. Pteridophyta: 220 (1970); Bull. Jard. Bot. Nat. Belg. 55: 176 (1985); Acta Botanica Barcinonensia 40: 35 (1991).

Habit: herb. **Habitat:** forest. **Distribution:** pantropical. **Chorology:** Montane. **Star:** excluded. **Alt. range:** 1-200m (1), 401-600m (1), 1801-2000m (1). **Fertile:** Oct (3). **Specimens:** - *Eastern slopes:* Thomas D. W. 9455. *Etinde:* Watts J. 523. *Mabeta-Moliwe:* Sidwell K. 115. *No locality:* Johnston H. H. 148

Dryopteris kilemensis (Kuhn) Kuntze

Fl. Zamb. Pteridophyta: 222 (1970); Bull. Jard. Bot. Nat. Belg. 55: 157 (1985).

Habit: herb. **Habitat:** forest-grassland transition. **Distribution:** Cameroon, Zambia, Zimbabwe, Malawi and East Africa. **Chorology:** Afromontane. **Star:** green. **Alt. range:** 1600-1800m (1). **Fertile:** Sep (1). **Specimens:** - *Etinde:* Thomas D. W. 9195

Dryopteris manniana (Hook.) C.Chr.

F.W.T.A. Suppl.: 70 (1959); Fl. Cameroun 3: 261 (1964); Acta Botanica Barcinonensia 40: 38 (1991).

Habit: herb. **Habitat:** forest. **Distribution:** tropical and subtropical Africa. **Chorology:** Afromontane. **Star:** green. **Alt. range:** 1600-1800m (1). **Fertile:** Oct (1). **Specimens:** - *Etinde:* Thomas D. W. 9527

Dryopteris pentheri (Krasser) C.Chr.

F.W.T.A. Suppl.: 70 (1959); Fl. Cameroun 3: 261 (1964); Bull. Jard. Bot. Nat. Belg. 55: 163 (1985); Acta Botanica Barcinonensia 40: 36 (1991).

Habit: herb. **Habitat:** forest-grassland transition. **Distribution:** tropical and subtropical Africa. **Chorology:** Afromontane. **Star:** green. **Alt. range:** 2000-2200m (2), 2201-2400m (1), 2401-2600m (1), 2801-3000m (1). **Fertile:** Oct (5). **Specimens:** - *Eastern slopes:* Mildbraed G. W. J. 20876; Sidwell K. 43; Tchouto P. 281, 297; Thomas D. W. 9329; Watts J. 455

Lastreopsis barteriana (Hook.) Tardieu

Fl. Cameroun 3: 279 (1964); Acta Botanica Barcinonensia 40: 53 (1991).
 Ctenitis barteriana (Hook.) Alston, F.W.T.A. Suppl.: 73 (1959).

Habit: herb. **Habitat:** forest. **Distribution:** Nigeria, Bioko and Cameroon. **Chorology:** Lower Guinea. **Star:** blue. **Alt. range:** 200-400m (1), 401-600m (1), 601-800m (1). **Fertile:** Sep (1), Oct (2). **Specimens:** - *Etinde:* Thomas D. W. 9113, 9124; Watts J. 547

Lastreopsis currori (Mett. ex Kuhn) Tindale

Bull. Jard. Bot. Nat. Belg. 55: 178 (1985); Acta Botanica Barcinonensia 40: 51 (1991).

Habit: herb. **Habitat:** forest. **Distribution:** Guinea to Bioko, Cameroon, Gabon, Congo (Brazzaville), Congo (Kinshasa), Angola, Uganda, Burundi and Madagascar. **Chorology:** Guineo-Congolian. **Star:** green. **Alt. range:** 200-400m (1), 401-600m (1), 601-800m (1). **Fertile:** May (1), Oct (1), Nov (1). **Specimens:** - *Etinde:* Cable S. 218; Thomas D. W. 9123. *Mokoko River F.R.:* Thomas D. W. 10164

Lastreopsis nigritiana (Baker) Tindale

Fl. Cameroun 3: 280 (1964); Fl. Cameroun 3: 280 (1964); Acta Botanica Barcinonensia 40: 51 (1991).
 Ctenitis pubigera Alston, F.W.T.A. Suppl.: 73 (1959).

Habit: herb. **Habitat:** forest. **Distribution:** Ivory Coast to Bioko, Cameroon, Gabon and Congo (Brazzaville). **Chorology:** Guineo-Congolian. **Star:** green. **Specimens:** - *Etinde:* Kalbreyer W. 20

Lastreopsis sp. aff. nigritiana (Baker) Tindale

Specimens: - *Mokoko River F.R.:* Thomas D. W. 10029, 10163

Lastreopsis subsimilis (Hook.) Tindale

Fl. Cameroun 3: 278 (1964); Fl. Cameroun 3: 278 (1964); Acta Botanica Barcinonensia 40: 50 (1991).
 Ctenitis subsimilis (Hook.) Tardieu, F.W.T.A. Suppl.: 73 (1959).

Habit: herb. **Habitat:** forest. **Distribution:** Liberia to Bioko, Cameroon and Gabon. **Chorology:** Guineo-Congolian. **Star:** green. **Specimens:** - *Etinde:* Kalbreyer W. 14

Polystichum fuscopaleaceum Alston

F.W.T.A. Suppl.: 70 (1959); Fl. Cameroun 3: 257 (1964); Bull. Jard. Bot. Nat. Belg. 55: 175 (1985); Acta Botanica Barcinonensia 40: 39 (1991).

Habit: herb. **Habitat:** forest-grassland transition. **Distribution:** Bioko, W. Cameroon, Mozambique, Malawi, Rwanda, Somalia, Ethiopia and East Africa. **Chorology:** Afromontane. **Star:** green. **Specimens:** - *No locality:* Keay R. W. J. FHI 28602

Polystichum transvaalense N.C.Anthony

Contr. Bolus Herb. 10: 146 (1982).

Habit: herb. **Habitat:** forest-grassland transition. **Distribution:** Mount Cameroon, Bioko, Congo (Kinshasa), Malawi, Zimbabwe, South Africa and East Africa. **Chorology:** Afromontane. **Star:** green. **Alt. range:** 2000-2200m (1), 2201-2400m (1). **Fertile:** Oct (2). **Specimens:** - *Eastern slopes:* Thomas D. W. 9350; Watts J. 434

Tectaria angelicifolia (Schumach.) Copel.

F.W.T.A. Suppl.: 74 (1959); Acta Botanica Barcinonensia 40: 55 (1991).

Habit: herb. **Habitat:** forest. **Distribution:** Guinea to Bioko, Cameroon, Gabon, Congo (Brazzaville), Congo (Kinshasa), Angola, Uganda and Sudan. **Chorology:** Guineo-Congolian. **Star:** green. **Alt. range:** 200-400m (1). **Fertile:** Nov (1). **Specimens:** - *Etinde:* Williams S. 66

Tectaria barteri (J.Sm.) C.Chr.

F.W.T.A. Suppl.: 74 (1959); Fl. Cameroun 3: 288 (1964); Acta Botanica Barcinonensia 40: 54 (1991).

Habit: herb. **Habitat:** forest, on rocky ground. **Distribution:** Nigeria, SW Cameroon and Bioko. **Chorology:** Western Cameroon. **Star:** gold. **Alt. range:** 1-200m (5), 201-400m (5). **Fertile:** May (4), Sep (1), Oct (1), Nov (3), Dec (1). **Specimens:** - *Etinde:* Cheek M. 5492, 5877; Thomas D. W. 9704; Wheatley J. I. 497. *Mabeta-Moliwe:* Wheatley J. I. 272. *Mokoko River F.R.:* Acworth J. M. 270; Fraser P. 435; Thomas D. W. 10070. *Onge:* Thomas D. W. 9900, 9901

Tectaria camerooniana (Hook.) Alston

F.W.T.A. Suppl.: 74 (1959); Fl. Cameroun 3: 290 (1964); Bull.

Jard. Bot. Nat. Belg. 55: 178 (1985); Acta Botanica Barcinonensia 40: 56 (1991).

Habit: herb. **Habitat:** forest. **Distribution:** Nigeria, Mount Cameroon, Bioko, S. Tomé, Congo (Kinshasa) and Uganda. **Chorology:** Guineo-Congolian (montane). **Star:** green. **Alt. range:** 200-400m (1), 401-600m (2), 801-1000m (2), 1601-1800m (1), 1801-2000m (1). **Fertile:** Mar (2), Oct (1), Nov (1), Dec (3). **Specimens:** - *Eastern slopes:* Viane R. 2828, 2836. *Etinde:* Cable S. 259; Cheek M. 5873; Faucher P. 50; Lighava M. 9; Watts J. 506. *No locality:* Mann G. 1362

Tectaria cf. camerooniana (Hook.) Alston

Specimens: - *Etinde:* Maitland T. D. 1136

Tectaria fernandensis (Baker) C.Chr.

F.W.T.A. Suppl.: 74 (1959); Fl. Cameroun 3: 291 (1964); Bull. Jard. Bot. Nat. Belg. 55: 178 (1985); Acta Botanica Barcinonensia 40: 57 (1991).

Habit: herb. **Habitat:** forest, on rocky ground. **Distribution:** Guinea to Bioko, Cameroon, Gabon, Congo (Kinshasa) and Rwanda. **Chorology:** Guineo-Congolian (montane). **Star:** green. **Alt. range:** 1-200m (1), 201-400m (1), 801-1000m (3), 1401-1600m (4), 1801-2000m (1). **Fertile:** Feb (1), Mar (1), Jun (1), Sep (1), Oct (3), Nov (1), Dec (1). **Specimens:** - *Eastern slopes:* Sidwell K. 99; Tchouto P. 380, 395; Thomas D. W. 9477; Viane R. 2839. *Etinde:* Cable S. 480; Maitland T. D. s.n.; Thomas D. W. 9224; Upson T. 184; Williams S. 65. *No locality:* Adams C. D. 1228

Tectaria sp. aff. fernandensis (Baker) C.Chr.

Specimens: - *Etinde:* Cable S. 286; Maitland T. D. 1131. *Onge:* Cheek M. 4984

Triplophyllum pilosissimum (T.Moore) Holttum

Kew Bull. 41: 246 (1986); Acta Botanica Barcinonensia 40: 44 (1991).
Ctenitis pilosissima (Moore) Alston, F.W.T.A. Suppl.: 71 (1959).

Habit: herb. **Guild:** shade-bearer. **Habitat:** forest, sometimes on rocks. **Distribution:** Guinea to Bioko, Cameroon, Gabon, Congo (Brazzaville), Congo (Kinshasa) and Angola. **Chorology:** Guineo-Congolian. **Star:** green. **Alt. range:** 1-200m (1). **Specimens:** - *Mokoko River F.R.:* Batoum A. 1. *Southern Bakundu F.R.:* Brenan J. P. M. 4064

Triplophyllum protensum (Sw.) Holttum

Kew Bull. 41: 247 (1986); Acta Botanica Barcinonensia 40: 45 (1991).
Ctenitis protensa (Afzel. ex Sw.) Ching, Fl. Cameroun 3: 272 (1964).

Habit: herb. **Guild:** shade-bearer. **Habitat:** forest and plantations, sometimes on rocks. **Distribution:** Senegal to Bioko, CAR, Gabon, Congo (Brazzaville), Congo (Kinshasa), Angola and Uganda. **Chorology:** Guineo-Congolian. **Star:** green. **Alt. range:** 1-200m (1). **Fertile:** May (1). **Specimens:** - *Etinde:* Schlechter F. R. R. 12371. *Mokoko River F.R.:* Ekema N. 872

Triplophyllum securidiforme (Hook.) Holttum var. nanum (Bonap.) Holttum

Kew Bull. 41: 243 (1986).
Ctenitis securidiformis (Hook.) Copel. var. *nana* (Bonap.) Tardieu, F.W.T.A. Suppl.: 73 (1959).

Habit: rheophyte. **Habitat:** on rocks in rivers. **Distribution:** Nigeria, Bioko, Cameroon, Gabon and Angola. **Chorology:** Guineo-Congolian. **Star:** green. **Alt. range:** 1-200m (2), 201-400m (1), 401-600m (1), 601-800m (1). **Fertile:** Mar (1), Jun (1), Oct (1), Nov (3). **Specimens:** - *Etinde:* Cable S. 217; Cheek M. 5463. *Mabeta-Moliwe:* Sunderland T. C. H. 1137. *Mokoko River F.R.:* Mbani J. M. 470. *Onge:* Tchouto P. 832; Thomas D. W. 9855

Triplophyllum securidiforme (Hook.) Holttum var. securidiforme

Kew Bull. 41: 242 (1986).
Ctenitis securidiformis (Hook.) Copel. var. *securidiformis*, F.W.T.A. Suppl.: 73 (1959).

Habit: herb. **Guild:** shade-bearer. **Habitat:** forest, on rocks and by streams. **Distribution:** Senegal to Bioko, Cameroon, Gabon, Congo (Kinshasa) and Angola. **Chorology:** Guineo-Congolian. **Star:** green. **Alt. range:** 1-200m (13), 201-400m (7), 401-600m (6), 601-800m (1), 1001-1200m (1). **Fertile:** Mar (1), Apr (4), May (6), Jun (6), Sep (1), Oct (9), Nov (5), Dec (3). **Specimens:** - *Etinde:* Cable S. 142, 161, 240; Cheek M. 5884; Kalbreyer W. 22; Lighava M. 50; Thomas D. W. 9126, 9127, 9163, 9706; Watts J. 495, 508; Wheatley J. I. 509. *Mabeta-Moliwe:* Cable S. 588; Nguembock F. 13c; Wheatley J. I. 88. *Mokoko River F.R.:* Acworth J. M. 209, 329; Akogo M. 290; Ekema N. 972, 1042; Fraser P. 426; Mbani J. M. 447, 448, 472, 473, 474; Ndam N. 1081, 1085, 1226; Thomas D. W. 10037, 10140. *Onge:* Cheek M. 4982; Ndam N. 605; Tchouto P. 833; Thomas D. W. 9902; Watts J. 960; Wheatley J. I. 696

Triplophyllum securidiforme (Hook.) Holttum var. nov.

Specimens: - *Mabeta-Moliwe:* Sunderland T. C. H. 1138

Triplophyllum varians (T.Moore) Holttum

Kew Bull. 41: 249 (1986).
Tectaria varians (Moore) C.Chr., F.W.T.A. Suppl.: 73 (1959).

Habit: herb. **Habitat:** forest, on rocky ground. **Distribution:** Guinea to Cameroon, Gabon and Angola (Cabinda). **Chorology:** Guineo-Congolian. **Star:** green. **Alt. range:** 1-200m (2). **Fertile:** Apr (1), May (1). **Specimens:** - *Mokoko River F.R.:* Akogo M. 270; Ekema N. 945

GLEICHENIACEAE

Dicranopteris linearis (Burm.f.) Underw. var. linearis

Fl. Zamb. Pteridophyta: 50 (1970); Acta Botanica Barcinonensia 31: 23 (1978); Bull. Jard. Bot. Nat. Belg. 53: 196 (1983).
Gleichenia linearis (Burm.f.) C.B.Clarke, F.W.T.A. Suppl.: 22 (1959).

Habit: scrambling herb. **Guild:** pioneer. **Habitat:** roadsides and forest clearings. **Distribution:** tropical and subtropical Africa.

Star: excluded. **Alt. range:** 1-200m (4). **Fertile:** Mar (1), Apr (1), Nov (2). **Specimens:** - *Mokoko River F.R.:* Acworth J. M. 206. *Onge:* Tchouto P. 517; Thomas D. W. 9872, 9927

GRAMMITIDACEAE

Xiphopteris cultrata (Bory ex Willd.) Schelpe
Fl. Zamb. Pteridophyta: 143 (1970).
Ctenopteris elastica (Bory ex Willd.) Copel., Fl. Cameroun 3: 330 (1964).
Xiphopteris elastica (Bory ex Willd.) Alston, F.W.T.A. Suppl.: 45 (1959).

Habit: epiphyte. **Habitat:** forest-grassland transition. **Distribution:** Bioko, Mount Cameroon, S. Tomé and Mascarene Islands. **Chorology:** Afromontane. **Star:** black. **Alt. range:** 2000-2200m (1). **Specimens:** - *Eastern slopes:* Richards P. W. 4381

Xiphopteris flabelliformis (Poir.) Schelpe
Bol. Soc. Brot. (ser.2) 41: 217 (1967); Fl. Zamb. Pteridophyta: 143 (1970); Acta Botanica Barcinonensia 33: 22 (1982).
Ctenopteris rigescens (Bory ex Willd.) J.Sm., Fl. Cameroun 3: 327 (1964).
Xiphopteris rigescens (Bory ex Willd.) Alston, F.W.T.A. Suppl.: 45 (1959).

Habit: epiphyte. **Habitat:** forest-grassland transition. **Distribution:** Bioko, Mount Cameroon, East Africa, Mascarene Islands and the Andes. **Chorology:** Montane. **Star:** green. **Alt. range:** 1600-1800m (1), 2201-2400m (1). **Fertile:** Nov (1). **Specimens:** - *Eastern slopes:* Cheek M. 5340. *No locality:* Adams C. D. 1325; Steele M. 56

Xiphopteris oosora (Baker) Alston var. oosora
F.W.T.A. Suppl.: 45 (1959); Fl. Cameroun 3: 325 (1964); Fl. Zamb. Pteridophyta: 143 (1970); Acta Botanica Barcinonensia 33: 21 (1982); Bull. Jard. Bot. Nat. Belg. 53: 199 (1983).

Habit: epiphyte. **Habitat:** forest and forest-grassland transition. **Distribution:** Guinea to Bioko, Cameroon, S. Tomé, Malawi, Tanzania and Madagascar. **Chorology:** Afromontane. **Star:** green. **Alt. range:** 1600-1800m (1), 1801-2000m (1). **Fertile:** Oct (1). **Specimens:** - *Eastern slopes:* Thomas D. W. 9447, 9503. *No locality:* Adams C. D. 1303

Xiphopteris villosissima (Hook.) Alston var. laticellulata Benl
Nova Hedwigia 27: 152 (1976).

Habit: herb. **Habitat:** forest. **Distribution:** Mount Cameroon. **Chorology:** Endemic. **Star:** black. **Alt. range:** 1000-1200m (2). **Specimens:** - *Eastern slopes:* Benl G. 70/27, 71/27

Xiphopteris villosissima (Hook.) Alston var. villosissima
F.W.T.A. Suppl.: 45 (1959); Fl. Zamb. Pteridophyta: 142 (1970); Acta Botanica Barcinonensia 33: 23 (1982).
Ctenopteris villosissima (Hook.) W.J.Harley, Fl. Cameroun 3: 328 (1964).

Habit: herb. **Habitat:** forest and forest-grassland transition. **Distribution:** tropical Africa. **Chorology:** Afromontane. **Star:** green. **Alt. range:** 1400-1600m (1), 1801-2000m (1). **Fertile:** Oct

HYMENOPHYLLACEAE

Hymenophyllum capillare Desv. var. capillare
F.W.T.A. Suppl.: 32 (1959); Fl. Cameroun 3: 75 (1964); Fl. Zamb. Pteridophyta: 80 (1970); Acta Botanica Barcinonensia 32: 7 (1980).

Habit: epiphyte. **Habitat:** forest, sometimes on rocks. **Distribution:** tropical Africa. **Chorology:** Afromontane. **Star:** green. **Specimens:** - *No locality:* Adams C. D. 1708

Hymenophyllum kuhnii C.Chr.
F.W.T.A. Suppl.: 32 (1959).
Hymenophyllum polyanthos Sw. var. *kuhnii* (C.Chr.) Schelpe, Fl. Zamb. Pteridophyta: 80 (1970); Acta Botanica Barcinonensia 32: 6 (1980).

Habit: epiphyte. **Habitat:** forest and forest-grassland transition. **Distribution:** tropical Africa. **Chorology:** Afromontane. **Star:** green. **Alt. range:** 600-800m (1), 1401-1600m (2), 1601-1800m (1), 1801-2000m (1). **Fertile:** Sep (1), Oct (3). **Specimens:** - *Eastern slopes:* Sidwell K. 89. *Etinde:* Cheek M. 3808, 3810; Kalbreyer W. 18716; Tekwe C. 238

Hymenophyllum splendidum Bosch
F.W.T.A. Suppl.: 32 (1959); Fl. Cameroun 3: 76 (1964); Acta Botanica Barcinonensia 32: 8 (1980).

Habit: epiphyte. **Habitat:** forest and forest-grassland transition. **Distribution:** Bioko, Cameroon, S. Tomé, Gabon, Congo (Kinshasa) and Tanzania. **Chorology:** Guineo-Congolian (montane). **Star:** green. **Alt. range:** 1200-1400m (1), 1601-1800m (1). **Fertile:** Sep (1). **Specimens:** - *Etinde:* Thomas D. W. 9220; Upson T. 41. *No locality:* Mann G. 1397

Hymenophyllum triangulare Baker
F.W.T.A. Suppl.: 32 (1959); Fl. Cameroun 3: 74 (1964); Acta Botanica Barcinonensia 32: 5 (1980); Bull. Jard. Bot. Nat. Belg. 53: 244 (1983).

Habit: epiphyte. **Habitat:** forest and Aframomum thicket. **Distribution:** Bioko, Cameroon, Rwanda and Tanzania. **Chorology:** Guineo-Congolian (montane). **Star:** green. **Alt. range:** 200-400m (1), 401-600m (1), 801-1000m (1), 1401-1600m (1), 1601-1800m (1). **Fertile:** Sep (1), Oct (2), Nov (1), Dec (1). **Specimens:** - *Etinde:* Cheek M. 5612; Thomas D. W. 9155, 9230, 9536; Williams S. 50

Microgonium benlii Pic.Serm.
Webbia 35: 254 (1982).
Trichomanes benlii (Pic.Serm.) Benl, Acta Botanica Barcinonensia 40: 79 (1991).

Habit: epiphyte. **Habitat:** forest. **Distribution:** Cameroon and Congo (Kinshasa). **Chorology:** Guineo-Congolian. **Star:** green. **Alt. range:** 1-200m (2), 401-600m (1). **Fertile:** Jan (1), Mar (1), Oct (1). **Specimens:** - *Etinde:* Thomas D. W. 9135. *Onge:* Benl G. Ka 74/6; Wheatley J. I. 849

(2). **Specimens:** - *Eastern slopes:* Thomas D. W. 9446, 9467. *No locality:* Adams C. D. 1310

Trichomanes africanum Christ

F.W.T.A. Suppl.: 31 (1959); Fl. Cameroun 3: 88 (1964); Acta Botanica Barcinonensia 32: 23 (1980).

Habit: epiphyte. **Habitat:** forest. **Distribution:** Guinea to Bioko, Cameroon, S. Tomé, CAR, Gabon, Congo (Brazzaville), Congo (Kinshasa), Angola and Sudan. **Chorology:** Guineo-Congolian. **Star:** green. **Specimens:** - *No locality:* Adams C. D. 1240

Trichomanes ballardianum Alston

F.W.T.A. Suppl.: 30 (1959); Acta Botanica Barcinonensia 40: 80 (1991).
  Microgonium ballardianum (Alston) Pic.Serm., Webbia 35: 255 (1982).

Habit: epiphyte. **Habitat:** forest, sometimes on rocks. **Distribution:** Nigeria, Cameroon, Gabon, Congo (Brazzaville) and Congo (Kinshasa). **Chorology:** Guineo-Congolian. **Star:** green. **Alt. range:** 400-600m (1). **Fertile:** Oct (1). **Specimens:** - *Etinde:* Thomas D. W. 9156

Trichomanes borbonicum Bosch

F.W.T.A. Suppl.: 31 (1959); Fl. Cameroun 3: 88 (1964); Fl. Zamb. Pteridophyta: 76 (1970); Acta Botanica Barcinonensia 32: 22 (1980).

Habit: epiphyte. **Habitat:** forest, sometimes on rocks. **Distribution:** tropical Africa. **Chorology:** Afromontane. **Star:** green. **Specimens:** - *No locality:* Mann G. 1400

Trichomanes clarenceanum Ballard

F.W.T.A. Suppl.: 30 (1959); Fl. Cameroun 3: 84 (1964); Acta Botanica Barcinonensia 32: 19 (1980).

Habit: epiphyte. **Habitat:** forest. **Distribution:** Liberia, Nigeria, Bioko and Cameroon. **Chorology:** Guineo-Congolian (montane). **Star:** green. **Specimens:** - *Eastern slopes:* Rosevear D. R. s.n.

Trichomanes erosum Willd. var. erosum

F.W.T.A. Suppl.: 30 (1959); Fl. Cameroun 3: 82 (1964); Acta Botanica Barcinonensia 32: 16 (1980).
  Trichomanes erosum Willd. var. *aerugineum* (Bosch) Bonap., Fl. Zamb. Pteridophyta: 76 (1970); Acta Botanica Barcinonensia 32: 17 (1980).
  Trichomanes aerugineum Bosch, F.W.T.A. Suppl.: 30 (1959).
  Trichomanes chamaedrys Taton, F.W.T.A. Suppl.: 30 (1959).
  Microgonium chamaedrys (Taton) Pic.Serm., Webbia 23: 181 (1968); Webbia 35: 254 (1982).

Habit: epiphyte. **Habitat:** forest, sometimes on rocks. **Distribution:** tropical and subtropical Africa. **Star:** excluded. **Alt. range:** 1-200m (1), 201-400m (7), 401-600m (1). **Fertile:** Oct (5), Nov (3). **Specimens:** - *Eastern slopes:* Box H. E. 3608; Richards P. W. 4317. *Etinde:* Cheek M. 5465, 5560, 5561; Kalbreyer W. 39; Thomas D. W. 9132, 9718, 9720. *Onge:* Cheek M. 5061; Tchouto P. 877; Thomas D. W. 9815. *Southern Bakundu F.R.:* Richards P. W. 4042

Trichomanes giganteum Bory ex Willd.

F.W.T.A. Suppl.: 31 (1959).

Habit: epiphyte. **Habitat:** forest-grassland transition. **Distribution:** Nigeria, Bioko, Cameroon, S. Tomé and Tanzania. **Chorology:** Guineo-Congolian (montane). **Star:** green. **Alt. range:** 2000-2200m (1). **Specimens:** - *Eastern slopes:* Adamseyer 1709

Trichomanes guineense Afzel. ex Sw.

F.W.T.A. Suppl.: 31 (1959); Fl. Cameroun 3: 91 (1964); Acta Botanica Barcinonensia 32: 13 (1980).

Habit: herb. **Guild:** shade-bearer. **Habitat:** forest. **Distribution:** Guinea to Bioko, Cameroon, Gabon and Congo (Brazzaville). **Chorology:** Guineo-Congolian. **Star:** green. **Specimens:** - *Southern Bakundu F.R.:* Richards P. W. 4041

Trichomanes liberiense Copel.

F.W.T.A. Suppl.: 30 (1959); Fl. Cameroun 3: 80 (1964).

Habit: epiphyte. **Habitat:** forest. **Distribution:** Liberia, Ivory Coast, Nigeria, Cameroon and Congo (Kinshasa). **Chorology:** Guineo-Congolian. **Star:** green. **Specimens:** - *Etinde:* Thorold C. A. 21

Trichomanes mannii Hook.

F.W.T.A. Suppl.: 30 (1959); Fl. Cameroun 3: 83 (1964); Acta Botanica Barcinonensia 32: 17 (1980).

Habit: epiphyte. **Habitat:** forest and Aframomum thicket. **Distribution:** tropical Africa. **Star:** excluded. **Alt. range:** 1000-1200m (2). **Fertile:** Sep (1), Oct (1). **Specimens:** - *Etinde:* Cheek M. 3786; Thomas D. W. 9231. *No locality:* Adams C. D. 1330

Trichomanes mettenii C.Chr.

F.W.T.A. Suppl.: 31 (1959); Fl. Cameroun 3: 85 (1964); Acta Botanica Barcinonensia 32: 19 (1980).

Habit: epiphyte. **Habitat:** forest, sometimes on rocks. **Distribution:** Sierra Leone to Bioko, Cameroon, CAR, Gabon, Congo (Brazzaville), Congo (Kinshasa) and Uganda. **Chorology:** Guineo-Congolian. **Star:** green. **Specimens:** - *Mabeta-Moliwe:* Dunlap 270

LOMARIOPSIDACEAE

Bolbitis acrostichoides (Afzel. ex Sw.) Ching

F.W.T.A. Suppl.: 68 (1959); Fl. Cameroun 3: 322 (1964); Hennipmann E., Monograph of Bolbitis: 149 (1977); Acta Botanica Barcinonensia 40: 61 (1991).

Habit: herb. **Guild:** shade-bearer. **Habitat:** forest, sometimes on rocks. **Distribution:** tropical Africa. **Star:** excluded. **Alt. range:** 1-200m (7), 201-400m (2), 401-600m (3). **Fertile:** Mar (1), Jun (1), Oct (4), Nov (4), Dec (3). **Specimens:** - *Etinde:* Cable S. 160, 275; Thomas D. W. 9120, 9137, 9722. *Mabeta-Moliwe:* Cable S. 537, 541, 589. *Mokoko River F.R.:* Mbani J. M. 442. *No locality:* Maitland T. D. 1131. *Onge:* Tchouto P. 943; Thomas D. W. 9905; Watts J. 1024; Wheatley J. I. 855

Bolbitis auriculata (Lam.) Alston

F.W.T.A. Suppl.: 68 (1959); Fl. Cameroun 3: 318 (1964); Hennipmann E., Monograph of Bolbitis: 136 (1977); Bull. Jard. Bot. Nat. Belg. 55: 179 (1985); Acta Botanica Barcinonensia 40: 59 (1991).

Habit: herb. **Guild:** shade-bearer. **Habitat:** forest, sometimes on rocks. **Distribution:** tropical Africa. **Star:** excluded. **Alt. range:** 200-400m (2), 401-600m (2). **Fertile:** May (1), Jun (2), Oct (2), Dec (1). **Specimens:** - *Etinde:* Cable S. 391; Kalbreyer W. 24; Thomas D. W. 9136, 9723. *Mokoko River F.R.:* Mbani J. M. 475, 493; Thomas D. W. 10038

Bolbitis fluviatilis (Hook.) Ching

F.W.T.A. Suppl.: 68 (1959); Fl. Cameroun 3: 316 (1964);
Hennipmann E., Monograph of Bolbitis: 142 (1977); Acta
Botanica Barcinonensia 40: 60 (1991).

Habit: herb. **_Iabitat:** forest, sometimes on rocks. **Distribution:**
Liberia to Bioko, Cameroon, Gabon, Congo (Brazzaville), Congo
(Kinshasa), S. Tomé and Principe. **Chorology:** Guineo-Congolian.
Star: green. **Alt. range:** 200-400m (1), 401-600m (2). **Fertile:** Sep
(1), Nov (1). **Specimens:** - *Etinde:* Cheek M. 5571; Thomas D. W.
9118, 9119. *Southern Bakundu F.R.:* Schlechter F. R. R. 12874

Bolbitis heudelotii (Bory ex Fée) Alston

F.W.T.A. Suppl.: 68 (1959); Fl. Cameroun 3: 319 (1964); Fl.
Zamb. Pteridophyta: 218 (1970); Hennipmann E., Monograph of
Bolbitis: 236 (1977); Bull. Jard. Bot. Nat. Belg. 55: 180 (1985).

Habit: rheophyte. **Habitat:** streams. **Distribution:** Senegal to
Congo (Brazzaville) and Angola. **Chorology:** Guineo-Congolian.
Star: green. **Specimens:** - *Etinde:* Richards P. W. 4020

Bolbitis sp.

Specimens: - *Mokoko River F.R.:* Mbani J. M. 445; Thomas D. W.
10036

Elaphoglossum acrostichoides (Hook. & Grev.) Schelpe

Acta Botanica Barcinonensia 40: 71 (1991).
   Elaphoglossum preussii Hieron., F.W.T.A. Suppl.: 66 (1959).

Habit: epiphyte. **Habitat:** forest. **Distribution:** Bioko and SW
Cameroon. **Chorology:** Western Cameroon. **Star:** black.
Specimens: - *No locality:* Johnston H. H. 132

Elaphoglossum aubertii (Desv.) T.Moore

F.W.T.A. Suppl.: 66 (1959); Fl. Cameroun 3: 303 (1964); Fl.
Zamb. Pteridophyta: 213 (1970); Bull. Jard. Bot. Nat. Belg. 55:
181 (1985); Acta Botanica Barcinonensia 40: 75 (1991).

Habit: epiphytic or terrestrial herb. **Habitat:** grassland.
Distribution: Bioko, Mount Cameroon, East and southern Africa.
Chorology: Afromontane. **Star:** green. **Alt. range:** 2000-2200m
(2). **Fertile:** Oct (1). **Specimens:** - *Eastern slopes:* Brenan J. P. M.
4213; Thomas D. W. 9319

Elaphoglossum barteri (Baker) C.Chr.

F.W.T.A. Suppl.: 66 (1959); Fl. Cameroun 3: 298 (1964).

Habit: epiphyte. **Habitat:** forest. **Distribution:** Guinea to SW
Cameroon. **Chorology:** Guineo-Congolian (montane). **Star:** green.
Alt. range: 1400-1600m (1), 1601-1800m (3). **Fertile:** Sep (1),
Oct (3). **Specimens:** - *Etinde:* Tchouto P. 407, 408; Tekwe C. 239;
Wheatley J. I. 615. *No locality:* Adams C. D. 1675

Elaphoglossum cinnamomeum (Baker) Diels

F.W.T.A. Suppl.: 66 (1959); Fl. Cameroun 3: 304 (1964); Acta
Botanica Barcinonensia 40: 77 (1991).

Habit: epiphyte. **Habitat:** forest, sometimes on rocks.
Distribution: Mount Cameroon and Bioko. **Chorology:** Western
Cameroon (montane). **Star:** black. **Alt. range:** 1000-1200m (1),
1401-1600m (1), 1601-1800m (2), 1801-2000m (1). **Fertile:** Sep
(1), Oct (4). **Specimens:** - *Eastern slopes:* Sidwell K. 88. *Etinde:*
Tchouto P. 409; Tekwe C. 240; Thomas D. W. 9162; Wheatley J. I.
617. *No locality:* Johnston H. H. 139

192

Elaphoglossum hybridum (Bory) Brack.

F.W.T.A. Suppl.: 66 (1959); Fl. Cameroun 3: 303 (1964); Fl.
Zamb. Pteridophyta: 213 (1970); Bull. Jard. Bot. Nat. Belg. 55:
181 (1985); Acta Botanica Barcinonensia 40: 74 (1991).

Habit: epiphyte. **Habitat:** forest, sometimes on rocks.
Distribution: tropical Africa and America. **Star:** excluded.
Specimens: - *No locality:* Mann G. 1377

Elaphoglossum isabelense Brause

F.W.T.A. Suppl.: 66 (1959); Acta Botanica Barcinonensia 40: 73
(1991).

Habit: epiphyte. **Habitat:** forest. **Distribution:** Guinea to Bioko,
Cameroon, and S. Tomé. **Chorology:** Guineo-Congolian
(montane). **Star:** green. **Fertile:** Jun (1). **Specimens:** - *Mokoko
River F.R.:* Mbani J. M. 483

Elaphoglossum kuhnii Hieron.

F.W.T.A. Suppl.: 66 (1959); Fl. Cameroun 3: 305 (1964); Fl.
Zamb. Pteridophyta: 215 (1970); Acta Botanica Barcinonensia 40:
76 (1991).

Habit: epiphyte. **Habitat:** forest. **Distribution:** Sierra Leone,
Liberia, Bioko and Cameroon. **Chorology:** Guineo-Congolian.
Star: blue. **Alt. range:** 400-600m (1). **Fertile:** Jun (1).
Specimens: - *Mokoko River F.R.:* Mbani J. M. 482. *No locality:*
Maitland T. D. 1051; Mann G. 1378

Elaphoglossum salicifolium (Willd. ex Kaulf.) Alston

F.W.T.A. Suppl.: 66 (1959); Fl. Cameroun 3: 302 (1964); Bull.
Jard. Bot. Nat. Belg. 55: 186 (1985); Acta Botanica Barcinonensia
40: 73 (1991).

Habit: epiphyte. **Habitat:** forest and forest-grassland transition.
Distribution: Guinea, Liberia, Ivory Coast, Bioko, SW Cameroon,
S. Tomé and Principe. **Chorology:** Guineo-Congolian (montane).
Star: green. **Alt. range:** 1600-1800m (1), 2401-2600m (1).
Fertile: May (1), Oct (1). **Specimens:** - *Eastern slopes:* Upson T.
101. *Etinde:* Thomas D. W. 9252. *No locality:* Adams C. D. 1211

Elaphoglossum subcinnamomeum (H.Christ) Hieron.

F.W.T.A. Suppl.: 67 (1959); Fl. Cameroun 3: 306 (1964).

Habit: lithophyte. **Habitat:** lava flows. **Distribution:** Mounts
Cameroon, Kenya, Elgon and Kilimanjaro. **Chorology:**
Afromontane. **Star:** black. **Alt. range:** 2800-3000m (1). **Fertile:**
Dec (1). **Specimens:** - *No locality:* Adams C. D. 1333a; Maitland
T. D. 846

Lomariopsis decrescens (Baker) Kuhn

F.W.T.A. Suppl.: 67 (1959); Fl. Cameroun 3: 312 (1964); Acta
Botanica Barcinonensia 40: 68 (1991).

Habit: epiphyte. **Habitat:** forest, sometimes on rocks.
Distribution: Nigeria, SW Cameroon and Bioko. **Chorology:**
Western Cameroon. **Star:** gold. **Alt. range:** 800-1000m (1).
Fertile: Dec (1). **Specimens:** - *Etinde:* Cable S. 502. *No locality:*
Mann G. 1391

Lomariopsis guineensis (Underw.) Alston

F.W.T.A. Suppl.: 67 (1959); Fl. Cameroun 3: 309 (1964); Acta
Botanica Barcinonensia 40: 64 (1991).

Habit: herbaceous climber. **Guild:** shade-bearer. **Habitat:** forest. **Distribution:** Guinea to Bioko, S. Tomé, CAR, Gabon, Congo (Brazzaville), Congo (Kinshasa), Angola and Sudan. **Chorology:** Guineo-Congolian. **Star:** green. **Alt. range:** 1-200m (2), 201-400m (3). **Fertile:** May (3), Jun (2), Nov (1). **Specimens:** - *Etinde:* Kalbreyer W. 25. *Mokoko River F.R.:* Acworth J. M. 319; Mbani J. M. 455; Thomas D. W. 10053, 10133, 10136. *Onge:* Thomas D. W. 9765

Lomariopsis mannii (Underw.) Alston

F.W.T.A. Suppl.: 67 (1959); Fl. Cameroun 3: 313 (1964); Acta Botanica Barcinonensia 40: 69 (1991).

Habit: epiphyte. **Habitat:** forest, sometimes on rocks. **Distribution:** Mount Cameroon and Bioko. **Chorology:** Western Cameroon (montane). **Star:** black. **Specimens:** - *No locality:* Johnston H. H. 146

Lomariopsis muriculata Holttum

F.W.T.A. Suppl.: 67 (1959); Acta Botanica Barcinonensia 40: 65 (1991).

Habit: epiphyte. **Habitat:** forest. **Distribution:** Sierra Leone, Nigeria, Bioko, Cameroon and Congo (Kinshasa). **Chorology:** Guineo-Congolian. **Star:** green. **Alt. range:** 200-400m (2). **Fertile:** May (2). **Specimens:** - *Mokoko River F.R.:* Thomas D. W. 10134, 10135

Lomariopsis cf. muriculata Holttum

Specimens: - *Etinde:* Cable S. 486

Lomariopsis palustris (Hook.) Mett. ex Kuhn

F.W.T.A. Suppl.: 67 (1959); Fl. Cameroun 3: 309 (1964).

Habit: epiphyte. **Habitat:** forest, sometimes on rocks. **Distribution:** Guinea to Angola. **Chorology:** Guineo-Congolian. **Star:** green. **Alt. range:** 200-400m (1). **Fertile:** Jun (1), Oct (1). **Specimens:** - *Mabeta-Moliwe:* Mann G. 785. *Mokoko River F.R.:* Mbani J. M. 454. *Onge:* Tchouto P. 892

Lomariopsis rossii Holttum

F.W.T.A. Suppl.: 67 (1959); Acta Botanica Barcinonensia 40: 67 (1991).

Habit: herbaceous climber. **Habitat:** forest, sometimes on rocks. **Distribution:** Guinea to Bioko, Cameroon, Gabon, Congo (Brazzaville) and Uganda. **Chorology:** Guineo-Congolian. **Star:** green. **Alt. range:** 1-200m (2). **Fertile:** May (2). **Specimens:** - *Mokoko River F.R.:* Ekema N. 949; Watts J. 1176

Lomariopsis cf. warneckei

Specimens: - *Mokoko River F.R.:* Mbani J. M. 456

Lomariopsis sp.

Specimens: - *Etinde:* Thomas D. W. 9141. *Mokoko River F.R.:* Mbani J. M. 450

MARATTIACEAE

Marattia fraxinea J.Sm. var. fraxinea

F.W.T.A. Suppl.: 20 (1959); Fl. Cameroun 3: 50 (1964); Fl. Zamb. Pteridophyta: 40 (1970); Acta Botanica Barcinonensia 32: 3 (1980); Bull. Jard. Bot. Nat. Belg. 53: 193 (1983).

Habit: herb. **Habitat:** forest. **Distribution:** palaeotropical. **Chorology:** Montane. **Star:** excluded. **Alt. range:** 1-200m (3), 201-400m (3), 801-1000m (1), 1201-1400m (1), 1401-1600m (1). **Fertile:** Mar (2), Apr (1), May (4), Jun (4), Sep (1), Oct (1), Nov (1), Dec (1). **Specimens:** - *Bambuko F.R.:* Watts J. 626. *Eastern slopes:* Mbani J. M. 120; Tchouto P. 393. *Etinde:* Lighava M. 55; Tekwe C. 178. *Mabeta-Moliwe:* Plot Voucher W 467; Sunderland T. C. H. 1112; Watts J. 388; Wheatley J. I. 119. *Mokoko River F.R.:* Mbani J. M. 476, 479; Ndam N. 1227, 1233, 1236. *No locality:* Johnston H. H. 149. *Onge:* Harris D. J. 3743

OLEANDRACEAE

Arthropteris cameroonensis Alston

F.W.T.A. Suppl.: 52 (1959); Fl. Cameroun 3: 115 (1964); Acta Botanica Barcinonensia 33: 33 (1982).

Habit: lithophyte. **Habitat:** lava flows. **Distribution:** Bioko and SW Cameroon. **Chorology:** Western Cameroon. **Star:** black. **Alt. range:** 600-800m (1), 1201-1400m (1). **Fertile:** Mar (1), Sep (1). **Specimens:** - *Etinde:* Tekwe C. 168; Upson T. 7. *No locality:* Mann G. 1395

Arthropteris monocarpa (Cordem.) C.Chr.

F.W.T.A. Suppl.: 52 (1959); Fl. Cameroun 3: 117 (1964); Fl. Zamb. Pteridophyta: 163 (1970); Acta Botanica Barcinonensia 33: 35 (1982).

Habit: epiphyte. **Habitat:** forest, stream banks and Aframomum thicket. **Distribution:** tropical Africa. **Chorology:** Afromontane. **Star:** green. **Alt. range:** 1-200m (2), 201-400m (3), 401-600m (3), 601-800m (1), 801-1000m (1), 1601-1800m (1), 1801-2000m (1), 2001-2200m (1). **Fertile:** Mar (1), May (1), Oct (7), Nov (2), Dec (2). **Specimens:** - *Eastern slopes:* Cheek M. 3707; Upson T. 73. *Etinde:* Cable S. 300, 435; Thomas D. W. 9153, 9533; Watts J. 534; Williams S. 90. *No locality:* Maitland T. D. 1102. *Onge:* Cheek M. 5046, 5064; Tchouto P. 870; Thomas D. W. 9793; Wheatley J. I. 724

NB. Lawralrée, in Bull. Jard. Bot. Nat. Belg. 60: 317-324 (1990) states that *A. monocarpa* only occurs on Reunion! Two names (new) for two allied taxa in Africa: *A. anniana* and *A. antungupffertiae.*

Cites:- Mt. Cameroon alt. 1950m. Breteler et al 263 (P) as *A. anniana* Lawralrée

Arthropteris palisoti (Desv.) Alston

F.W.T.A. Suppl.: 52 (1959); Fl. Cameroun 3: 114 (1964); Acta Botanica Barcinonensia 33: 31 (1982).

Habit: herbaceous climber. **Habitat:** forest. **Distribution:** Ivory Coast to Cameroon, Gabon, Congo (Brazzaville), Congo (Kinshasa), Angola, Madagascar and Asia. **Star:** excluded. **Alt. range:** 1-200m (2), 201-400m (1). **Fertile:** May (1), Oct (2). **Specimens:** - *Mabeta-Moliwe:* Wheatley J. I. 240. *Onge:* Cheek

M. 5144; Tchouto P. 834. *Southern Bakundu F.R.:* Richards P. W. 4045

Nephrolepis biserrata (Sw.) Schott
F.W.T.A. Suppl.: 50 (1959); Fl. Cameroun 3: 110 (1964); Fl. Zamb. Pteridophyta: 160 (1970); Acta Botanica Barcinonensia 33: 36 (1982); Bull. Jard. Bot. Nat. Belg. 55: 187 (1985).

Habit: epiphytic or terrestrial herb. **Habitat:** forest, plantations and lava flows. **Distribution:** pantropical. **Star:** excluded. **Alt. range:** 1-200m (4), 201-400m (1). **Fertile:** Apr (1), May (1), Oct (1). **Specimens: -** *Etinde:* Thorold C. A. 7. *Mabeta-Moliwe:* Cheek M. 5746. *Mokoko River F.R.:* Acworth J. M. 207; Ekema N. 1044. *Onge:* Cheek M. 5073

Nephrolepis cf. biserrata (Sw.) Schott
Specimens: - *Mokoko River F.R.:* Mbani J. M. 485

Nephrolepis pumicicola Ballard
F.W.T.A. Suppl.: 50 (1959); Fl. Cameroun 3: 108 (1964); Acta Botanica Barcinonensia 33: 37 (1982).

Habit: lithophyte. **Habitat:** lava flows. **Distribution:** Mount Cameroon, Bioko and S. Tomé **Chorology:** Western Cameroon (montane). **Star:** black. **Alt. range:** 600-800m (2), 1201-1400m (1), 1601-1800m (1), 1801-2000m (1). **Fertile:** Mar (1), May (1), Nov (1). **Specimens: -** *Eastern slopes:* Keay R. W. J. FHI 37500; Sunderland T. C. H. 1442. *Etinde:* Cable S. 225; Keay R. W. J. FHI 28665; Upson T. 5. *No locality:* Maitland T. D. 1091

Nephrolepis undulata (Afzel. ex Sw.) J.Sm. var. undulata
F.W.T.A. Suppl.: 50 (1959); Fl. Cameroun 3: 109 (1964); Fl. Zamb. Pteridophyta: 162 (1970); Acta Botanica Barcinonensia 33: 39 (1982); Bull. Jard. Bot. Nat. Belg. 55: 188 (1985).

Habit: epiphytic or terrestrial herb. **Habitat:** forest, plantations and roadsides. **Distribution:** tropical and subtropical Africa. **Star:** excluded. **Alt. range:** 1-200m (3), 401-600m (1), 601-800m (2), 1001-1200m (1), 1401-1600m (1), 1601-1800m (1), 2001-2200m (1). **Fertile:** May (1), Sep (1), Oct (4), Nov (4). **Specimens: -** *Eastern slopes:* Richards P. W. 4122; Upson T. 57. *Etinde:* Cable S. 157, 239; Thomas D. W. 9207, 9528; Wheatley J. I. 642; Williams S. 89. *Onge:* Cheek M. 5006; Tchouto P. 1013; Thomas D. W. 9898

Nephrolepis cf. undulata (Afzel. ex Sw.) J.Sm.
Specimens: - *Mokoko River F.R.:* Mbani J. M. 441

Oleandra distenta Kunze var. distenta
F.W.T.A. Suppl.: 52 (1959); Fl. Cameroun 3: 106 (1964); Fl. Zamb. Pteridophyta: 165 (1970); Acta Botanica Barcinonensia 33: 28 (1982); Bull. Jard. Bot. Nat. Belg. 55: 189 (1985).

Habit: epiphyte. **Habitat:** forest and plantations. **Distribution:** tropical and subtropical Africa. **Chorology:** Afromontane. **Star:** green. **Alt. range:** 1800-2000m (3). **Fertile:** Sep (1), Oct (1). **Specimens: -** *Eastern slopes:* Sidwell K. 87. *Etinde:* Tekwe C. 266, 298. *No locality:* Adams C. D. 1231

OPHIOGLOSSACEAE

Botrychium chamaeconium Bitter & Hieron.
F.W.T.A. Suppl.: 19 (1959); Fl. Cameroun 3: 47 (1964).

Habit: herb. **Habitat:** grassland. **Distribution:** W. Cameroon, Uganda and Sudan. **Chorology:** Guineo-Congolian (montane). **Star:** blue. **Specimens: -** *Eastern slopes:* Preuss P. R. 1037

Ophioglossum gomezianum Welw. ex A.Br.
F.W.T.A. Suppl.: 19 (1959); Fl. Zamb. Pteridophyta: 35 (1970); Bull. Jard. Bot. Nat. Belg. 53: 192 (1983).
    Ophioglossum gomezianum Welw. ex A.Br. var. *latifolum* Prantl, F.W.T.A. Suppl.: 19 (1959).

Habit: herb. **Habitat:** forest-grassland transition. **Distribution:** Sierra Leone, Ghana, Nigeria, Cameroon and Angola. **Chorology:** Guineo-Congolian (montane). **Star:** green. **Specimens: -** *No locality:* Mildbraed G. W. J. 3406

Ophioglossum reticulatum L.
F.W.T.A. Suppl.: 19 (1959); Fl. Cameroun 3: 45 (1964); Fl. Zamb. Pteridophyta: 37 (1970); Acta Botanica Barcinonensia 33: 3 (1982); Bull. Jard. Bot. Nat. Belg. 53: 193 (1983).

Habit: herb. **Guild:** pioneer. **Habitat:** lava flows. **Distribution:** pantropical. **Star:** excluded. **Specimens: -** *No locality:* Mann G. 2061

POLYPODIACEAE

Anapeltis lycopodioides (L.) J.Sm.
Acta Botanica Barcinonensia 33: 18 (1982).
    Microgramma owariensis (Desv.) Alston, F.W.T.A. Suppl.: 49 (1959).
    Anapeltis owariensis (Desv.) J.Sm., Bull. Jard. Bot. Nat. Belg. 53: 206 (1983).

Habit: epiphyte. **Habitat:** forest. **Distribution:** Guinea to Bioko, Cameroon, CAR, Rio Muni and Gabon. **Chorology:** Guineo-Congolian. **Star:** green. **Alt. range:** 1-200m (10), 201-400m (1). **Fertile:** Mar (1), Apr (2), May (3), Jun (2), Jul (1), Oct (1), Nov (2), Dec (1). **Specimens: -** *Etinde:* Cable S. 153; Rosevear D. R. 37/37; Tchouto P. 244. *Mabeta-Moliwe:* Cable S. 609; Sunderland T. C. H. 1332; Wheatley J. I. 156. *Mokoko River F.R.:* Acworth J. M. 181; Mbani J. M. 453, 510; Tchouto P. 1179; Watts J. 1149. *Onge:* Cheek M. 5091; Watts J. 1037; Wheatley J. I. 844

Drynaria laurentii (H.Christ ex De Wild. & Durand) Hieron.
F.W.T.A. Suppl.: 48 (1959); Fl. Cameroun 3: 337 (1964); Acta Botanica Barcinonensia 33: 10 (1982).

Habit: epiphyte. **Habitat:** forest. **Distribution:** tropical Africa. **Star:** excluded. **Alt. range:** 1-200m (1). **Fertile:** Apr (1). **Specimens: -** *Mokoko River F.R.:* Acworth J. M. 223

Drynaria volkensii Hieron.
F.W.T.A. Suppl.: 48 (1959); Fl. Cameroun 3: 336 (1964); Fl. Zamb. Pteridophyta: 149 (1970); Acta Botanica Barcinonensia 33: 11 (1982); Bull. Jard. Bot. Nat. Belg. 53: 209 (1983).

Habit: epiphyte. Habitat: forest and forest-grassland transition. Distribution: Bioko and Cameroon to East Africa. Chorology: Afromontane. Star: green. Alt. range: 1600-1800m (1), 2001-2200m (1). Fertile: Oct (1). Specimens: - *Eastern slopes:* Richards P. W. 4377; Thomas D. W. 9469

Lepisorus excavatus (Bory ex Willd.) Ching

Zink M., Systematics of Lepisorus: 37 (1993).
Pleopeltis excavata (Bory ex Willd.) Moore, Acta Botanica Barcinonensia 33: 16 (1982).

Habit: epiphyte. Habitat: forest-grassland transition. Distribution: Guinea to Bioko and Cameroon. Chorology: Guineo-Congolian (montane). Star: green. Alt. range: 1200-1400m (1). Fertile: Apr (1). Specimens: - *No locality:* Maitland T. D. 1092

Lepisorus preussii (Hieron.) Pic.Serm.

Zink M., Systematics of Lepisorus: 45 (1993).
Pleopeltis preussii (Hieron.) Tardieu, F.W.T.A. Suppl.: 49 (1959).
Habit: epiphyte. Habitat: forest. Distribution: Guinea to Bioko, Cameroon, Gabon and Congo (Kinshasa). Chorology: Guineo-Congolian (montane). Star: green. Specimens: - *Eastern slopes:* Migeod F. 34

Loxogramme abyssinica (Baker) M.G.Price

Amer. Fern. J. 74(2): 61 (1984).
Loxogramme lanceolata (Sw.) C.Presl, F.W.T.A. Suppl.: 48 (1959); Acta Botanica Barcinonensia 33: 24 (1982); Acta Botanica Barcinonensia 33: 26 (1982).

Habit: epiphyte. Habitat: forest-grassland transition. Distribution: tropical Africa. Chorology: Afromontane. Star: green. Alt. range: 1-200m (1), 401-600m (1), 1201-1400m (1), 1401-1600m (2), 2001-2200m (4). Fertile: Apr (1), Sep (3), Oct (4), Dec (1). Specimens: - *Eastern slopes:* Migeod F. 142; Sidwell K. 32; Tchouto P. 298, 312; Watts J. 437. *Etinde:* Cable S. 364; Tekwe C. 147; Thomas D. W. 9197, 9198. *Mokoko River F.R.:* Acworth J. M. 194

Microsorum punctatum (L.) Copel.

F.W.T.A. Suppl.: 49 (1959); Fl. Cameroun 3: 350 (1964); Fl. Zamb. Pteridophyta: 156 (1970); Acta Botanica Barcinonensia 33: 14 (1982); Bull. Jard. Bot. Nat. Belg. 53: 208 (1983).

Habit: epiphyte. Habitat: forest. Distribution: palaeotropical. Star: excluded. Alt. range: 1-200m (4), 201-400m (3). Fertile: Apr (1), May (2), Oct (2), Nov (1), Dec (1). Specimens: - *Etinde:* Brodie S. 2; Rosevear D. R. 39/37. *Mabeta-Moliwe:* Cheek M. 5749; Wheatley J. I. 120. *Mokoko River F.R.:* Tchouto P. 1178; Thomas D. W. 10024. *Onge:* Cheek M. 5025; Thomas D. W. 9908

Microsorum scolopendria (Burm.f.) Copel.

Univ. Calif. Publ. Bot. 16: 112 (1929).
Phymatodes scolopendria (Burm.f.) Ching, Fl. Cameroun 3: 352 (1964).
Phymatosorus scolopendria (Burm.f) Pic.Serm., Acta Botanica Barcinonensia 33: 13 (1982); Bull. Jard. Bot. Nat. Belg. 53: 207 (1983).

Habit: epiphyte. Habitat: forest. Distribution: palaeotropical. Star: excluded. Alt. range: 1-200m (3). Fertile: Apr (1), May (1), Jul (1). Specimens: - *Etinde:* Box H. E. 3624; Tchouto P. 243. *Mabeta-Moliwe:* Wheatley J. I. 277. *Mokoko River F.R.:* Acworth J. M. 195

Platycerium stemaria (P.Beauv.) Desv.

F.W.T.A. Suppl.: 46 (1959); Fl. Cameroun 3: 335 (1964); Acta Botanica Barcinonensia 33: 9 (1982).

Habit: epiphyte. Habitat: forest and plantations. Distribution: Senegal to Bioko, Cameroon, Gabon, Congo (Kinshasa), Angola, Sudan and Comoros. Chorology: Guineo-Congolian. Star: green. Alt. range: 1-200m (2). Fertile: May (1), Jun (1). Specimens: - *Etinde:* Kalbreyer W. 198. *Limbe Botanic Garden:* Upson T. 109. *Mokoko River F.R.:* Acworth J. M. 272

Pleopeltis macrocarpa (Bory ex Willd.) Kaulf. var. macrocarpa

Fl. Zamb. Pteridophyta: 152 (1970); Acta Botanica Barcinonensia 33: 15 (1982); Bull. Jard. Bot. Nat. Belg. 53: 203 (1983).
Pleopeltis lanceolata (L.) Kaulf., F.W.T.A. Suppl.: 49 (1959).

Habit: epiphyte. Habitat: forest-grassland transition, sometimes on rocks. Distribution: pantropical. Chorology: Montane. Star: excluded. Alt. range: 2200-2400m (1), 2601-2800m (1). Fertile: Oct (1). Specimens: - *Eastern slopes:* Sidwell K. 66. *No locality:* Maitland T. D. 1236; Mann G. 1374

Pyrrosia lanceolata (L.) Farw.

F.W.T.A. Suppl.: 46 (1959); Fl. Cameroun 3: 340 (1964); Fl. Zamb. Pteridophyta: 146 (1970).

Habit: epiphyte. Habitat: forest and plantations. Distribution: Cameroon to Congo (Kinshasa), East Africa and Asia. Star: excluded. Specimens: - *Mabeta-Moliwe:* Dunlap 179

Pyrrosia schimperiana (Mett.) Alston var. schimperiana

Fl. Zamb. Pteridophyta: 147 (1970).
Pyrrosia mechowii (Hieron.) Alston, F.W.T.A. Suppl.: 46 (1959).

Habit: epiphyte. Habitat: forest. Distribution: tropical Africa. Star: excluded. Alt. range: 1-200m (1). Fertile: Jun (1). Specimens: - *Eastern slopes:* Brenan J. P. M. 4391. *Mabeta-Moliwe:* Wheatley J. I. 336

PTERIDACEAE

Acrostichum aureum L.

F.W.T.A. Suppl.: 36 (1959); Fl. Cameroun 3: 130 (1964); Fl. Zamb. Pteridophyta: 99 (1970); Acta Botanica Barcinonensia 38: 22 (1988).

Habit: herb. Habitat: mangrove forest and saline mudflats. Distribution: pantropical. Star: excluded. Specimens: - *Etinde:* Kalbreyer W. 226. *Mabeta-Moliwe:* Maitland T. D. 976a

Pteris burtoni Baker

F.W.T.A. Suppl.: 42 (1959); Fl. Cameroun 3: 158 (1964); Bull. Jard. Bot. Nat. Belg. 53: 223 (1983); Acta Botanica Barcinonensia 38: 7 (1988).

Habit: herb. Habitat: stream banks in forest. Distribution: Guinea to Bioko, Cameroon, CAR, Gabon, Congo (Brazzaville), Congo (Kinshasa), Angola, Burundi and Tanzania. Chorology: Guineo-Congolian (montane). Star: green. Alt. range: 1-200m (1). Fertile: Dec (1). Specimens: - *Mabeta-Moliwe:* Cable S. 536

Pteris ekemae Benl
Nova Hedwigia 27: 147 (1976).

Habit: herb. **Habitat:** forest. **Distribution:** Mount Cameroon. **Chorology:** Endemic (montane). **Star:** black. **Alt. range:** 1000-1200m (1), 1201-1400m (1). **Fertile:** Feb (2). **Specimens:** - *Eastern Slopes:* Benl G. Ka 75/49, Ka 75/69

Pteris hamulosa (H.Christ) H.Christ
Fl. Zamb. Pteridophyta: 120 (1970); Bull. Jard. Bot. Nat. Belg. 53: 226 (1983); Acta Botanica Barcinonensia 38: 19 (1988).

Habit: herb. **Habitat:** wet areas in forest. **Distribution:** tropical Africa. **Star:** excluded. **Alt. range:** 200-400m (1). **Fertile:** May (1). **Specimens:** - *Mokoko River F.R.:* Thomas D. W. 10073

Pteris linearis Poir.
F.W.T.A. Suppl.: 42 (1959); Fl. Cameroun 3: 163 (1964); Bull. Jard. Bot. Nat. Belg. 53: 227 (1983); Acta Botanica Barcinonensia 38: 18 (1988).

Habit: herb. **Guild:** shade-bearer. **Habitat:** forest. **Distribution:** Guinea to southern Africa and Sudan. **Star:** green. **Alt. range:** 2200-2400m (1). **Specimens:** - *No locality:* Dunlap 219

Pteris manniana Mett. ex Kuhn
Acta Botanica Barcinonensia 38: 6 (1988).
   Pteris camerooniana Kuhn, F.W.T.A. Suppl.: 40 (1959).

Habit: herb. **Habitat:** forest and stream banks. **Distribution:** Ivory Coast, Nigeria, Bioko, SW Cameroon and CAR. **Chorology:** Guineo-Congolian. **Star:** green. **Alt. range:** 1-200m (4), 201-400m (3), 401-600m (1), 801-1000m (1), 1001-1200m (1). **Fertile:** Apr (1), Jun (1), Sep (1), Oct (2), Nov (4), Dec (1). **Specimens:** - *Etinde:* Cable S. 281, 440; Cheek M. 3779, 5458; Thomas D. W. 9112, 9714. *Mabeta-Moliwe:* Wheatley J. I. 110. *Mokoko River F.R.:* Acworth J. M. 330. *No locality:* Mann G. 1385. *Onge:* Thomas D. W. 9812; Watts J. 978

Pteris mildbraedii Hieron.
F.W.T.A. Suppl.: 42 (1959); Fl. Cameroun 3: 165 (1964); Acta Botanica Barcinonensia 38: 20 (1988).

Habit: herb. **Guild:** shade-bearer. **Habitat:** forest, often on stream banks. **Distribution:** tropical Africa. **Star:** excluded. **Alt. range:** 1-200m (5), 401-600m (1). **Fertile:** Jun (1), Sep (2), Oct (3), Nov (1). **Specimens:** - *Etinde:* Tchouto P. 804; Thomas D. W. 9122, 9707. *Mokoko River F.R.:* Mbani J. M. 481. *Onge:* Tchouto P. 1004; Thomas D. W. 9840; Watts J. 715

Pteris sp. aff. mildbraedii Hieron.

Specimens: - *Mokoko River F.R.:* Acworth J. M. 163; Thomas D. W. 10023

Pteris preussii Hieron.
F.W.T.A. Suppl.: 40 (1959); Fl. Cameroun 3: 160 (1964); Bull. Jard. Bot. Nat. Belg. 53: 231 (1983); Acta Botanica Barcinonensia 38: 10 (1988).

Habit: herb. **Habitat:** forest. **Distribution:** Bioko, Mount Cameroon, Rwanda, Burundi and East Africa. **Chorology:** Afromontane. **Star:** green. **Specimens:** - *Eastern slopes:* Preuss P. R. 585

Pteris prolifera Hieron.
F.W.T.A. Suppl.: 40 (1959); Fl. Cameroun 3: 158 (1964); Bull. Jard. Bot. Nat. Belg. 53: 232 (1983); Acta Botanica Barcinonensia 38: 11 (1988).

Habit: herb. **Habitat:** forest and forest-grassland transition. **Distribution:** Liberia, SW Cameroon, Bioko, CAR, Congo (Brazzaville), Congo (Kinshasa), Uganda and Sudan. **Chorology:** Guineo-Congolian (montane). **Star:** green. **Alt. range:** 1400-1600m (1), 1601-1800m (1). **Fertile:** Sep (1), Oct (1). **Specimens:** - *Eastern slopes:* Thomas D. W. 9474. *Etinde:* Thomas D. W. 9199. *No locality:* Adams C. D. GC 4100

Pteris pteridioides (Hook.) Ballard
F.W.T.A. Suppl.: 42 (1959); Fl. Cameroun 3: 170 (1964); Fl. Zamb. Pteridophyta: 117 (1970); Bull. Jard. Bot. Nat. Belg. 53: 233 (1983); Acta Botanica Barcinonensia 38: 9 (1988).
Habit: herb. **Habitat:** forest-grassland transition. **Distribution:** tropical and subtropical Africa. **Chorology:** Afromontane. **Star:** green. **Alt. range:** 1400-1600m (1), 2001-2200m (3). **Fertile:** May (1), Sep (1), Oct (1). **Specimens:** - *Eastern slopes:* Kalbreyer W. 135; Thomas D. W. 9312; Upson T. 61. *Etinde:* Thomas D. W. 9191

Pteris similis Kuhn
F.W.T.A. Suppl.: 42 (1959); Fl. Cameroun 3: 168 (1964); Acta Botanica Barcinonensia 38: 18 (1988).

Habit: herb. **Habitat:** forest, often on stream banks. **Distribution:** Guinea to Bioko, Cameroon, Gabon, Congo (Kinshasa), Angola, Uganda, Sudan and Tanzania. **Chorology:** Guineo-Congolian. **Star:** green. **Specimens:** - *Mabeta-Moliwe:* Mann G. 786

Pteris togoensis Hieron.
F.W.T.A. Suppl.: 40 (1959); Fl. Cameroun 3: 160 (1964); Acta Botanica Barcinonensia 38: 13 (1988).

Habit: herb. **Guild:** shade-bearer. **Habitat:** forest-grassland transition. **Distribution:** Guinea to Bioko, Cameroon, Gabon, Congo (Brazzaville), Angola, CAR, Malawi, Sudan, Tanzania and Kenya. **Chorology:** Guineo-Congolian. **Star:** green. **Alt. range:** 2000-2200m (3), 2201-2400m (1). **Fertile:** May (1), Oct (2). **Specimens:** - *Eastern slopes:* Tchouto P. 296; Thomas D. W. 9390, 9494; Upson T. 62. *No locality:* Maitland T. D. 1049

Pteris tripartita Sw.
Acta Botanica Barcinonensia 38: 15 (1988).
   Pteris marginata Bory, F.W.T.A. Suppl.: 42 (1959).

Habit: herb. **Guild:** shade-bearer. **Habitat:** forest. **Distribution:** tropical Africa. **Star:** excluded. **Specimens:** - *Mabeta-Moliwe:* Mann G. 789

Pteris sp.

Specimens: - *Eastern slopes:* Sidwell K. 26

SCHIZAEACEAE

Lygodium smithianum C.Presl. ex Kuhn
F.W.T.A. Suppl.: 22 (1959); Fl. Cameroun 3: 64 (1964); Acta Botanica Barcinonensia 31: 22 (1978).

Habit: herbaceous climber. Guild: pioneer. Habitat: forest clearings. Distribution: Guinea to Bioko, Cameroon, CAR, Gabon, Congo (Brazzaville), Congo (Kinshasa) and Angola. Chorology: Guineo-Congolian. Star: green. Alt. range: 1-200m (4). Fertile: Nov (1), Dec (3). Specimens: - *Mabeta-Moliwe:* Cable S. 561; Cheek M. 5755, 5783; Schlechter F. R. R. 12399. *Onge:* Thomas D. W. 9929

THELYPTERIDACEAE

Amauropelta bergiana (Schltdl.) Holttum
Acta Botanica Barcinonensia 38: 48 (1988).
Thelypteris bergiana (Schltr.) Ching, F.W.T.A. Suppl.: 61 (1959).
Habit: herb. Habitat: forest-grassland transition. Distribution: tropical and temperate Africa. Chorology: Afromontane. Star: green. Alt. range: 2000-2200m (1). Specimens: - *Eastern slopes:* Richards P. W. 4246. *No locality:* Johnston H. H. 131

Ampelopteris prolifera (Retz.) Copel.
F.W.T.A. Suppl.: 63 (1959); Fl. Zamb. Pteridophyta: 200 (1970); Bull. Jard. Bot. Nat. Belg. 53: 284 (1983).

Habit: herb. Habitat: stream banks. Distribution: palaeotropical. Star: excluded. Alt. range: 1-200m (1). Fertile: Nov (1). Specimens: - *Onge:* Thomas D. W. 9785

Christella dentata (Forssk.) Brownsey & Jermy
Acta Botanica Barcinonensia 38: 53 (1988).
Cyclosorus dentatus (Forssk.) Ching, F.W.T.A. Suppl.: 62 (1959).
Thelypteris dentata (Forssk.) E.P.St.John, Fl. Zamb. Pteridophyta: 197 (1970).

Habit: herb. Habitat: forest clearings and roadsides. Distribution: palaeotropical. Star: excluded. Specimens: - *Eastern slopes:* Fraser J. 30

Cyclosorus interuptus (Willd.) H.Itô
F.W.T.A. Suppl.: 62 (1959); Bull. Jard. Bot. Nat. Belg. 53: 280 (1983); Acta Botanica Barcinonensia 38: 56 (1988).

Habit: herb. Habitat: swampy areas. Distribution: tropical and subtropical Africa, Asia and the Pacific region. Star: excluded. Alt. range: 1-200m (1). Fertile: Nov (1). Specimens: - *Onge:* Thomas D. W. 9786

Cyclosorus molundensis (Brause) Pic.Serm.
Bull. Jard. Bot. Nat. Belg. 53: 282 (1983).

Habit: herb. Habitat: forest clearings and river banks. Distribution: Sierra Leone, Cameroon, CAR, Congo (Kinshasa) and Burundi. Chorology: Guineo-Congolian. Star: green. Alt. range: 200-400m (1). Fertile: May (1). Specimens: - *Mokoko River F.R.:* Thomas D. W. 10165

Cyclosorus striatus (Schumach.) Ching
F.W.T.A. Suppl.: 62 (1959); Fl. Cameroun 3: 247 (1964); Bull. Jard. Bot. Nat. Belg. 53: 282 (1983); Acta Botanica Barcinonensia 38: 55 (1988).
Thelypteris striata (Schumach.) Schelpe, Fl. Zamb. Pteridophyta: 199 (1970).

Habit: herb. Habitat: forest margins and swampy ground. Distribution: tropical Africa. Star: excluded. Specimens: - *Eastern slopes:* Fraser J. 41

Cyclosorus tottus (Thunb.) Pic.Serm.
Webbia 23: 173 (1968).

Habit: herb. Habitat: river banks. Distribution: Cameroon, Rwanda, Burundi and South Africa. Chorology: Afromontane. Star: green. Alt. range: 1-200m (1). Fertile: Mar (1). Specimens: - *Onge:* Wheatley J. I. 765

Pneumatopteris afra (H.Christ) Holttum
Bull. Jard. Bot. Nat. Belg. 53: 283 (1983); Acta Botanica Barcinonensia 38: 57 (1988).
Cyclosorus afer (Christ) Ching, F.W.T.A. Suppl.: 63 (1959).
Habit: herb. Habitat: plantations and open areas in forest. Distribution: tropical Africa. Star: excluded. Alt. range: 1-200m (10), 201-400m (2), 401-600m (1). Fertile: Mar (1), Apr (3), May (3), Jun (1), Aug (1), Oct (3), Nov (3). Specimens: - *Eastern slopes:* Fraser J. 38. *Etinde:* Thomas D. W. 9716. *Mabeta-Moliwe:* Sunderland T. C. H. 1139; Watts J. 194. *Mokoko River F.R.:* Acworth J. M. 79; Akogo M. 291; Fraser P. 470; Ndam N. 1342, 1354; Thomas D. W. 10139. *Onge:* Cheek M. 4983, 5131; Thomas D. W. 9787; Watts J. 691, 988, 1006

Pneumatopteris unita (Kunze) Holttum
Bull. Jard. Bot. Nat. Belg. 53: 283 (1983); Acta Botanica Barcinonensia 38: 61 (1988).
Cyclosorus patens (Fée) Copel., F.W.T.A. Suppl.: 62 (1959).
Thelypteris madagascariensis (Fée) Schelpe, Fl. Zamb. Pteridophyta: 196 (1970).

Habit: herb. Habitat: forest-grassland transition. Distribution: tropical Africa. Chorology: Afromontane. Star: green. Alt. range: 1400-1600m (1). Fertile: Sep (1). Specimens: - *Etinde:* Thomas D. W. 9193

Stegnogramma pozoi (Lagasca) Iwatsuki var. pozoi
Acta Botanica Barcinonensia 38: 49 (1988).
Leptogramma pilosiuscula (Wikstr.) Alston, F.W.T.A. Suppl.: 63 (1959).
Thelypteris pozoi (Lagasca) Morton, Fl. Cameroun 3: 239 (1964).

Habit: herb. Habitat: forest-grassland transition. Distribution: Bioko, Mount Cameroon, Sudan and East to South Africa. Chorology: Afromontane. Star: green. Specimens: - *Eastern slopes:* Richards P. W. 4246A. *No locality:* Mann G. 1375

VITTARIACEAE

Antrophyum mannianum Hook.
F.W.T.A. Suppl.: 35 (1959); Fl. Cameroun 3: 122 (1964); Fl. Zamb. Pteridophyta: 96 (1970); Acta Botanica Barcinonensia 33: 6 (1982); Bull. Jard. Bot. Nat. Belg. 53: 241 (1983).

Habit: epiphyte. Habitat: forest, sometimes on rocks. Distribution: tropical Africa. Chorology: Afromontane. Star: green. Alt. range: 200-400m (1), 1401-1600m (1). Fertile: Sep (1). Specimens: - *Etinde:* Tekwe C. 206; Thomas D. W. 9110. *No locality:* Mann G. 1364

Vittaria guineensis Desv. var. **camerooniana** Schelpe

Acta Botanica Barcinonensia 33: 8 (1982).

Habit: epiphyte. **Habitat:** forest-grassland transition. **Distribution:** Cameroon and Bioko. **Chorology:** Lower Guinea (montane). **Star:** blue. **Alt. range:** 1400-1600m (1). **Fertile:** Sep (1). **Specimens:** - *Etinde:* Thomas D. W. 9214

Vittaria guineensis Desv. var. **guineensis**

F.W.T.A. Suppl.: 35 (1959); Fl. Cameroun 3: 124 (1964); Acta Botanica Barcinonensia 33: 7 (1982).

Habit: epiphyte. **Habitat:** forest. **Distribution:** Guinea to Bioko, Cameroon, S. Tomé, Gabon, Congo (Brazzaville), Congo (Kinshasa), Angola and Uganda. **Chorology:** Guineo-Congolian (montane). **Star:** green. **Alt. range:** 1-200m (1). **Fertile:** Jun (1), Dec (1). **Specimens:** - *Mabeta-Moliwe:* Cable S. 614. *Mokoko River F.R.:* Mbani J. M. 484. *No locality:* Mann G. 1366

Vittaria owariensis Fée

F.W.T.A. Suppl.: 35 (1959).

Habit: epiphyte. **Habitat:** plantations and forest. **Distribution:** Liberia to Cameroon and S. Tomé. **Chorology:** Guineo-Congolian. **Star:** green. **Alt. range:** 1-200m (1). **Fertile:** Apr (1). **Specimens:** - *Mabeta-Moliwe:* Wheatley J. I. 86

WOODSIACEAE

Athyrium ammifolium (Mett.) C.Chr.

F.W.T.A. Suppl.: 64 (1959); Fl. Cameroun 3: 230 (1964); Acta Botanica Barcinonensia 38: 39 (1988).

Habit: herb. **Habitat:** forest and forest-grassland transition. **Distribution:** Nigeria, Bioko and W. Cameroon. **Chorology:** Western Cameroon (montane). **Star:** gold. **Alt. range:** 1400-1600m (1), 1801-2000m (2). **Fertile:** May (1), Sep (1), Oct (1). **Specimens:** - *Eastern slopes:* Thomas D. W. 9457; Upson T. 87. *Etinde:* Thomas D. W. 9196. *No locality:* Migeod F. 247

Athyrium schimperi Moug. ex Fée

F.W.T.A. Suppl.: 64 (1959); Fl. Cameroun 3: 229 (1964); Fl. Zamb. Pteridophyta: 202 (1970); Bull. Jard. Bot. Nat. Belg. 55: 155 (1985).

Habit: herb. **Habitat:** grassland and forest-grassland transition. **Distribution:** Cameroon, East Africa and Himalayas. **Chorology:** Montane. **Star:** excluded. **Alt. range:** 2000-2200m (1), 2201-2400m (2), 2601-2800m (1). **Fertile:** Oct (3), Nov (1). **Specimens:** - *Eastern slopes:* Cheek M. 3674, 5353; Tchouto P. 278; Thomas D. W. 9380. *No locality:* Adams C. D. 1219

Diplazium proliferum (Lam.) Kaulf.

F.W.T.A. Suppl.: 65 (1959); Fl. Cameroun 3: 237 (1964).
   Callipteris prolifera (Lam.) Bory, Acta Botanica Barcinonensia 38: 37 (1988).

Habit: herb. **Habitat:** forest. **Distribution:** tropical Africa. **Star:** excluded. **Alt. range:** 1-200m (3), 201-400m (2). **Fertile:** May (1), Jun (2), Nov (1). **Specimens:** - *Etinde:* Cheek M. 5461, 5864.

Mabeta-Moliwe: Watts J. 355. *Mokoko River F.R.:* Acworth J. M. 334; Ekema N. 1055

Diplazium sammatii (Kuhn) C.Chr.

F.W.T.A. Suppl.: 64 (1959); Fl. Cameroun 3: 233 (1964).

Habit: herb. **Habitat:** river banks in forest. **Distribution:** Guinea to Cameroon, Congo (Kinshasa) and Sudan. **Chorology:** Guineo-Congolian. **Star:** green. **Alt. range:** 1-200m (1), 201-400m (1). **Fertile:** Apr (1), May (1). **Specimens:** - *Mokoko River F.R.:* Acworth J. M. 87; Thomas D. W. 10162

Diplazium velaminosum (Diels) Pic.Serm.

Webbia 27: 443 (1973).
   Diplazium zanzibaricum sens. auct., F.W.T.A. Suppl.: 65 (1959); Fl. Cameroun 3: 236 (1964).

Habit: herb. **Habitat:** forest and forest-grassland transition. **Distribution:** tropical Africa. **Chorology:** Afromontane. **Star:** green. **Alt. range:** 800-1000m (1), 2001-2200m (1). **Fertile:** Dec (1). **Specimens:** - *Eastern slopes:* Richards P. W. 4313. *Etinde:* Faucher P. 53

Diplazium welwitschii (Hook.) Diels

F.W.T.A. Suppl.: 65 (1959); Fl. Cameroun 3: 234 (1964); Acta Botanica Barcinonensia 38: 41 (1988).

Habit: herb. **Habitat:** forest. **Distribution:** Sierra Leone to Bioko, Cameroon, Gabon, Congo (Brazzaville), Congo (Kinshasa), Angola, Uganda and Tanzania. **Chorology:** Guineo-Congolian. **Star:** green. **Alt. range:** 200-400m (1), 401-600m (1). **Fertile:** Apr (1), Oct (1), Dec (1). **Specimens:** - *Etinde:* Lighava M. 49; Thomas D. W. 9152. *Mokoko River F.R.:* Ndam N. 1084

Lunathyrium boryanum (Willd.) H.Ohba

Yokosuka City Mus. 11: 53 (1965).
   Athyrium glabratum (Mett.) Alston, F.W.T.A. Suppl.: 64 (1959).
   Dryoathyrium boryanum (Willd.) Ching, Acta Botanica Barcinonensia 38: 38 (1988).
   Deparia boryanum (Willd.) M.Kato, Bot. Mag. Tokyo 90: 36 (1977).

Habit: herb. **Habitat:** forest and forest-grassland transition. **Distribution:** Ivory Coast, Nigeria, Bioko and SW Cameroon. **Chorology:** Guineo-Congolian (montane). **Star:** green. **Alt. range:** 200-400m (1), 1001-1200m (1), 2001-2200m (1). **Fertile:** Oct (1), Dec (1). **Specimens:** - *Eastern slopes:* Adams C. D. 1634. *Etinde:* Lighava M. 51; Thomas D. W. 9521